CHAOTIC, FRACTAL, AND NONLINEAR SIGNAL PROCESSING

Mystic, CT July 10–14, 1995

EDITOR
Richard A. Katz
Naval Undersea Warfare Center
Newport, RI

AIP PRESS

AIP CONFERENCE PROCEEDINGS 375

American Institute of Physics

Woodbury, New York

L.C. Catalog Card No. 96-85356
ISBN 1-56396-443-0
DOE CONF- 950730

Printed in the United States of America

CONTENTS

3.1 APPLICATIONS IN ACOUSTICS

4. ADVANCED APPLIED SIGNAL PROCESSING METHODS

5. STOCHASTIC RESONANCE

6. MACHINERY DIAGNOSTICS

7. TURBULENCE

8. GEOPHYSICS

9. MEDICINE

10. NEW METHODS FOR MODELING NONLINEAR SYSTEMS

FOREWORD

The contents of this book are a compilation of technical topics from the Office of Naval Research (ONR)/Naval Undersea Warfare Center (NUWC) *Third Technical Conference on Nonlinear Dynamics and Full-Spectrum Processing.* The conference was held in Mystic, Connecticut from July 10 to 14, 1995 at the Seamen's Inne Conference Center.

The conference focus was on the latest advances in chaotic, fractal, and nonlinear signal processing methods leading toward practical problem solutions in engineering and the physical sciences. Methods using advanced linear techniques are also covered (see, for example, paper by A. Nuttall). Topics of discussion covered in the book include: mathematical frontiers; predictability and control of chaos; detection and classification with applications in acoustics; advanced applied signal processing methods (linear and nonlinear); stochastic resonance; machinery diagnostics; turbulence; geophysics; medicine; and recent novel approaches to modeling nonlinear systems.

In the book, we make the general distinction between linear versus nonlinear processes and associated methods. Linear processes may be considered those in which the additive property of superposition holds up. For nonlinear processes, this property breaks down. Hence, a process such as an undamped periodically oscillating pendulum is considered linear, while a chaotic process, such as a turbulent waterfall, falls into the category of nonlinear. The categorization of physical processes can sometimes be a challenging endeavor; hence, the optimal processing method must be weighed in light of the perceived physical mechanism(s) generating the particular "signal-of-interest" under investigation. For our purposes, a signal-of-interest may be considered any time sequence (or series) measurement of at least potential practical engineering import.

Sponsors for the conference included: the ONR, Chemistry and Physics S&T Division (Dr. Michael F. Shlesinger); the ONR Science Opportunities Program (Director, Dr. Albert J. Faulstich, Jr.); the ONR European Office, London (Director, Dr. John Silva); the Naval Undersea Warfare Center, Science and Technology Directorate (Dr. Richard H. Nadolink); and the International Science Foundation (Program Director, Ms. Eleanor A. Gorman). Conference endorsements were also provided by the American Institute of Physics and the American Physical Society.

There were several special events during the conference. An evening banquet appearance in the Mansion at the Inn at Mystic was made and remarks were given by Rear Admiral (RADM) Scott L. Sears, Commanding Officer of NUWC. Banquet entertainment included a lighthearted theatrical performance, with musical arrangements in the "Science of Chaos" subject theme, written by Ms. Jacqueline Princevalle of Science Applications International Corporation and directed by Ms. Lynn Britt of Britt Productions, Old Lyme, Connecticut. Special banquet awards were given to Dr. Robert Cawley for his support and contributions in the field of Nonlinear Dynamics (Chaos) and Full Spectrum Processing, and to RADM Sears in acknowledgment from the Naval Scientific Community of his leadership and community support in this important technical area.

The conference organizational team members were Joan Cembrola, Thomas Galib, and Richard Katz of NUWC; Theodore Frison of Randle, Inc.; James Kadtke of the University of California, San Diego; and Adi Bulsara of the Naval Command Control & Ocean Surveillance Center. Ted Frison provided particularly valuable program guidance. James Kadtke and Adi Bulsara served as international co-chairs, and were largely responsible for bringing together the assembly of international presenters for which we are most indebted. Conference planning and administrative support was provided by Science Applications International Corporation (Annette Zimmermann, Douglas Brown, Fernand Deltgen, and Cecelia Sullivan), and publication support was provided by Mr. Michael Hennelly (now with McGraw Hill) and Ms. Anita Hargrove of AIP Press.

Finally, we wish to express our gratitude to Henry D. I. Abarbanel and Francis C. Moon for providing the preface, *What's the Future for Chaos: Experiment and Theory?* Both have written prolifically on the subject. Professor Abarbanel has recently completed his new book *Analysis of*

Observed Chaotic Data, and Professor Moon has published an updated edition of his well known *Chaotic Vibrations*. Both were active participants in our last (2nd) conference held in 1993, and both were instrumental in its success.

The reader interested in most recent advances in signal processing may wish to begin with Chapters 3 and 4, starting with the paper by Stephen Hammel and Bo Hammer on *System Identification in Experimental Data*, or with any of the specialization topics covered in Chapters 5 through 9. The first two chapters, "Mathematical Frontiers" and "Predictability and Control of Chaos," are intended to provide a formal grounding and deeper insight into some of the later chapters. These chapters also discuss new directions in the field. Finally, Chapter 10 concludes with some new approaches to nonlinear modeling.

It has been just two short years since the publication of our last *Proceedings* (AIP Series No. 296) and, upon reflection of the past and the present, it is now abundantly clear that nonlinear methods of analysis are generally well established in the mainstream of scientific thought and engineering development. It is with this view in mind that this book was compiled. It is intended for use by linear and nonlinear researchers alike as a seedbed for expansion of one's own ideas and new discovery.

Richard A. Katz
March 12, 1996

PREFACE

What's the Future for Chaos: Experiment and Theory?

Before we attempt to address this large issue, it is well to review the past progress in nonlinear dynamics. The early successes in the field in the 1980's were in mechanics, electronics, fluid dynamics, and chemistry, and now in the 1990's, we have seen the theory applied to new problems in optics, communications, and biology. In engineering, chaos theory has been applied to more complex machine systems; exciting progress has been made using the ideas about controlling chaos in the areas of electronics, lasers, and biological systems.

If the study of chaos were another contemporary research fad, one would have thought that 18 years after Feigenbaum's period doubling papers, the field might be taking its place next to catastrophe theory. On the contrary, each year more books and journals devoted to this subject appear, conferences like this one continue to attract new novitiates, and the basic ideas of nonlinear dynamics and chaos theory are being introduced in the science and engineering curricula of undergraduates. The only downside about chaos theory is the stagnation or decrease in research support at a time when potential payoffs seem more fruitful than ever. To the extent that realization of chaos as an essential part of nonlinear systems represents a major paradigm change from the dual view — regular or noise — promoted by the study of linear systems, one might have expected extensive private and government support to pursue its depth and create applications of value to society.

As for the future, new ideas enter science and technology through young minds trained in this new thinking. The extremely successful effort in nonlinear dynamics created in the former Soviet Union and focused around the Institute for Applied Physics in Nizhny Novgorod (former Gorky) always had an annual Winter School devoted to training young students, and those students became the leaders in that effort over the years. Conferences such as this one play a partial role in that regard, and more focus on education and training would augment this effort in a healthy fashion. Already a new generation of dynamists have entered the research and productivity fields with modern methods of analysis and experiments in dynamics. One such area is the new set of nonlinear signal processing tools such as fractal dimension, false nearest neighbors, Lyapunov exponents, etc. These new tools have created a new methodology to build low order models for dynamical systems. These tools provide the analogue of the power of the FFT for analyzing linear systems. They are applicable to data from whatever source and have proven themselves in matters ranging from boundary layer flows of interest to the Navy and climatic fluctuations of large watersheds to the oscillations of individual neurons. One can diagnose a nonlinear system using them and build predictive models of great value for practical applications.

As chaos "looks like" noise without these tools, the initial reaction of engineers has been to banish to other parts of state space and to seek robust controls to suppress it. A new direction in this subject is the recognition that chaotic dynamics may be beneficial or useful in both natural and man-made systems. Recent work by the authors of this preface, for example, suggest that chaotic dynamics in material processing, such as machining of metals, may be desirable to produce a high quality surface. One can spoeculate that a little chaos is good in the operation of some machines, such as internal composition engines or complex mechanisms. Such a concept, if proved true, would invite new design methods for such systems. The use of chaos in communications systems starts from the recognition that it is a signal structured in state space, and then seeks to exploit that structure rather than suppress it and return to previously solved problems. In biological systems where chaotic motions abound, one must ask how animals use chaotic behavior for functional benefit. Evolution could well have banished chaos to the nether reaches of phase space, but it persists, and one can speculate that it is actually a necessary ingredient in the reliability and robustness biological systems (including the authors and the participants of this conference) required to achieve evolution.

Another new concept in chaos theory is the idea of "anti-control." In contrast to using control to force a chaotic system into a periodic orbit, "anti-control" schemes seek to push a periodic system into chaotic or transient chaos behavior, which may be more desirable for the health of the system. The idea that this might be useful is clear when thinking of mixing in chemical reactors, but the concept will lead far beyond that.

Another application area for nonlinear dynamics is in material processing and manufacturing. Almost all processing methods involve dynamics, yet practitioners in these fields usually view the methods as steady. Use of new diagnostic tools and control methods based on nonlinear dynamics concepts may bring the new theory into mainstream manufacturing.

As for theory, spatio-temporal problems continue to be a frontier. But the lack of simple paradigms, such as period doubling and the Lorenz attractor familiar in temporal chaos, leaves the field of spatial complexity without an organizing center.

Another area is complex adaptive systems or CAS. These new ideas are exciting but lack the kind of hard experimental basis that propelled chaos theory in the early 1980's. We shall have to wait for more progress in this field to predict its useful applications.

The success of the new dynamics and chaos theory can be measured in many ways, not the least of which is the frequency with which these new concepts have entered the English language and literature. But unlike the discovery of a new chemical or micro-electronics device, new ideas propagate by diffusion into the scientific and technical workplaces through education of newly trained scientists and engineers.

In order to continue to make progress in nonlinear dynamics, funding agencies and commercial ventures must realize that support of new students is the best vehicle for the propagation of new ideas and applications. From this support will arise the new theoretical ideas needed to drive the next phase of the dynamics revolution, and the academic and industrial leaders needed to make it happen.

H. D. I. Abarbanel *F. C. Moon*

SCIENCE ET METHODE

....Si nous connaissions exactement les lois de la nature et la situation de l'univers à l'instant initial, nous pourrions prédire exactement la situation de ce meme univers à un instant ultérieur. Mais, lors meme que les lois naturelles n'auraient plus de secret pour nous, nous ne pourrons connaitre la situation initiale qu'*approximativement.* Si cela nous permet de prévoir la situation ultérieure *avec la meme approximation,* c'est tout ce qu'il nous faut, nous disons que le phénomène a été prévu, qu'il est régi par des lois; mais il n'en est pas toujours ainsi, il peut arriver que de petites différences dans les conditions initiales en engendrent de très grandes dans phénomènes finaux; une petite erreur sur les premières produirait une erreur énorme sur les derniers. La prédiction devient impossible et nous avons le phénomène fortuit.

Henri Poincaré 1854-1912

SCIENCE AND METHOD

....If we could know exactly the laws of nature and the situation of the universe at the initial moment, we should be able to predict exactly the situation of that same universe at a succeeding moment. But, even when the natural laws should have no longer any secret for us, we could know the initial situation only *approximately.* If that permitted us to predict the succeeding situation *with the same approximation,* that is all we require, and we should say that the phenomenon had been predicted, that it is governed by laws; but this is not always so, since it may happen that small differences in the initial conditions produce very great ones in the final phenomena; a small error in the former will produce an enormous error in the latter. Prediction becomes impossible, and we have the fortuitous phenomenon.

(Translation)

CHAOS: MAKING A NEW SCIENCE

....The modern study of Chaos began with the creeping realization in the 1960's that quite simple mathematical equations could model systems every bit as violent as a waterfall. Tiny differences in input could quickly become overwhelming differences in output—a phenomenon given the name "sensitive dependence on initial conditions." In weather, for example, this translates into what is only half-jokingly known as the Butterfly Effect—the notion that a butterfly stirring the air today in Peking can transform storm systems next month in New York.

James Gleick 1987

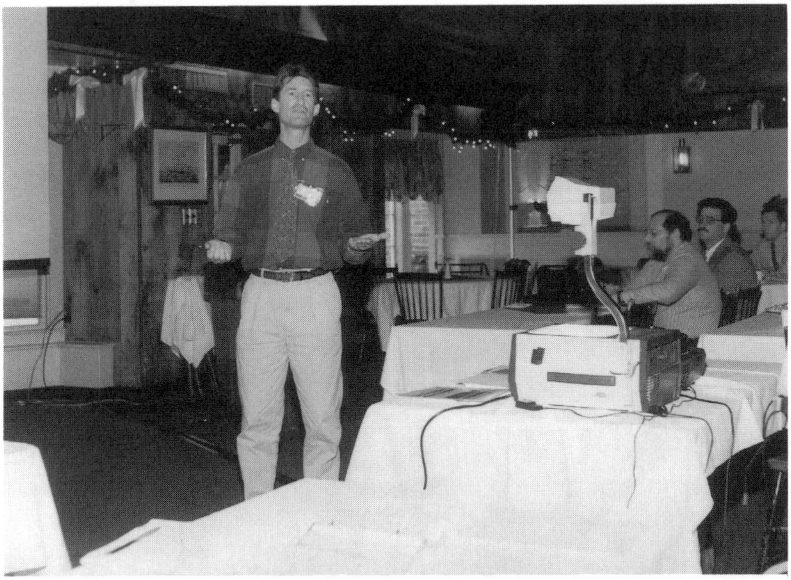

Dr. Stephen Hammel leads off technical discussions on the detection and classification of chaotic signals.

Dr. James Kadtke lectures on his global approach to modeling system dynamics.

Michelle Bright of NASA discusses her research on detecting stall events in high speed compressors, using chaotic diagostic methods.

Commodore Charles Munns, Commander Submarine Development Squadron Twelve, was one of a number of dignitaries attending the conference.

Professor Vadim Anishchenko of the former Soviet Union lectures on the stochastic resonance in passive and active electronic circuits.

Professor Ed Ott discusses chaos in physical experiments and the Ott-Grebogi-Yorke (OGY) method for controlling chaos

Dr. Bill Ditto addresses the audience on the Nonlinear Dynamics of a Physiologically viable model neuron!

Dr. Adi Bulsara, International Co-chair relaxes during social activities in the Mansion at the Inn at Mystic.

Banquet attendees enjoy festivities held on the terrace at the Mansion at the Inn at Mystic.

Rear Admiral (RADM) Scott L. Sears, Commanding Officer of NUWC addresses banquet attendees and offers his congratulations and best wishes.

Professor Yuri Kravtsov of the former Soviet Union makes some heartfelt remarks to banquet guests.

Jeffrey Brush shown in background helps out while Professor Kravtsov addresses dinner guests.

Chairman Dr. Richard Katz presents prestigious NUWC award to Dr. Robert Cawley for his significant contributions in the field of Chaos and Nonlinear Dynamics.

1. MATHEMATICAL FRONTIERS

Numerical Analysis of Bifurcations

John Guckenheimer

Mathematics Department, Cornell University, Ithaca, NY 14853

Abstract: This paper is a brief survey of numerical methods for computing bifurcations of generic families of dynamical systems. Emphasis is placed upon algorithms that reflect the structure of the underlying mathematical theory while retaining numerical efficiency. Significant improvements in the computational analysis of dynamical systems are to be expected from more reliance of geometric insight coming from dynamical systems theory.

Finite dimensional dynamical systems arise as models in many settings, including electrical networks, chemical reactors, neuronal networks and multibody mechanical systems. In all of these examples models are often assembled from component subsystems that are well characterized. Nonetheless, determining the behavior of such a system from is still a substantial challenge. Depending upon context this challenge manifests itself as one of two different problems. If we are building artificial systems, we have the problem of *synthesis* or design. A set of specifications are presented to us along with a kit of components from which systems can be assembled. Our goal is to build a system that performs according to the specifications. If we are studying natural systems, then we have the analog of the engineer's problem of *system identification.* We want to determine which components the whole system was assembled from by observing it. We assert that the insights coming from advances in dynamical systems theory during the past decades provide an intellectual substrate for enhancing our ability to solve these two problems. The realization of this opportunity requires that we develop better computational tools that allow us to explore the dynamics of nonlinear systems more fully. This paper discusses one important aspect of such computations, the location of bifurcations in multiparameter families of dynamical systems.

Modern developments in dynamical systems theory have relied heavily upon numerical investigations for motivation and intuition. Most of these studies have been performed in the context of low dimensional, often discrete systems. Here the errors associated with numerical computations are primarily due to round-off. On the other hand, continuous time dynamical systems are defined by systems of ordinary differential equations. These systems seldom have explicit solutions, any numerical methods are essential to compute even approximate solutions. Numerical integration is a classical subject, but it forms an incomplete substrate for the computation of properties of dynamical systems. Additional aspects of the numerical analysis of dynamical systems have drawn increasing attention in recent years. We review the work that is devoted to the computation of bifurcations, placing emphasis upon our own work.

We give a more formal statement of the problems that we seek to solve. Let $\dot{x} = f(x, \lambda)$ be a smooth (C^∞) vector field that depends upon parameters. The state space can be R^n or an n dimensional manifold, and we assume $\lambda \in R^k$. For each λ, the trajectories of the vector field partition the phase space. The geometry of this partitioning can be very complicated. To produce a more manageable context for understanding the geometry, attention is focussed upon generic fields and families. Within a designated "universe" of allowable systems, one seeks classification of phenomena that persist when a system is perturbed within the set of allowed systems. Mildly degenerate behavior is studied by embedding vector fields in multiparameter families so that the object of interest persists under perturbation of the family.

This viewpoint for studying bifurcation problems is based upon developments in singularity theory. The basic problem of singularity theory is to classify mappings $F : R^m \to R^n$ up to coordinate changes. The results of singularity theory yield a satisfying picture that we would like to adapt to the context of dynamical systems. The portion of singularity theory that we shall use most extensively deals with stratifications of the parameter space for a family of mappings. Call two mappings (topologically) equivalent if there are (continuous) coordinate changes that transform one map to the other. A map is stable if all perturbations of the map are equivalent to it. These definitions of equivalence and stability can be generalized to families of maps. Given a family of maps F with parameter space Λ, divide the space Λ into sets of mappings that are equivalent to one another. In families that are stable with respect to this equivalence relation, these sets are submanifolds that fit together to form a *stratification.* each stratum corresponds to a particular type of singularity. The singularity types can be classified by systematic procedures, and representatives of each equivalence class chosen. These representatives are called normal forms. The geometry of perturbations of each singularity can be studied by constructing a universal unfolding of its normal form. The normal form is embedded in a stable family of minimal dimension (the *codimension* of the singularity), and the stratification of this family is studied as an archetype. In studying other families of maps, the unfoldings of singularities serve as templates to guide the analysis.

We seek to transplant the strategy described above for the study of singularities to the study of bifurcations of dynamical systems. However, the bifurcation theory is only partially susceptible to this type of analysis, and attempts to build a comprehensive theory have floundered over technical difficulties and counterexamples. Pragmatic use of singularity theory approaches have produced significant progress, but we remain far from any sensible classification of "almost all" dynamical systems based upon topological equivalence. Lists of bifurcations have been constructed and analyzed with varying degrees of completeness. As with singularity theory, the bifurcations come with a set of *defining equations* and *nondegeneracy conditions*. The defining equations specify a closed subset in the universe of vector fields under investigation, most of which are presumed to have the type of the desired bifurcation. The nondegeneracy conditions are inequalities that determine that the bifurcations are no more degenerate than the type desired. There is also a distinction made between *local* and *global* bifurcations: the defining equations and nondegeneracy conditions of local bifurcations are expressed directly in terms of the analytic formulas of the vector field and do not depend upon numerical integration. If the set of bifurcations of a particular type is a submanifold of the universe of vector fields, the codimension of the bifurcation is the codimension of this submanifold.

We use two examples to illustrate this formulation of bifurcation problems. The saddle node bifurcation of an equilibrium is a codimension one local bifurcation. The defining equations of the saddle node bifurcation of an equilibrium are $f = 0$ and $\det(Df) = 0$. There are two nondegeneracy conditions, the first of which expresses that a particular second derivative of f does not vanish. The second nondegeneracy condition states that a first derivative of f with respect to the parameters does not vanish [15]. Homoclinic bifurcations of equilibria are codimension one bifurcations. If p is a saddle equilibrium point of the vector field f, then a homoclinic orbit is a trajectory that lies in the intersection of the stable and unstable manifolds of p. The stable and unstable manifolds can only be found through numerical integration for most vector fields, so homoclinic bifurcation is a global phenomenon. Because the stable and unstable manifolds have complementary dimensions in the phase space, but intersections that are generically one dimensional, the bifurcation has codimension one.

To apply bifurcation theory in the context of example systems, effective numerical algorithms are required. Pursuing the strategy suggested by singularity theory, we seek algorithms that are based upon (1) the analysis of normal forms for different types of bifurcations, and (2) the computation of smooth manifolds by solving regular systems of equations. The second requirement appears paradoxical at first glance because we seek to find singular objects. However, the procedures associated with singularity theory give us the means to reduce these computations that work with stratified sets as collections of submanifolds. The program that we outline is based upon work that is still very much in progress, so some of the comments are tentative.

The codimension one bifurcations of generic dynamical systems include

1. Equilibrium points
   (a) Saddle node bifurcation
   (b) Hopf bifurcation
2. Periodic orbits
   (a) Saddle node bifurcation
   (b) Period doubling bifurcation
   (c) Hopf bifurcation
3. Homoclinic bifurcation of equilibria

This list is not exhaustive. There are phenomena involving chaotic behavior such as the presence of homoclinic tangencies of periodic orbits that are not on the list. There are other phenomena such as the presence of invariant tori with a fixed winding vector that occur in codimension one but are not usually regarded as bifurcations.

Let us proceed to review what has been done to develop algorithms for the computation of each of these classes of bifurcations. The equilibrium bifurcations are local. As summarized above, defining equations for saddle node bifurcations are readily specified. One adds to the equilibrium equations $f(x,\lambda) = 0$ the equation $\det(Df)(x,\lambda) = 0$. The nondegeneracy conditions for a saddle node bifurcation to occur with respect to variations of the parameter λ_i are precisely that this system of equations should have a regular zero with variations of x and λ_i. In addition to the well known problems associated with solving systems of equations, there is one additional algorithmic difficulty in computing saddle node bifurcations. Computing the determinant of a square matrix directly from its definition is a complex and frequently unreliable procedure. Effective bifurcation algorithms incorporate ideas from numerical linear algebra for determining the singularity of a matrix that avoid explicit calculation of the determinant. One approach is the construction of *bordered matrices* to determine the singularity of a matrix. If A is an $n \times n$ matrix of rank $n-1$, then there are n vectors B and C so that the $(n+1) \times (n+1)$ matrix

$$E = \begin{pmatrix} A & B \\ C^T & 0 \end{pmatrix}$$

is nonsingular. For example the choices $B = \ker(A^T)$ and $C = \ker(A)$ satisfy this criterion. If

$$E^{-1} = \begin{pmatrix} K & L \\ M & N \end{pmatrix}$$

, then the 1×1 matrix N vanishes. If one perturbs A, then the vanishing of the perturbed matrix N is a criterion for the singularity of the perturbed matrix A [10]. Moreover, the computation of N can be accomplished by solving a system of linear equations by Gaussian eliminatiau with partial pivoting - an algorithm that is very reliable [9]. Thus the introduction of a bordering for the singular or near singular matrix A leads to an effective algorithm for the computation of a function that detects the presence of a saddle node bifurcation.

The computation of Hopf bifurcations presents additional numerical challenges. The defining criterion for a Hopf bifurcation of an equilibrium point is that its Jacobian matrix have a pair of pure imaginary eigenvalues. If A is a square matrix, then determining whether it has a pair of pure imaginary eigenvalues is more complex than determining whether it has a zero eigenvalue. Several approaches have been used during the past decade [11, 12, 18, 20, 21]. In our own work [12], we have employed an old algebraic construction [22] that produces a larger matrix, called the biproduct, whose eigenvalues are the pairwise sums of eigenvalues of A. Replacing the Jacobian matrix at an equilibrium point by its biproduct leads to an algorithm for detection of Hopf bifurcation that is a direct analog of the algorithm for computing saddle node bifurcations.

From a theoretical point of view, there is little difference in computing the bifurcations of periodic orbits and equilibrium points. Introduction of a cross-section yields a return map that is a discrete time dynamical system with a fixed point at the intersection of the cross-section with the periodic orbit. The techniques described above can be used to find defining equations that determine whether a matrix A has an eigenvalue of 1 (singularity of $I - A$), -1 (singularity of $I + A$) or a pair of complex eigenvalues of modulus one (singularity of the skew symmetric part of $I - A \otimes A$). However, computation of the return map relies upon numerical integration. With the simplest numerical integration methods, the continuous time flow is replaced by a discrete time map ψ that is an approximation to the time h map of the flow, for small h. The periodic orbit of the vector field may be close to an invariant curve of the map

h, but invariant curves of discrete maps are less stable than periodic orbits of continuous systems. The analysis of Chenciner [4] indicates that that there will arbitrarily small step sizes h for which the discretization of a vector field with a periodic orbit undergoing a saddle node bifurcation will *not* have an invariant curve. Instead, resonances of the discretization along the periodic orbit will produce more complex dynamical structures reminiscent of the resonance zones of symplectic maps [4]. Genericity arguments lead one to expect that small scale chaotic behavior will occur for some step sizes. There are still few studies that indicate how serious the problems in computing periodic orbit bifurcations associated with discretization of a flow are in practice. One of the examples in the next section suggests that the resonance phenomena are not a major hurdle in computing bifurcations.

In addition to methods based upon the computation of return maps, the detection of saddle nodes for periodic orbits has been approached by other methods, including "naive" topological approaches and the solution of boundary value problems. Let us use an example to illustrate why the naive approaches encounter substantial difficulty in producing definitive results. Consider the planar cubic vector field

$$\begin{aligned} \dot{x} &= y \\ \dot{y} &= m + nx + px^2 - x^3 + ry + sx^2y \end{aligned}$$

studied by Dangelmayr and Guckenheimer [5]. Their analysis suggested that there is a region of the parameter space for this system in which there is a single equilibrium point and four concentric limit cycles. Computations of Malo [19] located such a region. In a parameter plane obtained by fixing (m, p, s) and varying (n, r), this region is a strip whose width was of order 10^{-10}. Detection of the strip requires that the limit cycles be computed with very high accuracy. The boundary of the strip consists of two curves of saddle node bifurcations of periodic orbits. The search for concentric limit cycles in this example has prompted our attempts to reexamine more closely algorithms for computing periodic orbits.

One naive method for computing saddle node bifurcations of limit cycles in planar systems uses a divide and conquer strategy or "brute force" parameter searches. If A is an annulus containing the periodic orbit at the saddle node bifurcation, then on one side of the bifurcation locus in the parameter space, trajectories will cross A. On the other side of the bifurcation locus in the parameter space, trajectories will not cross A. Thus one can search for the bifurcation value of the parameter by testing whether trajectories from a suitably chosen initial point cross A. Using this test with a bisection method gives an algorithm for computing the bifurcation point that is thoroughly reliable and as accurate as the numerical integration. However, it suffers from two difficulties. First, it is subject to the problems associated with discretization of the flow by numerical integration creating resonances. For a fixed step size in the numerical integrator, the bifurcation boundary determined by the algorithm will not be a smooth curve for the reasons that are explained above. Indeed, theoretical results are not adequate to describe what we expect the algorithm to produce, though we do know a lot about the complexity of the dynamical behavior to be found.

The second difficulty with this naive algorithms is that it is very slow. Crude estimates of the computational time required for this algorithm are as follows. The diameter of the limit cycles we are computing and the magnitude of the vector field along the limit cycle are quantities that are comparable to 1. Suppose one uses a fourth order integration scheme for computing trajectories. To produce solutions that are accurate to 10^{-12} will require roughly 10^3 time steps per period. The normal form for the transverse motion of the vector field near the bifurcating limit cycle is $\dot{r} = \lambda + r^2$, with λ proportional to the distance form the bifurcation. For small positive λ, the time required to pass through the annulus A scales like $\sqrt{\lambda}$. Thus accuracy of 10^{-10} in locating the bifurcation requires integration times of the order of 10^5. Each step of the integrator requires several function evaluations, yielding roughly 100 flops per step. Thus the computation of the longest trajectory segments for our bifurcation test are likely to require roughly 10^{11} flops. On a workstation that achieves 10 megaflops, the algorithm requires hours of computation to test a single trajectory with the accuracy required to resolve important bifurcation structures for even this simple test problem. Clearly, algorithms with better performance on this problem are needed for practical use.

The example described above clearly demonstrates the need for algorithms based upon boundary value solvers to find bifurcations of periodic orbits. Some tools for this purpose are available, For example, the program AUTO [6] has been widely used for following families of periodic orbits using continuation methods coupled to collocation methods for the solution of boundary value problems. The program LOCBIF [16, 17] uses root finding algorithms applied to return maps for similar purposes. Further refinement of these techniques are warranted. In the collocation approach of AUTO, large systems of equations are solved, and the domain of convergence for root finding algorithms can be quite small. Thus establishing suitable initial conditions for the computations requires skill. On the other hand, shooting methods are subject to well known problems of ill-conditioning and numerical stability. Since the relative motion of trajectories in dynamical systems often involves exponential convergence or divergence, the return t map of a flow can appear to be nearly singular.

AUTO has also been used in the computation of homoclinic (and heteroclinic) orbits of dynamical systems [8]. This has been done in two ways: by following periodic orbit branches as their period diverges and by formulating two point boundary value problems whose solutions give approximations to homoclinic orbits [7]. These boundary value problems rely upon theoretical analysis by Beyn [3]. One searches for solutions of the differential equation that connect the linear unstable manifold of an equilibrium to its linear stable manifold in a specified time T. If T is large and the middle of the trajectory (with respect to time) is far from the equilibrium, then the trajectory can be expected to approximate a homoclinic orbit. Beyn estimates the asymptotic convergence properties of solutions to this boundary value problem as $T \to \infty$. Applying this strategy to the computation of homoclinic orbits is somewhat harder than finding periodic orbits. The local flow near an equilibrium point has "boundary layers" associated with different exponential rates of convergence to the equilibrium, and these are difficult to resolve with collocation methods based upon piecewise polynomial interpolation. Experimentation with example problems [8] suggests that further work at developing robust methods for computing homoclinic orbits is warranted.

The algorithms outlined above form a substrate for the computation of bifurcations in specific dynamical systems. One would like to achieve truly robust implementations of these algorithms and extend them to work with bifurcations in multiparameter families, including those of higher codimension. The extension to multiparameter families has been pursued mainly through the use of *continuation methods.* These algorithms compute curves of solutions for the zeros of regular maps $G : R^{m+1} \to R^m$ [1]. The continuation algorithms that have been used for bifurcation problems use the implicit function theorem to compute implicitly the tangent vector to the solution curve. This is incorporated into a marching algorithm that specifies a candidate point along the curve which is then refined through the application of a root finding algorithm. Gong beyond the calculation of one dimensional branches of codimension one bifurcations in two parameter families is currently a very active area of research on several fronts, but there are also a long list of obstacles that remain. In particular, we call attention to the following problems:

1. The theory of codimension two bifurcations of homoclinic and heteroclinic orbits is far from complete.

2. Multiparameter continuation methods need further development.

3. The defining equations and nondegeneracy conditions of higher codimension bifurcations involve large expressions in derivatives of a vector field. Accurate computation of these derivative is necessary for robust calculation of the bifurcations.

4. The unfoldings of higher codimension bifurcations often contain regions of dynamical behavior in which substantial qualitative changes occur in very small regions of a parameter space.

5. High precision calculations of bifurcations of periodic orbits remain a challenge.

While this list appears daunting, the computing power of modern workstations is adequate to support significant progress on these questions. The problems of synthesis and system identification for dynamical systems with several parameters require a much more powerful toolkit than those that have been created to date. Integrating the information learned from geometric studies of dynamical systems with sound numerical analysis is a path for progress in this area.

ACKNOWLEDGMENTS

This research was partially sponsored by the Department of Energy, National Science Foundation and the Office of Naval Research.

REFERENCES

[1] Allgower, E.L. and K. Georg, *Numerical Continuation Methods*, New York: Springer-Verlag, 1990.

[2] Back, A., J. Guckenheimer, M. Myers, F. Wicklin, and P. Worfolk., dstool: Computer assisted exploration of dynamical systems., *Notices Amer. Math. Soc.*, **39**,303-309(1992).

[3] Beyn, W. J., The numerical computation of connecting orbits in dynamical systems. *IMA J. Numer. Anal.*, **9**,169–181, 1990.

[4] Chencine, A., Bifurcations de points fixes elliptiques III. - Orbites périodiques de petites périodes et élimination résonnante des couples de courbes invariantes, *Publ.Math. I.H.E.S.* **66**,5-91(1988).

[5] Dangelmayr, G. and J. Guckenheimer, On a four parameter family of planar vector fields, *Arch. Rat. Mech. Anal.* **97**,321-352(1987)

[6] Doedel, E. *AUTO: Software for Continuation and Bifurcation Problems in Ordinary Differential Equations*, CIT Press: Pasadena, 1986.

[7] Doedel, E. and M. Friedman, Numerical computation of heteroclinic orbits, *J. Comput. and Appl. Math.*, **26**, 159-170, 1989.

[8] Doedel, E., M. Friedman and J. Guckenheimer, On computing connecting orbits: general algorithms and applications to the Sine-Gordon and Hodgkin-Huxley equations, *IEICE Trans. fundamentals*, **E77-A**, no. 11, 1994.

[9] Golub, G.H. and C.F. Van Loan, *Matrix Computations*, Second Edition, Maryland: The John Hopkins University Press, 1984.

[10] Govaerts, W. and Pryce, J.D., A singular value inequality for block matrices, *Lin. Alg. Appl.*, **125**, 141 - 148, 1989.

[11] Griewank, A. and G. Reddien, The Calculation of Hopf Points by a Direct Method, *IMA Journal of Numerical Analysis*, **3**, 295-303, 1983.

[12] Guckenheimer, J.G, Myers, M.R. and B. Sturmfels, Computing Hopf Bifurcations, *SIAM J. Num. Anal*, in press.

[13] Holodniok, M. and M. Kubiaček, New Algorithms for the Evaluation of Complex Bifurcation Points in Ordinary Differential Equations. A comparative Numerical Study, *Applied Mathematics and Computation*, **15**, 261-274, 1984.

[14] Guckenheimer, J. and P. J. Holmes, *Nonlinear Oscillations, Dynamical Systems, and Bifurcations of Vector Fields*, New York: Springer-Verlag,1983.

[15] Guckenheimer, J. and P. Worfolk, Dynamical Systems: Some Computational Problems, in *Bifurcations and Periodic Orbits of Vector Fields*, D. Schlomiuk, ed., Kluwer Academic Publishers, 1993, pp. 241-278.

[16] Khibnik, A.I., LINLBF: A Program for Continuation and Bifurcation Analysis of Equilibria Up to Codimension Three, in *Continuation and Bifurcation: Numerical Techniques and Applications*, D. Roose, ed., Kluwer Academic Publishers, 1990, pp. 283-296.

[17] Khibnik, A., Kuznetsov, Y., Levitin, V. and Nikolaev, E., *LOCBIF, version 2 : Interactive LOCal BIFurcation Analyzer*, Amsterdam: CAN Expertise Centre, 1992.

[18] Kubiček, M., Algorithm for Evaluation of Complex Bifurcation Points in Ordinary Differential Equations, *SIAM Journal of Applied Mathematics*, **38**, 103-107, 1984.

[19] S. Malo, *Rigorous Computer Verification of Planar Vector Field Structure*, Thesis, Cornell University, 1993.

[20] Moore, G., T.J. Garrett and A. Spence, The Numerical Detection of Hopf Bifurcation Points, in *Continuation and Bifurcation: Numerical Techniques and Applications*, D. Roose, ed., Kluwer Academic Publishers, pp. 227-246, 1990.

[21] Roose, D. and V. Hlavacek, A Direct Method for the Computation of Hopf Bifurcation Points, *SIAM Journal of Applied Mathematics*, **45**, 879-894, 1985.

[22] Stéphanos, C., Sur une extension du calcul des substitutions linéaires, *Journal de Mathématiques Pures et Appliqués*, **6**, 73-128, 1900.

Approximating Markov Chains: What and Why

Steve Pincus

990 Moose Hill Road, Guilford, CT. 06437

Abstract. Much of the current study of dynamical systems is focused on geometry (e.g., chaos and bifurcations) and ergodic theory. Yet dynamical systems were originally motivated by an attempt to "solve", or at least understand, a discrete-time analogue of differential equations. As such, numerical, analytical solution techniques for dynamical systems would seem desirable. We discuss an approach that provides such techniques, the approximation of dynamical systems by suitable finite state Markov Chains. Steady state distributions for these Markov Chains, a straightforward calculation, will converge to the true dynamical system steady state distribution, with appropriate limit theorems indicated. Thus (i) approximation by a computable, linear map holds the promise of vastly faster steady state solutions for nonlinear, multidimensional differential equations; (ii) the solution procedure is unaffected by the presence or absence of a probability density function for the *attractor*, entirely skirting singularity, fractal/multifractal, and renormalization considerations. The theoretical machinery underpinning this development also implies that under very general conditions, steady state measures are weakly continuous with control parameter evolution. This means that even though a system may change periodicity, or become chaotic in its limiting behavior, such statistical parameters as the mean, standard deviation, and tail probabilities change continuously, not abruptly with system evolution.

1. PERSPECTIVE: STEADY STATE CALCULATIONS AND STATISTICAL PROPERTIES

The aim here is to discuss the main ideas, and associated consequences, from the paper "Approximating Markov Chains" (1). The focus is to provide solution techniques for steady state behavior for dynamical systems and differential equations that avoid some crucial problems encountered in classical approaches. We illustrate this focus by two questions for a simply defined dynamical system, the logistic map $x_{n+1}=3.6x_n(1-x_n)$. **Question 1**: How would one compute the long-term time-average (and standard deviation) of this map? **Question 2**: How can one obtain correlation data for successive increments?

The classical means of obtaining steady state information for differential equations and dynamical systems are accompanied by at least two very different sets of difficulties. **First**, numerical procedures for steady state information proceed via the aggregation of transient information, where the form of the approximating system has been sufficiently restricted to assure that it shadows the true solution throughout time (Liapunov stability, ref. 2). The conditions to assure shadowing pose some restrictions (Lax-Richtmyer equivalence theorem, ref. 3), but much more computationally demanding is the need to acquire vast amounts of intermediate time information to estimate the steady state behavior. The straightforward response to Questions 1 and 2, time averaging, is subject to this difficulty. **Second**, analytic techniques for both transient and steady state solutions are generally nonexistent for nonlinear systems. Compounding analytic characterization difficulties, in general there is a lack of existence of a probability density function for the steady state measure, especially in discrete time. This singularity of the Radon-Nikodym derivative has spawned considerable interest in both a multifractal formalism (4-6) and renormalization techniques (7,8) as alternate means of analytic characterization of the limiting dynamics, but again, general explicit computational techniques from this perspective to date have been elusive. Furthermore, as a third difficulty, for chaotic maps, transient calculations in a classical sense are inappropriate, since some neighboring orbits ultimately diverge, given arbitrarily well-specified initial data.

Below, we circumvent these difficulties. After defining approximating Markov Chains to a dynamical system, we show that under general conditions, steady state (a.k.a., stationary or invariant) measures for each of two distinct versions of these chains converge weakly to true system steady state measures as the defining partition width goes to 0. For the purpose of obtaining steady state information, we are thus approximating deterministic nonlinear systems by probabilistic linear systems, which contrasts fundamentally with classical methods (9, 10). Classically, for a finite difference approximation, one solves for grid values at time $t+\Delta t$, *deterministically*, in terms of grid values at time t, Δt, the mesh dimensions, and nonlinear differential operators; in the present approach, grid values at time $t+\Delta t$ are *probabilistically specified* from the aforementioned data.

The point of the approach is severalfold: for the approximating discrete state space Markov Chains (i) calculations of the steady state measures are straightforward; (ii) these finite state calculations necessarily converge to the dynamical system steady state measure independent of any singularity of the latter; (iii) calculation of the steady state requires a single procedure, the solution of an eigenvector problem for the Markov Chain transition matrix- intermediate transient solutions are not required for this linear operator; (iv) via the Kolmogorov-Chapman equations, one also gets estimates of n-step steady state information, thus directly accounting for correlation in successive iterates or time-periods. In Computations 1 and 2 under *Convergence of Approximating Chains* (section 3.1), we illustrate these points, addressing Questions 1 and 2 by explicit arithmetic calculations.

At one level, such an approach may seem absurd, since the probabilistic make-up of the Markov chain insures eventual divergence of transient solutions to the true process transient. We comment on this a bit more below, in Section 5, *Stability Issues, Revisited*, where we motivate the sensibility of this approach by considering a simply answered, yet analogously posed question for 2 x 2 state Markov chains.

As well, formally, m^{th} order dynamical systems are m^{th} order Markov processes, *continuous* state space. The idea here is to approximate these systems by m^{th} order *discrete* state space Markov Chains, which are very well understood, with straightforward procedures to calculate stationary measures and rates of convergence to steady state (11). Of course, in attempting to provide straightforward, relatively general solution techniques, we sacrifice the beauty and elegance of the occasional, clever and insightful solution to a (nonlinear) problem, e.g., the exact characterization of the steady state measure for the dynamical system $x_{n+1} = 4\, x_n(1-x_n)$ via a sine-squared transformation (12), for the practicality of procedural generality.

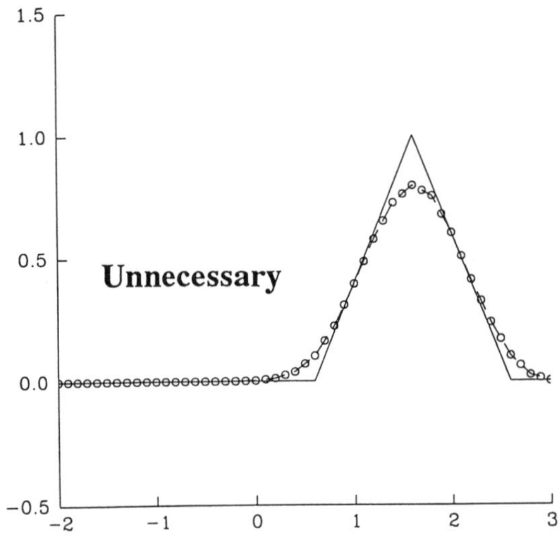

FIGURE 1. Schematic figure, indicating a primary philosophic difference in the Approximating Markov Chain approach compared to classical methods, insofar as obtaining steady state information. With Markov Chain approximations to dynamical systems, we do not require close shadowing of a true transient solution throughout time.

The focus on steady state behavior is partially motivated by applications. Many classical questions for systems described by differential equations, especially in design and performance, in effect ask about steady state or longterm dynamics. Representative questions, e.g., in aerodynamics, might concern moments of system variables (mean velocity); the percentage of time spent in prescribed domains (how often will pressure exceed 1.2 units ?); and measures of correlation between contiguous observations (will the pressure ever remain above 1.2 units for 10 minutes?) Mathematically, answers to these questions are given explicitly as "weak integrals" against the steady state measure, assuming ergodicity, which in practice almost always is satisfied. Hence we are interested not only that approximations to a true steady state measure converge, but moreover, that weak integrals against those approximating measures also converge to the true system value, insuring that straightforward statistical calculations for the "approximating" systems are indeed true estimates of actual system performance. Weak convergence is precisely the technical notion required to guarantee this behavior, and the theorems discussed below are weak convergence theorems.

We emphasize that we are not approximating transient information, e.g., where is the system at t=37.2, in the classical sense of this notion. Accurate transient approximations are often less important in many contexts, including those stated above, though quite obviously, in some contexts they are the primary questions of interest, e.g., what will the weather be in 5 days? It is in the determination of (the extent of) transient stability that parameters such as the largest Liapunov exponent (13,14) and the K-S entropy (15) address directly, and in such settings, a thorough understanding of these parameters remains paramount.

2. APPROXIMATING CHAINS: DESCRIPTION AND DEFINITIONS

The Markov chain approximations are the classical, finite (n) state objects, represented by an n x n matrix of transition probabilities. Think of them as collapsing the behavior of a continuous state dynamical system or stochastic process with values on some interval [A,B] onto some process approximation on the finite point set {A+r/2, A+3r/2, ... , B-r/2}. In effect, we encapsulate the local action of the map on intervals of width r into probabilistic information on a representative point (e.g., midpoint) from each interval. We focus on two distinct types of approximating chains. The <u>first type</u>, specified by Definition 2, is based on stationary conditional probabilities (time-averaging in the case of usual interest). This type is a *canonical* choice, and will give close approximation of the invariant distributions, but the cell transition probabilities must be numerically calculated from time-averages. For a deterministic map f, we call the <u>second type</u> a *transient* Markov chain; these are entirely distinct from the first type, with transition probabilities defined by the <u>fraction</u> of the Lebesgue measure <u>of the image</u> of the conditioning set that intersects a given interval (Definition 3). This approximation has one very appealing property: the transition probabilities can be calculated arithmetically from f, without any time-averaging, and hence with effectively 0 computational time expense. The trade-off is that steady state approximations from this transient approximating chain will be less accurate than from the first type, though still pretty good, as seen below (and convergent in the limit).

A motivating thrust here is that much general information is known about steady state computations for finite state Markov Chains, which are linear operators. Steady state calculations there are straightforward, given by a solution to an eigenvector problem; as well, rates of convergence to steady state are easily computable. The primary results proved in ref. 1 (stated by Theorems 3 and 4 below) establish that under general conditions, stationary measures for approximating chains and *transient* chains converge weakly to stationary measures for the given defining process as the mesh size converges to 0. We can thus estimate much about a dynamical system's behavior via straightforward Markov chain computations.

For deterministic and stochastic discrete time processes, support on some interval [A,B], we define *approximating (m,r) Markov Chains* in ref. 1. These chains are m^{th} order, state space {A+r/2, A+3r/2, ... , B-r/2}. Herein, we concentrate on the first-order case (m=1), to establish the primary ideas -- the higher order case definitions and theorems follow similarly. Recall

<u>Definition 1</u>. The Markov chain (or process) $\{X_n\}$ is of order m if the conditional probability $P\{X_n \in A_n \parallel X_k = a_k , k < n\}$ is independent of values a_k for $k < n-m$.

Notice that a dynamical system such as $x_{n+1}= f(x_n) =3.6x_n(1-x_n)$ is formally a first-order continuous state Markov process. We now approximate this by a first-order discrete state Markov chain. To maintain primary focus, we illustrate the basic case in Definition 2, assuming the steady state measure is a *physical* (Kolmogorov) stationary measure (ref.16, p.626). For a wide class of deterministic processes, this measure is unique, given by well-defined time-averages. In ref. 1, we indicate how to form the basic approximating Markov chains for other (non-physical) stationary measures, as well, e.g., those given by repelling fixed points. Therein, we also define approximating Markov chains in the far more general setting of stationary stochastic processes, for which the main weak convergence theorem also applies.

Definition 2. A. Take a discrete time, deterministic map f: [A,B] → [A,B].
B. Divide [A,B] into (B-A)/r cells; the ith cell C(i) = [x,x+r), x= A+(i-1)r.
C. Define mid(i) as the midpoint of C(i).
D. Define $p_{i,j}$ for all pairs (i,j) by limits of time averages:
$p_{i,j}$ = lim[# occurrences of $f^n \in C(j)$ & $f^{n-1} \in C(i)$] / [# occurrences of $f^{n-1} \in C(i)$].
When C(i) is a *transient* cell, conventionally define $p_{i,j} = 0$.
E. Define the approximating (1,r,A,B) Markov Chain Y_i as a 1^{rst} order chain, given by transition probabilities $p_{i,j}$ from state mid(i) to state mid(j).

Again to illustrate the main thrust, for the transient Markov chain definition, we focus here on the m=1 setting -- ref. 1 covers m^{th} order maps, which would correspond naturally to m^{th} order differential equations approximated in discrete time. We assume the minimal restriction on f that there exists no cells C on which f(C) is constant. This assures non-zero denominators, and hence well-defined $p_{i,j}$'s, in the following.

Definition 3. Transient (1,r,A,B) Markov Chain for a deterministic map f of order 1. This is defined on the same state space as for Definitions 2; the terminology and definitions (A-C) and (E) of Definition 2 apply. We form the conditional probabilities differently: **(D)** $p_{i,j} = \lambda(C(j) \cap f(C(i))) / \lambda(f(C(i)))$, where λ is Lebesgue measure on **R**.

Example 1. The map f(x)= 3.6x(1-x) on [0,1]. Transition probability matrices for both the (a) approximating (1,1/10) Markov chain, MAT, and the (b) transient (1,1/10) Markov chain, TMAT, for f(x) are shown below, with the $(i,j)^{th}$ entry corresponding to a transition from C(i) to C(j).

Here the interval [0,1] is divided in a coarse partition of 10 cells of width 0.1. We calculate the (1,1) entry $p_{1,1}$ of TMAT to indicate the arithmetic simplicity of this matrix generation. From Definition 3, $p_{1,1} = \lambda([0,.1] \cap f([0,.1])) / \lambda(f([0,.1]))$ $= \lambda([0,.1] \cap [0,.324]) / \lambda([0,.324]) = .1/.324 = .309$. In this calculation, λ is Lebesgue measure, thus interval length, and $f([0,.1]) = [f(0), f(.1)] = [0, 3.6(0.1)(1.0 - 0.1)] = [0, .324]$. The matrix TMAT is filled out in similar fashion, from which the steady state probabilities are computed via direct, classical techniques.

MAT

0	0	0	0	0	0	0	0	0	0
0	0	0	0	0	0	0	0	0	0
0	0	0	0	0	0	0	0	0	0
0	0	0	0	0	0	0	.223	.777	0
0	0	0	0	0	0	0	0	1.0	0
0	0	0	0	0	0	0	0	1.0	0
0	0	0	0	0	0	0	0	1.0	0
0	0	0	0	0	.864	.136	0	0	0
0	0	0	.480	.327	.193	0	0	0	0
0	0	0	0	0	0	0	0	0	0

TMAT

.309	.309	.309	.073	0	0	0	0	0	0
0	0	0	.302	.396	.302	0	0	0	0
0	0	0	0	0	.133	.556	.311	0	0
0	0	0	0	0	0	0	.407	.593	0
0	0	0	0	0	0	0	0	1.0	0
0	0	0	0	0	0	0	0	1.0	0
0	0	0	0	0	0	0	.407	.593	0
0	0	0	0	0	.133	.556	.311	0	0
0	0	0	.302	.396	.302	0	0	0	0
.309	.309	.309	.073	0	0	0	0	0	0

Stationary probabilities are (0 0 0 .217 .147 .129 .007 .048 .452 0) for MAT and (0 0 0 .123 .162 .138 .060 .108 .409 0) for TMAT, both for the point set $\{p_i\}_{i=1,10} = \{.05, .15, .25, \ldots, .95\}$. By Theorem 3, stationary probabilities of {mid(i)} for MAT agree with time-average probabilities for the {C(i)} given by iterations of f(x). To insure that all rows have probabilities that sum to 1 in MAT, we should delete cells from the state space with 0 stationary probability.

3. CONVERGENCE OF APPROXIMATING CHAINS

The (m,r) approximating chains can be used to "solve" deterministic and stochastic m^{th} order difference equations. Our orientation is computational; we are interested in moments of system variables, the percentage of time spent in prescribed domains, and measures of correlation between contiguous observations. Stationary measures provide this information, so they become the objects of study. In ref. 1, we show that under general conditions, stationary measures for the approximating (m,r) chains and the transient (m,r) chains converge weakly to stationary measures for a given m^{th} order process as $r \to 0$. We can thus estimate much about a dynamical system's behavior via straightforward approximating chain computations. Weak convergence results are most interesting in chaotic settings, where some neighboring orbits ultimately diverge. Transient information does not make sense for such systems, but we can still inquire about the quantities mentioned in the second sentence of this paragraph. Below, we discuss the theoretical underpinnings here. Now, we illustrate the focus of this approach via 2 computations.

3.1 Two Computations

Computation 1: Estimate the mean for $f(x) = 3.6x(1-x)$.

Solution: We utilize the transient (1, 1/10) Markov Chain TMAT indicated above, by solving for its steady state. This is a "routine" eigenvector problem, specified by $\pi\,TMAT = \pi$. As stated above, the solution vector: $\pi = \{\pi_i\}_{i=1,10} = [0, 0, 0, .123, .162, .138, .06, .108, .409, 0]$ for the point set $\{p_i\}_{i=1,10} = \{.05, .15,$

.25, ..., .95}. Hence the steady state mean = $\Sigma_{i=1,10}$ p_i π_i = .6595. The weak convergence theorem stated below insures that the true time-average mean for the nonlinear logistic map f given above ≈.6595. By numerical experiment with 10^6 points the true time-average mean for f(x) = 3.6x(1-x) ≈ .6461, about a 2% error from the estimate given by the computation using TMAT. Similarly, we could estimate other moments of f(x) from the {π_i} of TMAT, e.g, the variance, which would be approximately ($\Sigma_{i=1,10}$ p_i^2 π_i) - ($\Sigma_{i=1,10}$ p_i π_i)^2 Note once again that we have avoided some very difficult problems -- exact chacterization of the invariant measure of f (possibly via renormalization); the likely singularity (lack of a density function) of this measure; and especially, the need for "transient tracking" i.e., close approximation of a true solution by an approximation throughout time.

Computation 2: Correlation in 3 contiguous observations, f(x) = 3.6x(1-x).

Not only can we obtain a reasonable approximation of the true steady state distribution from the approximating chains, but we can ask for and realize more. Steady state probabilities do not tell the whole story, ignoring correlation between values at successive time points. We now indicate an analytical method for estimating e.g., π_3(Z), where Z:={$X_n \in A_1$& $X_{n+1} \in A_2$ & $X_{n+2} \in A_3$}, the steady state probability that three successive observations fall into prescribed sets for the map f(x) = 3.6x(1-x), for A_1=[0.8, 0.9], A_2=[0.4, 0.5], and A_3=[0.8, 0.9]. For Markov chains, such multiple time-step probabilities are calculated in explicit, straightforward fashion via the Chapman-Kolmogorov equations (sums over all intermediate possible paths). By computer experiment based on 10^6 points for f, π_3(Z) = .146. For the transient (1,.1) chain approximation TMAT, π_{TR}(Z) = Pr($X_{n+2} \in$ [.8,.9] || $X_{n+1} \in$ [.4,.5] & $X_n \in$ [.8,.9]) Pr($X_{n+1} \in$ [.4,.5] || $X_n \in$ [.8,.9]) π_{TR}($X_n \in$ [.8,.9]), where Pr and π_{TR} refer to the approximating chain. Since [.4,.5] and [.8,.9] are single atoms in the (1,.1) partition, π_{TR}(Z) = $p_{5,9}$ $p_{9,5}$ π_{TR}(9), with cell i corresponding to [.1(i-1), .1i). From TMAT and **π** above, we conclude that π_{TR}(Z) = (1.0) (.396) (.409) = .162. Note that π_{TR}(Z) is much closer to π_3(Z) than $\pi_1(A_1)\pi_1(A_2)\pi_1(A_3)$ = (.452)(.147)(.452) =.030 is, the latter the product of exact steady state probabilities, which would equal π_3(Z) if successive observations were uncorrelated. So not only do we have a good steady state estimate, we also realize only about 10% error in calculating a 3-in-row probability, where the only computational cost (very minimal) is the steady state solution for the 10 x 10 transient Markov chain. Theorem 6, ref. 1 establishes the convergence of this estimate to the true value as the mesh → 0. To analytically approximate the measure of an n-time unit event, (i) calculate all $p_{ivect,\ j}$ for a transient chain approximation TR for the given system (ii) compute all π_{ivect} by raising TR to a high power; (iii) derive all n-time unit probabilities via the Chapman-Kolmogorov equations.

3.2 Theoretical Underpinnings

The machinery to establish the applications indicated above falls under the aegis of weak convergence theorems. We recall this concept (17), herein applied to steady state measures. For the present discussion, we presume that our space, denoted by A, is a compact subset of **R**. In ref. 1, we handle the more general setting of $\mathbf{R}^m$ with no loss of generality. ν_n and ν will refer to probability measures (total probability measure 1).

Definition 4: Measures ν_n converges weakly to ν on A if for all continuous, bounded f, $\int_A f(x)\, d\nu_n(x) \to \int_A f(x)\, d\nu(x)$.

Weak convergence generally implies steady state convergence on all subintervals B (choose $f(x) = 1_B$; one needs a generally valid assumption of no atoms, i.e., points of positive stationary mass for ν). Furthermore, on a compact space, weak convergence implies convergence of all moments (choose $f(x) = x^n$), including the mean, standard deviation, and many other statistical parameters. Given the "n-in-row" weak convergence theorem (Theorem 6), convergence of many autocorrelation statistical parameters directly follows, including the autocorrelation function, and calculations such as Computation 2 above. However, it is imperative to note that weak convergence does not *a priori* imply "finer" modes of convergence, e.g., the Liapunov spectra or K-S entropy.

We wish to compare stochastic processes, particularly Markov processes, to one another and to deterministic maps, and so consider a space of transition probabilities on A, Tr(A).

Definition 5: Define $t \in Tr(A)$, t = {probability measures t_a on A for all $a \in A$, such that for all Borel-measurable subsets B of A, the map $a \to t(a,B)$ is a Borel-measurable function}.

For a deterministic map $f: A \to A$, $(t_f)_a$ will be the point mass $\delta_{f(a)}$ for all a.

Definition 6: Action of Tr(A). For $t \in Tr(A)$, and μ, a measure on A, define $t * \mu$ as a measure on A given by $\int f(z)\, d(t * \mu)(z) = \int [\int f(y) t(z,dy)]\, d\mu(z)$, for all Borel-measurable functions f.

Definition 7. A probability measure μ on A is stationary for $t \in Tr(A)$ if $t * \mu = \mu$.

Definition 7 agrees with the standard notion for deterministic f and for Markov Chains. By Lemma 1.2, ref. 14, there exists at least one such stationary measure for $t \in Tr(A)$. Below, **we do not presume absolute continuity** of stationary measures with respect to Lebesgue measure; typically these measures are singular. For $\{t_n\}, t \in Tr(A)$, we say that t_n converges to t ($t_n \to t$) if whenever μ_n converges to μ weakly on A, then $t_n\, \mu_n$ converges to $t\, \mu$ weakly on A.

The precise statement to insure that we can estimate a stationary measure for a dynamical system by finding the stationary measure for a ("regular") approximating chain, with small r is:

Theorem 3. Given $f: [A,B] \to [A,B]$, select a stationary measure μ for f, and define a (1,r) approximating chain A_r on {mid(i)} given by Definition 2. Suppose there exists r_0 such that for all $r < r_0$, A_r is irreducible, when restricted to those cells with positive μ measure. Then ν_r, the unique stationary measure for A_r, converges weakly to μ as $r \to 0$.

This theorem [1] requires minimal technical machinery for its proof. It has numerous extensions, to m^{th} order Markov chains, compact subsets of R^n, stationary stochastic processes, non-uniform partitions, and interestingly, to the non-physical or unstable stationary measures of f as well (this last extension does not apply in the transient approximating chain setting). In practice, the irreducibility assumption in the theorem statement is very generally satisfied. Irreducibility also follows from the geometric notion of topological transitivity.

The theorem of greater interest, the weak convergence theorem for transient approximating chains that justifies Computations 1 and 2 above, requires that a reasonable *physical measure* exists. For many dynamical systems, including irreducible Axiom-A systems (18), unique physical measures exist (19), and agree with SRB measures(20). If no physical measure exists, there is no ergodic behavior (16), a terrible state of affairs (21). Fortunately, in both computer experiments and the physical world, a small, "uncertain" perturbation is a component in system evolution. The next, primary result verifies the convergence of transient approximating chains to the true process on Tr(A). Steady state measure convergence to the physical measure of f follows, assuming the existence of such a measure.

Theorem 4. Given $f:[A,B] \to [A,B]$ continuous, and A_r a transient Markov chain for f, define $t_r \in Tr([A,B])$, recalling the notation of Definition 3: for $a \in [A,B]$, choose the unique i with $a \in C(i)$. Define $(t_r)_a$ to agree with $p_{i,j}$ for A_r, i.e., $(t_r)_a$ has atomic probability distribution $\Sigma_j \, p_{i,j} \, \delta_{mid(j)}$. Then t_r converges to t_f on Tr[A,B] as $r \to 0$.

Fig. 2 indicates that even coarse mesh approximating and transient chains can produce good estimates for the physical measure of a dynamical system. Stationary probability distribution functions are shown for the physical measure for the map $x_{n+1} = 3.6 \, x_n \, (1 - x_n)$, for (1,.1) and (1,.04) approximating Markov Chains, and for the (1,.1) transient Markov Chain. The (1,.04) approximating chain produces a very accurate stationary measure approximation; also observe that the (1,.1) approximating Markov Chain produces a more accurate estimate of the true process stationary measure than does the (1,.1) transient chain, as expected.

Theorem 4 is proved as an operator (weak-*) convergence result, which is generally stronger than weak (stationary measure) convergence, and immediately implies stationary measure convergence when the limit operator has a unique stationary measure. The proof requires in effect that 2 integral ($\int$) differences be driven to 0, which in turn follows from the continuity of f and the compactness of [A,B].

---

1) In the three theorems that are precisely stated here, we retain the theorem numbering from ref. 1, to maintain a consistency of reference. We have omitted all proofs from the current exposition, hopefully to facilitate a more easily read manuscript. Again, see ref. 1 for technical details.

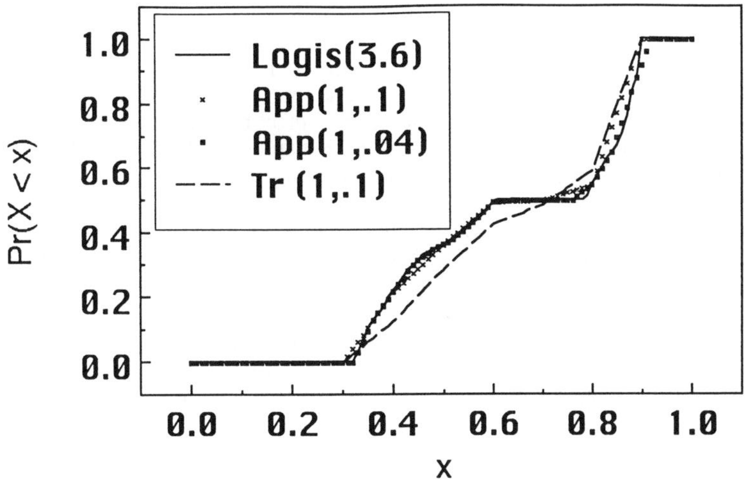

FIGURE 2. Stationary probability distribution functions for (i) physical measure for logistic map, $x_{n+1} = 3.6\, x_n\, (1 - x_n)$ [Logis (3.6)] ; (ii) (1,.1) and (iii) (1,.04) approximating Markov Chains [App(1, .1) and App(1, .04), respectively] ; (iv) (1,.1) transient Markov Chain [Tr (1, .1)] .

As for Theorem 3, there are numerous generalizations (e.g., Theorems 5 and 6, ref. 1). Similar steady state weak convergence results for transient approximating chains follow, with virtually identical proofs, for arbitrary compact subsets of R^m; arbitrary partitions, so long as sup mesh width → 0; and k^{th} order Markov processes on R^m (e.g., Henon map). This last generalization justifies applications of approximating Markov Chains to discrete time solutions of multidimensional ODE's.

In particular, the n-in-row, contiguous observation weak-convergence theorem (see Theorem 6, ref. 1 for a precise statement) that justifies Computation 2 above also follows as a direct generalization of Theorem 4, via a straightforward recursive argument.

4. TOWARDS DIFFERENTIAL EQUATIONS

This paper discusses discrete time results, for dynamical systems; an important extension of this approach is in application to the continuous-time setting for (systems of partial) differential equations (DE's). The same conceptual approach would be followed: (i) approximate the DE's by a discrete-time analogue; (ii) estimate steady state behavior for the discrete-time analogue by discrete-time, discrete state approximating Markov Chains. The added technical machinery to insure the overall validity of this approach, i.e., step (i), given the results herein, is in effect already completed by Kushner and Dupuis (22), though their problem orientation is very different -- they are developing techniques to solve optimal and adaptive stochastic control problems. The formalism in ref. 22 establishes conditions to insure weak convergence of steady state solutions for the general PDE setting, of classes of discrete-time and state approximating Markov Chains to the continuous-time and state defining problem. Kushner and Dupuis basically require a notion of "local consistency", which both types of approximating chains discussed in ref. 1 satisfy.

Probably the most important practical issue here is the possibility of dramatic computational time reduction in steady state solutions to PDE's, in utilizing an approximating Markov Chain approach in lieu of classical methods. As stated above, to obtain steady state data for PDE's via the classical approach, one determines intermediate solutions at each of times t_0, $t_0 + \Delta t$, $t_0 + 2\Delta t, \ldots,$ then aggregates these data from a suitably long time-interval. Using finite difference methods, e.g., at each time-step, one solves for the state variables of the system at specified grid points at time $t_0 + (n+1)\Delta t$ in terms of grid values in the recent past, e.g., $t_0 + n\Delta t_0$, $t_0 + (n-1)\Delta t$, typically via one of a variety of large, sparse matrix methods (23). The computational expense of this approach is often forbidding, especially for 3-dimensional problems. Currently, solutions to 3-dimensional viscous PDE's (physical reality) are often derived via partial solutions to 2-dimensional reduced problems, followed by some degree of system overlay. Unfortunately, there are well-known major problems with such an approach, at times yielding qualitatively incorrect "solutions", thus affecting numerous critical applications of such PDE models to e.g., aerodynamic and hydrodynamic design. In contrast, via the approximating Markov Chain approach, one still must solve a large, sparse matrix linear algebra (here an eigenvector) problem, but only once. Thus the potential speed-up attained by approximating Markov Chains may allow both qualitative and quantitative gains. Fine grids (meshes) would still be required for fine accuracy, but first order steady state approximations should be computable from coarse mesh chains, which as above, can produce quite respectable approximations, at relatively minimal computational cost. A substantial amount of existing machinery in numerical computer science should be directly applicable to the eigenvector problem here, via (i) linear algebra solution methods for Markov Chains (24, 25), and (ii) clever choices for sequences of grids, e.g., non-uniform partitions to exploit large gradients, adaptive meshes, and alternating fine/coarse (multigrid) schemes to save computational time (26, 27).

Furthermore, the local consistency required to insure steady state convergence via approximating Markov Chains is not so strict as classical stability conditions that insure Liapunov (transient) stability. Thus currently discarded unstable finite difference schemes (9), which are often faster and simpler to compute than their stable counterparts, may be shown to be valid for estimating steady state information for PDE's. Additionally, the capability to explicitly estimate the steady state distribution of state variables will allow one to evaluate assumptions of distributional form of these distributions that are often made in practice. For instance, given the mean and standard deviation of a velocity flow-field, one sometimes sees an associated assumption that the steady state velocity distribution is Gaussian, with the specified moments. For a general Markov chain, the steady state mass distribution is typically asymmetric, hence not an approximation of a Gaussian distribution.

5. STABILITY ISSUES, REVISITED

Classically, for the numerical solution of differential equations, we have been trained to tie an approximate solution to the true solution throughout time, that otherwise, we could say nothing about solutions to the equations. In approximating herein by a probabilistic object, we purposefully violate this approach. The question arises: is there a simple example to illustrate the sensibility of "guaranteed" close steady state measures, despite certain transient divergence? I like the following example to this end. Compare two 2x2 Markov Chains, that could e.g., represent the probability of (no) rain tomorrow given (no) rain today.

$$A = \begin{matrix} .5 & .5 \\ .33 & .67 \end{matrix} \qquad \text{vs.} \qquad B = \begin{matrix} .5 & .5 \\ .34 & .66 \end{matrix}$$

We compute steady state probabilities for $\begin{matrix} .5 & .5 \\ x & 1\text{-}x \end{matrix}$ to be (a 1-a), with a = x/.5+x.

Therefore, A and B have close stationary measures: A has steady state probability of (rain of) .33/.83, B has steady state probability of (rain of) .34/.84. This is assured independent of coupling or transient information; there will certainly be days for which it would rain in system A, not so in B, thus providing a "divergent" transient calculation.

The technical point is that to insure valid approximation of steady state behavior, one only needs a measure of <u>uniformity</u> in approximating local behavior; this guarantees the steady (global) result. Lemma 1 and Theorem 2, ref. 1 provide the technical machinery; they are, in effect, weak-* topology (28) convergence arguments performed in explicit fashion. More generally, the uniform continuity of continuous maps on compact spaces assures significant extension of the results in ref. 1 to approximations of compact operators on Banach spaces (29). Epistemologically, here we are approximating nonlinear dynamical systems by linear systems, to the level of steady state convergence.

Weak convergence, the natural measure of closeness of measures, thus provides a distinct notion of process stability. Classically, we consider three types of stability: Laplace (the solution remains bounded), Liapunov (usual, time-dependent closeness of the perturbed system to the true dynamics), and Poincare (orbital or tubular, close trajectories as geometric objects) (2). Here, in considering the stability of steady state solutions, we allow transient divergence forbidden by Liapunov's notion. As well, the analyst's weak convergence of measures (here) gives a different view of closeness of steady state solution than does the topologist's (Poincare's) view. Poincare's notion questions whether the supports of steady state measures are close, as <u>geometric</u> objects, without concern of the closeness of the probabilistic weights of the respective measures. To illustrate this point simply, consider 2 probability (e.g., steady state) measures that have values on the two points {0,1}. Measure # 1 has probability .9 at {0} and probability .1 at {1}. Measure # 2 has probability .1 at {0} and probability .9 at {1}. These measures have identical support {0,1}, and thus are identical from Poincare's perspective, yet are far apart from the analyst's viewpoint (of weak-distance). Weak continuity of steady state measures with system perturbation would thus be this notion of stability, not so restrictive as Liapunov's, yet more stringent than Poincare's, since the steady state measures must be close analytically, which implies geometric (Poincare) closeness, but does not follow from it.

Thematically, this type of stability is consistent with Ornstein's d-bar approximations of stationary "B-processes" by Markov processes (30, 31), Geman's result that hidden Markov models are dense among essentially all finite-state discrete-time stationary processes (32), and Kurtz' (finite-horizon) approximations of ODE's by pure-jump Markov processes (33,34). However, each of these three settings employ technically distinct means of measuring closeness of processes, and more pointedly, each works with canonical choices of Markov approximations, which would correspond here to the (regular) approximating Markov Chains. This stability is also consistent with real analytic evolution of the largest Liapunov exponent as a function of transition probabilities for random positive matrices (35), and with the stability of the Liapunov exponents for classes of perturbed dynamical systems (36).

6. SEQUENTIAL STATIONARY DATA APPROXIMABLE by MARKOV CHAINS

The n-in-row weak convergence Theorem 6 also bears directly on some claims of chaos derived from time-series analysis. While chaotic models for time-series are sensible when underlying system behavior is derived from differential equations, we can rarely, if ever infer chaos or determinism solely from data considerations. A finite value of the correlation dimension, an algorithm frequently applied to time-series, does not necessarily imply determinism or chaos (37). Furthermore, process determinism can not be established from (either one- or multi-dimensional) steady state information. This is especially germane to claims associated with *phase space plots* (e.g., Fig. 3). Such plots are graphs of a fixed number (n) of contiguous observations taken according to some prescribed sampling procedure; usually n=2 or 3 and the sampling is at constant time increments, thus the plots are of the form $x(t+\Delta t)$ vs. $x(t)$ or $x(t+2\Delta t)$ vs. $x(t+\Delta t)$ vs. $x(t)$. Phase space plots often display sequential correlation, suggesting that sequential independence is highly unlikely.

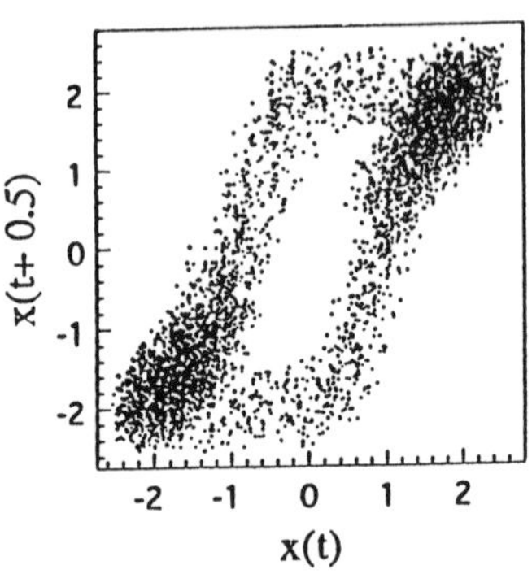

FIGURE 3. Representative phase space plot of x(t+0.5) vs. x(t) from the stochastic "perturbed van der Pol model", ref. 38.

Occasionally, such plots are accompanied by a discussion indicating that since there is no evidence of periodicity, the underlying data are chaotic. In effect, the labelling of bounded, aperiodic, yet correlated output as deterministic chaos is a false cognate.There is a technical point, here, that must be considered -- phase space plots display n-dimensional steady state output, and thus Theorem 6 applies to the n-dimensional data which define such plots. The n-in-row weak convergence implies that any n-dimensional steady state measure arising from a deterministic dynamical system model can be approximated to arbitrary accuracy by that from a stochastic Markov Chain. Thus it is invalid to infer deterministic chaos from n-dimensional steady state information, displayed numerically or graphically.

The correlation seen in such phase-space plots is typically real, as is geometric change reflecting a shift in ensemble process autocorrelation in some comparisons. However, these observations are entirely distinct from any claims regarding underlying model form (chaos vs. stochastic process). Similarly, in power spectra, decreasing power with increasing frequency (oftentimes labelled 1/f decay) is also a property of process correlation, rather than underlying determinism or chaos (38).[2]

One would expect that a transient test could be developed to establish determinism; such a test would necessarily require analysis of arbitrarily long strings of data. Furthermore, it is sensible that it is much easier to reject an i.i.d. or independence assumption with high probability (given finite data input) than it is to reject the presence of any (possibly correlated) stochastic behavior.

7. ORTHOGONALITY of GEOMETRY and ANALYSIS

One characteristic of chaotic models that has generated a great deal of interest is qualitative, geometric change in the limit set, i.e. support of the physical measure with evolving control parameter. For example, consider the limit set of the logistic map $f_r(x) = rx(1-x)$ on [0,1]. As r increases, the limit set changes from 1 point (steady state) to periodic (2 point set) to 4 point set, ..., until ultimately the support is infinite: the system is aperiodic. The Feigenbaum theorem (41) assures not only a period-doubling of the limit set until aperiodicity is achieved, it provides a constant (4.669...) for the limiting rate of doubling prior to chaos that is the same for a very large class of systems. In particular, from the topological and geometric viewpoint, the steady state evolves in a discontinuous manner with control parameter r.

However, one generally computes statistics, or more generally, estimates analysis parameters, from serial data: the mean, standard deviation, and proportion of time that $X > 1.2$ all are statistical parameters of a process, and can be written as integrals with respect to the invariant measure, as discussed above. We ask: in settings where there is geometric discontinuity in the point set with changing control parameter, is there corresponding statistical or analytic discontinuity? For example, in the logistic map near r=3.0, where the system goes from steady state to period 2, is the mean discontinuous? What about other weak integrals?

---

2) As well, a *strange attractor* or *fractal* steady state structure does not necessarily imply a deterministic model, see e.g., ref. 39 for fractal behavior arising from stochastic random matrices. Furthermore, "nice" steady state, e.g., support the entire space, can coincide with "pathologic" analysis (40), namely, the absence of a probability density function (Radon-Nikodym derivative).

We prove (Theorem 7) that under very general conditions, all moments of dynamical systems, indeed all weak integrals are continuous with control parameter evolution. Thus when the topologist sees structural change and instability with control parameter, through bifurcations and into chaotic regions, the analyst sees no discontinuities.

Precisely, we prove a weak continuity theorem for stationary measures of perturbed dynamical systems. First, for continuous $f:A\to A$, define $f^\varepsilon \in Tr(A)$, a uniform perturbation of magnitude $\leq \varepsilon$ (we can choose a different perturbation as well, so long as we are consistent in our choice for all f_r below). Pick $a \in \mathbf{R}^m$; define the measure $f^\varepsilon{}_a$ by the density function $\varphi_a(z)= K_a$ for $z \in A$ such that $\|z-f(a)\| < \varepsilon$, $\varphi_a(z)= 0$ otherwise, with $\|.\|$ the Euclidean metric on $\mathbf{R}^m$, and $K_a = 1/$(the m-dimensional volume of ($A \cap$ the ε-ball around a)).

Theorem 7. Assume a family of dynamical systems is given by continuous f_x: $A\to A$, where f_s converges uniformly on A to f_r. Choose ε, and assume that $f_r{}^\varepsilon$ has unique stationary measure ν_r. For each s, choose a stationary measure ν_s for $f_s{}^\varepsilon$; then ν_s converges weakly to ν_r.

Most familiar dynamical systems satisfy the uniform convergence assumption (e.g. Henon, logistic maps), hence computed perturbations of these systems generally fall under the scope of Theorem 7. Since these calculations are performed by computer, they yield statistics for a slightly perturbed version of the logistic map, which fits the context of Theorem 7.

To illustrate this result, we contrast (**A**) the continuity of the time-averaged mean and standard deviation as a function of the control parameter for the logistic map $x_{n+1} = rx_n (1- x_n)$ [Fig. 4] with (**B**) the bifurcation diagram, which displays the changing geometry of the steady state measure.

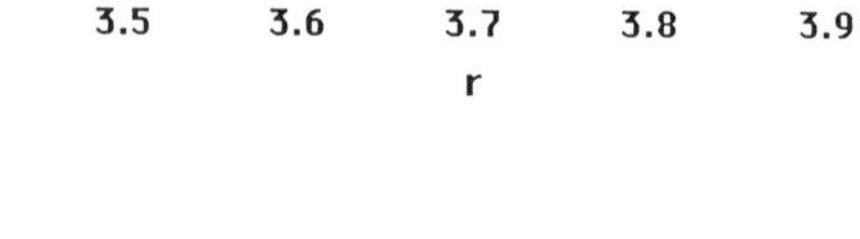

Bifurcation diagram of logistic map for $2.9 < a < 3.9$

(**A**) (**B**)

FIGURE 4. Contrast between continuity of statistical parameters mean and standard deviation **(A)** with discontinuity of steady state geometry **(B)**, as a function of a control parameter "r", for logistic map $x_{n+1} = rx_n (1- x_n)$.

Although explicit arithmetic verification of this theorem is in general difficult, in one very illuminating setting, such a computation is straightforward. We now establish weak continuity (hence all moment continuity) at a bifurcation point for the map $x_{n+1}= f(x_n)= rx_n(1-x_n)$ near r=3.0, where the system changes from a single limit point to a period 2 limit (Fig. 5). The pictorial intuition can be gleaned from the bifurcation diagram: weak continuity follows from the connectivity of the graph of the physical measure in the multiply periodic domain.

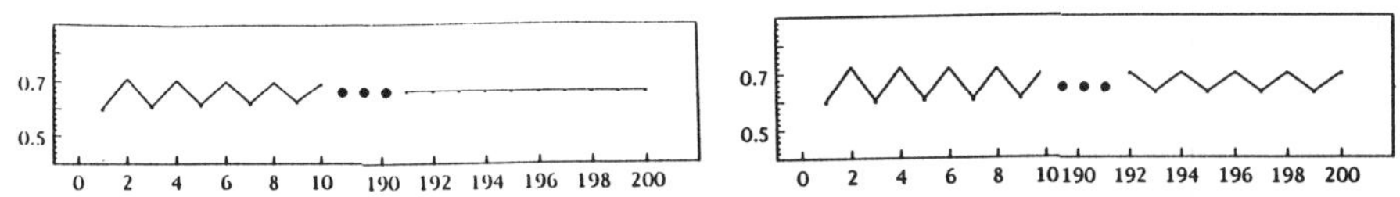

$x_{n+1}= 2.96\ x_n(1-x_n)$ $\qquad$ $x_{n+1} = 3.01\ x_n(1-x_n)$

FIGURE 5. Comparison of steady state dynamics near first bifurcation point for logistic map, $f(x_n)= rx_n(1-x_n)$ near r=3.0. For r=2.96, (*physical*) steady state is a single limit point, whereas for r=3.01, (*physical*) steady state is a period 2 oscillation. Yet these two steady state measures are close weakly.

For $r < 3.0$, $s(r) = 1-1/r$ is an attracting fixed point with physical measure $\nu_{phys} = \delta_{s(r)}$. Thus for a function g, for $r < 3.0$, $\int g(x)\, d\nu_{phys} = g(s(r))$. For $3.0<r<3.2$, $s(r) = 1-1/r$ is a repelling fixed point. The other fixed points of f^2, $t(r)$ and $u(r) = [(1+1/r)+/- \sqrt{(1-2/r - 3/r^2)}]/2$ are attracting, and thus the physical measure $= 1/2\ (\delta_{t(r)}+ \delta_{u(r)})$. Hence for $3.0<r<3.2$, $\int g(x)\, d\nu_{phys} = 1/2[g(t(r)) + g(u(r))]$ for a function g. Since t(r) and u(r) both converge to $s(r) = 2/3$ at r=3.0, weak convergence follows: $\lim_{r \to 3} \int g(x)\, d\nu_{phys} = g(2/3)$ for any continuous g.

This weak continuity should not be considered as a foregone conclusion. Parameters of dynamical system behavior that ascertain more delicate information are in general badly discontinuous with control parameter evolution, e.g. K-S entropy and largest Liapunov exponent. Fig. 12, Eckmann and Ruelle (16) illustrates such marked discontinuity for the logistic map.

Geometric changes in attractors appear to be manifested in the differentiability of a weak integral as a function of the control parameter. Thus, e.g., the mean for the logistic map is piecewise smooth in the multiply periodic domain, and non-differentiable at bifurcations and at returns to periodicity from chaos, most blatantly realized near 3.828, where the logistic map changes from chaotic to periodic, period 3 (Fig. 6). That period 3 gives a most extreme behavior is not at all surprising -- this is consistent with Sarkovskii's Theorem (42, 43), which provides an ordering of which periods imply other periodicities, with period 3 maximal -- it implies periods of all orders.

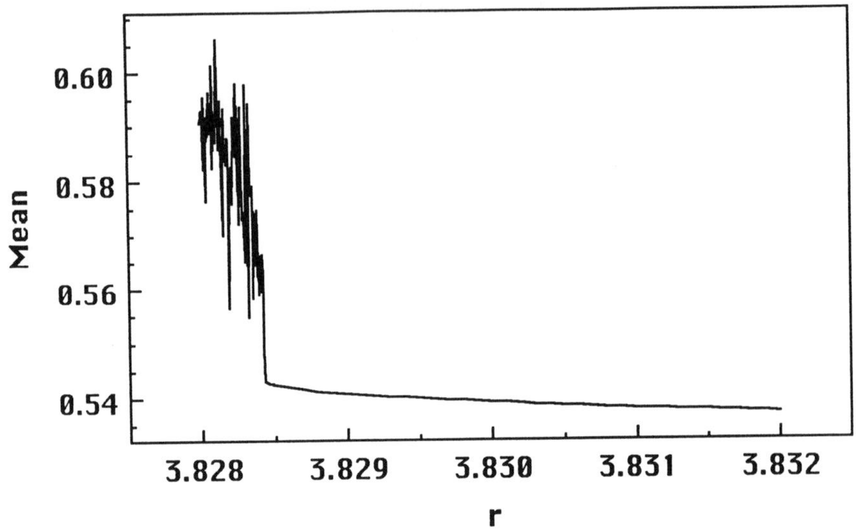

FIGURE 6. Close-up of the mean vs. control parameter evolution for logistic map $f(x_n) = rx_n(1-x_n)$ near period-three return, $r \approx 3.828$.

8. FUTURE DIRECTION

Three complementary directions appear to be natural next steps to the results discussed above. First is implementation towards 2-dimensional and 3-dimensional PDE's, following the discussion of Section 4, *Towards Differential Equations.* One would expect to work closely with numerical computer scientists here, to efficiently solve the large, sparse matrix eigenvector problem. A classical system, the (multidimensional) Burgers' equation $u_t + uu_x - \varepsilon u_{xx} = 0$ (44), seems a natural test problem for this method. This paradigm for modelling nonlinear and viscous effects has been extensively studied, both numerically and analytically (Hopf transformed the problem to a linear diffusion equation (45), affording exact analytical solutions).

A second direction is the establishment of rates of convergence of the error in estimating true dynamical system statistical parameters via approximating Markov Chains, as a function of the mesh width r.

A third, more theoretical direction is determining conditions that insure unique physical measures for dynamical systems and differential equations. Both the transient approximating chain steady state convergence, and the weak continuity with control parameter (Section 7) results are valid presuming that such a measure exists, but while this existence of a natural or preferred measure appears to be ubiquitous, a general theory to determine when a unique physical measure exists is very incomplete. Eckmann and Ruelle (16) note that Kolmogorov never formally proposed a definition of such a measure, indicating a strong sense of some of the difficulties in doing so, although this topic is clearly of significant interest to both Kolmogorov and his numerous disciples. Probably the most complete general characterizations to guarantee such measures are by Kifer (46); once one goes beyond initial classes of results, conditions are often quite technical. A spiritually related book that also indicates some of the technical difficulties in this area is that of Freidlin and Wentzell (47), which studies effects of large deviations type for diffusion perturbations of systems with usually relatively simply dynamics. However, in ref. 47, the outcome of the studies generally depends on perturbations (diffusion coefficients), so it does not precisely bear on the question of stability of dynamical systems. As a final thought to further complicate theoretical progress here, existence and uniqueness invariant measure questions are very different in discrete versus continuous time. Notably, in discrete time, fixed points of maps of all orders generate a vast number of invariant measures, but there is no direct analogue in continuous time. To develop firm technology to approximate (continuous time- and state-) PDE's, we will ultimately need to address the existence of physical measures in this latter setting. Regrettably, a paucity of general results exists in this context, with the vast body of current work on physical measures taking place in the discrete time setting.

REFERENCES

1. Pincus, S.M., *Proc. Natl. Acad. Sci. USA.* **89**, 4432-4436 (1992).

2. Struble, R.A., *Nonlinear Differential Equations*, New York: McGraw-Hill, 1962, ch. 5, pp. 129-169.

3. Strikwerda, J.C., *Finite Difference Schemes and Partial Differential Equations*, Belmont, CA: Wadsworth & Brooks, 1989, pp. 26, 222-226.

4. Tel, T., *Zeit. Naturforsch.* **43A**, 1154-1174 (1988).

5. Falconer, K., *Fractal Geometry: Mathematical Foundations and Applications*, Chichester, England: Wiley, 1990.

6. Brown, G., Michon, G., and Peyriere, J., *J. Stat. Phys.* **66**, 775-790 (1992).

7. Collet, P., and Eckmann, J.-P., *Iterated Maps of the Interval as Dynamical Systems*, Boston: Birkhauser, 1980.

8. Bedford, T., and Fisher, A.M., *Proc. London Math. Soc.* **64**, 95-124 (1992).

9. Lapidus, L., and Pinder, G.F. *Numerical Solution of Partial Differential Equations in Science and Engineering*, New York: Wiley, 1982.

10. Bender, C.M., and Orszag, S.A. *Advanced Mathematical Methods for Scientists and Engineers*, New York: McGraw-Hill, 1978, pp. 61-318.

11. Parzen, E., *Stochastic Processes*, San Francisco: Holden-Day, 1964, ch. 6, pp. 188-275.

12. Lauwerier, H.A., "One-dimensional iterative maps," in *Chaos*, ed. A.V. Holden, Princeton: Princeton University Press, 1986, pp. 52-56.

13. Oseledec, V.I., *Trans. Moscow Math. Soc.* **19**, 197-231 (1968).

14. Furstenberg, H., *Trans. Amer. Math. Soc.* **108**, 377-428 (1963).

15. Kolmogorov, A.N., *Dokl. Akad. Nauk. SSSR.* **119**, 861-864 (1958).

16. Eckmann, J.P., and Ruelle, D., *Rev. Mod. Phys.* **57**, 617-656 (1985).

17. Billingsley, P., *Convergence of Probability Measures*, New York: Wiley, 1968.

18. Smale, S., *Bull. Am. Math. Soc.* **73**, 747-817 (1967).
19. Kifer, Y.I., *Math. USSR Izv.* **8**, 1083-1107 (1974).
20. Young, L.S., *Trans. Amer. Math. Soc.* **318**, 525-543 (1990).
21. Newhouse, S., *Publ. Math. IHES* **50**, 102-151 (1979).
22. Kushner, H.J., and P.G. Dupuis, P.G., *Numerical Methods for Stochastic Control Problems in Continuous Time*, New York: Springer-Verlag, 1992, ch. 4-5, pp. 67-133.
23. Smith, B.T. *et. al.*, *Matrix Eigensystem Routines -- EISPACK Guide, 2nd* ed., vol. 6 of Lecture Notes in Computer Science, New York: Springer-Verlag, 1976.
24. Philippe, B., Saad, Y.,Stewart, W.J., *Numerical methods in Markov chain modeling*, Publication interne no. **495**, INRIA (1989).
25. Meyer, C.D., Plemmons, R.J., *Linear Algebra, Markov Chains, and Queueing Models*, New York: Springer-Verlag, 1993.
26. McCormack, S.F., *Multigrid Methods*, Philadelphia: SIAM, 1987.
27. McCormack, S.F., *Multilevel Adaptive Methods for Partial Differential Equations*, Philadelphia: SIAM, 1989.
28. Reed, M., and Simon, B., *Methods of Modern Mathematical Physics, Vol 1: Functional Analysis* , San Diego: Academic Press, 1980, pp. 111-114.
29. Rudin, W., *Functional Analysis*, New York: McGraw-Hill, 1973, pp. 97-105.
30. Ornstein, D., *Ergodic Theory, Randomness, and Dynamical Systems*, New Haven: Yale University Press, 1974.
31. Ornstein, D.S., and Weiss, B., *Ann. Prob.* **18**, 905-930 (1990).
32. Kunsch, H., Geman, S., and Kehagias, A., *Hidden Markov Random Fields*, preprint, 1994.
33. Kurtz, T.G., *J. Appl. Prob.* **7**, 49-58 (1970).
34. Kurtz, T.G., *J. Appl. Prob.* **8**, 344-356 (1971).
35. Peres, Y., *Ann. Inst. Henri Poincare* **28**, 131-148 (1992).
36. Ledrappier, F., and Young, L.S., *Ergod. Theor. Dyn. Syst.* **11**, 469-484 (1991).
37. Pincus, S.M., *Proc. Natl. Acad. Sci. USA* **88**, 2297-2301 (1991).
38. Pincus, S.M., *Math. Biosci.* **122**, 161-181 (1994).
39. Pincus, S.M., *Ann. Prob.* **11**, 931-938 (1983).
40. Pincus, S.M., *J Theor. Prob.* **7**,199-208 (1994).
41. Feigenbaum, M.J., *J. Stat. Phys.* **19**, 25-52 (1978).
42. Devaney, R.L., *An Introduction to Chaotic Dynamical Systems*, Menlo Park: Adddison-Wesley, 1985, pp. 60-67.
43. Li., T.-Y., and Yorke, J.A., *Am. Math. Monthly* **82**, 985-992 (1975).
44. Burgers, J. M., *Adv. Appl. Mech.* **1**,171-199 (1948).
45. Hopf, E., *Comm. Pure Appl. Math.* **3**, 201-230 (1950).
46. Kifer, Y.I., *Random Perturbations of Dynamical Systems*, Boston: Birkhauser, 1988.
47. Freidlin, M.I., and Wentzell, A.D., *Random Perturbations of Dynamical Systems*, New York: Springer-Verlag, 1984.

Measures of Complexity in Signal Analysis

J. Kurths[†], U. Schwarz[†], A. Witt[†], R. Th. Krampe[‡] and M. Abel[†]

[†]*Arbeitsgruppe Nichtlineare Dynamik der Max-Planck-Gesellschaft an der Universität Potsdam*, Am Neuen Palais, D-14415 Potsdam, PSF 601553, Germany*
[‡]*Institut für Psychologie, Universität Potsdam, Am Neuen Palais, D-14415 Potsdam, PSF 601553, Germany*

Observational data of natural systems, as measured in astrophysical, geophysical or physiological experiments are typically quite different from those obtained in laboratories. Due to the peculiarities with these data, well-known characteristics processes, such as periodicities or fractal dimension, often do not provide a suitable description. To study such data, we present here the use of measures of complexity, which are mainly basing on symbolic dynamics. We distinguish two types of such quantities: traditional measures (e.g. algorithmic complexity) which are measures of randomness and alternative measures (e.g. ϵ-complexity) which relate highest complexity to some critical points. It is important to note that there is no optimum measure of complexity. Its choice should depend on the context. Mostly, a combination of some such quantities is appropriate. Applying this concept to three examples in astrophysics, cardiology and cognitive psychology, we show that it can be helpful also in cases where other tools of data analysis fail.

I. INTRODUCTION

The study of natural systems is different from those in laboratories. In contrast to laboratory experiments, it is generally not possible to influence the medium under consideration or to repeat such observations under the same conditions. This often leads to rather short and noisy observational data which are sometimes irregularly sampled or show a non-stationary behavior. Due to these difficulties, well-known tools of linear data analysis, such as correlation and spectral analysis as well as popular methods from nonlinear dynamics, especially, estimates of fractal dimensions or Lyapunov exponents, have to be taken with very much caution.

To study such data, we have, therefore, to look for other approaches. In the case of inhomogeneous behavior, wavelets, recurrence plots or structure

*E-Mail: JKurths@agnld.Uni-Potsdam.de, WWW-Page: http://www.agnld.uni-potsdam.de/

functions are appropriate tools (Farge *et al.* (6), Koebbe and Mayer-Kress (13)). They enable to find, on the one hand, long-range correlations inherent the underlying dynamics, as was shown e.g. for heart-rate-variability (Peng *et al.* (20)) or for solar and stellar activity (Kurths *et al.* (16)). But they are, on the other hand, also helpful to detect some coherent structures which rather seldom occur, e.g. large eddies in turbulent motion (Arneodo *et al.* (1)) or grand minima in long-term solar variability (Kurths *et al.* (17)).

Here, we report another approach which is based on some concepts of complexity (Sect. 2). Presently a bunch of different definitions of complexity and corresponding measures of complexity exists. They have their origins in quite different fields, as coding theory or nonlinear dynamics. It is important to note that the most popular of these measures, as Shannon entropy or algorithmic complexity, are measures of randomness, i.e. they assign highest complexity for white noise-like behavior, where past and future are uncorrelated. Such a characterization is, however, not sufficient for many systems, especially in nonlinear dynamics. We, therefore, present another kind: non-traditional or alternative measures of complexity which relate highest complexity at phase transitions, e.g. the onset of chaos. The properties of both kinds of measures are compared by applying them to rather simple nonlinear systems. An important point is to discuss some artifacts which may occur if these measures are estimated from rather short experimental data. In Section 3 it is demonstrated how this concept can help in the analysis of special systems. We present examples from cardiology (sudden cardic death), cognitive psychology (synchronization of movements), and astrophysics (fine structures in solar flares). The results are summarized in Section 4.

II. COMPLEXITY MEASURES AND SYMBOLIC DYNAMICS

There are quite different approaches toward a definition of complexity. Some of them are related to well-known characteristics in nonlinear dynamics, such as approximative entropy (*ApEn*) to Kolmogorov-Sinai entropy. Most of them are, however, based on symbolic dynamics. Applying such complexity measures, a symbolic representation instead of the original trajectories of the system has to be considered. Before presenting some special measures of complexity, we, therefore, firstly discuss how to transform data into symbolic strings.

A. Transformation into Symbolic Strings

We have developed and used different techniques of symbolic dynamics for data analysis.

In a first step, the observational record has to be transformed into a symbolic string. In doing so one loses a certain amount of detailed information, but some of the invariant, robust properties of the dynamics are retained.

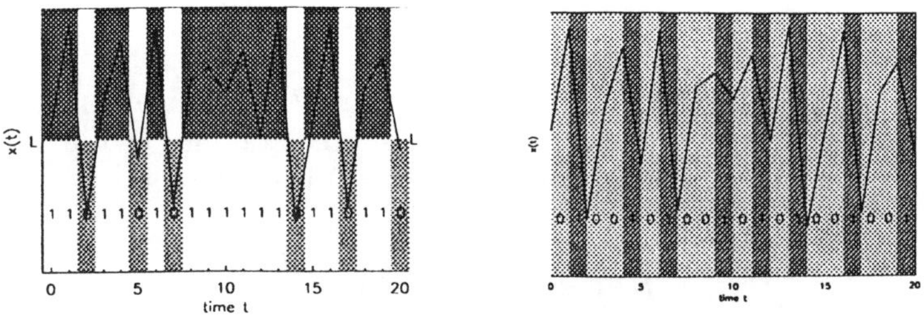

FIG. 1. Scheme of transformation of time series into symbolic string consisting of two different symbols (1 - dark, 0 - bright). Left: Static transformation – the 1D phase space is separated by the given threshold L into two cells. Right: Dynamical transformation – a step-to-step increase is denoted by a '0', a decrease by '1'.

In the best case, such a transformation generates a Markov partition. However, in most examples of natural systems we know neither the existence of such a partition nor their construction. Therefore, more pragmatic transformations have to be used which may be not Markovian ones. For simplicity we explain them for the case, in which only two symbols, say 0 and 1, are used. We distinguish

i) static transformations, where the phase space is separated into cells (denoted by symbols). These symbols are related to the system states x_i in dependence on a certain level X as follows (cf. Fig. 1 left)

$$s_i = \begin{cases} 1 & \text{if } x_i > X \\ 0 & \text{otherwise ,} \end{cases} \tag{1}$$

ii) dynamical transformations, where the step-to-step-differences in the sequence are taken into account. They consider the kind of difference between two adjacent measurement values (cf. Fig. 1 right)

$$s_i = \begin{cases} 0 & \text{if } x_i > x_{i-1} \\ 1 & \text{otherwise} \end{cases} \tag{2}$$

and

iii) a mixing of both kinds of transformation.

It is no difficulty to refine these symbolic descriptions by including more symbols (i.e. more levels resp. triplets etc. instead of pairs). The choice of the kind of transformation is of particular importance. In case of rather small data records, as it typically occur in many applications, a transformation into only few symbols is to recommend. In the following we use only two symbols. This manner, the transformation is context-dependent. If possible, such a coarse-graining should, hence, be based on some physical motivation. Otherwise, we highly recommend to compare different transformations into symbols. Such symbol sequences can be characterized by complexity measures, as discussed in the next paragraph.

B. Mutual Information

The mutual information $I(k)$ (cf. Farmer (7)) is a generalization of the correlation function. It measures the state predictability or the memory of a symbol sequence. In case of two symbols $I(k)$ is defined by

$$I(k) = \sum_{i,j=0}^{1} p_{ij}(k) \operatorname{ld} \frac{p_{ij}(k)}{p_i p_j} \tag{3}$$

It quantifies the average dependence of two symbols over k time steps, where p_i is the probability of the symbol $s_i \in \{0,1\}$ of a symbol string, and $p_{ij}(k)$ is the joint probability that the symbol s_i and, k steps later, the symbol s_j occurs. In particular, peaks in $I(k)$ exhibit the levels of memory quantitatively. This description is especially appropriate for nonlinear systems because the joint probabilities $p_{ij}(k)$ reflect more general dependences within a symbolic string than the autocorrelation function.

In the case of white noise, mutual information $I(k)$ vanishes for all $k \neq 0$, reflecting that white noise is a process without any memory. Periodic processes are characterized by peaks at the multiples of the period in k. Chaotic regimes as well as colored noise yield a decrease of mutual information $I(k)$ with growing k. Hence, mutual information gives an insight into the nature of the underlying system which is similar but more general than the correlation function.

C. Traditional Complexity Measures

Among the bunch of complexity measures, the Shannon entropy and the algorithmic complexity are most popular and often used. It is important to note that both are maximum in case of purely stochastic behavior (white noise), i.e. they are **measures of randomness**. We call this type traditional measures of complexity.

1. Shannon Entropy

The symbol sequence S is regarded as a set of realizations of a discrete random variable whose outputs are elements of the alphabet A. This leads to the Shannon information, a classical parameter of data analysis. The Shannon information of k-th order H_k is defined by:

$$H_k = - \sum_{s^k \in A^k, p(s^k)>0} p(s^k) \operatorname{ld} p(s^k) \tag{4}$$

s^k is a word of length k. It is easily checked that

- $H_k \geq 0$, and $H_k = 0$ only for constant symbol sequences.

• In case of periodic symbol sequences S with a prime period $m, m < k$ the Shannon information takes the value $H_k = \text{ld } m$.
• The maximum of H_k is taken if the probabilities of all words $s^k \in A^k$ are equal, then $H_k = k$.

This rather abstract quantity H_k is simply to interpret: it measures the average number of bits needed to specify an arbitrary word of length k in the sequence S. It has been proven that for symbol sequences generated by Markov processes, the algorithmic complexity agrees with Shannon entropy (Chaitin (3)). From our experiences we find that these both quantities more generally agree (Wackerbauer *et al.* (31)). Renyi (Balatoni & Renyi (2)) introduced a generalization of Shannon's approach

$$H_k^{(q)} = (1-q)^{-1} \text{ ld} \left(\sum_{s^k \in A^k} p(s^k)^q \right) \tag{5}$$

where q is a real number and $q \neq 1$. This measure converges to Shannon entropy as $q \to 1$. Due to different averaging it additionally reflects some inhomogeneity of the underlying probability distribution. If $q > 1$ those words of length k with large probability dominantly influence the Renyi information. For $0 < q < 1$ we have the reverse behavior.

2. Algorithmic Complexity

This complexity measure of a string is defined as the number of bits of the shortest computer program for a universal computer to regenerate this string (Kolmogorov (14)). To make this definition operational, Lempel and Ziv ((32)) used a procedure working with only two operations: to copy and to insert. Kaspar and Schuster ((11)) suggested an algorithm that calculates a normalized measure $AC(N)$ of this kind of complexity of a string, where N is the string length.

In the limit $N \to \infty$ this complexity takes the maximum for white noise and is normalized to $\max\{AC\} = 1$. It is a minimum ($= 0$) for a constant or simple, periodic functions. Estimation of algorithmic complexity works without any use of substring probabilities. Therefore, this measure includes also long words, i.e. this measure also takes into account rather long range effects.

3. Approximative Entropy

The approximative entropy ($ApEn$) is based on the Kolmogorov–Sinai entropy H_{KS}, defined as

$$H_{KS} = -\sup_{\mathcal{P}} \lim_{w \to 0} w^{-1} \sum_{w,j} \mu_{w,j} \text{ ld } \mu_{w,j} \tag{6}$$

where $\mathcal{P}$ is a partition, w denotes the "grain" of the used partition $\mathcal{P}$ and j is the index of the components in $\mathcal{P}$. μ is an invariant measure and $\mu_{w,j} = \mu(\mathcal{P}_{w,j})$. Because H_{KS} is invariant under isomorphic transformations, it is one of the most important quantities in ergodic theory (Ott (19)). Under some assumptions, H_{KS} can be estimated from a time series of length N (Grassberger and Procaccia (8)) as

$$H_{KS} = \frac{1}{N-m+1} \lim_{\epsilon \to 0} \lim_{m \to \infty} \lim_{N \to \infty} \sum_{i=1}^{N-m+1} \mathrm{ld} \frac{c_{m,\epsilon,i,\tau}}{c_{m+1,\epsilon,i,\tau}} \tag{7}$$

Here $c_{m,\epsilon,i,\tau}$ represents the conditional probability needed to calculate H_{KS}. To calculate them, we use the embedding procedure (Takens (28))

$$\vec{x}_j = (x_j, x_{j+\tau}, ..., x_{j+m\tau}) \tag{8}$$

where τ is the delay time and m the embedding dimension. From these embedded vectors $\vec{x}_j$, $c_{m,\epsilon,i,\tau}$ is determined:

$$c_{m,\epsilon,i,\tau} = \sum_{j=1}^{N-m+1} \theta(\| \vec{x}_i - \vec{x}_j \| - \epsilon), \tag{9}$$

where θ is the Heavyside function and ϵ represents the diameter of the phase space partition (grain). Without taking the limit process in Eq. (7), one has a dependency on the three parameters ϵ, m and N. This leads to the definition of the family of *ApEn* **statistics** (Pincus (21)):

$$ApEn(m, \epsilon, N, \tau) = \frac{1}{N-m+1} \sum_{i=1}^{N-m+1} \mathrm{ld} \frac{c_{m,\epsilon,N,\tau}}{c_{m+1,\epsilon,N,\tau}} \tag{10}$$

We have extended this *ApEn*–concept to include as additional parameter the delay time τ. This has shown useful, e.g. in the case of the noise–free logistic map (see Sect. II E 1).

There is a transitivity relation between *ApEn* and H_{KS} in the case of noise–free processes (Pincus *et al.* (22)). The *ApEn* is proper for the comparison of two states of one system which correspond to different parameter values. In this sense *ApEn* is a *relative* measure, which has the following basic properties:

• $ApEn \geq 0$

• *ApEn* is *not* necessarily zero for periodic cases, although in most cases it is. With appropriate chosen parameters it is minimized for periodic processes

• *ApEn* takes it maximum value for white noise (cf. Sect. II E 1). *ApEn* is finite. Applied to stochastic processes it can thus provide information to distinguish different stochastic processes (although H_{KS} tends to infinity).

D. Alternative Complexity Measures

Next, we describe two kinds of alternative complexity measures. In contrast to the traditional measures of complexity, they are not maximum for white noise.

One of the most interesting complexity measures is the **renormalized entropy**, originally introduced by Klimontovich (12) in thermodynamics. It takes into account that the energy of an open system changes with its control parameter, which makes a direct comparison of Shannon entropies impossible. The main idea is that the Shannon entropy for different system states is normalized to a fixed value of **mean effective energy**. This approach, losely speaking, renormalizes the entropy obtained from a time series $x(t)$ of a certain system state in such a manner that the mean effective energy coincides with that of a reference state $x_r(t)$.

Starting from these two time series, we can easily estimate the corresponding probability distributions $f(x)$ and $f_r(x)$. By using formal arguments from thermodynamics the effective energy is defined as:

$$h_{eff}(x) = -\log f_r(x) \tag{11}$$

The renormalization of f_r into $\tilde{f}_r$ is constructed such that the mean effective energies $< h_{eff} >$ of f and $\tilde{f}_r$ are equal. To make this idea operational, we first represent the distribution in terms of the canonical Gibbs distribution

$$\tilde{f}_r(x) = e^{\frac{\Phi(T_{eff})-h_{eff}(x)}{T_{eff}}} \tag{12}$$

which can be rewritten as

$$\tilde{f}_r(x) \quad = \quad C(T_{eff}) \cdot e^{-\frac{h_{eff}(x)}{T_{eff}}}, \tag{13}$$

where T_{eff} and $\Phi(T_{eff})$ are the effective temperature resp. the free effective energy. Because h_{eff} can be calculated from Eq. (11), there are two unknowns in Eq. (13): $C(T_{eff})$ and T_{eff}. They are determined from the following two conditions.

a) Normalization:

$$\int \tilde{f}_r(x)\; dx = 1$$

b) Equality of mean effective energy:

$$\int h_{eff}(x)\tilde{f}_r(x)\; dx = \int h_{eff}(x) f(x)\; dx.$$

Hence, $\tilde{f}_r$ fulfills the properties wanted. Consequently, we can compare the Shannon entropies of f and $\tilde{f}_r$

$$H = -\int f(x)\log f(x)\; dx \qquad \text{and} \qquad \tilde{H}_r = -\int \tilde{f}_r(x)\log \tilde{f}_r(x)\; dx \tag{14}$$

For that the renormalized entropy difference

$$\Delta\tilde{H} = H - \tilde{H}_r \tag{15}$$

is introduced. It is important to note that $\Delta\tilde{H}$ is a relative measure that depends on the reference state chosen.

If this reference state is chosen suitable, then $\Delta\tilde{H}$ relates smaller values to periodic and random behavior than to chaotic dynamics (Saparin *et al.* (23)).

2. ϵ-Complexity

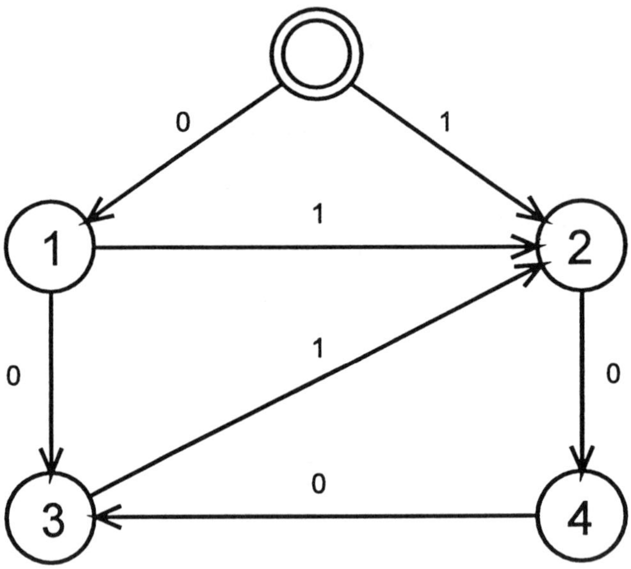

FIG. 2. Automaton for a period-3-sequence.

FIG. 3. Automaton for white noise.

The idea to use automata for a definition of complexity goes back to Kolmogorov (14) and led to the concept of algorithmic complexity. It is based on a deterministic automaton. Crutchfield and Young (5) generalized this approach and suggested to apply stochastic automata, which they call ϵ-machines. The determination of the complexity C_ϵ from a given sequence of an ϵ-machine can be divided into four main steps.

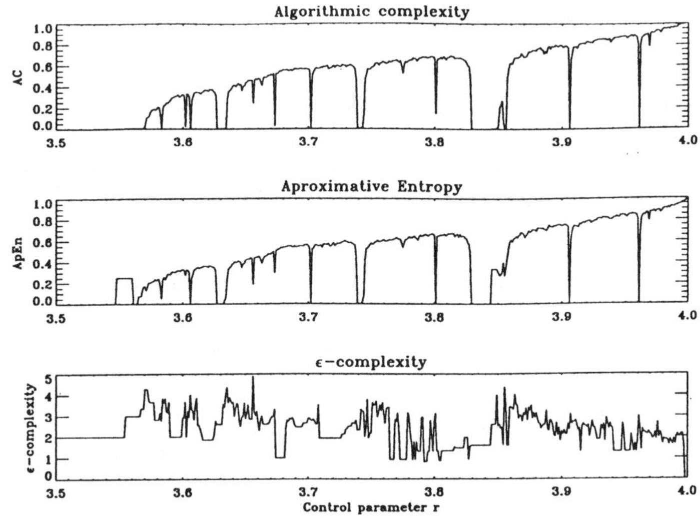

FIG. 4. Top: Algorithmic complexity of logistic map as an example of traditional complexity measures calculated for the symbol sequences of length $N = 10^5$. Middle: *ApEn* for $N = 5000, \tau = 1, \epsilon = 0.1\sigma, m = 2$. Bottom: ϵ-complexity of logistic map for $l_2 = 16$, $l_1 = 8$.

1. Construction of a tree: A binary tree $T = (V, E, a)$ of length l_1, with a finite set V of vertices, a set $E \subset V \times V$ of edges, and an origin $a \in V$ of T, is assigned to a given symbol sequence. All conditional probabilities $p_e = p_{v_1 \mapsto v_2}$ are denoted on the edges $e = (v_1, v_2), \quad v_1, v_2 \in V$, of T.

2. Search for equivalent subtrees: From this tree T of length l_1 all subtrees of length $l_2, (l_2 < l_1)$ are considered. Within the set of subtrees an equivalence relation (ϵ–similarity) is defined such that any two subtrees are equivalent, if the difference of the probabilities assigned to their edges e_1, e_2 is smaller than ϵ, $|p_{e_1} - p_{e_2}| < \epsilon$.

3. Construction of random automaton: The condition $|p_{e_1} - p_{e_2}| < \epsilon$ generates a classification in the set of subtrees. Each equivalence class is regarded as a state of the automaton, and each edge between the origins of two subtrees represents an edge between two states of the automaton. The transition probabilities between different states in this automaton are determined such that the original sequence is reconstructed in an ϵ-similar manner. Increasing l_1 and l_2 results in an increasing resolution of the dynamics.

4. ϵ-complexity: The ϵ-*complexity* C_ϵ is defined as the Shannon information of the state probabilities of the automaton.

The ϵ-complexity C_ϵ has the following basic properties:

- In case of periodic behavior with period $\wp$: $C_\epsilon = \log \wp$, if $\wp \leq l_2$. The corresponding automaton for the period-3-sequence $S =$ 00100100100100.... consists of a closed loop composed of 3 automaton states, thus providing a complexity $C_\epsilon = \log 3$ (see Fig. 2).
- In case of stochastic behavior the corresponding automaton has only one state (Fig. 3). Therefore, the complexity vanishes: $C_\epsilon = 0$.

E. Comparison of Measures of Complexity

1. Application to the Logistic Map

To understand how these different measures of complexity characterize various kinds of dynamical features, we apply them to the simple logistic map

$$x_{i+1} = rx_i(1 - x_i) \tag{16}$$

with $x_i \in [0, 1]$ and $r \in [0, 4]$. It exhibits a rich variety of dynamics depending on the control parameter r. We, especially, focus several critical points of Eq. (16), e.g. accumulation and band merging points, or interior and exterior crisis (Wackerbauer *et al.* (31)).

To analyze the logistic map we have chosen the following transformation into symbolic strings:

$$s_i = \begin{cases} 1 & \text{if } x_i > 0.5 \\ 0 & \text{otherwise} \end{cases} \tag{17}$$

The general behavior is as expected (Fig. 4): The envelope increases with r for the traditional measures (e.g. algorithmic complexity or approximative entropy), but goes to zero as $r \to 4$ for the ϵ-complexity, an alternative measure. This last measure relates highest complexity to accumulation and band merging points. We find that all these critical points are marked by different measures of complexity (Wackerbauer *et al.* (31)). It is, however, important to note that the renormalized entropy is the only one which indicates all these critical points in the logistic map (Saparin *et al.* (23)).

In case of the ϵ-machine we find a behavior far from that of Shannon entropy, algorithmic complexity or approximative entropy. Minima are considered for constant ($r < 3$) or pure stochastic symbol sequences (Fig. 4). There the automata have only one state. Surprisingly simple automata are also given in the case of band merging points. Therefore, the complexity yields local minima there. Maximum values occur for the accumulation points of the period doubling cascade asymptotically with growing l_1 and l_2.

Applying the renormalized entropy $\Delta\tilde{H}$ to the logistic map and using $r = 4$ as reference state we have found: The accumulation point of period doubling takes the minimum value of complexity. At the band-merging points as well as at inner crisis $\Delta\tilde{H}$ increases rapidly.

With respect to challenges from observations of natural systems, our special field of interest is to study the influence of the **finite observational length** N on the reliability of the estimated complexity measures. So, we have derived error estimates and have developed suitable concepts, how in this special situation the notion of complexity measures could be applied.

i) Mutual information

A remarkable advantage of $I(k)$ is that for each k only 4 probabilities (n^2 for n different symbols) have to be calculated. For short sequences, artifacts may occur in the calculation of $I(k)$. Therefore, we propose a method to judge the statistical significance of peaks in $I(k)$ by a statistical, semi-empiric method, called randomization (e.g. Chorafas (4)). From Monte Carlo simulation of white noise we have calculated a distribution of the estimated mutual information $I_{\text{noise}}^{\text{est}}(k)$. Chosen the 95% level of significance, a correspondent threshold g is given by the probability

$$P\left(\max_{0<k\leq k_{max}} \{I_{\text{noise}}^{\text{est}}(k)\} < g\right) = 0.95, \tag{18}$$

where $I_{\text{noise}}^{\text{est}}(k)$ is the estimated mutual information of white noise. From many simulations we found that g behaves reciprocal to the sequence length N (Fig. 6):

$$g(N, k_{\max}) \approx \frac{3.16}{N} k_{\max}^{\frac{1}{4}}. \tag{19}$$

In all the calculations of mutual information we have taken into account Eq. (19) to suppress the short series effects. It is important to note that this test is especially addressed to judge peaks in $I(k)$, but for checking a decay of $I(k)$ other tests are necessary.

ii) Shannon entropy and algorithmic complexity

The probabilities $p(s^k)$ to calculate H_k (Eq. 4) are in practical applications usually approximated by their relative frequencies. These estimates give rise to serious problems due to two facts:

A) The number of different substrings s^k grows exponentially with increasing length k. Using a rather short time series of length, say $N = 1024$, each word of length $k = 10$ has the mean frequency $2^k/N = 1$. Therefore, we can only analyze a short-range memory to get reliable results.

B) The entropies from such relative frequencies lead to systematic underestimates of the theoretical entropies, of course in dependence on the data length N. To reduce this effect, some corrections have been proposed (Grassberger (9), Schmitt *et al.* (25)), among them

$$\hat{H}_k^{(corr)} = \hat{H}_k^{(fr)} + \frac{L(k)}{2\ln 2N} \tag{20}$$

($\hat{H}_k^{(fr)}$ – estimate from relative frequencies, $L(k)$ – number of different words of length k).

An essential advantage of the algorithmic complexity is that it gives reliable results already for rather short data sets. If the process is periodic (period length ≤ 20) or a low-order Markovian one, then a time series length of $N = 500$ can be sufficient.

iii) Approximative entropy

The main problem is to determine appropriate values for the parameters ϵ, m, N and τ, which are generally system dependent. We have found that results are best if ϵ varies between 0.1σ and 0.2σ (σ = standard–deviation of the underlying time series). For the embedding dimension m, it turns out that already low dimensions produce satisfying results. Already rather short data records of length of about 1000 data points can produce satisfying results in distinguishing systems clearly (Fig. 5). The idea in using *ApEn* is to compare two systems with different parameter values. Thus, we look for the best combination of ϵ, m, N and τ to produce maximal differences between different parameter values. To be systematic one can follow the way to use $m = 2$,

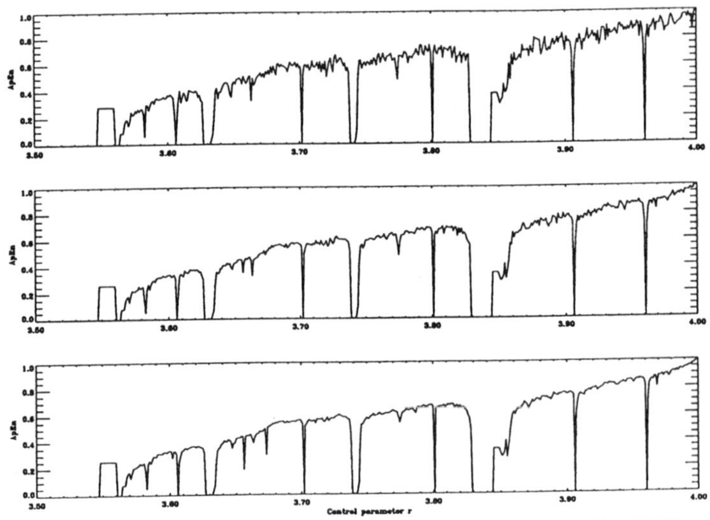

FIG. 5. *ApEn* for different data lengths, for N = 300, 1000, 3000 and 5000, $\tau = 1, \epsilon = 0.1\sigma, m = 2$.

N fixed (mostly because of experimental conditions, which imply often small data sets) and determine $\epsilon = \epsilon(m, N)$ such that the "resolution" of the differences in this measure is best. As is well known from correlation dimension, ϵ must be larger than any noise–level in the experimental data. Furthermore the time series should be comparable in the range of their distribution, otherwise one just compares essentially the σ's of the time series.

For the logistic map, the best choice for any data length is $\epsilon = 0.1\sigma, m = 2, \tau = 2$. Especially setting $\tau = 2$ removes numerical artifacts in the periodic windows, which otherwise can easily be misinterpreted as states of higher complexity. Here, the general ability of $ApEn$ to provide larger numbers for more developed chaos is not valid because of the special distribution of the time series. On the other hand, $ApEn$ produces the expected curve for chaotic processes well. Another way to get rid of these effects is a transformation of the distribution of the values to an equal distribution using a rank list to preserve the dynamics of the system (Pompe (24)). Going to higher values of m has not improved the ability to distinguish between different states of the system. With the above parameters, $ApEn$ distinguishes well between periodic and chaotic cases of motion, but the transitions are somewhat crude, especially at the end of the period–2 region (using $\tau = 2$). Summing up one can say, that $ApEn$ may be sensitive for changes in the parameters, especially ϵ. In the case of periodicity, one has to be careful of systematic errors (see Sect. II E 1)

iv) Renormalized entropy

For calculating the renormalized entropy a data set length of about 1000 point is sufficient to distinguish simple periodic from different types of complex behavior (Saparin *et al.* (23)). Artifacts may occur if both the reference state and the considered state have the same probability density. Then, the complexity $\Delta\tilde{H}$ vanishes.

We recommend to chose that reference state, which is the most complex one, i.e. the renormalized entropy $\Delta\tilde{H}$ is negative for all other states.

v) ϵ-complexity

The computation of the ϵ-complexity calls for huge sequence lengths, which we have chosen as larger than $1000 \times 2^{l_1 AC}$, where AC denotes the algorithmic complexity. To determine optimum values of ϵ, l_1 and l_2 for a given symbol sequence, we have recently proposed a robust procedure (31). Especially, we have found for the logistic map with $r = 4$ and $l_1 = 16$ that the minimum sequence length is about 50,000, in case of $r = 3.7$ about 500,000. Unfortunately, this is seldom fulfilled in observational data. Note that artifacts can occur if the values l_1 and l_2 are not large enough.

III. APPLICATIONS

We have applied the concept of complexity measures in quite different areas ranging from astrophysics to cognitive psychology (Hempelman and Kurths (10), Kurths and Schwarz (16), Kurths *et al.* (18), Saparin *et al.* (23), Schwarz *et al.* (26), Witt *et al.* (30)). Here, three examples are presented.

A. Cardiology: Detecting High-Risk-Patients for the Sudden Cardiac Death

Every year several 100,000 persons die due to the sudden cardiac death that is caused by cardiac fibrillation. Without any warning it even can occur by subjects who are up to this time apparently healthy or medically inconspicuous. By use of conventional methods only 30% of all affected persons are

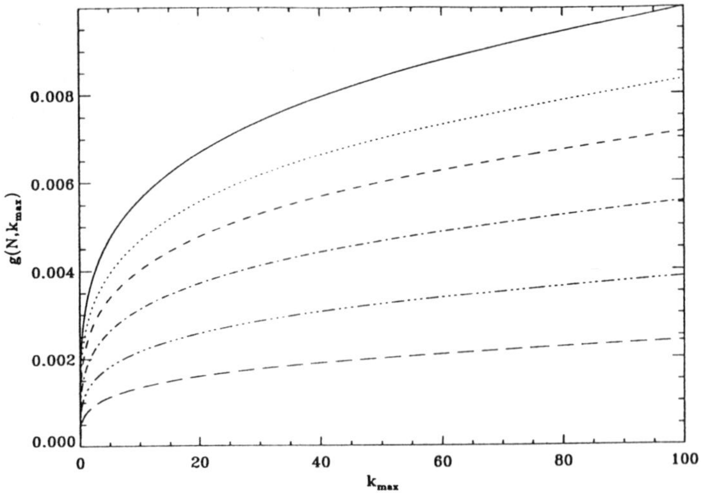

FIG. 6. Threshold $g(m, k_{\max})$ of mutual information depending on the memory k_{max} and symbol string length m (=1000(solid), 1200(dotted), 1400(dashed), 1800(dash dot), 2600(dash dot dot dot) and 4200(long dashes)) according to Eqs. (18) and (19).

diagnosed as high-risk-patients. This is an important challenge to nonlinear dynamics.

The aim of our investigation is to get by means of nonlinear dynamics a clear improvement of the detection rate of persons with a high risk for the sudden cardiac death. An essential point is the finding of new parameters, which describe the complex processes and their interactions for detecting those high-risk-patients, who could not recognized by traditional - mostly linear methods.

The basis of our analysis is the heart-rate-variability (HRV) which is yielded from 30 to 60 minutes ECG-measurements, i.e. from non-invasive methods. Even in case of healthy volunteers, the variability shows a broad variety of structures (cf. Fig. 7). This is essentially caused by the fact that the system which generates the HRV has to by considered as an open one whose energy changes temporally as well as from person to person.

From a lot of possible attempts we find the following two approaches as especially effective for a classification in healthy volunteers and high-risk-patients (Kurths *et al.* (18)).

A) Symbolic Dynamics

From various tests it comes out that for our purpose at least 4 different symbols are necessary. The most appropriate transformation is a static one:

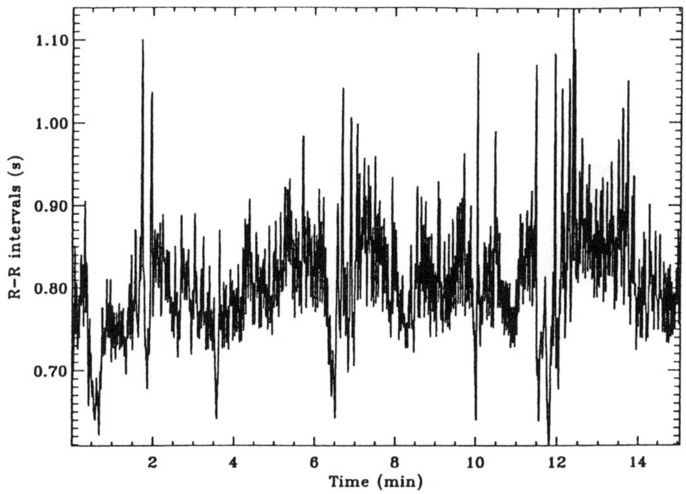

FIG. 7. Tachogram of a healthy person.

$$s_i = \begin{cases} 0 & \text{if} \quad t_i > (1+a)\mu \\ 1 & \text{if} \quad \mu < t_i \leq (1+a)\mu \\ 2 & \text{if} \quad (1-a)\mu < t_i \leq \mu \\ 3 & \text{if} \quad t_i \leq (1-a)\mu \end{cases} \tag{21}$$

where t_i are the RR-intervals, μ is their mean value and $a = 0.1$.

To analyze the symbolic strings, Shannon and Renyi entropies of length-3 words are calculated. As expected, the Shannon entropy (Eq. 4) is not so useful as the generalized Renyi entropies (Eq. 5). We use in particular the $H_k^{(q)}$ for $q = 0.25$ and for $q = 4$ to describe the complexity. It is interesting to note that already the distribution of length-3 words yields a criterion for a distinction of both groups: For persons with cardiac risk, this distribution is mainly concentrated on about 10 words (of 64 possible ones), whereas healthy persons are characterized by a more uniform distribution.

B) Renormalized Entropy

This relative measure is especially related to compare different states of one system. An important problem of its application is to choose a suitable reference state, i.e. special person in our case. We choose that healthy proband as reference person who has the largest renormalized entropy. Note that this choice does not sensitively influence the results. From this, we indeed get an indication for high cardiac risk in two directions. If $\Delta\tilde{H}$ (Eq. 15) is very low, a strongly reduced variability is expressed and, on the other side, if $\Delta\tilde{H}$ is rather large, an exceptional variability is indicated.

It is important to note that none of these measures of complexity alone is sufficient to describe the risk. Therefore, we combine these 4 parameters mentioned above. In a pilot study we have applied this integrated measure to the data of 85 persons. In comparison with standard methods, we could achieve a surprisingly improved detection rate for high-risk-patients. These results have to be validated in a larger study.

B. Cognitive Psychology: Synchronization and Coordination of Movements

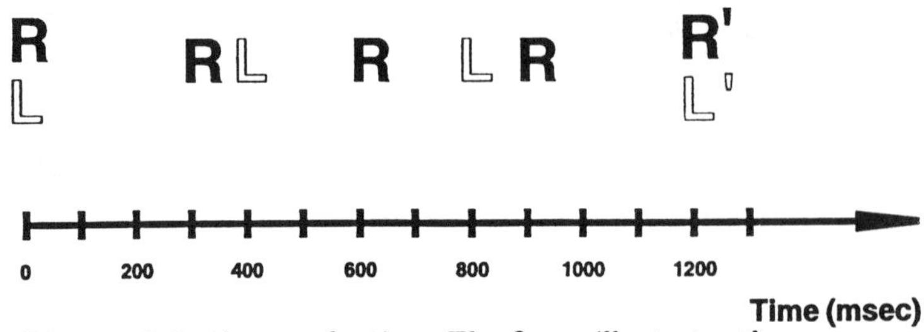

FIG. 8. Bimanual rhythm production: The figure illustrates the sequence of finger movements and their ideal timing for a polyrhythm (3 beats per cycle in the left hand versus 4 beats per cycle in the right hand) at a cycle duration of 1200 ms. R refers to the right, L to the left index finger. A cycle starts with simultaneous strokes of the two fingers.

The abilities to perform precise movements, coordinate movements between different limps, or adjust them to external performance constraints, constitute general, but highly complex human capacities. Biologists and psychologists have for a long time been interested in these capacities in order to gain insights into the functionality of the central nervous system. More recently, the dynamic systems perspective has been applied to a number of phenomena related to motor control (Stelmach and Requin (27)) and human development of cognition and action (Thelen and Smith (29)).

In a recent study, Krampe, Kliegl, and Mayr (15) investigated the production of bimanual rhythms in a large number of subjects differing in musical (pianist) skills, and also age. One task in these experiments required polyrhythmic performance, that is, the combination of different rhythms in the two hands (see Fig. 8).

The task was performed on two keys of an electronic piano connected to a computer which measured the time intervals between successive keystrokes. The external control parameter, that is the prescribed tempo, was experimentally varied across trials for each subject between 800 ms and 8200 ms cycle duration; for highly skilled subjects, performance was assessed at speeds as fast as 500 ms per cycle. After a short synchronization phase during which subjects could play along with the rhythm generated by the computer and adjust to the prescribed tempo, participants had to continue their performance without external support for another 12 cycles. A complete continuation

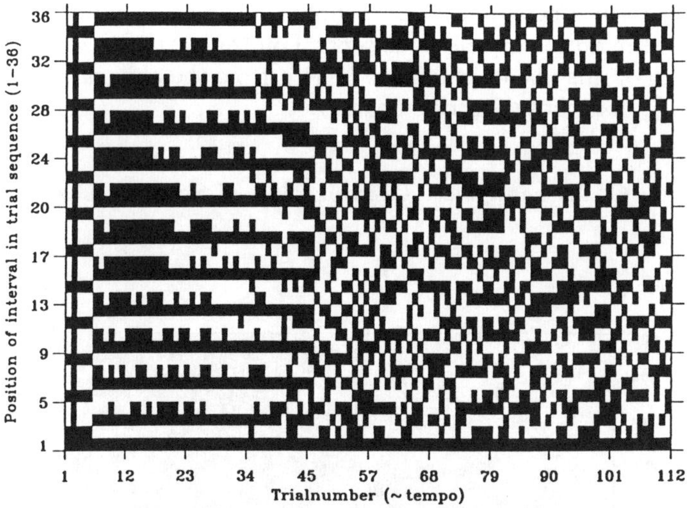

FIG. 9. Phase shift in the performance of a 3 against 4 polyrhythm as a function of tempo in one subject. Trials have been sorted by tempo for this illustration. The trial number is provided by the x-axis and corresponds to the external control parameter (tempo); trial number 1 refers to a cycle duration of 800 ms, trial number 112 to a tempo of 8200 ms per cycle. The position of a single interval within a given trial is provided by the y-axis. Only intervals from the left hand (1-36) are shown. Black pixels indicate intervals which are longer than their immediate predecessor, otherwise a white pixel is set. A change of the pattern can be observed in a region around trial number 47, which corresponds to a performance tempo of 1400 ms per cycle.

trial consists of 36 intervals between left, and 48 intervals between right hand keystrokes, which should ideally be of equal duration for a given hand.

Methods from symbolic dynamics permit to investigate whether the empirically observed variation in the duration of these intervals can be described as an orderly sequence of violations of the prescribed duration within trials, and whether these systematic patterns emerge or dissolve as a function of the external control parameter performance tempo. We have chosen one particular coding scheme to get symbolic strings (Fig. 9). The coding scheme used here transforms continuous measures of interval duration to a dichotomous variable based on local comparisons between successive keystrokes performed with the same hand. A given interval terminated by a left hand keystroke (only these data are considered here for illustration) receives a value of '1', if its duration exceeds the duration of the preceding interval between left-hand keystrokes, otherwise the value is '0'. Values of '1' are indicated by black pixels in Figure 9.

Already this simple transformation clearly suggests a shift in strategies for the realization of the polyrhythmic task as a function of performance tempo.

We use the term strategy in a more global sense here, without committing ourselves to the notion of conscious control on the part of the subject. Future investigation of this phenomenon will benefit from the application of measures of complexity like those described earlier.

C. Astrophysics: Organization of Solar Spikes

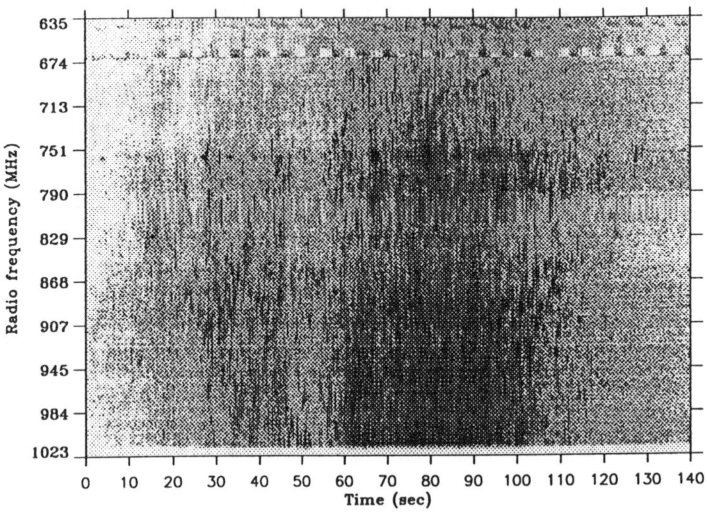

FIG. 10. Dynamic spectrograms of millisecond spikes. The data was recorded by the frequency-agile solar radio spectrometer in ETH Zürich on 6 June, 1983. Low flux: white. High flux: black. The resolution in frequency is 1 MHz and in time 0.2 sec.

The variation of solar electromagnetic radiation and the particle emission is mainly caused by solar activity. Flares are the most violent manifestation of solar activity. They are caused by a rapid release of energy stored in the coronal magnetic field. Understanding the flare phenomenon requires to identify and model a large variety of physical processes involved. From observations of solar radio emission we know that the impulsive phase of this primary energy release in flares is fragmented into a multitude of substructures, called spikes (Fig. 10) which are triggered almost simultaneously. Therefore, properties of spikes give some detailed insight into the nature of this impulsive phase.

Depending on the assumed energy release and emission processes, two types of fragmentation are now under discussion: a scenario of global organization (spikes are emitted in a succession of similar events by the same system) or a scenario of local organization (many systems are triggered by an initial event).

We have searched for interrelations of spikes emitted simultaneously at different frequencies during the impulsive phase of a flare event (26). To

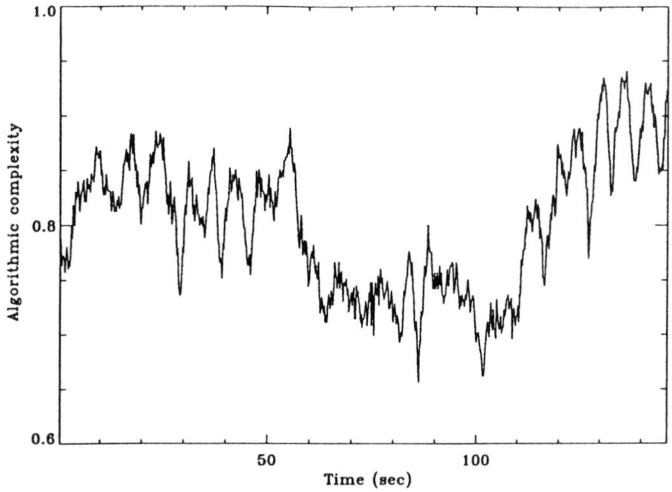

FIG. 11. Algorithmic complexity calculated from the irregular spike pattern presented in Fig. 10.

characterize such complex spatio-temporal patterns, such as dynamic spectra (Fig. 10) measured in solar radio astronomy, we use quantities of symbolic dynamics, such as mutual information $I(k)$, Shannon information and algorithmic complexity. This approach is appropriate to characterize these patterns, whereas the popular estimate of fractal dimensions and related techniques fail here.

In the case of the analyzed dynamic spectra the length of the symbolic strings is about 400. To improve the statistics, we concatenate up to ten symbol strings of successive scans of the dynamic spectrum. We observe in all cases that the only effect of such concatenation is a smoothing in mutual information, Shannon information and algorithmic complexity. Some artifacts concerning the mutual information occur from this concatenation if the values of k exceeds the length of the single symbolic string. However, we limit k here to values smaller than one half of the symbol string length.

By means of mutual information, we have found a memory over frequency intervals, especially in the range from 20 to 40 MHz (Fig. 11), which is indicative of spatial correlations, since different frequencies are emitted at different heights in the atmosphere.

The Shannon information and algorithmic complexity of the symbol sequence are used for comparing the considered observation with others, or with models (surrogates) and for characterizing the observation. These measures concern the whole frequency region, i.e. the global source region. This way, we find out that global organization is also apparent in quasi-periodic changes of these Shannon information and algorithmic complexity in the range of 2 –

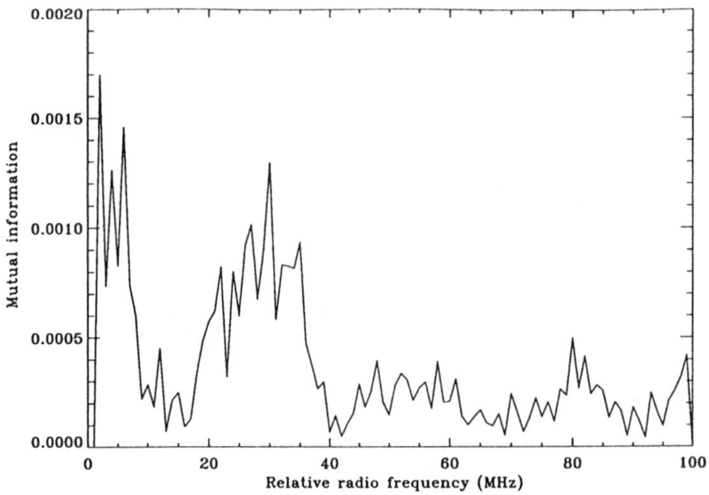

FIG. 12. Average of the mutual information obtained from the symbol string pattern of the irregular spike pattern presented in Fig. 10. There are two broad bands: The first one about 5 MHz reflects the bandwidth of the spikes and the second one at about 30 MHz reveals a spatial correlation between different sources of energy release.

8 seconds (Fig. 12).

Our analysis of four spike events suggests that the structure in frequency is not stochastic but a process in which spikes at nearby locations are simultaneously triggered by a common exciter.

IV. CONCLUSIONS

In this paper we have discussed several measures of complexity for the analysis of data obtained from natural systems. It has been demonstrated that this approach can be indeed helpful in cases where other tools of data analysis fail.

From a comparison of such measures, we have proposed a distinction into two types:

a) Traditional measures of complexity, such as algorithmic complexity or Shannon entropy. These quantities are measures of randomness, i.e. they are maximum for white noise.

b) Alternative measures of complexity which relate highest complexity to critical points. Examples are ϵ-complexity or renormalized entropy. This type seems to be especially suitable to characterize nonlinear dynamical systems.

It is interesting to note that the approximative entropy ($ApEn$) is in this classification also a traditional measure.

We would like to stress that there is no optimum measure of complexity because they are context-dependent. In many applications a combination of some such quantities is useful, e.g. to describe the individual risk for some heart diseases from ECG.

The concept of measures of complexity seems to be a promising tool in signal analysis. Nevertheless, it must be mentioned that this approach is in its infancy; there are many open problems on the theoretical foundation and on the estimate from experimental data.

ACKNOWLEDGEMENTS

We thank A. Voß and A.O. Benz for submitting the data. The authors are greatly indebted to Stefan Schmidt for assistance in preparing some figures and programs used in this paper.

REFERENCES

1. A. Arneodo, F. Argoul, E. Bacry, J. Elezgaray, E. Freysz, G. Grasseau, J.F. Muzy and B. Pouligny, In: Y. Meyer (ed.) 1992 Wavelets and Applications. Proceedings of the International Conference, Marseille, France, May 1989, Masson, Paris und Springer, Berlin 1992.
2. J. Balatoni and A. Renyi, in: Selected Papers of A. Renyi, Vol. I, Akademiai, Budapest, 1976, p. 558.
3. G.J. Chaitin, Algorithmic Information Theory, Cambridge University Press, 1987.
4. D.N. Chorafas, Statistical Processes and Reliability Engineering, Van Nostrand, Princeton, 1960.
5. J.P. Crutchfield and K. Young, Phys. Rev. Lett. **63**, 105 (1989).
6. M. Farge, J.C.R. Hunt and J.C. Vassilicos (eds.) Wavelets, Fractals, and Fourier Transforms, Clarendon Press, Oxford, 1993.
7. J.D. Farmer, Z. Naturforsch. **37a**, 1304 (1982).
8. P. Grassberger and I. Procaccia, Phys. Rev. Lett. **50**, 346 (1983).
9. P. Grassberger, Phys. Lett. A **128**, 369 (1988).
10. A. Hempelmann and J. Kurths, Astron. & Astrophys. **232**, 356 (1990).
11. F. Kaspar and H.G. Schuster, Phys. Rev. A **36**, 842 (1987).
12. Yu. L. Klimontovich, Turbulent motion and the Structure of Chaos, Kluwer Academic Publishers, Dordrecht, 1991.
13. M. Koebbe and G. Mayer-Kress, Use of Recurrence Plots in the Analysis of Time-Series Data, in: Nonlinear Modeling and Forecasting, SFI Studies in the Sciences of Complexity, Proc. Vol. XXI, (eds.: M. Casdagli & S. Eubank) Addison-Wesley, Santa Fe, 1992, pp. 361-378.
14. A.N. Kolmogorov, Probl. of Inform. Theory **1**, 3 (1965).
15. R. Krampe, R. Kliegl and U. Mayr, Research Report. Max Planck Institute for Human Development, Berlin, Germany, 1993.

16. J. Kurths and U. Schwarz, Space Science Reviews **68**, 171 (1994).
17. J. Kurths, U. Schwarz, C.P. Sonett and U. Parlitz, Nonlinear Processes in Geophysics **1**, 72 (1994).
18. J. Kurths, A. Voß, P. Saparin, A. Witt, H.J. Kleiner and N. Wessel, Chaos **5**, 88 (1995).
19. E. Ott, Chaos in Dynamical Systems, Cambridge UP, 1993.
20. C.-K. Peng, J. Mietus, J.M. Hausdorff, S. Havlin, H.E. Stanley and A.L. Goldberger, Phys. Rev. Lett. **70**, 1343 (1993).
21. S.M. Pincus, Proc. Natl. Acad. Sci. USA **88**, 2297 (1991).
22. S.M. Pincus, I.M. Gladstone and R.A. Ehrenkranz, J. Clin. Monit. **7**, 335 (1991).
23. P. Saparin, A. Witt, J. Kurths and V. Anishenko, Chaos, Solitons and Fractals **4**, 1907 (1994).
24. B. Pompe, J. Stat. Phys. **73**, 587 (1993).
25. A. Schmitt, H. Herzel and W. Ebeling, Europhys. Lett. **23**, 303 (1993).
26. U. Schwarz, J. Kurths, A. Witt and A.O. Benz, Astron. & Astrophys. **277**, 215 (1993).
27. G.E. Stelmach and J. Requin (Eds.), Tutorials in Motor Behavior II, North Holland, Amsterdam, 1992.
28. F. Takens, Lecture Notes in Mathematics **898**, Springer-Verlag, Berlin, 1981.
29. E. Thelen and L.B. Smith, A Dynamic Systems Approach to the Development of Cognition and Action, MIT Press, Cambridge, MA, 1994.
30. A. Witt, J. Kurths, F. Krause and K. Fischer, Geophysical and Astrophysical Fluid Dynamics, (1995, in press).
31. R. Wackerbauer, A. Witt, H. Atmanspacher, J. Kurths and H. Scheingraber, Chaos, Solitons & Fractals **4**, 133 (1994).
32. A. Lempel and J. Ziv, IEEE Trans. Inf. Theory IT-**22**, 75 (1976).

Detecting smoothness in noisy time series

Robert Cawley, Guan–Hsong Hsu*, and Liming W. Salvino†
Information Sciences and Systems Branch
Naval Surface Warfare Center, Dahlgren Division, White Oak
10901 New Hampshire Avenue, Silver Spring, MD 20903–5640

Abstract

We describe the role of chaotic noise reduction in detecting an underlying smoothness in a dataset. We have described elsewhere a general method for assessing the presence of determinism in a time series, which is to test against the class of datasets producing smoothness (*i.e.*, the null hypothesis is determinism). In order to reduce the likelihood of a false call, we recommend this kind of analysis be applied first to a time series whose deterministic origin is at question. We believe this step should be taken before implementing other methods of dynamical analysis and measurement, such as correlation dimension or Lyapounov spectrum.

INTRODUCTION.

It has long been observed that nonlinear deterministic systems can exhibit seemingly random behavior. An apocryphal example is smoke rising in still air from a cigarette. Although the time evolution obeys strict deterministic laws, still at relatively shallow heights in the smoke column complicated oscillations develop that seem so irregular as to be virtually random. Before the advent of nonlinear dynamics, commonly studied deterministic behaviors consisted of stationary, periodic, and quasi–periodic regular motions, or perhaps at best combinations and perturbations of these. The irregular oscillations we now would

---

*Physics Department, The American University

†ONR Postdoctoral Fellow

characterize as chaotic usually were ignored, or perhaps modeled as random or very high dimensionally quasi-periodic (Landau's turbulence picture).[1]

From a nonlinear dynamics viewpoint, there are, however, various geometrical and dynamical quantities such as fractal dimensions, entropy, Lyapunov exponents, prediction properties, *etc.*, that have been used to discriminate between deterministic and random systems. Devising effective tests of determinism for time series data based on these theories, however, has not been easy.[a]

Sufficient smoothness of a reconstructed phase portrait implies determinism, and this fact can be exploited to devise a quantitative index of determinism.[4] An important experimental fact in chaotic data analysis, however, is the ubiquitous presence of noise in time series. Noise, whether measurement noise, which is merely additive, or dynamical noise, whose origin lies in the dynamical process itself, can obscure the smoothness feature of a phase portrait. It is therefore important to incorporate noise reduction to make a smoothness test for determinism more robust.

Before we go any further, we need to deliver a couple of *caveats.* Some deterministic systems may be high dimensional, and look pretty noisy. If they have too many degrees of freedom, such as the cigarette smoke behavior at sufficient heights, it may be necessary, or even best, to regard them as effectively random for practical purposes. Despite our opening illustrative remarks, this fact is, of course, not very surprising. Indeed the effort to use the data analysis methods developed for nonlinear dynamics makes the best sense when the system is effectively low dimensional.

In addition, any test for determinism, any time series measurement or dynamical analysis, including ours, is necessarily performed in relation to some, generally low, chosen dimension, such as may be used to construct the phase portraits. Nevertheless, as a matter of principle, any conclusion asserting the absence of determinism is *inherently limited by this chosen dimension.* Evidently, an undetected finite higher dimensional determinism yet might be present.

We refer the reader to Ref.[4] for description of the method for detecting determinism, and to Ref.[3] for a hypothesis testing approach, where we seek to discriminate against determinism. Focusing in the present article on the noise

---

[a]We do not discuss the various problems of data analysis here. We refer the reader to Refs.[2] and[3] for discussion of some of the issues and some of the history of time series data analysis.

problem, we describe the effect of chaotic noise reduction on the index of determinism introduced in Ref.[4] The noise reduction algorithm we use is the local geometric projection (LGP) method introduced by Cawley and Hsu.[5]

It will be apparent from our results that chaotic noise reduction does not create smooth phase portraits, willy-nilly. There must be an underlying dynamics present in the first place.

After a brief introduction to establish some notation and describe the properties of our determinism index, we review the LGP method and then describe the effects of noise reduction on the determinism index for a few examples.

I. DETERMINISM PRELIMINARIES.

We consider time series that are both chaotic and noisy. We assume the noisy part is some form of randomness. We denote the given dataset by $v(t), t = 1, 2, \ldots, N_D$; without loss of generality, we may write

$$v(t) = V(t) + \epsilon\eta(t),\ t = 1, \ldots, N_D, \tag{1}$$

where $V(t)$ represents the underlying deterministic part. We have written the random part of the time series as $\epsilon\eta(t)$, where $\eta(t)$ is a zero mean process having unit variance. Although Eq.(1) is in the form of a dataset having additive noise, which models instrumental noise, the dynamical noise case also is represented as long as there is a shadowing property. The idea is that the noisy dataset, $v(t)$, is to represent an ϵ-pseudo-orbit, and this is guaranteed to be δ-shadowed by a true orbit for some δ. A chaotic true orbit is then represented by the observation coordinate $V(t)$. Indeed for a given $\epsilon > 0$ one generally expects uncountably many δ-shadowing true orbits. So $V(t)$ is just not necessarily unique.[b]

The delay vector time series,

$$x(t) = (v(t), v(t + \Delta), ..., v(t + (d - 1)\Delta)), \tag{2}$$

where Δ is the delay, and $t = 1, ..., N = N_D - (d - 1)\Delta$, specifies a phase space reconstruction of the given time series $v(t)$. When $v(t)$ represents a noise–free dynamical system, *i.e.* for the case $\epsilon = 0$, the tip of the delay vector, $x(t)$, traces

---

[b]Typically, one does not expect to have shadowing over infinitely long times unless the system is uniformly hyperbolic. But shadowing over finitely long segments of time is much more likely to occur.

out a geometrical object, the embedded image of the attractor in $\mathbb{R}^d$. For a chaotic time series this object is a fractal, but the flow lines, unstable manifolds, and dynamics all are smooth.

Precisely, for this noise free case, let an observed time series, $v(t) : t = 1, ..., N_D$, be the output of a differentiable dynamical system f^t on an m-dimensional manifold M; i.e., $f^t : M \rightarrow M$, where f^t is the tth iterate of f, is the nonlinear dynamics underlying the data and $v(t)$ is the measured time history of one of the coordinates for the orbit in M, $v(t) = S(\mathbf{x}_t), \mathbf{x}_t \in M$. By Takens' theorem[6], when delays are introduced an embedding of M into $\mathbb{R}^d$ typically results as long as the number of components, d, is made large enough. Smoothness properties of the dynamical system are now reproduced in the embedded image M' of M in $\mathbb{R}^d$.

Moreover, the vector time series, $x(t)$, now lives in that image. Thus the noise–free attractor image is contained in (and thus samples) the m-dimensional embedded image, $M' \subseteq \mathbb{R}^d$, of the original true space for the dynamics, *viz.* M. But when the data are noisy, as in Eq.(1) with $\epsilon > 0$, the orbit points in $\mathbb{R}^d$ fluctuate out of M', resulting in a "fuzzy" and "less smooth" image of M.

Properties of Determinism Index.

Our determinism index measures the smoothness of a phase portrait. It is defined from a quantity which is theoretically independent of the embedding, as we might certainly hope, but whose computed values, $W_M = W_M(\Delta)$, in practice nevertheless depend on the delay. But we have found a method to deal with the finite numerics responsible for this unwanted effect.

A mitigating fact about this embedding dependence is that graphs of $W_M(\Delta)$ settle down to approximately constant stable upper envelope values out to *very* high values of delay, *viz.* $\Delta \sim N_D/d$. For deterministic chaotic systems upper envelope values are slightly less than one (Figure 1). For datasets produced by nondeterministic systems the plots look the same but the corresponding upper envelope values are decisively less than one, *e.g.* 0.4 for white noise. We denote these upper envelope values simply by W_M, and this is what we call the determinism index.

For details see Refs.[4,3]. Summarizing material therein, we note three points:

(i) Our method provides a basis for assessment of whether a phase portrait is in the smooth class or not. If it belongs to the smooth class, then the time series that generates it is "not random at all;" that is, there is no need for surrogate

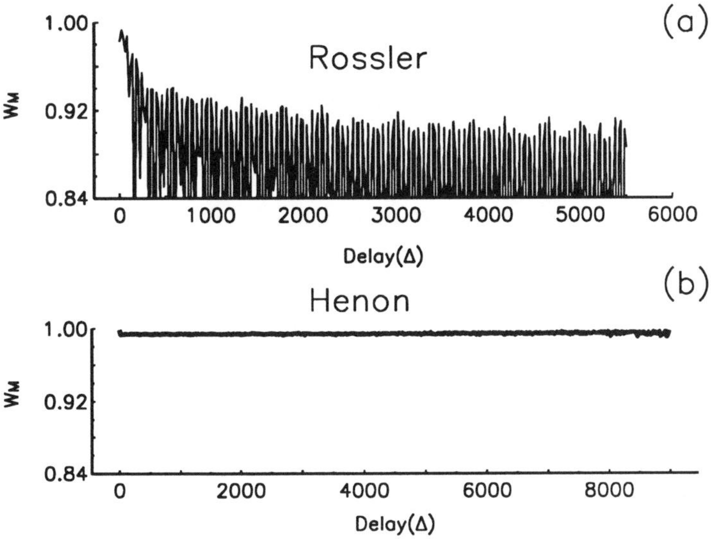

FIG. 1. Chaotic determinism case studies of x-coordinate time series from (a) Rössler system [$\dot{x} = -(y+z), \dot{y} = x+0.2y, \dot{z} = 0.2+zx-5.7z$, sampling time $\Delta t = 0.1$] and (b) Hénon map [$x' = 1 - 1.4x^2 + y$, $y' = 0.3x$]. Computed values of $W_M(\Delta)$ span a very large range of delays, out to $\Delta \sim N_D/d$.

comparisons to assess whether the given dataset can be distinguished from one or another surrogate class.[7] Smoothness implies determinism.

(ii) From its definition (see Ref.[4]) it follows immediately that our task in assessing the presence of determinism is that of deciding whether $W_M = 1$ or $W_M < 1$.

(iii) To decide which of these holds, we have introduced an informal determinism tolerance criterion, which accepts $W_M = 1$ if $0.9 \leq W_M < 1.0$; in this case we write $W_M \simeq 1$ and we label the given dataset as "not random at all." Also, we regard $W_M < 1$ if $0 < W_M < 0.7$. In this case we conclude there is no evidence for determinism.

The intermediate case, $0.7 \leq W_M < 0.9$ and for which we write $W_M \sim 1$, is known sometimes to arise from a deterministic time series. Since datasets in this category fail our determinism tolerance test, we have developed a separate method to deal with them. We analyze such data by means of a surrogate comparison test, the outcome of which, however, cannot be "not random at all" as in the cases where $W_M \simeq 1$, but instead only consistent or inconsistent with membership in classes from a chosen family of specific random surrogates. Thus we may decide here to accept $W_M = 1$ if one or more random surrogate classes are clearly distinguishable from the given dataset. Since here the conclusion $W_M = 1$ is based upon rejection of a specific set of surrogate classes, it is weaker than for the case where $W_M \simeq 1$ holds.

The discriminating values of W_M, 0.7 and 0.9, which set off our determinism tolerance criteria, as we have indicated, are informal. We have chosen these high values in order to implement our main requirement: as much strictness as we can practically manage against accepting a determinism call — we seek to zero out our "false alarm" rate. But, by incorporating chaotic noise reduction, we can increase our "detection" rate, without compromising this objective, as we now show.

II. CENTRAL ROLE OF CHAOTIC NOISE REDUCTION

"The use of noise–reduction methods is strongly recommended in any analysis of chaotic time–series data."[8] This is an integral part of our approach also, although now for new reasons as well as the salutary reasons of Ref.[8].

We have found in controlled numerical studies that even small amounts of noisy contamination of a chaotic time series can quickly lead to $W_M < 1$. Thus even if we have ruled clearly against determinism in some example, or have only been able to secure $W_M \sim 1$, the time series still might possess a deterministic part. That is, we might suppose that after a successful chaotic noise reduction, the output time series might no longer be distinguishable from the smooth class, and we would have then a strong deterministic call, $W_M \simeq 1$. We'd like to catch these in our "not random at all" net, too.

This is actually a radical program because we will be willing to say we have a chaos call on the basis of a time series derived from the given dataset, rather than just the initial dataset itself. The definition of the chaotic part issues from the mathematics of the chaotic noise reduction algorithm used. We also have the makings of an exactly similar kind of separation using a novel and different method, one based on predictability analysis[9], but which we cannot discuss here.[c]

In other words, integration of the use of chaotic noise reduction into our

---

[c]The question to which we have been able to give precise formulation is, "What is the sense in which one can say that a time series has a predictable component?" In principle this can be answered, and the answer suggests a way of isolating that component.

approach provides a way for us to hold more strictly to our tolerances against deterministic calls (fewer false alarms) and, at the same time, give the matter our best shot (more detections). This is why chaotic noise reduction is central to the problem of detecting determinism.

The Noise Reduction Process.

Evidently, when a raw dataset is given we do not have the decomposition in Eq. (1) available. For a noise reduction method to be meaningful (and successful!), it must correctly exploit some aspect of dynamical systems theory to provide a basis of identification of a deterministic part.

Reviews of a variety of different noise reduction methods that exist are in Refs.[8,10] These differ from one another according to which aspect is exploited. Typically, they proceed through an identifiable sequence of four basic steps.

Step 1. Embed; *i.e.* replace the given scalar dataset, $v(t)$, by a data–state vector, such as $x(t)$ in Eq. (2). This provides a phase portrait in $\mathbb{R}^d$ for Step 2 manipulations.

Step 2. Adjust the positions of the points of the phase portrait. This is what is normally regarded as the key noise reduction step, with the vector time series $x(t)$ replaced by $\hat{x}(t)$.

Step 3. Disembed; *i.e.* replace the altered data–state vector time series, $\hat{x}(t)$, by a new scalar dataset, $\hat{v}(t)$. This step, the necessity of which was first realized in Ref.[5], can also remove further noise.

Step 4. Iterate the foregoing, inputting $\hat{v}(t)$ in Step 1.

Sometimes the order of these steps is shuffled. For instance, in the trajectory adjustment step (Step 2) Kostelich and Yorke[11] iterate before disembedding. In their method one can do it either way in principle, although that is not always the case in other methods.

The method used in Step 3 is quite important. Sometimes Step 3 is avoided entirely, with resulting loss in performance. The issue is this. The altered data–state vector time series, $\hat{x}(t)$, is almost surely not itself a DCC, but the right answer, namely the DCC from $V(t)$, of course, must be. To disembed we construct the scalar time series $\hat{v}(t)$ having the following property: namely, that the vector time series given by its DCC is as close as possible to the altered vector time series, $\hat{x}(t)$.

This provides us our first guess at the noise-free scalar dataset, $V(t)$. For t-values not too close to endpoints of the time series, a simple least squares criterion gives

$$\hat{v}(t) = d^{-1}\Sigma_{j=1}^{d}\hat{x}_j(t-(j-1)\Delta), \tag{3}$$

where $\hat{x}_j$ denotes the jth component of the vector $\hat{x}$. Similar expressions result for end-point range t-values. We note the smoothing effect of the averaging that occurs in Eq. (3).

Of course, different disembedding norms besides the least squares (L^2), are possible. The once popular choice in the community, $\hat{v}(t) = \hat{x}_1$, namely that of the first component of $\hat{x}$, is a poor relation of the L^p disembedding with $0 < p < 1$. This also gives a single component, but usually not the first component.

We remark that iteration is an essential part of chaotic noise reduction since all the methods so far proposed in the research literature are based on successive approximations. Otherwise improvements are very small. This circumstance has significant and useful consequences involving attractor stability and instability under iteration[12], and quantitative measures of such effects[13].

Local Geometric Projection Method.

The term local geometric projection refers to the procedure employed in Step 2. In this noise reduction method we locate the embedded image M' of a manifold M for an assumed dynamics by estimating its tangent spaces. This is done by using a family of local neighborhoods of points from the reconstructed phase portrait. The family of neighborhoods is constructed so as to achieve a covering of the underlying attractor image. The resulting effect is to approximate the sampled part of the embedded image of M by a polytope. The dimension of the estimate, $k < d$, is that of the local spaces we use to make the approximation. If we already know m we can simply choose $k = m$, if not, we may examine properties of the output time series for increasing values of k, regarding it as a parameter of the method.[d]

For the noise reduction action of Step 2, we move the points of the phase portrait some fraction, typically half, the distances to the various local hyper-

[d]Naturally, the issue of whether the value of m is known or unknown is not peculiar to the LGP noise reduction procedure.

planes specifying the polytope. Hence the term local geometric projection. For precise details of the LGP method see Ref.[5].

Of course, the polytope as well as the phase portrait changes a little with each iteration; the changes that are realized with each iterate depend on the scalar time series, $\hat{v}(t)$, input from the corresponding iteration. Thus the entire process is highly nonlinear, any operator representation of the algorithm's procedure necessarily varying from one iteration step to the next.

As the noise reduction iterations proceed, phase portraits from a noisy chaotic dataset become less and less ragged owing to the smoothing action of local projection. So, we might expect this noise reduction method, in particular, to result in smoother phase portraits.

III. NOISE REDUCTION AND SMOOTHNESS.

We present some results of application of the LGP method to Lorenz system data. For the delay embedding construction we use the x-coordinate, with standard parameter choice: $\dot{x} = 10(y-x)$, $\dot{y} = 28x - y - xz$, $\dot{z} = -\frac{8}{3}z + xy$, sampling time $\Delta t = 0.05$ (~ 17 points/cycle).[e] The time series length is $N_D = 10,000$, and $W_M(\Delta)$-calculations have been performed in dimension $d = 3$ with a $20 \times 20 \times 20$ grid. Parameters for the noise reduction were $(d, k, \nu, \Delta) = (15, k, 40, 1)$, where $\nu + 1$ is the number of points chosen for the local neighborhood covering elements for the phase portraits.[f] We took $k = 2$ in one set of runs and $k = 3$ in another; in the Lorenz example, $m = 3$.

Smoothness after Noise Reduction: the Baseline Noise Case.

We first present results for the baseline noise case, *i.e.* for applications of LGP to noise-free data (see Ref.[5]). Owing to inevitable numerical imperfections, the action of any chaotic noise reduction algorithm introduces noise, the so-called baseline noise. For $k = 3$, the signal-to-noise ratio (SNR) after a single iterate, $n = 1$, is $\sim 70dB$ (0.03% noise), and falls slowly, approximately linearly in $\log n$. The rate of falloff is numerically about $4.4dB$ for each doubling of n; for $k = 2$ the

---

[e] The solution time step used to generate the sampled dataset was 0.001 and the resulting time series was not low passed before sampling.

[f] The choice $\Delta = 1$ always gives the maximum noise reduction (SNR gain).

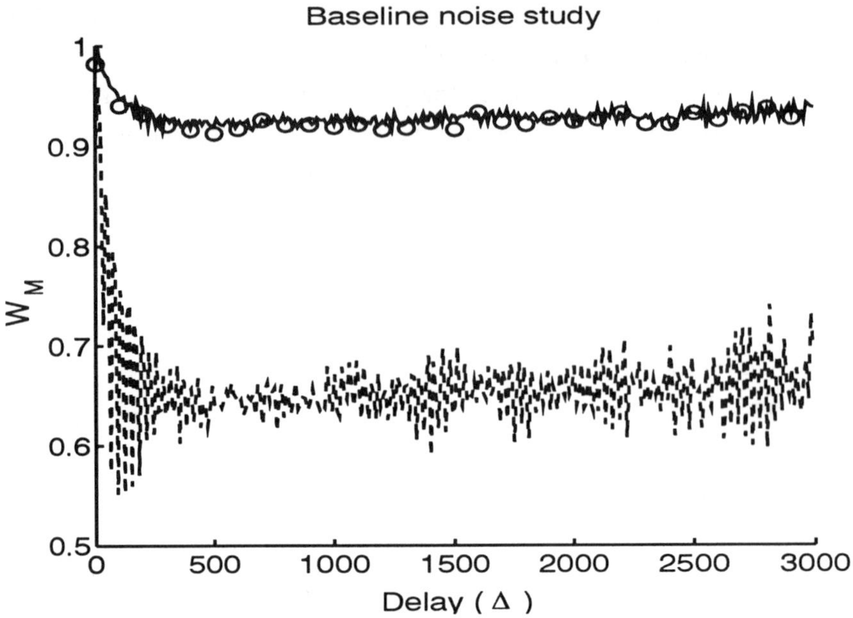

FIG. 2. Determinism index plot, $W_M(\Delta)$, for baseline noise Lorenz x-coordinate time series. Solid line: $n = 0$ (original time series before noise reduction); dashed line: $k = 2$; circles: $k = 3$.

rate is the same, but the starting point is now $60dB$ (0.1% noise).[g] Graphically, the attractor in the $k = 3$ case is hardly distinguishable from the original attractor before application of the algorithm, even after 256 iterates. But the $k = 2$ case is somewhat different, one begins to see attractor breakup even at 32 and 64 iterates.[12]

The behavior of the determinism index for these baseline noise cases is shown in Fig. 2. The solid line is for the original dataset, before application of the noise reduction algorithm ($n = 0$); the open circles are for $k = 3$ and $n = 256$. The two cases are nearly indistinguishable, and in each one we clearly have $W_M \simeq 1$. The dashed line is for $k = 2$ and $n = 256$, we now have $W_M < 1$. Thus the determinism index is stable under application of the LGP algorithm as long as $k > 2$. In addition, we have found a lack of determinism in the fractured attractor resulting in the $k = 2$ case.

---

[g]Extensive numerical experiments reveal that the initial *SNR* increases monotonically with k, and the falloff rate is sensibly independent of k.

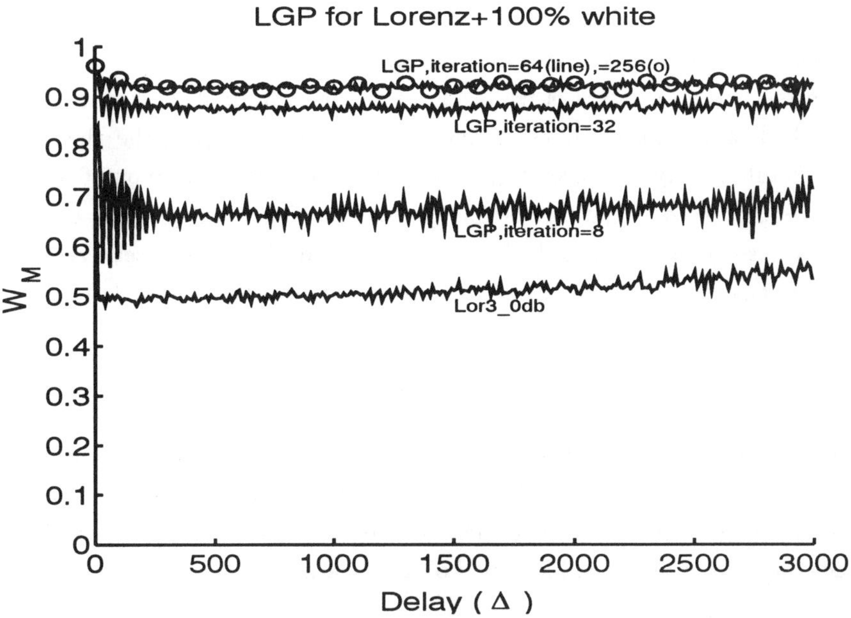

FIG. 3. Determinism index plot, $W_M(\Delta)$, for 100% noise ($0dB$) Lorenz x-coordinate time series. Upper part of the figure (data lying above $W_M = 0.9$) — Solid line: $n = 64$; circles: $n = 256$.

Smoothness after Noise Reduction: the $0dB$ (100%) Noise Case.

Even low levels of noise *can* interfere with our determinism test as we have already asserted. We have seen this in our analysis of experimental data from the chaotic oscillations of a magnetoelastic ribbon, where the initial SNR was about $35dB$.[4] There we found $W_M = 0.85$, and therefore only $W_M \sim 1$. On the other hand, note that 256 iterates for the case $k = 3$ on the Lorenz system dataset above resulted in $SNR = 70 - 8 \times 4.4 \simeq 35dB$, also. But here W_M remained high, well within our determinism tolerance interval, $0.9 \leq W_N < 1$. So, the SNR alone is not the complete determinant of whether $W_M \simeq 1$.

The behavior of the determinism index under application of LGP for the case of 100% white noise is displayed in Fig. 3. Algorithm parameters are the same as above, and with $k = 3$. W_M starts around 0.5 ($W_M < 1$) and increases with n until, by $n = 64$ iterates, $W_M \simeq 1$ has been recovered. Further iteration out to $n = 256$ has little further effect on its value, and the last held true out to $n = 1024$, as well.

When the LGP algorithm was applied to the ribbon data mentioned above, W_M increased from 0.85 to 0.91, which is the same effect we have just described for the Lorenz datasets. The fact that in the former case the noise was additive and in the latter case it was dynamical did not matter.

The method also passes other tests one would require of it. For instance, when applied to a (Fourier) phase randomized Lorenz surrogate, the algorithm does not result in any increase of W_M from an initial value of about 0.4, and $W_M < 1$ is preserved. That is, the LGP algorithm's smoothing effect did not result in a false determinism call. The same was true of ambient ocean acoustic data, where again noise reduction fails to recover an abysmal $W_M < 1$.

SUMMARY

We have described the interplay of chaotic noise reduction and determinism. Noise reduction can recover a "lost" determinism call. Its use allows us to retain an unforgiving posture in our strict acceptance criterion for a chaos call. This approach can help reduce the likelihood of false alarms.

Acknowledgments. This work was supported by the Office of Naval Research (Chemistry and Physics S & T Division) and by the Independent Research Programs of the Dahlgren and Carderock Divisions of the Naval Surface Warfare Center.

REFERENCES

[1] L. D. Landau, Fluid Mechanics, Addison-Wesley, 1960.

[2] James Theiler, Physics Letters A **196**, 335, 1995.

[3] Robert Cawley, Guan-Hsong Hsu and Liming W. Salvino, "Method to discriminate against determinism in time series data," in *Predictability of Complex Dynamical Systems*, James B. Kadtke and Yurii Alexandrovich Kravtsov, Eds., Springer-Verlag, Berlin (to appear).

[4] Liming W. Salvino and Robert Cawley, Phys. Rev. Letters **73**, 1091 (1994).

[5] Robert Cawley and Guan–Hsong Hsu, Phys. Rev. A **46**, 3057 (1992).

[6] F. Takens, "Detecting Strange Attractors in Turbulence" in *Dynamical Systems and Turbulence, Warwick, 1980*, Lecture Notes in Mathematics, No. 898, D. A. Rand and L. -S. Young, eds., Springer-Verlag, Berlin 1981, p. 366.

[7] James Theiler, Stephen Eubank, André Longtin, Bryan Galdrikian and J. Doyne Farmer,Physica D **58**, 77 (1992).

[8] Eric J. Kostelich and Thomas Schreiber, Phys. Rev. E **48**, 1752 (1993).

[9] Liming W. Salvino, Robert Cawley, Celso Grebogi and James A. Yorke, Phys. Lett. A **209**, 327 (1995).

[10] Peter Grassberger, Rainer Hegger, Holger Kantz, Carsten Schaffrath and Thomas Schreiber, Chaos **2**, 127 (1993).

[11] Eric Kostelich and James A. Yorke, Phys. Rev. A **38**, 1649 (1988).

[12] Robert Cawley and Guan–Hsong Hsu, "Another new chaotic noise reduction algorithm," in *Proceedings of the first experimental chaos conference, Arlington, VA, October 1–3, 1991*, Edited by Sandeep Vohra, Mark Spano, Michael Shlesinger, Lou Pecora and William Ditto, World Scientific Press, Singapore, 1992, p 38.

[13] Robert Cawley, Guan-Hsong Hsu and Liming W. Salvino, "Detection and diagnosis of dynamics in time series data: theory of noise reduction," in *The chaos paradigm: developments and applications in engineering science*, AIP Conference Proceedings Volume 296, American Institute of Physics, New York, 1994, p. 182.

2. PREDICTABILITY AND CONTROL OF CHAOS

Predictable Nonlinear Dynamics: Advances and Limitations

L.A.Anosov[†], O.Ya.Butkovskii[††], Yu.A.Kravtsov[*], E.D.Surovyatkina[**]

[†] *Vladimir Cardiological Center, Vladimir 600023, Russia;* [††] *Vladimir Technical University, Vladimir, 600029, Russia;* [*] *Space Research Inst, Russian Acad.Sci., Moscow 117810, Russia and* [**] *Moscow State Pedagogical Univ., Moscow, 119485, Russia*

Abstract. Methods for reconstruction chaotic dynamical system structure directly from experimental time series are described. Effectiveness of general methods is illustrated with the results of numerical simulation. It is of common interest that from the single time series it is possible to reconstruct a set of interconnected variables. Predictive power of dynamical models, provided by the nonlinear dynamics inverse problem solution, is limited firstly by the noise level in the system under study and is characterized by the horizon of predictability.

New physical results are presented, concerning nonstationary and bifurcation nonlinear systems: 1) algorithms for revealing of nonstationarity in random-like chaotic time-series are suggested based on discriminant analysis with nonlinear discriminant function; 2) an opportunity is established to predict the final state in bifurcation system with quickly varying control parameters; 3) hysteresis is founded out in bifurcation system with quickly varying parameters; 4) delayed correlation ‹noise-prediction error› in chaotic systemsis revealed.

1. HORIZON OF PREDICTABILITY FOR CHAOTIC PROCESSES

The ability to forecast a process evolving in time, or, as researchers in natural sciences would say, time series, is very much called for in different areas of human endeavour. In this work we will try to outline some approaches to the problem of dynamical forecasting, that is, forecasting based on certain dynamical laws.

Let us consider a process recorded from a sensing element and represented in Fig. 1.

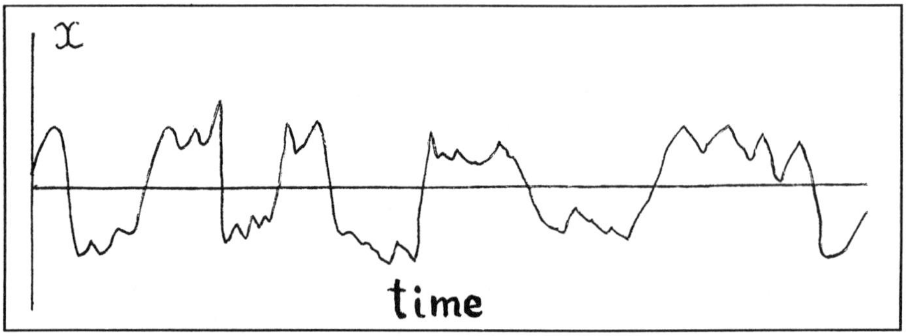

FIGURE 1. Example of a random-like process of dynamical nature. Processes of that kind exhibit both stochastic (in a sense) and deterministic properties.

The first question that arises as we look at the figure is about the nature of the process, that is, whether it is random or deterministic. Even at this early stage of our research we encounter a theoretically profound and practically essential problem of how to distinguish between randomness and determinateness. It is not until the nature of the process is identified that we can select a forecasting method. In fact, we have to do with methods to construct predictive models, since forecasting is always based on a distinct model.

If a process is judged to be random, it is only natural to turn to statistical methods of prediction originally worked out by N.Wiener [1-3].

If there are signs or suspicions that the process has a dynamical origin, it is more expedient to try a dynamical forecast based on relevant regularities (attempts to apply such an approach are reported in Refs.[4-8]). The difference between the statistical and dynamical forecasts can be truly significant.

The statistical forecast starting from probabilistic and correlation regularities is usually limited by relatively short intervals of time comparable with the autocorrelation time τ_c of the process. For cases where the dynamical origin is known, it seems logical to use dynamical forecasting. By way of example, we can refer to weather forecasting in which dynamical methods based on hydrodynamical equations for air flows are used to advantage. Admittedly, the dynamical methods, too, fail to predict a process' behavior far into the future. One of the main reasons for this is local instability inherent in most of the phenomena around us and governing such processes as the dynamical chaos.

A peculiarity of the equations giving the dynamical chaos is that they fail to provide reliable long-term forecasts. The point is that errors in initial conditions, along with other hindering occurring in model equations, a slight nonstationarity of a system etc.), tend to grow exponentially under local instability conditions, so that after some finite time all the workers's efforts would go down the drain.

Taking A to be the swing (typical magnitude) of oscillations under study and δ the perturbation in the system occurring due to one or other factor, the time of predictable behavior τ_{pred} can be written as (see, for instance, Ref.[9]):

$$\tau_{pred} = \frac{1}{2\lambda^+} \ln\frac{A}{\delta} , \qquad (1)$$

where λ^+ is maximum Lyapunov exponent giving the perturbation growth rate.

The maximal value of τ_{pred} which corresponds to minimal achievable value of inaccuracy δ (low instrumental noise, low ambient noise, good model for predictability) was suggested to be named as horizon of predictability [10].

Although the prediction time is limited, as follows from (1), it can exceed and often does exceed!) the auticorrelation time τ_c. It is the ratio $\tau_{\text{pred}}/\tau_c$ that characterizes the gain provided by the dynamical forecast as against the statistical prediction. The equations governing a systems's behavior are not known a priori. To reconstruct these equations while analyzing a time series is the inverse problem of nonlinear dynamics.

It is this problem that we will address here. We will show schematically how experimental data, treat some specific reconstruction algorithms and demonstrate examples.

When using the dynamical approach, elements of the statistical analysis and data processing need volens nolens to be employed. Moreover, apart from purely statistical (probabilistical) and consistently dynamical approaches, there exist a number of "intermediate" methods, based partly on probabilistic, partly on dynamical algorithms. Without going into such combined approaches, we only mention that the forecasts they provide are of intermediate nature, too.

In choosing between the statistical and dynamical forecasting strategy the worker must realize that with different test methods one and the same time series may at times be found to be random and at other times deterministic. The matter is that several agreements on how to define randomness coexist in modern science, and therefore a process satisfying a randomness test under one agreement can be judged to be deterministic under another agreement.

The problem of those different agreements on randomness has been addressed by Kravtsov [11-13]. The most widespread agreement is the one underlying the modern set-theoretical approach and the theory of probability. Within its frame work, anything having a probabilistic measure is deemed to be random. Unlike it, the algorithmical approach identifies randomness with algorithmical complexity.

Physicists generally hold to the "set-theoretical" agreement, while entertaining in some cases their own views of randomness, such as the criterion of decaying (uncoupling) correlations or the agreement commonly though implicitly adopted by the experimental physical community, according to which randomness is identified with unpredictability. The relations developing under this agreement have been formalized only recently [11-13].

Note that processes like dynamical chaos, predictable for times τ_{pred} and thus seemingly deterministic, appear to be random in the context of many other agreements (by the criterions of algorithmical complexity, decaying correlations etc.). That is why in tacking the inverse problem of nonlinear dynamics one can completely neglect the results of many randomness tests if they are useless in reconstructing the governing dynamical equations. Of course, the most important predictability test is correlation between what is observed and what was predicted. A suitable qualitative indicator of such a correlation is the degree of predictability (that is, of determinism) $D(\tau)$, which was introduced in [11] as a normalized function of correlation between the observed, y(t), and the predicted, *z(t)*, processes:

$$D(\tau) = \frac{\langle y(t), z(t)\rangle}{\sqrt{\langle y^2(t)\rangle\langle z^2(t)\rangle}}, \quad \tau = t - t^0 . \tag{2}$$

Here $\tau = t - t^0$ is a time interval after observations started.

A typical time in which such a correlation decays is called predictability time. For the dynamical chaos, the predictability time is estimated according to equation (1).

2. INVERSE PROBLEM OF NONLINEAR DYNAMICS AS APPLIED TO CHAOTIC PROCESSES

General approaches to the nonlinear dynamics inverse problem solution were developed due to A.N.Kolmogorov and D.Gabor a few decades ago. The present state-of-the art is outlined, say, in Ref.[14]. Algorithms for nonlinear dynamics inverse problem solution are generally applied to chaotic system too, but time of predictable behavior usually becomes shorter due to local instability of chaotic processes.

To predict using the dynamical approach, it is expedient to narrow the class of equations used to describe a given process. Otherwise the problem of dynamical forecasting becomes as indefinite as it is difficult to solve. Therefore, number of restricting assumptions is reasonable to make.

First, we exclude from our consideration those spatio-temporal processes whose description requires partial differential equations to be employed. Instead, we will do with dynamical systems given by ordinary differential equations. Second, we restrict ourselves to autonomous models involving no external forcing. Third, we assume those equations to be stationary. Whenever autonomy and stationarity are not assumed, the reconstruction algorithm gets dramatically complicated. The only exception is the ordinary nonstationarity systems, say, those with a stepwise or smooth change of parameters, for which the reconstruction algorithm is not unduly complicated (see below, sect.5).

Forth, in elaborating particular algorithms, we will be oriented to polynomial approximations of nonlinear functions appearing in differential equations under reconstruction.

Even under the said constraints, reconstruction remains to be an extremely indeterminate and tedious task. For example, when taking the exhaustive search strategy and trying equations from a rather limited model class, one by one, to fit the experimental data, we will see that no modern data processing techniques are powerful enough to meet this challenge. Fortunately, there are some algorithms available for such a purpose that are more effective than the exhaustive search method. Their elaboration was possible due to the following favorable factors.

In the first place, we point out the experimental fact that a major part of nonlinear systems exhibiting chaotic behavior have a relatively low dimension $n \approx 3 \div 5$. The abundance of low-dimensional systems is what obviously makes one believe that the inverse problem of nonlinear dynamics is solvable.

In the second place, there is much evidence that the evolution of dynamical processes, even in widely different systems, generally follows the same scenarios. The number of typical scenarios is relatively not large — not more than ten for low-dimensional systems. This is where the dynamical isomorphism principle reveals itself. It permits the class of tested systems of differential equations to be further confined.

Furthermore, one must be aware of the possibility, unexpected at first glance, to reconstruct the variables hidden from observation. Such a possibility only seems to be realizable in highly nonlinear systems where different variables are strongly interconnected, each bearing the information about the whole system.

Another favorable factor, even though unable of confining the class of differential equations, allows a considerable saving of the processing time. The matter is that increasing the length of time series under exanimation beyond a limiting value τ_{pred} does not necessarily lead to a better accuracy of the coefficients of equations determined under reconstruction, but invariably extends the calculation time. As a result, one can reduce the data processing time without sacrificing the accuracy of the model equation. Unfortunately, it is impossible to determine the limiting time τ_{pred} a priori, since it is to be defined in the course of data procession. One can only guess that the bottom limit of τ_{lim} is the time of prediction carried several time steps into the future.

Now, let us consider, following the earlier works [7,8], some specific procedures whereby predictive differential equations can be obtained. We will give most attention to revealing the hidden (unobservable) variables.

3. RECOVERING DYNAMICS FROM TIME SERIES

For reconstructing dynamical equations directly from time series a few specific techniques are usually applied.

3.1. Filtration

Filtering a given time series $x(t)$ implies a separation of fairly smooth "dynamical" component $X(t)$ and remainder $\xi(t)$ that, with some conventionality can be called "noise":

$$x(t) = X(t) + \xi(t) \ . \tag{3}$$

The conventionality about dividing x(t) by the "dynamical" and "noisy" components stems from the fact that the "noisy" component invariably incorporates the high-frequency portion of the dynamical process, while the "dynamical" portion contains the low-frequency noise component. This is where the worker is to demonstrate his sense of proportion, for too deep filtering can cause unacceptable distortions of the dynamics, while a superficial filtration fails to provide the required degree of differentiation.

The high-frequency noise filtration is effected by different ways. In Ref.[7,8] a multiple (up to 50-80 times) $x(t)$ filtering using the least-squares algorithm [15] was used with the weight coefficient of about 0.5. Sometimes such an algorithm was applied together with the method described in [6].

Fig.2 demonstrates an example of a noise-polluted process (Fig. 2,b) resulting from addition of a 15% short-correlated noise to the originally Lorenz process shown in Fig. 2,a. The result of filtration by the said algorithm is very close to that of the original process (see Fig. 2, the smooth curve $X(t)$).

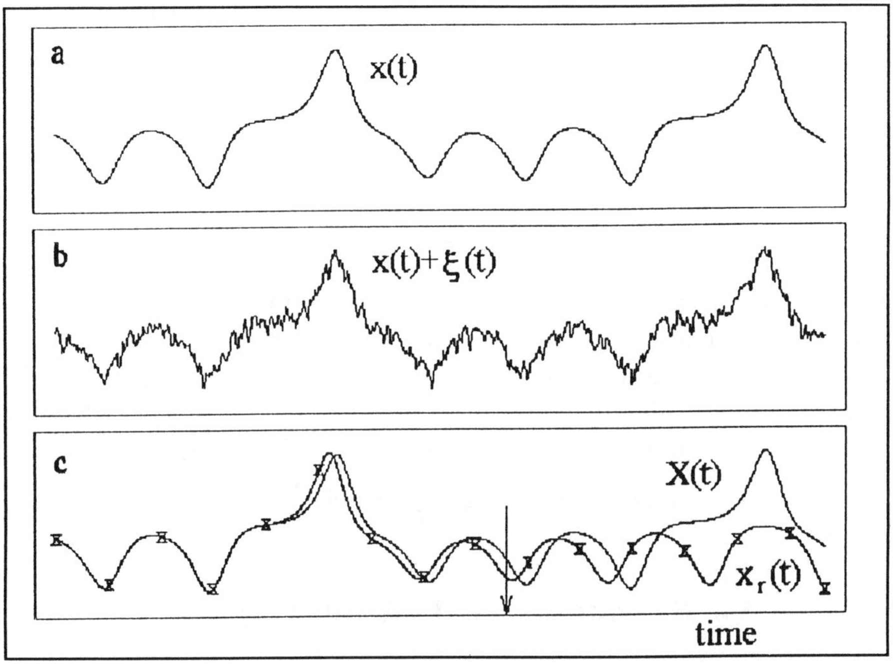

FIGURE 2. The effect of noise $\xi(t)$ added to the Lorenz system process is that (a) becomes (b). In (c) a filtered process $X(t)$ is shown by the unmarked curve, whereas the reconstructed process $x_r(t)$ is marked. The predictability time designated by arrows corresponds to a 15% divergence of the curves.

3.2. Determining the Dimension

To find out about the dynamical nature of the time series $X(t)$ one usually employs Takens' delayed variables [17] $X_1(t) = X(t)$, $X_2(t) = X(t+\tau)$,

..., $X_n(t)=X[t+(n-1)\tau]$, that can be combined together in an n-component state vector $X=\{X_1,...,X_n\}$. Here time delay τ is chosen close to the sampling rate Δt of the observed time series. We can determine the dimension of the embedding space using one of conventional procedures [18-19].

3.3. Choosing an Optimal Basis Set

Taking a sufficient small τ, we may expand the components of the state vector X^k into a Taylor series, and then to express them through the eigenvectors of correlation matrix [1]. It leads to a specific form of equations which in 3D case take the form:

$$\dot{X}_1 = \alpha_1 X_2 ,$$
$$\dot{X}_2 = \alpha_2 X_3 , \qquad (4)$$
$$\dot{X}_3 = \frac{F(X_1,X_2,X_3)}{f(X_1,X_2,X_3)} .$$

Here X_1, X_2, X_3 are new variables, connected with the former ones by a linear transformation, the dot denoting time differentiation, α_1, α_2 being constants, and f one of monomials of the form 1, X_i, X_j, X_iX_j (i=j=1,2,3). F is a polynomial with indeterminate coefficients that can be represented as a sum of all possible monomials G_j composed of the variables X_1, X_2, X_3:

$$f(X_1,X_2,X_3) = \sum A_j G_j(X_1,X_2,X_3) . \qquad (5)$$

3.4. Reconstruction of Differential Equations.

The task of seeking the undetermined hitherto coefficients A_j can be reduced to searching for the minimum of the functional

$$\Phi = \sum_{k=1}^{N} [g^k - F^k]^2 = \min , \qquad g^k = f^k \dot{X}^k_3, \qquad (6)$$

where f^k and F^k are values of the functions f and F at instant t^k. All the values of the variables $X^k_{1,2,3}$ from the initial time series in (6) must be recalculated in view of the performed substitution.

The minimum of the functional ϕ is achieved under the condition:

$$g = \hat{G}A, \tag{7}$$

where $g=\{g^k\}$, $g^k = f^k X^k_3$ and $A = \{A_j\}$ is a coefficient vector. The "nonlinearity matrix" G with elements G^k_j is a set of monomials G_j, their values at $t=t^k$, k=1,...,N being:

$$\hat{G} = \begin{pmatrix} 1 & x_1^{\,1} & x_2^{\,1} & x_3^{\,1} & (x_1^{\,1})^2 & x_1^{\,1}x_2^{\,1} & \ldots \\ & & & \cdot & & & \\ & & & \cdot & & & \\ & & & \cdot & & & \\ 1 & x_1^{\,N} & x_2^{\,N} & x_3^{\,N} & (x_1^{\,N})^2 & x_1^{\,N}x_2^{\,N} & \ldots \end{pmatrix}$$

The system of equations (7) is formally solved as

$$A = (\hat{G}^T \hat{G})^{-1} \hat{G}^T g . \tag{8}$$

With these coefficient values we obtain the reconstructed dynamical equations (4) which now have a predictive power. However this is not where the reconstruction procedure ends.

3.5. Selection of Coefficients

We can reduce the complexity of equations (4) with the coefficients (8) by reserving A_i that exhibit small variations δA_i within a time series. Thus, we only preserve coefficients fitting the inequality $A_i \gg \delta A_j$ and o,it (assume equaling zero) those coefficients that show unacceptable large variations $A_i \gg \delta A_j$.

3.6. Providing Global Stability

A reconstructed system is more tolerant of small disturbances once it has the property of dissipativeness [7,8]. In a dissipative system, the phase volume element gets smaller with time, corresponding to the negative value of div V, where $V_i = \dot{X}_i$.

By imposing the dissipativeness condition we can drop dubious, that is, small or highly fluctuating coefficients A_i, provided the condition div V<0 is met for the reduced model, even if, perhaps, sacrificing local predictability.

3.7. Example: Noise-polluted Lorenz system

There has been a number of reported attempts to make a consistent dynamical forecast basing on differential equations reconstructed from experimental time series. Breeden and Hubler [4] treated an idealized case of a noiseless system, reconstructing equations from experimental data with three hidden variables. Gouesbet [5] practiced an even more general reconstruction scheme than the system of equations (8): he represented the function f as a series with undetermined coefficients, similar to the function F. Reconstruction of equations in [5] was carried out in the absence of noise, from experimental data with three variables. The experimental data with three variables were also used by Brush and Kadtke [6].

In Ref.[7,8] a more complex task of reconstructing equations from a single noisy component was realized. We get back to Fig. 2. The initial realization of the variable x(t) shown in Fig. 2,a fits the Lorenz system:

$$\begin{aligned} \dot{x} &= -\delta\,(x-y), \\ \dot{y} &= rx-y-xz, \\ \dot{z} &= -bz+xy, \end{aligned} \qquad (9)$$

where σ=10, r=45, b=8/3. The $x(t)$ realization encompasses 2500 points spaced at 0.01. Each realization was added up with a wideband artificial noise of the form

$$\xi(t) = A_0\,\gamma(t)\cos p(t) ,$$

where $\gamma(t)$ and $p(t)$ are random values uniformly distributed over the intervals 0 to 1 and 0 to 2π, respectively. The amplitude of noise A_0 was 15% of that of the signal. The noisy Lorenz process is shown in Fig. 2,b.

After filtration, the noisy process $X+\xi$ was smoothed to $X(t)$, the process represented on Fig. 2,c by the unmarked curve.

The marked curve is the result of restored equations integration. Reconstruction of governing equations was performed following the above procedure. After returning to the initial variables, the reconstructed system of equations appeared to very much resemble the Lorenz system (9). At 1% noise, the coefficients of the nonlinear terms missing in the initial system (9) were on the average smaller than 0.02. Of the same order were relative errors of the reconstructed coefficients as against the initial coefficients. At 40% noise, the coefficients errors grew up to 5÷10%.

The example under discussion, Ref.[7], demonstrates effectiveness of hidden variables $y(t)$ and $z(t)$ reconstruction from single time series x(t). It was shown in Ref.[7], that values y_r and z_r follow the initial values y and z throughout the predictability time common for all three variables. This is another proof of a strong interconnection between the variables x, y and z: deviation of one variable under some forcing immediately causes departure of the other variables.

4. LIMITATIONS FOR AUTOREGRESSION MODELS

An autoregression prediction z_{n+1} is made using formulae of the type:

$$z_{n+1} = g(x_n, \dots, x_{n-m}) \ , \tag{10}$$

where $g(x_n, \dots, x_{n-m})$ is a polynomial in $x_n, \dots, x_{n-m}$ variables to the p-th degree (m is an order of autoregression). Linear autoregression (LAR) model for a given time series implies using only dependence on first degree of $x_n, \dots, x_{n-m}$:

$$z_{n+1} = a_0 x_n + a_1 x_{n-1} + \dots + a_m x_{n-m}. \tag{11}$$

Polynomial coefficients in (10) and coefficients a_k in linear model (11) are usually selected on the basis of (10) or (11) best fitting the experimental data. The autoregressive method of prediction is widely used in practice.

Our experience suggests that at fairly small noise levels (not more than 5% in amplitude), the time for which a dynamical system can be predicted by a dynamical algorithm τ^{n}_{pred} is usually several times (say 2-5 times) greater than the time τ^{NLAR}_{pred} corresponding to the algorithm of nonlinear autoregresion (NLAR), which in turn may be a little bit greater than predictability time τ^{LAR}_{pred} obtained by the linear autoregresion algorithm (LAR). The latter does not go beyond the limit τ_c of the autocorrelation of the process under review. Thus, the predictability times meet a set of inequalities:

$$\tau^{dyn}_{pred} > \tau^{NLAR}_{pred} > \tau^{LAR}_{pred} \sim \tau_c. \tag{12}$$

The statement that horizon of predictability for LAR models is a correlation interval τ_c of the observed process may be proved on the basis of degree of predictability analysis and illustrated with a diversity of examples, when degree of predictability may be defined analytically or numerically: the first- and second-order LAR-models; random processes with exponential and Gaussian correlation functions and processes, similar to them; series of dynamical nature, say, a logistic map.

Now we shall concentrate on recently studied new phenomena concerning properties of nonstationary and bifurcation nonlinear systems.

5. REVEALING OF NONSTATIONARITY IN CHAOTIC SYSTEMS

Procedure for nonstationarity retrieval in an observed process $x(t)$ are identical to discriminating the characteristics of $x(t)$, measured in two adjacent time intervals, say, in time-windows $[t_1,t_1+T_1]$ and $[t_2,t_2+T_2]$, where t_2 usual equals t_1+T_1 and time intervals T_1 and T_2 are usually equal each to other.

Distinction between characteristics of observed process $x(t)$ in two neighboring intervals may be detected by the analysis of so named discriminant function $d[x(t),a]$, where $a=(a_1,a_2,...,a_m)$ is a parameters set of a system under consideration.

Standard discriminant analysis deals with statistical characteristics of the process $x(t)$ under study and treats discriminant function values d_1 and d_2 as a moving averages of x or as variances of x inside time windows 1 and 2 correspondingly. Unfortunately statistical analysis is practically useless when the system under consideration manifests chaotic behavior. Statistical characteristics of the process $y(t)$ change only insignificantly even if system parameters change noticeably. That is why we suggest to use discriminant procedure (comparison of discriminant function values d_1 and d_2 in two neighboring windows), but for quite different type of discriminant function.

We consider it reasonable to use for d a nonlinear function of $x(t)$ and its derivatives which turns out to be zero when $x(t)$ is an exact solution of nonlinear differential (or difference) equations describing the system under consideration in noiseless situation. In order words equation $d[x(t),a]=0$ is simply a form of the governing equation, which the system obeys hypothetically to. The equation of this type contains indefinite parameters a_k, which should be determined. In this case time behavior of the coefficients set $a=(a_1,a_2,...\ a_m)$ testifies to nonstaionary properties of the system.

There is an other approach which uses Fisher criterium:

$$H = \frac{[M_1(t) - M_2(t)]^2}{[S_1^2(t) + S^2_2(t)]} . \qquad (13)$$

Here $M_1(t)$ and $M_2(t)$ are average values of a discriminant function $d[x(t),a]$ within two adjacent windows, and $S^2_1(t)$ and $S^2_2(t)$ are corresponding variations of $d[x(t),a]$. The new point is that nonlinear discriminant function $d[x(t),a]$ above described enters in Fisher criterium instead of statistical characteristics of the process $x(t)$ itself.

One may expect the faster parameters change the large criterium H. That is why we have applied such a procedure to revealing firstly a step-wise change in parameters of nonlinear system.

By the example let us consider a sequence $x(n)$ which obeys nonlinear map

$$x(n+1) = F[x(n),b], \tag{14}$$

where $b=\{b_1,b_2,...,b_{2m}\}$ is a system parameters set. Integer number n plays a role of discrete time here. As a model discriminant function we choose a difference between left and right sides of Eqn. (14):

$$d[x(n),a] = x(n+1) - F[x(n),a] \ , \tag{15}$$

where a is a set of so far indefinite parameters.

Discriminant function d, Eqn. (15), turns out to be zero in ideal, noiseless conditions, when parameters a_k to be fitted coincide with parameters b_k of real map Eqn. (14): $a_k=b_k$. Parameters a_k and b_k may be thought as coefficients of polynomial, approximating nonlinear map:

$$F[x(n),a] \approx a_0 + a_1 x(n) + a_2 x^2(n) + ... + a_m x^m(n) \ .$$

When parameter b_k of real system changes abruptly, the values $x(t)$ obey different equations (14) before and later the moment of parameter change. Resulting change in governing equation (14) may be detected by analysis of H criterium, Eqn. (13). Specific algorithms for the moment of parameter change detection implies not minimization discriminant function as above, but maximization the Fisher criterium by a proper parameters a_k choice. The efficiency of algorithm suggested we demonstrate by the example of logistic map

$$x(n+1) = r(n)x(n)[1-x(n)] \ , \tag{16}$$

where $r(n)$ is a time varying control parameters.

The Fig. 3,b shows a sequence $x(t)$ generated by a logistic map (16) with varying parameter $r(n)$. We suppose that control parameter changes abruptly at the moment $n=n_a$ from the initial value $r_a=3.80$ to the value $r_b=3.82$. At the moment $n=n_b$ control parameter $r(n)$ returns to initial value $r=r_a$. Increment of r is as small as 0.02. Thus $r(n)$ changes according to low

$$r(n) = \begin{cases} r_a, & when \quad n<n_a \\ r_b, & when \quad n_a \le n \le n_b \\ r_a, & when \quad n \ge n_b \end{cases},$$

shown at Fig. 3,a. Both values r_a and r_b succeed the critical value r_{cr}=3.5966 corresponding to chaos onset, so all sequence $x(n)$ at Fig. 3,b is of chaotic nature.

Fig. 3,b does not manifest any visible change in behavior. Average value $M_x(n)$ of $x(n)$ and variation $S^2_x(n)$ do not demonstrate any noticeably change as it may be seen from Fig. 3,c and Fig. 3,d. Practically the windows duration $T_1=T_2$ was as large as 20.

Meanwhile discriminant analysis is capable to reveal abrupt change in parameters. Let us consider quadratic discriminant function

$$d[x(n),a] = x(n+1) - a_1 x(n) - a_2 x^2(n), \tag{17}$$

which coincides with logistic map Eqn. (16) when a_1=-a_2=r. Results of the Fisher criterium computation with window length T=20 are shown at Fig. 3,e. Fisher criterium demonstrates well expressed peaks approximately as high as 5.7, though increment of control parameter was as small as 0.5%(!).

When 0.3% noise was added peaks became wider and lower ($H_{max} \approx 1.0$), as it may be seen from Fig. 3,f.

Dicriminant function of other type give much worse results. Fig. 3,g contains plot of Fisher criterium for truncated (linear) discriminant function

$$d_{tr} = x(n+1) - a_1 x(n) \ .$$

Maximal value H_{max} happens to be as small as 0.2. "Mixed" quadratic discriminant function

$$d_{mix} = x(n+1) - a_1 x(n) - a_2 x(n) x(n+1) \tag{19}$$

with mixed product x(n)x(n+1) instead of $x^2(n)$ does not provide maximal value higher than 0.5, Fig. 3,h. Fisher criterium acquires spontaneous maxima which are not related with the moments n_a and n_b of parameter r(n) jumps. Thus the results of discriminant analysis are very sensitive to the form of nonlinear discriminant function: for truncated, Eqn. (8), or "mixed", Eqn. (9), functions Fisher criterium H turns out to be correspondingly 25 or 10 times smaller as compared with adequate discriminant function (17).

Other examples of the nonlinear discriminant procedure applications (Rössler attractor, logistic map with sinusoidally varying control parameter) are given in Ref[21] and [22].

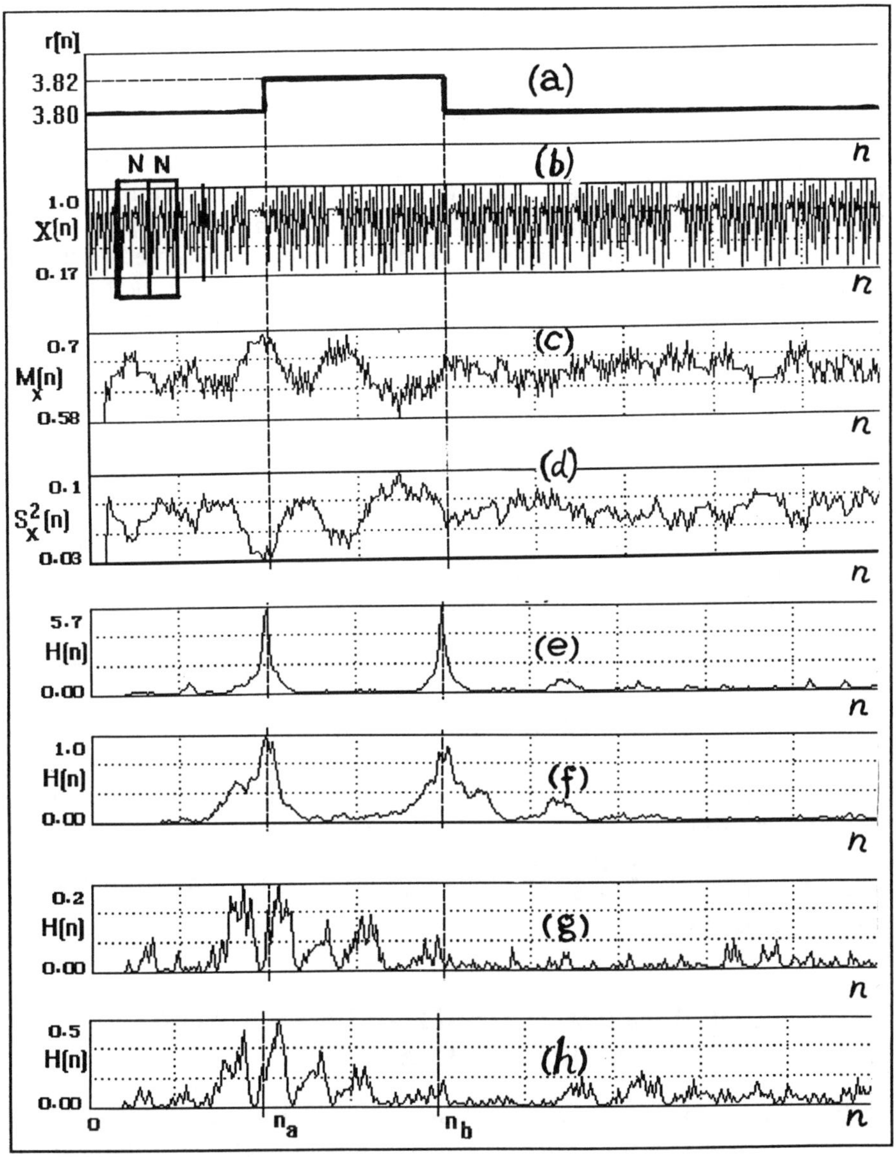

FIGURE 3. The sequence x(n) (Fig. 3,b), generated by logistic map with step-wise control parameter r(n) change (Fig. 3,a). Moving average value $M_x(n)$ and variation $S_x^2(n)$ (Figs. 1,c and 1,d) are not able to reveal 0.5% parameter r(n) change, whereas Fisher criterium H(n) with adequate nonlinear discriminant function (17) shows sharp peaks at the moments of parameter abrupt change (Fig. 3,e). Adding 0.3% noise makes the peaks wider and lower (Fig. 3,f). Nonadequate discriminant functions, truncated, Eqn. (18), and "mixed", Eqn. (19), provide much lower value of H(n) (Fig. 3,g and 3,h correspondingly).

6. PREDICTABILITY OF THE FINAL STATE IN BIFURCATION SYSTEMS WITH QUICKLY VARYING PARAMETERS

Quite recently two of us (O.L.Butkovskii and Yu.A.Kravtsov) together with J.S.Brush [23] revealed quite new property of bifurcation systems. When nonlinear systems undergo bifurcation they may wind up in any of two new stable equilibrium states. Which of states is observed during a particular trial has been considered a random variable, since the behavior of the system at the instant of bifurcation is determined by the influence of noise. The probabilities p_1 and p_2 of transition into either of two final states conventionally are considered to be equal: $p_1=p_2=0.5$.

However, in real system the final state is determined not only by noise, but also by speed of transition through the bifurcation point. When the system parameters are changed quickly enough, the noise process simply has no time to influence the resulting state of the system and transition occurs as if there were no noise at all. A noiseless bifurcation transition with finite speed turns out to be regular and fully predictable. In this case the probabilities p_1 and p_2 of the transition in two final states 1 and 2 are solely depends on the initial conditions and are either $p_1=0$, $p_2=1$ or $p_1=1$, $p_2=0$. Both these cases correspond to symmetry breaking in bifurcation systems.

It is shown in Ref. [23], that transition with symmetry breaking is separated from the probabilyticaly symmetric states by a rather narrow boundary which is described by the relationship between the transition speed and noise level. These general conclusions are illustrated in [23] by the example of the first period doubling bifurcation in logistic map and by a number of other physical situations with symmetry breaking in bifurcation systems.

7. NEW EFFECT: HYSTERESIS IN A NOISY BIFURCATION SYSTEMS

If a rate $s=dr/dt$ of control parameter r change is large enough, a specific hysteresis may be observed. By the example let us consider the first period doubling bifurcation in a logistic map. After transition through bifurcation value r_{bif}, the trajectory, instead of going to stable final states 1 or 2 (Fig. 4), continues to stay near instable state 3 for a long time. Time of locating nearby the instable state 3 depends on the noise level: the stronger noise, the quicker trajectory leaves the instable state 3 and goes to the stable states 1 or 2.

When control parameter decreases, trajectories return to the initial stable state after some time.

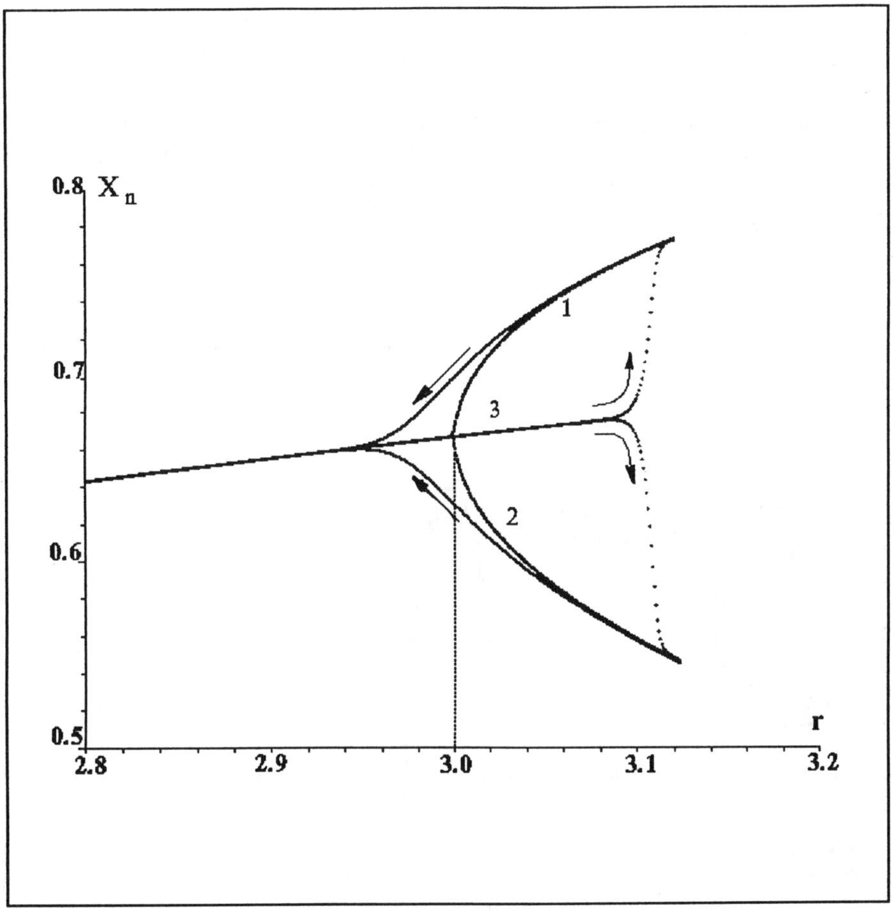

FIGURE 4. Hysteresis in the logistic map with changing control parameter r(t). Rate of control parameter change is 0.0064.

Measuring time delay $\Delta t = t_{real} - t_b$ between time t_0 of formal bifurcation transition (t_b corresponds to critical value of control parameter r_{bif}) and time t_{real} of real bifurcation, one may estimate noise level in a bifurcation system. It is important for many applications.

8. DELAYED CORRELATION <NOISE-FORECASTING ERROR>

The prevailing reason for prognosis error growth is internal (inherent) noise in the system under study. The question we would like it discuss here is to establish interrelationship between prediction error at given time moment and inherent noise, driving the system BEFORE current time.

We studied noisy logistic map and numerically derived correlation function $R_{uf}(\tau)$ between noise $f(t-\tau)$, acting time τ before t, and forecasting error $u(t)=y(t)-z(t)$, where y(t) is observed and z(t) is predicted process. Thus,

$$R_{uf}(\tau) = \langle u(t) f(t-\tau) \rangle \ . \tag{20}$$

Numerically calculated correlation function (20) is presented at Fig. 5. At $\tau=0$ correlation function equals zero due to initial condition $u(t_0) = y(t^0) - z(t^0) = 0$.

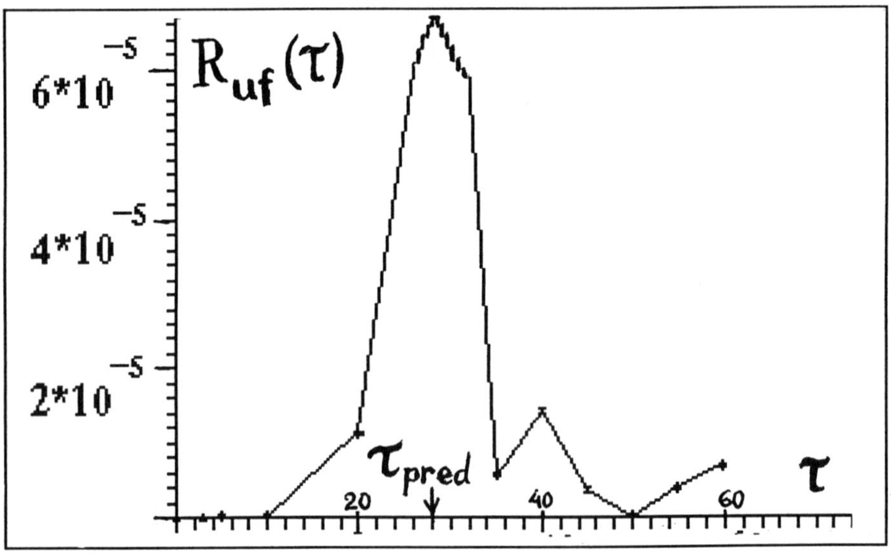

FIGURE 5. Delayed correlation <noise-prediction error> in a logistic map: control parameter r=3.80, noise level 10^{-3}. Maximum of correlation function corresponds to predictability time.

After period of exponential growth with Lyapunov exponent, $R_{uf} \sim \exp(\lambda_+ \tau)$, correlation function saturates at level $(R_{uf})_{\max} \sim A_{\sigma_f}$, where A is an attractor magnitude, and σ_f is r.m.s. value of noise. Time of saturation happens to be comparable with time of predictability (1), so that measurements of correlation function R_{uf} provide us with Lyapunov exponent λ_+ and with time of predictability (1). That is why such measurements are of practical interest.

9. POSSIBLE APPLICATION AREAS FOR DYNAMICAL PREDICTIVE ALGORITHMS

The technique for reconstructing dynamical equations from experimental time series has a wide application potential. Among the possible fields are medicine, economics and information processing and transfer.

In medicine, it can be used for revealing dynamical characteristics of heart and brain activity. The coefficients of reconstructed equations can serve as diagnostic indicators of a new type. The technique is promising as a groundwork for new methods of diagnosing heart muscle ischemia and dystrophy and as a way to facilitate the search for "dynamical" precursors of the arrhythmia.

In economics, the use of the dynamical prediction methods opens up a radically new possibility for going beyond the limitations inherent in the standard autoregression methods of prediction of the type (10). A dynamical prediction is far from an all-purpose tool in economics. It is only applicable to low-dimensional, close to autonomous, systems. Unfortunately, we have no list of such systems at hand, but suppose that macroeconom conglomerates wherein the dynamics of subsystems interaction prevails over the contribution of individual economic subjects are well suited for being analyzed and predicted by the dynamical algorithms. Another possible application filed in economics is isolation of a "noisy" nondynamical component. In truth, the notion of noise in economics still has to be defined. In our opinion, the philosophy of handling the inverse problems of nonlinear economics forms the right methodical groundwork for meeting this task. The reconstruction of dynamical systems in the area of signal processing has been addressed by J.S.Brush and J.B.Kadtke in their contributions to this conference. Similar algorithms hold much promise for reconstructing the dynamics of chemical, biological and ecological systems.

In conclusion we mention that the discussed reconstruction algorithms have underline an applications software designed to reconstruct governing equations directly from experimental data. It consists of several blocks, running which requires certain skills. The authors are ready to help those wishing to acquire such skills. Our experience tells us that more good use is made of the software application to specific fields of knowledge when potential users work closely with the developers. Along these lines, new problem-oriented versions of the software are to be developed.

REFERENCES

1. N.Winer. *The extrapolation, interpolation and smoothing of stationary time series.* N.Y.1950.
2. N.Wiener, P.Masani. *The prediction theory of multivariate stochastic processes.* Acta Math, 98, 111-150 (1957); 99, 93-137 (1958).
3. V.Mandrekar, H.Saleki, Eds. *Prediction theory and harmonic analysis.* North-Holland Publ.Co., Amst., 1983
4. J.Breeden and A.Hubler. *Reconstructing equations of motion from experimental data with unobserved variables.* Phys.Rev.A, 1990, 43, (10), 5817-5826
5. G.Gouesbet. *Reconstruction of the vector fields of continuous dynamical systems from numerical scalar time series.* Phys.Rev.A., 1991, 43 (10), 5321-5331
6. J.S.Brush and J.B.Kadtke. *Nonlinear signal processing using empirical global dynamical equations.* Proceeding of the ICASSP-92, San-Francisco, March 1992, 321-325.
7. D.A.Gribkov, V.V.Gribkova, Yu.A.Kravtsov, Yu.I.Kuznetsov, A.G.Rzhanov. *Reconstruction of dynamical system structure from time series.* Radiotekhnika i Elektronika, 1994, 39(2), 241-250 [Sov.J.Communication Technology Electronics, 1994, 39 (2)]
8. D.A.Gribkov, V.V.Gribkova, Yu.A.Kravtsov, Yu.I.Kuznetsov, A.G.Rzhanov. *Reconstruction of differential equations for autostochastic systems using single variable time series.* Zhurnal Technich.Fiziki, 1994, 64(3), 1-12 [Sov.Phys.-JTP, 1994, 64(3)]
9. R.S.Shaw. Zeitschr. Fur Naturfoschung, 36a (1), 80 (1981)
10. J.Lighthill. Proc.Royal Soc., London A407, 35 (1986)
11. Yu.A.Kravtsov. *Randomness, determinateness and predictability.* Sov.Phys.-Uspekhi, 1989, 32 (5), 434-449
12. Yu.A.Kravtsov. *Randomness and predictability in Dynamic Chaos.* In: Nonlinear Waves. 2. Dynamics and Evolution. A.V.Gaponov-Grekhov, M.I.Rabinovich, J.Engelbrecht, Eds.Springer Verlag, Berlin, Heidelberg, 1989, pp.44-56.
13. Yu.A.Kravtsov. *Fundamental and practical limits of predictability.* In: *Limits of Predictability*, Yu.A.Kravtsov, Ed., Springer Series in Synergetics, v.60, Springer Verlag, Berlin, Heidelberg, 1993.
14. L.Ljung. *System identification: theory for the user.* Prentice-Hall, 1987
15. D.Teodorescu. *Time series decomposition and forecasting.* Int J.Control, 1989, 50 (5), 1577-1596
16. P.Grassberger, R.Hegger, H.Kantz, C.Schaffrath, T.Schreiber. *On noise reduction methods for chaotic data.* Preprint WVB 92-13, Phys.Dept., Univ. of Wuppertal, Germany. Apr. 15, 1992
17. F.Takens. *Detecting strange attractor in turbulence.* Lecture Notes in Math. 898, Berlin: Springer-Verlag, 1981, p.366

18. P.Grassberger and I.Procaccia. Phys.Rev.Lett., 1983, 50 (5), 346-349
19. D.S.Broomhead and G.P.King. *Extracting qualitative dynamics from experimental data.* Physica D, 1986, 20 (2), 217-236.
20. D.A.Gribkov, V.V.Gribkova, Yu.A.Kravtsov, Yu.I.Kuznetsov, A.G.Rzhanov, O.L.Anosov, O.Ya. Butkovskii. *Dynamic equation reconstruction from the observed one-dimensional time series.* In: Dynamical Systems and Chaos (Proc.International Conf. on Dynam.Systems and Chaos, Tokyo, May 23-27, 1994) v.2, World Scientific, Singapore, 1994, pp.378-381.
21. O.L.Anosov, O.Ya.Butkovskii, D.A.Gribkov, V.V.Gribkova, Yu.A.Kravtsov, Yu.I.Kuznetsov, A.G.Rzhanov. *Discriminant analysis as applied to revealing of nonstationarity in chaotic systems.* In: Dynamical Systems and Chaos (Proc.Internal.Conf. on Dynam.System and Chaos, Tokyo, May 23-27, 1994) v.2, World Scientific, Singapore, 1994, pp.370-377.
22. O.L.Anösov, O.Ya.Butkovskii, V.V.Isakevich, Yu.A.Kravtsov. *Revealing nonstationarity in random-like signals of dynamical nature.* Radiotechnika i Elektronika, 40 (2), 255-260 (1995) [Transl: Sov.J.Comminic.Technol. Electron., 40 (2) 1995].
23. O.Ya.Butkovskii, J.S.Brush, Yu.A.Kravtsov. *Bifurcation paradox: the final state is predictable, if the transition is fast enough.* In: Predictability of Complex Dynamical Systems, Yu.A.Kravtsov and J.B.Kadtke, Eds., Springer Verlag, Berlin, Heidelberg (to be published in 1996).

CONTROLLING CHAOS

Edward Ott
Department of Physics, Department of Electrical Engineering, Institute for Plasma Research, and Institute for Systems Research, University of Maryland, College Park, Maryland
and
Mark Spano
Naval Surface Warfare Center, Silver Spring, Maryland and Applied Chaos Laboratory, Georgia Institute of Technology, Atlanta, Georgia

Abstract. The orbital complexity and exponential sensitivity of chaotic systems has the consequence that such systems offer the possibility of being feedback controlled by use of only small perturbations. The potential consequences of this recent realization are being investigated in a broad range of applications.

In the last decade scientists in many fields have increasingly found that the systems they study often exhibit a characteristic type of time evolution known as chaos. Its hallmark is wild, unpredictable behavior, a state often perplexing and unwelcome to those who encounter it. Not surprisingly, until recently, chaotic behavior was typically viewed as something to be avoided in most applications (an exception is the situation in which mixing of fluids is desired). More recently, it has been realized that chaos can actually be advantageous in many situations, and that when it is unavoidably present it can often be controlled and manipulated to obtain desired results. This article will present some of the basic ideas in feedback controlling chaos, review a few illustrative experimental results, and assess the status and future promise of the field[1]. We begin by presenting some background.

Dynamical Systems

A dynamical system is, in general, some mathematical rule for evolving the state of a system forward in time, given information on its present state and past history. Here time can be a continuous variable, in which case we denote it t; or else it can be a discrete integer-valued variable, in which case we denote it n. The case of a continuous-time dynamical system is exemplified by a system of d coupled first order autonomous ordinary differential equations,

$$d\mathbf{x}/dt = \mathbf{F}(\mathbf{x}(t)), \tag{1}$$

where $\mathbf{x}$ is a d-dimensional vector $\mathbf{x} = (x^{(1)}, x^{(2)}, \ldots, x^{(d)})$ representing the system state. Eq. (1) can in principle be solved for the orbit $\mathbf{x}(t)$ for all $t > 0$

given the initial condition $\mathbf{x}(0)$. The case of a discrete time dynamical system is exemplified by a d-dimensional map,

$$\mathbf{y}_{n+1} = \mathbf{G}(\mathbf{y}_n), \tag{2}$$

where again $\mathbf{y}$ is d-dimensional. Given an initial condition $\mathbf{y}_0$ at time $n = 0$ the system state at time $n = 1$ is $\mathbf{y}_1 = \mathbf{G}(\mathbf{y}_0)$; at time $n = 2$, it is $\mathbf{y}_2 = \mathbf{G}(\mathbf{y}_1)$; and so on.

Physicists are perhaps most used to dealing with continuous time systems. Newton's equations of motion, Maxwell's equations and Schrödinger's equation are all formulated in continuous time. We note, however, that continuous time dynamical systems can often profitably be reduced to discrete time systems by a *Poincaré surface of section.*

Chaos: Sensitivity and Orbit Complexity

The two most often used ways of characterizing chaos are (1) *exponential sensitivity* and (2) *orbit complexity.* These two are not independent properties but rather should be viewed as two sides of the same chaos coin. Either may be viewed as a way of defining chaos.

Exponential sensitivity refers to the fact that if we consider two chaotic orbits that initially are displaced only slightly from each other, then, as time evolves, the displacement between the two orbits grows, and this growth is *exponential.* Eventually, what might have been a microscopic separation between the orbits, relatively rapidly (because the growth is exponential) becomes a large displacement (i.e., of the order of the attractor size). Thus small unavoidable errors eventually defeat any attempt to predict the exact long-time evolution of a chaotic system.

By orbit complexity, we mean that many different kinds of motion are possible on a chaotic attractor. One manifestation of this is that chaotic attractors typically have embedded within them an *infinite number of unstable periodic orbits*[2].

The periodic orbits embedded in a chaotic attractor are all unstable in the sense that, if one displaces the system state slightly from a periodic orbit, this displacement grows exponentially in time. Thus periodic orbits are typically not observed in a free running chaotic system. Very infrequently a chaotic orbit may, in its ergodic wandering, approach close to a given periodic orbit. If this happens, the orbit may approximately follow the periodic orbit for a few cycles, but it subsequently moves away from it, resuming its wandering over the chaotic attractor.

Control of Chaos Based on Unstable Periodic Orbits

Both exponential sensitivity and orbit complexity provide points of view relevant to controlling motion on a chaotic attractor. In most of what follows we concentrate on orbit complexity, in particular, the presence of embedded unstable periodic orbits. (Later, when we briefly discuss the idea of *targeting control*, the exponential sensitivity view of chaos assumes center stage.)

Since we wish to accomplish control using only small perturbations, we do not expect to be able to create orbits very different from those already allowed by the dynamics on the strange attractor. Thus, we will seek to exploit the already existing unstable periodic orbits embedded in the attractor.

The approach is as follows[3]: We first determine some of the low period unstable periodic orbits that are embedded in the attractor. We then examine these periodic orbits. For each such orbit we determine the system performance that would result if that periodic orbit were actually followed by the system. (The exact measure of performance depends on the specific situation. If we use a laser as an example, the relevant measure of performance might be its output power at a given wavelength.) Typically, some of the periodic orbits will yield improved performance as compared to the free running chaotic motion, and some will not. (For instance, our hypothetical chaotic laser might have lower output power than the same laser controlled on a given periodic orbit.) We choose one of the unstable periodic orbits that yields sufficiently improved performance. Finally, we tailor our small time-dependent controls in such a way as to stabilize the chosen unstable periodic orbit. There are several methods of programming of the controls to achieve stabilization, and we describe some of these subsequently.

Loosely speaking, we can envision the control as small kicks putting the actual orbit back on the desired unstable periodic orbit whenever we sense that the actual orbit has wandered (e.g., due to noise) slightly away from the desired periodic orbit. The required control kicks are small if the actual orbit is near the desired orbit. Since chaotic orbits are ergodic on their attractor, they eventually wander close to any desired periodic orbit on the attractor, and can then be captured by a small control. Once captured, the required controls remain small (on the order of the inherent system noise).

If the situation is such that chaos control is practical, then the presence of chaos can be an advantage. The point is that any one of a number of different orbits can be stabilized, and the choice can be made to achieve the best system performance among those orbits. If, on the other hand, the attractor is not chaotic but is, say, periodic, then small controls typically can only change the orbit slightly. Basically we are stuck with whatever system performance the stable periodic orbit gives, and we have no option for substantial improvement, short of making large alterations in the system. To be concrete, one might wish to build chaos into a system for the same reason that the newer fly-by-wire fighter planes are designed to be unstable. The built-in instability gives the planes more maneuverability and lets them respond quickly to the pilots commands. In a similar fashion building chaos into a system might provide it with more flexibility than would otherwise be possible.

Furthermore, one may want a system to be used for different purposes or under different conditions at different times. Thus, depending on the use, different requirements are made of the system. If the system is chaotic, this type of adaptive behavior might be achieved without alteration of the gross system configuration. In particular, depending on the use desired, the system behavior could be changed by switching the temporal programming of the small controls used to stabilize the different orbits. In contrast, in the absence of chaos, completely separate systems might be required for each use. The flexibility provided by building chaos into devices may allow the development of smaller and lighter packages. Thus, when designing a system intended for multiple uses, purposely building chaotic dynamics into the system may provide the desired flexibility.

Methods for Stabilizing Periodic Orbits

In order to illustrate the possibility of control, we first consider the case of a d-dimensional map and rewrite Eq. (2) as

$$\mathbf{y}_{n+1} = \mathbf{G}(\mathbf{y}_n, c_n), \tag{3}$$

where c_n is a scalar control variable that we can adjust on each iterate. The free-running orbit on the chaotic attractor results if control is not applied, $c_n \equiv 0$. We assume that practical considerations constrain c_n to satisfy $|c_n| < c_{\max}$, where $c_{\max}$ is small.

For simplicity, assume that the unstable periodic orbit to be stabilized is a period one orbit of the map. We denote the location of this orbit by $\mathbf{y}_*$, where $\mathbf{y}_* = \mathbf{G}(\mathbf{y}_*, 0)$. Since we are interested in stabilization when $\mathbf{y}_n$ is near $\mathbf{y}_*$, and since c_n is small, we can expand Eq. (3) to linear order in c_n and in the deviation of the state from the periodic orbit $\mathbf{Y}_n = \mathbf{y}_n - \mathbf{y}_*$. Equation (3) becomes

$$\mathbf{Y}_{n+1} \cong \mathbf{A}\mathbf{Y}_n + \mathbf{B}c_n, \tag{4}$$

where $\mathbf{A}$ is the $d \times d$ matrix of partial derivatives of $\mathbf{G}$ with respect to $\mathbf{y}$ evaluated at $\mathbf{y} = \mathbf{y}_*$, and $\mathbf{B}$ is the d-vector of partial derivatives of $\mathbf{G}$ with respect to c_n. $\mathbf{Y}$ and $\mathbf{B}$ are regarded as column vectors.

The question now is how to program the time variation of the small controls c_n in order to drive $\mathbf{Y}$ to zero (i.e., drive $\mathbf{y}$ to $\mathbf{y}_*$). Since we decide the value to which we set c_n after observing the deviation $\mathbf{Y}_n$ from the periodic orbit, we can regard c_n as a function of $\mathbf{Y}_n$. The choice of this function specifies our chosen control scheme. Consistent with the linearity of Eq. (4), we assume a linear dependence of c_n on the deviation,

$$c_n = \mathbf{K}\mathbf{Y}_n, \tag{5}$$

where $\mathbf{K}$ is a row vector. The choice of $\mathbf{K}$ now specifies the control scheme. Equation (4) thus becomes

$$\mathbf{Y}_{n+1} = (\mathbf{A} + \mathbf{B}\mathbf{K})\mathbf{Y}_n. \tag{6}$$

[Note that, since $\mathbf{B}$ is a column vector and $\mathbf{K}$ a row vector, the quantity $\mathbf{BK}$ is (like $\mathbf{A}$) a $d \times d$ matrix.] In the absence of control, we would have $\mathbf{Y}_{n+1} = \mathbf{A}\mathbf{Y}_n$. Since the period one orbit, $\mathbf{y} = \mathbf{y}_*$, is unstable in the absence of control, the matrix $\mathbf{A}$ has at least one eigenvalue whose magnitude is greater than one. Our problem is solved (i.e., stabilization is achieved) if we can find a $\mathbf{K}$ such that the eigenvalues of the matrix $(\mathbf{A} + \mathbf{BK})$ all have magnitudes less than one. One can approach this problem purely empirically: simply choose a $\mathbf{K}$ and see if it works in the experiment; if it does not, try again. If one wants to be sophisticated, we note that the basic problem of choosing $\mathbf{K}$ is standard in control theory. In fact, if $\mathbf{A}$ and $\mathbf{B}$ satisfy a certain weak condition (called controllability), then a procedure exists such that the matrix $(\mathbf{A} + \mathbf{BK})$ can be made to have *any* desired set of d eigenvalues[4,5].

To illustrate the stabilization of the period one orbit $\mathbf{y} = \mathbf{y}_*$ in a more geometrical manner, consider the case of a two dimensional map, and assume that $\mathbf{A}$ has one eigenvalue larger than one (the "unstable eigenvalue") and one eigenvalue of magnitude less than one (the "stable eigenvalue"). Correspondingly, as shown in Fig. 1, the point $\mathbf{y}_*$ has a *stable manifold* and an *unstable manifold* such that points on the stable manifold approach $\mathbf{y}_*$ as time increases, $n \to +\infty$, while points on the unstable manifold diverge from $\mathbf{y}_*$ (or approach $\mathbf{y}_*$ going backwards in time, $n \to -\infty$). The location of an orbit point $\mathbf{y}_n$ at time n is shown in the figure. Without control this point would go to some other point, denoted $\tilde{y}_{n+1}$ in the figure. If we chose c_n so that $\mathbf{B}c_n$ displaces the orbit to the stable manifold (the point denoted $\mathbf{y}_{n+1}$ in the figure), then the control can subsequently be set to zero, and the orbit will naturally approach $\mathbf{y}_*$. (This works so long as $\mathbf{B}$ is not precisely parallel to the stable manifold[6].)

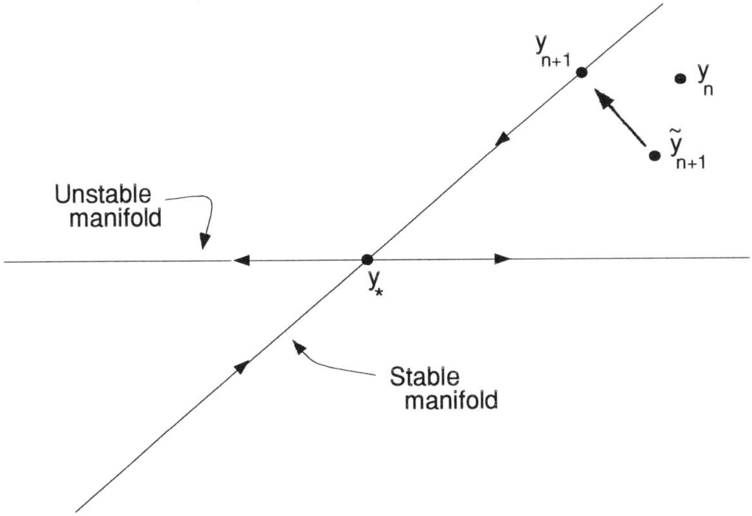

Figure 1. Control puts $\mathbf{y}_{n+1}$ on the stable manifold of y_*.

We emphasize that other means of periodic orbit stabilization have also been successfully implemented. In particular, K. Pyragas[7] considers a continuous time system of the form of Eq. (1) with a continuous time control $c(t)$, where Eq. (1) is replaced by $d\mathbf{x}/dt = \mathbf{F}(\mathbf{x}(t), c(t))$. For $c(t)$, Pyragas chooses $c(t) = \tilde{\mathbf{K}} \cdot (\mathbf{x}(t) - \mathbf{x}(t-T))$, where T is the period of the unstable periodic orbit that he wishes to stabilize. (Thus, if $\mathbf{x}$ were to follow the periodic orbit, then $c(t) = 0$.) Empirically, it is found that in interesting cases, stabilization can be achieved by proper choice of $\tilde{\mathbf{K}}$. Still another method, called occasional proportional feedback, has been proposed and implemented by E. R. Hunt[8]. This method has the advantage of not requiring detailed computations and can be carried out with comparatively little effort, although the aspect of choice of the periodic orbit is somewhat mitigated.

Real experiments are always subject to noise, both in the system itself, as well as in the measurement process. The effect of noise can be modeled in linear order by adding an additional stochastic vector to the right side of Eq. (4). The basic effect is that noise increases the minimum size of the allowed control perturbations required for effective operation[3].

Another situation often encountered is that in which the system undergoes some slow change with time, either intentionally or unintentionally (sometimes called "drift" or "nonstationarity"). Recent methods accommodate such slow changes by tracking the location of the controlled unstable orbit and its stability properties[9].

Implementation From Data

In many situations knowledge of a precise mathematical model of the system is lacking, or else there may be parameters characterizing the system that are unknown. One of the strengths of controlling chaos by means of unstable periodic orbits is that, by using data from experimental observations of the system dynamics, it has proven possible to implement control in the absence of a system model. Imagine that we set the control to zero ($c_n = 0$) and collect a long string of data $\mathbf{x}_1$, $\mathbf{x}_2 \ldots$, where $\mathbf{x}_n$ denotes the nth piercing of a surface of section. If two successive $\mathbf{x}$'s, say $\mathbf{x}_{21}$ and $\mathbf{x}_{22}$, are close, there will typically be a period one fixed point $\mathbf{x}_*$ of the surface of section map nearby. Having found one such close return we can search our data for other close return pairs $(\mathbf{x}_n, \mathbf{x}_{n+1})$ in the small region of the first observed pair. Because of the ergodicity of the chaotic orbit, we can, if desired, accumulate many such close returns, provided we observe long enough. We can then determine a best fit of the pairs data to a local linear relationship $\mathbf{x}_{n+1} = \hat{\mathbf{A}}\mathbf{x}_n + \mathbf{C}$. The fitted matrix $\hat{\mathbf{A}}$ then provides an approximation to the matrix $\mathbf{A}$ of Eq. (4), and the location of $\mathbf{x}_*$ is approximately by $\mathbf{x}_* = (1 - \hat{\mathbf{A}})^{-1}\mathbf{C}$. The vector $\mathbf{B}$ can then be determined by setting control to a constant value $c_n = c_0$ and repeating the above procedure and observing the resulting displacement Δ of the fixed point $\mathbf{x}_*$. The vector $\mathbf{B}$ is then approximated from $(1 - \hat{\mathbf{A}})^{-1}\Delta/c_0$.

In experiments it is sometimes difficult to measure the full state vector $\mathbf{x}$. As an extreme example of this situation, assume that one can only determine a single scalar function of the system state. In this case, it has been shown that, for a number of purposes, *delay coordinate embedding* provides a useful representation of the system state[10]. In particular, one can consider the delay coordinate vector,

$$\mathbf{Z}(t) = [z(t), z(t - T_D), z(t - 2T_D), \ldots, z(t - (Q - 1)T_D)] \tag{7}$$

where T_D is some conveniently chosen delay time and the number of delays Q should be large enough. Similarly, for a discrete time system (i.e., a map), if the observed scalar is w_n, then the delay coordinate vector is

$$W_n = [w_n, w_{n-1}, w_{n-2}, \ldots, w_{n-(Q-1)}], \tag{8}$$

U. Dressler and G. Nitsche[11] have shown how control can be implemented using a delay coordinate representation (see also Ref. 5). Thus control of chaos can often be achieved without full measurement of the system state and without knowledge of a global system model.

Controlling Chaos in Experiments

Chaos control techniques of the type we have just described have been implemented in a large number of experiments across a wide variety of fields. We now give a few representative examples chosen to illustrate the generality of these techniques.

A Simple Mechanical System

The first experiment was implemented by Ditto et al.[12] in a simple mechanical system consisting of a magnetoelastic metal beam clamped at its lower end. The Young's modulus of the beam could be decreased by over an order of magnitude by applying a magnetic field parallel to the beam, thereby causing the beam to buckle. This highly nonlinear system was placed in a biasing dc magnetic field and driven by a periodic ac field with a frequency on the order of 1 Hz. The position of one point on the beam was monitored by an optical sensor located at a spot near the beam's base. With appropriate choice of the dc and ac field amplitudes, the motion of the beam is observed to be chaotic. The position of the beam was measured once every drive period, experimentally taking a surface of section. Using these discrete time measurements, a simple delay coordinate embedding was constructed by plotting the current position of the beam versus its previous position ($Q = 2$ in Eq. (8)). The time evolution of the system reflected in the embedding identified the stable and unstable manifolds, as shown in Fig. 1. The application of small perturbations to the dc magnetic field, according to the scheme described above, controlled the system onto an unstable period 1 motion embedded in the attractor. By embedding the data with delays of 2, 3, 4..., higher period orbits were also found and controlled.

Electronics

A more practical application was devised by E. R. Hunt[8] of Ohio University. A driven, chaotic diode resonator circuit was controlled using a method related to that described above. This method recognizes that, for low dimensional chaos with a single control parameter, all of the feedback is controlled by a single constant, which may be determined empirically by varying the control parameter until control is achieved. The simplicity of the method entails little overhead and enables control at frequencies exceeding 50 kHz. Using this method the diode resonator has been successfully controlled at periods exceeding 80 times the drive period.

Lasers

An experiment of great practical interest was performed by R. Roy et al.[13] of the Georgia Institute of Technology. They examined the output of a solid-state Nd-doped yttrium aluminum garnet laser with a KTP doubling crystal in the laser cavity. As the pumping power of the laser was increased, the output became chaotic. The system was observed to undergo chaotic relaxation oscillations with a characteristic time scale of roughly 10^{-5} sec. Using an embedding constructed with this natural time as the delay, the researchers were able to employ the occasional proportional feedback method[8] to achieve control of this system.

Chemical Systems

The oscillatory Belousov-Zhabotinsky chemical reaction become chaotic for suitable values of the flow rates of reactants into the tank. The system may be monitored electrically by measuring the voltage of a bromide electrode placed in the tank. K. Showalter and his group at the University of West Virginia were able to stabilize[14] both period 1 and period 2 oscillations in the reactor by adjusting one of the flow rates using the same method as was used on the buckling beam.

Rollins et al.[15] of Ohio University applied chaos control to an electrochemical cell. Their approach, based on the work of Dressler and Nitsche[11], uses a recursive method to make available a wider choice of accessible system parameter for use in the control.

Biology - Heart and Brain Tissue

An experiment[16] on an *in vitro* rabbit heart septum used the drug ouabain to induce arrhythmias in the autonomous beating of the heart tissue. A discrete time embedding was constructed using the intervals between heartbeats as the system variable of interest. For this induced arrhythmia, the presence of deterministic chaos was confirmed by the observation of repeated approaches of the system state to a period 1 orbit with each approach along the same stable direction and with corresponding departures along the same unstable direction. In this case however it was not possible to find a system parameter that would move the system state point onto the stable manifold. However it was possible to directly intervene in the system by injecting a premature heartbeat with an interval timed to place the system state point onto the stable manifold. Then the dynamics of the system naturally tended to carry it toward the (unstable) period 1 motion. However because it was only possible to shorten the interbeat interval and not to lengthen it, the control achieved was, at best, period 3. Further control experiments are underway on an artificially perfused canine heart under ventricular fibrillation[17].

Tracking

As mentioned previously, drift or nonstationarity are usually present in experimental systems. Once chaos control has been established by the methods described above, an important concern is whether it will be possible to maintain control despite the drift. To deal with this question the technique known as *tracking* developed by Schwartz et al.[9] was implemented by Carroll et al.[18] in an electronic circuit and by In et al. in the magnetoelastic ribbon[19].

Targeting

As we have seen, stabilizing a system by small controls can be extremely effective once the system comes close to the desired unstable orbit. On the other hand, it might be the case that it takes an unacceptably long time for the ergodically wandering uncontrolled system to approach close enough to the desired orbit to be captured by a small control. This brings up the general problem of targeting: Given an initial condition A and a small target region B, how can we apply *small* controlling perturbations in such a way as to bring the orbit *quickly* to B? One suspects that, in the presence of chaos, this should be possible. The reason is the exponential sensitivity of chaotic orbits. Small changes eventually cause large changes. Thus, if the small changes are carefully chosen, then one might hope to effect a desired large change. This intuitive idea has been pursued[20], and algorithms for determining the appropriate controls have been developed and experimentally implemented[21]. An example predating these considerations, but nevertheless illustrating the general principal, is provided by an achievement of NASA scientists who, utilizing only small amounts of residual fuel, were able to send the ISEE-3/ICE spacecraft from an orbit near earth to an encounter with a comet half-way across the solar system[22]. This first- ever cometary encounter was effectively made possible by utilizing the chaotic exponential sensitivity of the three-body problem of celestial mechanics (the bodies being the Earth, the Moon and the spacecraft).

Conclusion

In the last few years there has been substantial theoretical, numerical, and experimental research demonstrating the feasibility of controlling systems that behave chaotically. The remaining question is whether it will prove possible to move from laboratory demonstrations on model systems to real- world situations of economic, engineering and societal importance. The prevalence of chaotic dynamics leads us to suspect that this will indeed be the case.

References

1. Some recent review articles on controlling chaos are the following: Shinbrot, T., Grebogi, C., Ott, E. and Yorke, J. A., *Nature* **363**, 411 (1993); Ditto, W. L. and Pecora, L. *Sci. Am.* **269** (2), 78 (August, 1993); Chen, G. and Dong, X., *Int. J. Bif. and Chaos* **3**, 1363 (1993); Hunt, E., *IEEE Spectrum* (1993).

2. See for example, Ott, E., Chaos in Dynamical Systems (Cambridge University Press, New York, 1993).

3. Ott, E., Grebogi, C. and Yorke, James A., *Phys. Rev. Lett.* **64**, 1996 (1990).

4. Ogata, K., Modern Control Engineering, 2nd Edition (Prentice-Hall, Englewood Cliffs, 1990).

5. Romeiras, F. J., Grebogi, C., Ott, E. and Dayawansa, W. P., *Physica D* **58**, 165 (1992).

6. In all of the previous discussions, for illustrative purposes, we have fixed attention on the case of a period one orbit of a map (possibly obtained from a surface of section of a continuous time system). We emphasize, however, that generalization to higher periods is straightforward. See Ref. 5; and Ott, E., Grebogi, C. and Yorke, J. A. in Chaos: Soviet-American Perspectives on Nonlinear Science, edited by D. K. Campbell (Am. Inst. of Phys., New York, 1990).

7. Pyragas, K., *Phys. Lett.* **A170**, 421 (1992).

8. Hunt, E. R., *Phys. Rev. Lett.* **67**, 1953 (1991); Rollins, R. W., Permanda, P. and Sherard, P., *Phys. Rev.* **E47**, R780 (1993).

9. Schwartz, I. and Triandaf, I., *Phys. Rev.* **A46**, 7439 (1992).

10. Takens, F., in Dynamical Systems and Turbulence, edited by D. A. Rand and L. S. Young, Springer Lecture Notes in Mathematics, **898** (Springer-Verlag, New York, 1980); Sauer, T., Yorke, J. A., and Casdagli, M., *J. Stat. Phys.* **65**, 579 (1991); a review article containing much information on this optic is H. D. I. Abarbanel, R. Brown, J. J. Sidorowich and L. S. Tsimiring, *Rev. Mod. Phys.* **65**, 1331 (1993); see also a the book Coping with Chaos, edited by E. Ott, T. Sauer and J. A. Yorke (John Wiley and Sons, New York, 1994).

11. Dressler, U. and Nitsche, G., *Phys. Rev. Lett.* **68**, 1 (1992).

12. Ditto, W. L., Rauseo, S. N. and Spano, M. L., *Phys. Rev. Lett.* **65**, 3211 (1990).

13. Roy, R., Murphy, T. W., Jr., Maier, T. D. and Gills, Z., *Phys. Rev. Lett.* **68**, 1259 (1992).

14. Petrov, V., Gaspar, V., Masere, J., and Showalter, K., *Nature* **361**, 240 (1993).

15. Rollins, R. W., Parmanda, P. and Sherard, P., *Phys. Rev.* **E47**, R780 (1993).

16. Garfinkel, A., Spano, M. L., Ditto, W. L. and Weiss, J., *Science* **257**, 1230 (1992).

17. Witkowski, F. X., Kavanagh, Katherine M., Penkoske, P. A., Plonsey, R., Spano, M. L., Ditto, W. L. and Kaplan, D. T., in press.

18. Carroll, T. L., Triandaf, I., Schwartz, I. and Pecora, L., *Phys. Rev.* **A46**, 6189 (1992).

19. Adaptive Control and Tracking of Chaos in a Magnetoelastic Ribbon, In, V., Ditto, W. L., and Spano, M. L., in press.

20. Shinbrot, T. et al., *Phys. Rev. Lett.* **65**, 3215 (1990); Kostelich, E. et al., *Phys. Rev.* **E47**, 305 (1993); Bradley, E. in Lecture Notes on Control and Information Services, No. 165, edited by G. Jacob and F. Lamnabhi-Lagarrigue (Springer Berlin, 1991).

21. Shinbrot, T., Ditto, W. L., Grebogi, C., Ott, E., Spano, M. L. and Yorke, J. A., *Phys. Rev. Lett.* **68**, 2863 (1992).

22. Farquhar, R., Muhonen, D. and Church, L. C., *J. Astronaut. Sci.* **33**, 235 (1985).

Controlling Spatiotemporal Chaos in One- and Two-Dimensional Coupled Logistic Map Lattices

Vladimir V. Astakhov, Vadim S. Anishchenko,
Galina I. Strelkova, Alexey V. Shabunin

Laboratory of Nonlinear Dynamics,
Saratov State University, Saratov, Russia

Abstract. A method of control of spatiotemporal chaos in lattices of coupled maps is proposed in this work. Forms of spatiotemporal perturbations of a system parameter are analytically determined for one- and two-dimensional logistic map lattices with different kinds of coupling to stabilize chosen spatiotemporal states previously unstable. The results are illustrated by numerical simulation. Controlled transition from the regime of spatiotemporal chaos to the previously chosen regular spatiotemporal patterns is demonstrated.

INTRODUCTION

Recently, in the frames of the dynamical chaos theory there has appeared a new branch. It is the control of chaos, e.g. transformation of initial chaotic behaviour of the system to the desired regular one by means of small influences on the system. This problem is connected with different fields of science and attracts a growing interest due to its potential applications (1-13).

Now, the possibility of effective control of temporal chaos has been investigated rather well. A number of works devoted to the important problem of controlling spatiotemporal chaos in high-dimensional and distributed systems has also appeared (10-13), which has suggested some methods of controlling chaos in coupled map chains.

In this work we propose a way of spatiotemporal chaos control in coupled map lattices. Forms of spatiotemporal perturbations of a system parameter are analytically determined for one- and two-dimensional coupled logistic map lattices to convert the systems to the chosen regular spatiotemporal states of the lattice. Controlled transitions of the system from the regime of spatiotemporal chaos to the spatio-homogeneous regimes with temporal periods one, two, four and to more complicated regular regimes with different spatial and temporal periods have been demonstrated by numerical simulations.

CONTROL OF SPATIOTEMPORAL CHAOS IN ONE-DIMENSIONAL COUPLED MAP LATTICES

We shall consider the coupled logistic map chains with two types of coupling.

$$x_{n+1}(i) = \alpha - x_n^2(i) + \gamma(2x_n^2(i) - x_n^2(i-1) - x_n^2(i+1)) \tag{1}$$

$$x_{n+1}(i) = \alpha - x_n^2(i) + \gamma(2x_n(i) - x_n(i-1) - x_n(i+1)) \tag{2}$$

with periodic boundary conditions $x_n(1) = x_n(N+1)$, where α is the parameter of nonlinearity, γ is the coefficient of coupling and $x_n(i)$ is a value of the dynamical variable in the point with discrete coordinate i and in discrete time moment n ($i = 1, 2, ..., N$; $n = 0, 1, 2, ...$, where N is the number of elements in the chain). With varying parameters α and γ, in the systems (1) and (2) there exists complex spatiotemporal dynamics. For example, the formation of domain structures, different spatio-periodic regimes as well as the regime of spatial intermittency of competitive structures can occur in system (1). (14–17). As the parameter α is further increased, the system transits to the regime of developed spatiotemporal chaos. In the chain with linear coupling (2) before the critical parameters values the zigzag unstability prevails. Then, the stable spatio-periodic regime with wavelength $\lambda = 2$ and temporal period $T = 2$ is formed on the basis of the zigzag unstability. While α is increased, it is changed by quasi-periodic motion, after which the transition to the spatiotemporal chaos takes place. On the basis of zigzag patterns appearance of defects is also possible as well as Brownian motion of the defects and the defect turbulence (14–17). A variety of different unstable regular spatiotemporal states co-exists on the chaotic attractor of these systems. There is an important question, how to stabilize a chaotic trajectory near these unstable structures by applying the perturbations on a system parameter.

Suppose, that unstable spatio-homogeneous states with temporal period $T = s$ and spatio-periodic structures with wavelength $\lambda = 2$ and temporal period $T = 2$, $\lambda = 3$ and $T = 2$, $\lambda = 5$ and $T = 4$ co-exist in the regime of chaos. Consider the simplest case. It is the spatio-homogeneous regime with some temporal period $T = s$. Suppose, the element of the chain (1) with coordinate i enters a small neighbourhood of the fixed-point $\bar{x}_k$ of the periodic orbit with period s. Then we present the dynamical variable in the form:

$$\begin{aligned} x_{n+k-1}(i) &= \bar{x}_k + \tilde{x}_{n+k-1}(i), \\ x_{n+s}(i) &= \bar{x}_1 + \tilde{x}_{n+s}(i), \\ (k &= 1, ..., s) \end{aligned} \tag{3}$$

(where $\tilde{x}_{n+k-1}(i)$ is a small perturbation of the dynamical variable near the fixed-point $\bar{x}_k$). We shall consider the parameter of nonlinearity α as available for control. Write it in the form:

$$\alpha = \alpha_{n+k-1}(i) = \alpha_0 + \tilde{\alpha}_{n+k-1}(i), \quad (k = 1, ..., s), \tag{4}$$

(where $\tilde{\alpha}_{n+k-1}(i)$ is a small perturbation of the parameter near its nominal value α_0).

Rewrite the system (1), taking into account (3) and (4), as follows:

$$\bar{x}_{k+1} + \tilde{x}_{n+k}(i) = \alpha_0 + \tilde{\alpha}_{n+k-1}(i) - (\bar{x}_k + \tilde{x}_{n+k-1}(i))^2 + \gamma(2(\bar{x}_k + \tilde{x}_{n+k-1}(i))^2 - 2\bar{x}_k^2) - \gamma((x^2_{n+k-1}(i+1) - \bar{x}_k^2) + (x^2_{n+k-1}(i-1) - \bar{x}_k^2)), \tag{5}$$

where $k = 1, ..., s$.

Suppose, the values of the neighbouring elements with coordinates $i-1$ and $i+1$ differ from the value of the fixed point $\bar{x}_k$ by such a magnitude, that the last term in the equation (5) can be comparable with $\tilde{x}_{n+k-1}(i)$. Then, we write the equations for the fixed point and for the evolution of the dynamical variable perturbation of the i-th element:

$$\bar{x}_{k+1} = \alpha_0 - \bar{x}_k^2, \ \bar{x}_{s+1} = \bar{x}_1, \tag{6}$$

$$\tilde{x}_{n+k}(i) = \tilde{\alpha}_{n+k-1}(i) - 2\bar{x}_k(1-2\gamma)\tilde{x}_{n+k-1}(i) - \gamma((x^2_{n+k-1}(i+1) - \bar{x}_k^2) + (x^2_{n+k-1}(i-1) - \bar{x}_k^2)), \tag{7}$$

where $k = 1, ..., s$.

Choose the parameter perturbations of the i-th element depending on its variable and on the variables of the neigbouring elements as follows:

$$\tilde{\alpha}_{n+k-1}(i) = a_1\tilde{x}_{n+k-1}(i) + a_2((x^2_{n+k-1}(i+1) - \bar{x}_k^2) + (x^2_{n+k-1}(i-1) - \bar{x}_k^2)). \tag{8}$$

Substituting (8) into (7), we obtain coefficients a_1 and a_2 at which the perturbations of the dynamical variable of the i-th element decrease in time. The form of the parameter perturbation of the system (1) to convert the system (1) from the regime of spatiotemporal chaos to the spatio-homogeneous regime with temporal period $T = s$ is as follows:

$$\tilde{\alpha}_{n+k-1}(i) = 2(1-2\gamma)\bar{x}_k(x_{n+k-1}(i) - \bar{x}_k) + \gamma((x^2_{n+k-1}(i+1) - \bar{x}_k^2) + (x^2_{n+k-1}(i-1) - \bar{x}_k^2)), \tag{9}$$

where $k = 1, ..., s$.

To test the obtained results, numerical simulation has been carried out with the coupled logistic map chain (1). Let our task be to convert chaotic behaviour to the spatio-homogeneous one with $T = 1$. At first, we determine from (6) the value of the fixed point $\bar{x}_1$. Then, having set the values of the parameters $\alpha = 1.6$ and $\gamma = 0.01$, which correspond to the regime of spatiotemporal chaos, we wait for the moment in the evolution of the system when the value of one of elements falls within the ϵ-neighbourhood of the $\bar{x}_1$. After that, the magnitude $\gamma((x_n^2(i+1) - \bar{x}_1^2) + (x_n^2(i-1) - \bar{x}_1^2))$ is evaluated. If it is less than the given ϵ, the parameter perturbation in the form (9) is introduced to influence this element. When other elements enter the given neighbourhood, the perturbations begin to act on their parameters, if the mentioned above condition is satisfied. Gradual transition of the elements to the chosen state is observed. At first, there appear small randomly distributed spatio-homogeneous patterns. Then, they gradually expand and as the result occupy the whole chain and a stationary spatio-homogeneous regime is observed in the system. The process of the control is consistently illustrated in the fig. 1a. The parameter perturbations having been turned off, the system naturally returns to the regime of spatiotemporal chaos.

Transformation of chaotic behaviour into a spatio-homogeneous one with larger temporal period is performed in the similar way. In this case the form of the parameter perturbation gets complicated as seen from the expression (9). Thus, to transit to the regime with temporal period two it is necessary to use two kinds of perturbations. One of them acts when the system elements enter the neighbourhood of the point $\bar{x}_1$, while the other perturbation acts at the next iteration when they enter the neighbourhood of the point $\bar{x}_2$. For the regime with temporal period four, there are four similar expressions of the perturbations, every of which influences the system parameter at the determined time moment and so on. The transition of the system from the regime of spatiotemporal chaos to the spatio-homogeneous regime with temporal periods $T = 2$ and $T = 4$ is illustrated in fig. 1b,c.

It should be mentioned, that both in our investigations and in the work (13), the small patterns of unhomogeneous behaviour remained uncontrollable when the system has been converted to spatio-homogeneous regimes with temporal periods larger than one. To eliminate these dislocations we have applied a small random influence on them. As the result the whole chain has converted to the regimes of spatio-homogeneous motions.

The more complicated situations are controlled transitions of the chain to spatio-periodic structures with $\lambda = 2$ and $T = 2$, $\lambda = 3$ and $T = 2$, and $\lambda = 5$ and $T = 4$. In this case our approach is as follows. We decompose all elements of the lattice on λ groups. The elements of every such group, located at the distance $p\lambda$ $(p = 1, 2, ...)$ from each other, have equal values of the dynamical variable $x_n(i) = x_n(i + p\lambda)$. Having written the equation (1) for every group of the elements we obtain the system of equations:

$$\begin{aligned} x_{n+1}(\lambda m + 1) &= \alpha - x_n^2(\lambda m + 1) + \gamma(2x_n^2(\lambda m + 1) - x_n^2(\lambda m + 2) - \\ &\quad - x_n^2(\lambda m)), \end{aligned}$$

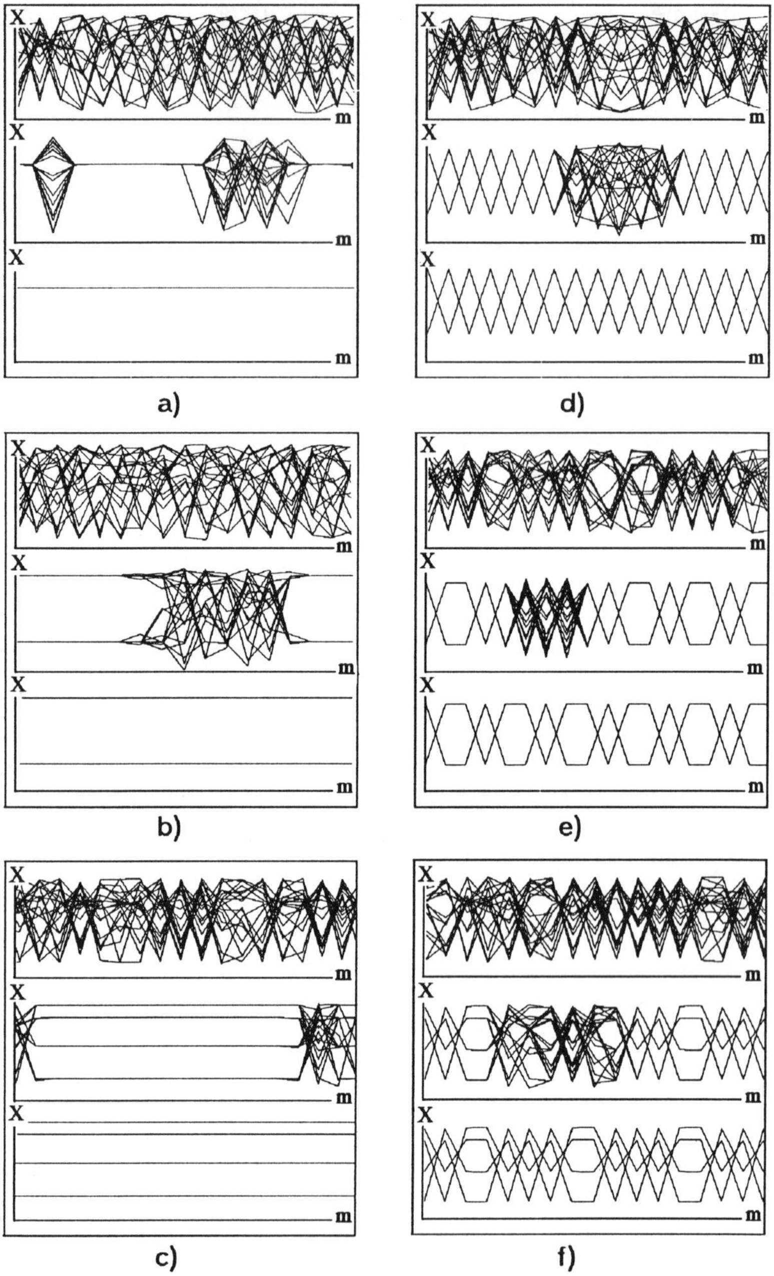

FIGURE 1. Controlled transition from the spatiotemporal chaos to the spatio-homogeneous (a - c) and spatio-periodic (d - f) regimes.

$$
\begin{aligned}
x_{n+1}(\lambda m+2) &= \alpha - x_n^2(\lambda m+2) + \gamma(2x_n^2(\lambda m+2) - x_n^2(\lambda m+3) - \\
&\quad -x_n^2(\lambda m+1)), \\
&\quad \ldots \\
x_{n+1}(\lambda m+\lambda) &= \alpha - x_n^2(\lambda m+\lambda) + \gamma(2x_n^2(\lambda m+\lambda) - x_n^2(\lambda m+\lambda+1) - \\
&\quad -x_n^2(\lambda m+\lambda-1)),
\end{aligned} \tag{10}
$$

($m = 0, 1, \ldots, (N/\lambda) - 1$). Then, using the system (10) and applying the approach considered for spatio-homogeneous regimes to every such group of the elements, we obtain expressions for the parameter perturbations to convert the coupled map chain (1) from the regime of spatiotemporal chaos to the spatio-periodic structures mentioned above.

For example, for the controlled transition to the state with $\lambda = 2$ and $T = 2$ the following form of perturbations has been used:

$$
\begin{aligned}
\tilde{\alpha}_n(i) &= 2p_1(i)(1-2\gamma)(x_n(i) - p_1(i)) + \gamma((x_n^2(i+1) - \\
&\quad -p_2^2(i)) + (x_n^2(i-1) - p_2^2(i))),
\end{aligned} \tag{11}
$$

$$
\begin{aligned}
\tilde{\alpha}_{n+1}(i) &= 2p_{11}(i)(1-2\gamma)(x_{n+1}(i) - p_{11}(i)) + \gamma((x_{n+1}^2(i+1) - \\
&\quad -p_{21}^2(i)) + (x_{n+1}^2(i-1) - p_{21}^2(i))),
\end{aligned} \tag{12}
$$

$$
p_1(i) = \begin{cases} \bar{x}_1, \\ \bar{x}_2, \end{cases} \quad p_2(i) = \begin{cases} \bar{x}_2, & \text{if } i = 2m+1, \\ \bar{x}_1, & \text{if } i = 2m+2, \end{cases}
$$

$$
p_{11}(i) = \begin{cases} \bar{x}_{11}, \\ \bar{x}_{21}, \end{cases} \quad p_{21}(i) = \begin{cases} \bar{x}_{21}, & \text{if } i = 2m+1, \\ \bar{x}_{11}, & \text{if } i = 2m+2, \end{cases}
$$

where $\bar{x}_1, \bar{x}_2, \bar{x}_{11}, \bar{x}_{21}$ are values of the fixed points (see fig.1d), $m = 0, 1, 2 \ldots, (N/2) - 1$.

To convert the lattice to the state with $\lambda = 3$ and $T = 2$, the character of the perturbations is as follows:

$$
\begin{aligned}
\tilde{\alpha}_n(i) &= 2p_1(i)(1-2\gamma)(x_n(i) - p_1(i)) + \gamma((x_n^2(i+1) - \\
&\quad -p_2^2(i)) + (x_n^2(i-1) - p_3^2(i))),
\end{aligned} \tag{13}
$$

$$
\begin{aligned}
\tilde{\alpha}_{n+1}(i) &= 2p_{11}(i)(1-2\gamma)(x_{n+1}(i) - p_{11}(i)) + \gamma((x_{n+1}^2(i+1) - \\
&\quad -p_{21}^2(i)) + (x_{n+1}^2(i-1) - p_{31}^2(i))),
\end{aligned} \tag{14}
$$

$$p_1(i)=\begin{cases}\bar{x}_1,\\ \bar{x}_2,\\ \bar{x}_3,\end{cases}\quad p_2(i)=\begin{cases}\bar{x}_2,\\ \bar{x}_3,\\ \bar{x}_1,\end{cases}\quad p_3(i)=\begin{cases}\bar{x}_3, & \text{if } i=3m+1,\\ \bar{x}_1, & \text{if } i=3m+2,\\ \bar{x}_2, & \text{if } i=3m+3,\end{cases}$$

$$p_{11}(i)=\begin{cases}\bar{x}_{11},\\ \bar{x}_{21},\\ \bar{x}_{31},\end{cases}\quad p_{21}(i)=\begin{cases}\bar{x}_{21},\\ \bar{x}_{31},\\ \bar{x}_{11},\end{cases}\quad p_{31}(i)=\begin{cases}\bar{x}_{31}, & \text{if } i=3m+1,\\ \bar{x}_{11}, & \text{if } i=3m+2,\\ \bar{x}_{21}, & \text{if } i=3m+3,\end{cases}$$

where $\bar{x}_1, \bar{x}_2, \bar{x}_3, \bar{x}_{11}, \bar{x}_{21}, \bar{x}_{31}$ are the values of the fixed points (see fig. 1e), $m = 0, 1, 2, \ldots, (N/3) - 1$.

Forms of the parameter perturbations to convert the system to the spatio-periodic structure with $\lambda = 5$ and $T = 4$ can be obtained in the similar way. The figure 1d–f demonstrates the results of numerical simulation for the controlled transition from the spatiotemporal chaos to spatio-periodic structure with $\lambda = 2, T = 2$ (fig. 1d), $\lambda = 3, T = 2$ (fig. 1e), $\lambda = 5, T = 4$ (fig. 1f). In principle, the procedure of control is the same as in the case of spatio-homogeneous states. Only the form of the parameter perturbations gets complicated: the form of the chosen structure is more complicated, the character of the perturbation is more complicated. It depends on both a coordinate i and an iteration n, when a value of the dynamical variable falls within the neighbourhood of the fixed point.

In considered situations if we use not all but a part of conditions on the parameter perturbations (11–14), the effect of control is reached also. However, in this case the time of transient process extremely grows.

Suggested approach for spatiotemporal chaos control can be applied to chains with other kinds of coupling. It is easy to get expressions for the parameter α perturbations to convert system (2) to different spatio-homogeneous and spatio-periodic regimes. For example, for controlled transition to the spatio-homogeneous regime with temporal period $T = 1$ the expression of the parameter perturbations is as follows:

$$\tilde{\alpha}_n(i) = 2(\bar{x}_1 - \gamma)(x_n(i) - \bar{x}_1) + \gamma(x_n(i-1) - \bar{x}_1) + (x_n(i+1) - \bar{x}_1)). \quad (15)$$

The parameter perturbations ensuring transition to the spatio-periodic structure with $\lambda = 2$ and $T = 2$ in the chain (2) are as follows:

$$\begin{aligned}\tilde{\alpha}_n(i) &= 2(p_1(i) - \gamma)(x_n(i) - p_1(i)) + +\gamma(x_n(i+1) - p_2(i)) \\ &\quad +\gamma(x_n(i-1) - p_2(i)), \end{aligned} \quad (16)$$

$$\begin{aligned}\tilde{\alpha}_{n+1}(i) &= 2(p_{11}(i) - \gamma)(x_{n+1}(i) - p_{11}(i)) + \gamma(x_{n+1}(i+1) - p_{21}(i)) + \\ &\quad +\gamma(x_{n+1}(i-1) - p_{21}(i)), \end{aligned} \quad (17)$$

$$p_1(i) = \begin{cases} \bar{x}_1, \\ \bar{x}_2, \end{cases} \quad p_2(i) = \begin{cases} \bar{x}_2, & \text{if } i = 2m+1, \\ \bar{x}_1, & \text{if } i = 2m+2, \end{cases}$$

$$p_{11}(i) = \begin{cases} \bar{x}_{11}, \\ \bar{x}_{21}, \end{cases} \quad p_{21}(i) = \begin{cases} \bar{x}_{21}, & \text{if } i = 2m+1, \\ \bar{x}_{11}, & \text{if } i = 2m+2, \end{cases}$$

Numerical investigations have shown that spatiotemporal chaos in the linear coupled logistic map chain can be convert to the same spatio-homogeneous and spatio-periodic states which have been considered in system (1). The transient process of system (2) from the regime of spatiotemporal chaos to the chosen states does not differ qualitatively from the corresponding processes in system (1), which have been described and illustrated in fig. 1.

Efficiency of controlling chaos in the coupled map chains (1) and (2) depends on magnitude of the parameter of nonlinearity α and the coefficient of coupling γ. For example, preliminary investigations have shown that transformation of system (1) to the spatio-homogeneous regime with temporal period one in the interval of parameters values $1.6 < \alpha < 1.9$ and $0 < \gamma < 0.4$ during 15000 iterations at $N = 20$ is reached only in the range of the parameters values allocated in the fig. 2 by hatching. We could not transfer the system from chaotic regime to the spatio-homogeneous one during 15000 iterations if γ was more than 0.04. However, transition of the system to spatio-periodic regimes has been feasible outside the allocated domain.

It is evident that the possibility to convert the system from the regime of spatiotemporal chaos to the desired spatiotemporal structure depends on the probability of the chain element to enter the ϵ-neighbourhood of the corresponding fixed point. Fig. 3 demonstrates how the probability density of the existence of a chain element variable in the fixed point $\bar{x}_1$ of the spatio-homogeneous regime with $T = 1$ depends on the parameter values. One can see that the largest magnitudes of the probability density are located in the domain, where the controlled transition to the mentioned spatiotemporal state is observed (see fig. 2).

Both our preliminary investigations and the results of work (13) don't present comprehensive answer to the question about the dependence of the parameter values on the effect of chaos control. It requires further investigations.

CONTROL OF SPATIOTEMPORAL CHAOS IN TWO-DIMENSIONAL COUPLED MAP LATTICES

The suggested approach of controlling spatiotemporal chaos can be used to lattices of higher dimension.

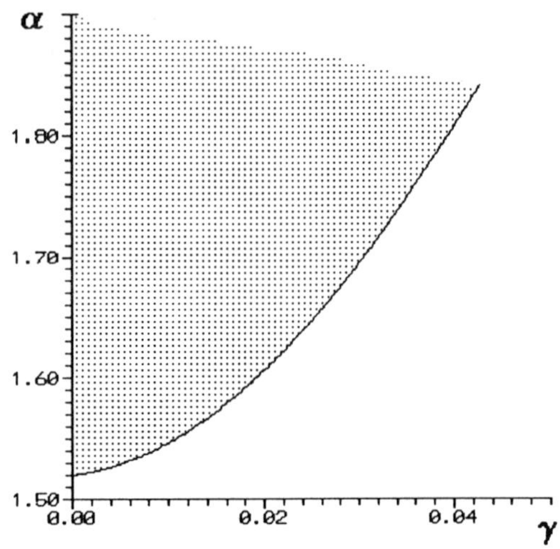

FIGURE 2. The area of parameters values where controlled transition to the spatio-homogeneous state with T = 1 is observed.

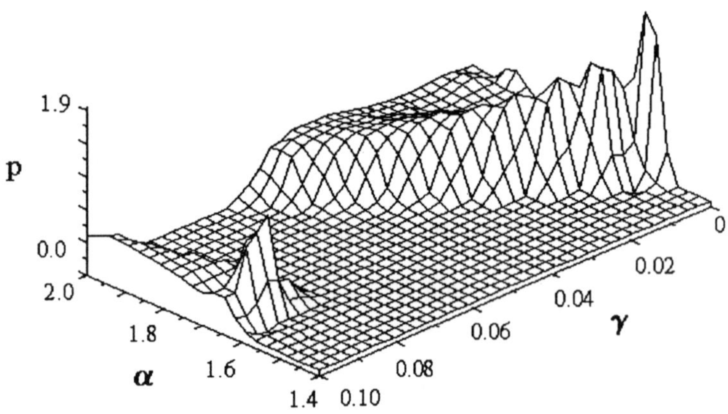

FIGURE 3. The dependence of the probability density of the chain element variable existence in the unstable fixed point $\mathbf{x}_1$ on the values of the parameters.

Consider two-dimensional coupled logistic map lattice:

$$\begin{aligned} x_{n+1}(i,j) &= \alpha - x_n^2(i,j) + \gamma(4x_n^2(i,j) - x_n^2(i-1,j) - \\ &\quad -x_n^2(i+1,j) - x_n^2(i,j-1) - x_n^2(i,j+1)); \qquad (18) \end{aligned}$$

with periodic boundary conditions. $x_n(1,j) = x_n(N+1,j), x_n(i,1) = x_n(i,N+1)$, $(i,j = 1,2, \ ..., \ N; N \times N$ is the size of the lattice).

When the parameters α and γ are varied, complex spatiotemporal dynamics is observed in the lattice. Without detailed discussion, we shall only note that there exist parameters values areas with developed spatiotemporal chaos. It is preceded by complicated bifurcational transitions between different spatiotemporal regimes (16,17).

To convert the lattice from the regime of spatiotemporal chaos to regular states, we determine the form of the parameter perturbations in the same way as for the chains. As the example, we shall write the forms of perturbations for controlled transition to spatio-homogeneous regimes with temporal period $T = s$ and to so-called "checkerboard structure" (17–19). In this structure neighbouring elements of the lattice oscillate out-phase with temporal period $T = 2$. The prototype of this regime in the one-dimensional lattices (1) and (2) is zigzag structure, which is spatio-periodic regime with $\lambda = 2$ and $T = 2$.

From the analysis of the evolution equations of the dynamical variable perturbation of an element in the neighbourhood of the noted unstable regimes, it is seen that to convert the lattice (18) to the previously unstable spatio-homogeneous regimes with $T = s$ we have to add the perturbation in the form:

$$\begin{aligned} \tilde{\alpha}_{n+k-1}(i,j) &= 2\bar{x}_k(1-4\gamma)(x_{n+k-1}(i,j) - \bar{x}_k) + \gamma((x_{n+k-1}^2(i-1,j) - \bar{x}_k^2) + \\ &\quad (x_{n+k-1}^2(i+1,j) - \bar{x}_k^2) + (x_{n+k-1}^2(i,j-1) - \bar{x}_k^2) + \\ &\quad +(x_{n+k-1}^2(i,j+1) - \bar{x}_k^2)), \qquad (19) \end{aligned}$$

(where $k = 1,2, \ ..., \ s$, $\bar{x}_k$ is the fixed point of the orbit with $T = s$) to the parameter of nonlinearity of the system.

If the perturbation has the form:

$$\begin{aligned} \tilde{\alpha}_n(i,j) &= 2p_1(i,j)(1-4\gamma)(x_n(i,j) - p_1(i,j)) + \gamma((x_n^2(i+1,j) - \\ &\quad -p_2^2(i,j)) + (x_n^2(i-1,j) - p_2^2(i,j)) + (x_n^2(i,j-1) - \\ &\quad -p_2^2(i,j)) + (x_n^2(i,j+1) - p_2^2(i,j))), \qquad (20) \end{aligned}$$

$$\begin{aligned} \tilde{\alpha}_{n+1}(i,j) &= 2p_{11}(i,j)(1-4\gamma)(x_{n+1}(i,j) - p_{11}(i,j)) + \gamma((x_{n+1}^2(i+1,j) - \\ &\quad -p_{21}^2(i,j)) + (x_{n+1}^2(i-1,j) - p_{21}^2(i,j)) + (x_{n+1}^2(i,j-1) - \\ &\quad -p_{21}^2(i,j)) + (x_{n+1}^2(i,j+1) - p_{21}^2(i,j)) \qquad (21) \end{aligned}$$

$$p_1(i,j) = \begin{cases} \bar{x}_1, \\ \bar{x}_2, \end{cases} \quad p_2(i,j) = \begin{cases} \bar{x}_2, & \text{if } (i+j) - \text{even}, \\ \bar{x}_1, & \text{if } (i+j) - \text{odd}, \end{cases}$$

$$p_{11}(i,j) = \begin{cases} \bar{x}_{11}, \\ \bar{x}_{21}, \end{cases} \quad p_{21}(i,j) = \begin{cases} \bar{x}_{21}, & \text{if } (i+j) - \text{even}, \\ \bar{x}_{11}, & \text{if } (i+j) - \text{odd}, \end{cases}$$

(where $\bar{x}_1, \bar{x}_n, \bar{x}_{11}, \bar{x}_{12}$ are the fixed points) one can stabilize the checkerboard structure.

Figures 4–6 show the results of numerical simulation of spatiotemporal chaos control in the two-dimensional coupled logistic map lattice. The lattice with 20×20 elements has been used in the experiments. We have chosen the values of the parameters $\alpha = 1.7$ and $\gamma = 0.01$ when there is a regime of spatiotemporal chaos in the lattice (fig. 4a, 5a, 6a). The procedure of control in the lattice have been the similar to that in the chain. Following the value of the variable of every element in the lattice we wait for the moment when it enters the chosen ϵ-neighbourhood. Then, the corresponding perturbations of the system's parameter are introduced if the conditions on the amplitude of the influence are satisfied. The figures consistently illustrate the controlled transitions from the regime of spatiotemporal chaos to the spatio-homogeneous ones with $T = 1$ (fig. 4), $T = 2$ (fig. 5) and to the checkerboard structure (fig. 6).

The transient process is the same as in the chains. At first, there appear small randomly distributed patterns, expanding step-by-step and as the result occupying the whole lattice (see fig. 4–6).

CONCLUSIONS

This work presents a method of spatiotemporal chaos control in one- and two-dimensional coupled map lattices. The controlled transitions from the regime of spatiotemporal chaos to different chosen regular spatiotemporal states have been demonstrated following the example of the logistic map lattices with two kinds of coupling.

Duration of the transient process is rather small. Unfortunately, to apply this method it is necessary to know the obvious form of the dynamical system for determining structure of the available system parameter. Nevertheless, we suppose the obtained results can be useful for study of processes of spatiotemporal chaos control in more complicated lattices with unknown view of a partial element.

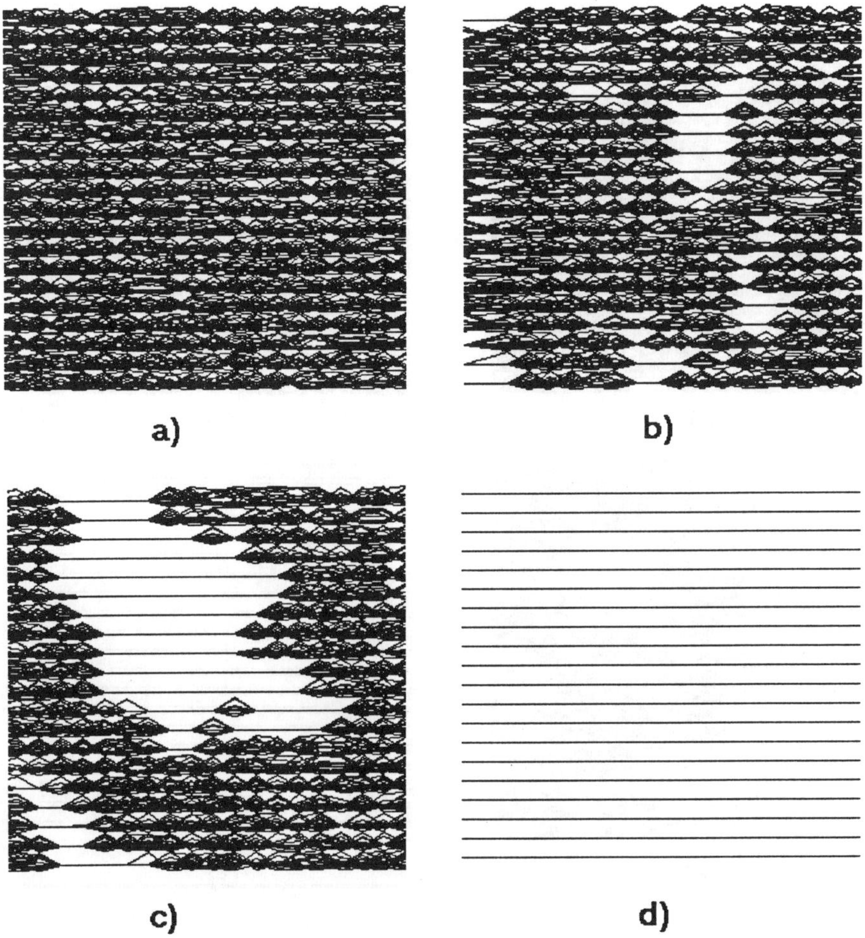

FIGURE 4. Controlled transition to the spatio-homogeneous regime with T = 1 in two-dimensional coupled map lattice. $\alpha = 1.7$, $\gamma = 0.01$.
(a) - the regime of spatiotemporal chaos, d) - the spatio-homogeneous regime with T=1. The values of the dynamical variable of every lattice element are presented during 20 iterations.)

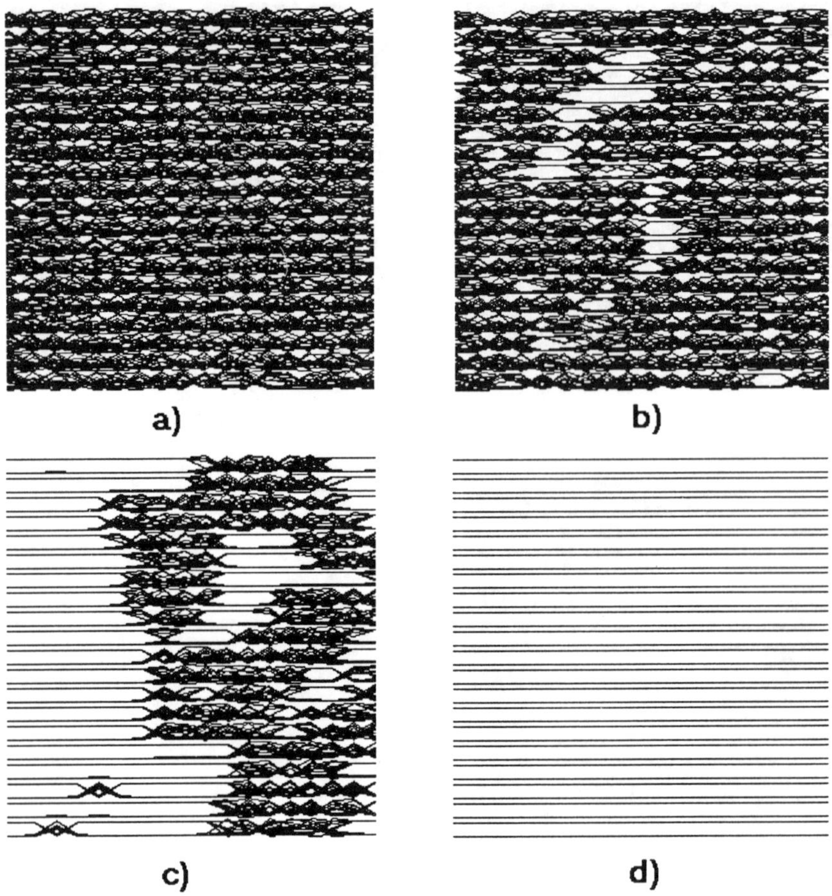

FIGURE 5. Controlled transition to the spatio-homogeneous regime with T = 2 in two-dimensional coupled map lattice. $\alpha = 1.7$, $\gamma = 0.01$.

(a) - the regime of spatiotemporal chaos, d) - the spatio-homogeneous regime with T=2. The values of the dynamical variable of every lattice element are presented during 20 iterations.)

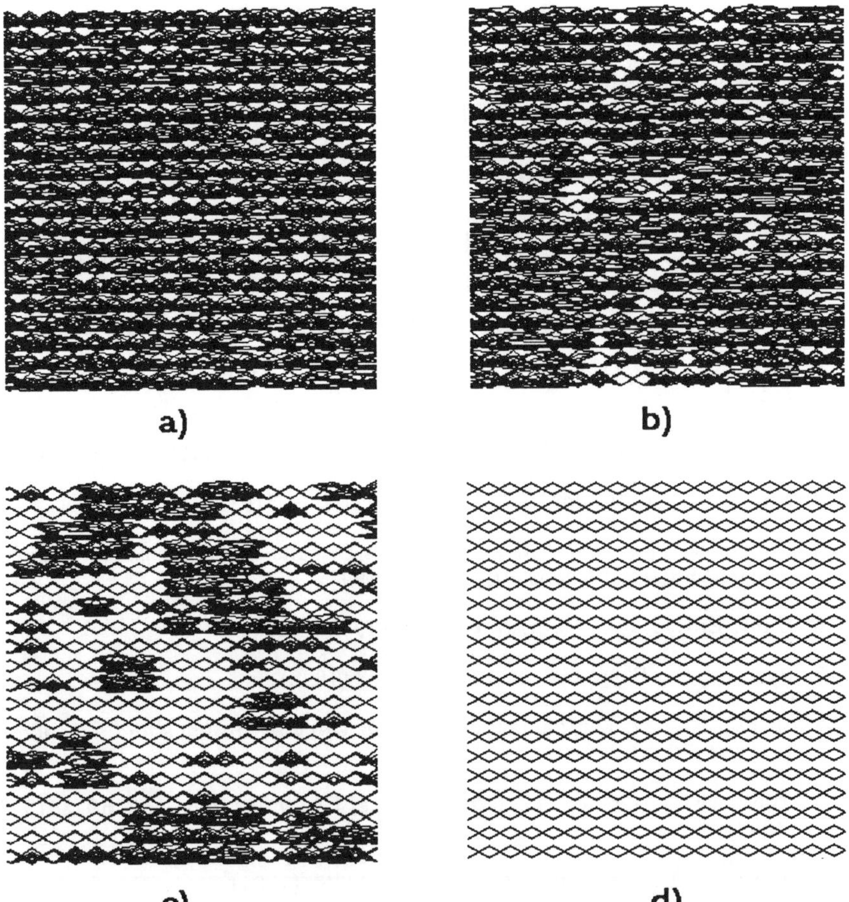

FIGURE 6. Controlled transition to the "checkerboard structure" in two-dimensional coupled map lattice. $\alpha = 1.7,\ \gamma = 0.01.$ (a) - the regime of spatiotemporal chaos, d) - the "checkerboard structure". The values of the dynamical variable of every lattice element are presented during 20 iterations.)

ACKNOWLEDGMENTS

The work was supported by International Science Foundation (grant RNO 000) and Russian Foundation of Fundamental Natural Sciences (grant 93-8.2-10)

REFERENCES

1. Ott, E., Grebogy, C., Yorke, J.A., *Phys. Rev. Lett.* **64**, 1196–1199 (1990).
2. Romeiras, F.J., Grebogi, C., Ott, E., Dayawansa, W.P., *Physica* **D 58**, 165–192 (1992).
3. Braiman, J., and Goldhirsch, J., *Phys. Rev. Lett.* **66**, 2545–2548 (1990).
4. Kapitaniak, T., *Chaos, Solitons and Fractals* **2**, 519–530 (1992).
5. Ogorzalek, M.J., *IEEE Transactions on Circuits and Systems – Fundamental Theory and Applications* **40**, 700–706 (1993).
6. Hunt, E.R., *Phys. Rev. Lett.* **67**, 1953–1955 (1991).
7. Pyragas, K., *Phys. Lett.* **A 170**, 421–428 (1992).
8. Murali, K., and Lakshamanan, M., *J. of Circuits, Systems and Computers* **3**, 125–137 (1993).
9. Auerbach, D., Grebogi, C., Ott, E., Yorke, J.A., *Phys. Rev. Lett.* **69**, 3479–3482 (1992).
10. Auerbach, D., *Phys. Rev. Lett.* **72**, 1184–1187 (1994).
11. Hu Gang, and Qu Zhilin, *Phys. Rev. Lett.* **72**, 68–71 (1994).
12. Lai, Y.C., and Grebogi, C., *Phys. Rev.* **E 50**, 1894–1899 (1994).
13. Youssefmir, M., and Huberman, B.A., submitted to *Physical Review E.*
14. Waller, I., Kapral, R., *Phys. Rev.* **A 30**, 2047–2055 (1984).
15. Kaneko, K., *Physica* **D 34**, 1–41 (1989).
16. Crutchfield, J.P., Kaneko, K., "Phenomenology of Spatio - Temporal Chaos", in *Directions in Chaos*, Singapure: World Scientific Publ., 1987, pp. 272.
17. Kuznetsov, A.P., and Kuznetsov, S.P., *Radiophysics, Transactions of higher schools* **34**, 1079–1115 (1991).

3. DETECTION AND CLASSIFICATION

System Identification in Experimental Data

Stephen Hammel and P. W. Bo Hammer

Nonlinear Dynamics and Wavelets Group, B44
Naval Surface Warfare Center
White Oak, Maryland 20903-5640

Abstract

A technique to identify the state of a dynamical system is proposed. The technique is based upon an identification of all period-one orbits present in the system. These orbits can then be classified in a way that permits an organization into a hierarchical ordering. The scheme is applied to time-series data gathered from a carefully constructed damped driven pendulum.

INTRODUCTION

Over the last twenty years, there have been numerous fruitful applications of the techniques developed in the study of chaotic dynamical systems. Tools which had been developed to advance the theory of nonlinear dynamics have been found to generate powerful data analysis techniques. In fact, it is part of the nature of novel experimental directions that novel forms of data analysis often must be co-developed.

One noteworthy aspect of some of the new techniques is the utilization of the complexity of the signal as an asset. A first example is the method of embedding, which has been used extensively on experimental data. There are several embedding procedures, all of which seek to re-constitute the full higher-dimensional nature of the background systems, given a one-dimensional time-series record. These techniques work well when the signal samples a reasonably substantial portion of the phase space. A second application of chaotic dynamics to data analysis is the signal-noise separation problem. Given an irregular signal, the goal is to separate the deterministic portion from the random, uncorrelated portion. Again, a number of practical techniques explicitly utilize the "chaotic" nature of the signal as an asset in the noise reduction process.

The classification procedure discussed in this paper was initiated to extend current techniques to characterize the state of a complex system. In this paper a signal is complex if it has broad band components in its power spectrum, and if it deviates substantially from a simple harmonic motion. We are not necessarily making a statement about the nature of the background system creating the signal. The initial premise is that we are given several different time series records, each of modest size and each recorded at a different occasion. We seek to understand how these records compare, and thence to determine if the background system has changed significantly.

A classical answer to such questions is to measure the power spectral density of each time series, and compare them. In certain cases, when strong harmonic components are present, this procedure can work well. However, there are many possible changes in system state that are difficult to detect in the power spectrum.

Our interest is systems for which simple periodic motion is somewhat exceptional, and chaotic motion is the more prevalent condition. The system should also exhibit transitions between periodic motions and chaotic motions within the same time series. The classification problem is restricted a search for some characteristic such that a change of state of the system can be recognized. It is not expected that we can develop a model for the system, or fully decompose it into constituent modes.

We will report here on preliminary progress in the classification problem. Our objective is to develop some new methods of classifying the behavior of a "simple" nonlinear dynamical physical system. We use a driven, damped, rigid pendulum as a typical case, since systems which are oscillatory are nearly ubiquitous. It should be emphasized that a mature classifier should operate without any dependence on the availability of a good model.

We define classification as a map from the space of time-series to a *hierarchical parameter space.* This means that an order relation has been defined for a multi-dimensional set of parameters. Thus a successful classification scheme will allow any two time series to be ordered in relation to each other, and to other time series, by insertion into the the hierarchy. We are building a relative descriptor rather than an absolute one.

The method is based upon the periodic orbits exhibited by a system. These periodic orbits can be attracting, or they can be periodic saddle orbits. The importance and utility of periodic orbits as hierarchical classifiers was noted by Cvitanovic [1]. We follow his usage primarily for the sake of brevity, and use the term cycles for periodic orbits from this point. Several different groups have analyzed experimental data by means of the topological organization of cycles [2], [3]. These investigations use cycles of periods greater than one as an integral part of the analysis. This stands in contrast with our work: we base the entire classification scheme upon the one-cycles (period-one orbits) that can be located within the attractor.

THE DAMPED DRIVEN PENDULUM

The equation for the motion of the rigid driven pendulum is:

$$\ddot{\theta} + \lambda\dot{\theta} + \omega_0^2 \sin\theta = G\sin(\omega t). \tag{1}$$

This can be rewritten as an autonomous system:

$$\begin{aligned} \dot{\theta} &= \alpha \\ \dot{\alpha} &= -\lambda\alpha - \omega_0^2 \sin\theta + G\sin(\omega\tau) \\ \dot{\tau} &= 1 \end{aligned} \tag{2}$$

with $(\theta, \alpha, \tau) \in \mathbf{R}^3$. Since θ is 2πperiodic, and the driving term is $(2\pi/\omega)$-periodic, we project system (2) onto $S^1 \times \mathbf{R} \times S^1$, and henceforth consider:

$$(\theta, \alpha, \tau) \in S^1 \times \mathbf{R} \times S^1$$

with $\theta = \theta \bmod 2\pi$ and $\tau = \tau \bmod (2\pi/\omega)$. However, in some of the subsequent definitions it is convenient to return to the formulation $(\theta, \alpha, \tau) \in \mathbf{R}^3$ which will be referred to as the covering space. A comparison with experiments using the physical pendulum generated the following values for the parameters: $\omega_0 \approx 8.98$, $\lambda = 1.85$, $\omega = \omega_0 1.1(\sqrt{5}-1)/2$.

The periodicity in τ is motivated by the definition of a Poincaré return map for equation (2). Using this map, a $(\theta, \dot{\theta})$-plane is defined for $\tau = 0 \bmod (2\pi/\omega)$. This means that for the flow ϕ^t which solves equation (2), the Poincaré map $P : x \mapsto \phi^{2\pi/\omega}x$ provides a cross-section to the flow. This cross-section generated by the Poincaré map gives a useful snapshot of the dynamics. The process is based upon one-cycles (period-one orbits) which are *fixed points* of the Poincaré map P.

The method described in the following sections utilizes the hyperbolicity and the dissipative nature of equation (2). The dissipation is readily determined by finding the divergence for system (2):

$$\nabla \cdot F = -\lambda$$

Hence phase space volumes contract as the system evolves, and under the Poincaré return map $P(V_0) = V_1$, with

$$V_1 = \exp(-\frac{2\pi\lambda}{\omega})V_0. \tag{3}$$

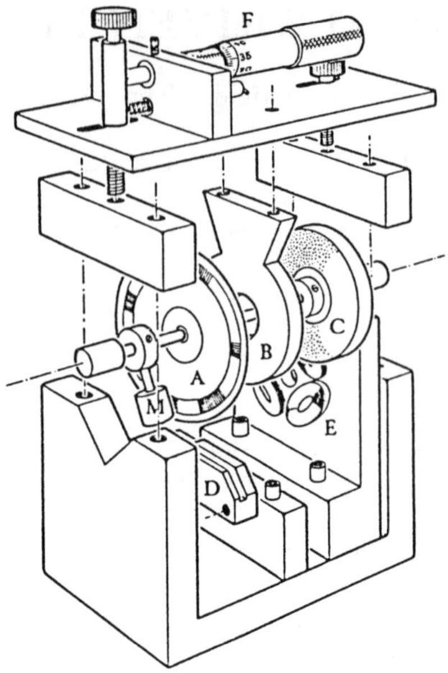

Figure 1: The physical pendulum available from Daedalon Corporation: a rigid bob M is fixed to a shaft driven by the brushless, slotless motor which consists of permanent magnets within disk C, driven by coils E. Eddy current damping is adjusted by moving the plate B; and time series data is converted from optical interruptions taken from the slotted disk A.

THE EXPERIMENTAL PENDULUM

The experimental device consisted of a small rigid pendulum with damping and driving, commercially available from Daedalon Corporation [4]. Most of the relevant features of the pendulum are readily apparent in Fig. 1. The rigid bob is free to move in the vertical plane, including rotation over the top.

The pendulum apparatus was carefully calibrated, and the relevant physical parameters were measured. The resonant frequency $\omega_0 \approx 8.98$, and we set the damping factor $\lambda = 1.85$. We chose a driving frequency which was not easily resonant with the natural frequency ω_0, namely $\omega = \omega_0 1.1(\sqrt{5}-1)/2$. The driving function $G\sin(\omega t)$ was generated by a function generator. The control parameter is the forcing amplitude G, which we investigated for the range $0 < G < 300$.

ISOLATING CYCLES

Before presenting a description of the classification process via one-cycles, it is appropriate to discuss some of the rationale behind this approach. We recorded a large amount of map data, generated by a Poincaré section. However, much of our data failed to settle into a long-term steady-state behavior. There are several reasons for this. For some system states, the embedded cycles may be attracting, but the system may not relax to that particular cycle. This is a manifestation of the common phenomenon of multiple basins of attraction. Furthermore, the basin boundaries may be fractal, and the system may persistently and yet intermittently switch between different sets of cycles.

Comparisons with our model system indicated that, for larger driving amplitudes G, equation (2) was overly simplistic, and other terms (such as damping effects due to motion of the pendulum bob in air) should be included, as well as nonlinearities in the motor drive mechanism. These effects, as well as experimental noise, could perturb orbits sufficiently to change anomalously the basins of attraction.

CYCLE CLASSIFICATION

The process of locating cycles within the system is actually a problem of *approximating* these cycles. This holds for both the numerical simulation and for the experimental data analysis. We seek segments of trajectories that are sufficiently close to the one-cycle to represent the cycle well. This is an effective procedure because the system is dissipative and hyperbolic (although not uniformly hyperbolic). The examples presented in the following sections are generated in this way.

In order to refine the classification of cycles, it is necessary to distinguish different one-cycles. We will define two quantities that are the basis of the one-cycle classification scheme—the winding number ρ, and the extent χ. The first one-cycle to be observed, visible for small $G > 0$, is a simple closed curve, with $-\pi < \theta < \pi$. As the forcing G is increased, eventually a first "running" mode is encountered: the pendulum swings over the top, and continues to swing over the top in the same direction, once for each period of the driving.

To define the *winding number* ρ for a cycle, we return to the original definition of the (θ, α, τ) from equation (2), with $(\theta, \alpha, \tau) \in \mathbf{R}^3$, the covering space. Then

$$\rho = \lim_{t \to \infty} \left(\frac{\theta(t)}{\omega \tau(t)} \right) \qquad (4)$$

A second identifier can distinguish cycles with the same winding number.

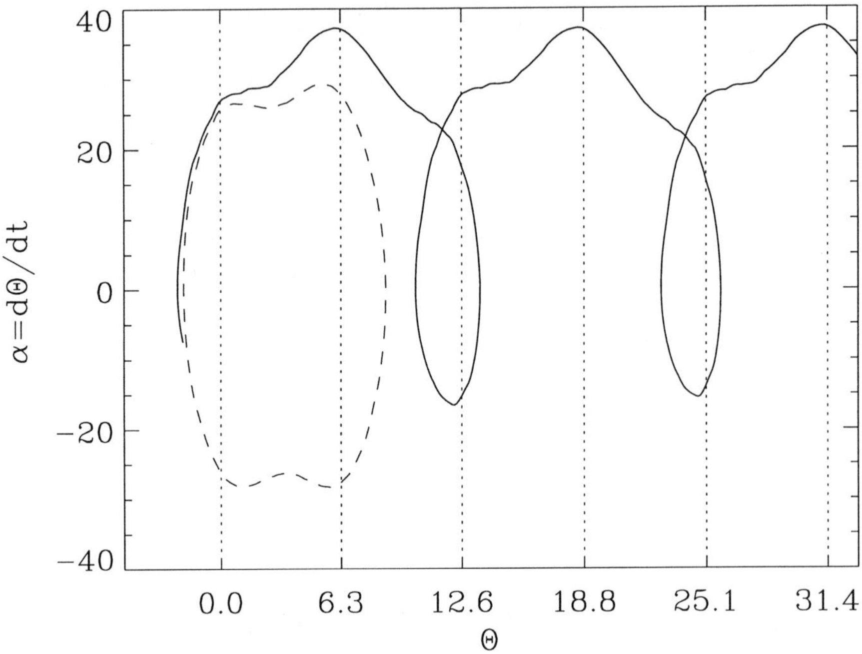

Figure 2: Two attracting orbits generated by the experimental pendulum. The vertical dotted lines indicate $2\pi-$periodic panels in this covering space view of the cylinder. One orbit is a closed loop, hence it has $\rho = 0$, and extent $\chi = 1.68$. The second orbit has $\chi = 2.51$ and winding number $\rho = +2$, which is clearly apparent in the figure, as the cycle advances through 2 of the $2\pi-$cylinders per period.

For a $k-$cycle (period k orbit), the *extent* χ is defined by:

$$\chi = \max_{T \in [0, 2\pi k/\omega]} \frac{1}{2\pi} \left[\max_{t \in T_1} \theta(t) - \min_{t \in T_1} \theta(t) \right] \tag{5}$$

where $T_1 = [T, T + 2\pi k/\omega]$. We want this quantity to be defined uniquely for all cycles, whether $\rho = 0$ or $\rho \neq 0$. When $\rho = 0$, the extent χ is simple and obvious to define. However, for $\rho \neq 0$, in order for the quantity to be well-defined, it is necessary to examine a length of the cycle over one period of the driving, and choose the initial point such that the cycle extends over a maximal region.

The model equation has the symmetry

$$(\theta, \alpha, \tau) \mapsto (-\theta, -\alpha, \tau + \pi/\omega) \tag{6}$$

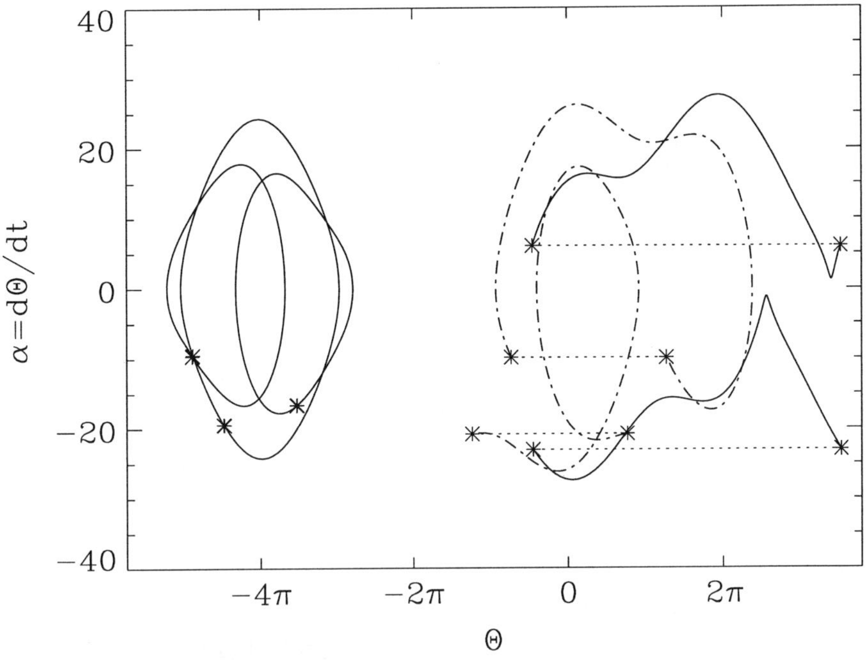

Figure 3: The one-cycles for the model with $G = 100$. There are three symmetric pairs of cycles shown, with $\rho = 0, \rho = \pm 1$, and $\rho = \pm 2$. The seventh cycle shown is self-symmetric. The offset of -4π for the three $\rho = 0$ cycles is simply for clarity.

This simplifies the classification problem, since every winding mode $\rho = +k$ and extent $\chi = d$ has a symmetric mode with $\rho = -k$ and extent $\chi = d$. If only one of a symmetric pair of cycles is located, the second can be readily found using the relation (6). Each cycle has a distinct, symmetric twin *except* for cycles which are self-symmetric.

For cycles with $\rho = 0$, there are two possibilities. First, the two points of the symmetry (6), (θ, α, τ) and $(-\theta, -\alpha, \tau + \pi/\omega)$, can lie on the *same* cycle. Or each point can exist on a different cycle, so that the pair together have the symmetry (6), but each alone is not self-symmetric. These features are illustrated in Figure 3. A self-symmetric cycle ($\rho = 0$) can undergo a pitchfork bifurcation and spawn an attracting pair of cycles which are asymmetric in themselves, but the pair satisfies (6).

We chose to illustrate the full set of identified cycles at $G = 100$ since there are only seven distinct cycles located by the algorithm. As G increases, more one-cycles appear, making the figure difficult to read. Three of the one-cycles have winding number $\rho = 0$; these appear as closed curves. For clarity in Figure 3 they have been displaced by -4π in the covering space for the cylinder $(\theta, \dot{\theta})$. A single self-symmetric cycle can be seen. Also a pair which satisfy the symmetry (6) overlap the single cycle. These two appear via a pitchfork bifurcation from the self-symmetric cycle, which is unstable here.

There are two more pairs of orbits, and each pair is coupled by the symmetry (6). One pair has $\rho = \pm 1$; this is the pair of cycles indicated by the dashed lines. The straight dotted lines simply show the 2π change in θ after one period of the driving, hence the non-zero winding number ρ. The second pair, for which the dotted straight line is twice as long, has $\rho = \pm 2$. Thus, the upper cycle winds $+4\pi$ in θ with every period of the driving. Similarly, the bottom cycle winds -4π per driving period. The cycles are symmetric under (6); the π/ω offset is responsible for the "asymmetry" of the endpoints.

The $\rho = \pm 1$ pair also nicely illustrates the importance of the maximization process in definition (5). As they are shown in Fig. 3, $\|\theta_{max} - \theta_{min}\|$ is not the same for the pair. One of the two has attained the maximum required by the definition. The endpoints for the second must be moved to verify that it has precisely the same χ-value.

A MULTI-PARAMETER BIFURCATION SPACE

The classification process is based upon a classification stable and unstable one-cycles. One-cycles are identified by seeking close returns in the period-one map. In practice, this involves a determination of a threshold near-return tolerance. Period-one segments of the flow are examined for close returns, and a set of candidates is generated. Each of the candidates can be checked, and redundant cases can be rejected. For example, near attracting cycles a number of trajectory segments will pass the close return test. The one with the closest return is retained, and the others are discarded. We hope to isolate at least one portion of the trajectory that is near enough to a one-cycle to yield a reasonable representation of that cycle. The fact that the system is dissipative and hyperbolic ensures that relatively small data sets will yield reasonable approximations to some of the one-cycles.

The construction of the multi-parameter bifurcation space can be accomplished for both the experimental and for the model system. For the model system, the differential equation solver is initialized with a grid of initial conditions, and the ensemble is integrated through two periods of the driving function. This generates the Poincaré return map described above. The initial grid of points is defined $(\theta, \dot{\theta}) \in S_{grid}$. The reference section S_0 is defined by $S_0 = P(P(S_{grid}))$. Next, one more full period of the integrator yields

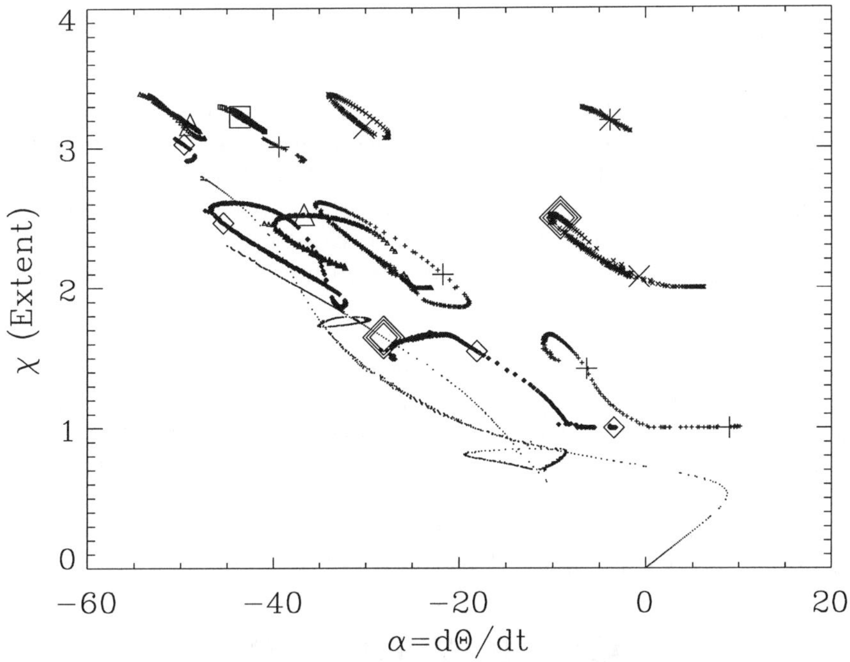

Figure 4: A projection of the five dimensional bifurcation space $(\theta, \dot{\theta}, G, \chi, \rho)$ for the pendulum one-cycles. $\mid \rho \mid$ increases with increasing χ. $\rho = 0$ for simple dots; for the other points, $\diamond \rightarrow \rho = -1$; $+ \rightarrow \rho = +1$; $\triangle \rightarrow \rho = -2$; $\times \rightarrow \rho = +2$; $\square \rightarrow \rho = -3$; $* \rightarrow \rho = +3$. The larger triple nested $\diamond$ locate two simulated one-cycles that are near experimental one-cycles.

$S_1 = P(S_0)$, defining the first-return map. The location of one-cycles is approximated by close returns in the map $P : S_0 \mapsto S_1$. A close return generates endpoints for an orbit segment that is close to a one-cycle.

Note that S_0 is generated from S_{grid} using only two iterations of the return map. This is because the dissipation in the return map is substantial (see eqn. 3), and roughly speaking, and volumes are reduced like $V_1 \approx e^{-\lambda} V_0$ for our choice of parameters, and $\lambda = 1.85$.

This definition of S_0 means that a set of initial conditions will have relaxed well onto a neighborhood of the unstable manifold set supporting the attractor. We want to cover the underlying unstable manifold with initial conditions, in order that *saddle* cycles can be located. Otherwise, if the system is allowed to relax completely onto the attractor, some of the saddle cycles will be stranded and therefore numerically invisible.

The driving amplitude G is the control parameter for our studies. To generate a full multi-parameter bifurcation space we increment G in small steps. (Figure 4 shows one of the projections of this space.) All of the one-cycles are determined at a small value of the driving. As G is increased from $G \approx 0$, a set of points is generated, for which $(\theta, \dot{\theta})$ is on a one-cycle at the Poincaré section define by $\tau = 2\pi n/\omega, n = 0, 1, 2, \ldots$ The extent χ and the winding number ρ can be measured for each of the one-cycles. Thus a discrete set of points is determined at this particular G-value. The coordinate axes of this multi-parameter space are $(\theta, \dot{\theta}, G, \chi, \rho)$, or as defined in eqn. (2), $(\theta, \alpha, G, \chi, \rho)$.

The locations of the one-cycle fixed point set are continued on to the new G value. The full multiple parameter bifurcation space can be constructed in this way. One of the goals of system identification is to determine the unknown G as a function of some of the remaining variables. The multi-factorial nature of this space should permit an identification even in cases when some of the variables cannot be measured.

The utility of these particular parameters can be seen by noting that in a rough sense, $\dot{\theta}$ and χ increase for larger values of G.

APPLICATION TO EXPERIMENTAL DATA

We locate the one-cycles for the experimental data in an analogous way. The search in this case is along a single long trajectory, which may either approach a one-cycle closely, or coincide with a one-cycle. Note that a first-return map with return-time equal to one driving period can be defined just as well for this sort of single trajectory, as for a large ensemble of initial conditions integrated forward one period. A long trajectory is simply snipped into one-period segments.

The application of these methods to experimental data generated by the driven pendulum can now be described. The apparatus is set at a particular value of the driving amplitude. Continuous data for the $(\theta, \dot{\theta})$ values for the pendulum position is sampled at $\approx$ 146 Hz. The resultant trajectory is regarded to be a short-time representative of the behavior of the system at that parameter value.

Severe limitations in the data acquisition apparatus meant that continuous flow data could be taken for only 36 driving periods before the buffer overflowed. The acquisition apparatus subsequently failed altogether, so we must describe experimental results from a true paucity of data. However, in Fig. 6 two approximations to the embedded one-cycles are shown. These were generated by setting the experimental driving amplitude to $G_{exp} = 137.2$.

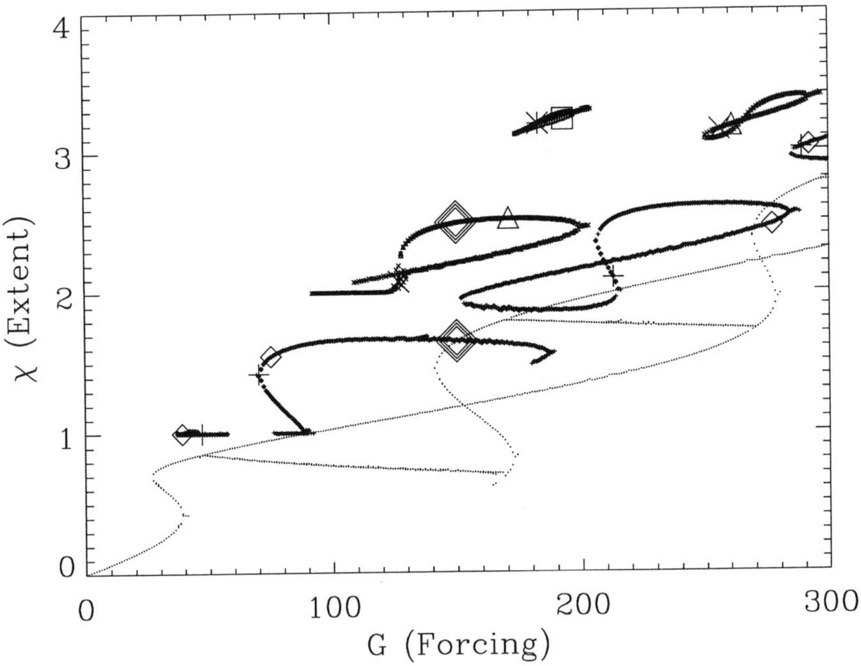

Figure 5: A different projection of the five dimensional bifurcation space $(\theta, \dot{\theta}, G, \chi, \rho)$ for the pendulum one-cycles, onto the $\dot{\theta} - \chi$ plane. The same correspondence for symbols and ρ−values as in the previous figure. The larger triple nested ◇ mark the same two simulated one-cycles mentioned in the previous figure that are near experimental one-cycles.

The utility of the multi-parameter bifurcation space derives from the possibility of measuring some or all of the values $(\theta, \dot{\theta}, \chi, \rho)$ at the Poincaré section. In Figure 6, two cycles for the experimental pendulum are shown as dotted lines. The system was observed to settle into a neighborhood of one of the cycles, and then meander, settling into the second, mildly attracting cycle. A traditional bifurcation diagram does not help locate the parameter value of the system. The input driving value for the experiment was set at $G_{exp} = 137.2$. At that value for G, the model is in a chaotic regime.

But it is easy to measure the values of $\theta, \dot{\theta}, \rho$, and χ for the experimental one-cycles. For one cycle, $\rho = 0, \chi = 1.68$; for the second, $\rho = +2, \chi = 2.51$. This measurement permits a comparison in the multi-dimensional bifurcation space. This particular realization of the system can be placed in a relative ordering.

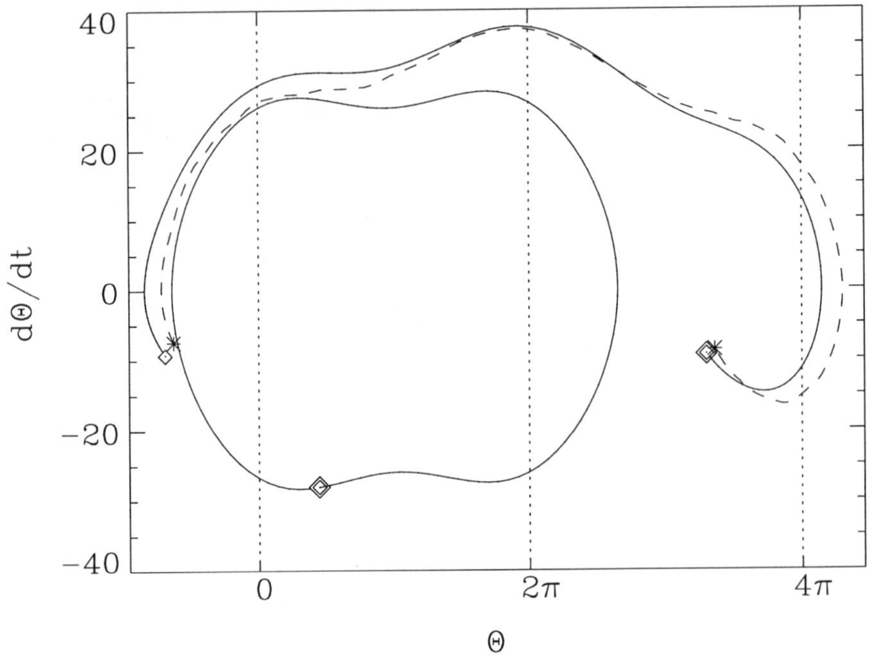

Figure 6: Two one-cycles extracted from the experimental pendulum are shown in dotted lines. These cycles are classified: $\rho = 0, \chi = 1.68$; and $\rho = +2, \chi = 2.51$. The solid lines show model orbits determined by matching points in the multi-dimensional bifurcation space, and are the two orbits indicated by triple-$\diamond$ in the previous 2 figures. These model orbits are: $\rho = 0, \chi = 1.65$, and $\rho = +2, \chi = 2.49$. Also note a close comparison between the Poincaré section $(\theta, \dot{\theta})$ values for model ($\diamond$) and experimental ($*$) one-cycles.

The values of ρ and χ for the two experimental cycles are shown in both projections of the space, figures 4 and 5. They correspond well with the model cycles at $G \approx 150$. A superposition of the model one-cycles at that $G-$value is shown in Figure 6. The model orbits (two of the eleven one-cycles existing at G=150) are classified as: $\rho = 0, \chi = 1.65$, and $\rho = +2, \chi = 2.49$, which is a close comparison. Also, as suggested by the figure, there is a good comparison between $(\theta, \dot{\theta})$ for the two sets.

CONCLUSION

We have presented a new classifier for a dynamical system that can be applied to experimentally generated time series data. The new classifier has been demonstrated on a particular model, a damped driven pendulum. The method is based upon locating the one-cycles for the system, and subsequent measurement of characteristics of the cycles, namely winding number ρ and extent χ.

From the one-dimensional time series data, a multi-dimensional bifurcation space can be constructed. Although a complete test has not yet been accomplished, this space shows potential as a hierarchical organization of system states. The restriction to only period-one orbits appears to increase the robustness of the classification scheme.

ACKNOWLEDGMENTS

This work was supported by Independent Research funding at the Naval Surface Warfare Center. The current affiliation of Bo Hammer: American Institute of Physics, One Physics Ellipse, College Park, MD 20740.

References

[1] Cvitanovic, P., Phys. Rev. Lett., **61**, 2729, (1988).

[2] Lathrop, D. P. and Kostelich, E. J., Phys. Rev A, **40**, 4028, (1989).

[3] Mindlin, G. B., Solari, H. G., Natiello, M., Gilmore, R., Hou, X-J., J. Nonlinear Sci., **1**, 147, (1991); Papoff, F., Fioretti, A.,Arimondo, E., Mindlin, G. B., Solari, H., Gilmore, R., Phys. Rev. Lett. **68**, 1128, (1992).

[4] Blackburn, J. A., Physics Dept. Wilfrid Laurier Univ., Waterloo, Ontario, Canada, and Smith, H. J. T., Physics Dept. Univ. of Waterloo, Waterloo, Ontario, Canada, designed the pendulum, available from Daedalon Corporation, 35 Congress Street, Salem MA 01970.

Bayesian Approach to Estimation and Detection of Chaotic/Nonlinear Signals

Maribeth Bozek-Kuzmicki (maribeth@mitre.org)
Paul Hriljac (hriljac@mitre.org)
Garry Jacyna (gjacyna@mitre.org)

The MITRE Corporation
7525 Colshire Drive
McLean Va. 22102

Abstract. An important issue in nonlinear dynamics is the optimal estimation and detection of the partially observed states of a system at low signal-to-noise ratios. In this paper, we will outline a Bayesian-based approach that allows for the optimal determination of the state probability density function in time as a function of the observations. This leads to optimal detector designs based on the notion of generalized innovation sequences. Here, the density functions are defined over a computational grid which is designed to capture the phase space dynamics of the nonlinear system. Partial measurements are used to update the projected system state and density function. Estimation and detection decisions are based on the propagated density functions.

INTRODUCTION

Many physical systems are best modeled as nonlinear dynamical systems. In fact, in many systems, the behavior is so complicated that it appears chaotic. In this paper, we will study deterministic-based systems that are seemingly chaotic in their responses.

Assume that we have a physical system that can be modeled as a nonlinear, possibly chaotic, process. Data is collected by discretely sampling the system. Noise is added to the process to adjust the resulting signal-to-noise ratio (SNR). From these noisy observations (SNR << 0dB) we want to estimate the system state and develop a detection decision. There are numerous techniques in the literature that address this problem. However, the majority of these papers assume some form of state and/or observation linearization in addition to restricting the number of significant statistical moments examined. (3,5,10,12) Other techniques use Taken's embedding approach (16) to exploit the geometric structure of the system. However, the use of linearization assumptions or geometric embedding arguments is inadequate at low SNRs.

The technique presented here differs from those techniques mentioned above in the following manner. A true nonlinear system model is used. No linearization is assumed. New observations are used to update the system state and past information is carried along in a probabilistic manner. The technique addressed in this paper is based on a Bayesian probabilistic formulation where past information is weighted with current information through the use of conditional probability density functions. The density functions are updated by new observations. Estimates of the system states are determined directly from the probability density function. An optimal detection statistic is determined from the ratio of conditional densities under hypotheses of signal and noise, respectively.

Here will we outline the proposed Bayesian-based estimation and detection techniques. We will give results for several well known nonlinear dynamical systems. Lastly, we will discuss future research in this area.

THE BAYESIAN-BASED ALGORITHM

The Bayesian-based nonlinear estimation and detection algorithm is a model based technique that is confined to a bounded region in the system phase space Ω. The bounded region should contain behavior representative of the nonlinear system model. See Figure 1. For this algorithm, the model must be a diffeomorphism of Ω (one-to-one and onto) with at least one observable state. It is assumed that a model of the appropriate diffeomorphism is given a priori. The initial system states are assumed to be Gaussian distributed unknown parameters. Lastly, we assume that the background noise is a Gaussian, bandlimited white zero-mean process and that a noise-only data cut is available for pre-processing. This pre-processing step allows one to calculate the distribution parameters, such as standard deviation σ, for the background noise environment.

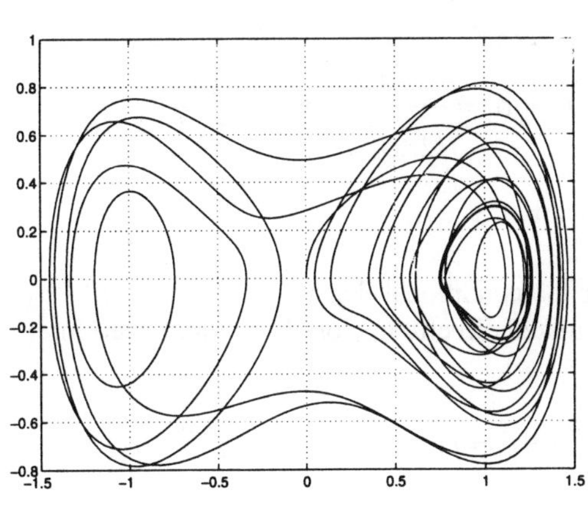

FIGURE 1. Example of the bounded phase space region for the Duffing Oscillator.

Consider a system of equations of the form $X_n = M(X_{n-1}), X_n = (x_n, y_n)$, where M is a (diffeomorphic) nonlinear model. At each time step, observations z_n are reported back to the system. Note that the form of z_n depends on the absence, H_0, or presence, H_1, of the signal, i.e. hypothesis H_0 implies $z_n = w_n$ and hypothesis H_1 implies $z_n = x_n + w_n$, where w_n is the sampled, bandlimited Gaussian white noise background process. The results presented in this paper will be restricted to two-dimensional systems with one observable state. The results are easily generalized for a d dimensional system with s observable states.

Let $z^n = \{z_n, z_{n-1}, ... z_n\}$ denote the sequence of observations up to and including time step n. We will build the detector by forming the classical likelihood ratio as shown in Equation (1).

$$l(z^n) = \frac{p(z^n \mid H^1)}{p(z^n \mid H^0)} \underset{H_0}{\overset{H_1}{\gtrless}} \eta \tag{1}$$

Since we assume the background noise is Gaussian, and we can obtain the standard deviation σ for the background noise, we can write the expression for the conditional density function $p(z^n \mid H_0)$ directly:

$$p(z^n \mid H_0) = p(z_n \mid z^{n-1}; H_0) = G(z_n, \sigma) \tag{2}$$

where

$$G(v, \sigma) = \frac{1}{(2\pi\sigma^2)^{\frac{m}{2}}} \exp(-\frac{|v|^2}{2\sigma^2}) \tag{3}$$

and m is the dimension of v.

The expression for the conditional density function, $p(z^n \mid H_1)$, under hypothesis H_1 can not be calculated directly in closed form and, therefore, must be recursively expanded as follows:

$$p(z^n \mid H_1) = p(z_n \mid z^{n-1}; H_1) p(z^{n-1} \mid H_1) \tag{4}$$

where

$$p(z_n \mid z^{n-1}; H_1) = \int_\Omega p_{\mathbf{w}_n}(z_n - \bar{x}_n \mid z^{n-1}; H_1) p_{\bar{\mathbf{x}}_n}(\bar{x}_n \mid z^{n-1}; H_1) d\bar{X}_n \tag{5}$$

and $\overline{X}_n = (\overline{x}_n, \overline{y}_n)$ is the set of all possible system states over Ω.

Since the current noise component w_n is independent of all past noise components $w_{n-1}, w_{n-2}, \ldots w_0$ (bandlimited white noise assumption) and of all observations $x_n, x_{n-1}, \ldots, x_0$ up to and including the current time n (causality), we can expand the right hand side of Equation (5):

$$p(z_n \mid z^{n-1}; H_1) = \int_\Omega G(z_n - \overline{x}_n, \sigma) p_{\overline{\mathbf{x}}_n}(\overline{x}_n \mid z^{n-1}; H_1) d\overline{X}_n \tag{6}$$

Using the fact that our model M is a diffeomorphism, we can further expand the right hand side of Equation (6) as follows:

$$p(z_n \mid z^{n-1}; H_1) = \int_\Omega G(z_n - \overline{x}_n, \sigma) \frac{p_{\overline{\mathbf{x}}_{n-1}}(M(\overline{x}_n) \mid z^{n-1}; H_1)}{\mid J(M(\overline{x}_n)) \mid} d\overline{X}_n \tag{7}$$

where $J(\bullet)$ is the Jacobian matrix of M and $\mid \bullet \mid$ denotes the matrix determinant. Thus Equations (2) and (7) provide the necessary components for the likelihood ratio in Equation (1). Detection decisions can now be made by comparing the likelihood ratio statistic to a pre-set threshold η.

The nonlinear estimator can be determined by calculating the conditional density function of the state given all available information, i.e. $p_{\overline{x}_n}(\overline{x}_n \mid z^n)$. Using Bayes rule, we can rewrite $p_{\overline{x}_n}(\overline{x}_n \mid z^n)$ for the hypothesis H_1 in the following manner:

$$p_{\overline{x}_n}(\overline{x}_n \mid z^n) = \frac{p_{z_n}(z_n \mid \overline{x}_n) p_{\overline{x}_n}(\overline{x}_n \mid z^{n-1})}{p_{z_n}(z_n \mid z^{n-1})} \tag{8}$$

If we treat each new observation as a new initial condition that is Gaussian distributed, we can write $p_{z_n}(z_n \mid \overline{x}_n)$ as a Gaussian density function of Ω centered at the observation z_n.

$$p_{z_n}(z_n \mid \overline{x}_n) = G(z_n - \overline{x}_n, \sigma) \tag{9}$$

As in the detector formulation, we can use the fact that the model M is a diffeomorphism and write $p_{\overline{x}_n}(\overline{x}_n \mid z^{n-1})$ in terms of a projection of the conditional density function via the model M and the Jacobian of the state equations at the current time:

$$p_{\bar{x}_n}(\bar{x}_n \mid z^{n-1}) = \frac{p_{\bar{x}_{n-1}}(M(\bar{x}_n) \mid z^{n-1}; H_1)}{\mid J(M(\bar{x}_n)) \mid} \tag{10}$$

Note that $p_{z_n}(z_n \mid z^{n-1})$ is independent of the estimates. Therefore, we choose the density to satisfy the normalization condition:

$$p_{\mathbf{z}_n}(z_n \mid z^{n-1}) = \int_{\Omega} p_{\mathbf{z}_n}(z_n \mid \bar{x}_n) p_{\bar{\mathbf{x}}_n}(\bar{x}_n \mid z^{n-1}) d\bar{X}_n \tag{11}$$

Once the density function in Equation (7) is determined, one can use the maximum aposterior probability (M.A.P) estimation criterion to determine the state estimate. This estimate corresponds to the maximum of the conditional density function.

Now that the foundation for the Bayesian-based nonlinear estimator and detector has been presented, we will briefly address implementation issues.

IMPLEMENTATION

Implementating the above procedure is computationally plausible if one defines a bounded grid as the support of the conditional density functions. Using the model M, one can define a bounded region in the phase space Ω, representative of the system behavior. See Figure 1. By dividing this bounded region into uniform squares (a uniform grid), we can treat the intersections of the grid points as centers of cells with width and height γ. See Figure 2.

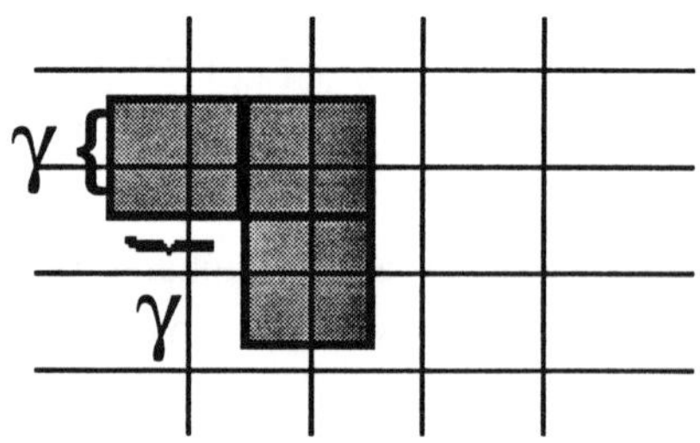

FIGURE 2. An example of how the grid vertex points become centers of cluster cells.

Using this grid as support, we will build the conditional density function in the third dimension. See Figure 3. The estimator and detector algorithms described above are executed over the cell centers to calculate propagation to the next time step. It is assumed that the entire cell is propagated with the cell center; thus the cell size must be sufficiently small. If the grid is not fine enough, spurious dynamics will be propagated via the cell center propagation.

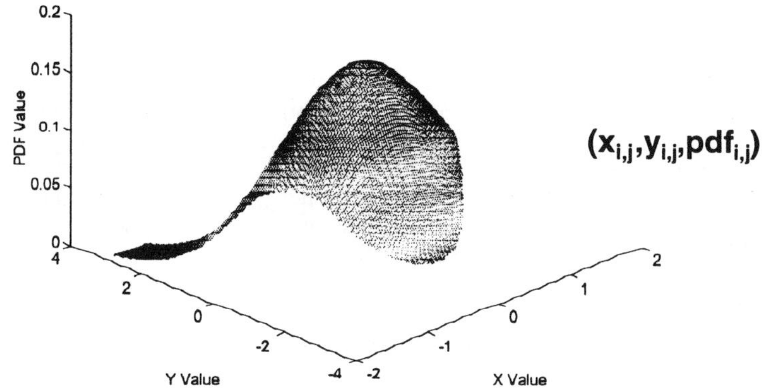

FIGURE 3. Graphical representation of how a density function is built on the bounded region using a grid structure.

The issue of integrating over a grid structure is an important step in the proposed algorithm. In order to carry out the estimation and detection routines, we must be able to integrate the density functions over the grid structure. Consider the following mapping:

$$T = \begin{cases} x = f(u,v) \\ y = g(u,v) \end{cases} \quad D \subset S \xrightarrow{T} D^* \subset S \tag{12}$$

and a function $F(x,y)$. The area over D^* can be approximated by

$$\iint_{x,y,D^*} F(x,y)dxdy \cong \sum_x \sum_y F(x,y)\frac{\mu(D^*)}{N} \tag{13}$$

where $\mu(D^*)$ is the measure of each cell and N is the number of cells in D^*. The measure of each cell is the area associated with each cell. Since we are operating over a uniform grid, we can calculate the total area in D^* and divide by the number of cells, N. Thus if we consider our model M as the function F above, we can effectively integrate over the grid structure by performing a double sum operation.

RESULTS

The theory presented above was tested on two well-known dynamical systems: the Henon map and the Duffing-Ueda map. By using two systems of equations that are prevalent throughout the literature, we are able to bench mark performance results against those sighted in the open literature. Here we will provide results for both systems relative to a bandwidth-matched energy detector. The Duffing-Ueda results will also be compared against that of a second order moment-based Extended Kalman Filter (EKF) approach (9).

The Henon Map

The following set of equations were used to generate data for the Henon map. These equations were also used as the model in the Bayesian-based estimation and detection algorithm:

$$\begin{aligned} x_{n+1} &= 1.4 + .3y_n - x_n^2 \\ y_{n+1} &= x_n \end{aligned} \qquad (14)$$

The receiver operating curve (ROC), relating detection probability to SNR at a fixed false alarm rate (PFA), in Figure (4) shows that the Bayesian-based algorithm performs slightly better than the energy detector at a PFA of 0.01. We are encouraged by these results because the Henon map, while generated from a simple set of dynamical equations, produces some of the most complicated behavior (comparable to bandlimited white noise) known in the field of nonlinear dynamics.

The computational grid used in the case of the Henon map has boundaries of (-2,2) in the x - direction and (-3,3) in the y -direction. Each cell measures 0.01 by 0.015 in the x and y directions, respectively. These boundaries result in a support grid composed of 400 x 400 cells.

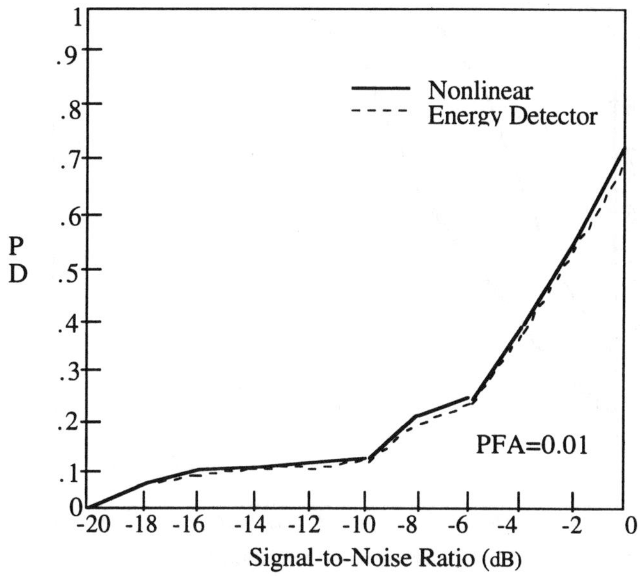

FIGURE 4. ROC curve results for the nonlinear detector and the energy detector at PFA=0.01 when applied to the Henon map.

The Duffing-Ueda Map

The set of equations used as the model for the Duffing-Ueda map are given as follows:

$$\begin{aligned} \dot{x} &= y \\ \dot{y} &= -.25y - x^3 + .275\cos(\omega t) \end{aligned} \tag{15}$$

where $\omega = 1$ and $\Delta t = \dfrac{2\pi}{100}$.

Continuous data is sampled at the rate Δt to produce a sequence of discrete points. The results obtained for the Duffing-Ueda map are very encouraging. As the ROC curve in Figure (5) shows, the nonlinear Bayesian-based detector results in a 10-12 dB gain over the energy detector and a 4-6 dB gain over that of the Extended Kalman Filter (EKF) detector at a fixed PFA of 0.01.

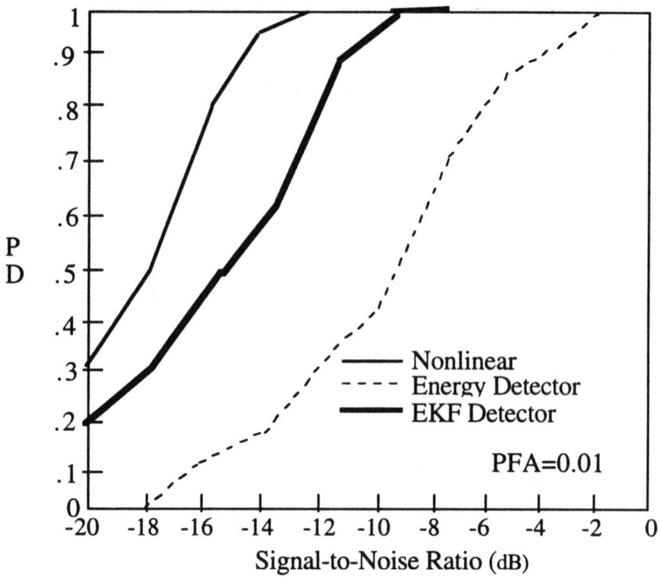

FIGURE 5. ROC curve results for the nonlinear detector, the energy detector and the Extended Kalman Filter (EKF) detector at PFA=0.01 applied to the Duffing-Ueda map.

The grid used in the Duffing-Ueda map had boundaries of (-3,3) in both the x and y directions. Slightly bigger cells are used for the Duffing map as compared to the Henon map. This is because the Duffing-Ueda map has dynamics that do not spread as rapidly as the Henon map: more neighboring points can be targeted as having similar dynamics. The cell size utilized was a square cell with side measurement 0.02. These parameters result in a 300x300 cell support grid for the density functions.

CONCLUSIONS/FUTURE WORK

We have developed a new method which can accurately estimate the current system state and detect the presence of a signal at very low SNRs (<<0dB). Initial tests show that performance is better than that of the energy detector and the second moment-based Extended Kalman Filter detectors for two known nonlinear system maps.

Since the proposed method is still in the early stages of development, there are several areas where further work may be possible. The use of a non-uniform grid may prove more computationally efficient. Perhaps it is more efficient to have larger cells defined over less dynamical regions (such as an attractor or well) and smaller cells defined more dynamical regions in the state space (such as transition trajectories). One can investigate the use of non-rectangular cells also as well. Also, since the Bayesian-based method assumes a given nonlinear diffeomorphic model, further research in nonlinear modeling will broaden the application range of our algorithm.

ACKNOWLEDGMENTS

The authors would like to thank NAWC for their continued support of this project. In particular, we thank Tony Passamante and group for their assistance in this research effort.

REFERENCES

1. Colella, David; Hriljac, Paul; Jacyna, Garry,"A Fokker-Planck Statistical Model for Neural Nets", MITRE Corporation Annual Publication, (1991).
2. Devaney Robert, *An Introduction tn Chaotic Dynamical Systems*, Second Edition, Addison-Wesley, New York, 1989.
3. Farmer, J Sidorwich, J,"Predicting Chaotic Time Series",*Physica Rev. Letters,* no 59, pp. 845-848, (1987).
4. Gelb, Arthur editor, *Applied Optimal Estimation* , The M.I.T. Press, Massachussetts, 1988.
5. Grebogi, Celso; Ott, Edward;Yorke,James,"Chaos, Strange Attractors and Fractal Basin Boundries in Nonlinear Dynamics",*Science (American Association for the Advancement of Science)*Vol 238, pp 633-637, (1987).
6. Guckenheimer, John; Holmes, Philip, *Nonlinear Ocsillations, Dynamical Systems, and Bifurcations of Vector Fields,* Applied Mathematical Sciences #42 Springer Verlag, New York, 1983.
7. Gulick, Denny, *Encounters with Chaos,* Mc Graw Hill, New York. pp 90-91
8. Hsu, C.S., *Cell to Cell Mapping: a Method of Global Analysis for Nonlinear Systems,* Applied Mathematical Sciences #64 Springer Verlag, New York, 1987.
9. Jacyna, Garry: Mueller, Jeff, "Near Optimal Detection and Estimation of Chaotic-Like Signals",*JUA*, vol.44, no. 2, pp. 631-652, (1994).
10. Passamante A; Farrell,M.;Hediger,T,"Characterizing Attractors Using Local Instrinsic Dimensions Calculated by SVD and Information-Theoretic Criteria", *Physical Review A,* vol 41, no.10, pp. 5325-32, (1990).
12. Passamante A; Gollub,M; Hediger,T,"Fractal Dimension and Local Intrinsic Dimension".*Physical Review A*,vol. 39, no. 7., pp. 3640-5, (1989).
13. Poor, Vincent, *An Introduction to Signal Detection and Estimation,* Springer-Verlag, New York, 1988.
14. Ruelle, David, *Chaotic Evolution and Strange Attractors,* Cambride University Press, New York, 1989.
15. Ruelle, David; Eckmann, J,"Ergodic Theory of Chaos and Strange Attractors", *Reviews of Modern Physics,* vol. 57, pp. 617, (1985).
16. Takens, F, "Detecting Strange Attractors in Turbulence", Lecture Notes in Math #898, Springer-Verlag, 1981.

17. Van Trees, Harry, *Detection, Estimation and Modulation Theory:Part 1,* John Wiley & Sons, New York, 1968.
18. Walpole, Ronald;Myers, Raymond,*Probability and Statistics for Engineers & Scientists,* The MacMillan Company, New York, 1972.
19. Whalen, Anthony, *Detection of Signals in Noise*, Academic Press, New York, 1971.
20. Wiggins, S., *Introduction to Applied Nonlinear Dynamical Systems and Chaos,* Texts in Applied Mathematics #2, Springer Verlag, New York.,1990.

Classifying Transient Signals with Nonlinear Dynamic Filter Banks

Jeffrey S. Brush

Cambridge Research Associates
1430 Spring Hill Road, Suite 200, McLean, VA 22102

Abstract: In recent years, several specific advances in the study of chaotic processes have been made which appear to have immediate applicability to signal processing. This paper describes two applications of one of these advances, nonlinear modeling, to signal detection & classification, in particular for short-lived or transient or signals. The first method uses the coefficients from an adaptively fit model as a set of features for signal detection and classification. In the second method, a library of predictive nonlinear dynamic equations is used as a filter bank, and statistics on the prediction residuals are used to form feature vectors for input data segments. These feature vectors provide a mechanism for detecting and classifying model transients at signal-to-noise ratios as low as -10 dB, even when the generating dynamics of the transient signals are not present in the filter bank. The second method and some validating experiments are described in detail.

INTRODUCTION

Detection and classification (D/C) of low level transient signals is of crucial importance in a number of arenas. In electronic warfare and SONAR applications, spread-spectrum methods and reverberation can obscure signals of interest (SOIs) so that relatively "clean" examples occur only rarely, and then at very low signal-to-noise ratios (SNRs) and short duration. Even at high SNRs, very brief impulsive transients such as cavitation behind a propeller blade, machinery failure modes, pump pulses, etc. can be hard to classify, and yet are frequently of crucial importance in distinguishing friend from foe (or either from nothing!). Methods of detecting these transient signals that rely on linear filtering or Fourier methods are severely hampered by the high noise and short time scales of these signals, and also by the broadband nature of both the SOIs and the background noise. Existing shortcomings have prompted investigation into the development of D/C methods based on advances in the analysis of chaotic and other nonlinear systems.

ADVANCES IN CHAOTIC SYSTEMS RESEARCH

Driven by developments in nonlinear mathematics and applications thereof to physics, particularly in the fields of fluid dynamics (turbulence) and solid state physics (semiconductor lasers), the field of nonlinear dynamics (NLD) has expanded to have ramifications from the physical to the social sciences. In the past 20 years or so, the field has grown to represent an entirely new paradigm for viewing variability in the natural world. Where in the past the concepts of stochasticity and randomness (i.e. noise) have been invoked to explain variations in data, we now routinely check for the presence of a nonlinear dynamical system -- a deterministic system whose steady state or asymptotic behavior may yield not fixed points or limit cycles but bounded nonperiodic trajectories (chaos). Systems operating in these modes typically exhibit broadband Fourier spectra, and may frequently satisfy many statistical tests for noise. The study of chaotic systems has led to the development of new analytical tools ranging from methods to detect the presence of chaos in data sets to characterizing the underlying generating system to explicitly modeling and exploiting that system. These new methods, rooted in time series analysis, suggest new concepts and techniques to exploit the unique characteristics of chaotic systems for the field of signal processing.

In recent years, several specific advances in the study of chaotic processes have been made which appear to have immediate applicability to signal processing. These advances include the following:

***Modeling nonlinear systems using only observed time series data*:** The ability to generate predictive models of underlying generating systems based only on time series observations of the output of the system is the culmination of a large body of work, beginning with the experimental work of Packard, Farmer, and others [1,2], corroborated by the embedding theorems of Takens and Ruelle [3,4], extended by the early work by Crutchfield & McNamara, Broomhead & King, and Kostelich & Yorke [5-7], and buttressed by all of the pre-existing curve-fitting and modeling literature for linear systems [e.g. 8-12]. Recent theoretical results by Sauer & Yorke and application oriented results by Cawley & Hsu, Breeden et al., Brush & Kadtke, and others [13-23] have firmly established the science of obtaining dynamic models from experimental data. As a whole, this body of work has resulted in the ability to generate detailed predictive models, in the form of systems of differential equations or of iterated maps, of a wide range of systems. These models can be global (i.e., stationary over time), adaptive (time varying), or atlas-base (state space varying) depending upon the nature of the underlying system and of the intended application of the model.

***Synchronization of chaotic systems*:** One defining characteristic of nonlinear (and therefore of chaotic) systems is a phenomenon known as "sensitive dependence on initial conditions." Simply stated, this means that two sets of initial conditions that differ only slightly will eventually result in uncorrelated future behavior, unlike the case for linear systems which have no mechanism for amplifying initial uncertainties (of course, linear systems with stochastic components can amplify initial uncertanties, here the discussion is restricted to deterministic systems). Similarly, if a given set of initial conditions are known precisely, and is used to initiate the behavior of two (or more) systems which differ only slightly, their future behavior will become uncorrelated. As a result of investigating mechanisms for generating uniformly nonlinear (i.e. replicable) electric circuits, Carroll & Pecora of the Naval Research Laboratories [24] developed a synchronization scheme for chaotic systems which directly builds on the special characteristics of nonlinear systems. Many researchers have expanded on these results, and now the study of chaotic synchronization is expanding rapidly.

Control of chaotic systems: Initial application-oriented investigations of chaotic behavior occurred in the physical sciences, particularly in fluid dynamics. Irregular behavior that had not been satisfactorily explained by the existing linearized models of fluid flow was observed (particularly in turbulent regimes), and was originally thought to be some uncontrolled noise source. As insight into the nature of nonlinear dynamical systems grew, the realization that the observed behavior was an integral feature of the systems under consideration spread. Several physicists, along with their colleagues in theoretical mathematics, began developing extensions to linear control theory in order to control or constrain the behavior of chaotic systems. Significant early results were obtained by Hubler et al. [25], and by Ott, Grebogi & Yorke [26], whose technique (now termed the OGY method) has been adapted to become the standard method of controlling chaotic systems [27-29]. Successful applications of this method in controlling irregular cardiac and hippocampal behavior in experimental settings (by Spano & Ditto et al. [29]) have justified years of research in these areas.

***Development of VLSI implementations of chaotic systems*:** As mentioned above, a key attribute of chaotic nonlinear systems is their sensitivity to errors in initial conditions and therefore to finite precision representations. Since all digital computer implementations of these systems are finite precision models of infinitely precise analog systems, they have inherently limited predictive utility. The capability to manufacture non-software devices which repeatably generate known dynamics will be a key to the eventual commercialization of any technology relying on rapid chaotic sequence generation. Analog circuit implementations of chaotic systems are difficult to replicate (as mentioned above in relation to the work of Carroll & Pecora), since nonlinear elements are not manufactured to the required precision. A group in Berkeley (Chua et al. [31]) has developed a VLSI (Very Large Scale Integrated) circuit implementation of a simple nonlinear system which, although digital, nonetheless generates extremely rich behavior (Chua's circuit has in excess of 50 distinct known attractors, depending upon parameter settings [32,33]). The inherent replicability of VLSI circuits has opened the way to near-term commercialization of a range of chaos-based devices.

The above list, which is by no means exhaustive, should give a sense of the state of research and development in nonlinear dynamics. Only a few years ago, the field was dominated by the development of software tools for extracting system invariants from time series and by phenomenological investigations. With the maturation of the science, real-world applications and devices are clearly possible. In the field of signal processing, potential applications include improved source coding, noise reduction, system identification, and signal detection & classification.

This paper describes an application of nonlinear modeling to signal detection & classification which exploits recent developments in NLD-based signal processing, in particular for transient or short-lived signals. The method presented here is based on prediction residuals, and has been demonstrated to be able to detect and classify some transient signals at broadband SNRs down to at least -10 dB (decibels). The prediction residual method is amenable to implementation in hardware (as part of a digital signal processing system, for example), and can augment existing signal processing and conditioning equipment and methodologies.

REVIEW OF NONLINEAR MODELING

Before describing the prediction residual method for detection and classification of transients, it is appropriate to briefly review some aspects of nonlinear modeling (the concepts presented here have been developed in detail in numerous papers [2,6,23,34]).

Global (Stationary) Modeling

Nonlinear state space modeling methods for chaotic systems differ somewhat from traditional linear methods. Typically, one observes scalar output from a multi-dimensional system, and so the first step is to reconstruct a vector space in which to perform the modeling. The most common method for this reconstruction is that of time delay embedding: an embedding dimension D and a time-delay parameter τ are estimated from the data $x[t]$ (using any of a variety of methods), and a new D-dimensional time series is generated from the scalars by constructing vectors $\overline{Y}[t] = \{ x[t], x[t-\tau], x[t-2\tau], ..., x[t-(D-1)\tau] \}$. The objective is to model the time evolution of $\overline{Y}$, or to find a vector function F such that $\overline{Y}[t+1] \cong F(\overline{Y}[t])$. The D functional components of F are denoted f_d $(1 \leq d \leq D)$. Noting that the only new information in $\overline{Y}[t+1]$ is $x[t+1]$, only f_1 is required and the modeling procedure reduces to finding a function F such that $x[t+1] \cong F(\overline{Y}[t])$. There is an unlimited choice of possible bases for F, including polynomial and exponential expansions, radial functions, and combinations thereof. In situations where some understanding of the underlying dynamical system allows one to postulate a model form, coefficients can be fit to that form directly. For data from unknown systems, however, it is useful to use a ratio of two polynomial expansions p_i and q_k, and write the composite function as a ratio of linear combinations of these bases. The models can either express $x[t+1]$ directly (iterated maps) or can specify $x'[t]$ (differential equations), as shown below:

$$x'[t] \cong F(\overline{Y}[t]) = \frac{\sum_j a_j p_j(\overline{Y}[t])}{1+\sum_k b_k q_k(\overline{Y}[t])}. \tag{1}$$

Now, finding the model is equivalent to finding the coefficients of the chosen combination of bases. For bases which are is linear in the coefficients, algebraic methods such as singular value decomposition can be used for coefficient estimation, otherwise more complicated methods involving gradient descent or nonlinear optimization must be used. Coefficient estimation procedures and the sensitivity to noise and data set size using global dynamical equations such as these have been explored extensively [20].

Adaptive (Nonstationary) Modeling

By allowing the coefficients in Equation 1 to vary over time, one creates an adaptive nonlinear model, and rewriting it as shown in Equation 2 allows the use of a gradient descent method to track the best fit solution over time [35]:

$$x'[t] = \sum_j a_j p_j(\overline{Y}[t]) - x'[t] \sum_k b_k q_k(\overline{Y}[t]). \tag{2}$$

The gradient terms (with respect to the coefficients) are:

$$\frac{\partial\, e[t]^2}{\partial\, a_j} = -2\mu\, e[t]\; p_j(\overline{Y}[t]) \tag{3a}$$

and

$$\frac{\partial\, e[t]^2}{\partial\, b_k} = 2\mu\, e[t]\, x'[t]\; q_k(\overline{Y}[t])\,, \tag{3b}$$

where $e[t]$ is the instantaneous error $x[t]$ - $\tilde{x}[t]$ ($\tilde{x}[t]$ is our estimate of $x[t]$). The Widrow-Hoff least mean square method is used to adapt the a_j and b_k, minimizing the instantaneous squared error $e[t]^2$ by adapting the model coefficients via a simple update scheme [36].

In Equations 1-3, the scalar signal components x (as part of the reconstructed vectors $\overline{Y}$) are modeled directly, rather than by using a reference signal as in the conventional adaptive filtering case [37]. By directly modeling x, the assumption is that unwanted signal components are either uncorrelated noise or have a significantly shorter time scale than the convergence time dictated by the control parameters in the coefficient update scheme. As for the global method, the adaptive method has a number of refinements and alternative approaches that could be and have been employed [38-40].

Adaptive Model Application to Transient D/C

The time-dependent coefficients which result from the adaptive modeling process have utility as the basis of a transient D/C method. The essential idea is to exploit the locations, in coefficient space, of estimated coefficient vectors as they evolve over time. Detection and classification could be performed by examining the distribution in the coefficient space of particular signal types. In the absence of *a priori* knowledge about given signals of interest, one would first observe the coefficient distribution of background noise, and define detection of a novel event as occurring when the coefficient vector evolves substantially outside of this distribution. If examples of particular signal types are available, then classification might be performed by observing the statistical coefficient distribution for each signal type, partitioning the coefficient space accordingly, and classifying the signal based on which region the coefficient trajectory visits.

An early version of this coefficient space partitioning method, based on a windowed temporal approach, was capable of nearly perfect discrimination among 5 model signal types at 0 dB (the average probability of detection vs. probability of false alarm -- p_D/p_{FA} -- was 0.998/0.002 at 0 dB SNR). The details of this early approach are described in [23].

A truly adaptive version of the coefficient space partition method was implemented, and was used to perform detection and classification simultaneously as the adapted coefficient vector trajectory evolved. As a demonstration of the method, a synthetic transient data set consisting of bursts of Gaussian noise and normalized x components of Ikeda and Pickover data (see Appendix) was generated. This data set was comprised of alternating segments (2000 points each) of noise, Ikeda, noise, Pickover, noise, etc., with the signal components at an SNR of +10 dB (32% noise). A three-dimensional, second-order polynomial map (with 10 free coefficients) was used as a notional model (this is the 'wrong' model for both signal types), and the coefficients of the model were adaptively fit to the data set. Figure 1 shows the noisy time series along with the time evolution of two of the coefficients, revealing clear deviations at the 2000 point intervals.

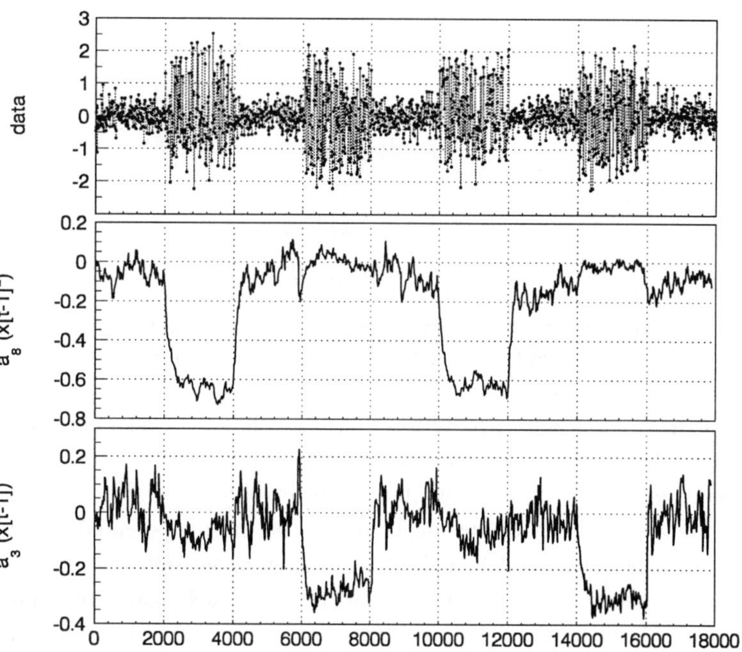

Figure 1. Coefficients from a model adapted to alternating Gaussian, Ikeda, and Pickover data at +10 dB SNR. The changes at 2000 point intervals reflect the transient boundaries.

In Figure 2, a two-dimensional projection of these coefficients is shown, with the regions corresponding to the noise and signal types indicated. It is clear that the method works well, at least for these signal types at this relatively high SNR.

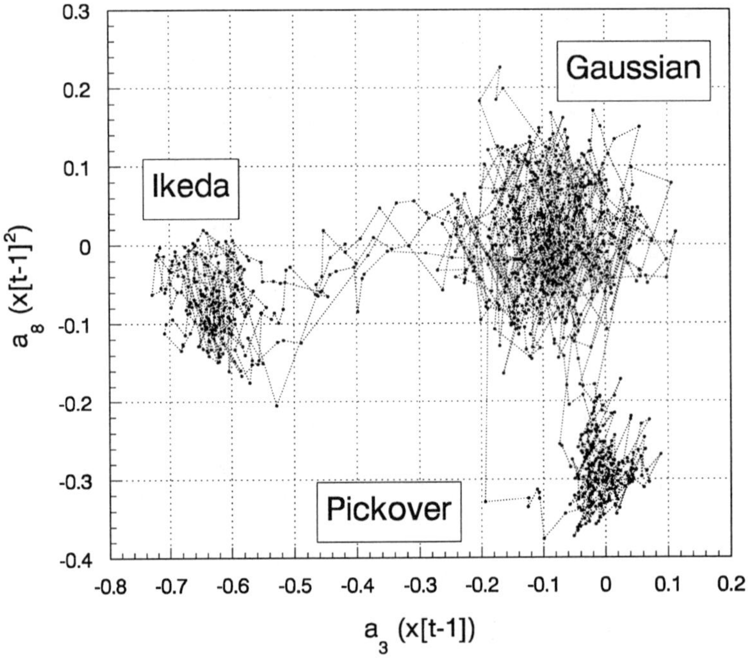

Figure 2. Projection of the coefficient trajectories from Figure 1 into two dimensions. The regions of coefficient space visited during the various signal types are quite distinct.

In spite of the encouraging results presented above, the adaptive coefficient space method for D/C is severely hampered by a number of issues, including convergence rates of the coefficients (the example above consisted of rather long 'transient' signals), stability of the adaptation algorithms, the potential difficulty in choosing useful notional models and applying appropriate statistical measures to the results. In the following sections, a related approach based on employing known dynamical models and using prediction residuals as the elements of feature vectors is discussed.

PREDICTION RESIDUALS FOR SIGNAL IDENTIFICATION

Original Concept

As originally described [34,41] the concept of prediction residuals for system identification was to be based on a library of models, one for each transient type of interest. These models were intended to be 'good' models of the dynamic behavior of the transients in the absence of noise or distortion (i.e. they were to be obtained by the methods outlined above and were expected to yield accurate short-term predictions of future behavior of the system). Given this library of models, an incoming data stream was fed into each model as a set of continuously changing initial conditions, and the model was used to predict the next point in the data stream. Through careful analysis of the residuals (the error between the prediction of each model and the actual next data value), statements could be made regarding the presence or absence of a particular signal type at a particular time.

To be more concise, assume a situation where a library of theoretical or derived models exists for each of the different signals of interest. That is, there is a library of N dynamical systems $F_n(\overline{Y}[t])$ $(1 \leq n \leq N)$, each with known coefficients and bases. These models could be obtained, for example, by modeling relatively clean data segments from known sources and saving the resulting models. Recall that a typical linear detection method is just a power detector, based on forming a windowed variance

$$E_0[t] = \sum_{i=-w}^{w} x[t+i]^2 , \tag{4}$$

for some data window of half-width w (assuming zero-mean data). Values of E_0 which exceed some threshold indicate statistical detection based purely on increased signal power. Note that $E_0[t]$ will be identical for data from two signals of equal power generated by two different dynamical systems. The utility of E_0 is extended by using a measure based on first defining a time series of one-step prediction residuals (multi-step residuals could be used with no loss of generality) for each dynamical system F_n, and denoting these residuals $R_n[t]$. Applying a windowed variance filter to each R_n yields the following one-step error variances:

$$E_n[t] = \sum_{i=-w}^{w} R_n[t+i]^2 = \sum_{i=-w}^{w} \left(x[t+1] - F_n(\overline{Y}[t])\right)^2 . \tag{5}$$

When the 'correct' model is used (generating $E_{correct}$) for a particular received transient signal, some of the contribution to E_0 is accounted for by the action of the model dynamics, so the residual variance is reduced ($E_{correct}$ will be less than E_0). When the 'wrong' model is used, the residual variance increases relative to E_0 due to the action of the nonlinearities in the model (E_{wrong} will be greater than E_0). Normalizing the contributions due to E_0 in R_n forms the discriminant series $D_n[t] = \log_{10}(E_n[t]/E_0[t])$. The measures D_n are sensitive to the amount of variance in the original signal accounted for by the particular nonlinear dynamical model of interest. Detection and classification of signal components occurs simultaneously, as the appropriate filter registers not only detection but 'anti-detection'. Anti-detection occurs because the filter is sensitive to the presence of deterministic structure in the signal, however it is of the incorrect type and hence magnifies prediction errors in combination with the filter dynamics.

Refined Approach

One potential limitation of the prediction residual method as outlined above is its reliance on good predictive models of the signals of interest. In an effort to mitigate this dependence, the use of relatively arbitrary models for the library $F_n(\bar{Y}[t])$ was explored, with encouraging results. At a given SNR, transients of a particular type will have some average value for each of the measures D_n. Taking these measures as the components of a vector yields, for each signal type/SNR combination, an N-vector whose components are the measures D_n for each model in the library. If these vectors are created from very long transients (or taken as an ensemble of very many short ones), then the result can be thought of as a 'reference vector' for that signal type and noise level. That is, the reference vector describes the expected location in 'residual space' of the D_n results from the particular signal type at a particular noise level. Given a family of reference vectors (one for each data type in the library), and a 'test vector' obtained from some actual incoming data, it should be possible to determine which one of the reference vectors the test vector was closest to, in a Euclidean sense. Ideally, this concept would form the basis of a classification scheme. Detection could be accomplished as well, by including reference vectors for the background noise. Classifying a short time window of the signal as a background noise transient would be akin to having no detection for that period. Thus, the refined approach is as follows:

1. For each signal type of interest, obtain a relatively noise- and interference-free set of time series.

2. Use these time series, along with samples of the noise environment of interest, to generate data sets with a wide range of SNRs.
3. At each SNR, generate the residuals $D_n[t]$ (using arbitrary models) and consider the set of residuals as a vector. Plotting these vectors generates one trajectory in residual space for each data type, and the trajectories are parametrized by SNR.
4. Given these reference trajectories, classify incoming signals by obtaining (from the incoming signal) a test vector of the $D_n[t]$ measurements and determining which reference trajectory the test vector is closest to. The point of closest approach along this trajectory can be used to estimate the SNR, or if known *a priori*, the SNR can be used to constrain which trajectory segments that are searched for a match.

A set of experiments to demonstrate this process was conducted, using transient signals from model data and synthetic noise. Details on the experimental protocol and results follow.

DEMONSTRATION EXPERIMENTS

The first step in the experiments was to create a set of model data types for use as transient signal sources. Model systems with broadband, non-periodic output were chosen, since signals of this type are the most difficult to detect using traditional methods. A number of iterated maps and their various components were chosen for sources of transient signals. In order to make the first- and second-order moments of each of the experimental transient types match, several of the generating equations were normalized so that the output was roughly zero mean and unit variance. Long (10^6 points) data sets were then generated and stored. For those generating systems that could not be transformed, long data sets were generated from the standard equations, and these data sets were normalized. The appendix contains a list of the 17 original (and, where available, normalized) functional forms used to generate data sets. The six functional forms that could be normalized (Equations 6-11) were used as elements of the filter bank (i.e. the F_n discussed above). Thus the library of transient types included six which were modeled explicitly by the filter banks, and 11 which were not.

The second step was to choose an SNR range. All of the validation experiments were performed with computer generated pseudo-Gaussian noise. The initial SNR range was +35 dB (2% noise) to -35 dB (5600% noise), in 5 dB increments. At each of these SNRs, a noisy data set was generated from each of the

17 signal types, as well as a "noisy" noise data set for comparison. Reference vectors for each signal/SNR combination were computed by using the noisy data set as a set of initial conditions for the six filter bank elements. One-step prediction residuals were generated, and single measures of E_0 and R_n were computed for the long files (that is, w in Equations 5 and 6 was taken to be 5×10^5 points). The results of this procedure was a list of reference vectors for each of the data types. Analysis of this list revealed that for SNRs in excess of about +15 dB, the vectors were significantly different for the different data types, and that below about -10 dB, all the reference vectors looked the same, since the output of the nonlinear filter bank was dominated by the high noise level. Based on these observations, reference vectors were recomputed, and all subsequent analyses performed, using an SNR range from +15 to -15 dB, in 2 dB steps.

Plotting the reference vectors for each signal/SNR combination results in a set of "reference trajectories", or traces through residual space. These traces or trajectories are parametrized by noise level. Figures 3 and 4 show 3-dimensional projections for each of the 18 (each of the transient types plus one for the noise) reference trajectories onto two separate subsets of the six filter elements.

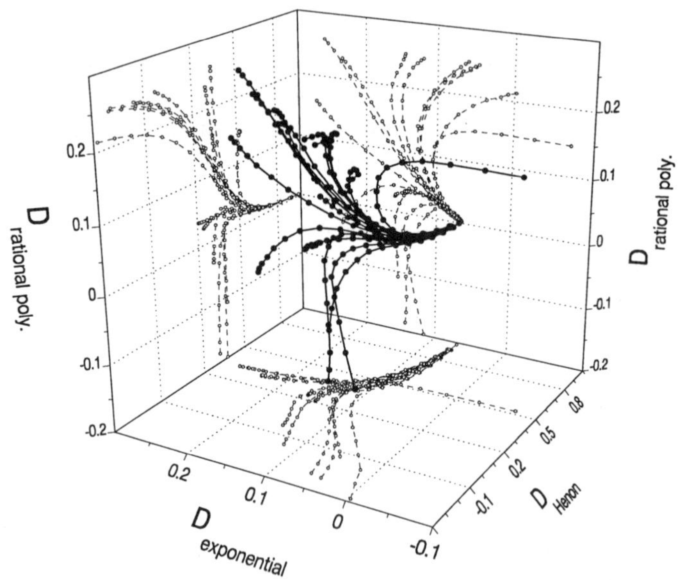

Figure 3. Reference trajectories from 18 different signals and from Gaussian noise, from +15 to -15 dB. The trajectories converge in the low SNR region, and the symbols are spaced 2 dB apart. Each trajectory is also projected onto the axis planes (dotted lines and open symbols).

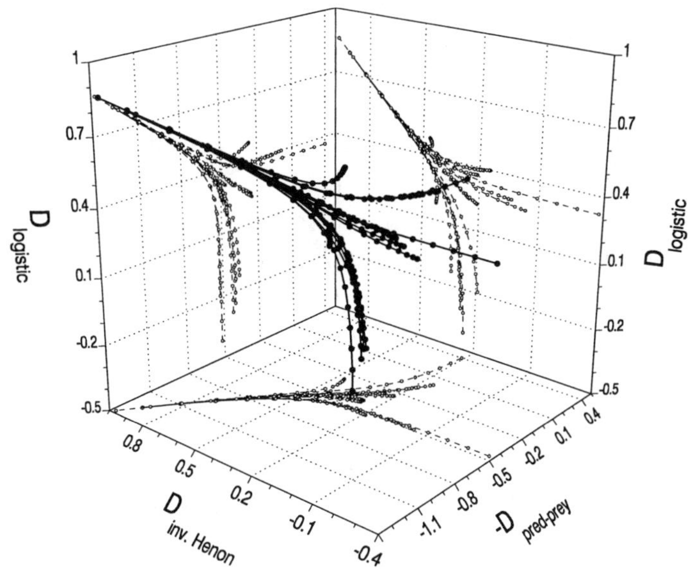

Figure 4. Reference trajectories projected onto the three axes not used in Figure 3.

Having obtained the reference trajectories, detection and classification of transient signal proceeds by computing a test vector for a short segment of data, and comparing the location of that test vector with the set of known reference trajectories. Assignment of an identity to the signal fragment for which the test vector was computed is based on measuring the distance between that test vector and the reference trajectories. The reference trajectory to which the test vector lies closest corresponds to the most likely data type, and the location along the trajectory corresponds to the most likely SNR. For the experiments performed to date, it was assumed that the instantaneous SNR was known, therefore simplifying the search of trajectories. Figure 5 shows three of the reference trajectories, along with 100 different test vectors from the corresponding transient types at an SNR of +9 dB, and demonstrates that the test vectors from the different signal types do indeed fall into different regions of residual space.

The performance of the filter bank was measured by computing the probabilities of detection (p_D) and of false alarm (p_{FA}) for each of the signal types. Probability of detection was defined as the fraction of transients of a given type that were correctly identified. Probability of false alarm for a given type was defined as the fraction of transients of all other types that were incorrectly identified as the specified type. Noise bursts at various levels were also used to generate test vectors, so that p_{FA} includes the probability that a particular transient type will be declared in the absence of any signal whatsoever.

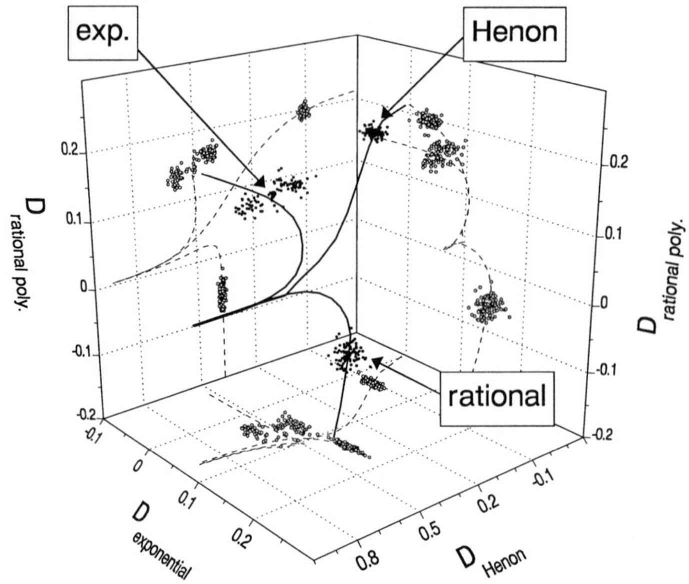

Figure 5. Perspective views of three of the reference trajectories, along with 100 test vectors from the corresponding signal type, at an SNR of +9 dB.

For each of the signal types, transient bursts of four different lengths (30, 100, 300, and 1000 points) were used to generate 1000 test vectors at each of the SNRs between -15 and +15 dB. For each transient, the distance between the test vectors and each reference vector at that SNR was computed. The data type corresponding to the reference vector closest to each test vector was used as the classification of each transient. Figure 6 shows a typical set of results for one of the signal types, and Figure 7 shows the results for bursts of Gaussian noise, presented as p_D and p_{FA} as functions of SNR and burst length.

The overall performance of the filter bank was measured by taking averages of curves such as those shown in Figures 6-7. The signal types were grouped into two categories for this averaging -- the results for the six transients whose dynamics were used as filter bank elements ('modeled' transients) were averaged separately from the remaining 11 ('non-modeled') transient types. Figure 8 shows the aggregate performance of the filter bank against the modelled transients, while Figure 9 shows the performance for the non-modelled signals.

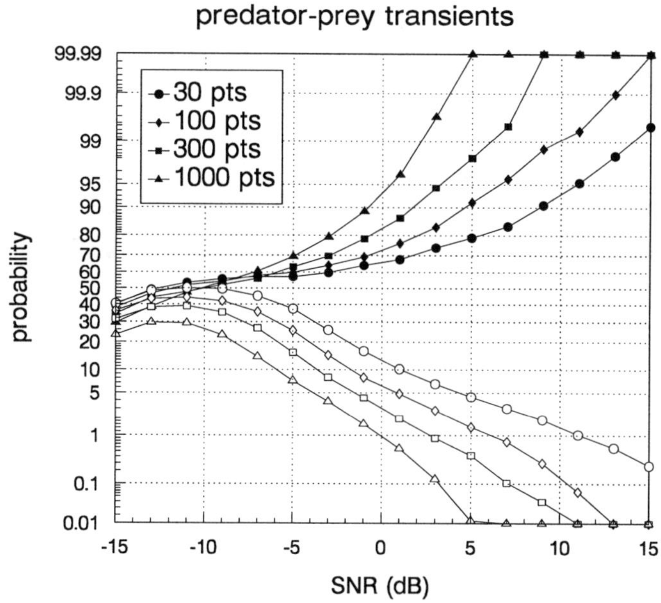

Figure 6. Probability of detection (p_D, solid symbols) and false alarm (open symbols, p_{FA}) vs. SNR for transients from the predator-prey system (Equation 9).

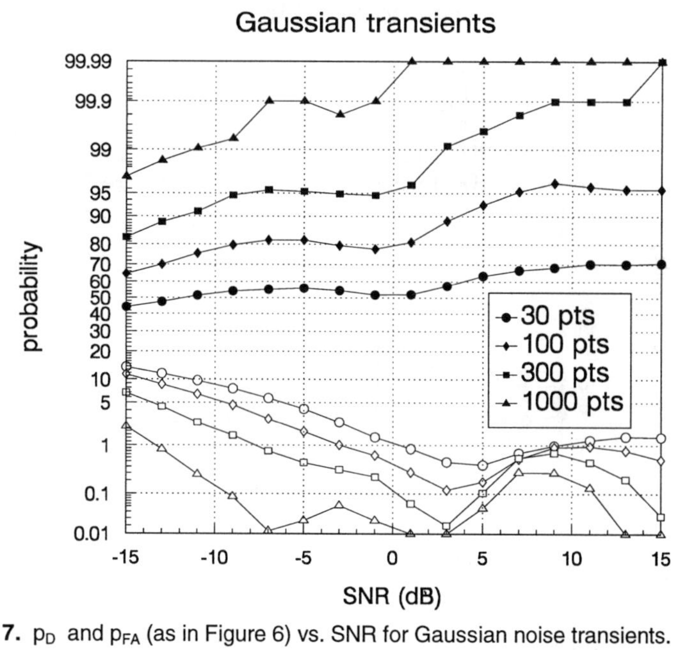

Figure 7. p_D and p_{FA} (as in Figure 6) vs. SNR for Gaussian noise transients.

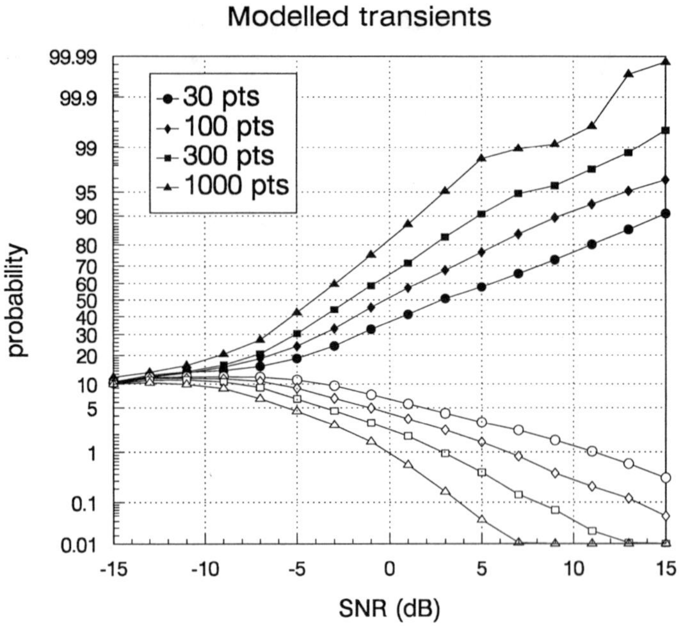

Figure 8. Average p_D and p_{FA} vs. SNR for transients from the six modeled systems.

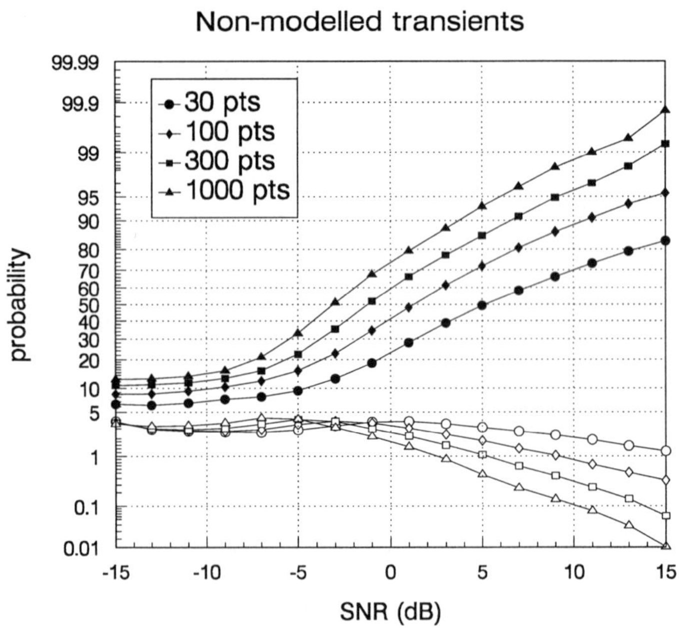

Figure 9. Average p_D and p_{FA} vs. SNR for the eleven non-modeled systems.

As can be seen in Figures 6-9, the nonlinear dynamic filter bank achieves significant detection probability with low false alarm rates at low SNRs even for data types whose dynamics are not explicitly represented in the filter. This implies that it should be possible to discriminate between a variety of real-world transients, providing that an appropriate set of filter elements can be chosen. Comparing Figures 8 and 9, it can be seen that the explicitly modeled transients have generally higher p_D values than those from the non-modeled signal types, but that they have slightly higher p_{FA} values (this is to be expected, since the p_{FA} from 'guessing' would be 1/6 = 0.17 for the modeled systems and 1/11 = 0.09 for the non-modeled systems).

FUTURE DIRECTIONS

Validation

Perhaps the most important remaining aspect of this investigation is the requirement for a method with which to compare the prediction residual (and any other technique) against traditional methods. Ideally, one would obtain an 'ideal' D/C method which is not feasible in reality but which would provide a standard benchmark against which to judge the algorithms. One possible ideal method is time domain matched filtering, in which it is assumed that one has a perfect copy of each noise-free transient, and the linear correlation between those ideal transients and the noisy data segments that we have for testing is used to perform classification. Another possibility is to use time-averaged power spectra for the correlations, which may be more realistic since one would normally never have detailed time histories of actual noise-free transients, but may well have detailed spectral estimates. Given some benchmark, one way to report the results of a given algorithm is in terms of an SNR reduction from the matched filter method. That is, if the prediction residual method at an SNR of -5 dB performs as well as the matched filter method at -20 dB, then there is a +15 dB reduction. If some other method (at -5 dB) performs as well as the matched filter at -24 dB, then that method has a +19 dB reduction, and the nonlinear filter bank method would outperform it by 4 dB. Choosing one or more benchmarks and characterizing the performance of the prediction residual method (as well as any other candidate D/C algorithm) needs to be done.

"Fuzzy" Classification

As can be seen in Figure 5, the actual transient test vectors form clusters in residual space that appear to be more akin to ellipsoids than to spheres. Quantifying the shapes of these clusters as functions of SNR and transient duration will allow a substantially more sophisticated method for declaring detections and classifications. In particular, understanding the distributions of the test vectors will allow the distances from reference trajectories to be expressed in terms of probabilities, rather than just Euclidean distance. This will allow the correct identification of a test vector that may be physically closest to the 'wrong' reference trajectory. The ability to attach significance values to the D/C declarations should improve the performance of the method dramatically.

Custom Reference Trajectories

In Figure 9 (non-modeled transients), the p_D performance is in general degraded compared to that shown in Figure 8 (modeled transients). This degradation is due to the relatively dense clustering of reference trajectories (Figures 3 and 4). This decrease in performance may possibly be mitigated by the probabilistic methods described above. Alternatively, since the reference trajectories created above are a reflection of the actual dynamics used in the nonlinear filter functions, it may be possible to design a set of filters with specific characteristics for particular signal and noise types. This may result in a set of reference trajectories that are relatively distant from one another in residual space. Investigation into just how the dynamics of the filter function map into the reference trajectories and the design of specific filters should also improve the performance of this method.

SUMMARY

The use of filter banks based on nonlinear mapping functions for transient detection and classification seems feasible. Methods involving monitoring the time evolution of adaptive filter ceofficients and of static filter residuals were presented along with analyses of their performance against model data sets. Several potential enhancements and future directions have been identified. The method capitalizes on several advances in the analysis of chaotic systems.

APPENDIX: GENERATING EQUATIONS

The six models below are normalized versions of the standard forms -- when iterated they generate time series with approximately zero mean and unit variance. They were used as the components of the dynamic filter bank:

Logistic:

$$\tilde{x}[t+1] = 0.7285 - 1.237\,\tilde{x}[t] - 0.7561\,\tilde{x}[t] \tag{6}$$

Hénon:

$$\tilde{x}[t+1] = 1.009 + 0.30\,\tilde{x}[t-1] - 0.7197\,\tilde{x}[t] - 1.009\,\tilde{x}[t]^2 \tag{7}$$

Inverted Hénon (upside down attractor):

$$\tilde{x}[t+1] = -1.009 + 0.30\,\tilde{x}[t-1] - 0.7197\,\tilde{x}[t] + 1.009\,\tilde{x}[t]^2 \tag{8}$$

Predator-Prey:

$$\begin{aligned}\tilde{x}[t+1] = & -0.3401 - 1.316\,\tilde{x}[t-1] + 0.8376\,\tilde{x}[t] + 0.6050\,\tilde{x}[t-1]^2 \\ & -0.6284\,\tilde{x}[t]\tilde{x}[t-1] + 0.2889\,\tilde{x}[t]\tilde{x}[t-1]^2\end{aligned} \tag{9}$$

Exponential Map:

$$\tilde{x}[t+1] = -1.014 - \frac{\log_e(|1 - 0.7340\,e^{-0.9877\,\tilde{x}[t]}|)}{0.9881} \tag{10}$$

Rational Polynomial:

$$\tilde{x}[t+1] = \frac{\begin{array}{c}0.2125 - 1.391\,\tilde{x}[t-1] + 1.641\,\tilde{x}[t] \\ -0.9087\,\tilde{x}[t-1]^2 + 0.2211\,\tilde{x}[t]\tilde{x}[t-1] + 0.6460\,\tilde{x}[t]^2\end{array}}{1 - 1.375\,\tilde{x}[t-1] + 1.077\,\tilde{x}[t] + 0.7892\,\tilde{x}[t-1]^2 + 0.7892\,\tilde{x}[t]^2} \tag{11}$$

The models below were used to generate data from which each component was normalized after it was generated. Each component of each model was used as a different test signal type:

Hyperchaos:

$$x[t+1] = 3.8x[t](1 - x[t]) - 0.05(y[t] + 0.35)(1 - 2z[t]) \tag{12a}$$

$$y[t+1] = 0.1[(y[t] + 0.35)(1 - 2z[t]) - 1](1 - 1.9x[t]) \tag{12b}$$

$$z[t+1] = 3.78z[t](1 - z[t]) + 0.2y[t] \tag{12c}$$

Pickover:

$$x[t+1] = \sin(2.24y[t]) - z[t]\cos(0.43x[t]) \quad (13a)$$

$$y[t+1] = z[t]\sin(-0.65x[t]) - \cos(-2.43y[t]) \quad (13b)$$

$$z[t+1] = \sin(x[t]) \quad (13c)$$

Ikeda:

$$x[t+1] = 1.0 + 0.7\left[x[t]\cos(\tau) - y[t]\sin(\tau)\right] \quad (14a)$$

$$y[t+1] = 0.7\left[x[t]\sin(\tau) + y[t]\cos(\tau)\right] \quad (14b)$$

$$\tau = 0.4 - \frac{6}{1 + x[t]^2 + y[t]^2} \quad (14c)$$

Lazelles-Ferriere:

$$x[t+1] = -9y[t]e^{-0.1(x[t] + y[t])} \quad (15a)$$

$$y[t+1] = 0.2x[t]e^{-0.1(x[t] + y[t])} + 0.8y[t]e^{-0.05y[t] - 0.025x[t]} \quad (15b)$$

ACKNOWLEDGMENTS

I wish to thank Dr. Michael Shlesinger of the Office of Naval Research for his support for this work (under Navy contract N00014-93-C-0242). I would also like to thank James Adams for his technical and analytical contributions, and the RTA Corporation, for supporting this effort and its precursors.

REFERENCES

1. Farmer, J.D. and Sidorowich, J.J. (1987) "Predicting chaotic time series." *Phys. Rev. Lett.*, **59(8)**:845-848.
2. Packard, N.H.; Crutchfield, J.P.; Farmer, J.D.; and Shaw, R.S. (1989). "Geometry from a Time Series." *Phys. Rev. Lett.*, **45(9)**:712.
3. Takens, F. (1981) "Detecting Strange Attractors in Fluid Turbulence." In Dynamical Systems and Turbulence. D. Rand and L.S. Young (Editors). Springer-Verlag:Berlin, New York.
4. Eckmann, J.P. and Ruelle, D. (1985) "Ergodic Theory of Chaos and Strange Attractors." *Reviews of Modern Physics*, **57(3)**:617-656.
5. Crutchfield, J.P. and McNamara, B.S. (1987) "Equations of Motion from a Data Series." *Complex Systems*, **1**:417-452.

6. Broomhead, D.S. and King, G.P. (1986) "Extracting Qualitative Dynamics from Experimental Data." *Physica D*, **20**:217-236.
7. Kostelich, E. and Yorke, J.A. (1990) "Noise Reduction: Finding the Simplest Dynamical System Consistent with the Data." *Physica D*, **41(2)**:183-196.
8. Shannon, C. (1948) "A Mathematical Theory of Communication." *Bell Tech. Jour.*, **27**:379-423, 623-656.
9. Wiener, The Extrapolation, Interpolation and Smoothing of Stationary Time Series with Engineering Applications (Wiley, N.Y., 1949).
10. Kolmogorov, A.N. (1983) "On Logical Foundations of Probability Theory." In: Probability Theory and Mathematical Statistics (K. Ito and J.V. Prokhorov, eds.). Springer-Verlag, Lecture Notes in Mathematics 1021, Berlin, New York.
11. Kalman, R.E. (1960) "A New Approach to Linear Filtering and Prediction Problems." *Trans. ASME J. Basic Eng. Ser. D*, **82**:35-45.
12. Press, W.H.; Teukolsky, S.A.; Vetterling, W.T.; and Flannery, B.P. Numerical Recipes (Cambridge University Press, 1988).
13. Sauer, T.; Yorke, J.A.; and Casdagli, M. (1992) "Embedology." *J. Stat. Phys.*, **65**:579-616.
14. Cawley, R. and Hsu, G.-H. (1992) "SNR performance of a noise reduction algorithm applied to coarsely sampled chaotic data." *Phys. Lett. A.*, **166**:185-196.
15. Cawley, R.; and Hsu, G.-H. (1991) "Local-Geometric-Projection Method for Noise Reduction in Chaotic Maps and Flows.", *Phys. Rev. A*, **46(6)**:3057.
16. Breeden, J.L. and Hubler, A. (1990) "Reconstructing Equations of Motion from Experimental Data with Unobserved Variables." *Phys. Rev. A*, **42(10)**:5817-4826.
17. Breeden, J.L. and Packard, N.H. (1992) "A Learning Algorithm for Optimal Representations of Experimental Data." *Center for Complex Systems Research Technical Report* CCSR-92-11, University of Illinois at Urbana-Champaign.
18. Baake, E.; Baake, M.; Bock, H.G. and Briggs, K.M. (1992) "Fitting Ordinary Differential Equations to Chaotic Data." *Phys. Rev. A*, **45(8)**:5524-5529.
19. Brown, R.; Rulkov, N.F.; and Tracy, E.R. (1994) "Modeling and Synchronizing Chaotic Systems from Time-Series Data." *Phys. Rev. E.*, **49(5)**:3784-3800.
20. Brush, J.S. and Kadtke, J.B. (1992) "Nonlinear Signal Processing Using Global Empirical Equations of Motion." *Proceedings of the ICASSP-92*, **V**:321-325.
21. Cremers, J. and Hubler, A. (1987) "Construction of Differential Equations from Experimental Data." *Z. Naturforschung*, **42(A)**:797.
22. Judd, K. and Mees, A. (1995) "On selecting models for nonlinear time series." *Physica D*, **82**:426-444.
23. Kadtke, J.B. and Brush, J.S. (1994) "Global Modelling of Chaotic Time Series with Applications to Signal Processing." In The Chaos Paradigm: Developments and Applications in Engineering and Science. *AIP Conference Proceedings*, **296**:205-232.
24. Carroll, T.L. and Pecora, L.M. (1991) "Synchronizing Chaotic Circuits." *IEEE Trans. on Circuits & Systems*, **38**:453-456.
25. Hübler, A. and Lüscher, E. (1989) "Resonant Stimulation and Control of Nonlinear Oscillators." *Naturwissenschaften*, **76**:67.

26. Ott, E.; Grebogi, C.; and Yorke, J.A. (1990) *Phys. Rev. Lett.*, **64**:1196.
27. Sinha, S.; Ramaswamy, R.; and Rao, J.S. (1990) "Adaptive Control in Nonlinear Dynamics." *Physica D*, **43**:118-128.
28. Sardar, H.M. and Ahmadian, M. (1992) "A Modified Model Referenced Adaptive Control Technique for Nonlinear Dynamic Systems." *Transactions of the ASME*, **114**:154-160.
29. Breeden, J.L. (1994) "Open-Loop Control of Nonlinear Systems." *Phys. Lett. A.*, **190**:264-272.
30. Garfinkel, A.; Spano, M.L.; and Ditto, W.L. (1992) "Controlling Cardiac Chaos." *Science*, **257**:1230-1235.
31. Chua, L.O.; Wu, C.W.; and Huang, A. (1993) "A Universal Circuit for Studying and Generating Chaos: Routes to Chaos." *IEEE Trans. on Circuits & Systems, Part I*, **40**:732-744.
32. Glover, J. and Mees, A.I. (1992) "Reconstructing the dynamics of Chua's circuit." *J. Circuits Syst. Comput.*, **3**:201-214.
33. Murali, K. and Lakshmanan, M. (1993) "Chaotic Dynamics of the Driven Chua's Circuit." *IEEE Trans. on Circuits & Systems*, **40**:836-840.
34. Brush, J.S. and Kadtke, J.B. (1994) "Adaptive Nonlinear Dynamical Processing for Time Series Analysis." In The Chaos Paradigm: Developments and Applications in Engineering and Science. *AIP Conference Proceedings*, **296**:159-181.
35. Leung, H. and Haykin, S. (1994) "Detection and Estimation Using an Adaptive Rational Function Filter." *IEEE Trans. on Signal Processing*, **42(12)**:366-3376.
36. Roy, R.J. and Sherman, J. (1967) "A Learning Technique for Volterra Series Representation." *IEEE Trans. on Automatic Control*, **1967(12)**:761-764.
37. Widrow, B.; Glover, J.R.; McCool, J.M.; Kaunitz, J.; Williams, C.S.; Hearn, R.H.; Zeidler, J.R.; Dong, E.; and Goodlin, R.C. (1975) *Proc. of the IEEE*, **63(12)**:1692-1716.
38. Koh, T. and Powers, E.J. (1983) "An Adaptive Nonlinear Digital Filter with Lattice Orthogonalization." *Proceedings of the ICASSP-83*, **1(10)**:37-40.
39. Orgren, A.C.; Malik, N.R.; and Youn, D.H. (1986) "Convergence of an Adaptive Echo Cancellation System with an Augmented Predictor." *Proceedings of the ICASSP-86*, **19.4.1**:961-964.
40. Bottomley, G.E. and Alexander, S.T. (1991) "A Novel Approach for Stabilizing Recursive Least Squares Filters." *IEEE Trans. on Signal Processing*, **39(8)**:1770-1779.
41. Kadtke, J.B. and Brush, J.S. (1993) "Adaptive Methods for Chaotic Communication Systems." In Chaos in Communications (L. Pecora, ed.), *Proceedings of the SPIE*, **2038**:182-193.

Time Series Analysis of Transient Chaos: Theory and Experiment

Imre M. Jánosi[1] and T. Tél[2]

[1]*Department of Atomic Physics, and* [2]*Institue for Theoretical Physics*
Eötvös University, Budapest, Puskin u. 5-7, H-1088, Hungary

Abstract A simple method is described how nonattracting chaotic sets can be reconstructed from time series by gluing those pieces of many transiently chaotic signals together that come close to this invariant set. The method is illustrated by both a map of well known dynamics, the Hénon map, and a signal obtained from an experiment, the NMR laser. The strange saddle responsible for the transient chaotic behavior is reconstructed and its characteristics like dimension, Lyapunov exponent, and correlation function are determined.

INTRODUCTION: TRANSIENT CHAOS

The appearance of chaos on finite time scales is known as *transient chaos* (for a review see [1]), and provides an example of a kind of "nonequilibrium process" that cannot be understood as an asymptotic behavior. In such cases one observes a moving around of the system in an apparently chaotic manner and then, rather suddenly, a settling down to a steady state which is either a periodic or a chaotic motion (but of different type than the transients). Studying only the asymptotic behavior of such systems means loosing the interesting, chaotic part contained in the transients.

In physical systems exhibiting transient chaos there exists in phase space a *nonattracting chaotic set: a chaotic saddle* [2, 3, 4, 1] that coexists in dissipative cases with an attractor, and in Hamiltonian systems with an exit channel leading to an asymptotic motion. Trajectories starting from randomly chosen initial points then approach the attractor or the asymptotic state with probability one. Because of the saddle's stable foliation, however, they might come close to the saddle and stay in its vicinity for a shorter or longer time. What is observable numerically or experimentally is not the chaotic saddle but rather a *small neighborhood* of it. This results in the appearance of a chaotic motion with a well defined average lifetime of $1/\kappa$ where κ is the escape rate, a basic characteristic of the chaotic saddle that truns out to be independent of the size of small neighborhood considered.

In dissipative systems, that we shall mainly focus on in this paper, a typical occurence of transient chaos is in the periodic windows inside the chaotic region. The positivity of the topological entropy then ensures the existence of a chaotic saddle. Such windows occur between two crisis configurations: a boundary crisis [2] (where a chaotic attractor ceases to exists) and an interior crisis [2] (where a chaotic attractor is suddenly enlarged). Since the total measure of windows is finite in the parameter space, just like that of strictly chaotic cases, we conclude that the probability to find transient chaos is *on the same order* as that of permanent chaos.

In the realm of transient chaotic signals, one can distinguish three different types:

(a) Signals settling down on a periodic attractor. This is the case in the period doubling regime inside a periodic window.

(b) Signals settling down on a multi-piece attractor. This happens beyond the period doubling regime in a periodic window. The illustrative examples of the present paper will belong to this second class.

(c) Signals settling down on an attractor lying far away from the chaotic saddle so that it can be considered to be at infinity. In Hamiltonian systems the attractor at infinity is replaced by an asymptotic motion.

Transient chaos is closely connected to many other aspects of nonlinear dynamics. In systems with fractal basin boundaries [5], trajectories staying for a long time around such boundaries exhibit transient chaos. The boundaries themselves turned out to be stable manifolds of chaotic saddles. The so called *noise induced chaos* [6], when a system with simple attractors turns chaotic in the presence of noise, has recently been explained as a consequence of transient chaos [7]. Transient chaos can also be a sign of permanent chaos to be born. Large attractors born at crises incorporate into themselves the chaotic saddles existing before. Consequently, dynamical properties of the saddle are partially inherited by the large attractor. Transient chaos can thus provide a *backbone* of the motion on attractors [8]. The OGY method of controlling chaos [9] converts the motion on a chaotic attractor before control sets in into a kind of transient chaos.

Interesting other phenomena are related to the Hamiltonian version of transient chaos. For scattering processes in open Hamiltonian systems chaos is inevitably of transient type and *chaotic scattering* [10] has attracted great recent interest. A particularly appealing application is the chaotic advection of tracers in open hydrodynamical flows [11, 12, 13]. *Diffusion* and other transport phenomena can also be interpreted as consequences of chaotic scattering [14].

The experimental investigation of transient chaos has received, in spite of its relevance, relatively little attention, and has mainly concentrated on determining the average chaotic lifetime. In this paper we illustrate that transient chaos can be analyzed from time series in very much the same manner as permanent chaos. Chaotic saddles underlying transient chaotic dynamics can be reconstructed and their characteristics taken with respect to their natural measure can be determined from time series.

TIME SERIES ANALYSIS: GENERAL CONSIDERATIONS

Suppose, we follow one single variable of the system and record the data as a time series. Because of escape from any neighborhood of the strange saddle, the input of our method has to be an *ensemble* of time series of the preselected variable with several initial conditions in some finite domain and with endpoints on the true attractor [3].

It is worth starting with a brief characterization of transient chaotic signals. The transient part of such time series consists typically of three regions

(I) transient from the initial point to a neighborhood of the chaotic saddle,

(II) motion in this neighborhood, and

(III) escape to the attractor.

In general, a simple visual observation of the complete signal clearly indicates when the attractor is reached, i.e., where the end point of region III is. Separating regions I and III from region II is less well defined. Region I (III) is associated with the part of trajectories which approach (escape) the saddle along its stable (unstable) manifold. The rate of this exponential appraoch (escape) is governed by the negative (positive) Lyapunov exponent of the saddle. The length of region I (III) has thus to be chosen as a few times the reciprocal value of the negative (positive) Lyapunov exponent's modulus.

For the reconstruction of the chaotic saddle, one only needs the points of region II. This part of the signal will be called hereafter the *truncated time series*. Thus, from the complete signal we have to delete those typically long parts which lie close to the attractor, as well as Regions I and III. Thus, an ensemble of truncated time series arises. The average length of these truncated signals is on the order of the average transient life time $1/\kappa$. One has to try different cuttings of the transient pieces so that the results do not depend on details of the truncating procedure.

For many purposes, like e.g., for reconstructing the chaotic saddle or for extracting its dimensions, it is sufficient to have a great number of typically short trajectories with points far away from the attractor. If, however, one is interested in the Lyapunov exponent on the chaotic saddle or in the power spectrum of the transient signal, a single long trajectory might be needed created via some appropriately chosen gluing procedure. This trajectory has to faithfully represent the motion around the chaotic saddle. The knowledge of such a long trajectory is also useful because it enables us to take over numerical routins worked out for analysing permanent chaotic time series.

In numerical simulations the so-called PIM triple method [15] introduced by Nusse and Yorke provides an algorithm for generating long trajectories being all the time close to the chaotic sdaddle. Without knowing the equations of motion this algorithm is obviously not applicable and cannot be of help for us in the time series analysis. In a similar spirit, however, we can construct a long artificial signal with points close to the saddle extracted from an ensemble of transient chaotic time series. The method, originally proposed in [16], provides thus an experimental analog of a PIM triple trajectory.

Our aim is thus to construct an *artificial time series* representing the long time motion in a close vicinity of the chaotic saddle. The most naive approach is to simply glue together all the truncated signals without taking care of adjusting the ending and starting points. This may, however, lead to the appearance of some unrealistic points on a Poincaré map. An effective way to avoid this can be the use of the *matching condition*: Choose a point in the truncated time series which has a value close enough to the endpoint of the previous segment, discard the intermediate points and glue the remaining part to the endpoint of the preceding segment. The gluing process is inevitably an excess noise source whose effect can, however, be kept under control (see [16]).

Next, one can apply a low-dimensional time-delay coordinate embedding, produce a Poincaré section, and plot the invariant set. Further, one can also use the time series analysis method (for a review see [17]) to determine quantitative characteristics of the chaotic saddle.

The most important steps of the method can thus be summarized as follows:

1. Take an ensemble of time series containing transients to a steady state behavior.

2. Locate the attractor of the dynamics in the variable investigated and separate the transient parts.

3. Construct truncated time series.

4. Glue the truncated signals together, apply a low-dimensional time-delay coordinate embedding, produce a Poincaré section, and plot the invariant set.

5. Apply the usual time series analysis method to determine quantitative characteristics of the chaotic saddle.

We shall carry our the method described in two cases, for the Hénon map [16], and for an experimental signal generated by an NMR laser [18].

TIME SERIES ANALYSIS: TWO EXAMPLES

Hénon map: We use the well-known recusion relation

$$x_{n+1} = 1 - ax_n^2 + by_n \qquad y_{n+1} = x_n \tag{1}$$

to generate a time series $\{x_i\}$. This recursion represents the most general form of invertible quadratic maps. As a particular parameter setting we take $a = 1.2715$ and $b = 0.3$ corresponding to a configuration where a seven piece chaotic attractor coexists with a chaotic saddle.

NMR laser: The signals of an NMR laser were investigated whose chaotic behavior has been studied for over a decade [19]. The detailed description of the experimental setup is available in the literature (see e.g. [20]), here we just mention the features most relevant for our subject. The laser output is a time-dependent voltage. Our investigations were carried out at a control parameter value just below a crisis point where a (small) chaotic attractor coexists with a chaotic saddle. Consequently, the permanently chaotic motion on the attractor can only be observed after chaotic transients. The laser transients were generated pushing out the system from its permanent chaotic state. Then, the laser output was recorded as a normalized, dimensionless scalar time series $\{\xi_1, \xi_2, \ldots \xi_n\}$, $0 < \xi_i < 1$ of length $n = 1024$. The sampling frequency was 2500 Hz.

In the remaining part of this section we briefly describe the main steps of the time series analysis of our examples and illustrate these steps graphically by plots obtained for one of the two examples at least.

1. Take an ensemble of time series containing transients to a steady state behavior

Hénon map: 4000 truncated trajectories were taken. Figure 1a shows one typical signal along with regions I, II, and III. The asymptotic motion is restricted to 6 intervals in the x-variable (shaded strips) due to an overlap of the projection of two different attractor pieces. Figure 1b displays the behavior of an ensemble of 200 trajectories. One clearly sees also how the gaps between the strips become with increasing time less and less populated by the transients. The length of the densely populated region in the gaps gives a fairly good estimate to the averaged transient chaotic lifetime.

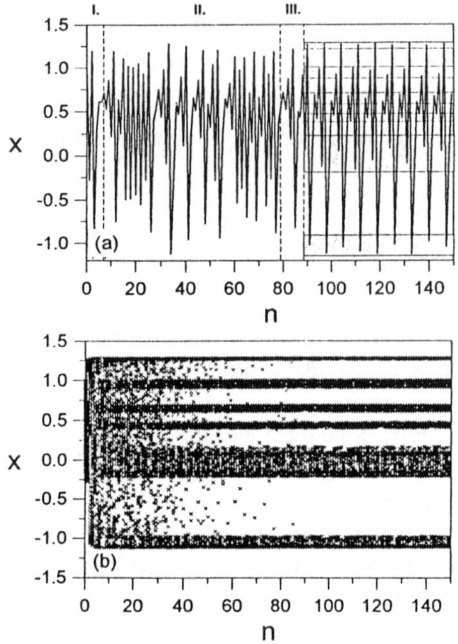

FIGURE 1. (a) Hénon signal settling down on a multi-piece chaotic attractor. The initial point is $x_0 = 0.06$, $y_0 = 0$. The shaded stripes represent the projection of the 7-piece chaotic attractor to the x-axis. (b) An ensemble of 200 transient chaotic trajectories of Eq. (1) belonging to uniformly distributed initial points in the interval $| x |< 0.5, y = 0$.

NMR laser: An ensemble of 9599 short records ending on the attractor has been generated. The length of these records (t = 0.4096s) was chosen to be about four times the average chaotic lifetime, so that we have just a few signals exhibiting transient chaos during the full record. Another, smaller ensemble of only several hundred longer (t = 1.9968 s) transient records was used to prove that the probability to find transients on longer time scales was negligible. To characterize the chaotic steady state, a long series of 256000 data was also recorded.

Fig. 2a shows a typical segment of the steady chaotic signal. The dotted line denotes the maximum observed peak height in this state. Fig. 2b illustrates a typical transient behavior. Switching back the quality factor at $t = 0$, a few peaks of relatively large amplitude arise after a delay time of approximately 0.02 s. The laser action between subsequent pulses is almost quenched. The various transient pieces will be glued together at this very low laser amplitude, the laser background level. The apparently chaotic motion outside the attractor then settles down to the steady state. Note that between the large transient impulses there are no peaks of intermediate amplitude characterizing the motion on the inner loop of the attractor (see Fig. 4a below).

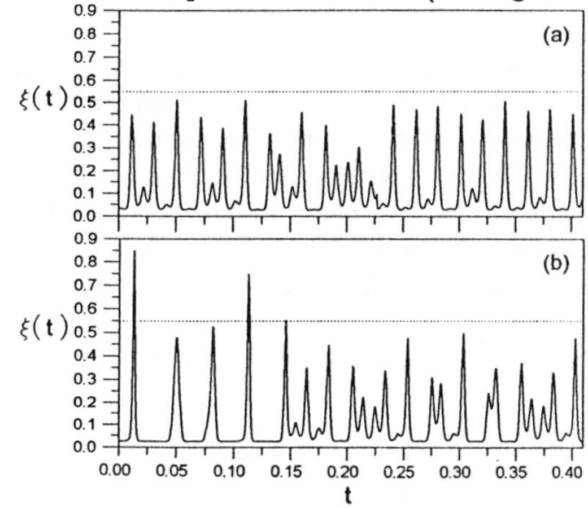

FIGURE 2. Typical NMR laser time series segments $\xi(t) \equiv \xi_i$ with $t = i \cdot 4 \cdot 10^{-4}$ s consisting of 1024 data points (a) representing the steady chaotic motion on the attractor, (b) illustrating a typical transient. The dotted line represents the maximum measured value of the laser signal in the chaotic steady state.

2. Locate the attractor of the dynamics in the variable investigated and separate the transient parts.

Hénon map: The chaotic attractor can easily be obtained by e.g., constructing a first return map from any trajectory after it has entered the shaded strips of Fig. 1. The result is shown in Fig. 3. The first parts of the signals ending with the last point outside of any of the strips were considered as the transient parts.

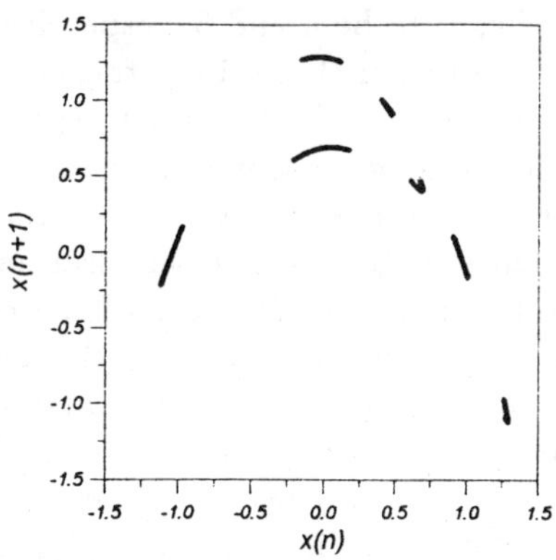

FIGURE 3. The seven piece chaotic attractor of the Hénon map. The attractor pieces are lines folded in a complicated manner so that their intersection with a straight line is a fractal (not seen in this resolution).

NMR laser: From the long permanently chaotic signal the attractor can be reconstructed by means of standard methods [17]. To separate the transient parts, we have applied the following empirical rule: Find the last peak of amplitude larger than the maximal observed amplitude at the steady state (dotted line in Fig. 2b) and keep the segment ending with this peak and a short piece of the laser background after. (The average distance between the peaks of the transient is ≈ 0.04 s with a random distribution, thus, the length of the short pieces added to the last peak was chosen to be 0.02 s.) Of course, some information has been lost in this way, because the last transient impulse may be smaller than the cutting level, but we have not found any better condition for the separation.

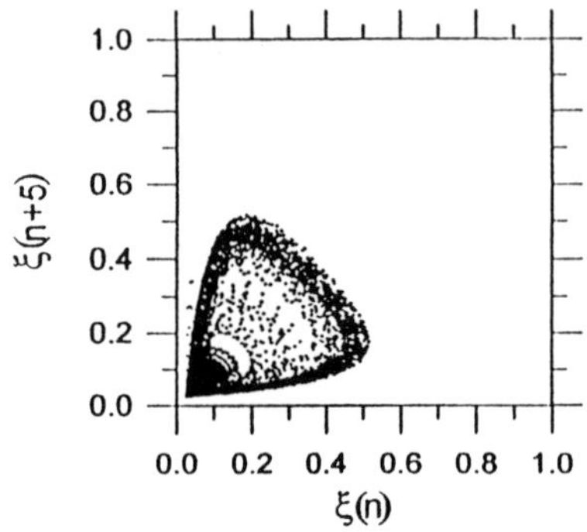

FIGURE 4. The chaotic attractor of the NMR laser signal on a return map

3. Construct truncated time series.

Hénon map: 4000 truncated signals weere used and the typical length of regions I and III was chosen to be 4 and 5, respectively. The average length of the truncated signals in region II was 13 iterations.

NMR laser: The empirical rule we used to separate the motion on the attractor also ensured discarding the transients from the vicinity of the saddle to the attractor. There was no need for an extra definition of region III here. In order to obtain truncated time series, the transients toward the saddle (regions I) are also to be discarded. Because of the rather large modulus of the first negative Lyapunov exponent governing this decay (see below), it was sufficient to drop just a little piece before the first large peak only. Time series truncated in this way had the property that probability to find a piece of length T decreased exponentially with T. The number $D(T)$ of truncated signals of lengths T is plotted in Fig. 5, after applying a local averaging. An exponential fit gives the escape rate $\kappa = 9.3$ 1/s with a relative error of less than 1%.

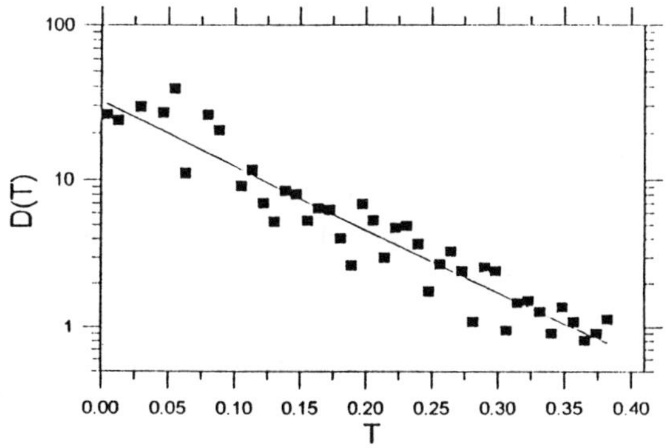

FIGURE 5. Coarse grained life time distribution of the NMR laser transients. The solid line shows the exponential fit: $D(T) = 34.23\exp(-T/0.1078s)$.

4. Glue the truncated signals together, apply a low-dimensional time-delay coordinate embedding, produce a Poincaré section, and plot the invariant set.

Hénon map: The matching procedure briefly sketched in the previous section has been used to glue pieces of truncated signals together to form an artificial long time series of total length 48502. The first return map of this series yields the points of Fig. 6 that are practically indistinguishable from the chaotid saddle obtained by direct computation, e.g., by means of the PIM triple algorithm.

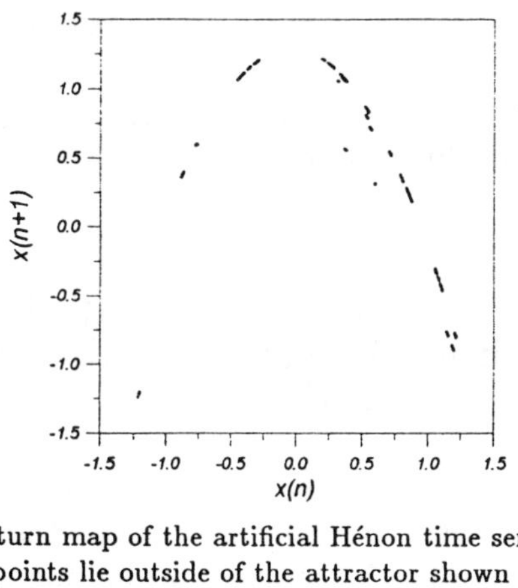

FIGURE 6. First return map of the artificial Hénon time series representing the chaotic saddle. Note that its points lie outside of the attractor shown in Fig. 3 and exhibit fractal features along the unstable direction, too.

NMR laser: In order to decrease the noise coming from the gluing procedure, extremely short truncated signals were discarded. Those longer than three effective periods of evolution (≈ 0.12 s), which is on the order of $1/\kappa$, have been glued together at the background signal level between the peaks. Fig. 7a shows a segment of the long artificial time series of 2037577 data points obtained from 5608 truncated signals in this way. The difference between the steady state and this motion is clear (cf. Fig. 2a). A single peak (between the rising and falling edges) in the artificial time series consists of typically 20–30 data points, thus the correlation is very high between consecutive data points. Therefore $\Delta n = 5$ was chosen in a two-dimensional time-delay embedding. Fig. 7b displays the plot of the return map of the time series representing the motion near the chaotic saddle. This is to compare with Fig. 3 in which the return map of the coexisting chaotic attractor is plotted. Careful inspection shows that the inner structures are not identical at all, however, the branches of the saddle are close to the attractor. The presence of a low dimensional invariant set is clear. A comparison of the Poincaré section of the saddle and the attractor, respectively, in a three-dimensional embedding (not shown here) leads to similar conclusion. Note that the chaotic saddle has a larger extent than the attractor.

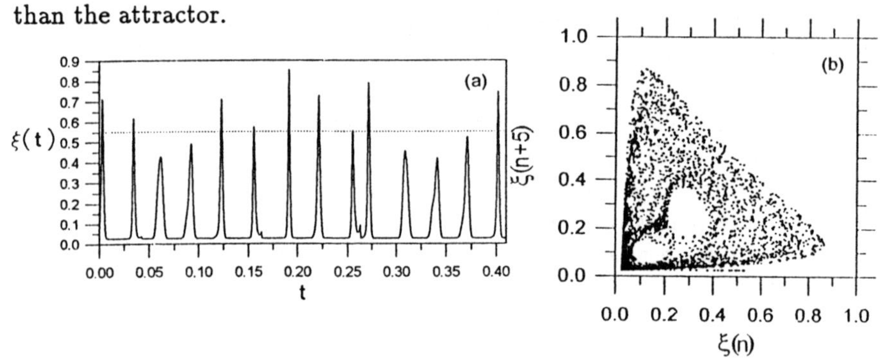

FIGURE 7. (a) A piece of the artificial long NMR laser time series representing the motion near the chaotic saddle. (b) Reconstruction of the chaotic saddle by the time delay embedding method.

5. Apply the usual time series analysis method to determine quantitative characteristics of the chaotic saddle.

Both long artificial time series have been analysed by using the the embedding technique [17, 21]. A phase point is defined a the d-dimensional time delay vector. The correlation sum is obtained by counting the number of neighboring phase points which are closer to each other than a distance r. In the scaling region $r \ll 1$, the correlation dimension D_c is given by the exponent of the best power fit to the correlation sum vs. r. Embedding dimensions up to $d = 6$ and $d = 10$ have been used in the Hénon and the NMR laser case, repectively. The correlation dimension extracted are

$$D_{c,H} = 1.07 \quad \text{and} \quad D_{c,NMR} = 2.3 \tag{2}$$

for these two cases, respectively, with a relative uncertainty of 5%. While the existence of a low-dimensional saddle is obvious for the Hénon map, it is much less trivial for the NMR laser. The saddle's dimension is on the same order as the attractor-dimensions investigated earlier [19, 20]. The results support the view that the transient NMR signal can be faithfully described in a three-dimensional phase space and the use of a two-dimensional return map is justified.

We also computed the Lyapunov exponents of the time series by applying well known methods [22, 23]. For the positive Lyapunov exponent on the chaotic saddle and on the coexisting attractor we obtained

$$\lambda_{s,H} = 0.58 \qquad \lambda_{a,H} = 0.06 \tag{3}$$

and

$$\lambda_{s,NMR} = 225 \ \ 1/\text{s} \qquad \lambda_{a,NMR} = 126.5 \ \ 1/\text{s}, \tag{4}$$

respectively. The relativ errors are here again less than 5%. The higher values on the saddle clearly show that the transient motion is more unstable, i.e. the divergence of the nearby trajectories is much faster there than on the attractor. The modulus of the negative Lyapunov exponent was in all the cases on the same order as λ_s.

There is a famous formula [3] expressing the information dimension $D_1^{(u)}$ of a chaotic saddle in a two-dimensional map along the unstable manifold in terms of the escape rate κ and the positive Lyapunov exponent λ_s as

$$D_1^{(u)} = 1 - \kappa/\lambda_s, \tag{5}$$

which is expected to be close to the partial fractal dimension. (Note that for chaotic attractors $\kappa = 0$ and one recovers $D_1^{(u)} = 1$.) With the values determined above, we obtain

$$D_{1,H}^{(u)} = 0.86 \quad \text{and} \quad D_{1,NMR}^{(u)} = 0.96, \tag{6}$$

respectively. The latter number is close to 1 which explains that the structure of the NMR laser saddle is not "airy", i.e. it does not contain large holes, in contrast to they Hénon saddle. This is due to the fact that the average chaotic life-time $1/\kappa$ is long on the time scale of predictability: $1/\lambda_s$.

As another illustration of the advantage of using long artificial time series, we computed their autocorrelation function in both examples. The results are very similar, therefore, in Fig. 8 we just show the correlation function of the

NMR laser signal. The most striking feature is that the correlation decay on the saddle is much faster than in the case of permanent chaos. In addition, one can observe a regular behaviour superimposed on the chaotic motion in the attractor case. This is so because the motion on the chaotic attractor is more coherent than on the saddle. In summary, the results strongly indicate that the motion on a chaotic saddle is, in general, *more chaotic* than the one on a coexisting strange attractor.

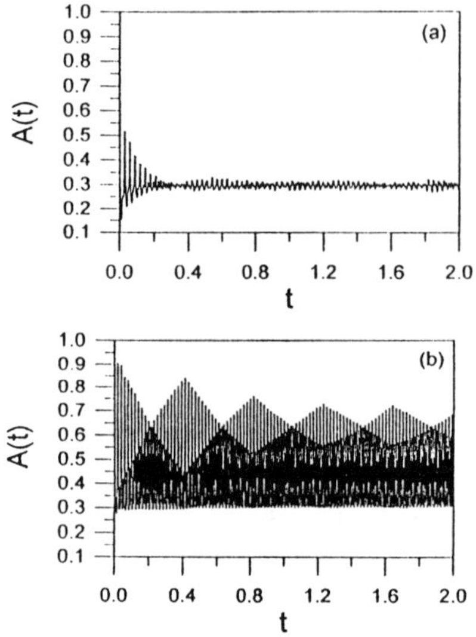

FIGURE 8. NMR laser autocorrelation function $A(t) = \langle \xi(t')\xi(t'+t) \rangle$ for the saddle (a), and for the attractor (b). Time is measured in seconds.

CONCLUDING REMARKS

With these studies we hope to contribute to the establishment of transient chaos as an experimentally accessible phenomenon. It is very likely that many experimentally observed transient chaotic signals were considered in the past to be uninterpretable and were therefore discarded. This might be due to the fact that the asymptotic feature of chaos has been overemphasized, and the use of long time averages has been more common in the definition of chaos than that of ensemble averages. As we have seen transient chaos is an example for a phenomenon where long time and ensemble averages can be different. An ensemble of trajectories staying around the chaotic saddle for long, yields averages characterizing the saddle, different from their long time asymptotics. The concept of nonattracting chaotic sets responsible for transient chaos, and their associated natural distributions helps, however, understanding the very strong similarities between transient and permanent chaos. We believe that transient chaos is an essential part of nonlinear dynamics also needed to the correct interpretation of many interesting phenomena.

Other examples of chaotic saddles reconstructed from time series would enrich our picture of transient chaos. It would be interesting to find experimental saddles exhibiting pronounced chaotic features also along the unstable direction, as at the Hénon map. An appealing application of the method presented above could be the time series analysis of chaotic tracer motion in open two-dimensional incompressible flows [11, 12, 13], a phenomenon which is a good candidate for being the first experimental investigation of classical chaotic scattering.

ACKNOWLEDGMENTS

It is a pleasure for the authors to acknowledge their fruitful collaboration with Leci Flepp on the transient NMR laser experiment. This work has been supported by the Hungarian National Science Foundation (OTKA) under Grant No. 2090, No. 2091, and No. T4439.

REFERENCES

[1] T. Tél, in *Directions in Chaos, Vol. 3*, ed. Hao Bai Lin (World Scientific, Singapore, 1990) p. 149.

[2] C. Grebogi, E. Ott and J. Yorke, Phys. Rev. Lett. **48**, 1507 (1982); Physica (Amsterdam) **D 7**, 181 (1983).

[3] H. Kantz and P. Grassberger, Physica **D 17**, 75 (1985).

[4] G. H. Hsu, E. Ott, and C. Grebogi, Phys. Lett. **127A**, 199 (1988).

[5] C. Grebogi et al, Phys. Rev. Lett. **50**, 935 (1983), E. Ott et al, Phys. Rev. Lett. **71**, 4134 (1993)

[6] M. Iansiti et al, Phys. Rev. Lett. **55**, 746 (1985)

[7] A. Hamm, T. Tél, and R. Graham, Phys. Lett. **185A**, 313 (1994).

[8] K. G. Szabó and T. Tél, Phys. Lett. A196, 173 (1994)

[9] E. Ott et al, Phys. Rev. Lett **64**, 1196 (1990)

[10] Focus issue on chaotic scattering, Chaos **3**, No.4 (1993)

[11] C. Jung and E. Ziemniak, J. Phys. **A 25**, 3929 (1992)

[12] J. B. Kadtke and E. A. Novikov, Chaos **3**, 543 (1993); H. N. Luithardt, J. B. Kadtke and G. Pedrizetti, Chaos **4**, 681 (1994)

[13] C. Jung, T. Tél, and E. Ziemniak, Chaos **3**, 555 (1993)

[14] P. Gaspard, G. Nicolis, Phys. Rev. Lett. **65**, 1693 (1990); P. Gaspard, J. Stat. Phys. **68**, 673 (1992)

[15] H. E. Nusse and J. Yorke, Physica **D 36**, 137 (1989).

[16] I. M. Jánosi, and T. Tél, Phys. Rev. **E49**, 2756 (1994)

[17] H. D. I. Abarbanel, R. Brown, J. J. Sidorovich, and L. Sh. Tsimring, Rev. Mod. Phys. **65**, 1331 (1993).

[18] I. M. Jánosi, and T. Tél, Phys. Rev. Lett. **73**, 529 (1994)

[19] G. Broggi et al., Phys. Rev. **A 39**, 434 (1989).

[20] H. Kantz et al., Phys. Rev. **E 48**, 1529 (1993).

[21] P. Grassberger and I. Procaccia, Physica **D 9**, 189 (1993)

[22] Th.M. Kruel, M. Eiswirth, and F. Schneider, Physica **D 63**, 117 (1993).

[23] H. Kantz, Phys. Lett. **A 185**, 77 (1994).

Global Dynamical Modeling of Time Series and Application to Restoration of Broadband Signal Characteristics

Dmitrii A.Gribkov, Valentina V.Gribkova, Yurii I.Kuznetsov and Alexei G.Rzhanov

Physical Department of Moscow State University, Moscow,119899, Russia

Abstract. We show the principle possibility of the external complex action restoring using the nonlinear dynamics inverse problem solution. It is shown that broadband signal can be restored from the time series of the process generated by nonstationary chaotic system using the reconstructed model as the nonlinear filter.

INTRODUCTION

The nonlinear dynamics analysis has a great number of applications in signal processing: noise reduction, signal detecting and classification , encoding and decoding of information.[1,2,3,4]

In this paper we are trying to solve the problem of regular signal restoring from the scalar time series of the process generated by chaotic dynamical system which experiences an external action. This task is closely connected with more general problem: dynamical equations construction using the single - variable time series of chaotic system.

Let we have one - dimensional time series of stationary chaotic process. Begining from the moment t_0 , the chaotic system is forced by broadband signal f(t). We propose the approach which allows to restore this signal. Our approach contains two main procedures :
1) Construction of the dynamic model in a given class of ordinary differential equations using the time series of autonomous chaotic system; 2) External signal restoring from the time series of nonstationary system using the constructed model as a nonlinear filter. The applicability of mentioned approach depends on following conditions: a) it is necessary to detect the region of time series

produced by autonomous system, b) we need apriory information about nonstationary system structure, i.e. how the forcing signal is involved in dynamical system.

MODEL EQUATIONS CONSTRUCTION FROM SCALAR TIME SERIES

In this section we give a brief description of procedure we use to construct a dynamical model of original autonomous system using its single - variable time series.[5,6]

Let $x_k = x(t_k)$ be a scalar time series registrated from autonomous dynamical system, where $k=1,\dots N_0$. We can determine a state vector $\vec{X}_k = \vec{X}(x_{k1},\dots x_{kN})$ of dynamical system, where $x_{kj} = x(t_k + j\tau)$ - is j-th Takens variable, $j=1,\dots N$, N - is an embedding dimension, τ - is time delay. Let n be the dimension of the phase space of the original dynamical system ($n < N$). We then construct a new basis via a singular system analysis[7], by first computing the eigenvectors $\vec{S}_i$ $(i = 1,\dots N)$ of the covariance matrix $\hat{C}$. These are obtained from the following equation:

$$\left(\hat{C} - \beta_i \cdot \hat{E}\right) \cdot \vec{S}_i = 0,$$

where

$$\vec{S}_i = \begin{pmatrix} \mu_{i1} \\ \vdots \\ \mu_{iN} \end{pmatrix}, \quad \hat{C} = \begin{pmatrix} \sigma_{11} & \cdots & \sigma_{1N} \\ & \vdots & \\ \sigma_{N1} & \cdots & \sigma_{NN} \end{pmatrix},$$

$\sigma_{jj} = \sum_{k=0}^{N_0} x^2(t_k + j\tau)$, $\sigma_{jl} = \sigma_{lj} = \sum_{k=0}^{N_0} x(t_k + j\tau)x(t_k + l\tau)$, μ_{ij} - are coordinates of i-th eigenvector, β_i - is the i-th eigenvalue, $\hat{E}$ - is a unit matrix. Now note that the first n vectors $\vec{S}_i$ determine the orthonormal basis for new variables:

$$w_{ik} = (\vec{X}_k \cdot \vec{S}_i). \tag{1}$$

If $\tau \to 0$ we can use a Taylor series for each Takens variable ($j=1,\dots N$) So, expression (1) can be rewritten with accuracy $O(\tau^n)$ as follows:

$$w_{ik} = \sum_{m=1}^{n} \left(\frac{d^{m-1}}{dt_k^{m-1}} x_{1k} \right) \cdot u_{im}, \tag{2}$$

where

$$u_{im} = \sum_{j=1}^{N} c_{imj} \cdot \mu_{ij},$$

and u_{im} - are the coordinates of the new basis vectors $\vec{S}_i^*$, and c_{imj} are constants (i=1,...n). Let us now denote derivatives in (2) as new variables

$$x_{mk}^* = \frac{d^{m-1}}{dt_k^{m-1}} x_{1k}.$$

According to (2) the transformation to the new variables can be obtained from equation:

$$\vec{X}_k^* = \hat{R}^{-1} \cdot \vec{W}_k, \tag{3}$$

where

$$\vec{X}_k^* = \begin{pmatrix} x_{1k}^* \\ \vdots \\ x_{nk}^* \end{pmatrix}, \quad \hat{R} = \begin{pmatrix} u_{11} & \cdots & u_{1n} \\ & \vdots & \\ u_{n1} & \cdots & u_{nn} \end{pmatrix}, \quad \vec{W}_k = \begin{pmatrix} w_{1k} \\ \vdots \\ w_{nk} \end{pmatrix}.$$

Variations of the embedding dimension N and time delay τ change the direction of vectors $\vec{S}_i$. So by varying N and τ, we can reduce not only the noise influence but minimize the error resulting from Taylor series.

Using new variables obtained from (3) we shall search for a dynamical model of 3-dimensional system in the form:

$$\frac{dx}{dt} = y, \quad \frac{dy}{dt} = z, \quad \frac{dz}{dt} = F_j(x, y, z, \vec{a}, \vec{b}), \tag{4}$$

where $f(x,y,z,\vec{a}) = a_1 + a_2x + a_3y + a_4z + a_5x^2 + a_6xy + \ldots$ - is a polynomial of I-th order; $\psi(x,y,z,\vec{b}) = b_1x + b_2y + b_3z + b_4x^2 + \ldots$ - is a polynomial of J - th order ($J < I$), $\vec{a}$, $\vec{b}$ - are the vectors of polynomial coefficients; $\varphi(x,y)$ - is one of the following functions: 1, x , y, x^2, xy .

Using this general functional form, we can now attempt to determine the specific model by fitting the coefficients of the model to the data. Calculation of the polynomial coefficients follows from the minimization of the cost function :

$$\Phi_j = \sum_{k=1}^{N_0} \left[\frac{dz}{dt_k} - F_{jk} \right]^2, \qquad (5)$$

where $F_{jk} = F_j(x_k, y_k, z_k)$, $j=1, 2$.

Least - squares minimization of function (5) yields a linear equation of form:

$$\vec{D}_Z = \hat{P}_F \cdot \vec{A}, \qquad (6)$$

where $\vec{D}_Z$ - is an N_0 - dimensional vector of derivatives dz/dt_k, $k= 1,\dots N_0$; $\vec{A}$ - is a M - dimensional vector of polynomial coefficients ($M= I+ J$, for $j=1$; $M= I + 1$, for $j=2$); and $\hat{P}_F$ - is a $N_0 \times M$ - dimensional matrix of polynomial nonlinear terms.

Solving (6) numerically, we obtain a set of polynomial coefficients ($\{a_l\}$, $\{b_m\}$, $l=1,\dots I$, $m= 1,\dots J$, for $j=1$; $\{a_l\}$, $l= 1,\dots I+1$, $\{b_m\}= \varnothing$, for $j=2$).

As an example we have applyed the methods mentioned above for constructing a system of dynamical equationes of form (4) for well known Lorenz system[8] using single - variable time series of one of its coordinates. The original equations of Lorenz system with definite parameter values are written as follows:

$$\begin{aligned} \frac{dX}{dt} &= -10 \cdot (X - Y), \\ \frac{dY}{dt} &= 50 \cdot X - Y - XZ, \\ \frac{dZ}{dt} &= -\frac{8}{3} \cdot Z + XY. \end{aligned} \qquad (7)$$

Using X - coordinate of (7) we have constructed a new , topologically equivalent system of form (4):

$$\frac{dx}{dt} = y,$$
$$\frac{dy}{dt} = z, \qquad (8)$$
$$x \cdot \frac{dz}{dt} = A_1x^2 + A_2xy + A_3xz + A_4y^2 + A_5yz + A_6x^4 + A_7x^3y,$$

where $A_1 = 3190.407$, $A_2 = -70.34543$, $A_3 = -21.38827$, $A_4 = 16.81685$, $A_5 = 1.000430$, $A_6 = -15651.33$, $A_7 = -943.9095$. These model coefficients were obtained for time step $\Delta t = 0.005$.

EXTERNAL SIGNAL RESTORING

In this section we show the possibility of broadband signal restoring from scalar time series of nonstationary Lorenz system. Equations (7) of previous section can be rewritten including external signal as follows:

$$\frac{dX}{dt} = -10 \cdot (X - Y),$$
$$\frac{dY}{dt} = 50 \cdot X - Y - XZ, \qquad (9)$$
$$\frac{dZ}{dt} = -\frac{8}{3} \cdot Z + XY + f(t),$$

where $$f(t) = \begin{cases} 0, & 0 < t \le t_0, \\ C \cdot \cos(p \cdot |\cos(\omega \cdot t)|) / |B + \cos(\omega \cdot t)|, & t > t_0, \end{cases}$$

and C, B, p, ω - parameters of external action. For this nonstationary case the third equation of (8) becomes:

$$x \cdot \frac{dz}{dt} = A_1x^2 + A_2xy + A_3xz + A_4y^2 + A_5yz + A_6x^4 + A_7x^3y + f(t) \cdot x^2. \qquad (10)$$

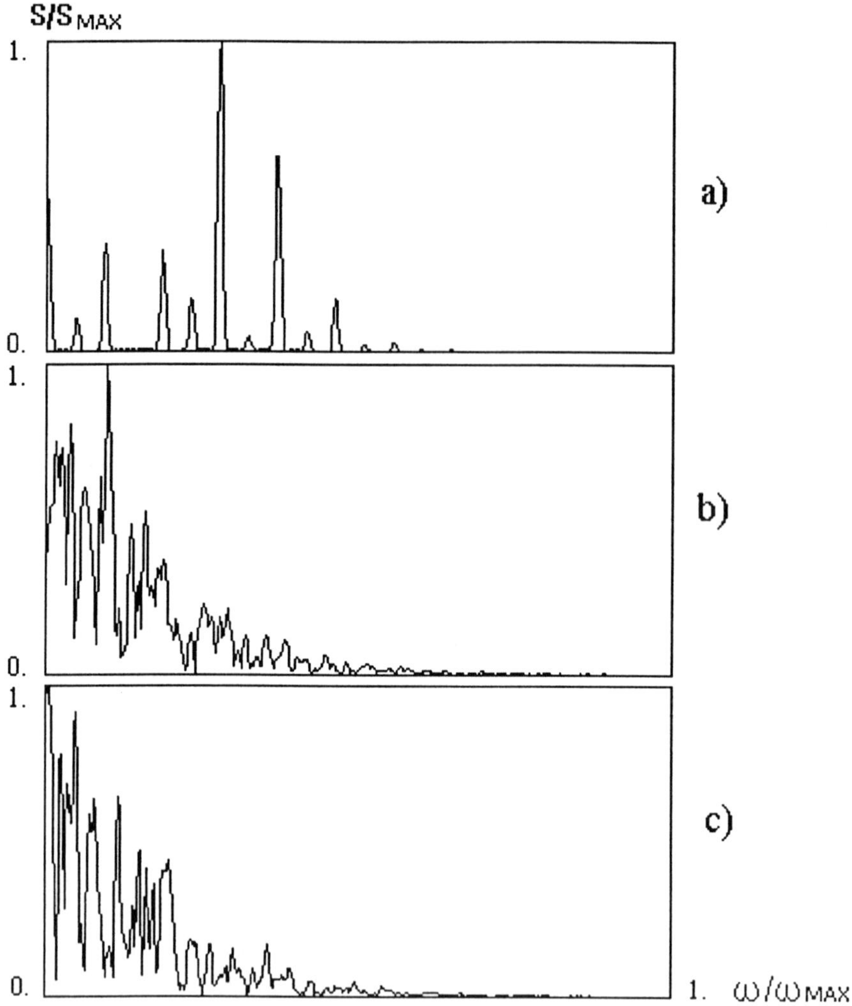

FIGURE 1.

For t $>t_0$, using (10) and taking time series of variable X(t) = x(t) of the nonstationary system (9) , we can reveal signal f(t) in the form:

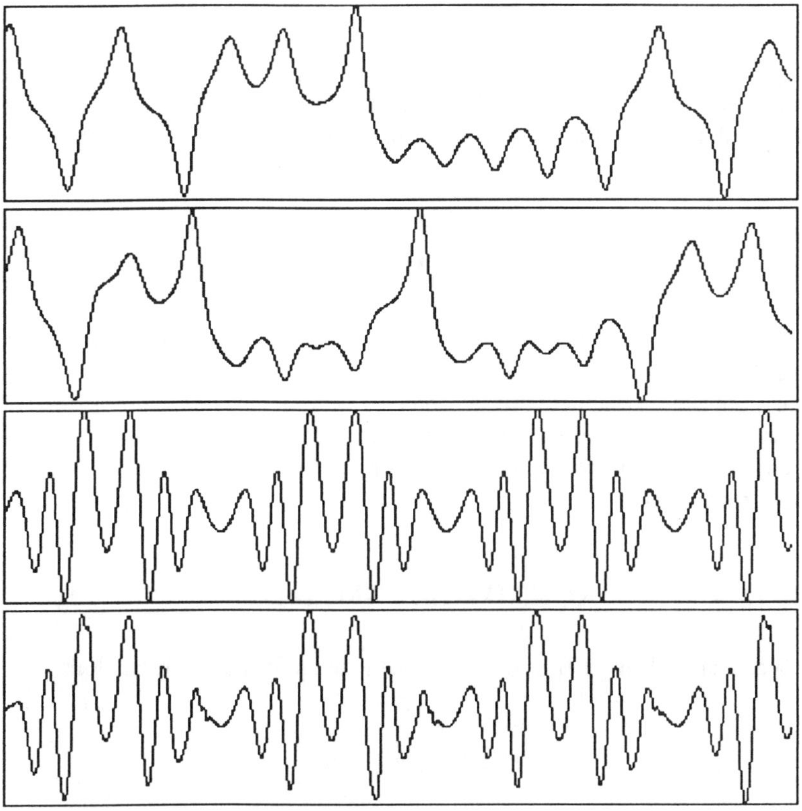

FIGURE 2.

$$f(t) = \frac{1}{x^2} \cdot \left(x \cdot \frac{dz}{dt} - F(x,y,z,\vec{a}) \right), \qquad (11)$$

where $F(x,y,z,\vec{a}) = A_1 x^2 + A_2 xy + A_3 xz + A_4 y^2 + A_5 yz + A_6 x^4 + A_7 x^3 y$.

For the numerical simulation we put the following parameters of external force: C= 30, B= 4, p= ω = 8. In fig. 1a),b),c) we present spectrums of the signal and variables X of autonomous and nonautonomous systems, correspondently. As it follows from figures 1b),c), spectrums of system (7) and (9) are, practically, the same. The nonlinear interaction of signal and chaotic process makes impossible to reveal signal components from the nonautonomous system by linear procedures. Using the constructed model of autonomous system (8) and expression (11) we restore f(t) completely. Figure 2 demonstrates the quality of this restoration. In figures 2a),b) we present the time series of autonomous and nonautonomous systems, correspondently. Figure 2c) shows the original signal and figure 2d) presents the restored signal. Forms of these signals are completely coinside.

CONCLUSION

Thus, we have showed the principle possibility of uknown broadband signal restoring from chaotic process generated by nonautonomous chaotic system using the nonlinear dynamics inverse problem solution. This becomes possible if: 1) we can registrate a piece of time series generated by autonomous dynamical system (without signal) and 2) we have information about the way of forcing signal action on system.

ACKNOWLEDGMENTS

We wish to thank J.B.Kadtke for useful and informative discussions.

REFERENCES

1.Teodorescu, D., *Int. J. Control* **50**, 1577- 1591 (1989).
2.Kadtke, J.B. and Brush, J.S. , " Noise Reduction Methods for Chaotic Signals using Empirical Equations of Motion ", in *Proceedings of the SPIE Conference on Signal Processing, Sensor Fusion, and Target Recognition*, 1992, pp. 338 - 349.
3. Ott, E., Grebogi, C. and Yorke, J.A. *Phys. Rev. Lett.* **64**, 1196 (1990).
4. Matthews, R. *Cryptologia* **13**, 29 - 42 (1989).
5. Gribkov, D.A., Gribkova, V.V., Kravtsov, Yu.A., Kuznetsov, Yu.I. and Rzhanov, A.G., *Radiotekhnika i electronica* **39**, 269-277 (1994).
6. Gribkov, D.A., Gribkova, V.V., Kravtsov, Yu.A., Kuznetsov, Yu.I. and Rzhanov, A.G., *Zhurnal tekhnicheskoi fiziki* **64**, 1-12 (1994).
7. Broomhead, D.S., King, G.P., *Physica D.* **20**, 217 (1986).
8. Lorenz, E.N., *J.Atmos.Sci.* **20**, 130 (1963).

Signal Classification Using Global Dynamical Models, Part I: Theory

James Kadtke and Michael Kremliovsky†

Institute for Pure and Applied Physical Sciences
University of California at San Diego
9500 Gilman Dr., La Jolla, CA 92037–0360

† *also at: RTA Corp., P.O. Box 5267,*
Springfield, VA 22150

Abstract. Detection and classification of signals is one of the principal areas of signal processing, and the utilization of nonlinear information has long been considered as a way of improving performance beyond standard linear (e.g. spectral) techniques. Here, we develop a method for using global models of chaotic dynamical systems theory to define a signal classification processing chain, which is sensitive to nonlinear correlations in the data. We use it to demonstrate classification in high noise regimes (negative SNR), and argue that classification probabilities can be directly computed from ensemble statistics in the model coefficient space. We also develop a modification for non-stationary signals (i.e. transients) using non-autonomous ODEs. In Part II of this paper, we demonstrate the analysis on actual open ocean acoustic data from marine biologics.

INTRODUCTION

For nearly a decade, time series analysis methods derived from nonlinear dynamical systems theory (NLD) have been applied to signal processing problems, which is a natural application of these ideas. Methods have been developed for prediction [1,2], noise reduction [3-5], and secure communications [6,7], among others, and these schemes have demonstrated significant improvement over conventional linear techniques in several contexts. Another major area of signal processing now receiving attention is signal detection (identifying the presence of a deterministic signal-of-interest) and classification (assignment of a detected signal to a particular class), particularly in the high-noise regime. Detection and classification (D/C) methods have widespread application to areas such as physics, Sonar/Radar processing, pattern recognition, voice recognition, remote sensing, image processing, and biomedical and geophysical data analysis.

From a signal processing point of view, the methods of NLD are innovative because they allow a compact representation of nonlinear signal information (i.e. nonlinear correlations) to all orders, by treating the signal as an evolution of a dynamical system in a (Takens) re-constructed state space. Although powerful, conventional linear methods (e.g. matched filtering) are largely insensitive to nonlinear correlations or can distort them [8], and hence do not make use of the additional nonlinear information which may be present in many natural data sets. The use of nonlinear correlations for detection and classification is not new (e.g. bi-spectral methods, first proposed in the 1960's); however, these methods often have exhorbitant data requirements or perform poorly in the high noise regime, as well as capturing only one moment of the nonlinear expansion. Nonlinear correlations can, however, be efficiently represented to all orders in the dynamical phase space, where they are interpreted as phase couplings between dynamical modes.

In recent years, a few techniques have appeared in the literature [9] which take steps toward using nonlinear dynamical information for classification in the signal processing sense. Typically, many of these methods require a priori exact signal models or clean (i.e. noise-free) signal realizations, or perform well only at modest noise levels. Also, few methods incorporate the dynamical information into a useful classification chain, yielding classification probabilities as an output (an exception is recent work by Fraser, 1995). Here, however, we develop a very general D/C method which has several advantages from a practical, operational point of view [10]. Even though we model signals as a dynamical system, this method does not assume signal stationarity or an 'attractor' over the measurement time scales, and so is equally applicable to linear (e.g. tonals) and nonlinear signal components, and stationary or transient signals. Because of the averaging characteristics of the model class (sets of ODEs), this technique can perform well in the high-noise regime. Here, we define the high noise regime to be the negative dB range, with $dB = 20\log_{10}(\sigma_{\mathrm{signal}}/\sigma_{\mathrm{noise}})$, where σ is standard deviation, since this is the range where most conventional D/C schemes have difficulty. This technique also does not assume the existence of any exact model for the signals-of-interest (SOI), nor does it initially require clean realizations of particular signal classes to be useful. It also provides a simple way of calculating the classification probabilities, and hence performance can be estimated in a straightforward fashion. Finally, the method is numerically efficient and the data requirements are minimal, and hence the scheme is likely applicable to real-time data processing.

The inclusion of a NLD technique into a consistent D/C processing scheme can often be confusing, and it is useful to consider the scheme as a sequence of distinct steps. Here, for conceptual organization, we divide the D/C process as follows: we choose first the type of data representation, e.g. scalar data, or time-delay reconstructed (vector) data; second, an information quantifier, such as a power spectrum, or the fitting of a dynamical model; third, an information metric, which is capable of assigning values to differences in quantifier realizations in a useful fashion; and finally a D/C statistic, which is typically a function of the metric. We have investigated overall D/C performance using combinations of different techniques for each step above, and in what follows we describe the algorithm we have found as most useful.

TECHNICAL APPROACH

We develop our D/C scheme by first discussing the algorithmic procedure, then discuss how it is operationally applied to a given data set. To begin, we assume the existence of a time series $x(t)$ of measurements of some generic physical observable of a system, or of an artificially generated signal. We use as our data representation the time-delay reconstructed state-space vectors $\mathbf{x} = (x(t),\ x(t-\tau), \ldots,\ x(t-(D-1)\tau)$ as per Takens [11], for some embedding dimension D and delay τ. Intuitively, this step constructs from the scalar data $x(t)$ a higher-dimensional dynamical state-space representation with potentially D degrees-of-freedom, isomorphic to the original phase space for sufficient D. As is well-known, a deterministic signal evolution (even for linear systems) in this re-constructed state-space can be described by a system of D first-order differential equations, which may be coupled and nonlinear. The signal evolution generates state-space trajectories whose time-asymptotic distribution can be characterized by different properties: global topology (the fixed-point structure and manifold geometry); determinism (existence of a dynamical "rule"); differentiability (for continuous systems); and distributional (e.g. invariant probability density). We attempt to develop a classification scheme which is sensitive to all these various types of state-space characteristics simultaneously, even though much of it is often obscured at high noise levels.

Therefore, we choose as our information quantifier an empirically-fit global dynamical model [12, 13]. That is, we fit a set of model differential equations $d\mathbf{x}(t)/dt = \mathbf{F}(\mathbf{x}, t)$, to the data evolution in the re-constructed state space vectors $\mathbf{x}$. As a fundamental point, we assert that if the original data generator was actually a continuous dynamical system, then a dynamical model is of the "correct" model class, and can be simultaneously sensitive to all the types of dynamical information mentioned above. That is, in the limit of vector observables, infinitely long time series, no noise, and the proper basis expansion, we can recover the original dynamical system to arbitrary accuracy. In practice, for D/C we are not concerned with the exact model structure, and we construct a model by assuming some general expansion for the D functions $\mathbf{F}(\mathbf{x}, t)$ in some basis set $\phi(\mathbf{x}(t))$. Typically, for the autonomous case $\mathbf{F}(\mathbf{x}, t) = \mathbf{F}(\mathbf{x})$, one can simply choose all polynomials up to some order P, with $\phi_j(\mathbf{x}(t)) = x(t)^k \cdot x^m(t-\tau) \cdot \ldots \cdot x^q(t-(D-1)\tau)$, where $0 \leq k,\ m,\ q \ldots \leq P$ and $k + m + q + \ldots \leq P$, writing for each of the $1 \leq d \leq D$ equations

$$F_d(x,t) = a_{0,d} + a_{1,d}x(t) + \ldots = \sum a_{j,d}\phi(\mathbf{x}(t)) \quad j = 0, \ldots N \tag{1}$$

and the expansion includes all $N = (P+D)!/P!D!$ cross terms. The unknown model coefficients $a_{j,d}$ are determined by constructing a general linear least-squares problem [14] for each F_d defined by the matrix equation $\mathbf{Y} = \mathbf{BA}$, where $\mathbf{Y}$ is the vector of time derivatives $dx(t)/dt$ estimated at successive observation times $x(1),\ x(2), \ldots$; $\mathbf{B}$ is a matrix of the basis terms $\phi_j(\mathbf{x}(t))$ evaluated at $x(1),\ x(2), \ldots$ (i.e. the "data matrix"); and $\mathbf{A}$ is the vector of unknown coefficients $a_{j,d}$. Each F_d is determined by least-squares solution of its matrix equation using, e.g. singular value decomposition. Many papers have appeared in the literature in recent years discussing various numerical approaches to this *inverse* problem, and the reader is referred there. For prediction and modeling purposes, nontrivial numerical issues are still outstanding [15], however it has been shown that recovery of a system's dynamical invariants can often be achieved even for significant noise [16, 17].

It is important to note that, for D/C purposes, our scheme diverges fundamentally from a modeling approach. For modeling purposes determination of a correct D, P, etc. is necessary to recover exact dynamical information about the original system, which is typically problematic. However for D/C purposes, we postulate that it is only necessary to incorporate sufficient D and P (and other parameter ranges) to distinguish the required signal classes to a particular performance level, regardless of the exact form of the original dynamical generators. Typically, we find that D and P can usually be small (e.g. 2 or 3), since signal power is usually distributed mostly in the lowest orders of the model coefficients (in analogy with spectral expansions). More importantly, we find it crucial to define a standard form (i.e. fixed P, D, τ) for the dynamical model (i.e. a fixed "dynamical filter") in a particular application, otherwise comparison of arbitrary signal classes which may have radically different forms for their generators is impossible.

We now assume that we have extracted estimated coefficient sets for a variety of signal observations α. There now remains the non-trivial issue of quantifying differences in coefficient distributions in a meaningful way (i.e. the information metric). We use a simple but useful scheme [18]: we construct a vector space $\mathcal{A}$ from the coefficients of all $\phi_j(\mathbf{x}(t))$ by forming vectors $\mathbf{A}_\alpha = (a_{0,1},\ a_{1,1}, \ldots,\ a_{j,d}, \ldots)$ for all α data realizations, where the $a_{j,d}$ includes all $D \cdot N$ available coefficients from the D model functions F_d. For each $\mathbf{A}_\alpha$, the $a_{j,d}$ determined from the fitting process is embedded in this high-dimensional coefficient space $\mathcal{A}$. Note that each $\mathbf{A}_\alpha$ in $\mathcal{A}$ defines a unique model dynamical system, describing a particular data realization, and the set of all $\mathbf{A}_\alpha$ for an ensemble of observations yields a corresponding distribution in $\mathcal{A}$. Qualitative differences between the different signal realizations can then be quantified using a Euclidean metric to measure differences in the $\mathbf{A}_\alpha$.

A central part of any D/C scheme is the calculation of probabilities and confidence estimates. Using the representation $\mathcal{A}$, classification probability estimates can be directly derived from the statistical properties of the ensemble distributions in $\mathcal{A}$. That is, the probability that the α-th data observation belongs to the m'th signal class can be estimated from the probability that a particular AM lies within the ensemble distribution C_m of that class in $\mathcal{A}$. For C_m which are Gaussian, this probability is directly estimated from the distance of $\mathbf{A}_\alpha$ to the centroid of C_m, normalized in units of the standard deviation of C_m. In practice, the C_m are typically non-Gaussian, hence we utilize non-parametric statistics to estimate probabilities. In this context, we note that pure noise must generate Gaussian C_m centered at the origin of $\mathcal{A}$, and that here the "detection" process consists simply of the statistical exclusion of a given $\mathbf{A}_\alpha$ from the C_m of the noise realization. Hence, we may perform detection and classification simultaneously, if we are given prior (noisy or clean) realizations of the SOI and the pure noise background (which may be colored).

The above discussion describes the algorithmic tools used in the scheme, and we now describe how they are used operationally. Typically, we will be interested in either transient signals which have a short, finite lifetime, or short observations of a potentially stationary signal, both of which can effectively limit the window-of-observation to a few characteristic timescales. This, coupled with the fact that the model basis will generically be incorrect, means that even noise-free signal observations will produce a statistical distribution in the coefficient estimates for the $\mathbf{A}_\alpha$. Hence, one must resign oneself to the idea that even repeated observation of an identical signal will generically produce a distribution in $\mathcal{A}$, and only a probabilistic estimate for classification can be made. These facts imply two things about the technique: first, the details of the observation procedure become important in the coefficient estimation scheme, i.e. the sampling rate, the data observation window length L_w (in samples), and number of observations, among others, must be accounted for and standardized in the definition of the "dynamical filter" used in the D/C process. Secondly, characterization of the statistical properties of the coefficient ensembles in $\mathcal{A}$ is a natural and useful way to estimate the probabilities of the finite-observation time effects, and provides a natural metric for studying the differences of signal classes. Therefore, we will think operationally in terms of collecting statistics on model estimates of many short windows of independent observations of signals and noise, then using these ensembles to perform D/C on future observations of arbitrary signals.

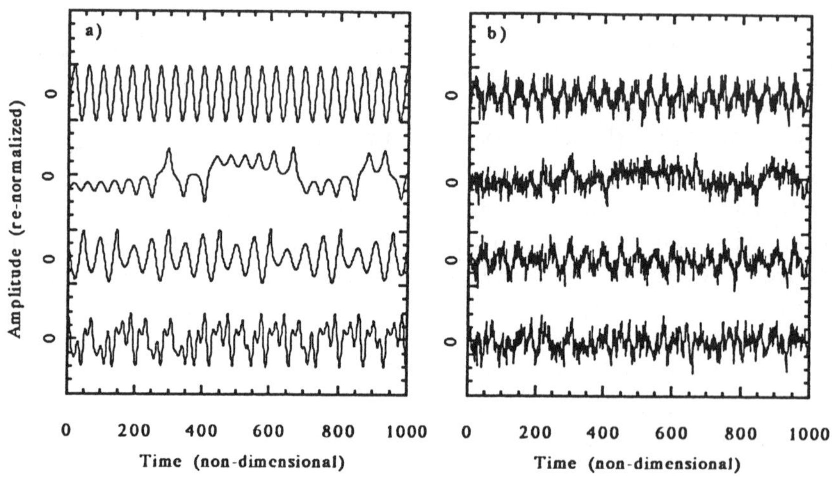

FIGURE 1. Segments of pure (left) and 0 dB noisy (right) time series, from sine wave (top), Lorenz system, Rössler system, and Pecora circuit (bottom).

NUMERICAL EXPERIMENTS

We now clarify the above ideas using several numerical examples. Throughout, we will explicitly construct our simulated data to minimize or remove amplitude and spectral information, hence D/C results are due principally to nonlinear correlation. We first simulate data observations from a stationary system by generating long time series $x(t)$ from three systems: a sine wave, and the Rossler and Lorenz chaotic dynamical systems; we then also include real data from the Pecora chaotic circuit [6]. All series simulate discrete sampling at 50 to 100 points per characteristic "orbital" time, which normalizes the time scales so that power spectral distributions for the different systems overlap. Simulated Gaussian-distributed noise was then added at different noise levels from +5 to −10 *dB*. All time series were then normalized to zero-mean and unit variance throughout (via a windowed average) to remove amplitude information and make conventional signal variance detectors insensitive. Segments of resulting pure and noisy signals are shown in Fig. 1(a) and 1(b), respectively. Power spectra for the broadband signals (not shown) show little structure for negative SNR, hence matched filter techniques are also not useful.

To simulate a D/C procedure, we choose a standard dynamical model with $D = 3$, $P = 2$, and $\tau = 12$ ($\approx 1/4$ cycle). This system has 30 total coefficients $a_{j,d}$ in $D = 3$ time-delayed variables, which we re-write for clarity as $x = x(t)$, $y = x(t-\tau)$, and $z = x(t-2\tau)$. We then divide each time series into $L_w = 500$ point windows (about 10 characteristic cycles), to simulate individual and independent short data observations. For each signal class and noise level (and additionally for pure noise data realizations), we fit the dynamical model to each data window, generating the $\mathbf{A}_\alpha$. In this sense, we "train" the D/C algorithm by accumulating observations of the four known signal classes, and a "library" of signal distributions C_m is stored consisting of the $\mathbf{A}_\alpha$. For the data described above, Fig. 2 shows the C_m for the 0 *dB* SNR cases, in the 3-space formed from only the linear basis term coefficients $a_{1,1}$, $a_{2,1}$, and $a_{3,1}$ of one equation F_1, for visualization purposes. Clear separation is evident between the C_m for different signal classes and noise, except for the Lorenz and Pecora distributions which can be separated using alternate axes. As an important point, note that both detection and classification are achieved simultaneously in this scheme.

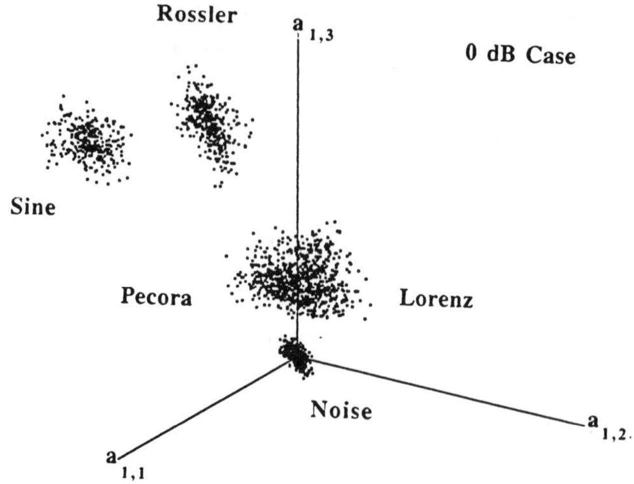

FIGURE 2. Three-variable coefficient space ($a_{1,1}$, $a_{2,1}$, $a_{3,1}$) showing embedded ensembles of model coefficient realizations, for signals of Fig. 1.

As mentioned previously, D/C probabilities can be assigned using the statistical properties of the C_m, such as those shown in Fig. 2. We have found that the C_m for deterministic signals are usually significantly non-Gaussian, and tightly clustered about the centroid. Often more than 90% of the $\mathbf{A}_\alpha$ lie within one equivalent σ of the mean, hence higher confidence estimates can be achieved than would be indicated by Gaussian statistics. To illustrate this, Fig. 3(a) shows a sample distribution of a C_m along only one coefficient axis, for the Rossler system at five representative SNR levels. As is evident, the distribution is tightly clustered at larger SNR, and becomes more Gaussian for lower SNR, while the centroid simultaneously moves toward the origin of $\mathcal{A}$.

The above analysis has assumed data at a constant, known SNR level. In practice, SNR levels are often unknown and individual data observations may appear at arbitrary SNR. Importantly, we have demonstrated that systematic scaling exists for the ensemble centroid positions with varying SNR level. This is demonstrated by numerical experiments in Fig. 3(b), showing the scaling with SNR of the centroids of several signal classes, projected along two axes of A for visualizaton. As the SNR varies, the centroid positions smoothly trace a line which moves to the origin of A as the SNR moves to $-\infty$ (note that these lines are significantly more separated when additional axes of $\mathcal{A}$ are used; also, all lie in the same quadrant due to artificial time normalization). These results indicate that it is possible to perform classification for arbitrary noise level, by having prior signal observations at only a few SNR, and then extrapolating the SNR-scaling of the centroid. Classification statistics are then measured from this SNR-parameterized line, rather than a single position at a specific noise level. Note that Fig. 3(b) indicates that D/C is possible in these cases even down to -10 *dB*.

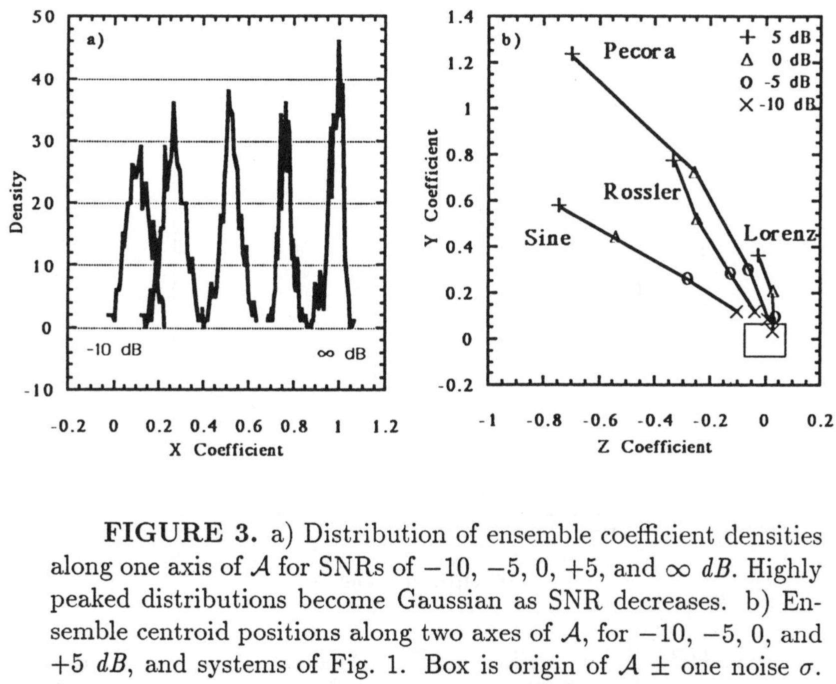

FIGURE 3. a) Distribution of ensemble coefficient densities along one axis of $\mathcal{A}$ for SNRs of -10, -5, 0, $+5$, and ∞ *dB*. Highly peaked distributions become Gaussian as SNR decreases. b) Ensemble centroid positions along two axes of $\mathcal{A}$, for -10, -5, 0, and $+5$ *dB*, and systems of Fig. 1. Box is origin of $\mathcal{A}$ $\pm$ one noise σ. Centroids trace smooth line to origin as SNR decreases.

We now make the point that in practice, reliable D/C can usually be achieved using relatively few axes of $\mathcal{A}$. Firstly, we note that for time-delay reconstructed data, the Toeplitz nature of the data matrix implies that all equations F_d contain qualitatively similar or redundant information, particularly in the high-noise regime, and hence we need only monitor the $a_{j,d}$ of one of them. Secondly, for any given signal class, we typically find that few coefficients contain statistically significant information, the rest reflecting only noise properties. We find that these significant coefficients can be identified as follows: we renormalize all coefficients $a_{j,d}$ by their standard deviations $\sigma_{j,d}$ estimated from the C_m of the given class. We then argue that the new variables $a_{j,d}/\sigma_{j,d}$ that lie within one or two $\sigma_{j,d}$ of zero contain little or no signal information, since they statistically include zero. We thus define the significant coefficients as those lying outside this threshhold, and these must scale with noise magnitude. As an example, Fig. 4 shows a plot of the lowest order $a_{j,d}/\sigma_{j,d}$ vrs. coefficient number for the Rossler signals of Fig. 1, for varying SNR value. The one-σ threshold is also indicated. Here, the two statistically significant coefficients characteristic of this signal class are easily identified. This property is generally valuable since restriction of the D/C statistic to the significant coefficients can decrease the probability of false identification.

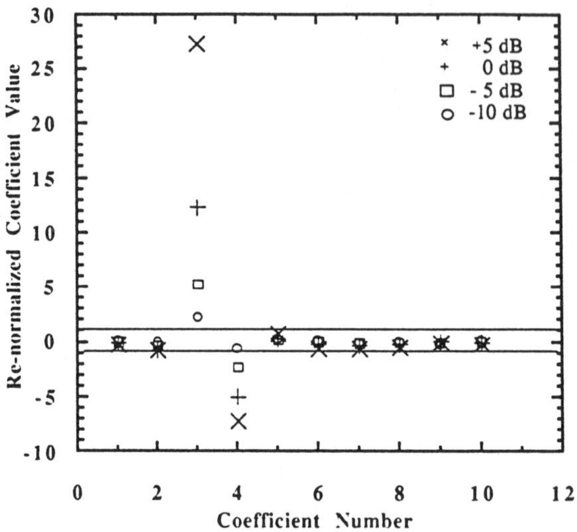

FIGURE 4. Normalized coefficient values $a_{j,d}/\sigma_{j,d}$ for Rossler data, at different SNR. Only two coefficients lie outside one-sigma threshold (indicated by lines) and are statistically significant.

CLASSIFYING TRANSIENT SIGNALS

The numerical experiments above demonstrate the general D/C procedure, by using short windows of otherwise stationary signals. A more difficult problem is encountered with non-stationary signals, where the useful observation length is limited by the characteristic signal time. In particular, D/C of short-lived, pulse-like signals (i.e. "transients") is important in many applications, and few reliable methods exist, especially for high noise regimes. We now demonstrate D/C of transients in high noise using a simple modification of the previous technique. We modify our technique solely by explicitly including time-dependent functions in the global model, i.e. we make the $\mathbf{F}(\mathbf{x})$ of Eq. 1 non-autonomous by including time-local basis terms. Coefficient estimation and representation in $\mathcal{A}$ are then performed exactly as before. We illustrate this by the following numerical experiment: we generate a simulated transient signal from Rossler data using exponential enveloping, i.e. $x(t) = \exp(-|t - t_0|/\lambda)R(t)$, with $R(t)$ a Rossler time series segment from Fig. 1. We choose λ so that the pulse is "short", i.e. it contains < 10 characteristic oscillations of the Rossler dynamics. We then add Gaussian-distributed simulated noise at differing SNR between $+5$ and -10 dB. Fig. 5 shows samples of pure and noisy (0 dB) single pulse realizations. We now define the $\mathbf{F}(\mathbf{x}, t)$ to be

$$\begin{aligned} dx(t)/dt &= a_{0,1}x(t)(t - t_0) + a_{1,1}x(t) + a_{2,1}y(t) + \dots \\ dy(t)/dt &= a_{0,2}y(t)(t - t_0) + a_{1,2}x(t) + a_{2,2}y(t) + \dots \end{aligned} \qquad (2)$$

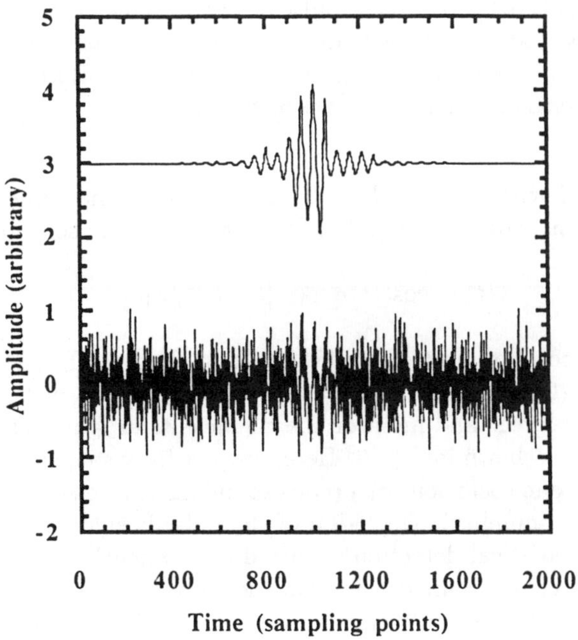

FIGURE 5. Time series of simulated Rossler chaotic "pulse" at ∞ dB (top) and 0 dB SNR (bottom).

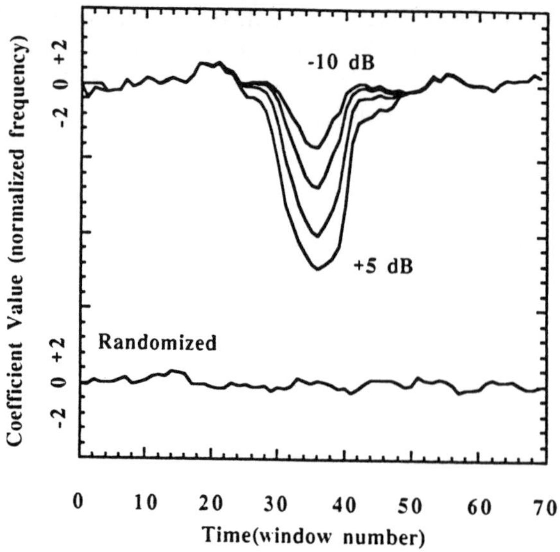

FIGURE 6. Detection of pulse of Fig. 5 using windowed non-autonomous model. Plot shows one coefficient (re-scaled frequency) vrs. time (i.e. window number), for four SNR levels. Bottom curve is same experiment for phase-randomized data.

i.e., a model with $D = 2$, $P = 2$ and one additional time-dependent term per equation. Note that Eq. 2 contains as a possible solution

$$x(t) = \cos(bt)\exp(-|t - t_0|^2/\lambda) , \tag{3}$$

i.e. a pulse-like solution, but of a different form than the Rossler envelope. We then fit Eq. (2) in a sliding $L_w = 150$ point window over the signal, at the various noise levels, and monitor the coefficients in $\mathcal{A}$ as before. $\mathcal{A}$ plot of typical results is shown in Fig. 6. Here, we plot for visualization purposes the value of only one coefficient $a_{1,2}$ (corresponding to a re-scaled frequency) vrs. time (window number). Detection of the pulse is evident even for SNR of -5 dB, and statistical detection occurs down to nearly -10 dB; slightly better performance is obtained using additional axes of $\mathcal{A}$. As a check, we phase randomize [19] the pulse data to destroy nonlinear correlations, and repeat the experiment. The resulting plot (bottom curve, Fig. 6) shows no variation over the pulse, hence we infer that the D/C ability is due principally to nonlinear correlations.

To summarize this experiment, we have demonstrated detection of a short-lived, highly non-stationary, strongly nonlinear signal in a high noise regime, even while using the "incorrect" dynamical model to perform coefficient estimation. By comparison, most conventional schemes show poor performance for these signals for 0 dB and lower.

CONCLUSION

In conclusion, we would like to note that the discussion and results presented above represent only the most basic D/C scheme. Many modifications can be made, with hopefully corresponding improvement in performance, by including such ideas as specialized model classes, optimization of τ for particular signal classes, and calculation of the D/C probabilities and confidence estimates by defining a variety of statistics. In addition, this method may have a natural modification to include spatio-temporal data sets, and hence be applicable to classification of fields or image processing. Preliminary experiments on the latter indicate that this method is capable of distinguishing patterns based on differences in nonlinear spatial correlations, at least in some instances. Finally, we point out that the numerical efficiency of these algorithms make it quite adaptable to real-time signal analyses, via dedicated processing chips.

ACKNOWLEDGMENTS

The authors would like to thank G. Mayer-Kress, A. Bulsara, and C.F. Driscoll for technical input, and M. Shlesinger for invaluable encouragement. Data was supplied by L. Pecora. This project was supported by a grant to the RTA Corp. from the Office of Naval Research, Number N00014-92-C-0045. The authors also acknowledges travel support from the NATO Office of Scientific Affairs, Grant Number 900088.

REFERENCES

1. Farmer, J.D., and Sidorowich, J.J. *Phys. Rev. Lett.* **59**, 845 (1987).
2. Casdagli, M., *Physica D* **35**, 335 (1989).
3. Hammel, S. *Phys. Lett. A* **148**, 421 (1990).
4. Kostelich, E., and Yorke, J.A. *Physica D* **41**, 183 (1990).
5. Cawley, R., and Hsu, G. *Phys. Rev. A* **46**, 3057 (1991).
6. Pecora, L., and Carroll, T. *Phys. Rev. Lett.* **64**, 821 (1990).

7. Hayes, S., Grebogi, C., Ott, E., and Mark, A. *Phys. Rev. Lett.* **73**, 1781 (1994).

8. Badii, R., and Politi, A., in *Dimensions and Entropies in Chaotic Systems* (ed. G. Mayer-Kress), Springer (1986), p. 67.

9. for example, see: Mindlin, G., and Gilmore, R. *Physica D* **58**, 229, (1992).

10. Kadtke, J. *Phys. Lett. A*, in press (May, 1995).

11. Takens, F., in *Dynamical Systems and Turbulence* (ed. D. Rand and L.S. Young), Springer (1981), p. 366.

12. Crutchfield, J., and McNamara, B. *Complex Systems 1*, 417 (1987).

13. Cremers, J., and Huebler, A. *Z. Naturforsch.* **42**a, 797, (1987).

14. for example, see: Press, et.al. *Numerical Recipes*, Cambridge (1986), p.509.

15. Kadtke, J., and Brush, J. *AIP Conf. Proc.* **296**, (ed. R. Katz), AIP Press (1993).

16. Kadtke, J., Brush, J., and Holzfuss, J. *Intl. J. Bif. and Chaos* **3**, 607 (1993).

17. Brown, R. *Phys. Rev. E* **47**(6), 3962, (1993).

18. Brush, J., and Kadtke, J. *SPIE Proc.* **2037**, (1993), p. 94.

19. Theiler, J. et.al. *Physica D* **58**, 77, (1992).

3.1 APPLICATIONS IN ACOUSTICS

Signal Classification Using Global Dynamical Models, Part II: SONAR Data Analysis

Michael Kremliovsky[†] and James Kadtke

Institute for Pure and Applied Physical Sciences
University of California at San Diego
9500 Gilman Dr., La Jolla, CA 92037–0360

[†] *also at: RTA Corp., P.O. Box 5267, Springfield, VA 22150*

Abstract. In Part I of this paper, we described a numerical method for nonlinear signal detection and classification which made use of techniques borrowed from dynamical systems theory. Here in Part II of the paper, we will describe an example of data analysis using this method, for data consisting of open ocean acoustic (SONAR) recordings of marine mammal transients, supplied from NUWC sources. The purpose here is two-fold: first to give a more operational description of the technique and provide rules-of-thumb for parameter choices; and second to discuss some new issues raised by the analysis of non-ideal (real-world) data sets. The particular data set considered here is quite non-stationary, relatively noisy, is not clearly localized in the background, and as such provides a difficult challenge for most detection/classification schemes.

INTRODUCTION

In Part I of this paper, we described a nonlinear detection and classification (D/C) scheme derived from the theory of nonlinear dynamics (NLD), which utilizes global dynamical models (systems of ODEs) as a signal quantifier [1,2]. We demonstrated that this method was able to detect and classify non-stationary, noisy (-10 dB), and typically short data observations in an algorithmically simple and numerically efficient way, at least on simulated data sets. Here, we turn to the analysis of a real-world data set, to examine its practical usefulness. Obviously, the analysis of real data presents new problems, often yielding ambiguous results, and we discuss these issues in some detail here. We also provide some recommendations for future modifications to the D/C approach, which should improve performance. In what follows, we first describe the data itself, then discuss the choice of typical parameters used to analyze the signals, then present results and draw conclusions.

THE DATA

The data of interest consists of un-classed open ocean SONAR recordings from magnetic tape, consisting of long stretches of background noise punctuated by occasional events consisting of various biologic acoustic transients. The data was generously supplied by T. Luginbuhl and M. Gouzie of NUWC, New London. This data was digitized at a sampling rate of 25 *KHz*, and the signal-to-noise ratio (SNR) of the transients above background is typically only a few *dB*. Samples of time series of the data are shown in Fig. 1, where we have included segments consisting of pure background noise, porpoise "whistles", and whale "cries". Event location of the biologic signals can usually be determined from amplitude variation, and hence detection is not a major difficulty, so that classification becomes the primary issue. An expanded time domain scale often shows some pulse structure, but not significantly so. Spectral analysis of the pulse regions are also difficult due to the low SNR and non-stationarity.

The principal question we will ask is this: can we construct a classification scheme which can determine the difference (in a statistical sense) between the noise, the porpoise "whistles", and whale "cries"? This data is a particularly difficult test, since the SNR is relatively low, and the signals-of-interest (SOI) are quite non-stationary, are not easily localized, and apparently vary in length and structure. Indeed, there is no reason to assume that such biologic signals can be modeled as a low-dimensional dynamical system at all, or hence that this particular method will prove sensitive to any information beyond linear. Although there is some precedent to indicate that biologic vocalizations can be modeled as a dynamical system [3], whether or not this dynamical structure is repeatable from individual to individual (or even from "word" to "word" of a particular individual) is an open question. Here, we proceed in a "black box" fashion, and simply perform the analysis without rigorous justification.

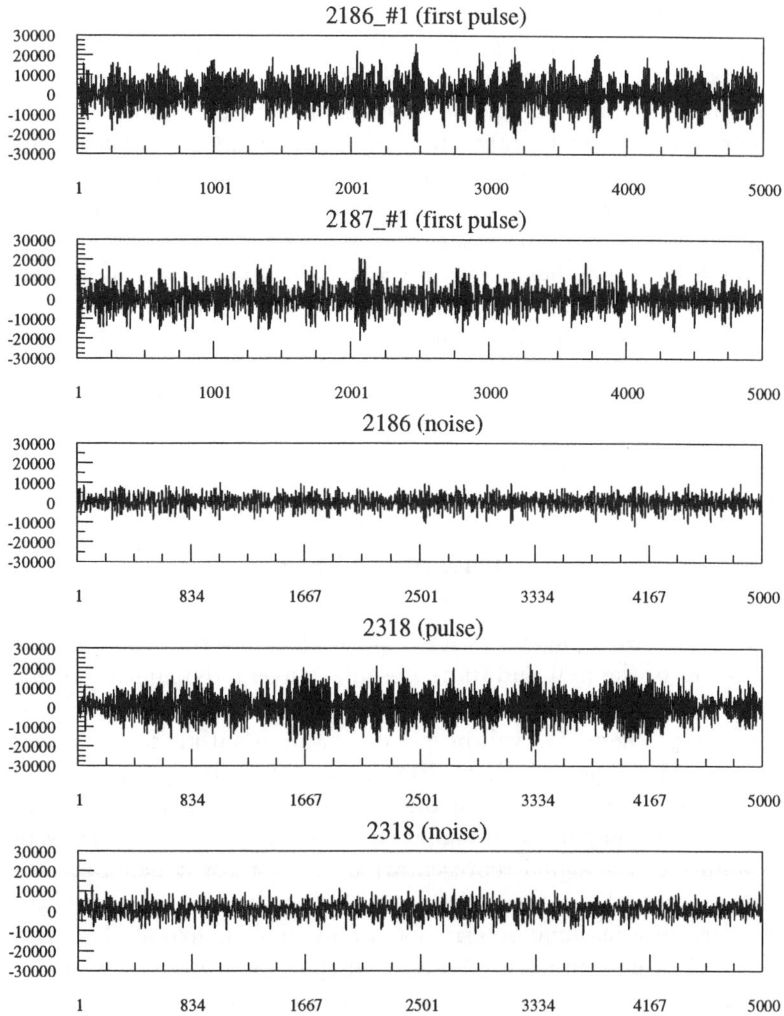

FIGURE 1. The samples of whale "cries" (2186 and 2187), porpoise "whistles" (2318) and noisy background from 2186 and 2318 data files.

The data files themselves consisted of rather long sequences (at least $3 \cdot 10^5$ points) of background signals, from which were manually extracted several segments containing samples of acoustic events. These consisted of four samples of whale "cries" of approximately 5000 points each, two samples of porpoise "whistles" of approximately 6000 points each, and various nearby segments of pure background noise, Fig. 1. Each data segment did not contain a pure stationary signal realization, but rather consisted typically of short transients of between 100 and 200 points in length, punctuated by short segments with no signal (pure noise) of 40 to 100 points in length. As such, some care had to be taken to isolate data windows which consisted of mostly SOI, otherwise windows considered to be signal which in actuality were mostly noise would give erroneous results. In an actual D/C processing chain however, this is not necessarily an issue, since one would typically look at the integrated time behaviour in the classification space to provide D/C statistics. Here, the sum of the individual data windows which contained mostly SOI provided the data ensemble for a given signal class. Finally, the extracted windows of data and noise were normalized to zero mean and unit variance, to remove amplitude information and standardize the model representation [1, 2]. We found that the typical time scale for the pulses were on the order of 40 points/characteristic period, and the pulses themselves seemed to overlap spectrally.

THE ANALYSIS

To proceed with the analysis, one must make several assumptions about the character of the SOI and its representation as a dynamical system, and also about the parameter ranges which may be relevant to the particular time scales of the problem. We will briefly cover, in an outline form, these issues below. One general point is that, although we attempt to perform D/C using a low-dimensional nonlinear dynamical model, we generally do not require this property of the particular signal class we analyze. We typically require only stability of the signal representation in the reconstructed phase space (i.e. non-divergent), otherwise the Takens representation is not well-defined. We of course also assume determinism, and to some extent this algorithm is a detector of determinism. Aside from this, we are certainly justified in representing a variety of signal classes using a dynamical representation, e.g. a tonal has a simple representaion as a set of linear, two-dimensional ODEs. In particular, in the low SNR regime, we expect only vestiges of any detailed phase space structure to survive, and hence we require a dynamical model for the classifier to be only sufficiently complex to be sensitive to this surviving information, and to deliver the required D/C perfomance.

A) General Classification Framework

Classification schemes generally assume a known set of signal classes to which an arbitrary signal is compared and possibly assigned. One valid question is how one obtains the 'canonical' set of signals used for comparison. One may consider cases where one is given exact models of the expected signals, or clean (zero noise) realizations of the signal classes available, or know the exact statistical properties of the background noise (white, colored), etc. Here, we assume no a priori fundamental knowledge of the signal types, or a detailed understanding of the characteristic noise spectrum. Hence, we attempt to perform only relative D/C, in the sense that we will observe an ensemble of realizations of noise and noise + SOI, hence building a "library" of reference signals. Note that, in a sense, this consists of "training" our classifier. We then attempt to classify new, arbitrary signal observations based on the structure (partitioning) in coefficient space of these known signal classes. In this paper, we will not present a detailed classification space analysis nor derive statistical estimates, which can be complicated issues, but rather we will simply demonstrate separation of the three signal classes in the model coefficient space.

B) Choice of a "Dynamical Filter"

As was discussed in Part I, we define a fixed "dynamical filter" to extract signal information, through which all signal classes are viewed. This filter is, in our case, a set of nonlinear ODEs defined in the re-constructed Takens state space, and hence the system is defined by a dimension D and an order P. Larger D can have an effect on the noise averaging characteristics (and "over-embedding" methods can even be implemented), and in general we need at least $D = 2$ to represent a tonal, so a typical choice is $1 < D < 6$. However for low SNR and a limited data window length, higher D can result in very sparse data and correspondingly poor numerical results. For the purposes of this data set, we chose $D = 3$ based on noise level. Similarly, a larger P can capture much more signal complexity and hence improve classification ability, however in the presence of strong noise this information is typically minimal, and higher P can actually increase the model instability, again because of data sparseness. Generally, we limit ourselves to $1 < P < 5$, and here, because of the low SNR and short window lengths, we chose $P = 2$.

C) Choice of Delay Time

The delay time τ used to construct the Takens vectors is an important parameter, although the results are typically stable within a significant range once this is determined. For clean data from well-defined dynamical generators, one can obtain a good estimate for τ using, e.g. mutual information. For very noisy data, and in particular for non-stationary data, the issue is significantly more complicated. One guideline which we have found useful is to simply use approximately 1/4 of the period of the characteristic time scale of the pulse oscillation, which generally gives good results. Since most of our data here lies between 40 to 50 points per characteristic period, we choose a $\tau = 12$. Generally, one must also insure that the chosen τ is longer than any correlation time which may be artificially induced by the analysis procedure, for example the smoothing procedure used to determine the local derivative.

D) Choice of Data Windows

Our general modeling technique consists of choosing a single window of data samples, then fitting a set of differential equations to this data by constructing a matrix equation and solving in a least-squares sense [1]. Although for a D-dimensional system, one requires D differential equations to completely describe the time evolution, one can argue that because of symmetries induced by the Takens construction, and also the loss of detailed information in the high noise regime, that in actuality each separate ODE contains qualitatively similar information from a D/C performance standpoint. Hence, we typically utilize only one equation, which comprises our dynamical filter, of the form

$$\begin{aligned}\frac{d}{dt}x(t+(d_e-1)\tau) =& c_0 + c_1x(t) + c_2x(t+\tau) + c_3x(t+2\tau) + \ldots + \\ &+ c_{d_e}x(t+(d_e-1)\tau) + \textit{high order terms}\end{aligned} \tag{1}$$

This filter effectively maps a single window of the data into the coefficient space constructed from the $\{c\}_i$. Note that if P and D are the order and dimension of the filter, respectively, then the total number of coefficients is $n = \frac{(P+D)!}{P!D!}$. Many observations of a given signal class are fit in this way, and the resulting coefficients from each window form an ensemble in the coefficient space. The variations in this ensemble are due to the noise component, the short obervation time, least-squares numerical variation, etc. For stationary signals, we can simulate many individual observations by sliding the data window along a single time series. Thus we have one and potentially two additional parameters to specify: the length of the window, L_w, and the shift of the window, L_s (in the latter case). Some guidelines can be made concerning choice of these two parameters: if L_w is too long in the case of a low *SNR* signal, then it is difficult for the filter (1) to "adjust" itself self-consistently over the whole data window. In other words, there are too many singular equations after singular value decomposition (SVD) of the matrix equation. In the case of a short window ($L_w \sim T$), there is not enough information to specify the topological pecularities. Typically, we find that the optimum value can be found in the range $2T \leq L_w \leq 6T$, where T is the characteristic time of the signal. In the case of sliding windows for a stationary data set, we put we generally put $L_s = L_w/2$, which makes the windows statistically independent, but usually allows us to increase the density of the ensemble in coefficient space.

E) Estimation of Local Derivatives

One of the most important elements of fitting the model dynamical equations is the estimation of the left-hand-side of the matrix equation, consisting of the local first-derivatives at some data point. Success of the D/C scheme strongly depends on whether we are able to extract continuous time derivatives from a badly corrupted (i.e. noisy) signal. To some extent, we assume and utilize the property that determinism is equal to differentiability [4]. It must be stressed however that for D/C, continuity of the derivative is more important than its true value, because the latter can be renormalized or even smoothly transformed without loss of generality. What is important is that random data produce random derivatives (on average), which in essense defines a null hypothesis for the technique. In practice, one must always define some smoothing procedure for noise corrupted (i.e. non-differentiable) data, and we utilize a least-squares interpolation procedure of quadratic order which averages over L_{sm} number of data points before and after the point in question. Since this smoothing procedure itself may induce artificial structuring of the data, one must avoid these problems by judicious choice of time scales. This is usually most easily accomplished by requiring that τ is at least as large as L_{sm}. Here, for this rather highly noisy data, we will use a L_{sm} of 8, and always choose τ at least this large.

F) Simple Statistical Estimate of Classification Probabilities

Any D/C procedure must be concerned with estimation of the probabilities and confidences associated with any class assignment, particularly in the high noise regime. Here, we find that our confidence estimates can be made directly by examining the statistics of the distribution of ensemble points in the coefficient space for a particular signal class. Simply, we aim at calculating the statistical probability that any new data observation is a member of a particular signal class, by calculating the probability that its coefficent space point lies within the coefficient space distribution defined by a given class. Since these distributions are typically non-Gaussian, sophisticated non-parameteric methods will likely provide the best estimators. At present, we only define a simple-minded statistic, which can yield rough preliminary results. We do this by assuming a Gaussian form for the coefficient space distribution, and then calculating the mean and standard deviation for the each particular signal class. An arbitrary point in coefficient space can then be assigned a probability of belonging to a particular distribution by calculating its distance from the distribution mean, normalized in units of its standard deviations.

In practice, not all coefficients of a given dynamical filter may be significant for a particular D/C purpose, and we can also define a criterion for choosing coefficients which is based on the coefficient ensemble distributions. We calculate the mean, $\langle c \rangle$, and the standard deviation σ_c in the usual way, for each coefficient separately of a given distribution. One can show that for pure noise, we must ideally have $\langle c \rangle \approx 0$ and $|\langle c \rangle|/\sigma_c \ll 1$. The latter property can be used to define a criteria for "significance" of any coefficient value: we consider that any coefficient for which $\langle c \rangle/\sigma \geq 1$ is statistically significant, the contrary implying that its value statistically includes zero. Hence, we may be able to reduce the number of coefficients to be compared during a particular classification application, thereby reducing the probabilities of "false alarms" due to additional axes which contain no information.

For determining whether an arbitrary data observation is a member of a particular class, we can similarly measure the distance between the coefficient space point and the mean of the given signal class distribution, using the significant coefficients. Hence, we can write the distance between centers of clouds from signal 1 and signal 2

$$r_c = \sqrt{\sum (\langle c^{(1)} \rangle_i - \langle c^{(2)} \rangle_i)^2} \tag{2}$$

is normalized by the sum of weighted variances :

$$\sigma_\Sigma = \sum q_i \sigma_i^{(1)} + \sum q_i \sigma_i^{(2)} , \tag{3}$$

where summation is done over significant coefficients and weights are determined by the projection

$$q_i = \frac{|\langle c^{(1)} \rangle_i - \langle c^{(2)} \rangle_i|}{r_c} . \tag{4}$$

Since we approximate using Gaussian statistics, this measure can be directly translated into probabilities that a given observation belongs to the given class. It is easy to see that, for example, for r_c/σ_Σ is greater than 1, the probability of a false alarm (when coefficients from a particular observation of Signal 1 are found inside distribution of the coefficients from Signal 2) in this case is less than 5%. Correspondingly, if $r_c/\sigma_\Sigma \geq 2$, then probability of the false alarm does not exceed 1%. Similar probability measures as these are quoted in the Results section to indicate D/C performance. We also note that utilization of more sophisticated statistical measures will significantly improve performance in marginal cases, since they can make use of the significantly non-Gaussian nature of the distributions (i.e. the distributions are typically tightly grouped).

RESULTS

Using the above guidelines, an analysis of the NUWC data was performed with the simple, autonomous D/C code described in Part I. As an example, we present the coefficient space results when the following code parameters were used: $D = 3$, $P = 2$, $\tau = 12$, $L_{sm} = 8$, $L_w = 300$, and $L_s = 150$. These parameters provide a good (although not optimal) distribution separation, although for clarity some outliers corresponding to data windows which contained mostly noise components were removed manually. Fig. 2 shows a $3 - D$ view of the coefficient ensembles derived for the two pulse classes (dolphin and whale), and the pure noise components, using three significant coefficients of the space (up to ten are available). This plot shows clear separation between the two signal types, and between the signals and pure noise (which is clustered about the origin). Qualitatively similar results are obtained using significant variations in the code parameter values, for example $150 \leq L_w \leq 600$, $7 \leq \tau \leq 15$, $4 \leq L_{sm} \leq 10$, hence the general performance is stable.

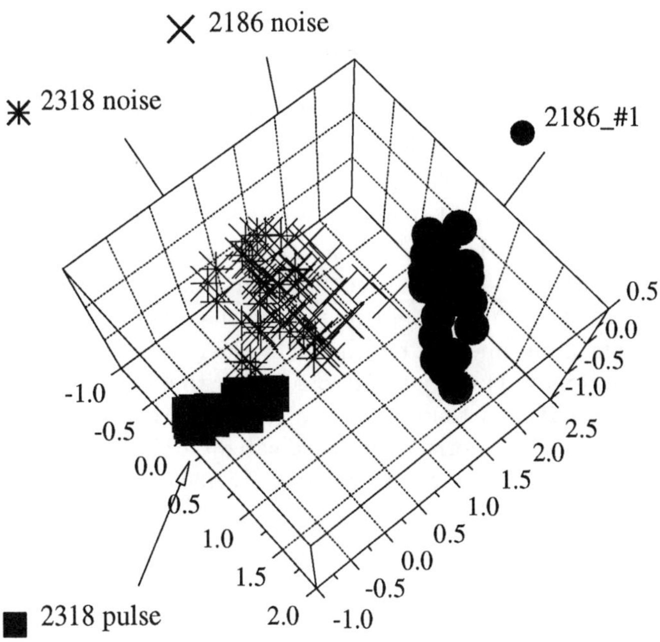

FIGURE 2. 3-dimensional coefficient space based on linear coefficients $c_{1,2,3}$. In this example we used data from 2186_#1 (first pulse) and 2318-pulse and noise was taken from both files, see Table I for the complete numerical presentation of the D/C scheme.

A quantitative measure of the classification performance is obtained using the simple statistical arguments discussed above. In Table I, we summarize the statistical measures $\langle c \rangle / \sigma$ and r_c / σ_Σ, indicating separation from noise (detection) and separation from other signal types (classification), computed for the various data samples analyzed. As the table indicates, the signal distributions are typically separated by 2 to 4 standard deviations, indicating a high confidence level as estimated by the Guassian approximation. As indicated above, one can roughly estimate Probability-of-Detect vrs. Probability-of-False-Alarm (Pd/Pfa) using the Gaussian arguments. In reality, the ensemble distributions are more tightly grouped than Gaussian, hence non-parametric methods likely result in even higher confidence levels.

One interesting point from the analysis is that the ensembles from the different whale pulses nearly overlap. At present, it is not clear whether these separate pulses came from the same whale emitting the same "word", the same whale emitting different "words", or different whales emitting the same or different "words". If the first case holds, then this would imply at least that the classifier is stable in its performance. If the second case holds, then it would imply that this D/C method may be capturing some fundamental property (nonlinear correlation) of a given single whale's vocalization cavity, which could be an interesting result. If the third case holds, then the same could be said for the vocalization cavities of the class of whales in general, and this result could be remarkable. Further analysis is underway at present to understand this result more fully.

It should also be noted that this method is numerically quite efficient. The bulk of the computational work (for the analysis of a single window) consists of the least-squares solution of the particular matrix equation, which essentially consists of the SVD of a matrix which is typically of size 10 by 300. Overall computation times are multiplied by the number of windows actually analyzed, which may typically be 50 to 100. For these typical numbers, total run time is usually less than 1 minute on an IBM 486 class PC. Hence, hardcoding of such an algorithm on a dedicated CPU, which would speed up the computation time significantly, could easily make the technique realizable for real-time analysis.

TABLE I. Classification conducted on whale "cries" (2186_#1, #2 and 2187_#1, #2) and porpoise "whistles" (2318-pulse and 2324-pulse) and noise (from 2186 and 2318 data files). If the normalized distance between clouds (right column) is more than 1, then separation of the signals is statistically significant.

Signal	Center of distribution	Significance $\langle c \rangle / \sigma$	Signal to compare with	r_c / σ_Σ
2186_#1	$c_1 = 0.77$	**2.08**	2186_#2	0.16
first pulse	$c_2 = 2.01$	**6.25**	2187_#1	0.44
	$c_3 = -0.33$	**1.1**	2187_#2	0.51
			2318-pulse	4.01
			2324-pulse	3.00
			2186-noise	1.61
			2318-noise	1.94
2186_#2	$c_1 = 0.66$	**2.13**	2187_#1	0.64
second pulse	$c_2 = 2.13$	**6.87**	2187_#2	0.57
	$c_3 = -0.28$	**1.04**	2318-pulse	4.54
			2324-pulse	3.53
			2186-noise	1.86
			2318-noise	2.18
2187_#1	$c_1 = 0.63$	**2.03**	2187_#2	0.33
first pulse	$c_2 = 1.58$	**3.06**	2318-pulse	1.28
	$c_3 = -0.15$	0.70	2324-pulse	2.02
			2186-noise	1.02
			2318-noise	1.36
2187_#2	$c_1 = 0.64$	**2.28**	2318-pulse	3.26
second pulse	$c_2 = 1.78$	**4.14**	2324-pulse	2.53
	$c_3 = 0.096$	0.38	2186-noise	1.28
			2318-noise	1.63
2318	$c_1 = 0.45$	**2.65**	2324-pulse	0.92
pulse	$c_2 = -0.64$	**2.67**	2186-noise	2.26
	$c_3 = -0.13$	0.60	2318-noise	1.50
2324	$c_1 = 0.17$	**1.76**	2186-noise	1.15
pulse	$c_2 = -0.31$	**3.87**	2318-noise	1.40
	$c_3 = 0.07$	0.70		
2186-noise			2318-noise	0.50

FUTURE WORK

The algorithm presented here clearly provides good relative classification capability, even for real-world data which are very non-stationary, short, noisy, and not well-localized. It should be noted, however, that only the simplest algorithm has been discussed here, and a variety of fundamental and technical improvements should be implemented. Choice of basis set (e.g. polynomials) [5], non-autonomous models, modification of the least-squares problem, and alternate normalization methods should be investigated. In terms of data analysis, optimal choice of time scales for non-stationary data, superior windowing techniques, and a more sophisticated statistical analysis are necessary to obtain the best classification results. We are currently developing several improvements, including an adaptive coefficient estimation procedure [6], which we plan to present in the future. In addition, we hope to continue working closely with NUWC personnel, and obtain more extensive data sets to benchmark the technique more fully.

ACKNOWLEDGEMENTS

The authors would like to thank D. Gribkov, M. Inchiosa, T. Luginbuhl, M. Shlesinger, and C. Swanson for valuable comments. This work was supported in large part by a grant to the RTA Corp. from the Office of Naval Research, Grant N00014-92-C-0045. JBK also wishes to acknowledge travel support from the NATO Office of Scientific Affairs, Grant 900088.

REFERENCES

1. see Kadtke, J. and Kremliovsky, M., this volume.
2. see also: Kadtke, J. *Phys. Lett. A*, in press, Spring, 1995.
3. Herzel, H., Berry, D., Titze, I.R., Saleh, M., *J. Speech and Hearing Res.* **37**(5), 1008, (1994)
4. Salvino, L., and Cawley, R. *Phys. Rev. Lett.* **73**(8), 1091, (1994).
5. Kadtke, J., and Brush, J. *AIP Conf. Proc.* **296**, (ed. R. Katz), AIP Press (1993), p.205.
6. Brush, J., and Kadtke, J. *AIP Conf. Proc.* **296**, (ed. R. Katz), AIP Press (1993), p.159.

Acoustic Chaos

Werner Lauterborn[§], Ulrich Parlitz[§], Joachim Holzfuss[‡], Andreas Billo[‡] and Iskander Akhatov[*]

[§]*Drittes Physikalisches Institut, Universität Göttingen, D-37073 Göttingen, Germany*
[‡]*Institut für Angewandte Physik, Technische Hochschule Darmstadt, D-64289 Darmstadt, Germany*
[*]*Department of Continuous Media Mechanics, Bashkir State University, Ufa 450074, Russia*

Abstract. Acoustic cavitation, a complex, spatio-temporal dynamical system, is investigated with respect to its chaotic properties. The sound output, the "noise", is subjected to time series analysis. The spatial dynamics of the bubble filaments is captured by high speed holographic cinematography and subsequent digital picture processing from the holograms. Theoretical models are put forward for describing the pattern formation.

INTRODUCTION

Acoustics, the science of sound, is an old science compared to chaos the now more and more recognized offspring of nonlinear dynamics with puzzling properties. Chaos forms the bridge between predictable deterministic systems and unpredictable nondeterministic noise filling the gap smoothly between these two antagonists. The new insight into the operation of nature is enormous. We now realize that determinism does not imply predictability.

Acoustic chaos has been among the first experiments to add to this insight (1 - 5) and continuously contributes to the field of nonlinear dynamics. Acoustics can be regarded as a section of the physics of vibration and waves. Seen in this slightly more abstract context the work on acoustic chaos can be expected to have a more general impact on physics because almost any measurement of a dynamical variable (pressure, voltage, etc.) yields a curve that increases and decreases on and on, in other words, is a vibration or oscillation of some sort. A large part of chaos physics therefore is devoted to chaotic vibrations (see, for instance, (6, 7)).

Quite a number of acoustic systems are known to show complex and chaotic dynamics. In musical acoustics a variety of musical instruments (8) have the potential to be played with chaotic sound emission although this is usually avoided. Among them are string instruments (9, 10), woodwind instruments (11 - 13) and gongs (14).

Also, nonlinear oscillators and their attractors may define new classes of sounds to compose with. Indeed, new musical instruments may be generated by systematically exploring nonlinear dynamical systems. For instance, Chua's circuit (15) with its large class of different attractors displays interesting features for composers (16). Sound may be generated by heat, a field covered by thermoacoustics (17, 18). A gas-filled tube, closed at one end and open at the other, emits sound when the closed end is heated and/or the open end is cooled. This system, when externally driven, has been investigated with respect to its bifurcation properties, routes to chaos and scaling laws (19). The thermoacoustic oscillator may also be used as a musical instrument (20).

Chaotic dynamics has also been found in connection with the speech production process. To produce voice sounds the vocal cords are set into vibration through the air flow between them. Models of vocal cords are presently investigated to learn about their bifurcation and chaos properties (21, 22), and bifurcations and chaos have been found in the cries of newborn infants (23).

When sound of high intensity is passed through a liquid, the liquid may rupture giving rise to the phenomenon of acoustic cavitation, whereby bubbles appear and are set into complicated motions. The analysis of this phenomenon has revealed that acoustic cavitation is a chaotic system (1 - 5). Moreover, it is a complex, spatio-temporal system deserving more investigations. It appears, for instance, in underwater acoustics, when sonar systems are driven too hard, in medical applications, where kidney stones are destructed by extracorporeal shock wave lithotripsy (24 - 26), and in ultrasonic cleaning (see, e. g., (27)). This type of cavitation originating from acoustic waves will be discussed in the sequel.

CAVITATION NOISE INVESTIGATIONS

Acoustic cavitation is accompanied by strong sound emission called acoustic cavitation noise, and a cloud of bubbles appears in the liquid forming dancing filaments. To investigate these phenomena an experimental arrangement has been set up as depicted in Fig. 1. A cylindrical transducer of piezoelectric material is used to set up a sound field in water. A typically used transducer was of 76 mm length, 76 mm inner diameter, and 5 mm wall thickness submerged in a water filled glass container. The cylinders were driven at different frequencies between 7 kHz and 23 kHz. A standing acoustic wave is set up inside the cylinder having its maximum pressure amplitude along the axis of the transducer. A hydrophone monitors the sound output of the liquid. Some electronics and a computer surround the experiment to drive the cylinder and to store the sampled pressure data from the hydrophone.

A spectral analysis of fully developed cavitation appearing at high driving levels reveals a broadband noise spectrum with lines superimposed at the driving frequency and its harmonics, but also at the subharmonics, notably the first subharmonic at one half the driving frequency and its harmonics. To learn about the onset of this

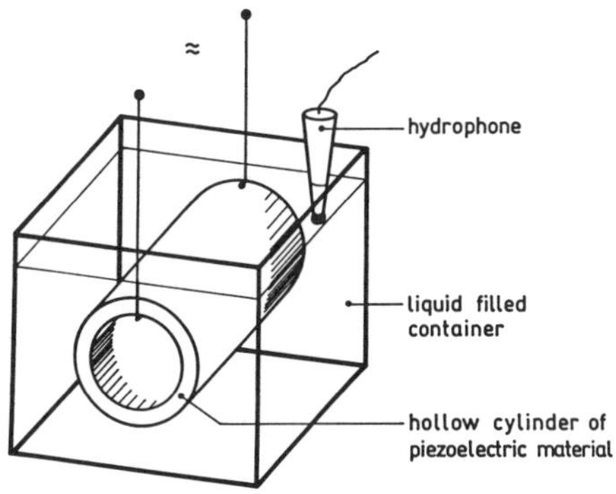

Figure 1. Basic elements of the acoustic cavitation experiment: a water filled container with a piezoelectric cylindrical transducer and a hydrophone.

spectrum experiments with increasing sound pressure amplitude of the driving sound field at, e. g., 23 kHz have been done. Conditions have been found, where a period-doubling cascade is observed leading to a low-dimensional strange attractor (1, 5, 28). Figure 2 shows an example of the spectrum of cavitation noise where two pe-

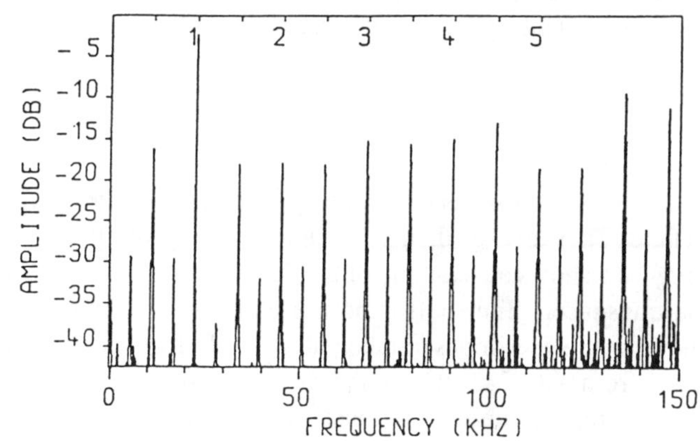

Figure 2. Power spectrum of acoustic cavitation noise after two period-doubling bifurcations.

riod doublings have taken place. The spectrum extends to high harmonics with almost equal amplitudes. The chaotic attractor (actually there seem to be different types) has been visualized by embedding the pressure samples into higher dimensional spaces (3, 28) and quantified by determining the fractal dimension (3) and the Lyapunov spectrum (4). An astonishingly low-dimensional strange attractor is obtained, having a dimension between two and three and just one positive Lyapunov exponent (Fig. 3). This fact is so astonishing, because the sound output is generated

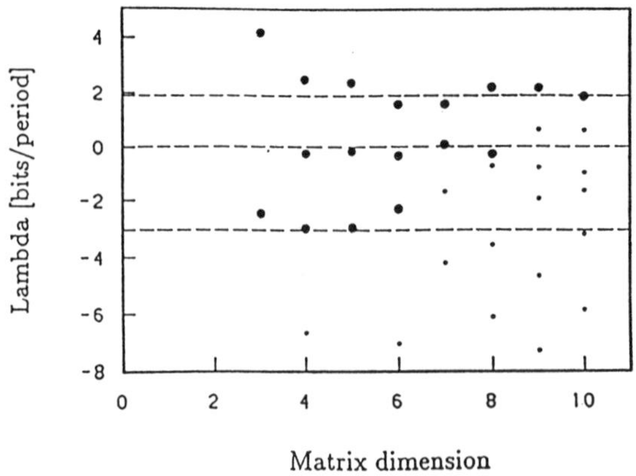

Figure 3. Lyapunov spectrum for a chaotic acoustic cavitation noise attractor.

by hundreds of bubbles of different size each being a nonlinear oscillator and each being capable of period doubling to chaotic oscillations (29, 30). Therefore some cooperative effects must take place synchronizing the bubble motions. A first approach to model bubble agglomeration in a sound field will be given below. First we turn to the measurement of the spatial dynamics.

FILAMENTARY BUBBLE PATTERN

The acoustic cavitation noise is produced by a cloud of bubbles oscillating in the driving sound field. They arrange themselves into filaments resembling Lichtenberg figures (31). Figure 4 shows an example of the dendritic filament pattern seen along the axis of the transducer. Ten short exposure photographs (15 ns from a copper vapor laser) have been superimposed taken at 1632 Hz from acoustic Lichtenberg figures produced from a 16.323 kHz sound field. As this mode of photographing enhances the structure, the bubbles obviously move along the branches. To clarify the dynamics of the spatial patterns and of the bubbles forming them time-resolved investigations have been initiated as described below.

High Speed Holographic Cinematography

The observation and the visualisation of the dynamics of three-dimensional motion is a difficult problem. The visualisation is presently approached by virtual reality with quickly improving results. The observation of real physical systems by measurement, however, has attracted much less attention and is intrinsically difficult

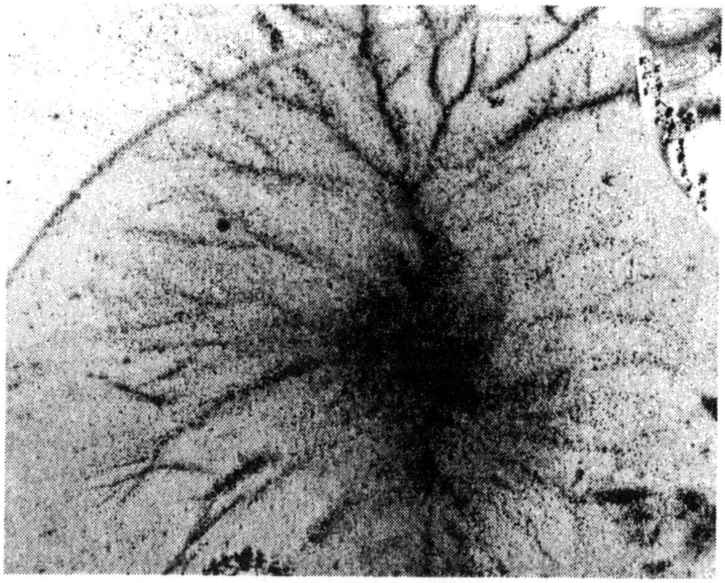

Figure 4. Example of the filamentary bubble pattern inside the cylinder in the projection along the axis of the transducer.

to perform in real time. Should the optical measurements be done with similar time resolution as the acoustic ones, for instance, at 1 MHz sampling rate, at least a few million GByte per second data rate would be needed, as a single three-dimensional image contains about 1 GByte of voxels (volume picture elements). While this cannot yet be done in real time, high speed holographic cinematography comes close to it off-line (32). By taking holograms at a rate of up to 69300 holograms per second it could be shown that the whole filamentary structure is undergoing period doubling simultaneously with period doubling of the sound output (33). The acoustic source has thus been traced back to bubble oscillations. For these investigations a cavity dumped argon ion laser has been used. Due to the limited power per light pulse for taking a hologram, however, only holograms of a few millimeter squared could be exposed sufficiently giving unsatisfactory resolution in the reconstructed three-dimensional images.

Therefore an improved arrangement has been developed utilizing a copper vapor laser pumped dye laser for producing a series of high power light pulses (28, 34). The device is capable of taking holograms at rates up to 20000 holograms per second with an area per hologram of up to one centimeter square and a total capacity of about 40 holograms or more depending on the actual size chosen for a hologram. The arrangement for taking in-line holograms with a rotating holographic plate for spatial separation of the individual holograms is depicted in Fig. 5.

The copper vapor laser is capable of emitting a quasi infinite series of short light

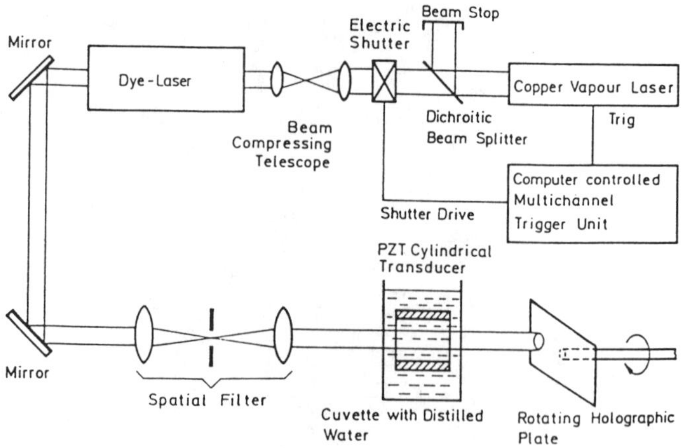

Figure 5. Arrangement for high speed holographic cinematography with a copper vapor laser pumped dye laser in an in-line geometry.

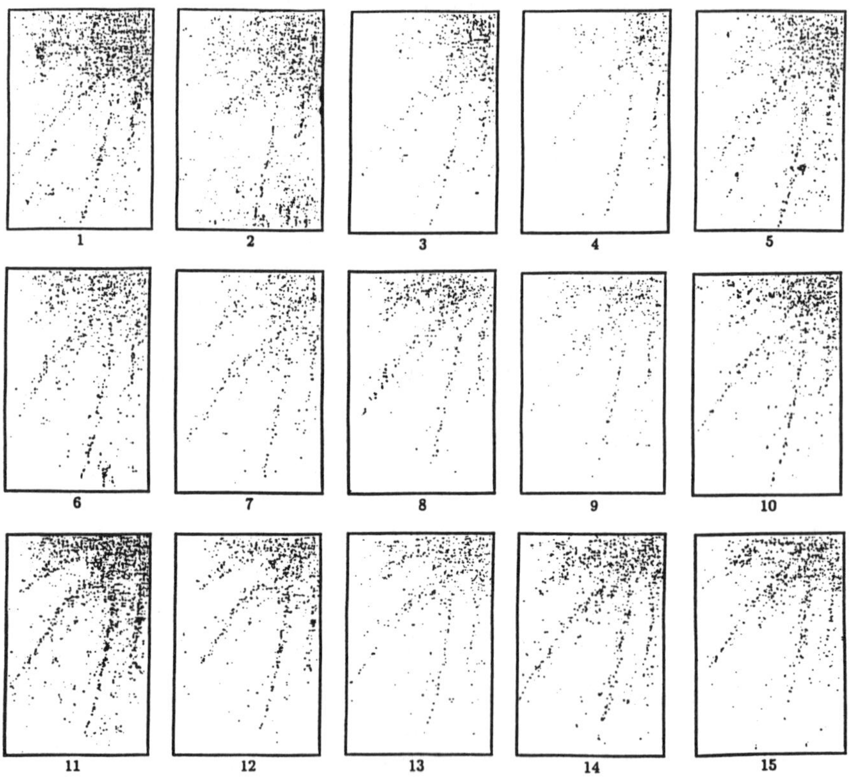

Figure 6. Reconstructed holographic images from a series of holograms taken at a rate of 5700 holograms per second of a filament bubble pattern (Courtesy of A. Judt).

pulses with a duration of about 15 ns and an energy of about 1 mJ each. The light consists of two spectral lines, green and yellow. The dichroic beam splitter selects the green line to pump the dye laser and dumps the yellow line into a beam stop. The electric shutter opens for the time of one revolution of the rotating holographic plate on which the series of holograms is recorded. The dye laser serves as a coherence transformer boosting the coherence length from the few millimeters of the copper vapor laser light pulses up to about 7 cm. Simultaneously the wavelength is changed from $\lambda = 510.6$ nm to $\lambda = 571$ nm (Rhodamine 6G dye). The spatial filter improves the beam quality by removing interference fringes collected by diffraction on dust particles on the mirrors. An in-line holographic arrangement is used. In this geometry the light passing undisturbed through the liquid serves as the reference beam for the light scattered at the bubbles. A series of 15 holographic images is shown in Fig. 6 taken photographically from a series of in-line holograms recorded at 5700 holograms per second. The driving sound field had a frequency of about 7 kHz in this case. The filaments to be seen are quite stable. They are mainly made up of bubbles with a typical size of 50 μm.

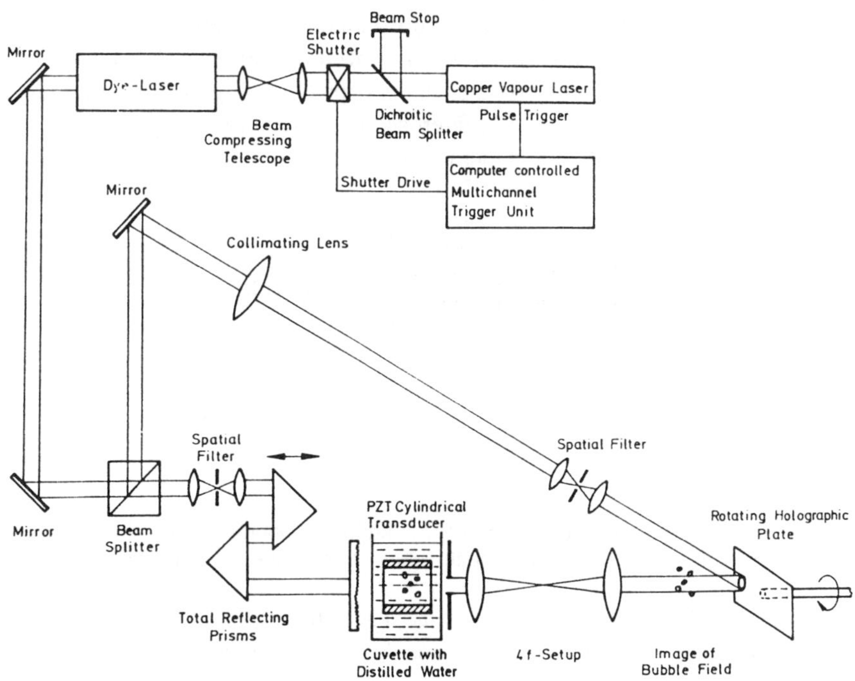

Figure 7. Arrangement for high speed holographic cinematography with a copper vapor laser pumped dye laser in an off-axis geometry.

In-line holography has the disadvantage that large and small objects cannot be recorded simultaneously with sufficient accuracy in depth location, because differently sized objects demand for a different optimum distance of the holographic plate from the objects. Therefore off-axis holography has been employed with diffuse illumination to spread the interference pattern all over the hologram size. Figure 7 shows the arrangement for off-axis holographic cinematography. This arrangement is more involved than the in-line case as a separate reference beam is needed and special precautions have to be taken to adjust the lenghts of reference and object beams to near equality. The two prisms serve as an adjustable delay line for that purpose. The two lenses in the 4f-arrangement image the bubble cloud in front of the rotating holographic plate for improved recording via an adjustable distance of the cloud to the holographic plate. The images of the holograms have been subjected to three-dimensional image processing as described next.

Digital Holographic-Image Processing

A series of holograms contains an enormous amount of information that cannot properly be analyzed manually. Therefore a digital image processing system for three-dimensional images from holograms has been developed to handle the large amount of data (35). The basic arrangement is depicted in Fig. 8. The hologram is illuminated with the phase conjugate reference beam for projecting the real image into space. An image dissector camera with 4096x4096 addressable pixels scans

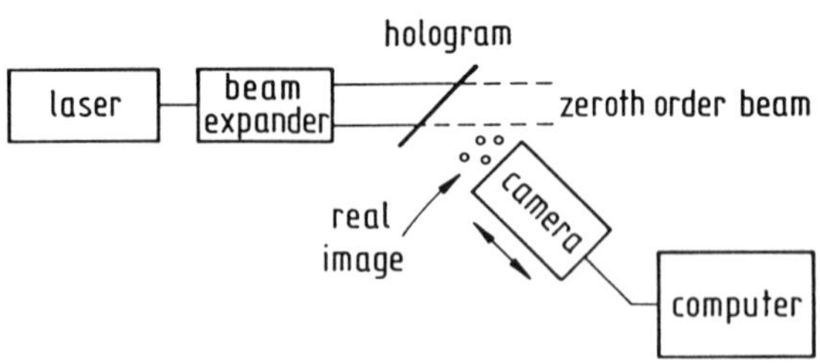

Figure 8. Arrangement for digital image processing from holograms using the phase conjugate image, high resolution image dissector camera and parallel computing.

the three-dimensional image plane by plane, typically 50 to 100 planes depending on the depth of the investigated volume, and feeds the data into a computer system. It mainly consists of a transputer net of up to 45 transputers T800 attached to a workstation with 128 MBytes RAM. The transputers do the preprocessing of filtering and condensing the data leaving a data set per hologram of about 100 MByte down from the about 1 GByte of raw data read in. A large software package has been written for detecting and locating bubbles in the three-dimensional real image. Figure 9 gives a three-dimensional view of the output of the holographic-image processing system for one hologram of a series. The filaments are best seen in the z-projection. Correlation studies between subsequent images are under way.

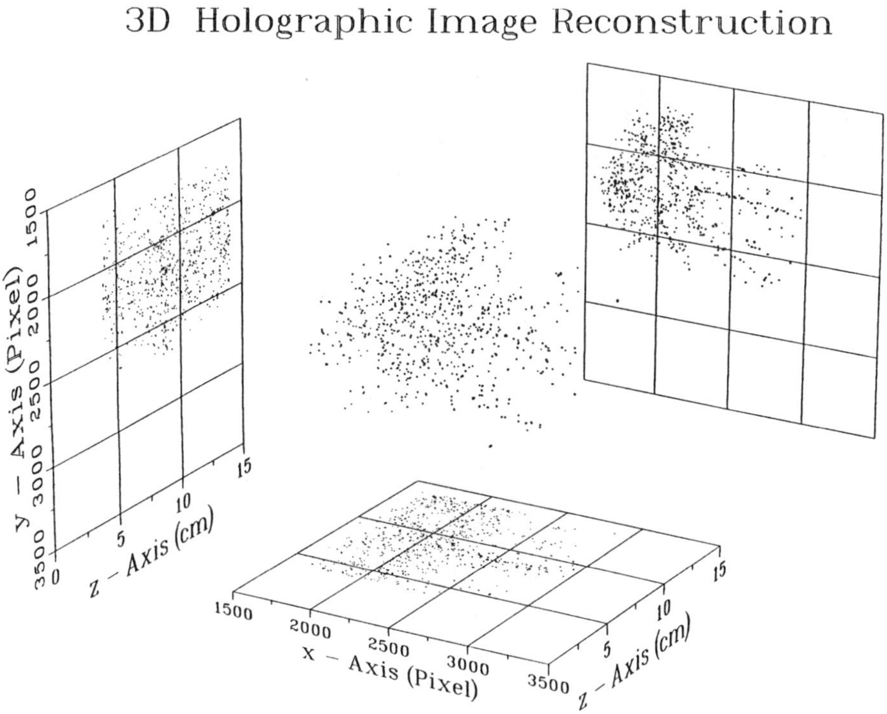

Figure 9. Plot of a bubble filament pattern with three projections as obtained with the digital holographic image processing system.

It is notoriously difficult to visualize three-dimensional point distributions. The solution, that, however, cannot be printed, is virtual reality where the observer can look around in the distribution. As a first step to implement a virtual reality system for this purpose we have stored a data base of three-dimensional images of bubble distributions obtained from a holographic series. One of the images is given in Fig. 10 with color coded depth information and bubbles plotted as spheres according to their size. They can be viewed on a computer monitor with stereo capabilities and shutter eyewear. The holographic-image processing system is continuously improved for better performance with the ultimate goal to be able to study space-time chaotic systems.

Figure 10. Color coded three-dimensional plot of a bubble distribution (Courtesy of H. Chodura). *(See Color Plate 1.)*

THEORY

The acoustic as well as the optical measurements have revealed a subharmonic route to chaos. For a theoretical understanding a model for the processes occurring in the cavitation bubble cloud must be formulated. In view of the thousands of interacting bubbles this is an involved task. As a first step single spherical bubbles were subjected to a sound field to look for period-doubling sequences (36). They abound even with spherically oscillating single bubbles (30, 37).

To approach the filament formation the spatial aspect must enter the considerations. Due to (primary) Bjerknes forces (38, 39) the bubbles move to specific locations in the sound field. This motion changes the spatial distribution of microbubbles that reacts back on the sound field due to the dependence of the speed of sound on the bubble concentration. A model incorporating these effects has been developed (40).

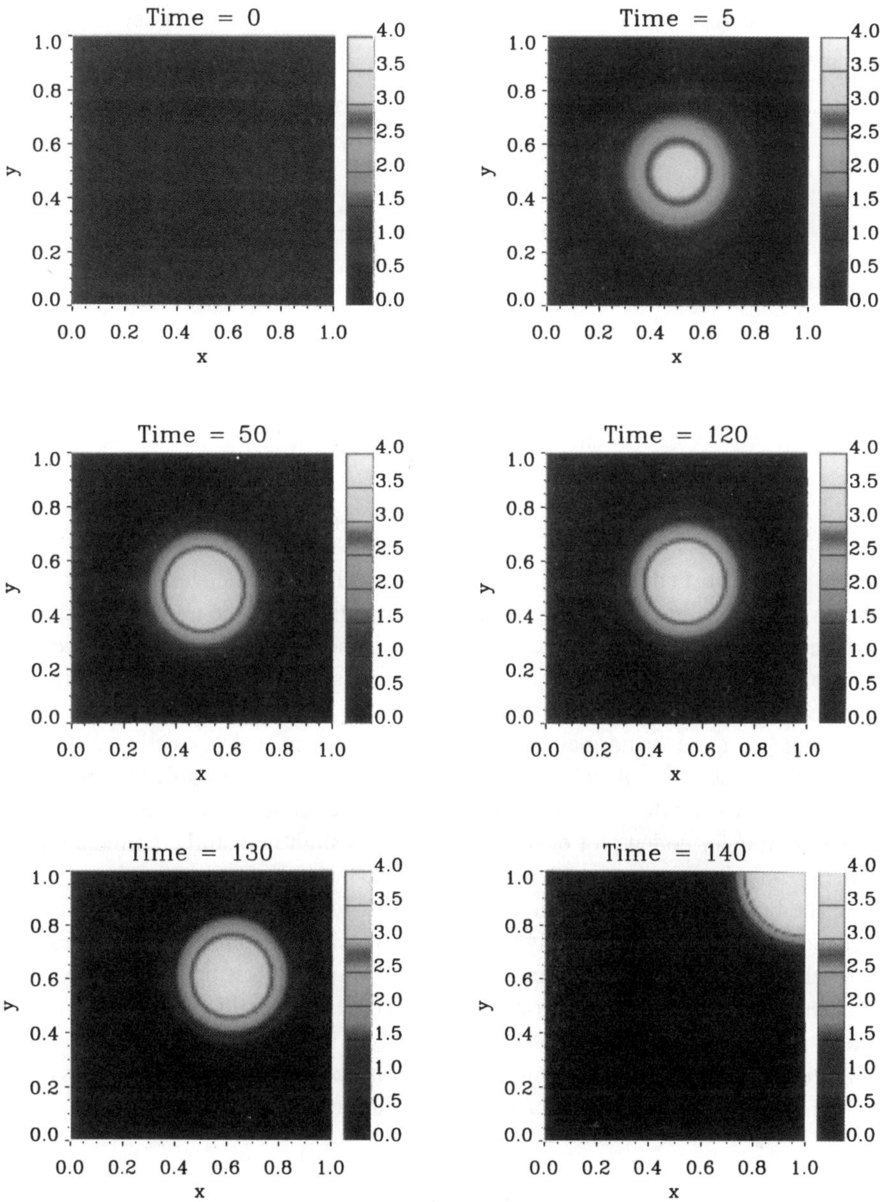

Figure 11. Color coded plot of the two-dimensional evolution of bubble concentration n of a slightly disturbed initial bubble distribution. Parameter values: $\tau_1 = 0.1$, $\tau_2 = 0.01$, $\gamma = 0.001$ $A_\infty^2 = 2$. *(See Color Plate 2.)*

It consists of a set of coupled partial differential equations describing the spatio-temporal evolution of the sound field amplitude and of the bubble concentration. In the case of a two-dimensional wavefront the equations read:

$$i\frac{\partial w}{\partial \xi} = \frac{\partial^2 w}{\partial \eta^2} + \frac{\partial^2 w}{\partial \zeta^2} + nw \tag{1}$$

$$\frac{\partial n}{\partial \xi} + \frac{\partial (n u_x)}{\partial \eta} + \frac{\partial (n u_y)}{\partial \zeta} = -\frac{n - f(|w|^2)}{\tau_1} \tag{2}$$

$$\tau_2 \frac{\partial u_x}{\partial \xi} + u_x = \gamma \frac{\partial}{\partial \eta}\left(|w|^2\right) \tag{3}$$

$$\tau_2 \frac{\partial u_y}{\partial \xi} + u_y = \gamma \frac{\partial}{\partial \zeta}\left(|w|^2\right) \tag{4}$$

$$f(|w|^2) = A_\infty^2[1 - \exp(-|w|^2/A_\infty^2)], \tag{5}$$

where w is the normalized sound field amplitude, n is the normalized bubble concentration, u_x and u_y are the bubble velocity components, τ_1 is the normalized characteristic time of dissolution of microbubbles, τ_2 is the normalized characteristic time for the relaxation of the velocity of the bubbles, γ is the normalized bubble drift parameter, A_∞^2 is the normalized saturation value for bubble production by the sound field, and ξ, η, ζ are the normalized time, x- and y-coordinates, respectively. This set of equations has been subject to a stability analysis and has been solved with appropriate boundary conditions (no bubble motion at the rectangular boundary). Figure 11 shows a color coded plot of the evolution of a slightly disturbed almost homogeneous bubble distribution (Time=0). In the sound field the distribution is unstable. The bubbles agglomerate first at the center (due to the initial condition of disturbance), until, at approximately Time=50, a quasistable concentrated distribution is reached. This state prevails until Time=120, where a positional instability takes over shifting the bubble concentration towards the boundary (Time=140). Thus this model shows a self-concentration effect of bubbles in a sound field. The sound field amplitude w behaves in a way similar to the bubble concentration, i. e., focuses down to higher amplitude where the bubble concentration is higher.

DISCUSSION

Chaotic dynamics in acoustics has grown to a large area of research. Most developed is the area of hydroacoustic chaos. Acoustic cavitation proves as a testbed for a spatio-temporal dynamical system that is accessible also experimentally. Self-organization phenomena can be studied here for developing an understanding of how structures may emerge from instabilities and nonlinearities. Moreover, bubbles are a microcosmos with puzzling properties, for instance light emission (41) that even may possess chaotic properties (42).

ACKNOWLEDGMENTS

This work is supported by the Deutsche Forschungsgemeinschaft (Sonderforschungsbereich 185 – Nonlinear Dynamics), the Stabsstelle für Internationale Beziehungen (Contract X222.31), the Ministry for Science, Higher Education and Technical Policy of the Russian Federation and the State Committee of the Russian Federation for Higher Education. Thanks go to the members of the Nonlinear Dynamics Group of the Drittes Physikalisches Institut, Universität Göttingen, in particular M. Wiesenfeldt and C. D. Ohl for help with the figures.

REFERENCES

1. Lauterborn, W., and Cramer, E., *Phys. Rev. Lett.* **47**, 1445–1448 (1981).
2. Smith, C. W., Tejwani, M. J., and Farris, D. A., *Phys. Rev. Lett.* **48**, 492–494 (1982).
3. Lauterborn, W., and Holzfuss, J., *Phys. Lett.* **115A**, 369–372 (1986).
4. Holzfuss, J., and Lauterborn, W., *Phys. Rev.* **A39**, 2146–2152 (1989).
5. Lauterborn, W., and Holzfuss, J., *Int. J. Bifurcation and Chaos* **1**, 13–26 (1991).
6. Thompson, J. M. T., and Stewart, H. B., *Nonlinear Dynamics and Chaos*, Chichester: Wiley, 1986.
7. Moon, F. C., *Chaotic Vibrations*, New York: Wiley, 1987.
8. McIntyre, M. E., Schumacher, R. E., and Woodhouse, J., *J. Acoust. Soc. Am.* **74**, 1325–1345 (1983).
9. Tuffillaro, N. B., *Am. J. Phys.* **57**, 406–414 (1989).
10. Müller, G., and Lauterborn, W., *Acustica*, in the press (1995).
11. Maganza, C., Caussé, R., and Laloë, F., *Europhys. Lett.* **1**, 295–302 (1986).
12. Gibiat, V., *J. Sound Vibr.* **123**, 529–536 (1988).
13. Idogawa, T., Kobata, T., Komuro, K., and Iwaki, M., *J. Acoust. Soc. Am.* **93**, 540–551 (1993).
14. Legge, K. A., and Fletcher, N. H., *J. Acoust. Soc. Am.* **86** 2439–2443 (1989).
15. Madan, R. N. (Ed.), *Chua's Circuit: A Paradigm for Chaos*, Singapore: World Scientific, 1993.
16. Mayer-Kress, G., Choi, I., Weber, N., Barger, R., and Hübler, A., *IEEE Trans. Circuits and Systems* **40**, 688–695 (1993).
17. Swift, G. W., *Physics Today* **48**, 22–28 (July 1995).
18. Swift, G. W., *J. Acoust. Soc. Am.* **84**, 1145–1180 (1988).
19. Yazaki, T., *Phys. Rev.* **E48**, 1806–1816 (1993).

20. Müller, G., and Lauterborn, W., "Experiments with the Thermoacoustic Oscillator," in *Proceedings of the International Symposium on Musical Acoustics, July 2–6, 1995, Le Normont, Dourdan, France* 1995, pp. 178–183.

21. Herzel, H., *Appl. Mech. Rev.* **46**, 399–413 (1993).

22. Steinecke, I., and Herzel, H., *J. Acoust. Soc. Am.* **97**, 1874–1884 (1995).

23. Mende, W., Herzel, H., and Wermke, K., *Phys. Lett.* **A145**, 418–424 (1990).

24. Coleman, A. J., Saunders, J. E., Crum, L. E., and Dyson, M., *Ultrasound in Med. Biol.* **13**, 69–76 (1987).

25. Philipp, A., Delius, D., Scheffczyk, C., Vogel, A., and Lauterborn, W., *J. Acoust. Soc. Am.* **93**, 2496–2509 (1993).

26. Carnell, M. T., Fiadeiro, P. T., and Emmony, D. C., *J. Acoust. Soc. Am.* **97**, 677–679 (1995).

27. Young, F. R., *Cavitation*, London: McGraw-Hill, 1989.

28. Lauterborn, W., Schmitz, E., and Judt, A., *Int. J. Bifurcation and Chaos* **3**, 635–642 (1993).

29. Lauterborn, W., *J. Acoust. Soc. Am.* **59**, 283–293.

30. Parlitz, U., Englisch, V., Scheffczyk, C., and Lauterborn, W., *J. Acoust. Soc. Am.* **88**, 1061–1077 (1990).

31. Lichtenberg, G. C., "De nova methodo naturam ac motum fluidi electrici investigandi," in *Novi Commentarii Societatis Regiae Scientiarum Gottingensis*, tomus VIII, ad annum 1777, Commentationes physicae et mathematicae classis, Gottingae: apud Joann. Christian Dieterich, 1777, pp. 168–180.

32. Hentschel, W., and Lauterborn, W., *Opt. Eng.* **24**, 687-691 (1985).

33. Lauterborn, W., and Koch, A., *Phys. Rev.* **A35**, 1774–1976 (1987).

34. Lauterborn, W., Judt, A., and Schmitz, E., *Opt. Lett.* **18**, 4–6 (1993).

35. Haussmann, G., and Lauterborn, W., *Appl. Opt.* **19**, 3529–3535 (1980).

36. Lauterborn, W., and Suchla, E., *Phys. Rev. Lett.* **53**, 2304–2307 (1984).

37. Lauterborn, W., and Parlitz, U., *J. Acoust. Soc. Am.* **84**, 1975–1993 (1988).

38. Blake, F. G., *J. Acoust. Soc. Am.* **21**, 551 (L) (1949).

39. Crum, L. A., *J. Acoust. Soc. Am.* **57**, 1363–1370 (1975).

40. Akhatov, I., Parlitz, U., and Lauterborn, W., *J. Acoust. Soc. Am.* **96**, 3627–3635 (1994).

41. Crum, L. A., *Physics Today* **47**, 22–29 (Sept. 1994).

42. Holt, R. G., Gaitan, D. F., Atchley, A. A., and Holzfuss, J., *Phys. Rev. Lett.* **72**, 1376–1379 (1994).

Ray Chaos of Underwater Sound in Long-Range Propagation

Jianguo Yan and Kang K. Yen

Dept. of Electrical and Computer Engineering
Florida International University
Miami, FL 33199

Abstract. Palmer, *et al.* (Geophysical Research Letters, Vol. 15, No. 6, 1988, 596-572) have shown that acoustic ray paths in a weakly range-dependent deterministic ocean model exhibit chaotic behavior. The objective of this paper is to investigate whether ray paths in a realistic ocean-acoustic problem are chaotic. To achieve this goal, measured sound speed data and Munk, *et al.*'s horizontal ray equations, which take into account earth's curvature and flattening, are used to study sound propagation from Heard Island to Ascension Island. Along with the integration of the ray equations, Lyapunov exponents are calculated to identify chaos. Numerical results show that all the horizontal ray paths we construct were chaotic. Predictability limitations due to the chaos are discussed.

INTRODUCTION

Palmer *et al.* (1) have shown that acoustic ray paths in a weakly range-dependent deterministic ocean model exhibit chaotic behavior, i.e., exponentially sensitive dependence on initial conditions. This implies that even if the sound speed structure were known with infinite precision, a limitation is imposed on one's ability to make deterministic predictions using ray theory. The Palmer *et al.*'s pioneering work created a new field of underwater acoustics: Acoustic ray chaos. Later, Brown *et al.* (2), Smith *et al.* (3,4), and Yan (5) further studied the ray chaos, and provided some new insights.

However, the previous studies (1-5) have a shortcoming: The ray equations used by those researchers are two-dimensional Hamiltonian equations. The equations are valid for, at most, about 100 km range, but they were integrated to thousands of kilometers (far beyond their range limitation) to construct Poincare sections and estimate Lyapunov exponents. This can lead to some unreasonable results. For example, a predictability horizon of 1000 to 2000 km has been estimated in ref. (3) using Lyapunov exponents, while the ray equations used are only valid for about 100 km. This suggests that predictability of ray acoustics is determined by ray equations, rather than the chaos. Thus, that predictability horizon is, of course, not useful.

The objective of this paper is to investigate whether ray paths in a realistic ocean-acoustic problem are chaotic. During January 1991, the Heard Island Feasibility Test (HIFT) was carried out to determine the feasibility of acoustically measuring the rate of ocean warming. Underwater acoustic signals were transmitted from a ship in the southern Indian Ocean and monitored at listening stations throughout the world (6,7). To interpret the received HIFT transmissions, we have constructed the horizontal ray paths between Heard Island and Ascension Island by using Munk et. al.'s horizontal ray equations (8) that take into account earth curvature, earth flattening, and horizontal refraction, and thus, are valid for very long ranges. We found that all the ray paths were chaotic, and present the numerical results in this paper.

METHODS

To construct horizontal ray paths, we adopt the following ray equations:

$$\frac{d\phi}{ds}=\frac{\cos\alpha}{\mu}, \tag{1}$$

$$\frac{d\lambda}{ds}=\frac{\sin\alpha}{\nu\cos\phi}, \tag{2}$$

and

$$\frac{d\alpha}{ds}=\frac{\sin\alpha\tan\phi}{\nu}-\left(\frac{\sin\alpha}{\mu}\frac{\partial}{\partial\phi}-\frac{\cos\alpha}{\nu\cos\phi}\frac{\partial}{\partial\lambda}\right)\ln C_n, \tag{3}$$

where

$$\mu=\frac{a(1-\epsilon^2)}{(1-\epsilon^2\sin^2\phi)^{3/2}}, \tag{4}$$

and

$$\nu=\frac{a}{(1-\epsilon^2\sin^2\phi)^{1/2}}. \tag{5}$$

Here, ϕ is the geographic latitude with north positive, λ is the longitude with east of Greenwich positive, α is the azimuth measured clockwise from north, C_n is the phase velocity of mode n, a is the semimajor radius of the reference ellipsoid, and ε is the eccentricity. These equations take into account both curvature and flattening of the earth so that they are particularly useful for studying global-scale transmissions. The derivation of these equations was first given by Munk *et al.*(8).

The phase velocity in Eqs. (1)-(3) can be obtained by solving the Sturm-Liouville problem:

$$\frac{d^2\psi_n}{dz}+(k^2-\kappa_n^{\;2})\psi_n=0, \tag{6}$$

with the boundary conditions

$$\psi_n(0)=0,\frac{d\psi_n(D)}{dz}=0, \tag{7}$$

where ψ_n is an eigenfunction, z is the depth of the ocean, k is propagation constant, _$_n$ is the horizontal component of k, and D is the depth of the ocean bottom. The boundary conditions in Eq. (7) imply that the ocean bottom is rigid. Here we use the rigid-bottom model for simplicity, assuming that for global-scale propagation, the interaction of sound waves with the bottom is relatively unimportant. By numerically solving Eq. (6), we obtain the eigenvalue _$_n$. Since

$$\kappa_n=\frac{\omega}{C_n}, \tag{8}$$

where ω is angular frequency, for a given frequency we can calculate the phase velocities from Eq. (8).

We normalize the eigenvector to obtain the group velocity:

$$U_n=1/[C_n\int_0^D\psi_n^{\;2}(z)/C^2(z)dz], \tag{9}$$

and compute the travel time for the n-th mode:

$$t_n=\int_0^S\frac{ds}{U_n}, \tag{10}$$

where S is the length of ray path.

For ray tracing, we use the National Oceanographic Data Center (NODC) archival sound speed data for the seasonal period December through March. First, we use the interpolated sound speed data to calculate phase and group velocities. The finite difference method, which has been developed by Porter and Reiss (9), is used in this calculation. Then, we integrate the ray equations to construct horizontal ray paths and calculate travel times. Eqs. (1)-(3) are integrated by using the fourth-order Runge-Kutta method with adaptive step size control (10). The calculation accuracy is controlled by specifying a maximum fractional error (10^{-6}) in any single integration step. We have tried 10^{-5}, and 10^{-6}, and have found by numerical experiments that using 10^{-6} as the maximum

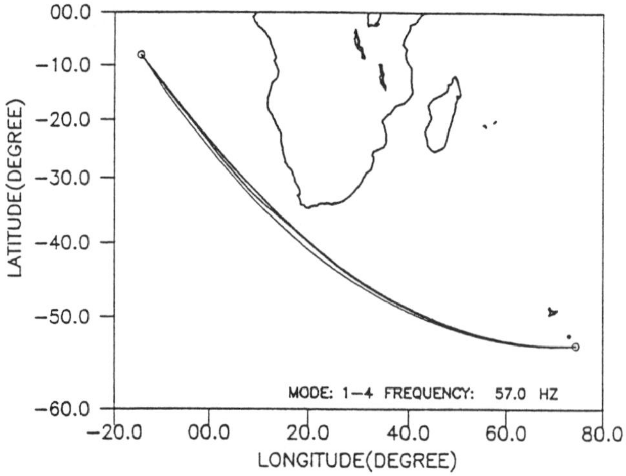

FIGURE 1. Horizontal ray paths between Heard Island and Ascension Island (Mode 1 through 4).

fractional error allows the computed travel times to converge in their sixth significant digit. Eigenrays are determined by using successive shooting method. A ray is determined as an eigenray, if it passes within 0.005° in both latitude and longitude of Ascension Island.

To identify chaos, Lyapunov exponents

$$\sigma_i = \lim_{s \to \infty} \frac{1}{s} \log_2 \frac{p_i(s)}{p_i(0)} \tag{11}$$

are estimated along ray paths using Wolf *et al.*'s FORTRAN code (11). Here p_i is the length of an ellipsoidal principle axis. For chaos the largest Lyapunov exponent is positive.

NUMERICAL RESULTS AND DISCUSSIONS

Sensitivity to Initial Conditions

Fig. 1 shows the horizontal ray paths between Heard Island (53°22'S, 74°30'E) and Ascension Island (7°57'S, 14°24'W). The corresponding ray parameters are listed in Table 1. We can see from Table 1 that to find the eigenrays (in Fig. 1) that satisfy the latitude/longitude tolerance criterion at Ascension Island (as mentioned in last section, <0.005°), the launch azimuths need to be specified with 13-16 digits. This phenomenon (extremely sensitive dependence on initial

TABLE 1. Ray parameters and Lyapunov exponents

Mode number	Launch Azimuth (Degree)	Arrival Longitude Error[a] (Degree)	Lyapunov exponents km^{-1}
1	266.2025639765	0.003	-0.28, 1.03, -7.16
2-1	265.7094041278737[b]	0.007	-0.15, 0.76, -7.11
2-2	266.9311828613281	0.002	-0.71, 1.36, -7.72
3-1	265.7386920991651[b]	0.008	-0.09, 0.60, -7.09
3-2	267.0751953125	0.002	-0.65, 1.15, -7.64
4	265.8042083741925[b]	0.005	-0.02, 0.47, -7.08

[a]This is obtained by using $|\lambda_{ending}-14.40|$.
[b]These rays lose predictability.

TABLE 2. An Example of Loss of Predictability (Mode 4)

Number of iteration	Launch azimuth (Degree)	Ending Longitude (Degree)
40	265.8042083741953	-14.390
41	265.8042083741939	-14.390
42	265.8042083741932	-14.390
43	265.8042083741929	-14.390
44	265.8042083741927	-14.390
45	265.8042083741926	-14.490

conditions) is a characteristic of chaos.

Loss of Predictability

There are three eigenrays (in Table 1) that we could not found for given latitude/longitude tolerance criterion (<0.005°). They are mode 2-1, mode 3-1, and mode 4. For these rays, the launch azimuths were specified with 16-digit accuracy (This is the maximum accuracy we can use when we use FORTRAN's double precision in VAX 8800,) but the ending longitude still did not satisfy the given precision criterion. In other words, we lose predictability in these cases. As seen from Table 2, at the 40-th step of shooting, we began to lose predictability (Change in launch azimuth did not significantly affect longitude.) Thus, we did not arrive at an ending longitude satisfying the tolerance criterion (<0.005°), although we know that the eigenray's launch azimuth is between 265.8042083741926 and 265.8042083741925. The later corresponds to ending longitude -14.405° (The exact ending latitude is 14.40°.)

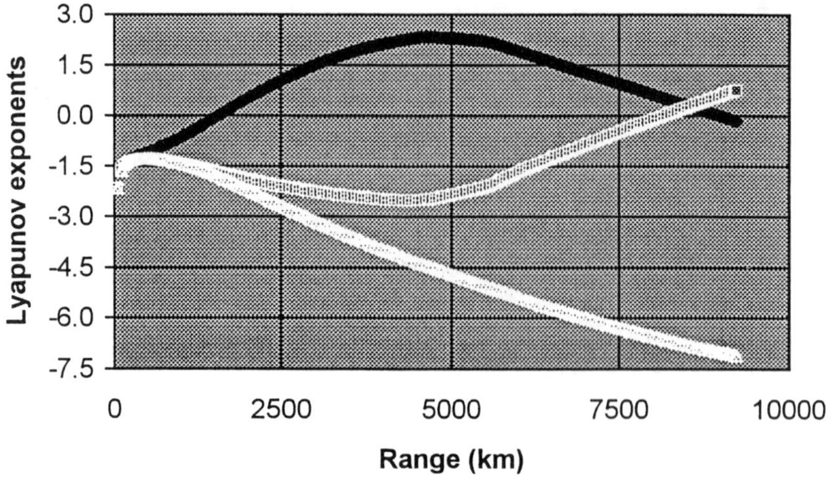

FIGURE 2. Lyapunov exponent estimation for horizontal ray (mode 1).

Lyapunov Exponents

As mentioned above, we have observed two characteristics of chaos, i.e., sensitivity to initial conditions, and loss of predictability. We tried to use Lyapunov exponents to quantify the predictability, but we had difficulty in estimating Lyapunov exponents defined by Eq. (11), as illustrated as follows.

Fig. 2 shows the Lyapunov exponents we estimated for eigenray mode 1. It can be seen that we could not obtain a Lyapunov exponent in the limit $s \rightarrow \infty$. In fact, we have no knowledge of behavior beyond s=9232.1 km in this case, because the ray path terminated at the receiver.

Let us take the values terminated at the receiver, and list them in Table 1. Strictly speaking, these values are not exactly the Lyapunov exponents conventionally defined, but for the limited ranges we considered, they should still provide some information on the exponential divergence or convergence of nearby ray trajectories. As seen from Table 1, corresponding to each ray, there is a positive exponent, indicating an exponential divergence of nearby ray paths. This is consistent with the two characteristics of chaos we have observed in Table 1 and 2.

However, the magnitude of these exponents cannot be used to determine the range scale on which the ray acoustics studied here become unpredictable. In fact, for the three rays with superscript *b* in Table 1, the exponents were smaller. Since these rays lose predictability, their exponents should be larger than other's, if the magnitude of these exponents is associated with predictability. The reason for this might be that we could not obtain Lyapunov exponents in the limit $s \to \infty$. In fact, for a realistic ocean acoustic problem, we are never able to estimate Lyapunov exponents defined by Eq. (11), because the range is always limited.

SUMMARY

The objective of this paper is to investigate whether ray paths in a realistic ocean-acoustic problem are chaotic. This was done by numerically integrating Munk *et al.*'s ray equations for long-rang propagation with measured sound speed data. Numerical results show that all the rays constructed were sensitive to initial condition (launch azimuth), and that some rays lose predictability. These are the characteristics of chaos. Corresponding to each chaotic ray, there was a positive Lyapunov exponent. We were unable to estimate Lyapunov exponents in the limit $s \to \infty$, therefore, the magnitude of these Lyapunov exponents cannot be used to quantify predictability. For predictability, no general conclusions can be made from this work, however, we have demonstrated that to produce a predicted receiver's location with error of 0.002-0.008°, we need to specify launch azimuth with 13-16 digits accuracy.

ACKNOWLEDGMENTS

We acknowledge funding from the Office of Naval Research (Grant N00014-95-1-0443). Part of this work was done at NOAA/Atlantic Oceanographic and Meteorological Laboratory, while J. Yan held a Resident Research Associateship of the National Research Council.

REFERENCES

1. Palmer, D. R., Brown, M. G., Tappert, F. D., and Bezdek, H. F., *Geophys. Res. Lett.* **15**, 569-572 (1988).
2. Brown, M. G., Tappert, F. D., and Goni, G. J., *Wave Motion* **14**, 93-99 (1991).
3. Smith, K. B., Brown, M. G., and Tappert, F. D., *J. Acoust. Soc. Am.* **91**, 1950-1959 (1992).
4. Smith, K. B., Brown, M. G., and Tappert, F. D., *J. Acoust. Soc. Am.* **91**, 1939-1949 (1992).
5. Yan, J., *J. Acoust. Soc. Am.* **94**, 2739-2745 (1993).
6. Baggeroer, A., and Munk, W., *Physics Today* **Sept.**, 22-30 (1992).
7. Palmer, D. R., Georges, T. M., Wilson, J. J., Weiner, L. D., Paisley, J. A., Mathiesen, R., Pleshek R. R., and Mabe, R. R., *J. Acoust. Soc. Am.* **90 Pt. 2**, 2330 (1991).

8. Munk, W. H., O'Reilly, W. C., and Reid, J. L., *J. Phys. Oceanogr.* **18**, 1876-1989 (1988).
9. Porter, M., and Reiss, E. L., *J. Acoust. Soc. Am.* **76**, 244-252 (1984).
10. Press, W. H., Flannery, B. P., Teukolsky, S. A., and Vetterling, W. T., *Numerical Recipes, The Art of Scientific Computing*, New York: Cambridge University Press, 1988, ch. 15, pp. 554.
11. Wolf, A., Swift, J. B., Swinney, H. L., and Vastano, J. A., *Physica* **16D**, 285-317 (1985).

Target Characterization Using Hidden Markov Models and Classifiers

David H. Kil
and
Frances B. Shin,

Lockheed Sanders, Inc.
Advanced Engineering & Technology Division
Signal Processing Center of Technology
PTP2-A001
P.O. Box 868
Nashua, NH 03061-0868

and

J. Robert Fricke

Department of Ocean Engineering
Massachusetts Institute of Technology
77 Massachusetts Avenue, Room 5-218
Cambridge, MA 02139

Abstract

We investigate various projection spaces and extract key parameters or features from each space to characterize low-frequency active (LFA) target returns in a low-dimensional space. The projection spaces encompass (1) time-embedded phase map, (2) segmented matched filter output, (3) various time-frequency distribution functions, such as Reduced Interference Distribution, to capture time-varying echo signatures, and (4) principal component inversion for signal cleaning and characterization. We utilize both dynamic and static features and parameterize them with a hybrid classification methodology consisting of hidden Markov models, classifiers, and data fusion. This clue identification and evaluation process is complemented by concurrent work on target physics to enhance our understanding of the target echo formation process. As a function of target aspect, we can observe (1) back scatter dominated by axial n=0 modes propagating back and forth along the length of the shell, (2) direct scatter from shell discontinuities, (3) helical or creeping waves from phase matching between the acoustic waves and membrane waves (both shear and compressional), and (4) the "array response" of the shell, with coherent superposition of elemental scattering sites along the shell leading to a peak response near broadside. As a function of target structures (the empty shell and the ribbed/complex shells), we see considerable complexity brought about by multiple reflections of the membrane waves between the rings. We show the merit of fusing parameters estimated from these projection spaces in characterizing LFA target returns using the MIT/NRL scaled model data. Our hybrid classifiers outperform the matched filter-based recognizer by an average of 5 to 25%. This improvement can be attributed to a combination of good features that maximize inter-class discrimination and appropriate classifier topologies that exploit the underlying multi-dimensional feature probability density function.

INTRODUCTION

Low-frequency active target echo characterization is of considerable interest from the perspective of target physics and signal processing because of complex, time-varying echo structures. In order to facilitate our understanding of target physics, the Massachusetts Institute of Technology (MIT) in collaboration with the Naval Research Laboratory (NRL) conducted a scaled model experiment in which monostatic and bistatic returns were recorded for three cylinder types along the 360 degree azimuthal sector. For $2 < ka < 10$, where k and a refer to the wavenumber and the radius of the cylinder, respectively, backscattered returns from finite cylinders consist of early and mid-to-late returns. The early returns are dominated by backscatter from target discontinuities for angles away from broadside and by the target "array response" at broadside generated by coherent superposition of elemental scattering along the length of the shell. The mid-to-late returns are dominated by supersonic helical or creeping waves produced by phase matching between the acoustic and membrane waves (both shear and compressional), and by slow flexural waves interacting with discontinuities in the shell. The echo return structure varies as a function of target type, aspect, ka, and time.

In order to take advantage of the variable echo structure, we investigate classifier topologies appropriate for echo characterization. Hidden Markov model (HMM) (1) is one of the most popular techniques to characterize time-varying patterns . The HMM models temporal variation in the feature space with a finite number of states, state transition probability matrices, observation probabilities for each state, and initial state occupancy probabilities. Features refer to parameters that capture essential target attributes useful for target characterization and discrimination. One drawback of the HMM-based recognition paradigm is that although the HMM maximizes the class likelihood ratio, it does not address the issue of inter-class discrimination, which is key to achieving good target recognition performance (2). On the other hand, both conventional and neural net classifiers emphasize discrimination when coupled with an appropriate feature optimization and rank order algorithm. However, they generally lack a mechanism to explicitly accommodate temporal variations.

Therefore, we develop a *reconfigurable classifier architecture* that combines the advantages of the HMM and classifiers. Our integrated target characterization paradigm can be succinctly described by:

1. low-dimensional data projection for feature extraction,
2. ranking of features and time segments in terms of their contribution to overall classification,
3. selection of an appropriate classifier architecture that is best mapped to the underlying multi-dimensional feature probability density function, and
4. recognition performance analysis in terms of rank order curves, confusion matrices, and classification receiver operating characteristic (ROC) curves.

Figure 1 depicts the integrated classification methodology that maximizes both discrimination and likelihood.

TARGET PHYSICS

In this paper we work exclusively with laboratory scale model backscatter data for finite cylindrical shells. To facilitate interpretation of features extracted by the low-dimensional classifiers, it is useful to have an underlying understanding of the target physics. In particular, features in the data can be attributed to specific physical mechanisms, which are linked to both the target type and the ensonification aspect. The next section briefly reviews the experimental design followed by a section reviewing the dominant signal physics and echo formation.

Experiment

The experimental data used in this paper was collected at NRL in collaboration

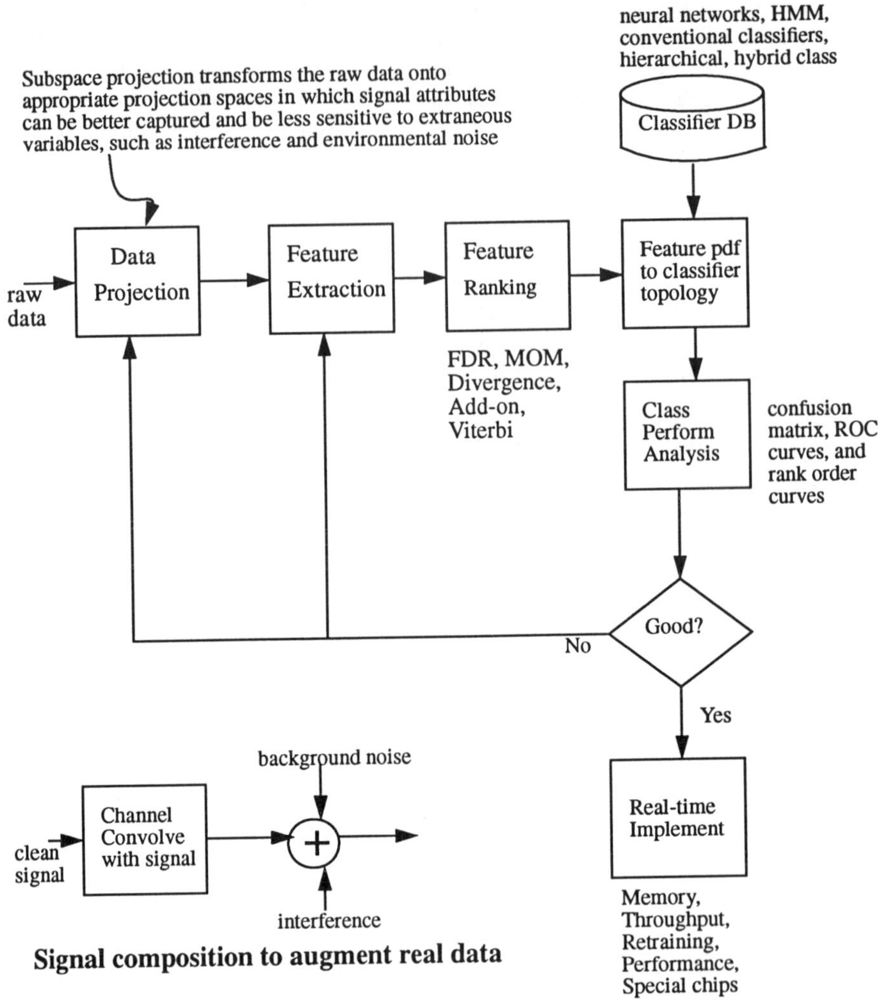

Figure 1: Our integrated classification paradigm consists of low-dimensional projection, feature optimization, matching the classifier topology to the underlying feature distribution, and performance analysis.

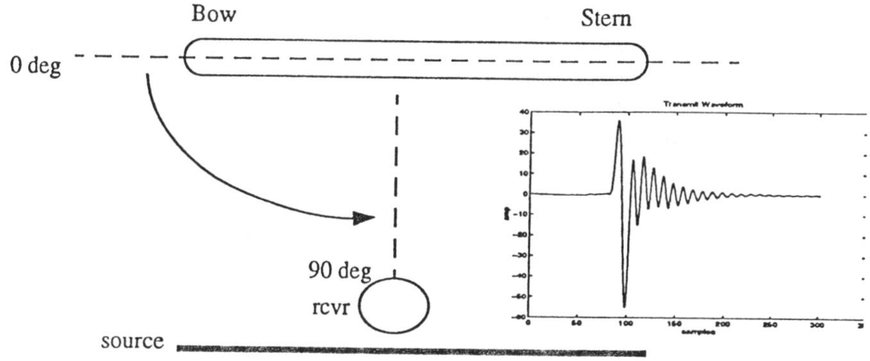

Figure 2: Experimental geometry & transmit waveform

with the ONR sponsored structural acoustics research program at MIT. Three finite cylinders were fabricated at varying levels of complexity. The external appearance of all the models is identical. They are 0.862 m in length with a diameter of 0.111 m. The skin is Ni-200 nickel with thickness of $0.532\text{x}10^{-3}$ m, which yields a 100:1 shell thickness to radius ratio. The shells differ in their internal configuration. The simplest has no internal structure, only the shell plating is present. The next level of complexity involves the placement of four unequally spaced deep-rings with an aggregate mass equal to that of the shell. The highest level of complexity is a shell with four suspended masses isolated from the rings with rubber bulkheads at each ring location and four delrin rods connecting the masses running the length of the shell. This model is not intended to mimic a full scale target, but rather to produce acoustic complexity comparable to that of a full scale target. The three models are known respectively as the "empty", "ribbed", and "complex" models. For the purposes of this paper scatter from the three models may be classified into two categories: (1) scatter from the empty model and (2) scatter from the ribbed and complex model. As may be inferred from this statement, the ribs play a dominant role in scattering process of the complex shell. Both monostatic and bistatic measurements were taken for each shell, but only the monostatic measurements are discussed in this paper. Further, since the ribbed and complex cylinders are so similar, of the two only the ribbed data will be discussed in comparison with the empty shell data.

The models were placed in an acoustic underwater test facility at NRL and ensonified by a plane wave source as shown in Figure 2. For the monostatic measurements the source array and single receiver hydrophone remained stationary. The models were rotated through a range of 360° in 1° steps. Since the problem is approximately quadrant symmetric only azimuth angles from 0 degrees (bow) to 90 degrees (beam) are considered here. This range shows the salient properties of scattering from finite cylinders. At each angle 100 pings were averaged to compute the backscattered return from the target. The source waveform was a wide-band pulse with useful energy in the range of 10-50 kHz, which corresponds to $2 < ka < 10$. To minimize the effects of clutter from the measurement tank a clutter subtraction process was used to clean the data. This process consisted of measuring the tank response with no target in place and subtracting this response from that measured with the target present. Details of the acquisition and initial data analysis are discussed by Corrado (3).

This paper presents an alternate analysis approach to the same data set, as is described below. To understand the analysis, however, we must first consider some of the fundamental physical processes associated with the target scattering. Much of this understanding is due to Corrado's work.

Physical Interpretation

The data may be divided into two major classification groupings as shown in Figure 3.

Namely,

1. time zones: early, mid, and late
2. azimuthal zones:
   (a) $\pm 20^o$ bow/stern
   (b) $20^o < \theta < 60^o$ low return sector
   (c) $60^o < \theta < 85^o$ helical wave sector
   (d) $85^o < \theta < 90^o$ beam sector

Of these 12 regions only six are of importance for this paper: all the early time regions, the mid-time membrane zone region, and the late-time bow zone region. Each of these regions is dominated by one of several physical processes, which are, in large part, individually determined by a given wave type: compressional, shear, and flexural. For the ribbed case, the boundary between the low-return zone and the helical zone is not distinct due to scattering from the rings.

In order to facilitate understanding of target echo formation process, we used the Reduced Interference Distribution (RID) (4) to project raw time series onto a high resolution time-frequency map as shown in Figure 4. The RID is particularly useful when the signal consists of a number of overlapping components in time and frequency which can cause "cross-term" interferences.

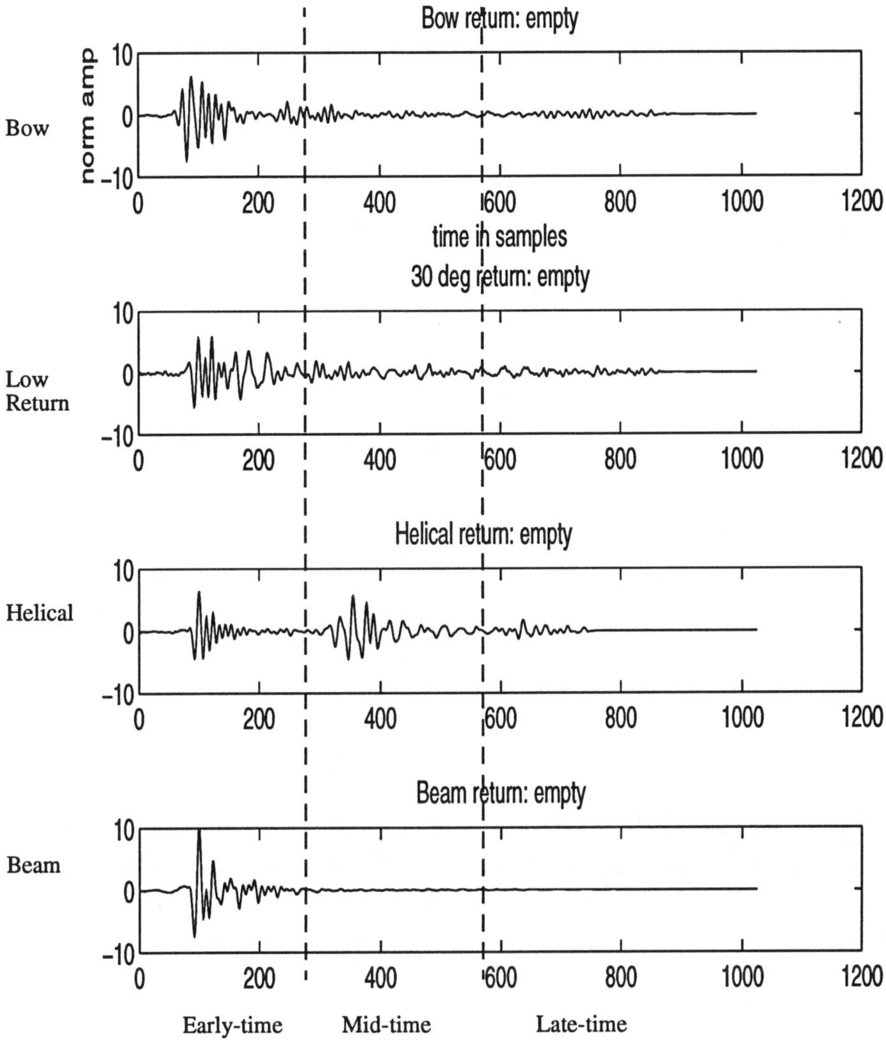

Figure 3: Echo groupings as a function of time and aspect for LFA return characterization. The sampling period is 4 microsec.

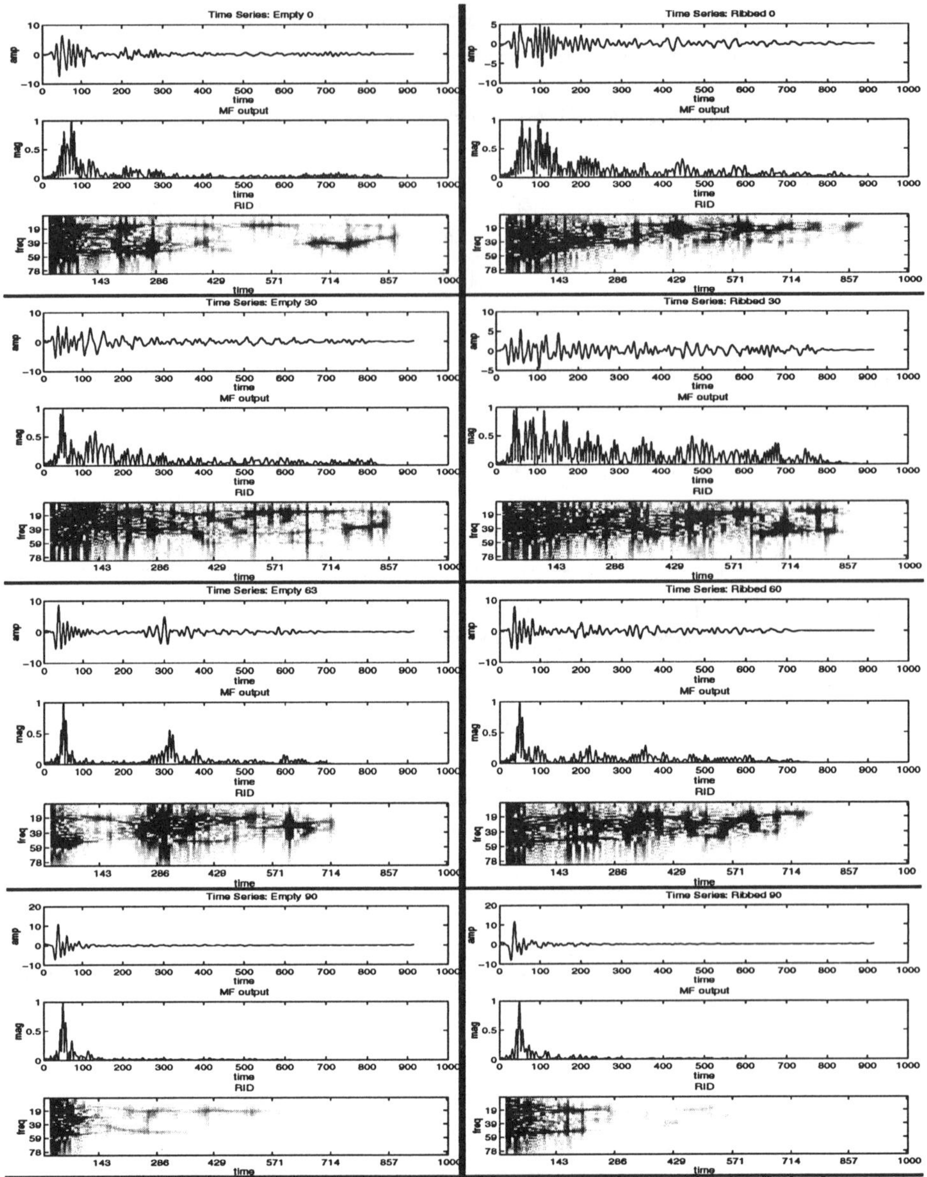

Figure 4: The RID spectrograms provide good time and frequency resolution without suffering from cross-interference terms. Time-series and matched filter outputs are also shown for comparison. The left plots show empty target echos with 0, 30, 60, and 90 deg aspect groups from top to bottom while the right plots show ribbed target returns in the same aspect group order. The time axis is in samples where the sampling period is 4 microsec. The frequency axis is in kHz covering ka of 0 to 17.

Let us first consider the four major azimuthal divisions, which are approximately quadrant symmetric. At bow incidence ($\pm 20^o$) n=0 modes dominate. The ability to excite higher circumferential modes is low due to the plane wave excitation which produces axisymmetric pressures around the shell. These n=0 modes propagate back and forth down the hull as though it were a transmission line (5) scattering from discontinuities as it goes. Next consider the helical zone, which is so called because here the incident acoustic wave is phase, or trace, matched to both compressional and shear helical waves. These waves are supersonic and have an axial component and a circumferential component, which results in a winding helical propagation path down the shell. The circumferential mode number for helical waves is greater than zero due to the unsymmetric forcing of the shell by the incident wave at these angles. Now, the low-return zone is simply an area where the axisymmetric forcing of n=0 modes is weak and trace matching to helical wave cannot take place due to geometric considerations. The main return here is direct scatter from the target discontinuities. Finally, the beam zone is dominated by direct backscatter from the entire shell length due to the coherent superposition of elemental scattering centers. These returns may be thought of as an "array response" for the shell, and the monostatic return is simply the peak response of the beam pattern near broadside.

Now consider the temporal divisions of the data. At early time the dominant backscatter is due to direct acoustic scatter from discontinuities in the shell including endcaps, slope discontinuities, and rings. At each discontinuity every wave type scatters into every other wave type to varying degrees. As an example, at bow incidence the earliest return is direct scatter off the endcap followed about 200 μs later by a return, which derives from direct excitation of a flexural wave in the endcap that then backscatters an acoustic wave when it encounters the slope discontinuity between the endcap and the cylindrical shell (6). At mid-time, we continue to see direct scatter from the ribs and far endcap over a broad range of angles. The more interesting feature, however, is the formation of the helical waves seen in a $\pm 30^o$ sector with respect to beam for the empty shell and $\pm 45^o$ for the ribbed shell. The difference between the shells is due to multiple scattering of the helical wave between the rings as it winds down the shell. All late time events are almost certainly due to flexural waves propagating slowly down the shell scattering into acoustic energy at discontinuities. Since the flexural wave is dispersive, we expect to see frequency down swept chirps for the late returns. For the empty shell, which produces particularly simple signal structure, the down-chirps are relatively distinct.

Comparing the empty and ring stiffened shell responses one sees considerable complexity brought about by multiple reflections of the helical waves between the rings. Early time response is due to compressional and shear waves in the skin scattering from the rings and the endcaps. Late time response derives from flexural waves slowly lumbering their way down the shell and scattering through mode conversion at the rings and endcaps. At bow incidence, one can see the result of multiple reflections bouncing back and forth in the first bay of the shell. In contrast, the empty shell response is dominated by scattering from the endcaps with a longer delay time between bounces. The observed acoustic energy derives both from direct scatter of compressional and shear waves and from mode conversion of flexural waves. For the ribbed shell the late time events between 30° and 60° are especially hard to interpret, since their nature is deterministic but very complicated in terms of multipath and mode conversion at discontinuities. The 0° and 90° cases are end members, and "relatively" simple.

PROJECTION SPACE INVESTIGATION

The conceptual framework of low-dimensional projection is based on data compression which uses as few bits of information as possible to convey the entire message with the lowest possible bit error rate. One simple example is to use an oversampled Gabor representation to characterize exponentially damped sinusoids. Advantages of extracting features from the low-dimensional projection space encompass (1) robustness to extraneous variables such as noise due to energy compaction, (2) computational efficiency in classification due to inherent data compression, and (3) facilitation of feature discriminant analysis.

After our thorough investigation of target physics, we explore the following four projection spaces for target characterization. We select them because they provide an accurate time "snapshot" of spectral contents. Furthermore, they perform adaptive smoothing and/or filtering to remove as much out-of-band noise as possible. In short, these projection spaces provide a temporal map where appropriate clues or features can be extracted at each time snapshot after noise filtering. The goal at this stage is to extract as many pertinent features as possible from each projection space for later feature optimization and fusion.

Reduced Interference Distribution (RID) Spectrogram

The RID allows us to construct a "time-frequency acoustic signature" of an echo. Instead of using the entire RID spectrogram, we implement an innovative image compression algorithm to extract features from the RID spectrogram. The image compression algorithm consists of two-dimensional transform for further coefficient compaction, transform coefficient encoding, encoded transform coefficient compression with the singular value decomposition (SVD) to overcome the "curse of dimensionality", vector quantization (VQ), and entropy or arithmetic encoding of the VQ codebook indexes (14). We perform the coefficient compression by using the signal subspace eigenvectors where we use the minimum description length (MDL) criterion (7) to determine the rank of the covariance matrix. Again a combination of transform and SVD-based subspace filtering results in data compaction and improved SNR for robust target characterization.

Segmented Matched Filter Output

We divide the matched filter output into a number of equal time segments. From each time segment, we extract shape and amplitude statistics (i.e., mean, standard deviation, skewness, and kurtosis). We also extract the same statistical parameters from the difference between the matched filter output and the raw energy detector output. This is done to evaluate the discrepancy between the correlated and uncorrelated components of the return energy with the transmit waveform, especially in the late arrival segment.

Principal Component Inversion (PCI) Output

The PCI methodology works as an adaptive Wiener filter in that it estimates the time-varying signal structure and utilizes the SVD to separate the data into signal subspace and alternate subspace components (8). From the signal subspace component, we can estimate the clean signal structure by using diagonal averaging. From each PCI output, we extract center frequency, bandwidth, linear predictive coding (LPC) coefficients, cepstral coefficients, delta-cepstral coefficients, state transition parameters, singular value distribution, and low-rank dimension. Figure 5 illustrates the noise reduction performance improvement with PCI in comparison to a conventional filtering matched to the transmit pulse bandwidth.

Compressed Phase Map with the SVD

In nonlinear dynamical modeling of chaotic structures, the time-embedded representation or phase map is often used. An embedding dimension of two is sufficient to characterize the Henon noise whose dynamics are governed by a second order differential equation (9). However, for the MIT/NRL tank data, we do not observe such a low-order, deterministic structure in the phase map. Therefore, we use an embedding dimension of 32, but perform data compression using the SVD so that only the first eight principal components are used for data analysis.

INTEGRATED TARGET CHARACTERIZATION PARADIGM

After feature extraction, we perform thorough feature discriminant analysis followed by target characterization performance assessment. For this analysis, we generate eight classes: two target types (empty and ribbed) and four aspect groups (near 0°, 30°, 60°, and 90°). Our goal is to find a good feature subset and an appropriate classifier topology matched to the underlying good feature distribution.

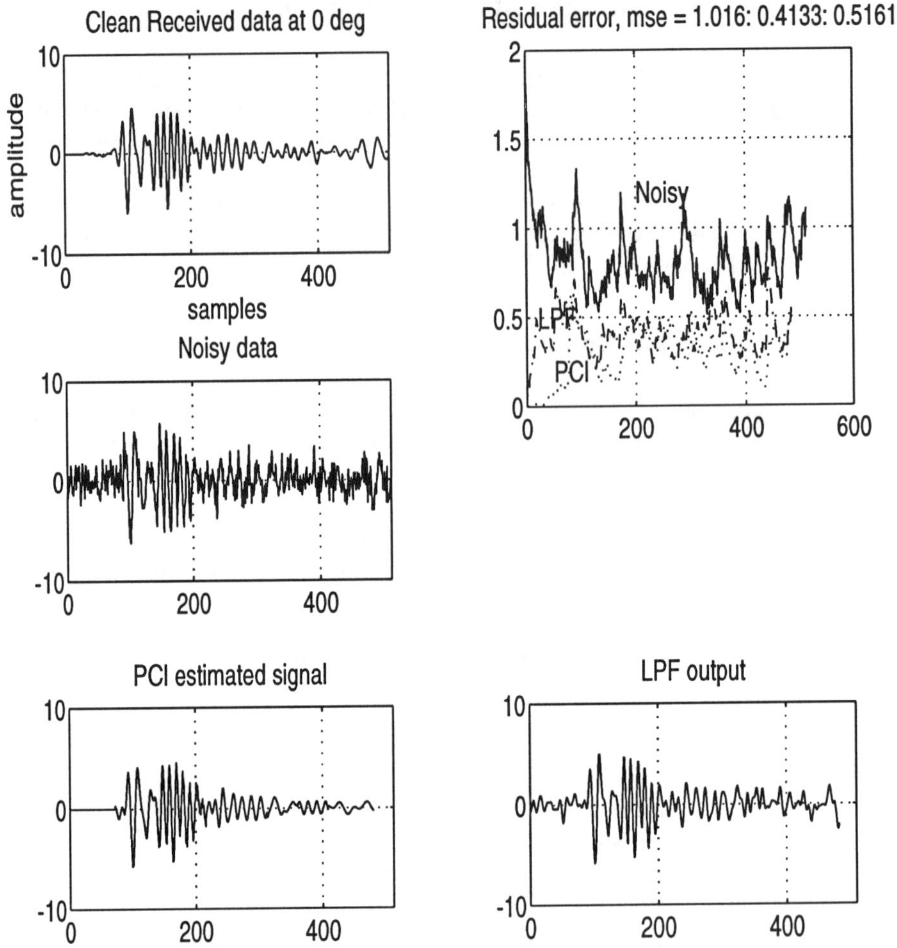

Figure 5: PCI performs time-adaptive filtering and yields better noise rejection performance than the low-pass filtering.

Feature Discriminant Analysis

In order to maximize class separability for robust recognition performance, we evaluate features in terms of class discrimination and the degree of feature correlation. Feature analysis algorithms consist of Fisher's discriminant ratio (FDR), Procrustes angle, multi-modal overlap measure (MOM), divergence, add-on/knock-out, Viterbi, and projection-based discriminant ratio tests. While one dimensional feature optimization algorithms, such as FDR, MOM, divergence, and Procrustes angle, are fast, they fail to take feature correlation into consideration. Further, when single dimensional feature distributions exhibit a high degree of overlap among classes, the one dimensional feature optimization algorithms yield inferior performance to that of multi-dimensional feature optimization algorithms. Unfortunately, the add-on or Viterbi algorithms are computationally expensive, especially for a large number of training tokens. In an attempt to combine the strengths of single and multi-dimensional feature ranking algorithms, projection-based algorithms perform feature compression followed by discriminant analysis in the compressed or reduced feature space.

Linear Fisher's Discriminant Ratio (FDR)

The FDR is a statistical rank order method which determines a feature priority by computing a detection index, $\frac{\Delta\mu}{\sigma}$, between classes where $\Delta\mu$ and σ refer to the mean difference between any pair of classes and feature standard deviation, respectively. Generally, FDR is ideal for unimodally distributed, Gaussian features and can be computed as follows:

$$FDR1(p) = \sum_{i=1}^{K}\sum_{i\neq j}^{K} \frac{\left(\mu_p^i - \mu_p^j\right)^2}{\left(\sigma_p^i\right)^2 + \left(\sigma_p^j\right)^2}, \tag{1}$$

$$FDR2(p) = \frac{\sum_{i=1}^{K}\sum_{j=1}^{K}\left(\mu_p^i - \mu_p^j\right)^2}{\sum_{i=1}^{K}\left(\sigma_p^i\right)^2}, \tag{2}$$

where

$$\begin{aligned} K &= \text{the number of classes,} \\ p &= \text{feature index,} \\ \mu_p^i &= \text{mean of the } p^{th} \text{ feature for the } i^{th} \text{ class, and} \\ \sigma_p^i &= \text{standard deviation of the } p^{th} \text{ feature for the } i^{th} \text{ class.} \end{aligned} \tag{3}$$

The main difference between FDR1 and FDR2 is that FDR1 tends to emphasize separation between any two classes while FDR2 averages over all the classes. Therefore, FDR2 is more appropriate for rank ordering features with more than two classes.

For problems involving a large number of classes, instead of summing detection indexes over all the classes, we can use the worst-case detection index to rank order individual features for more robust recognition performance. This concept of making the worst-case performance as favorable as possible is the backbone of the minimax algorithm.

Procrustes Angle

Procrustes angle (10,11) is closely related to the least squares approximation and measures the relationship between two given subspaces. Since the vector subspace defined by the eigenvectors corresponding to the significant eigenvalues of the Fisher covariance matrix is optimal in the least squares sense, it is intuitive that the angle between the kth feature and its orthogonal projection onto the Fisher projection subspace should be small for good features, and large for less useful ones. This formulation is conceptually similar to linear Fisher's discriminant analysis.

Multi-modal Overlap Measure

MOM is appropriate for features which exhibit multi-modal and non-Gaussian probability density functions. Feature rank is determined by integrating the area of overlap between class pdfs. As expected, features with the least degree of overlap are assigned the highest ranks.

Another discriminant measure based on the estimated feature pdfs is often referred to as divergence and can be computed as follows (12):

$$D_{ij}(k) = I_k(i,j) + I_k(j,i) \tag{4}$$

where

$$\begin{aligned} I_k(i,j) &= \int_x p_i^k(x) \ln\left(\frac{p_i^k(x)}{p_j^k(x)}\right) dx, \qquad (5) \\ k &= \text{feature index}, \\ p_i^k(x) &= \text{pdf of class i for feature k, and} \\ p_j^k(x) &= \text{pdf of class j for feature k.} \end{aligned}$$

For multi-class problems, following the minimax approach, we can attempt to maximize the worst-case performance by rank ordering features based on the minimum value of $D_{ij}(k)$ over i and j instead of summing $D_{ij}(k)$ over i and j (i.e., $D(k) = \sum_{i<j} \sum_j D_{ij}(k)$) (2). Although a little pessimistic in its philosophy, this minimax or "maximin" approach can yield robust recognition performance under certain situations.

For difficult problems with very complex class boundary functions and a substantial amount of overlap in the single dimensional feature space, it is advantageous to perform multi-dimensional feature optimization. Conceptually similar to its original use in convolutional encoding and decoding in communication, Viterbi algorithm (13) or dynamic programming considers many subsets in parallel to find the best M feature subset out of N candidate features ($M \leq N$). If the performance measure increases monotonically as a function of the feature subset size and the performance at any stage is a function of the previous feature subset and the current feature (i.e., Markov property), then this process will result in the optimum M feature subset.

The Viterbi rank order procedure is summarized below.

1. Evaluate the performance of N subsets, each consisting of one feature.
2. For each subset of one feature, append one of the remaining N-1 features, evaluate the performance of N-1 two-feature subsets, and select the two-feature subset that yields the best performance.
3. Now for each subset of two features, append one of the remaining N-2 features and select the three-feature subset that yields the best performance. Repeat the same procedure until each subset contains M features or performance degradation occurs.
4. Select the path that yields the best performance. Features that fall into the optimal path constitute the feature subset to be used for classification.

Although dynamic programming is more computationally tractable than the exhaustive search method, it is still very time-consuming. As a consequence, for all practical problems, we resort to suboptimal add-on or knock-out algorithms to find a "reasonably" good feature subset. The only difference between add-on/knock-out and Viterbi is that the former considers only one best path at any stage, thereby saving computational loading by a factor of N. Based on our extensive classification experiences, the performance difference between add-on and Viterbi is approximately 0 to 4 %.

In general, multi-dimensional feature optimization tends to yield optimistic recognition performance. The optimal feature subset composition is likely to change from run to run, provided that random cross validation is performed to independently assess the recognition performance. This means that the feature subset composition may not remain fixed during random cross validation. That is, the recognition performance averaged over multiple random runs may be based on different feature subsets. Therefore, the classification performance based on the Viterbi and add-on algorithms provides the theoretically attainable upper bound (i.e., similar to Cramer-Rao Lower Bounds).

Classifier Architecture

Conventional and neural network classifiers have been used extensively in pattern recognition. They can be divided into parametric, non-parametric, and boundary decision classifiers. Parametric classifiers, such as a multi-variate Gaussian classifier (MVG), makes a certain statistical assumption regarding the underlying feature pdf, resulting in recognition performance that depends on the goodness of the statistical fit. Non-parametric classifiers, such as a binary tree classifier, make no such assumption and generally require a large number of training tokens. Boundary decision classifiers, such as a backpropagation neural net (BPN), attempt to find a class boundary function that best separates classes based on some error criteria and typically suffer from long training time and possible convergence to one of the local minima. As a result of these differences among classifiers, it is crucial that we perform thorough feature discriminant analysis and match the classifier architecture to the underlying feature pdf.

HMMs are popular in modeling temporal variability. Feature tokens are computed at an appropriate frame rate. The HMM models feature variation in time by assigning tokens with similar statistical characteristics to a common state. Within each state, the observation probability of all the tokens is usually based on a Gaussian mixture model with a diagonal or full covariance matrix. In speech modeling, the left-to-right state model is widely used to closely follow speech articulation. In nonlinear dynamical modeling, each state typically represents a cluster in the time-embedded phase map. The HMM model parameters, consisting of initial state occupancy, state transition, and observation probabilities, are estimated using either segmental k-means or forward-backward algorithms. Since the transition matrix is full (i.e., no zero elements), we will denote such HMMs "ergodic".

Although the HMM is quite useful in statistical characterization of time-varying dynamic patterns, one potential drawback of the HMM is that it tends to maximize the likelihood of the correct class, but does not suppress the likelihood of other incorrect classes. That is, the HMM does not address an important issue of discrimination and robustness which are key to achieving good classification performance. Therefore, it makes sense to emphasize class discrimination during training. In order to design a classifier topology that combines the merits of classifiers with good class discrimination and HMMs with good temporal variability characterization, we present the three reconfigurable classifier architectures for the LFA target echo characterization as depicted in Figure 6.

Left-to-right HMM

This is the classic HMM used in speech recognition. The one exception in our implementation is that instead of using all the features in the likelihood ratio computation, we use only those that provide good class discrimination. Feature ranking is performed as a function of time (i.e., state in this case). Since the HMM generally

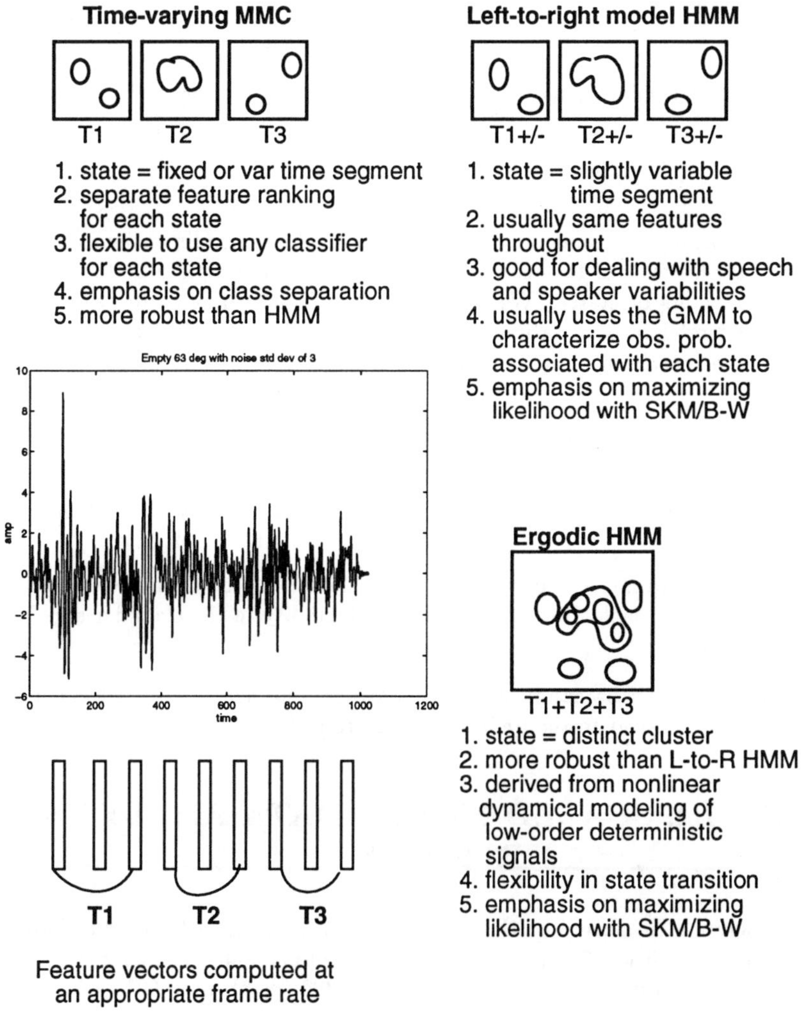

Figure 6: The three classifier topologies attempt to maximize both discrimination and likelihood by taking advantage of feature optimization and by accommodating temporal variability.

uses the same feature subset for the log-likelihood ratio (LLR) score computation, feature ranking is done globally over the entire observation period. Furthermore, we identify and exclude time segments that add to confusion. Within each state, we use the Gaussian mixture model for characterizing observation probabilities.

Ergodic HMM

After feature optimization and ranking, we populate the multi-dimensional feature space with all the feature tokens from each class. Now we perform the VQ to find distinct clusters or states. We can trade-off the number of states versus the complexity of modeling observation probability for each state. Unlike the left-to-right HMM, the transition probability matrix of this HMM is usually full.

Temporally Adaptive Classifier

The structure of this classifier is similar to the left-to-right HMM with the following two exceptions:

1. Features are optimized separately for each state (i.e., local feature optimization). This approach allows the maximum flexibility in the classifier and feature architecture.

2. Any classifier can be assigned to any state as long as the selected classifier provides the best fit to the underlying feature pdf.

MIT/NRL SCALED MODEL DATA ANALYSIS RESULTS

Due to the symmetric nature of the target echo structure, we focus our characterization efforts on 0 to 90 degree aspect for the empty and ribbed targets. We subdivided the quadrant into near 0° ($\pm 5^o$), near 30° ($\pm 5^o$), near 60° ($\pm 5^o$), and near 90° ($\pm 5^o$). Therefore, the task of a recognizer is to determine the aspect and target type of an echo corrupted in white Gaussian noise.

The training data consists of a clean data with the signal-to-reverberation ratio of at least 10 dB. We normalize the clean data so that its mean and standard deviation are 0 and 1, respectively. For testing at various SNR's, we corrupt the clean data with independent white Gaussian noise.

Recognition Performance With Matched Filters

At first, we pose the following question: what if we use the received target echo as a matched filter? By contrast, the matched filter projection space for feature extraction utilizes the transmit waveform, not the received waveform. In essence, classifiers use templates or features of known classes for pattern matching. Figure 7 shows confusion matrices along with the overall, and individual class recognition performances as a function of the noise standard deviation.

Confusion matrix, as the name implies, is a measure of how classifiers respond given an input signal with unknown identity. It can be generated by inputing a large number of test feature vectors and by grading the classifier outputs against the known ground truth. The correct recognition values can be read from the diagonal elements of the confusion matrix and off-diagonal elements indicate the degree of confusion with incorrect classes. Most confusion occurs for the empty 60° and empty/ribbed 90° classes because of a good deal of variability as a function of aspect and the lack of any late return structure, respectively. Our next task is to evaluate the three classifier candidates to determine if they can outperform the matched filter by a judicious combination of good features and appropriate classifier architecture.

Multi-dimensional Feature Distribution

After feature extraction and ranking, we look into the multi-dimensional feature pdf to derive the matching classifier architecture. Figure 8 illustrates the compressed feature scatter plot as a function of time. Note the highly non-Gaussian and multi-modal feature distribution which will make the MVG a poor choice for this problem. Moreover, there is a considerable amount of temporal variation in the feature pdf. This figure illustrates the importance of using the right classifier architecture that takes advantage of the time-varying feature pdf.

Feature Rank Order Curves

Rank order curves are quite useful in determining an appropriate feature dimension for classification. Initially, as we add good features, the recognition performance increases. It reaches a plateau after a while and may even degrade as we increase the feature dimension beyond what is necessary. In short, the rank order curves are a useful tool to detect the occurrence of underfitting (i.e., using less features than necessary) or overfitting (i.e., using too many features). Figure 9 illustrates the rank order curve for the LFA target characterization.

Recognition Performance Comparison

We initially train the three classifiers with clean data. Next we test them with noise corrupted data. Figure 10 illustrates their recognition performance as a function of noise standard deviation.

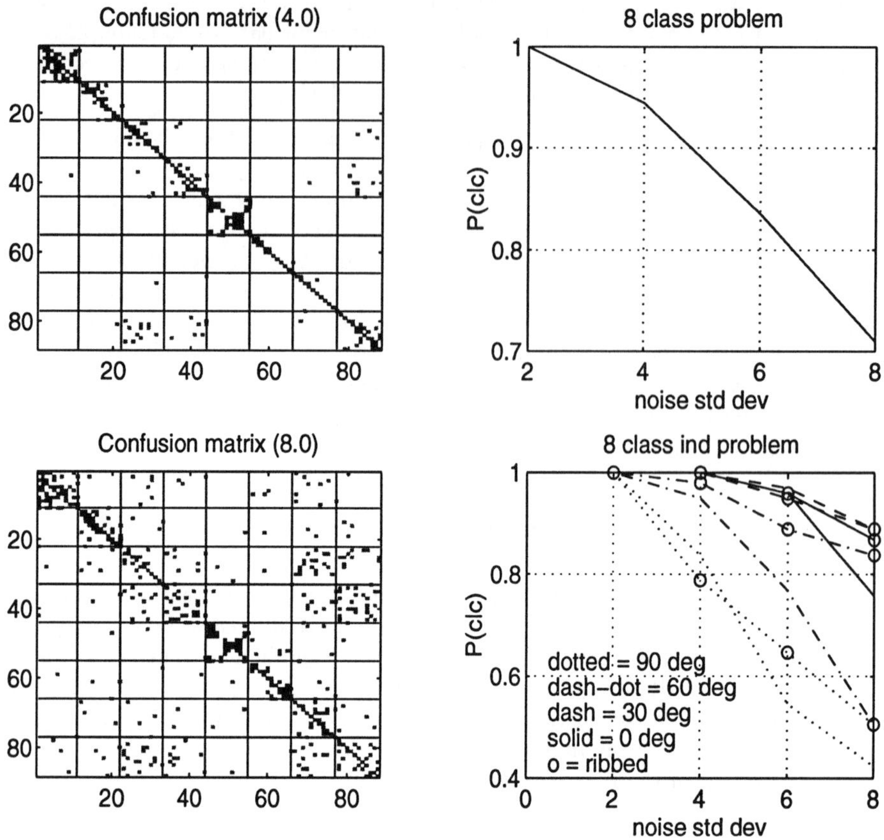

Figure 7: Recognition performance achieved by using the received waveforms as matched filters for aspect and target type variation. The 8-by-8 confusion matrices show the degree of confusion with other classes. The numbers in the parenthesis represent the noise standard deviation. The order of classes from top to bottom and left to right is as follows: empty0, empty30, empty60, empty90, ribbed0, ribbed30, ribbed60, and ribbed90. The y-axis represents the true class while the x-axis shows the classifier output.

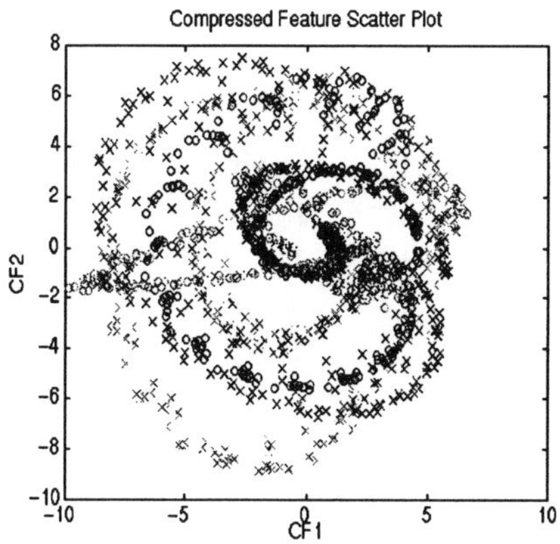

at T0

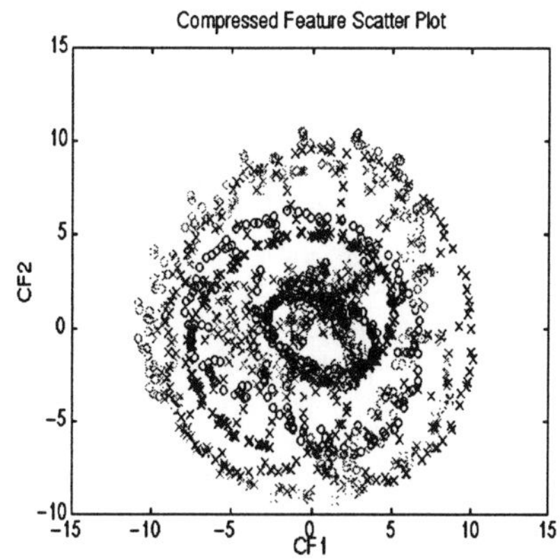

at T1

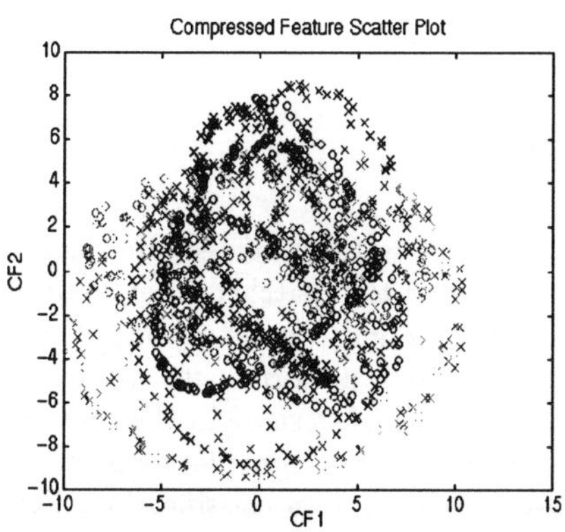

at T2

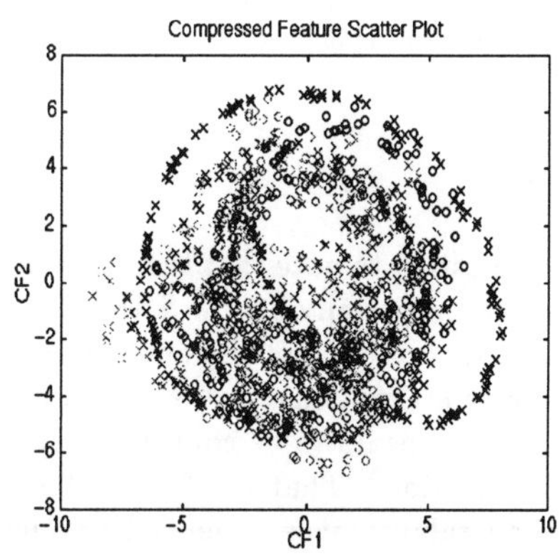

at T3

Figure 8: Compressed feature pdf plots show a highly non-Gaussian, multi-modal, and time-varying structure.

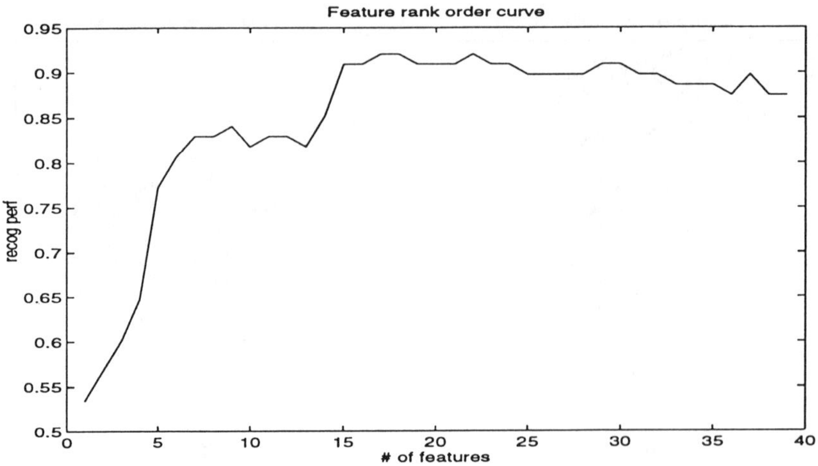

Figure 9: Rank order curves are useful in determining an appropriate number of features for maximum and robust recognition performance.

DISCUSSION

It is interesting to note that the three classifiers outperform the matched filter-based recognizer, especially at low SNR. Initially, we expected the matched filter to provide the theoretical upper bounds (i.e., similar to the Cramer-Rao lower bounds on parameter estimation) on the LFA target characterization performance. Although this result appears to be contradictory at first glance, it makes sense in the context of classification. That is, if some temporal segments do not provide good inter-class separability, it is beneficial to remove those segments from classification. For detection, such a strategy will yield suboptimal detection performance.

Furthermore, it is worthwhile to investigate as many pertinent projection spaces as possible, provided that each projection space yield orthogonal features. For instance, we achieved 69.3 %, 76.1 %, and 46.6 % recognition performance from RID, matched filter (transmit waveform), and PCI-derived features, respectively. When we fused all the good features, our recognition performance improved to 92.1 %. Due to the broad-band nature of the transmit pulse, the PCI-derived features do not work as well as those derived from the other projection spaces. PCI is more appropriate when the transmit waveform exhibits a time-varying narrow-band structure (i.e, LFM, HFM, or FSK).

In order to demonstrate the importance of matching the classifier architecture to the underlying feature pdfs, we replace the multi-modal classifier (MMC) with the MVG classifier in the temporally adaptive classifier topology. Figure 11 illustrates a dramatic performance difference between the two classifiers. The left plots show the individual and overall target characterization performance as a function of time.

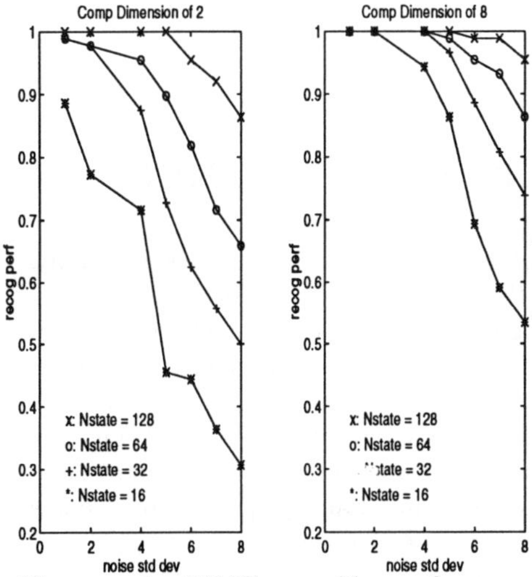

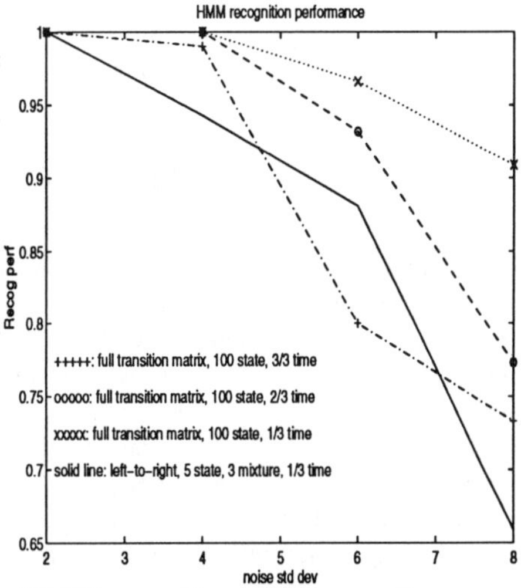

Time-varying MMC recognition performance with a different number of features as a function of the number of states and noise standard deviation.

HMM recognition performance after feature optimization as a function of the time segments and noise standard deviation.

Performance Summary of the Above

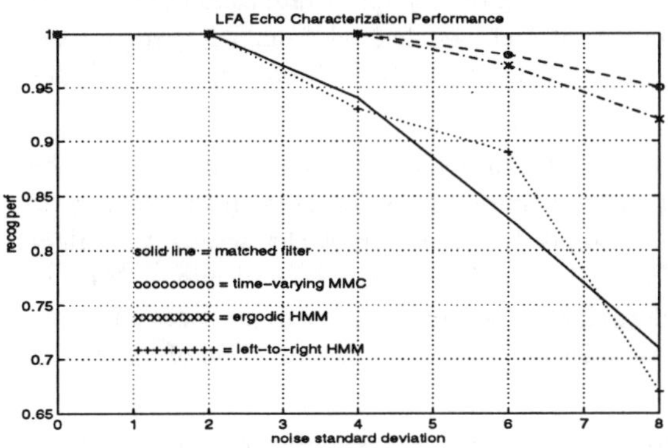

The key to achieving robust target characterization performance:
(a) energy compaction via low dimensional projection,
(b) thorough feature optimization,
(c) matching the classifier topology,
(d) and temporally adaptive LLR score integration.

Figure 10: Target characterization performance as a function of noise standard deviation, classifier topology, and feature.

The y-axis represents the recognition performance with an offset of 1.0 added for each case. That is, the overall recognition performance with the "+" legend should read 0.0 to 1.0 because of the offset of 4.0.

CONCLUSION & FUTURE DIRECTION

In this paper, we demonstrated the crucial link between target physics and signal processing with the scaled model data for the two cylinder types at the four aspect groups. We also developed the three recognizer topologies that exploited the underlying time-varying feature distributions. We also discussed LFA target echo formation process and the integrated classification paradigm that exploits an inherent relationship between features and classifiers. In short, a combination of (1) energy compaction via low-dimensional projection, (2) thorough feature discriminant analysis as a function of time, and (3) appropriate classifier topology generation is a key to achieving robust active target characterization performance.

With the noise standard deviation of 6, the matched filter-based classifier yields 83.5 % correct recognition performance of all eight classes while the time-varying MMC, left-to-right HMM, and ergodic HMM are able to achieve 98 %, 89 %, and 97 % correct echo characterization, respectively. This improved performance is attributable to feature optimization and selection of an appropriate classifier topology for this problem. Furthermore, fusion of features derived from RID, segmented matched filter, and PCI projection spaces results in 16 % improvement in target characterization performance.

Since we demonstrated an excellent target characterization performance with the scaled model tank data, the next natural extension is for more realistic targets. The first area of future research is to apply the same target characterization algorithms to real world threats, such as scaled model submarines and mines. We can characterize target recognition performance in terms of confusion matrices and classification receiver operating characteristics (ROC) curves as a function of frequency, bandwidth, aspect, and time.

Furthermore, it is crucial that we investigate the impacts of confusion factors, such as ambient noise, clutter, and rapidly fluctuating channel responses in shallow water, on the target characterization performance. We are currently looking into blind deconvolution, hypothesis-directed matched field processing for depth and channel estimation, and channel deconvolution using probe pulses as a potential means of deconvolving the medium effects out of the received waveform. The key to successful target characterization is to remove as much confusion as possible prior to feature extraction and classification.

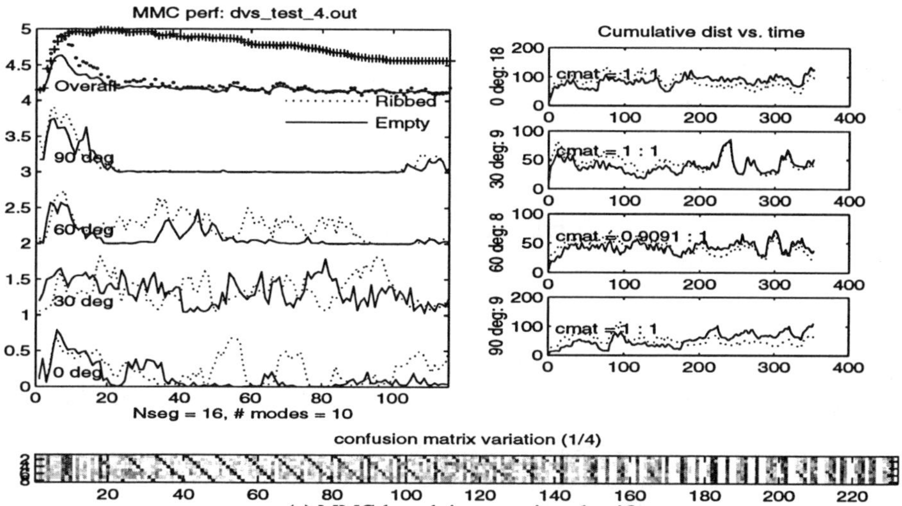

(a) MMC-based time-varying classifier

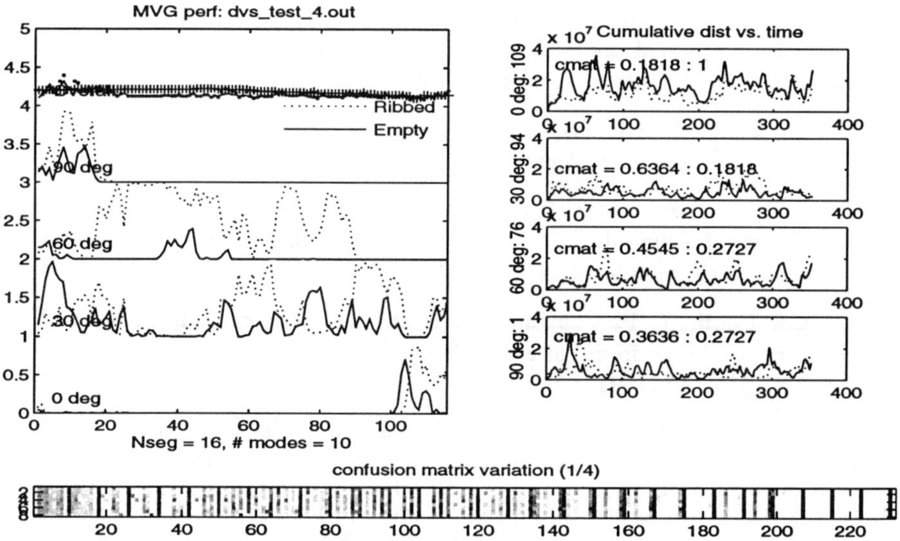

(b) MVG-based time-varying classifier

Figure 11: This example illustrates the importance of using the appropriate time segments that offer good class discriminability and the right classifier architecture that matches the underlying feature pdf. At low SNR, the late returns are highly corrupted by noise, thereby rendering them less than useful for discrimination. Furthermore, since the features exhibit non-Gaussian and multi-modal characteristics, MMC outperforms MVG by a wide margin.

ACKNOWLEDGEMENTS

This work was supported in part by the Office of Naval Research under contract N00014-91-C-0125. The technical reviewer for this program was Dr. Kam Ng of ONR, Code 122.

REFERENCES

[1] L.R. Rabiner, "A Tutorial on Hidden Markov Models and Selected Applications in Speech Recognition", *Proceedings of the IEEE*, vol.77, no.2, February 1989.

[2] K.-Y. Su and C.-H. Lee, "Speech Recognition Using Weighted HMM and Subspace Projection Approaches", *IEEE Trans. on Speech and Audio Processing*, Vol.2, No.1, Part 1, January 1994.

[3] Charles N. Corrado Jr., Mid-Frequency Acoustic Backscattering from Finite Cylindrical Shells and the Influence of Helical Membrane Waves, Ph.D. Thesis, Massachusetts Institute of Technology (1993).

[4] J. Jeong and W.J. Williams, "Kernel Design for Reduced Interference Distributions", *IEEE Trans. on Signal Processing*, Vol.40, No.2, February 1992.

[5] Joseph E. Bondaryk, Array Processing and Forward Modeling Methods for the Analysis of Stiffened, Fluid-Loaded Cylindrical Shells, Ph.D. Thesis, Massachusetts Institute of Technology (1994).

[6] Matthew Conti, Y. P. Guo, and Ira Dyer, "Influence of endcaps on mid-frequency scattering from finite cylinder shells at axial incidence", *J. Acoust. Soc. Am.*, 93, 2413 (1993).

[7] M. Wax and I. Ziskind, "Detection of the Number of Coherent Signals by the MDL Principle", *IEEE Trans. on ASSP*, Vol. 37, No. 8, August 1989.

[8] D.W. Tufts, D.H. Kil, and R.R. Slater, "Reverberation Suppression and Modeling", Ocean Reverberation edited by D.D. Ellis, J.R. Preston, and H.G. Urban, *Kluwer Academic Publishers*, Boston, MA, 1993.

[9] C.S. Myers, A. Singer, F. Shin, and E. Church, "Modeling Chaotic Systems with Hidden Markov Models", *Proceedings of ICAASP*, 1992.

[10] G.H. Golub and C.E. Van Loan, Matrix Computations, *Johns Hopkins Press*, Baltimore, MD, 1989.

[11] T.E. Luginbuhl, S.G. Greineder, and R.L. Streit, "Feature Set Reduction Algorithms for Probabilistic Neural Networks", *U.S. Navy Journal of Underwater Acoustics*, Vol. 43, No.2, April 1993.

[12] P.A. Devijver and J. Kittler, Pattern Recognition: A Statistical Approach, *London: Prentice-Hall*, 1982.

[13] A.J. Viterbi, "Error Bounds for Convolutional Codes and Asymptotically Optimum Decoding Algorithm", *IEEE Trans. on Information Theory*, Vol. IT-13, April 1967.

[14] D.H. Kil and F.B. Shin, "Reduced Dimension Image Compression for Remotely Distributed Underwater Signal Processing", submitted to the *1995 International Conference on Image Processing*, January 1995.

Active^3 Noise Reduction

Joachim Holzfuss

Institut für Angewandte Physik, TH Darmstadt
Schloßgartenstr. 7, 64289 Darmstadt, Germany

Abstract. Noise reduction is a problem being encountered in a variety of applications, such as environmental noise cancelation, signal recovery and separation. Passive noise reduction is done with the help of absorbers. Active noise reduction includes the transmission of phase inverted signals for the cancelation. This paper is about a threefold active approach to noise reduction. It includes the separation of a combined source, which consists of both a noise and a signal part. With the help of interaction with the source by scanning it and recording its response, modeling as a nonlinear dynamical system is achieved. The analysis includes phase space analysis and global radial basis functions as tools for the prediction used in a subsequent cancelation procedure. Examples are given which include noise reduction of speech.

INTRODUCTION

Noise is a term having a pretty negative image. Environmental noise like traffic noise or machinery noise noise is such an example. The ideal case would be to just get rid of it. A noise polluted signal or a "noisy" image is a more complex case, where one would like to cancel the noise completely but retain an interesting part, namely the "noise free" signal or sharp image. An example would be a conversation in the vicinity of a big noisy machine or the detection of a weak signal in a noisy environment. Noise reduction therefore is a reward winning but nevertheless difficult task.

Different classes of noise reducing or eliminating techniques can be created. The applicability of any of these has to be considered differently at each single case. The first method is the passive absorber, where certain damped resonating devices e.g. Helmholtz resonators transfer energy of an incoming sound wave into heat. With the help of this filtering effect sources occupying a small band in the frequency domain can be eliminated. A lot of different resonators or a lot of resonances have to be combined to reduce the noise of a broad band source. Examples are absorber walls in buildings or along traffic routes, sound absorbing headsets or an absorbing casing of a machine.

A more complex technique is the active noise reduction. Here a signal is picked up by a microphone, phase inverted and fed to an amplifier and loudspeaker. The radiated phase inverted signal thus cancels the noise. This works quite well, where areas of limited spatial extent and long wavelengths are considered. The problem is that in a 3 dimensional environment places may exist, where both, signal and source are in phase, thus increasing the noise locally. But real world applications have been demonstrated, where this kind of reduction works quite well, such as eliminating certain engine induced noises in a car. This class of noise reduction schemes may be classified as active^1, because one active process, the transmission of a phase inverted signal is involved.

An active^2 process would be used, where the source consists of both a noise a an interesting signal part. The active process additionally has to split the source into its parts and radiate the noise after a phase inversion or just the amplified signal part. While this can be done easily with the use of traditional filters, if the two parts are separated in the frequency domain, the method is of limited use, when noise and signal are indistinguishable in the frequency spectrum. In this case methods have been developed, that separate signals according to their dimensionality in state space. These methods have become of great interest in the area of research on nonlinear dynamical systems [1, 2, 3, 4]. One part of the source is being modelled as a low dimensional system and the other part as an additive, stochastic, high dimensional system. Separation is achieved, when the stochastic part, be it the signal of interest or not, is of limited, not too large amplitude.

In this paper an active^3 process is proposed, that follows the active^2 process for nonlinear dynamical systems. Additionally, this process interacts with the source through some kind of scanning or random perturbations, recording its response and model it as a low dimensional, nonlinear dynamical system. Subsequently after the scanning and modelling phase the separation of the two signals can be carried out. This method might be of use, when the nonlinear dynamical system has different coexisting attractors that can be reached from different initial conditions. More information is gathered this way to achieve a better modelling.

THEORY

We first describe an active[2] procedure, upon which active[3] is built. The source we wish to separate into noise and signal is introduced as an observable p of a nonlinear dynamical system. The flow $\Psi^t(\mathbf{z}_0)$ acts on an m-dimensional state space $\mathbb{R}^m$ and starts initially at $\mathbf{z}_0$. The observable maps the states of the system from the state space into a time dependent measurable value.

$$\begin{aligned} p : \mathbb{R}^m &\rightarrow \mathbb{R} \\ \Psi^t(\mathbf{z}_o) &\mapsto p\left(\Psi^t(\mathbf{z}_o)\right) \end{aligned} \tag{1}$$

The source we are dealing with is a noisy observable

$$\begin{aligned} p : \mathbb{R}^m \times \mathbb{R} &\rightarrow \mathbb{R} \\ \Psi^t(\mathbf{z}_o) &\mapsto p\left(\Psi^t(\mathbf{z}_o) + r(t)\right) \end{aligned} \tag{2}$$

$r(t)$ is considered here as additive Gaussian noise. We now wish to separate the source in eq. 2 into $\Psi^t(\mathbf{z}_o)$ and $r(t)$ by using the nonlinear dynamics of the flow. In this approach we are constructing an equivalent state space with the help of time shifted coordinates [5, 6]. The equivalent dynamics is described by a flow $\Phi^t(\mathbf{x}_0)$ constructed from n-tuples of the observable $p(t)$

$$\begin{aligned} \Phi^t : \mathbb{R}^n &\rightarrow \mathbb{R}^n \\ \mathbf{x}_o &\mapsto \Phi^t(\mathbf{x}_o) = \Big(p(t), p(t+T), \ldots, p(t+(n-1)T)\Big) \end{aligned} \tag{3}$$

T is the time shift and n the embedding dimension. Their values can be determined by different methods [7, 8, 9, 10].

A global flow can be approximated by different procedures [11, 12, 4, 13]. We use the method of global radial basis functions. The approximation is done with an approximating function f and random subsets $\left\{\mathbf{x}^k\right\}, \left\{\mathbf{y}^l\right\}$ of all points in state space

$$\begin{aligned} f : \mathbb{R}^n &\rightarrow \mathbb{R}^n \\ \Phi^{\Delta t}(\mathbf{x}^k) \approx f(\mathbf{x}^k) &= \sum_{l=1}^{N} \mathbf{c}_l \, \sigma(\mathbf{x}^k, \mathbf{y}^l), \quad \mathbf{c}_l \in \mathbb{R}^n \\ \sigma(\mathbf{x}^k, \mathbf{y}^l) &= \sqrt{r^2 + \|\mathbf{x}^k - \mathbf{y}^l\|^2}, \quad r \in \mathbb{R}^+ \end{aligned} \tag{4}$$

The σ are radial basis functions. r is a stiffness parameter. The special form of this function used here seems not to be of great importance, as a lot of these functions exist [14, 15, 16, 17]. The main idea behind this is, that the time evolution of a certain state is a function of its Euclidean distance to the other states.

The procedure leaves a matrix equation $\sigma C = F$, which can be solved for the coefficients of the matrix C. Because normally one takes less reference points $\mathbf{y}^l$ than data points $\mathbf{x}^k$, the system is over-determined. We use the singular value decomposition to find the solution with the least square error thus approximating a noise averaged flow.

In the next step we have to split off the noise from the original source. A simple method is to average a noisy state point with a predicted state from the noise averaged flow. Suppose the prediction time is Δt we get a "cleaner" state point by iterating the process

$$\mathbf{y}(t) = \alpha\Big[f\left(\mathbf{x}(t) + \mathbf{r}(t)\right)\Big] + (1-\alpha)\Big[\Phi^{\Delta t}\left(\mathbf{x}(t)\right) + \mathbf{r}(t+\Delta t)\Big], \quad \alpha \in (0,1) \quad (5)$$

A single time series can be extracted by either taking one coordinate $y_i(t)$ or by averaging over several coordinates, remembering the fact that they were originally constructed from time shifted coordinates [13].

So far all state points used in the analysis are located on an attractor $\mathcal{A} \subset \mathbb{R}^n$. The equivalent flow is constructed from a single trajectory initially originating from a single state $\mathbf{z}_0$ in the original state space or, equivalently, $\mathbf{x}_0$ in the time shift coordinate state space. As we have no requirements in the approximation procedure for the state points to be acquired sequentially in time, we now arrive at the proposal, that we don't have to be "on" an attractor. This means, we could take short time segments recorded from the evolution of different initial conditions. The present theories justifying the use of time delay embedding [5, 18] require that the data come from single (strange) attractors (ergodic theorem). We use our approximation procedure from a set of initial state points that are equally distributed in state space and got good predictability everywhere in the state space, including on the (possibly coexisting) attractors.

This opens the possibility of interaction with an unknown system to approximate its dynamics. This scanning procedure consists of random perturbations of the system and recording its response immediately after the perturbation has been shut off. Only short time segments have to be recorded, just enough to follow the evolution of the perturbed state in the time shift coordinate state space for a time Δt. the total recording time is then $t_{rec} = (n-1)T + \Delta t$. It is essential to have random perturbations that do not interfere with characteristic times of the scanned system. This way the full spectrum of the dynamics is being probed and made visible. Much more information about a system is acquired than just the dynamics on one attractor. Long term predictions for final states, short term predictions on strange attractors can be made and dynamical invariants can be calculated by predicting the flow for a long time and analyzing the statistics.

APPLICATIONS

The predicting abilities using the scanning procedure are demonstrated in Fig. 1. The global flow of a nonlinear oscillator [19] is approximated by disturbing the system such that the time evolution from different initial points are recorded. The x and $\dot{x}$ coordinates of all initial points lie within the little square in Fig. 1. Predictions of the final attractor for starting points lying in a Poincarè plane (black: period-1, white: chaotic) are shown in Fig. 1. Long term predictions within the inner square are very good, even predictions from the outside are possible.

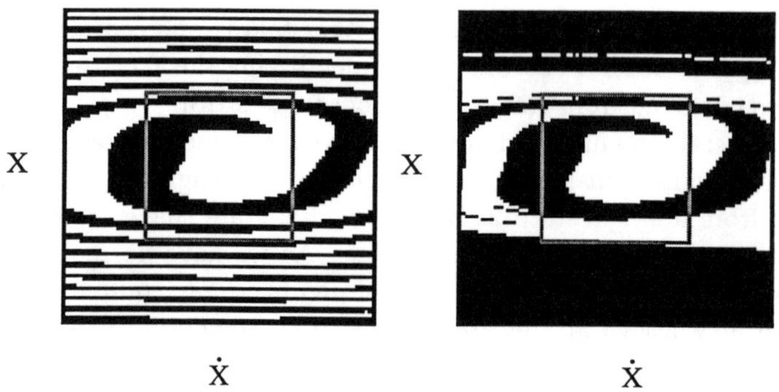

FIGURE 1. Long term predictions of the final attractor of a Duffing oscillator (black: period-1, white: chaotic). The x and $\dot{x}$ coordinates of the initial conditions used for flow approximations lie within the little square. Left: original equation. Right: approximation.

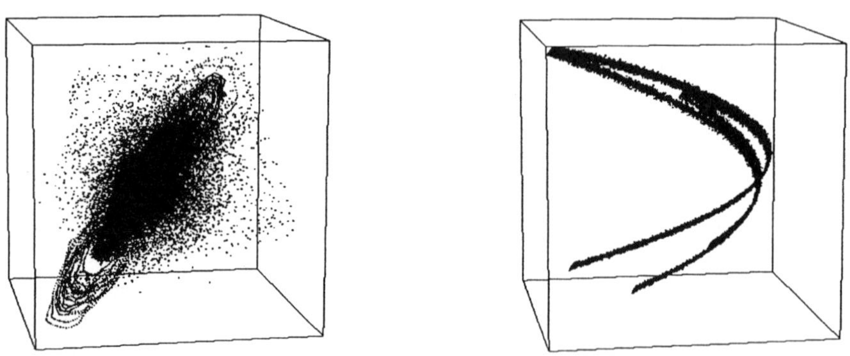

FIGURE 2. Attractors in the time shift coordinate state space. Left: voice signal. Right: voice+henon noise (-40dB SNR).

A human speech signal has been contaminated by the noise of the henon system [20]. The signal to noise ration was -40 dB ($\Delta s_{voice}/\Delta s_{noise}$). In the resulting signal no voice could be recognized. Fig. 2 a,b shows the signals in the time shift coordinate state space. The global flow has been approximated with the radial basis function approach. The henon noise is modelled as a dynamical system and the voice is treated as statistical noise. Deviations from the predicted behaviour of the henon system have been identified successfully. While playing these deviations, clearly the voice can be understood. To demonstrate the results, time series of the pure voice and the noise reduced voice signal is shown in Fig. 3. A spectrogram is presented for noise+voice, noise reduced noise+voice and pure voice. The similarity between the latter two is easily seen (Fig. 4 a,b,c).

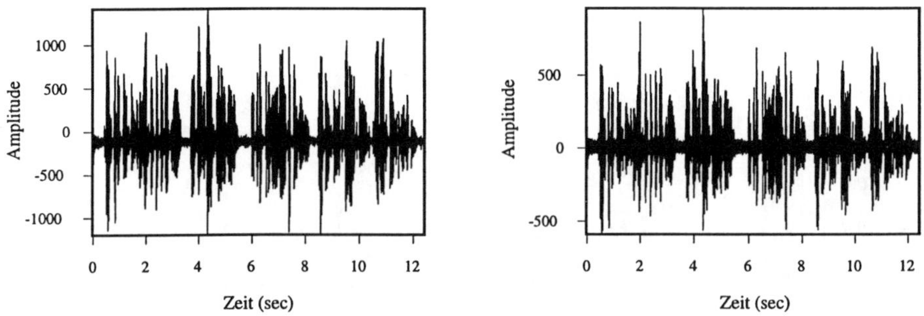

FIGURE 3. Time series of the pure voice and the voice signal after recovery.

SUMMARY

A method has been proposed, that has a threefold approach to active noise reduction. First, an unknown system is scanned and probed for its dynamics Second, the dynamics is reconstructed. Third, the recorded signal is separated into a nonlinear low dimensional part and a statistical part. The method of scanning can be applied to any of the known methods for the reconstruction of nonlinear dynamics. It can also be used to predict final states of multi-stable systems and for short term predictions on strange attractors, if no long time stationarity can be achieved. Examples of noise reduction of human speech show, that the method can also be applied to real world systems.

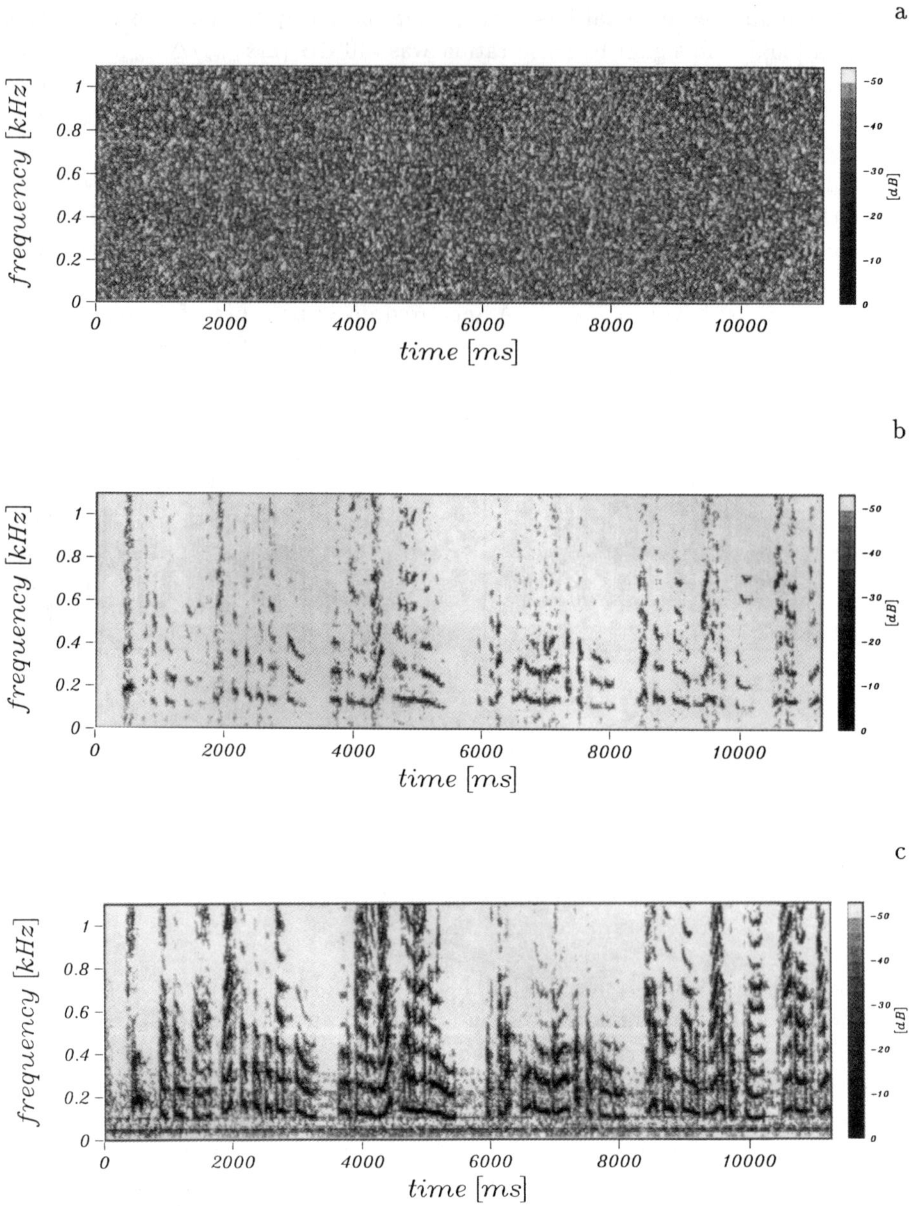

FIGURE 4. Spectrogram of a) voice + Henon noise, b) recovered voice, c) pure voice.

ACKNOWLEDGEMENTS

The author acknowledges support of the Sonderforschungsbereich 185 Nichtlineare Dynamik of the Deutsche Forschungsgemeinschaft and wishes to thank James B. Kadtke for cooperations concerning noise reduction.

REFERENCES

1. Farmer, D. and Sidorowicz, J., "Exploiting Chaos to predict the future and reduce noise," *Physica D47*, 373 (1988).

2. Grassberger, P., Hegger, R. Kantz, H., Schaffrath, C. and Schreiber, Th., "On Noise Reduction Methods for Chaotic Data," preprint Wuppertal (1992)

3. Kostelich, E.J. Yorke, J.A., "Noise Reduction: Finding the Simplest Dynamical System Consistent with the Data," *Physica D* **41** N2, 183–196 (1990).

4. Brush, J.S and Kadtke, J.B. "Nonlinear Signal Separation Using Global Empirical Dynamical Equations" Proceedings of the International Conference on Acoustic, Speech, and Signal Processing, San Francisco, Ca., USA, 1992 (5) pp. 321–325.

5. Takens, F., "Detecting strange attractors in turbulence," in *Dynamical systems and turbulence* ed. Rand, D.A. & Young L.-S., Lecture notes in mathematics, Vol. 898, Berlin: Springer, 1981, pp. 366 – 381.

6. Packard, N.H., Crutchfield, J.P., Farmer, J.D. and Shaw, R.S. "Geometry from a time series," *Phys. Rev. Lett.* **45**, 712 – 716, (1980).

7. Frazer, A.M., and Swinney, H.L., "Independent coordinates for strange attractors from mutual information," *Phys. Rev. A* **33**, 1134–1140, (1986).

8. Liebert, W., Pawelzik, K., and Schuster, H.G., "Optimal embeddings of chaotic attractors from topological consideration," *Europhys. Lett.* **14**, 6, 521–526, 1991.

9. Buzug, Th., Reimers, T., and Pfister, G., "Optimal Reconstructions of strange-attractors from purely geometrical arguments," *Europhys. Lett.* **13**, 7, 605–610, 1990.

10. Holzfuss, J. and Lauterborn, W., Nonlinear Dynamics of Acoustic Cavitation Noise, in: *Frontiers of Nonlinear Acoustics: Proceedings of the* 12th *ISNA* Eds.: M.F. Hamilton und D.T. Blackstock, London: Elsevier Science, Publishers Ltd., (1990) pp. 464–469.

11. Cremers, J. and Hubler, A. "Construction of differential equations from experimental data," *Z. Naturforsch.* **A42**,797–802, (1987).

12. Crutchfield, J., and MacNamara, B. "Equations of motion from a data series," *J. Complex Systems* **1**, 417–448, (1987).

13. Holzfuss, J., and Kadtke, Global nonlinear noise reduction using radial basis functions, *Int. J. of Bifurc. and Chaos* **3**, 10, 589–596, 1993

14. Franke, R., "Scattered Data Interpolation: Tests of some Methods," *Math. of comp.*, **38**, 157, 181 - 200, 1982.

15. Broomhead D.S., and Lowe, D. "Multivariable functional interpolation and adaptive networks," *Complex Systems* **2**, 321 – 355, 1988.

16. Casdagli, M., "Nonlinear Prediction of chaotic time series," *Physica D* **35**, 335 - 356, 1989.

17. Holzfuss, J., and Parlitz, U., "Lyapunov exponents from time series," in *Lyapunov Exponents* ed. Arnold, L., Crauel, H. & Eckmann, J.-P., Lecture Notes in Mathematics, Berlin: Springer, 1991, Vol. 1486, pp. 263–270, 1991

18. Sauer, T., Yorke, J.A., and Casdagli, M. "Embedology" *J. Stat. Phys.* 65, 395–400 (1991).

19. Duffing oscillator: $\ddot{x} + 0.2\dot{x} - x + x^3 = \cos(1.7\ t)$

20. Iteration: $x_{n+1} = 1 - 1.4x_n^2 + y_n,\ y_{n+1} = 0.3x_n$

Multifractal Nature of Unvoiced Speech Signals

Olufemi A. Adeyemi†,* K. Hartt†† and G. F. Boudreaux-Bartels†
†Department of Electrical Engineering, ††Department of Physics
University of Rhode Island
Kingston, RI 02881
adeyemi@ele.uri.edu

Abstract

A refinement is made in the nonlinear dynamic modeling of speech signals. Previous research successfully characterized speech signals as chaotic. Here, we analyze fricative speech signals using multifractal measures to determine various fractal regimes present in their chaotic attractors. Results support the hypothesis that speech signals have multifractal measures.

Introduction

Speech is a nonlinear phenomenon that has been qualitatively described in terms of fluid flow and turbulence [1]. Recently, however, speech has been found to exhibit low-dimensional chaos [2]. Strange attractors, positive Lyapunov exponents and low correlation dimensions have been found [2]. There exists an infinite hierarchy of generalized dimensions [3]. Typically, the generalized dimensions are not all equal. A chaotic attractor with unequal generalized dimensions is called a multifractal. The largest of those dimensions

*Supported in part by a Patricia Robert Harris Fellowship

also is an indicator of the minimum dimension required in the phase space of any realistic dynamical model of speech signals. Chaotic signals are characterized by broadband spectra, impulsive autocorrelation functions and sensitivity to initial conditions (exponential divergence of two closely spaced points on the chaotic orbit) [3]. Although some of these properties are reminiscent of random noise, it is now easy to distinguish between randomness and chaos [4].

Speech signals can be broadly classified into voiced and unvoiced speech. Voiced speech is produced by the vibration of the vocal tract whereas unvoiced sounds are generated by forming a constriction at some point in the vocal tract and forcing air through the constriction at high velocity [5]. In speech modeling literature, the word "turbulence" is often used in a qualitative sense to mean "nonlaminar flow". However, in the dynamical modeling of what is technically called turbulence, the dimension of the model is expected to be so high as to discourage such an approach. Yet the turbulent velocity field of turbulence has been extensively investigated to reveal a low-dimensioned multifractal structure [6, 7]. In the next section, multifractal will be defined. Pre-turbulent nonlaminar flow can sometimes be described in a low-dimensioned phase space [4]. The current investigation is motivated by a desire to find a more complete low-dimensioned dynamical description of speech signals. Chaotic systems typically have a multifractal orbital structure. The current work seeks to better define this structure.

Classical speech production techniques have modeled speech as the output of a quasi-stationary linear system consisting of an all-pole filter excited either by a periodic train of pulses for voiced speech or by broadband noise for unvoiced speech. This recursive model assumes that the vocal tract can be modeled over a short time interval by a linear system described by the transfer function

$$H(z) = \frac{A}{1 - \sum_{k=1}^{p} a_k z^{-k}} \qquad (1)$$

relating the air flowing into the glottis to the speech sound emanating from the mouth. The coefficients a_k are obtained by linear prediction of the present speech sample from the past p samples and are updated every 10-20 milliseconds or two to three pitch periods. The poles of $H(z)$ in (1), i.e., the roots of the denominator polynomial, are used to represent the three to five resonant frequencies which characterize most speech sounds [8].

In this paper, we present preliminary results of examining the multifractal measures of the acoustic waveforms generated during the production of voiced fricative speech signals. Our method is based on the estimation of generalized dimensions from the phase-space reconstructed vectors of a single scalar variable of speech signals. We test the technique on sounds generated by both male and female speakers. Our multifractal analysis of speech from reconstructed phase-space helps to determine whether speech signals exhibit low-dimensional chaos rather than turbulence [9]. This may lead to better speech modeling, synthesis and recognition.

Mathematical Background

State Space reconstruction for multifractal analysis

In time series analysis of experimental data, often only one component of an F-dimensional[1] vector is available. This is particularly crucial in speech production where any attempt to insert transducers in the vocal cavity will alter or impair the speech produced. For most speech analysis, that component is the acoustic pressure waveform recorded by a microphone in front of the mouth. The chaotic dynamics of the speech can be reconstructed from this single component by using the embedding space method [10]. If the embedding space is generated properly, the behavior of the trajectories in this embedding space will have the same geometrical and dynamical properties as those that characterize the actual trajectories in the full F-dimensional state space.

The reconstructed trajectory $\mathbf{X}$ from a single component can be expressed as a matrix where each column is a phase-space vector, i.e.,

$$\mathbf{X} = (X_1 \;\; X_2 \;\; \ldots \;\; X_P) \tag{2}$$

[1]F is the actual dimension of the dynamical system being analyzed. F is unknown in the case of speech data.

where X_i is the state of the system at discrete time i. Given an N-point time series, $\{x_1, x_2, x_3 \ldots x_N\}$, each X_i is given by

$$X_i = (x_i \;\; x_{i+\tau} \;\; \ldots \;\; x_{i+(D-1)\tau}) \tag{3}$$

where τ is a chosen time delay, and D is the embedding dimension. The reconstruction of the system phase space from such scalar measurements is theoretically supported by results from Takens' embedding theorem [10]. An optimum time delay often chosen is the first minimum of the mutual information [11]. The standard procedure is to increase the embedding dimension D incrementally until the calculated multifractal dimensions appear to have converged.

Multifractal measures

In this paper, the generalized dimension

$$D_q = \lim_{l \to 0} \frac{\ln C_D(l, q)}{\ln(l)} \tag{4}$$

was evaluated for different speech sounds using the generalized correlation integral $C_D(l, q)$ as defined in [12]. Note, theoretically in (4), if in the limit $l \to 0$, we plot $\ln C_D(l, q)$ as a function of $\ln(l)$, this log-log plot is linear with slope D_q. However, for real systems (with limited data points), only a portion of this curve is linear because the expression in (4) is an asymptotic result which requires an infinite number of data points. This portion will be referred to as the "linear region" below.

A numerical approximation to the generalized correlation integral is evaluated as

$$C_D(l, q) \approx \lim_{N \to \infty} \left\{ \frac{1}{N_{seed}} \sum_{k=1}^{N_{seed}} \left\{ \frac{1}{N} \sum_{\substack{j=1 \\ j \neq k}}^{N} U(l - \|\mathbf{X_k} - \mathbf{X_j}\|) \right\}^{q-1} \right\}^{\frac{1}{(q-1)}} \tag{5}$$

where U is the Heaviside unit step function, $\| \, . \, \|$ denotes maximum norm, j runs over the (N-1)-element fractal subset and k runs over N_{seed} equally-spaced points which form a subset of the N measured points [13]. We then determine the spectra of dimensions D_q in (4) for various q from the linear regions of the log-log plots of $\ln C_D(l,q)$ versus $\ln(l)$. The linear region in the log-log plot is selected by looking for a linear region in the plot. A good automated procedure for selecting the linear region is desirable. A systematic procedure for estimating the dimensions D_q is as follows:

For each value of q:

- Sample the range of $\ln(l)$ into 100 evenly spaced points.

- Specify a number M to represent the number of consecutive $\ln(l)$ values to be contained in the "linear region". This number should be large enough to avoid local straight runs of points and small enough to fit into the linear region.

- Estimate the approximate boundaries of the "linear region" by modeling the correlation integral curves with a least squares polynomial fit of order n (where n varies from three to five). The boundary is set equal to the innermost points p_1 and p_2 where the derivative of the polynomial changes sign. Start at $i = p_1$. Make a least squares linear fit of length M sequence of data points, $\ln C_D(l,q)$, $l = i,\ i+1, \ldots i+M-1$, in the estimated linear region of equation (4). Record the slope and error for each i. Increment the starting point $i = i+1$ until the last starting point is $i = p_2 - M + p_1$.

- Average the slopes of the linear fit corresponding to the five lowest errors to obtain the estimate of the dimension D_q.

The automated procedure described above is similar to the algorithm found in [13]. However, our algorithm efficiently estimates the linear region by modeling the correlation integral curves with a least squares polynomial fit and selecting points p_1 and p_2 described above. In order to assess the reliability of the results provided by the multifractal measures, we also analyzed strange attractors with known multifractal dimensions (e.g. Lorenz attractor and logistic map [3]). With 10,000 data points our algorithms gave the standard results found in the literature for the Lorenz attractor, i.e. $D_0 = 2.06$ and $D_2 = 2.05$ [14, 15]. The algorithm was also tested on the logistic map for which

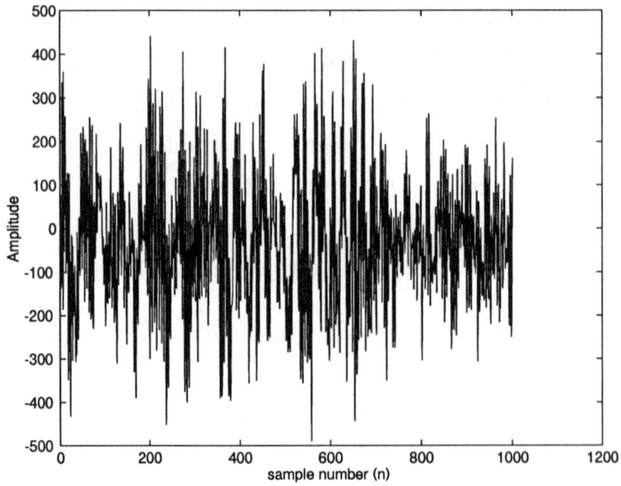

Figure 1: A segment of the voiced fricative sound "Z".

the analytic D_q values are known in closed form [16]. The estimated dimensions for the logistic map are close to the analytic values when the number of data points are on the order of 10^6 or more.

Speech Analysis

The time series used in this study consists of acoustic pressure waveforms of speech from the ISOLETE (standard) database [17]. ISOLETE is a database of letters of the English alphabet spoken in isolation. The database consists of 7800 spoken letters with two productions of each letter given by 150 speakers. The sampling rate is 16 kHz. Fig. 1 shows a segment of a voiced fricative with the noise-like characteristic clearly displayed. The voiced fricatives consist of two distinct components. One quasi-periodic part is due to the vibration of the vocal cord, whereas the other noise-like part is due to turbulent air flow in the neighborhood of the constriction in the vocal tract [5]. The multifractal analysis was performed on eight speakers (four males and four females) pronouncing the unvoiced fricative "Z". It is important to perform the multifractal analysis both on male and female speech signals because of the differences in their bandwidth and average pitch period. We show some of these results in the next section. In all cases, 10,000 data samples were used.

Results

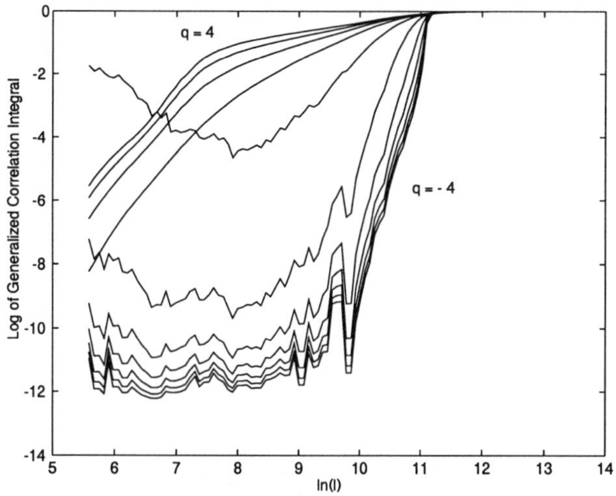

Figure 2: Log of the Generalized Correlation Integral, $\ln C_D(l, q)$, plotted from -4 $\leq$q $\leq$ 4 for one fricative speech sound, "Z", spoken by one male speaker.

Fig. 2 shows the values of the log of the generalized correlation integral in (5), $\ln C_D(l, q)$, plotted from -4 $\leq$q $\leq$ 4 for the voiced fricative sound "Z" spoken by a male speaker. The slopes of these curves in the "linear regions" correspond to the generalized dimensions D_q which are depicted in Fig. 3. Fig. 4 shows the values of $\ln C_D(l, q)$ versus $\ln(l)$ for the same sound spoken by a female speaker; the corresponding generalized dimensions are shown in Fig. 5. For the voiced fricative 'Z' speech signals, the dimensions at different values of q are not the same ($D_0 \neq D_1 \neq D_2$), which implies the presence of a multifractal.

Conclusion

We have analyzed voiced fricative speech sounds spoken by eight speakers (four

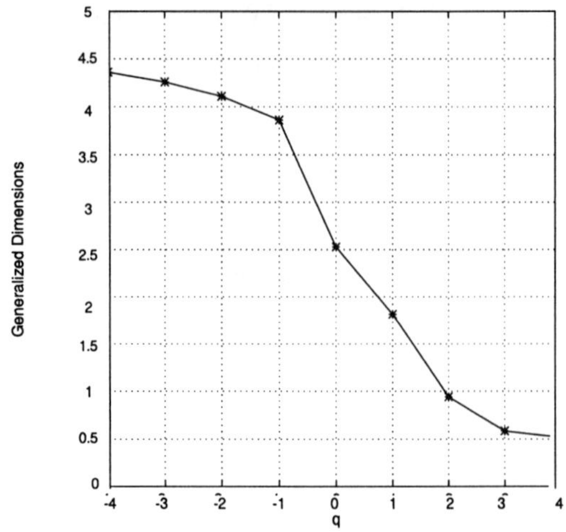

Figure 3: D_q in (4) versus q for a fricative sound "Z" spoken by a male speaker.

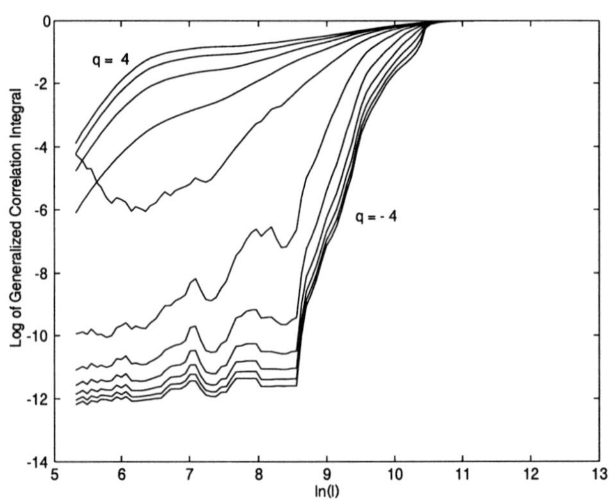

Figure 4: Log of the Generalized Correlation Integral, $\ln C_D(l, q)$, plotted from -4 $\leq$q $\leq$ 4 for one fricative speech sound, "Z", spoken by one female speaker.

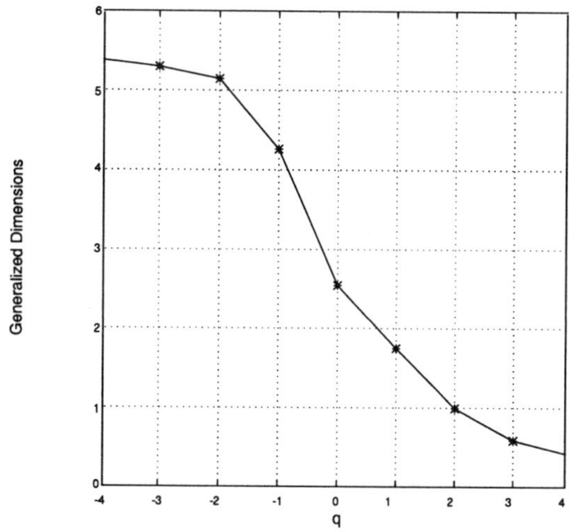

Figure 5: D_q in (4) versus q for fricative speech "Z" spoken by a female speaker.

males and four females) and found that they have non-constant generalized dimensions. This supports our hypothesis that fricatives and unvoiced speech signals are not only chaotic, but their attractors have multifractal measures. Some of the initial multifractal results give us a range of the number of components needed for a dynamical model of speech. It is impossible to obtain this range from estimating only the correlation dimension. The generalized dimensions for a male speaker are different from those of a female speaker. We hope this will be useful as a speaker identification criterion. We hope to perform multifractal analysis on existing biomechanical models of the vocal folds [18]. Finally, the goal is to model speech as a chaotic multifractal and quantitatively compare the performance of such a model with the classical all-pole model described earlier.

References

[1] H. M. Teager and S. M. Teager, "A phenomenological model for vowel production in the vocal tract," in *Speech Science: Recent advances* (R. D. Daniloff, ed.), California: College-Hill Press, 1985.

[2] S. S. Narayanan and A. A. Alwan, "A nonlinear dynamical systems analysis of fricative consonants," *J. Acoust. Soc. Am.*, vol. 97, no. 4, pp. 2511–2524, 1995.

[3] R. C. Hilborn, *Chaos and Nonlinear Dynamics: An Introduction for Scientist and Engineers.* New York: Oxford University Press, 1994.

[4] H. Schuster, *Deterministic Chaos.* VCH Publishers, 1988.

[5] L. R. Rabiner and R. W. Schafer, *Digital Processing of Speech Signals.* Prentice Hall, 1978.

[6] C. Meneveau and K. R. Sreenivasan, "Simple Multifractal Cascade Model for Fully Developed Turbulence," *Phys. Rev. A*, vol. 59, pp. 1424–1427, April 1987.

[7] C. Meneveau and K. R. Sreenivasan, "The multifractal spectrum of the dissipation field in turbulent flows," *Nuclear Physics B (Proc. Suppl)*, no. 2, pp. 49–76, 1987.

[8] B. S. Atal and N. David, "On synthesizing natural-sounding speech by linear prediction," *Proc. of Int. Conf. Acoust. Speech, Signal Processing*, 1979.

[9] F. Heslot, B. Castaing, and A. Libchaber, "Transitions to turbulence in helium gas," *Physical Review A*, vol. 36, no. 12, pp. 5870–5873, 1987.

[10] F. Takens, *Detecting strange attractors in turbulence*, vol. 898 of *Lecture Notes in Mathematics*, pp. 365–381. Berlin: Springer, 1981.

[11] W. Liebert and H. G. Schuster, "Proper choice of time delay for the analysis of chaotic time series," *Phys. Lett. A*, vol. 142, pp. 107–111, December 1989.

[12] K. Pawelzik and H. G. Schuster, "Generalized dimensions and entropies from a measured time series," *Phys. Rev. A*, vol. 35, pp. 481–484, January 1987.

[13] L. V. Meisel and M. A. Johnson, "Multifractal analysis of imprecise data: Baddi-Politi and correlation integral approaches," *Physical Review E*, vol. 50, pp. 4214–4219, 1994.

[14] H. Mori, "Fractal Dimensions of Chaotic Flows of Autonomous Dissipative Systems," *Prog. Theor. Phys.*, vol. 63, no. 3, pp. 1044–1047, 1980.

[15] P. Grassberger and I. Procaccia, "Measuring the strangeness of strange attractors," *Physica D*, vol. 9, pp. 189–208, 1983.

[16] E. Ott, *Chaos in Dynamical Systems.* New York: Cambridge University Press, 1993.

[17] R. Cole, Y. Muthusamy, and M. Fanty, "The ISOLETE Spoken Letter Database," *Technical Report*, Feb. 1991.

[18] H. Herzel, D. Berry, I. Titze, and I. Steinecke, "Nonlinear dynamics of the voice: Signal analysis and biomechanical modeling," *Chaos*, vol. 5, no. 1, pp. 30–34, 1995.

[19] H. D. I. Abarbanel, R. Brown, J. L. Sidorowich, and L. S. Tsimring, "The analysis of observed chaotic data in physical systems," *Rev. Mod. Phys.*, vol. 65, pp. 1331–1392, October 1993.

4. ADVANCED APPLIED SIGNAL PROCESSING METHODS

Channel Equalization For Communication with Chaotic Signals‡

Alan V. Oppenheim*, Kevin M. Cuomo†, Richard J. Barron*, and Alan E. Freedman*

**Research Laboratory of Electronics, Massachusetts Institute of Technology, Cambridge, MA 02139*

†*Lincoln Laboratory, Massachusetts Institute of Technology, 244 Wood Street, Lexington, MA 02173*

Abstract. Chaotic signals and synchronized systems are potentially useful as modulation or masking waveforms for communications systems. This paper explores the effect on synchronization of frequency dependent channel gain, *i.e.* linear filtering. Several approaches to compensating for these effects are also discussed.

I. INTRODUCTION

Over the past several years there has been considerable interest in utilizing chaotic signals for spread-spectrum communications. Most strategies that have been proposed exploit the self-synchronization property of a class of chaotic systems. By necessity, synchronization of the receiver requires that the received drive signal be undistorted or that it first be appropriately equalized in amplitude, spectral content and phase. Specifically, it can be anticipated that a realistic transmission channel will introduce a time varying attenuation due to fading, scattering, etc., will modify the spectral characteristics of the transmitted signal due to channel filtering and multipath and will introduce additive noise. The effects of additive noise on synchronization have been discussed in [1]. In this paper we propose a number of techniques for estimating and compensating for the effects of the channel on amplitude, spectral content

---

‡This work was supported in part by the Department of the Navy, Office of the Chief of Naval Research, contract number N00014-93-1-0686 as part of the Advanced Research Projects Agency's RASSP program and in part by the U.S. Air Force Office of Scientific Research under grant AFOSR-91-0034-C. The views expressed are those of the authors and do not reflect the official policy or position of the U.S. Government.

and phase of the synchronizing drive signal. In Section II we briefly review the various approaches that have been proposed for exploiting synchronized chaotic systems for communications. In Section III we incorporate into the system description the presence of a channel. Section IV describes the effect of channel gain on synchronization error and our approach and results for gain compensation. Section V illustrates the effect of channel filtering and proposes an approach to estimating the frequency response of an equalizing filter.

II. COMMUNICATIONS USING SYNCHRONIZED CHAOTIC SYSTEMS

Since the discovery that two chaotic systems can be made to synchronize when the second system (receiver) is driven by the first (transmitter) [2,3], researchers have been actively exploring how this characteristic might be used for practical applications. Many techniques have now been proposed for combining information-bearing signals with a chaotic carrier and for recovering the information with a self-synchronizing receiver system. One of the first demonstrations of how these concepts might be used for private communications is given in [4]. In that paper, the notion of cascaded chaotic subsystems is used to numerically demonstrate that low-level information-bearing signals can be added to a more powerful chaotic drive signal and recovered at the receiver by exploiting the robust synchronization and signal recovery properties of the Lorenz system. Later, it was shown that this chaotic signal masking and recovery concept could be implemented with standard analog hardware [5–7] and could be made to work with other self-synchronizing chaotic systems [8].

Recognizing some potential drawbacks of this approach in realistic communication scenarios, researchers developed alternative ways for privately communicating information with chaotic signals and systems. For example, by modulating a transmitter parameter with a binary-valued bit stream, it was demonstrated that digital signals can be privately transmitted and recovered with a self-synchronizing receiver system [5,6,9]. This concept has been recently extended to allow multiple independent messages to be privately communicated [10]. Other somewhat more sophisticated modulation techniques are based on chaotic "spread-spectrum" concepts, *i.e.* the information signal is directly multiplied by the broadband chaotic drive signal [11,12]. Note that each of these communication strategies have one important feature in common—they utilize the self-synchronization property of properly matched chaotic receivers to readily recover the information.

Chaotic communication methods have also been proposed that relax the requirement for synchronization at the intended receivers to readily recover the information-bearing signals. One approach utilizes the information signal to modulate the parameter(s) of a chaotic map; the information is then recovered at the intended receivers by inverting the map with respect to the modulation parameter(s) [13]. Another approach is based on "chaotic switching," *i.e.* information is transmitted by switching between different chaotic systems in the transmitter and detected at the receiver via maximum likelihood state estimation techniques [14]. Information can also be privately communicated with chaotic maps by altering the known unstable periodic orbits in a way that depends on whether a "zero" or "one" is to be transmitted [15]. In order for any of these chaotic communication methods to work well in practice, however, undesirable channel effects must be equalized in order to accurately recover the information signals at the intended receivers. Below, we address this issue by developing some straightforward channel equalization methods for chaotic signals.

III. SYSTEM DESCRIPTION INCORPORATING CHANNEL EFFECTS

The general structure for a synchronizing chaotic transmitter and receiver with channel effects and channel compensation is shown in Fig. 1. In the system of Fig. 1, $x(t)$ is referred to as the chaotic drive signal and $x_r(t)$ as the reconstructed drive signal. In the absence of transmission channel effects, $\tilde{x}(t)$ and $x(t)$ are identical and no compensator is required. In that case $s(t) = x(t)$ and for systems that have the self-synchronizing property the reconstructed drive signal $x_r(t)$ will equal $s(t)$ after an initial transient. In the presence of channel effects and inexact channel compensation the receiver won't synchronize and $x_r(t)$ will not equal $s(t)$, resulting in synchronization error, defined as

$$e_x(t) \triangleq s(t) - x_r(t) \quad . \tag{1}$$

In this paper we propose several approaches to utilizing $s(t)$ or the synchronization error to adaptively determine an appropriate compensator. As a measure of synchronization error we will make reference to P_x, $P_{\tilde{x}}$, and P_{ex} representing the time-average power over some appropriately chosen time window in the chaotic drive signal $x(t)$ the received signal $\tilde{x}(t)$ and the synchronization error, respectively. We assume that for any chaotic system exhibiting steady state behavior on the attractor, P_x is a quantity that can be determined apriori either empirically or analytically.

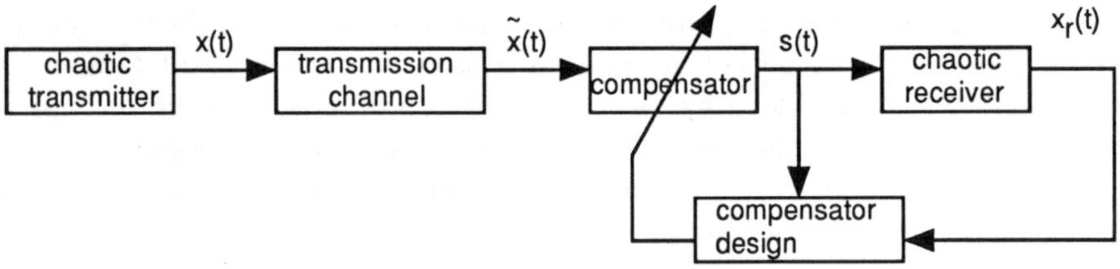

FIG. 1. General structure for a synchronizing chaotic transmitter and receiver with channel effects and channel compensation.

In the system of Fig. 1 and throughout our discussion in this paper we are not incorporating any specific modulation strategy utilizing the chaotic transmitter and receiver. Our specific focus here is on determining a compensator assuming only that the chaotic drive signal has been modified during transmission through a channel that can be represented by a linear system whose characteristics may be slowly time varying.

In this paper we propose and describe a number of approaches to determining the channel compensation. While the approaches are not specific to any particular chaotic system, we have used the Lorenz transmitter-receiver system as the prototype for this study. Within the Lorenz framework, the transmitter and receiver equations used are

$$\begin{aligned} \dot{x} &= \sigma(y-x) \\ \dot{y} &= rx - y - xz \\ \dot{z} &= xy - bz \quad , \end{aligned} \tag{2}$$

$$\begin{aligned} \dot{x}_r &= \sigma(y_r - x_r) \\ \dot{y}_r &= rs(t) - y_r - s(t)z_r \\ \dot{z}_r &= s(t)y_r - bz_r \quad , \end{aligned} \tag{3}$$

$$\begin{aligned} \text{where } s(t) &= \text{drive signal in the receiver equations,} \\ x, y, z &= \text{transmitter state variables,} \\ x_r, y_r, z_r &= \text{receiver state variables.} \end{aligned}$$

The analysis of these equations with regard to synchronization of the transmitter and receiver is given in [5–7].

The specific parameter values used were $\sigma = 16, r = 45.6$, and $b = 4$. For the simulations presented in this paper, the Lorenz equations were numerically integrated using a fourth-order Runge-Kutta algorithm with a time step of .005.

IV. CHANNEL GAIN COMPENSATION

In this section we assume that the channel imposes only a constant or time-varying gain on the transmitted drive signal $x(t)$, *i.e.* the received signal is $\tilde{x}(t) = G(t)x(t)$. If $G(t)$ differs from unity, then synchronization deteriorates, as illustrated Fig. 2. Specifically, Fig. 2 shows the synchronization error power P_{ex} with no channel compensation, *i.e.* $s(t) = \tilde{x}(t)$, and for constant channel gain or attenuation for a range of channel gain variation. As we expect, P_{ex} is essentially zero for $G = 1$ and rapidly increases as the channel introduces non-unity gain or attenuation.

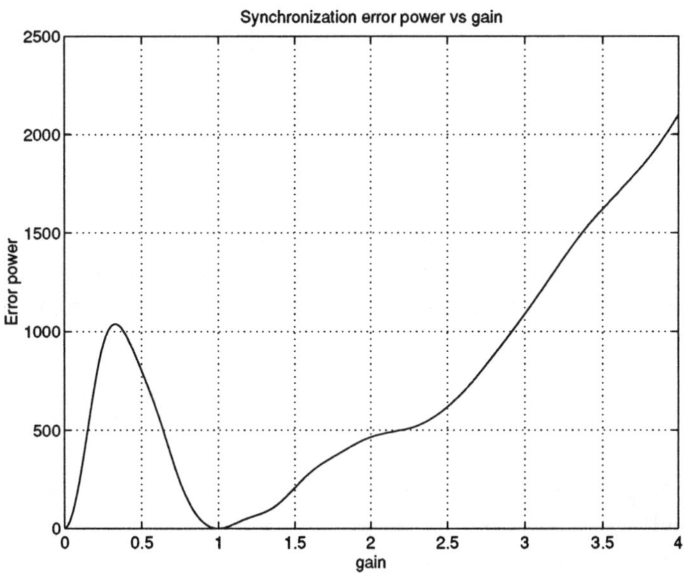

FIG. 2. Synchronization error power vs. gain.

There are several potential approaches to estimating the channel gain or attenuation so that a compensating gain can be applied. From Fig. 2 we see that the synchronization error power is unimodal for a range of gains around unity. Consequently, one approach is to iteratively adjust the compensating gain to minimize P_{ex}. This assumes that an initial estimate of the gain is within the range in Fig. 2 which is unimodal and includes $G = 1$ and for which the slope of the curve is sufficiently high to permit a reasonable gradient search for the gain.

A reasonable initial estimate of the gain can be obtained by utilizing the fact that P_x, the power in the chaotic drive signal at the input to the channel, can be expected to be independent of the specific sample path. Empirical measurement of P_x over a range of initial conditions in the Lorenz system with

500 independent trials and a time window of 800 sec resulted in an average value of 159.78, a variance of .018, a maximum of 160.16 and a minimum of 159.45. Based on this, a reasonable initial estimate of $\hat{G}(t)$ is then taken as

$$\hat{G} = \sqrt{\frac{P_{\tilde{x}}(t)}{P_x}} = \sqrt{\frac{P_{\tilde{x}}(t)}{159.78}} \ . \tag{4}$$

A rectangular sliding window of duration Δ is used to estimate $P_{\tilde{x}}$, *i.e.*

$$\hat{P}_{\tilde{x}}(t) = \frac{1}{\Delta}\int_{t-\frac{\Delta}{2}}^{t+\frac{\Delta}{2}} \tilde{x}^2(\tau)d\tau \tag{5}$$

In Fig. 3 we show a representative example of compensation for a constant gain $G(t) = 3$. In this figure we show the synchronization error when the compensating gain is the reciprocal of the initial gain estimate $\hat{G}(t)$ in Eq. (4). We also show on the same graph the synchronization error when for each time window, a gradient search is applied to minimize P_{ex}, starting with the initial estimate in Eq. (4). A 20-second sliding window was used to estimate $P_{\tilde{x}}$. For time scale comparison, a representative 20-second segment of $x(t)$ is shown in Fig. 4. In Fig. 3(a) the initial error represents the error transient prior to synchronization. In Fig. 3(b) we show the error on an expanded scale for a 10-second interval after the initial transient. Clearly the gradient search following the initial gain estimate results in significant reduction in synchronization error.

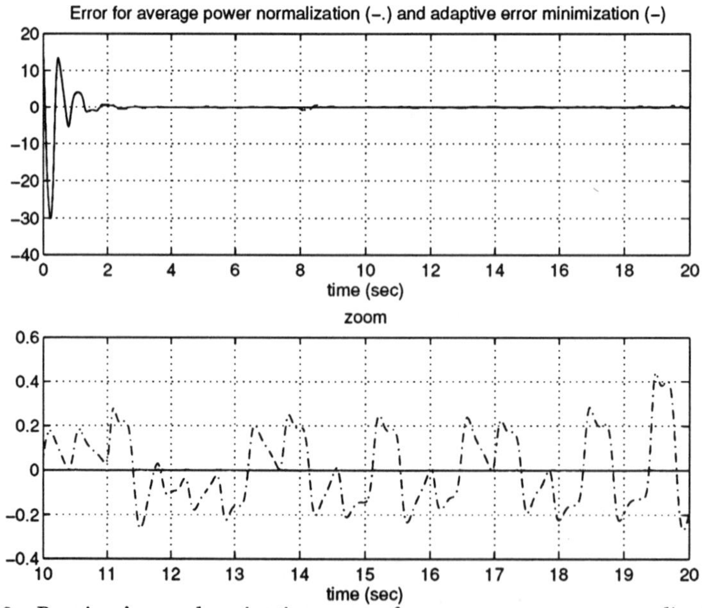

FIG. 3. Receiver's synchronization error for average power normalization and adaptive error normalization.

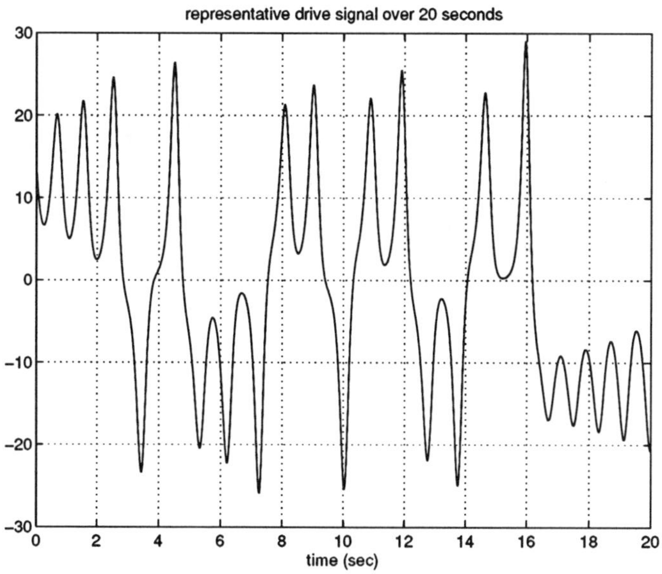

FIG. 4. Representative drive signal over a 20-second time interval.

In the example represented by Fig. 3, the channel gain was constant. In Fig. 5 we show corresponding results with a linearly time varying gain

$$G(t) = 0.95 + (t/2000) \quad . \tag{6}$$

A 20-second window is again used to estimate $P_{\tilde{x}}(t)$. Superimposed in Fig. 5 are the instantaneous error using only the initial gain estimate and using a gain estimate based on applying a gradient search to minimize P_{ex} within each window. In Fig. 5, the smaller instantaneous error curve represents the compensation resulting from adaptive error minimization and the larger error curve the compensation utilizing only the initial estimate of Eq. (4). While for this particular example there is some modest improvement with the adaptive error minimization it is not as significant as the case of constant channel gain. The difficulty in using adaptive error minimization when the gain is time varying lies in the fact that the procedure relies on Fig. 1 which assumes constant gain and an accurate estimate of synchronization error power. With the gain time varying, P_{ex} must be estimated over a finite duration window and consequently will inherently have some variance. Furthermore the gain varies over the window duration. Empirically we've found that the adaptive error minimization algorithm is sensitive to time varying channel gain and reduces the synchronization error over the use of the initial gain estimate in Eq. (4) only if the variance in gain over the window is small relative to the variance in the estimate of $P_{ex}(t)$. We are currently also exploring a number of other ways of estimating and compensating for time varying channel gain.

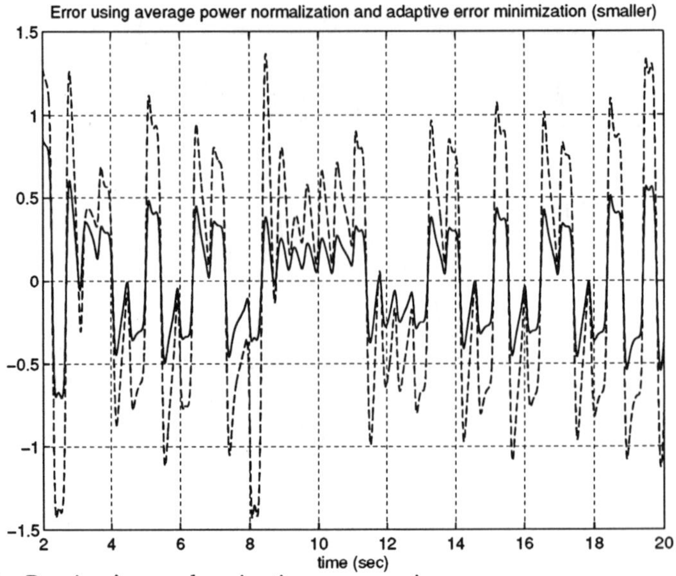

FIG. 5. Receiver's synchronization error using average power normalization (dashed) and adaptive error minimization (solid) for a time varying gain channel.

V. THE EFFECTS OF FILTERING

A realistic channel can also be expected to modify the spectral characteristics of the drive signal due to such effects as multipath, frequency-dependent attenuation etc. As with gain compensation, our proposed approach to compensating for channel filtering is to assume that the power density spectrum and higher-order power spectra of the chaotic drive signal are essentially independent of the specific sample path. The channel equalizer is then designed so that the spectral characteristics of $s(t)$ match those expected for the chaotic drive. For example, let $S_{\tilde{x}}(f)$ denote the power density spectrum of the received drive signal and $S_x(f)$ denote the power density spectrum of a representative drive signal for the transmitter. As with the calculation in Eq. (4), we assume that the power density spectrum for the specific drive signal that generated the received $\tilde{x}(t)$ will be $S_x(f)$, *i.e.* that $S_x(f)$ is independent of the specific sample path $x(t)$. With $H(f)$ denoting the frequency response of the channel,

$$S_{\tilde{x}}(f) = |H(f)|^2 S_x(f) \quad . \tag{7}$$

Since $S_x(f)$ is known and $S_{\tilde{x}}(f)$ is estimated from the received data, $|H(f)|^2$ can be estimated. If the channel is assumed to be minimum phase, then $H(f)$ can be constructed from $|H(f)|^2$ using any of a variety of techniques such as spectral factorization, the complex cepstrum or the Hilbert transform.

To illustrate the loss of synchronization due to filtering and estimation of $|H(f)|^2$, consider the minimum phase first order filter with transfer function

$$H(s) = \frac{2.3}{s+1} \quad . \tag{8}$$

In Fig. 6 we show the frequency response magnitude and group delay for the filter. As indicated in Fig. 7, the significant spectral content of the Lorenz drive signal $x(t)$ extends well beyond 2 Hz, *i.e.* the one pole filter alters the spectral content significantly. In Fig. 8 we illustrate the effect of filtering on the synchronization. Specifically, Fig. 8(b) shows the instantaneous synchronization error when the channel is represented by the single pole filter and no compensation is applied. For comparison, in Fig. 8(a) we show the synchronization error with no channel distortion.

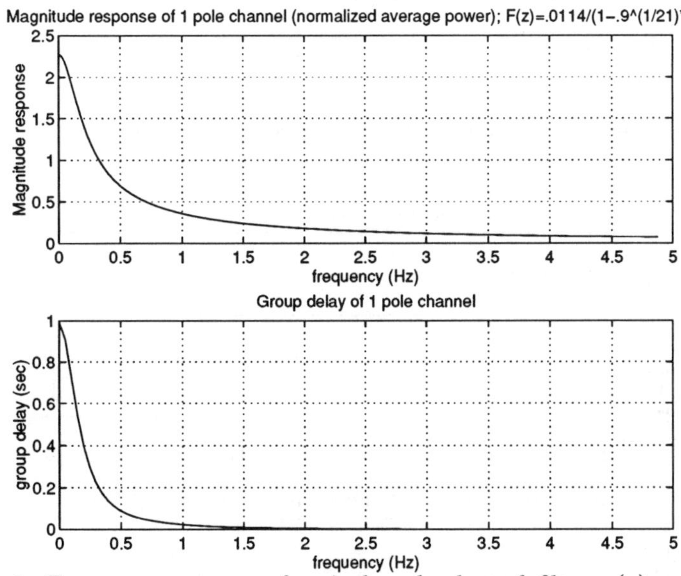

FIG. 6. Frequency response of a single pole channel filter. (a) magnitude response. (b) group delay.

For this example, $|H(f)|$ was estimated through the use of Eq. (7). The result is shown in Fig. 9 overlaid with the exact frequency response magnitude corresponding to Eq. (8). We see that for this simple and noise-free case, we are able to recover $|H(f)|$ over a much wider frequency range than would be required. We also note from Fig. 7 that since $S_x(f)$ is down by 100 dB at 10 Hz from its peak, any realistic noise in the channel would preclude estimating $H(f)$ out to even this frequency.

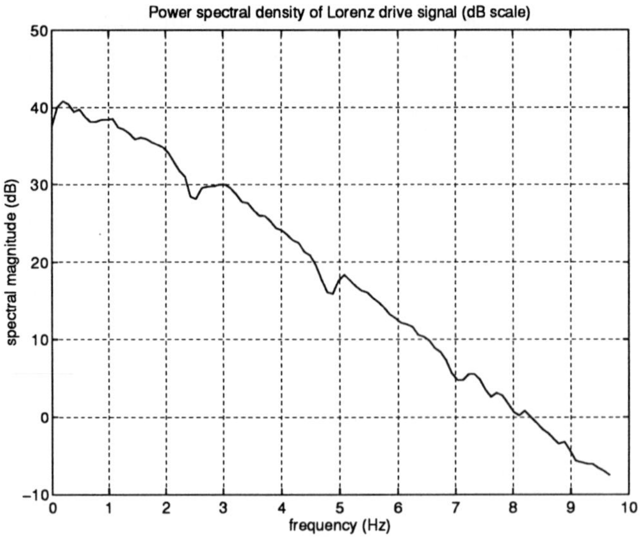

FIG. 7. Power spectral density of the Lorenz drive signal (dB scale).

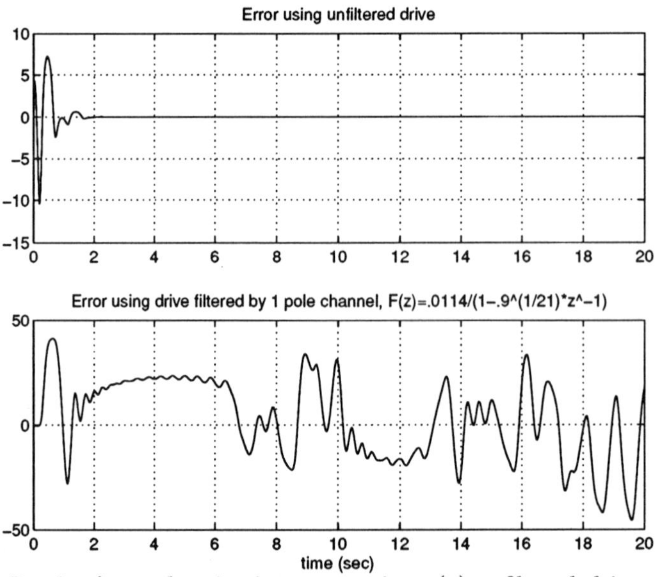

FIG. 8. Receiver's synchronization error using: (a) unfiltered drive signal; and (b) drive signal filtered with a single pole channel filter.

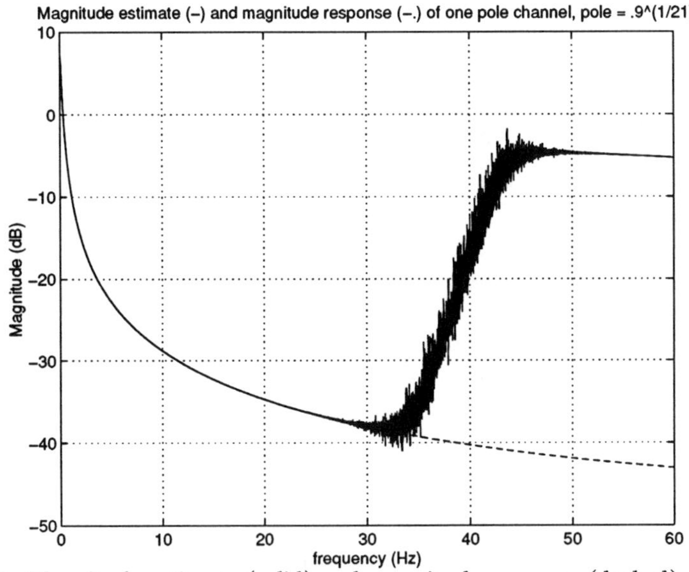

FIG. 9. Magnitude estimate (solid) and magnitude response (dashed) of a single pole channel filter.

As a second example of the effect on synchronization of channel filtering, we consider an all-pass channel with transfer function

$$H(s) = \frac{s + .995}{s - .995} \quad . \tag{9}$$

For this example, $|H(f)|^2 = 1$. The group delay for this channel is shown in Fig. 10. The instantaneous synchronization error is shown in Fig. 11(b). Again, for comparison, the synchronization error with the correct drive signal is shown in Fig. 11(a). As this example clearly demonstrates, exact compensation of the magnitude of the frequency response is not sufficient to eliminate synchronization error since the error in Fig. 11(b) is due entirely to distortion introduced in the drive signal by the group delay. We are currently developing and exploring several approaches to estimating and compensating for the group delay characteristics of the channel as well as the frequency response magnitude.

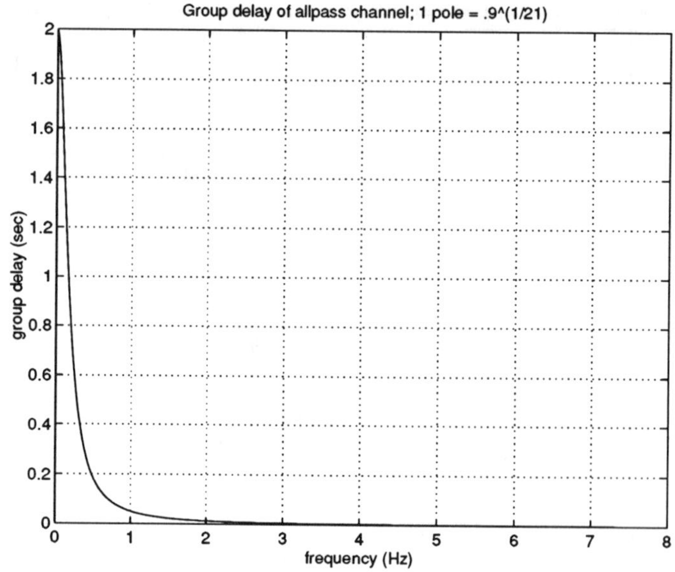

FIG. 10. Group delay of a single pole-zero allpass channel.

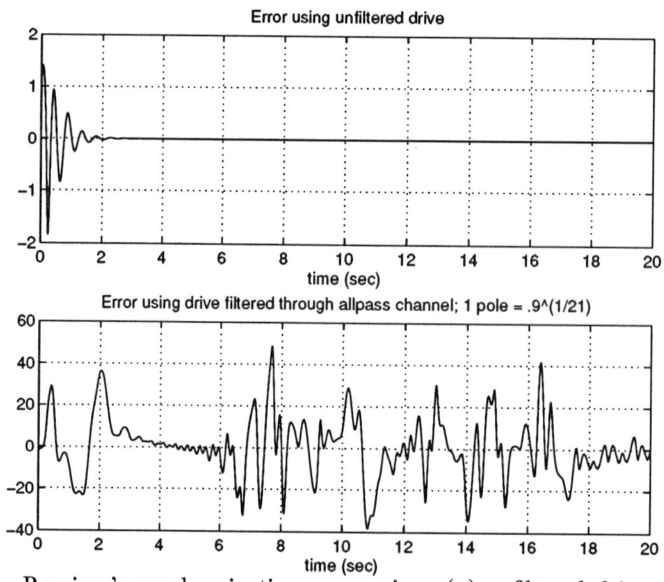

FIG. 11. Receiver's synchronization error using: (a) unfiltered drive signal; and (b) drive signal filtered with a single pole-zero allpass channel filter.

[1] Cuomo, K.M., Oppenheim, A.V., and Strogatz, S.H., "Robustness and signal recovery in a synchronized chaotic system," *Int. J. Bif. Chaos* **3**(6), pp. 1629-1638, 1993.

[2] Pecora, L.M., and Carroll, T.L., "Synchronization in chaotic systems," *Phys. Rev. Lett.* **64**(8), pp. 821-823, 1990.

[3] Carroll, T.L., and Pecora, L.M., "Synchronizing chaotic circuits," *IEEE Trans. Circuits and Syst.* **38**(4), pp. 453-456, 1991.

[4] Oppenheim, A.V., Wornell, G.W., Isabelle, S.H., and Cuomo, K.M., "Signal processing in the context of chaotic signals," in *Proc. 1992 IEEE ICASSP* **IV**, pp. 117-120, 1992.

[5] Cuomo, K.M., and Oppenheim, A.V., "Synchronized chaotic circuits and systems for communications," *Technical Report 575*, MIT Research Laboratory of Electronics, 1992.

[6] Cuomo, K.M., and Oppenheim, A.V., "Circuit implementation of synchronized chaos with applications to communications," *Phys. Rev. Lett.* **71**(1), pp. 65-68, 1993.

[7] Cuomo, K.M., Oppenheim, A.V., and Strogatz, S.H., "Synchronization of Lorenz-based chaotic circuits with applications to communications," *IEEE Trans. Circuits and Syst.* **40**(10), pp. 626-633, 1993.

[8] Kocarev, Lj., Halle, K.S., Eckert, K., and Chua, L.O., "Experimental demonstration of secure communications via chaotic synchronization," *Int. J. Bif. Chaos* **2**(3), pp. 709-713, 1992.

[9] Parlitz, U., Chua, L.O., Kocarev, Lj., Halle, K.S., and Shang, A., "Transmission of digital signals by chaotic synchronization," *Int. J. Bif. Chaos* **2**(4), pp. 973-977, 1992.

[10] Barda, A., and Laufer, S., "Multiple user receivers for CDMA networks based on chaotic signals," submitted to *IEEE Journal on Selected Areas in Communications*, special issue on *Wireless Local Communications*, 1995.

[11] Cuomo, K.M., Oppenheim, A.V., and Isabelle, S.H., "Spread spectrum modulation and signal masking using synchronized chaotic systems," *Technical Report 570*, MIT Research Laboratory of Electronics, 1992.

[12] Halle, K.S., Wu, C.W., Itoh, M., and Chua, L.O., "Spread spectrum communication through modulation of chaos," *Int. J. Bif. Chaos* **3**(2), pp. 469-477, 1993.

[13] Elmirghani, J.M.H., and Cryan, R.A., "New chaotic-based communication technique with multi-user provision," *Elect. Lett.* **30**(15), pp. 1206-1207, 1994.

[14] Papadopoulos, H.C., and Wornell, G.W., "Optimal Detection of a Class of Chaotic Signals," in *Proc. 1993 IEEE ICASSP* **III**, pp. 117-120, 1993.

[15] Abarbanel, H.D.I., and Linsay, P.S., "Secure communications and unstable periodic orbits of strange attractors," *IEEE Trans. Circuits and Syst.* **40**(10), pp. 643-645, 1993.

Detection Performance of Power-Law Processors for Random Signals of Unknown Location, Structure, Extent, and Strength

Albert H. Nuttall

Naval Undersea Warfare Center, New London, CT 06320

Abstract. A signal (if present) is located somewhere in a band of frequencies characterized by a total of N search bins. The signal occupies an arbitrary set of $\underline{M}$ of these bins, where not only is $\underline{M}$ unknown, but also, the locations of the particular $\underline{M}$ occupied bins are unknown. Also, the signal strength is unknown.

A class of processors, called the power-law processors, is investigated, in which the available data is raised to the ν-th power prior to summation over all data values. The receiver operating characteristics have been determined for values of power $\nu = 1, 2, 2.5, 3, \infty$ for a wide range of values of $\underline{M}$. These results allow for accurate extraction of required signal-to-noise ratios to achieve a specified level of performance, as measured by the false alarm and detection probabilities, P_f and P_d. One of the most surprising and useful results of this study is the discovery that the power-law processor with $\nu = 2.4$ performs near the absolute optimum, even without any knowledge of the number of occupied bins $\underline{M}$ or the signal-to-noise ratio.

INTRODUCTION

The reliable detection of a weak signal of unknown structure is a recurring problem in underwater acoustics and other fields. When the extent and location of the signal are also unknown, the difficulty is further compounded. The goal of this investigation is to determine a class of processors that function at or near the optimum level of performance, and that can be easily realized and analyzed, even in situations of scant knowledge about the detailed signal characteristics. We will derive the form of the optimum processor in this environment, and then simplify it to the point of realizing an alternative simple robust processor, namely the class of power-law processors. This power-law class will not require or use information such as the average signal-to-noise ratio per bin, $\underline{S}$, or the number of bins, $\underline{M}$, occupied by signal (when present); this is consistent with the fact that such information is not available in practical applications anyway.

Some previous results on the performance of power-law processors have been reported in [1] and [2]. However, they considered detection of a Gaussian burst signal in Gaussian noise, whereas our input data have exponential probability density functions. Also, they did not consider very small values of $\underline{M}$/N, where N is the total number of search bins in which the signal could lie; we will address cases of $\underline{M}$ varying all the way from $\underline{M} = 1$ to the large value $\underline{M} = N = 1024$. Their processor was arrived at in an ad hoc fashion, whereas ours is derived as an approximation to the optimum processor. Finally, they did not compare the performance of their power-law processors with any absolute bound on performance; we have undertaken that latter task and derived a very tight upper bound [3].

Although the previous authors couched their signal as a contiguous burst of $\underline{M}$ samples in time [2; page 210, above (1)], they never used this fact for deriving an optimum processor, nor in the ad hoc processor they adopted. Thus, their results actually apply to the broader class of signals whose occupancy pattern in time (or frequency) is completely unknown. The optimum processor for contiguous bursts in time (or frequency) would utilize this contiguous knowledge and perform significantly better, by virtue of being able to reduce the size of the search problem considerably. This case of increased signal location information is not addressed here.

PROBLEM DEFINITION

The search space consists of N (frequency) bins, each containing independent identically-distributed noises of unit power. This situation is presumed to be accomplished by an earlier normalization procedure. The number N is under our control and is always a known quantity. When signal is absent, hypothesis H_0, the probability density function of each of the bin output noises is completely known.

When signal is present, hypothesis H_1, the quantity $\underline{M}$ is the actual number of bins occupied by the signal; this is frequently an unknown parameter. The quantity $\underline{L}$ is the actual set of bins occupied by signal, when signal is present; for example, if $\underline{M} = 4$, then we might have for the occupied set, $\underline{L} = \{2,3,7,29\}$, meaning that bins 2,3,7,29 have signal in them. This quantity $\underline{L}$ is always unknown in our investigation. Finally, the quantities $\{\underline{S}_m\}$ are the actual average signal powers in the m-th bin in occupied set $\underline{L}$, when signal is present; these average signal powers are unknown. We shall presume here that all the actual signal powers per bin are equal to a common (unknown) value $\underline{S}$ in the occupied set of bins, $\underline{L}$, and zero elsewhere.

Probability Density Functions of Individual Bin Outputs

We now specify the detailed character of the probability density functions of the input data, namely p_0 and p_1, under hypotheses H_0 and H_1, respectively. In both hypotheses, the bin outputs or observations $\{\mathbf{x}_n\}$ are taken as the squared envelopes of the outputs of (disjoint) narrowband filters subject to a Gaussian input random process; alternatively, the observations can be interpreted as the magnitude-squared outputs of a fast Fourier transform subject to a Gaussian input process. It is assumed that these bin outputs (random variables) $\{\mathbf{x}_n\}$ are statistically independent of each other, which is consistent with a frequency-disjoint requirement.

Since the bin output noises have been normalized at unit level, the probability density function of the n-th observation $\mathbf{x}_n$ is, under hypothesis H_0, an exponential of the form

$$q_0(u_n) = \exp(-u_n) \quad \text{for } u_n > 0 \,, \quad 1 \le n \le N \,. \tag{1}$$

On the other hand, when signal is present, hypothesis H_1, with signal power $\underline{S}$ per bin, the density of $\mathbf{x}_n$ is changed to

$$q_1(u_n) = \underline{a} \exp(-\underline{a}\, u_n) \quad \text{for } u_n > 0 \ , \quad n \in \underline{L} \ , \tag{2}$$

where we have defined the strength parameter

$$\underline{a} = \frac{1}{1 + \underline{S}} \ . \tag{3}$$

Observe that the actual signal power per bin, $\underline{S}$, can also be interpreted as the actual signal-to-noise power ratio per bin, since the noise power per bin has been normalized at unity.

Joint Probability Density Functions of the Set of Observations

The joint probability density function governing the complete observation $\{\mathbf{x}_n\}$ under hypothesis H_0 follows from (1) and the statistical independence of $\{\mathbf{x}_n\}$ as

$$p_0(u_1,\ldots,u_N) = \prod_{n=1}^{N} \{\exp(-u_n)\} \ . \tag{4}$$

Under hypothesis H_1, the signal can land in any set of $\underline{M}$ disjoint bins, out of the total of N search bins. This results in a total number of possibilities $K = (N|\underline{M})$, where the latter quantity is the binomial coefficient. Furthermore, each set occurs with equal probability 1/K. Thus, there are K occupancy sets, each of size $\underline{M}$, namely $\{L_k\}$ for $1 \le k \le K$. The joint probability density function governing the observation $\{\mathbf{x}_n\}$ is therefore

$$p_1(u_1,\ldots,u_N) = \sum_{k=1}^{K} \left[\frac{1}{K} \prod_{n \in L_k} \{\underline{a} \exp(-\underline{a} u_n)\} \prod_{n \notin L_k} \{\exp(-u_n)\}\right] \ , \tag{5}$$

where we used (1) and (2).

DERIVATION OF OPTIMUM PROCESSOR

We initiate the investigation of optimum signal processing and detection by presuming, for the time being, that the number of bins, $\underline{M}$, occupied by signal (when present), and the common average signal power per bin, $\underline{S}$, are both known, but that the specific locations, $\underline{L}$, of the bins occupied by signal are completely unknown. This will enable us to derive the optimum (unachievable) processor in this environment.

Then, we will simplify this complicated time-consuming likelihood ratio procedure, eliminating the dependence on the parameters $\underline{M}$ and $\underline{S}$ in the process; this end result is consistent with the practical situation, where these parameters are unknown. The resulting approximations will turn out to yield the class of power-law processors, which will occupy most of the succeeding analysis and performance prediction.

The likelihood ratio for observation $\{\mathbf{x}_n\}$ is, from (4) and (5), the random variable

$$LR \equiv \frac{p_1(x_1,\ldots,x_N)}{p_0(x_1,\ldots,x_N)} = \frac{1}{K}\, \underline{a}^{\underline{M}} \sum_{k=1}^{K} \left[\prod_{n \varepsilon L_k} \{\exp(x_n[1 - \underline{a}])\}\right] =$$

$$= \frac{1}{K}\, \underline{a}^{\underline{M}} \sum_{k=1}^{K} \exp\left(\underline{w}\; X_k\right) , \tag{6}$$

where we have defined

$$\underline{w} \equiv 1 - \underline{a} = \frac{\underline{S}}{1 + \underline{S}} , \qquad X_k \equiv \sum_{n \varepsilon L_k} x_n \quad \text{for } 1 \le k \le K = \binom{N}{\underline{M}} . \tag{7}$$

The likelihood ratio test is therefore

$$\sum_{k=1}^{K} \exp\left(\underline{w}\; X_k\right) \gtrless v , \tag{8}$$

where v is a fixed threshold. The likelihood ratio test indicates to compute <u>all</u> K possible linear sums $\{X_k\}$ in (7), each of size $\underline{M}$, weight each by common value $\underline{w}$, exponentiate each term, and sum over all the possible K sets. For large N, the integer K is very large; then, (8) is a very time-consuming impractical prescription and requires knowledge of $\underline{S}$, which is very unlikely.

Special Cases

As a special case, for $\underline{M} = 1$, then $K = (N|1) = N$, and (7) and (8) reduce to

$$\sum_{n=1}^{N} \exp(\underline{w}\; x_n) \gtrless v . \tag{9}$$

This optimum processor can be easily realized if $\underline{S}$ is known; it can also be readily simulated in order to determine its performance.

At the other extreme, for the special case of $\underline{M} = N$, then $K = (N|N) = 1$, and (7) and (8) become

$$\exp\left(\underline{w} \sum_{n=1}^{N} x_n\right) \gtrless v , \quad \text{or equivalently,} \quad \sum_{n=1}^{N} x_n \gtrless v , \tag{10}$$

where the latter v is a different threshold value, of course. This processor, called the energy detector, does not require knowledge of $\underline{S}$ or $\underline{M}$. It has already been analyzed, and receiver operating characteristics have been obtained [4; pages 21 - 22 and 81 - 90].

Quadratic Simplification of Likelihood Ratio Test

In an attempt to simplify the likelihood ratio test (8), suppose we approximate exponential function exp(x) by quadratic function $\alpha + \beta\, x^2$ over the range of typical arguments x likely to be encountered. The quadratic does not emphasize the very large arguments as much as the exponential does, but it does give more importance to larger arguments than a linear function would. The cubic alternative $\alpha + \beta x^3$ will be considered below.

When we substitute the quadratic approximation into (8), there follows

$$\sum_{k=1}^{K} \exp\left(\underline{w}\, X_k\right) \cong \sum_{k=1}^{K} \left(\alpha + \beta\, \underline{w}^2\, X_k^2\right) = K\,\alpha + \beta\, \underline{w}^2 \sum_{k=1}^{K} X_k^2 \,. \qquad (11)$$

The corresponding approximate likelihood ratio test is therefore simply

$$Q_2 \equiv \sum_{k=1}^{K} X_k^2 \gtrless v \,, \qquad (12)$$

where we have discarded all the data-independent factors that we can. A very important feature of test (12) is that knowledge of average signal power per bin, $\underline{S}$, is not required for its realization. However, knowledge of $\underline{M}$ is still required for test (12). Also, K can be too large to make test (12) practical.

In order to simplify (12) further, we expand each and every linear sum X_k in (12) in terms of its particular $\underline{M}$ components $\mathbf{x}_n$ according to set L_k in (7). We also define the data power sum

$$\mathbf{T}_\nu \equiv \sum_{n=1}^{N} \mathbf{x}_n^\nu \,, \qquad (13)$$

where ν need not be integer. Then, after some manipulation, expansion of (12) yields the identity

$$Q_2 = \binom{N-1}{\underline{M}-1} \left(a\, \mathbf{T}_2 + b\, \mathbf{T}_1^2\right) , \qquad (14)$$

where coefficients

$$a = \frac{N-\underline{M}}{N-1} \,, \qquad b = \frac{\underline{M}-1}{N-1} \,; \qquad a + b = 1 \,. \qquad (15)$$

For $\underline{M} = 1$, a is 1, while b is zero; thus, Q_2 in (14) is then composed solely of the <u>quadratic</u> sums of the data $\{\mathbf{x}_n\}$, namely $\mathbf{T}_2$; see (13). Conversely, for $\underline{M} = N$, a is zero, while b is 1; that is, Q_2 in (14) is then converted over to the square of a sum of <u>linear</u> terms in the data $\{\mathbf{x}_n\}$, namely $\mathbf{T}_1$. In between these two extreme values of $\underline{M}$, the changeover in the data-dependent part of Q_2 is linear in $\underline{M}$ for both coefficients a and b, as indicated by (15). Thus, there is a smooth transition in Q_2 in (14), from using only squared terms to using only linear terms in the data $\{\mathbf{x}_n\}$, as $\underline{M}$ varies from 1 to N.

On the other hand, if we use the cubic function $\alpha + \beta\, x^3$ as an approximation to the exponential function exp(x), then (8) can be developed as follows:

$$\sum_{k=1}^{K} \exp\left(\underline{w}\, X_k\right) \cong \sum_{k=1}^{K} \left(\alpha + \beta\, \underline{w}^3\, X_k^3\right) = K\,\alpha + \beta\, \underline{w}^3 \sum_{k=1}^{K} X_k^3 \,. \qquad (16)$$

The corresponding approximate likelihood ratio test is then

$$Q_3 \equiv \sum_{k=1}^{K} X_k^3 \gtrless v \,. \qquad (17)$$

Although simpler than original test (8), test (17) still requires too many terms to make it practical.

Again, as above, we substitute for each term X_k in (17), its linear composition (7) in terms of data $\{x_n\}$, and expand out all the cubic terms. The result is, after considerable manipulation, the identity

$$Q_3 = \binom{N-1}{\underline{M}-1} \left(a\, T_3 + b\, T_2\, T_1 + c\, T_1^3\right) , \qquad (18)$$

where coefficients

$$a = \frac{N-\underline{M}}{N-1}\,\frac{N-2\underline{M}}{N-2} \,, \qquad b = 3\,\frac{\underline{M}-1}{N-1}\,\frac{N-\underline{M}}{N-2} \,, \qquad c = \frac{\underline{M}-1}{N-1}\,\frac{\underline{M}-2}{N-2} \,. \qquad (19)$$

For $\underline{M} = 1$, we have a = 1, b = 0, c = 0; thus, sum Q_3 in (18) is then composed solely of <u>cubic</u> terms in the data $\{x_n\}$, according to T_3 in (13). Conversely, for $\underline{M} = N$, we find a = 0, b = 0, c = 1; that is, sum Q_3 is then converted over to the cube of a sum T_1 involving only <u>linear</u> terms in the data $\{x_n\}$. In between these two extreme values of $\underline{M}$, the changeover in the data-dependent part of Q_3 is quadratic in $\underline{M}$ for all three coefficients a, b, c. Coefficient b reaches its peak at $\underline{M} = (N+1)/2$, with a value approximately 3/4, while c is about 1/4 there. A list of values of the coefficients a, b, c at selected values of $\underline{M}$ is given below. We always have a + b + c = 1, as may be verified from (19).

Values of Coefficients

$\underline{M}$	a	b	c
1	1	0	0
2	$\frac{N-4}{N-1}$	$\frac{3}{N-1}$	0
$\frac{N}{2}$	0	$\frac{3}{4}\frac{N}{N-1}$	$\frac{1}{4}\frac{N-4}{N-1}$
$\frac{N+1}{2}$	$\frac{1}{2}\frac{-1}{N-2}$	$\frac{3}{4}\frac{N-1}{N-2}$ (max)	$\frac{1}{4}\frac{N-3}{N-2}$
$\frac{3}{4}N$	$\frac{-N^2/8}{(N-1)(N-2)}$ (min)	$\frac{9}{16}\frac{N(N-4/3)}{(N-1)(N-2)}$	$\frac{9}{16}\frac{(N-4/3)(N-8/3)}{(N-1)(N-2)}$
N	0	0	1

Observations on Data Processing

The cubic-approximation results in (18) and (19) reveal that the optimum processor indicates a gradual transition from emphasis on cubed data, through squared data, to linear data, as $\underline{M}$ is increased from 1 to N. Coupled with the similar result for the quadratic approximation in (14) and (15), this suggests that one should consider the use of sum T_ν in (13), for general ν, as the decision variable and as a possible candidate for near-optimum processing as $\underline{M}$ varies.

It should be observed that some rather drastic approximations have been adopted in the extraction of T_ν as a candidate decision variable. First, the exponential function in (8) has been replaced by either a quadratic or a cubic function. Then, the resulting terms have been expanded out and collected in a particular power form. Finally, the behaviors of the coefficients of these power forms have been used in a heuristic fashion to deduce the power-law processor.

The best choice of ν is an open question at this point, as is the specific degradation incurred by resorting to the approximation T_ν as the decision variable, instead of using the optimum likelihood ratio test (8). Analytical and/or simulation comparisons of (8) and T_ν must be conducted in order to ascertain the exact degradations in performance that accompany the various alternative approximations to the optimum likelihood ratio test, and to determine the best value(s) of ν to use for minimum degradation.

The use of power sum T_ν in (13) as the decision variable is very attractive from a computational viewpoint. It involves only N calculations, not K, which are extremely simple if ν is an integer, like 1 or 2 or 3. Other intermediate (non-integer) values of ν would take longer. Also, no knowledge of $\underline{M}$ or $\underline{S}$ is required to realize the approximate likelihood ratio test

$$T_\nu = \sum_{n=1}^{N} x_n^\nu \gtrless v . \qquad (20)$$

This is called the class of power-law processors, and will be the major item of numerical investigation here.

For $\nu = 1$, the power-law processor corresponds to the energy detector [4; pages 21 - 22]. On the other hand, as $\nu \to \infty$, power-law processor (20) tends towards the maximum processor; that is, the performance of (20) tends towards that of $\max\{x_1,\ldots,x_N\}$. This maximum processor has already been thoroughly analyzed and quantified in a study on the sum-of-M-largest processor in [5], when M = 1 there.

The power-law nature of test (20) emphasizes the stronger samples in the given data $\{x_n\}$ over the weaker ones, in making its decision about signal presence or absence when the occupied set of bins, $\underline{L}$, is completely unknown; recall that the data $\{x_n\}$ can be interpreted themselves as power quantities, being the envelope-squared outputs of narrowband filters. Also, in keeping with optimum test (8), approximate test (20) reaches its decision based upon a combination of all the data values $\{x_n\}$, and not upon a maximum of some subsets of the data.

ANALYSIS OF POWER-LAW PROCESSORS

We are interested in the performance of power-law processor (20) when the statistics of the input data $\{x_n\}$ are given by (1) and (2). The specific results for $\nu = 1$ are given by [4; pages 21 - 22 and 81 - 90], while the case of $\nu = \infty$, that is, the maximum processor, $\max\{x_1,\ldots,x_N\}$, is treated in [5; (30) - (33) and tables 1 and 2].

Quadratic-Law Processor

In this subsection, we consider the special case of $\nu = 2$, for which we have the test

$$z \equiv T_2 = \sum_{n=1}^{N} x_n^2 \equiv \sum_{n=1}^{N} y_n \gtrless v . \qquad (21)$$

Under hypothesis H_1, the characteristic function of the individual auxiliary random variable y_n is

$$f_y(\xi) \equiv \overline{\exp(i\xi y_n)} = \overline{\exp\left(i\xi x_n^2\right)} = \int_0^\infty du\ \underline{a}\ \exp(-\underline{a}u + i\xi u^2) =$$

$$= (1+i)\ \underline{a}\left(\frac{\pi}{8\xi}\right)^{\frac{1}{2}} w\left(\underline{a}\,\frac{-1+i}{(8\xi)^{\frac{1}{2}}}\right) \quad \text{for } \xi > 0 , \qquad (22)$$

where we used (2) and the error function w( ) of complex argument [6; chapter 7]. An alternative expression for this characteristic function, in terms of the real auxiliary functions f and g defined in [6; 7.3.5 & 7.4.22 and 7.3.6 & 7.4.23], is given by

$$f_y(\xi) = \underline{a}\left(\frac{\pi}{2\xi}\right)^{\frac{1}{2}}\left[f\left(\frac{\underline{a}}{(2\pi\xi)^{\frac{1}{2}}}\right) + i\, g\left(\frac{\underline{a}}{(2\pi\xi)^{\frac{1}{2}}}\right)\right] \quad \text{for } \xi > 0 . \tag{23}$$

To complete the characteristic function, we have $f_y(0) = 1$ and $f_y(-\xi) = f_y(\xi)^*$, to complement the results in (22) and (23).

The characteristic function of $\mathbf{y}_n$ under hypothesis H_0 is obtained from (22) or (23) by setting average signal power $\underline{S}$ equal to 0, that is, by setting strength parameter $\underline{a}$ equal to 1; denote it by $f_y^o(\xi)$. Then, using the independence of input data $\{\mathbf{x}_n\}$ in sum (21), the characteristic function of decision variable z under H_0 is

$$f_z^o(\xi) = \left(f_y^o(\xi)\right)^N . \tag{24}$$

On the other hand, the characteristic function of z under H_1 is, regardless of which $\underline{M}$ bins are occupied, given by

$$f_z(\xi) = \left(f_y^o(\xi)\right)^{N-\underline{M}} \left(f_y(\xi)\right)^{\underline{M}} , \tag{25}$$

where $\underline{M}$ is the actual number of bins containing signal. At this point, we have all the results we need in order to compute false alarm and detection probabilities of the quadratic-law processor in (21). The numerical method we employ is the accurate and efficient fast Fourier transform procedure for going directly from a given characteristic function to its exceedance distribution function, as given in [7].

The false alarm probability,

$$P_f = \text{Prob}(z > v | H_0) , \tag{26}$$

obtained through use of (22) and (24), is plotted versus threshold v in figure 1, for total search sizes N = 64, 128, 256, 512, 1024. The detection probability, defined as

$$P_d = \text{Prob}(z > v | H_1) , \tag{27}$$

is obtained by use of (22) and (25). A sample receiver operating characteristic for N = 1024 and $\underline{M}$ = 16, with average signal power per bin $\underline{S}$ (in decibels) as a parameter, is given in figure 2. A complete set of receiver operating characteristics for N = 1024 and

$$\underline{M} = 1, 2, 3, 4, 8, 16, 32, 64, 128, 256, 512, 1024, \tag{28}$$

is given in [8; appendix A].

Since these results used a very accurate computer routine for the complex error function w( ) in (22), the false alarm probabilities in the 1E-6 range are accurate, and the receiver

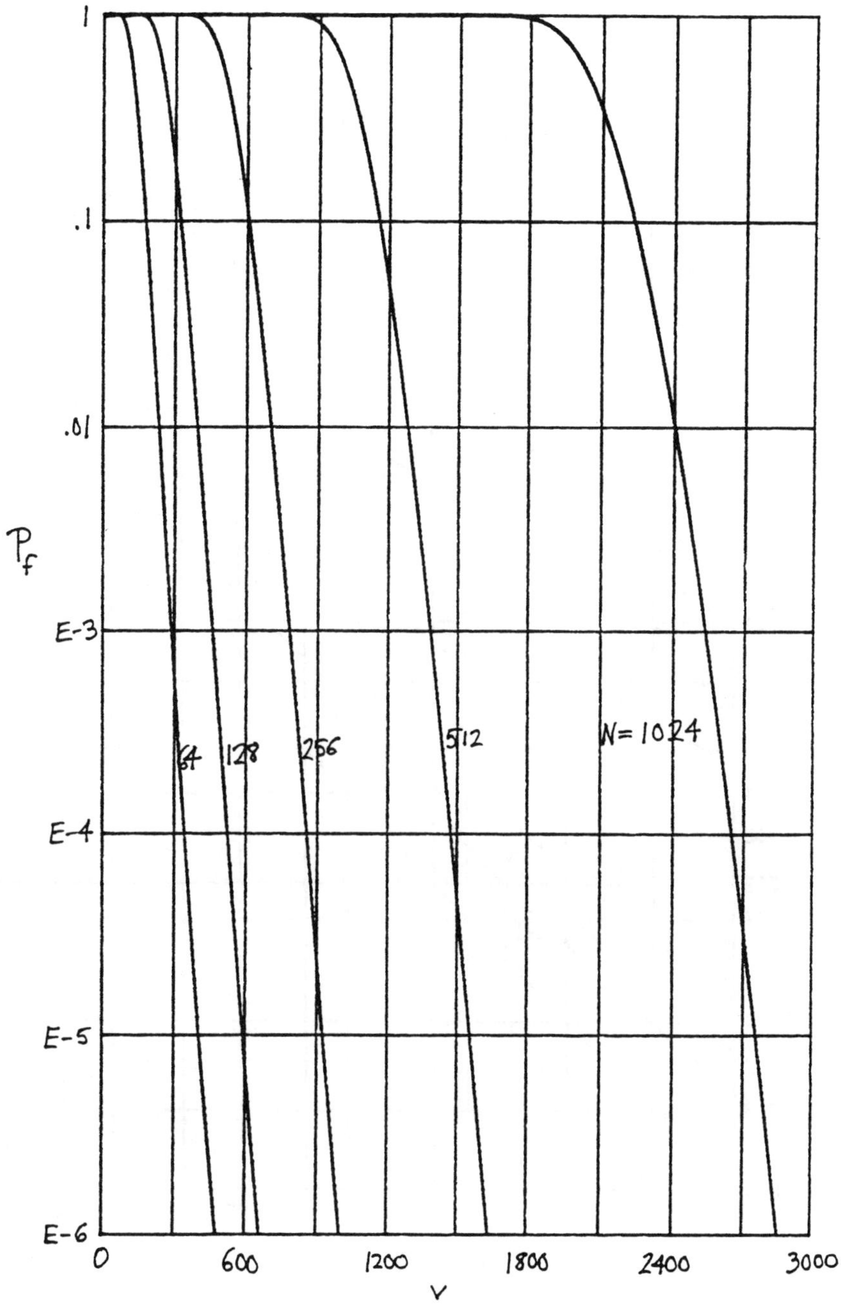

Fig. 1. False Alarm Probability for Quadratic-Law Processor

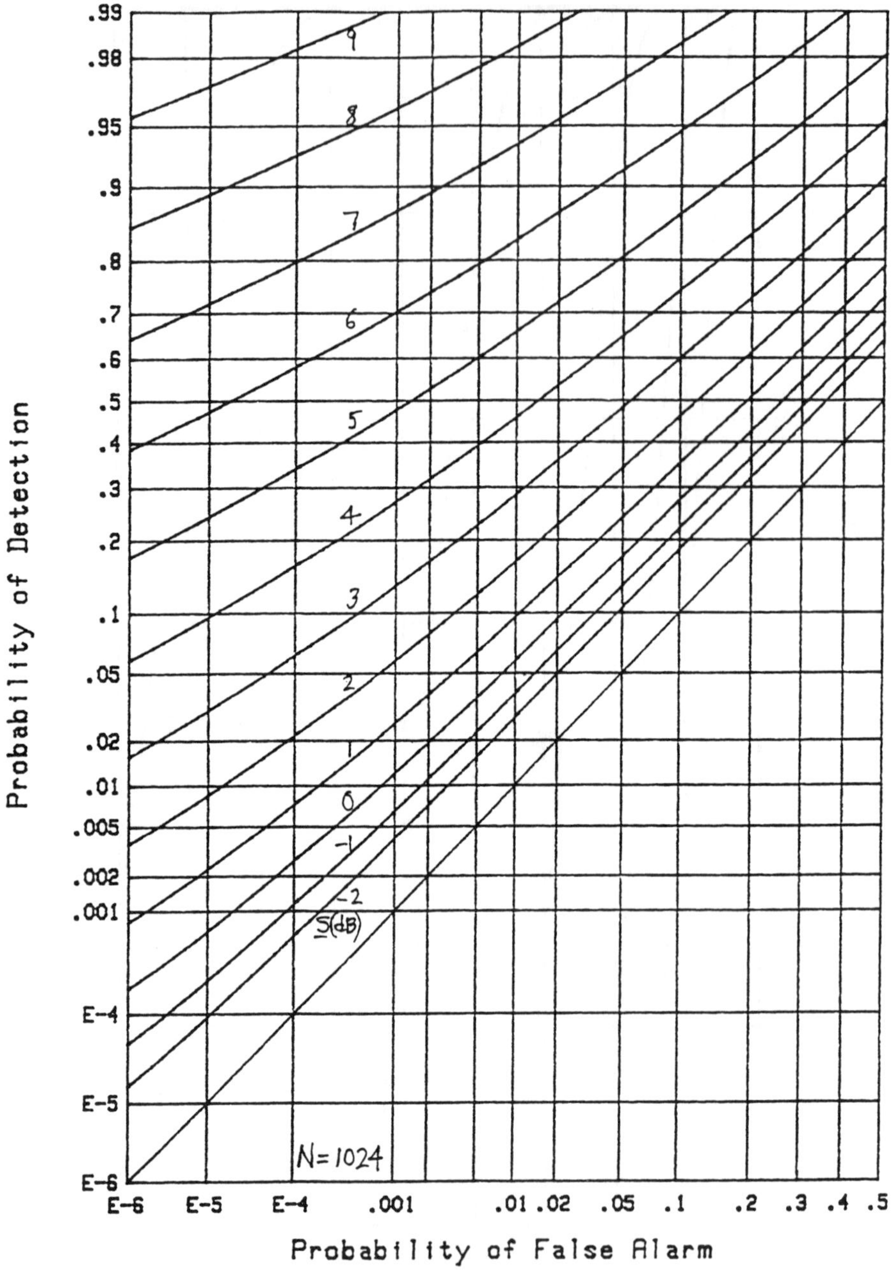

Fig. 2. Operating Characteristic for ν = 2, M = 16

operating characteristics are accurate over the complete range plotted. These curves enable us to extract the required signal-to-noise ratio per bin to realize a specified level of performance, such as $P_f = 1E-3$, $P_d = 0.5$. For example, for $N = 1024$ and $\underline{M} = 1$, we find that we must have the large value $\underline{S} = 14.8$ dB, whereas for $N = 1024$ and $\underline{M} = 1024$, we requires the much smaller value $\underline{S} = -9.4$ dB per bin, since the number of occupied bins, $\underline{M}$, is so much larger.

Cubic-Law Processor

Here, we consider the case of power law $\nu = 3$ in test (20), namely

$$z \equiv T_3 = \sum_{n=1}^{N} \mathbf{x}_n^3 \equiv \sum_{n=1}^{N} \mathbf{y}_n \underset{<}{>} v \ . \qquad (29)$$

Under hypothesis H_1, the characteristic function of an individual auxiliary random variable $\mathbf{y}_n$ is

$$f_y(\xi) \equiv \overline{\exp(i\xi\mathbf{y}_n)} = \overline{\exp\left(i\xi\mathbf{x}_n^3\right)} = \int_0^\infty du\ \underline{a}\exp(-\underline{a}u + i\xi u^3) =$$

$$= \underline{a}\,\frac{\sqrt{3}+i}{2}\int_0^\infty dr\ \exp\left(-\xi r^3 - \underline{a}\frac{\sqrt{3}}{2}r - i\,\underline{a}\frac{r}{2}\right) \quad \text{for } \xi \geq 0 \ , \qquad (30)$$

where we used (2). In order to obtain the latter integral form, we moved the contour in the u-plane to the radial line with angle $\pi/6$ radians and then made the change of variable $u = r\exp(i\pi/6)$.

The characteristic function of $\mathbf{y}_n$ under H_0, denoted by $f_y^o(\xi)$, follows from (30) by setting $\underline{a} = 1$ ($\underline{S} = 0$), and the characteristic function of decision variable $\mathbf{z}$ in (29) is again given by form (24). The lack of any computer routine for the complex function of ξ in (30) caused us to adopt the following numerical procedure. Integral (30) with $\underline{a} = 1$ ($\underline{S} = 0$) was accurately evaluated on the fine grid $\xi = \overline{0(.0001)}3$ and stored. These values were then used in (24) to evaluate the characteristic function of z under hypothesis H_0, for values of N = 64, 128, 256, 512, 1024. The numerical procedure in [7] was then employed to compute the false alarm probability, with the results given by figure 3. The curvatures in the results at the 1E-6 level are accurate, and are due to the cubic rule in (29).

The detection probability P_d could also have been obtained in a similar fashion, by using (30) with strength parameter $\underline{a} < 1$ ($\underline{S} > 0$). However, since we are generally interested in detection probability values in the neighborhood of 0.5 to 0.9, it was decided to avoid the bookkeeping, and to instead directly simulate test (29) for signal powers $\underline{S} > 0$. The results for $N = 1024$ and the example of $\underline{M} = 16$ again is given in figure 4. This plot utilized at least 10,000 independent trials of random variable z in (29); thus, the receiver operating characteristics

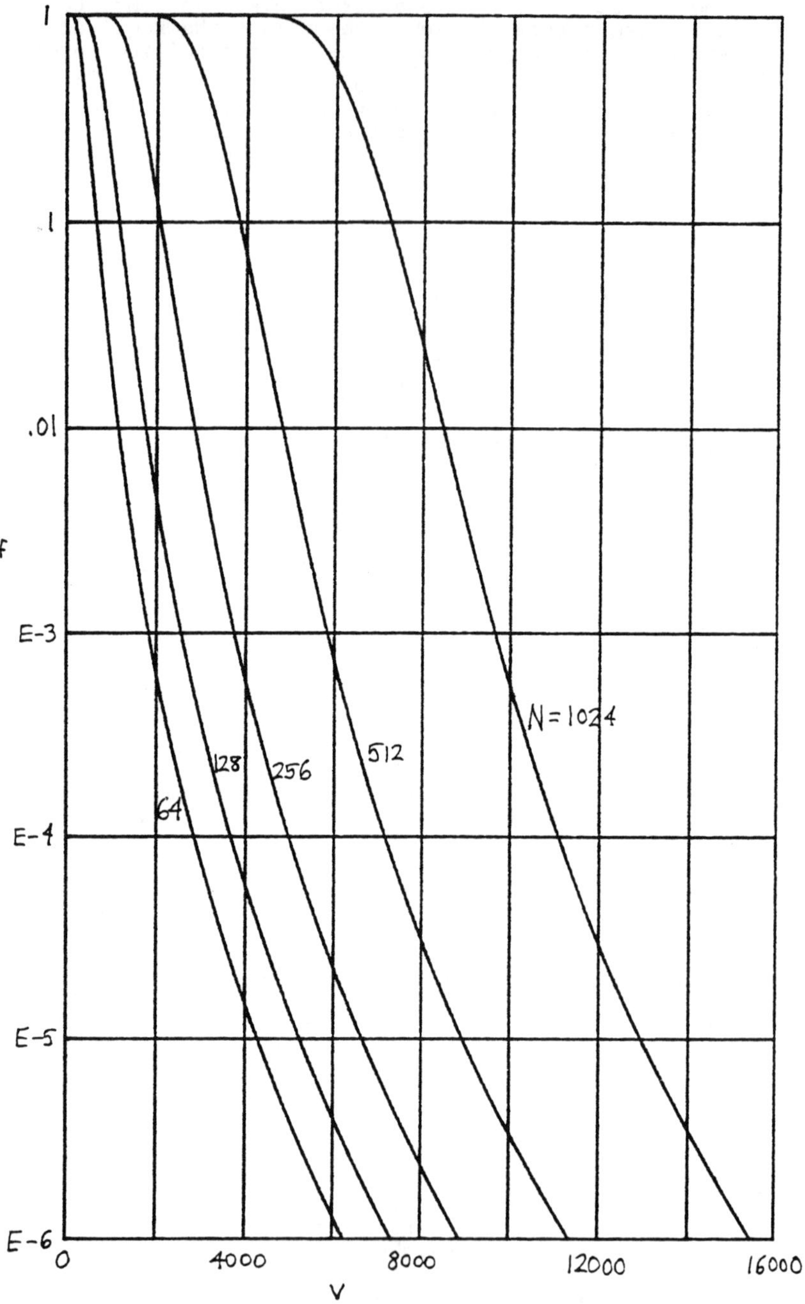

Fig. 3. False Alarm Probability for Cubic-Law Processor

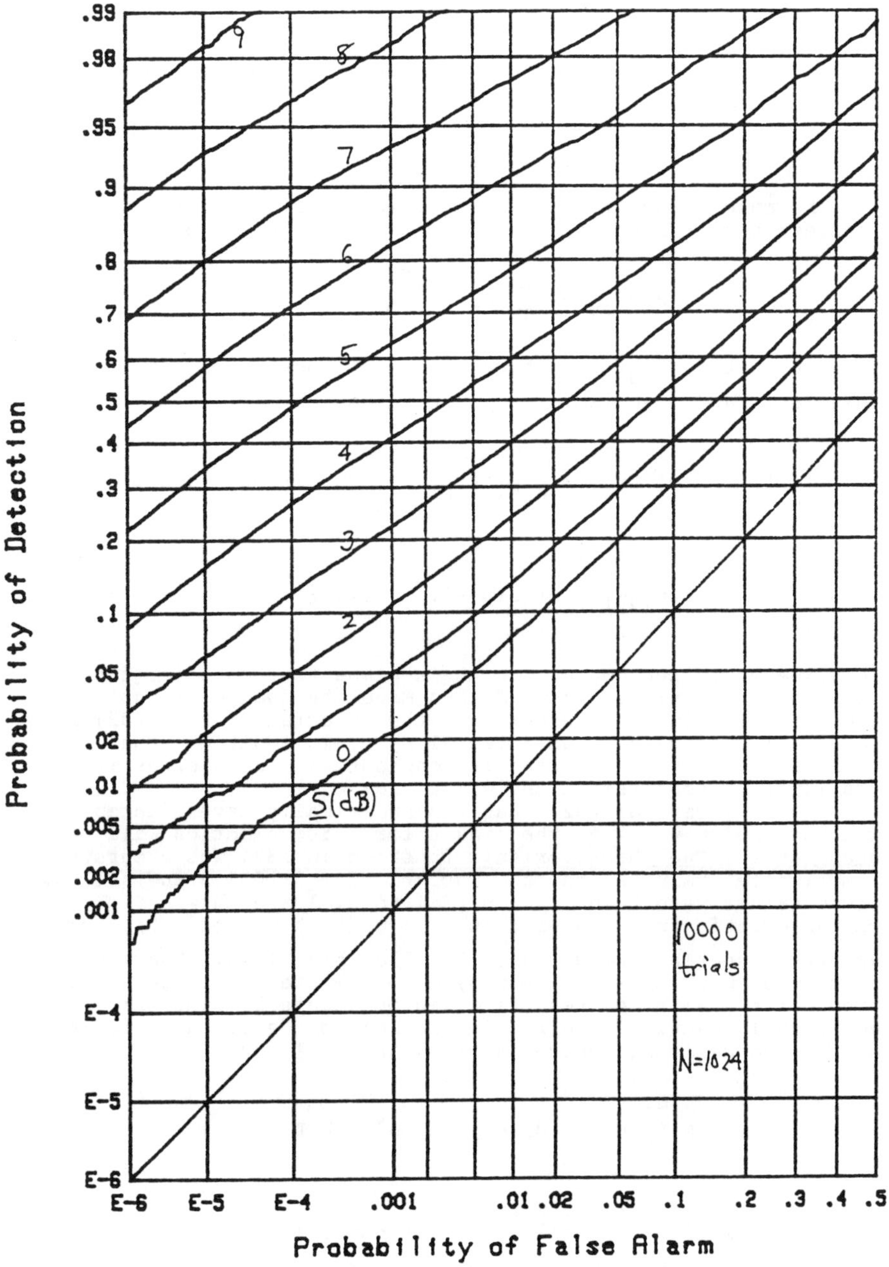

Fig. 4. Operating Characteristic for ν = 3, $\underline{M}$ = 16

are very stable except for very small P_d values of no practical interest. There is sufficient stability to be able to accurately determine required signal-to-noise ratios to operate in the desired 0.5 to 0.9 range of detection probabilities.

General Power-Law Processor

The case of general power $\nu > 1$ in power-law test (20) is of interest here. That is, the decision variable is now

$$\mathbf{z} \equiv \mathbf{T}_\nu = \sum_{n=1}^{N} \mathbf{x}_n^\nu \equiv \sum_{n=1}^{N} \mathbf{y}_n \gtrless v \ . \tag{31}$$

Under hypothesis H_0, the characteristic function of individual auxiliary random variable $\mathbf{y}_n$ is, upon use of (1),

$$f_y^o(\xi) \equiv \overline{\exp(i\xi \mathbf{y}_n)} = \overline{\exp\left(i\xi \mathbf{x}_n^\nu\right)} = \int_0^\infty du \exp(-u + i\xi u^\nu) =$$

$$= \exp\left(\frac{i\pi}{2\nu}\right) \int_0^\infty dr \exp\left[-\xi r^\nu - \exp\left(\frac{i\pi}{2\nu}\right) r\right] \ . \tag{32}$$

Here, we moved the contour in the u-plane to the radial line with angle $\pi/(2\nu)$ radians and then made the change of variable $u = r \exp(i\pi/(2\nu))$. Since $\nu > 1$, the integral on r in (32) has more rapid decay than the integral on u; in addition, the oscillation of the integrand is constant with r, whereas it continually increases in the u integral.

The only case of (32) that was numerically investigated here was the choice $\nu = 2.5$; the reason for this selection will be seen later. The characteristic function in (32) was accurately numerically evaluated on the fine grid $\xi = 0(.001)3$ and stored. These values were then used in (24) to evaluate the characteristic function of $\mathbf{z}$ in (31) under hypothesis H_0, for values of search size N = 64, 128, 256, 512, 1024. The numerical procedure in [7] was then employed to compute the false alarm probability, with the results given by figure 5. The only additional quantity required is the mean of decision variable $\mathbf{z}$ under hypothesis H_0, which is given generally by

$$\overline{\mathbf{z}} = N \ \overline{\mathbf{x}_n^\nu} = N \int_0^\infty du \exp(-u) \ u^\nu = N \ \Gamma(\nu+1) \ , \tag{33}$$

and which equals $N \sqrt{\pi} \ 15/8$ for $\nu = 2.5$.

The detection probability P_d for power-law test (31), with $\nu = 2.5$, was simulated for signal powers $\underline{S} > 0$. The results for $N = 1024$ and for the same value $\underline{M} = 16$ is given in figure 6. This plot also utilized at least 10,000 independent trials of random variable $\mathbf{z}$ in (31); thus, the receiver operating characteristics are very stable except for very small P_d values of no practical interest.

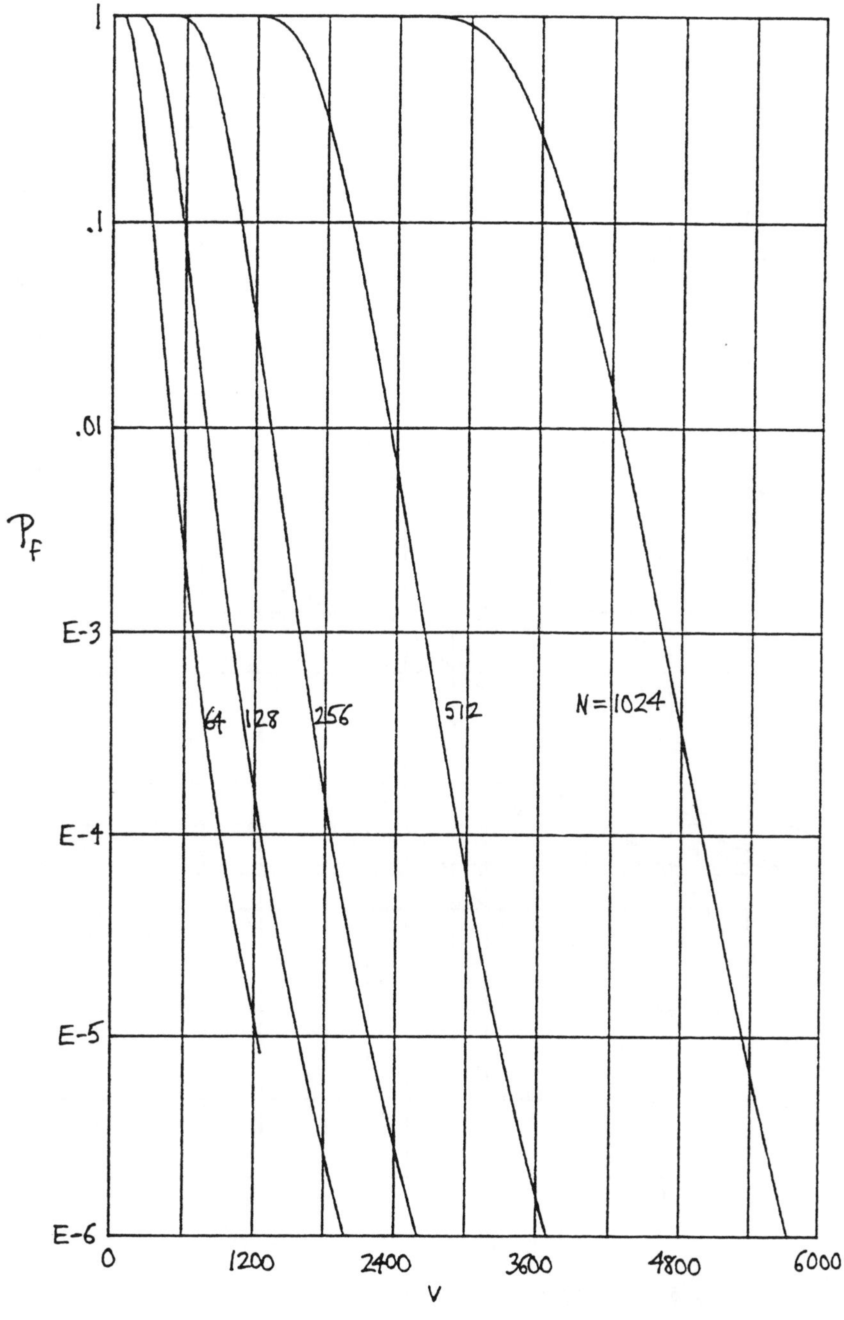

Fig. 5. False Alarm Probability for Power-Law Processor ν=2.5

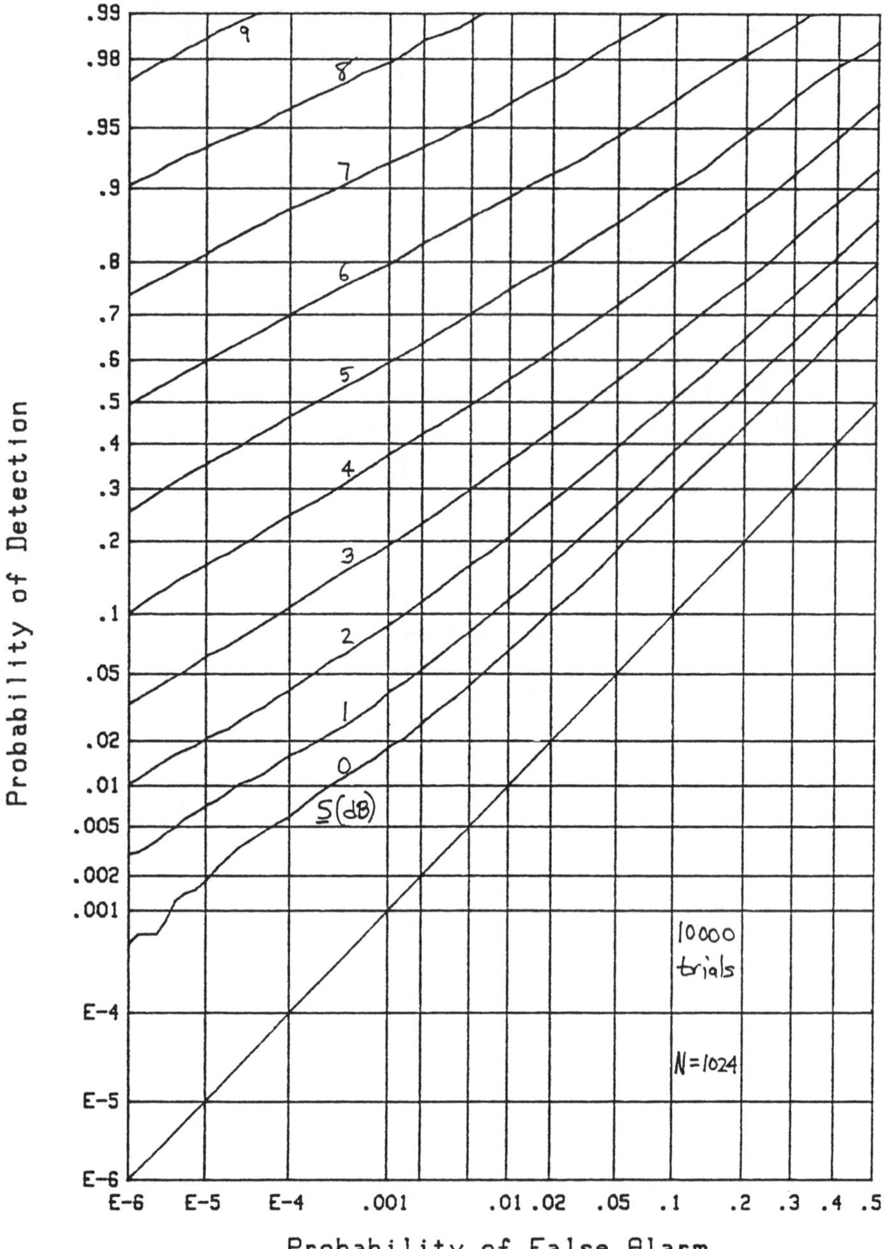

Fig. 6. Operating Characteristic for ν = 2.5, M = 16

COMPARISON OF POWER-LAW PROCESSORS

The compilation of 36 receiver operating characteristics in [8] for $\nu = 2, 2.5$, and 3 prompts us to condense this information for easier interpretation and accessibility. To accomplish this, we define a low-quality operating point $P_f = 1E-3$, $P_d = 0.5$, and a high-quality operating point $P_f = 1E-6$, $P_d = 0.9$. We then read off those curves the values of signal power per bin, $\underline{S}$(dB), which are required to realize these two levels of performance. These results are listed in tables 1 and 2, and are plotted in figures 7 and 8 for the low-quality and high-quality operating points, respectively, for $\underline{M}$ ranging over the set of values 1, 2, 3, 4, 8, 16, 32, 64, 128, 256, 512, 1024. The values of the power-law considered are $\nu = 1, 2, 2.5, 3, \infty$. The results for $\nu = 1$ and $\nu = \infty$ come from [5; pages 40 - 41]. The ordinate, $\underline{M}\,\underline{S}$ in decibels, in figures 7 and 8 is actually the total signal power required in order to achieve the specified performance level. This total quantity is more meaningful and it condenses the range of ordinate values to a more manageable regime.

It is immediately seen that the best value of ν, in order to realize the minimum average signal power per bin $\underline{S}$, varies with the number of occupied signal bins, $\underline{M}$. For convenience, we confine the following discussion to the low-quality operating point, presented in figure 7. For example, if $\underline{M} = 1$, the best value for ν is ∞, although $\nu = 3$ is only 0.4 dB poorer. On the other hand, if $\underline{M} = 1024$, the best ν is 1, but $\nu = 2$ is only 0.5 dB poorer. In between these extreme values of $\underline{M}$, the best ν sweeps through all the intermediate values as $\underline{M}$ changes. This behavior is consistent with the earlier observations based upon approximations Q_2 in (14) and Q_3 in (18). Figures 7 and 8 constitute a numerical confirmation of the tenuous approximations that were employed in deriving (14) and (18), and in arriving at power-law test (20).

The reason for considering power-law value $\nu = 2.5$ is now clear from looking at the low-quality operating point results in figure 7. When $\underline{M}$, the number of bins occupied by signal, is completely unknown, the best compromise value for ν is 2.4. This particular power-law processor performs about 1.2 dB poorer than the best in its class at the extreme values $\underline{M} = 1$ and $\underline{M} = 1024$; however, for intermediate values of $\underline{M}$, it is close to the best in this class of processors.

On the other hand, for the high-quality operating point results in figure 8, perhaps the best power-law value is near $\nu = 2.25$. Numerical results for this last case were not run; they can be found upon use of (32) and (33). However, for analytic simplicity and ease of practical realization, the quadratic-law processor $\nu = 2$ is recommended.

When the results in figures 7 and 8 are compared with the corresponding results for the modified generalized likelihood ratio processor in [4] and the sum-of-M-largest processor in [5; pages 42 - 48], the corresponding lower envelopes of required signal power are very close over the entire range of $\underline{M}$ from 1 to N = 1024. That is, the best performer in each class of processors requires about the same amount of signal power to achieve the same level of performance.

However, there is one outstanding attribute of the power-law class that highly recommends it over the earlier two classes of processors, namely the modified generalized likelihood ratio processors [4] and the sum-of-M-largest processors [5]. The

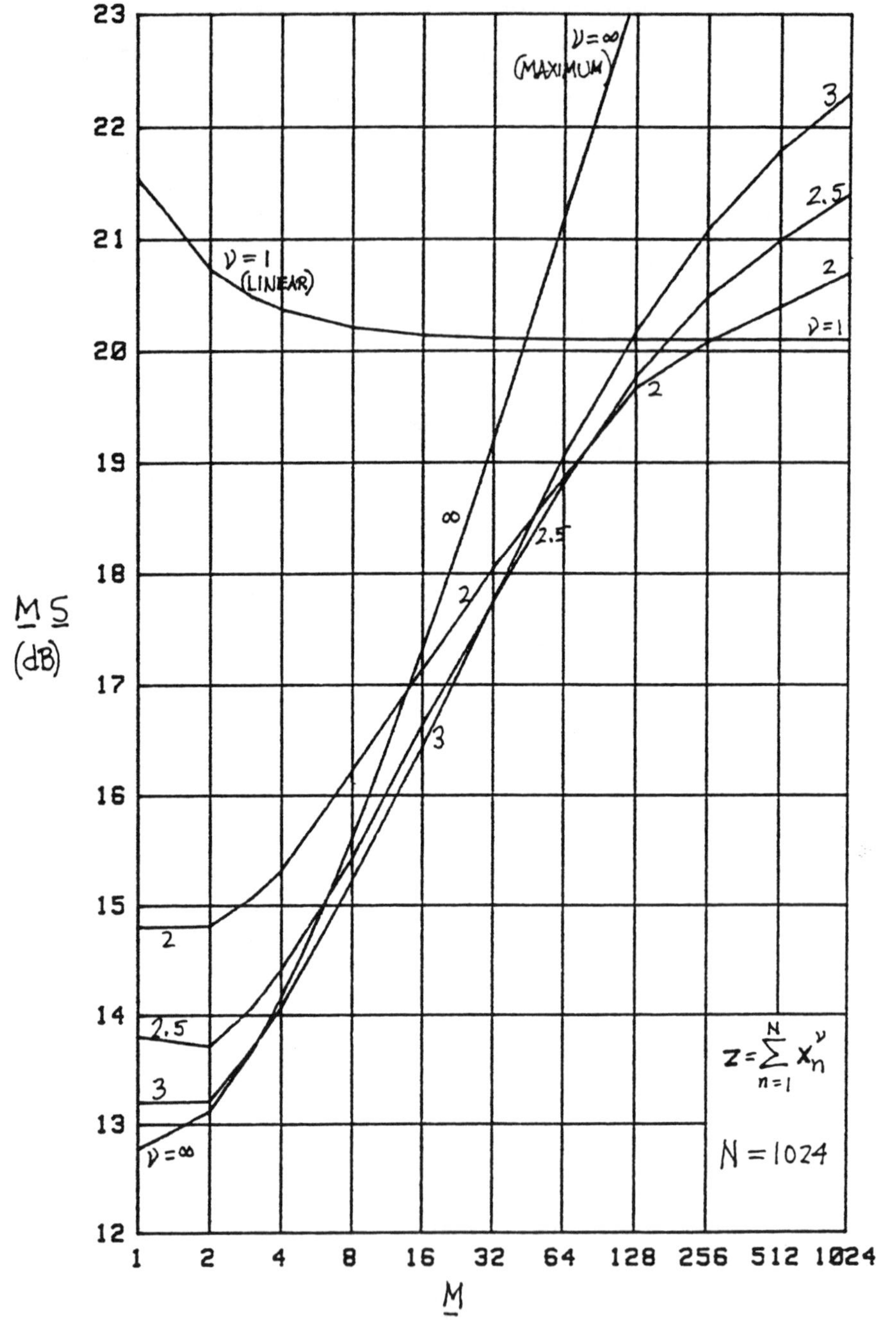

Fig. 7. Required Total Signal Power for $P_f = 10^{-3}$, $P_d = 0.5$

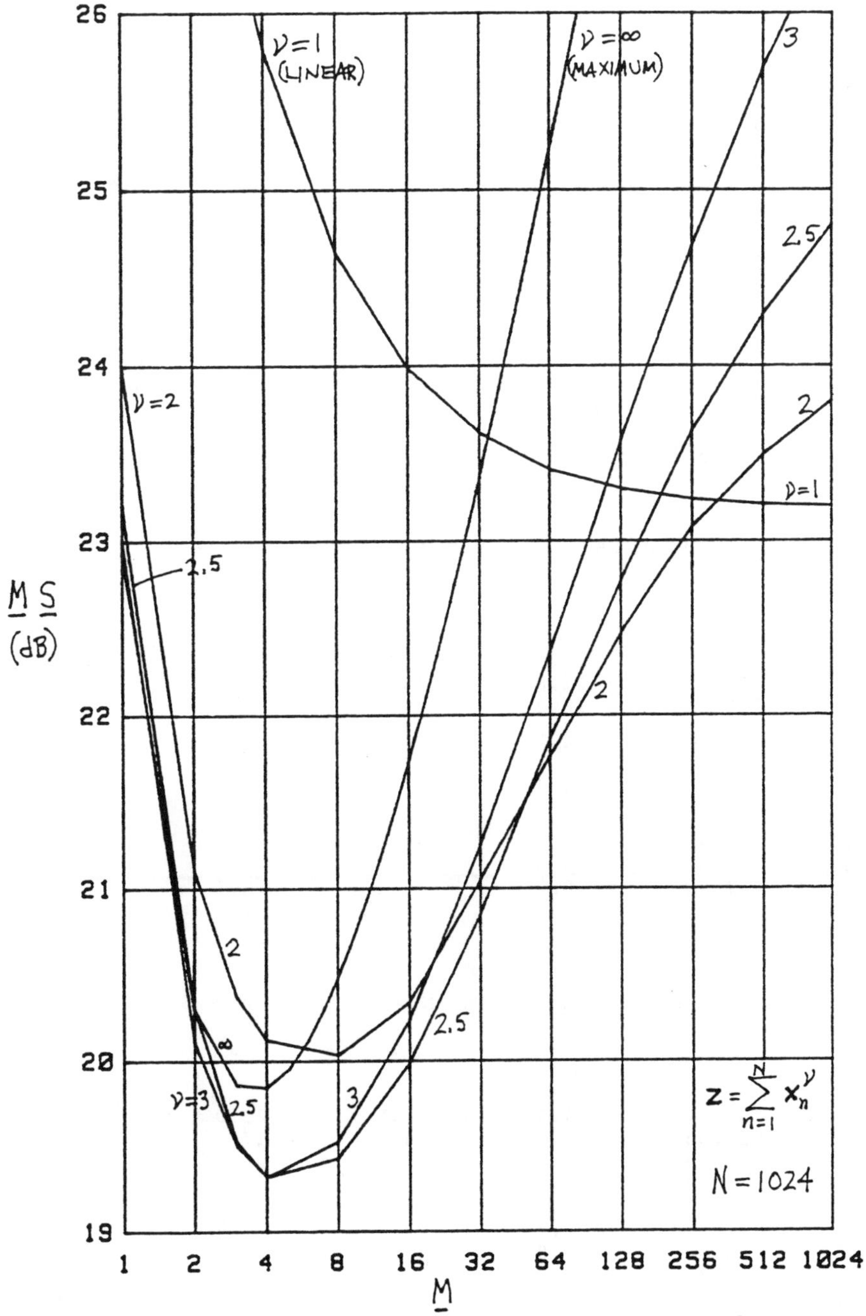

Fig. 8. Required Total Signal Power for $P_f = 10^{-6}$, $P_d = 0.9$

power-law processor, with ν fixed at value 2.4, does not need to know the value of $\underline{M}$ in order to perform near its best level. At the low-quality operating point, for example, it loses no more than 1.2 dB with regard to the best level found thus far, regardless of the unknown value of $\underline{M}$, the number of occupied signal bins. There is no need to guess at a breakpoint, x_o in [4], or a number of terms, M in [5], to use, as there was with the earlier processors; these choices could never be made intelligently without knowledge of $\underline{M}$ in the earlier processors.

This relative independence of the power-law class on $\underline{M}$ is a remarkable result and certainly could not have been anticipated from the debatable manipulations employed in arriving at the power-law class of processors. However, the question is still open as to the ultimate level of performance that can be attained by the optimum processor in this environment. That is, we would like to have an absolute lower bound on the curves in figures 7 and 8. This would tell us how good or bad the power-law class is, and whether we should bother to try to find yet another class of processors that performs still better. That problem has been solved and published [3].

Table 1. Required $\underline{S}$(dB) for N = 1024, P_f = 1E-3, P_d = 0.5

$\underline{M}$ \ ν	1	2	2.5	3	∞
1	21.53	14.8	13.8	13.2	12.77
2	17.71	11.8	10.7	10.2	10.11
3	15.70	10.3	9.3	8.9	8.90
4	14.34	9.3	8.4	8.05	8.14
8	11.17	7.2	6.4	6.2	6.59
16	8.09	5.1	4.6	4.4	5.28
32	5.05	3.0	2.7	2.7	4.15
64	2.03	0.8	0.75	1.0	3.12
128	-0.98	-1.4	-1.3	-0.9	2.17
256	-3.99	-4.0	-3.6	-3.0	1.27
512	-7.00	-6.7	-6.1	-5.3	0.39
1024	-10.01	-9.4	-8.7	-7.8	-0.48

Table 2. Required $\underline{S}$(dB) for N = 1024, P_f = 1E-6, P_d = 0.9

$\underline{M}$ \ ν	1	2	2.5	3	∞
1	31.81	24.0	23.2	23.0	22.92
2	24.85	18.1	17.3	17.1	17.29
3	21.73	15.6	14.75	14.7	15.09
4	19.77	14.1	13.3	13.3	13.82
8	15.61	11.0	10.4	10.5	11.45
16	11.94	8.3	7.95	8.2	9.70
32	8.56	6.0	5.8	6.2	8.31
64	5.34	3.7	3.8	4.3	7.17
128	2.22	1.4	1.7	2.5	6.18
256	-0.85	-1.0	-0.45	0.6	5.31
512	-3.89	-3.6	-2.8	-1.4	4.53
1024	-6.91	-6.3	-5.3	-3.6	3.81

SUMMARY

The class of power-law processors is characterized by raising each observed data point, $\mathbf{x}_n$, to the ν-th power and summing over <u>all</u> the data points from 1 to N, regardless of the (unknown) value of $\underline{M}$, the number of bins occupied by signal. The power-law processor can be regarded as a rough approximation to the optimum processor operating in this environment, which is attempting to detect a weak signal without any structure.

The required threshold settings for achieving false alarm probabilities in the range down to 1E-6 have been determined in figures 1, 3, 5 for power values ν = 2, 3, 2.5, respectively. The receiver operating characteristics have been determined and plotted in [8], for these same values of power ν, for a wide range of values of $\underline{M}$. These results allow for accurate extraction of required signal-to-noise ratios to achieve a specified level of performance, as measured by the false alarm and detection probabilities.

One of the most surprising and useful results of this study is the discovery that the power-law processor with ν = 2.4 performs near the absolute optimum, even without any knowledge of the number of occupied bins $\underline{M}$, or the average signal power per bin, $\underline{S}$. This conclusion has been drawn based only upon the numerical example of N = 1024, and for probabilities P_f, P_d in the range between the low-quality operating point 1E-3,0.5 and the high-quality operating point 1E-6,0.9. Additional ranges of numerical values have yet to be investigated.

The results in figure 8, for the required total signal power to achieve a specified level of performance at the high-quality operating point, indicate an interesting behavior. Namely, there is a best division of the total power into approximately $\underline{M}$ = 4 or 5 bins, at which point the power-law processors with ν = 2.5 or 3 will achieve the specified performance with the minimum value of $\underline{S}$ = 19.3 dB per bin. However, this situation may not be achievable in the typical practical passive application where $\underline{M}$, the number of occupied bins, is not under the receiver's control.

The greatest shortcoming of the results here is that the signal powers per bin have all been assumed equal in the occupied signal bins, with common value $\underline{S}$. An extension to allow for unequal arbitrary signal powers per bin, $\{\underline{S}_m\}$, is clearly necessary and is underway. A bound on performance, based on an argument analogous to that in [3], will be required in order to determine how close the power-law class is to the absolute optimum.

REFERENCES

[1] B. H. Maranda and J. A. Fawcett, "The Performance Analysis of a Fourth-Moment Detector", Proceedings IEEE International Conference, **Acoustics, Speech, and Signal Processing,** Albuquerque, NM, pages 1357 - 1360, 3 - 6 April 1990.

[2] J. A. Fawcett and B. H. Maranda, "The Optimal Power Law for the Detection of a Gaussian Burst in a Background of Gaussian Noise", **IEEE Transactions on Information Theory**, volume 37, number 1, pages 209 - 214, January 1991.

[3] A. H. Nuttall, **Limiting Detection Performance for Random Signals of Unknown Location, Structure, Extent, and Strength,** NUWC-NPT Technical Report 10839, Naval Undersea Warfare Center Division, Newport, RI, 2 March 1995.

[4] A. H. Nuttall, **Detection Performance of a Modified Generalized Likelihood Ratio Processor for Random Signals of Unknown Location,** NUWC-NPT Technical Report 10539, Naval Undersea Warfare Center Detachment, New London, CT, 23 November 1993.

[5] A. H. Nuttall, **Detection Performance of Generalized Likelihood Ratio Processors for Random Signals of Unknown Location, Structure, Extent, and Strength,** NUWC-NPT Technical Report 10739, Naval Undersea Warfare Center Division, Newport, RI, 25 August 1994.

[6] **Handbook of Mathematical Functions**, U. S. Department of Commerce, National Bureau of Standards, Applied Mathematics Series, number 55, U. S. Government Printing Office, Washington, DC, June 1964.

[7] A. H. Nuttall, **Accurate Efficient Evaluation of Cumulative or Exceedance Probability Distributions Directly From Characteristic Functions,** NUSC Technical Report 7023, Naval Underwater Systems Center, New London, CT, 1 October 1983.

[8] A. H. Nuttall, **Detection Performance of Power-Law Processors for Random Signals of Unknown Location, Structure, Extent, and Strength,** NUWC-NPT Technical Report 10751, Naval Undersea Warfare Center Division, Newport, RI, 16 September 1994.

NOVEL SONAR SIGNAL PROCESSING TOOL USING SHANNON ENTROPY

Azizul H. Quazi, Code 3122
Naval Undersea Warfare Center, Detachment
New London, CT 06320 USA

ABSTRACT

Traditionally, conventional signal processing extracts information from sonar signals using amplitude, signal energy or frequency domain quantities obtained using spectral analysis techniques. The object is to investigate an alternate approach which is entirely different than that of traditional signal processing. This alternate approach is to utilize the Shannon entropy as a tool for the processing of sonar signals with emphasis on detection, classification, and localization leading to superior sonar system performance.

Traditionally, sonar signals are processed coherently, semi-coherently, and incoherently, depending upon the a priori knowledge of the signals and noise. Here, the detection, classification, and localization technique will be based on the concept of the entropy of the random process. Under a constant energy constraint, the entropy of a received process bearing finite number of sample points is maximum when hypothesis H_0 (that the received process consists of noise alone) is true and decreases when correlated signal is present (H_1). Therefore, the strategy used for detection is:

(I) calculate the entropy of the received data; then,
(II) compare the entropy with the maximum value; and, finally,
(III) make decision: H_1 is assumed if the difference is large compared to pre-assigned threshold and H_0 is otherwise assumed.

The test statistics will be different between entropies under H_0 and H_1.

Here, we shall show the simulated results for detecting stationary and non-stationary signals in noise, and results on detection of defects in a Plexiglas bar using an ultrasonic experiment conducted by Hughes.[1]

INTRODUCTION

Sonar signals, using passive sonars, or echoes from active sonars, must almost always be observed in a noise background and/or in reverberation. Before any other system function is performed, the sonar system must first detect the presence of signals in this background. In other words, the total received sonar signal content consists of the desired signal and noise and/or reverberation. The desired signal may be deterministic, chaotic, stationary random, or non-stationary random. Background noise is generally random, either stationary or non-stationary, and also could be chaotic in nature.

Traditional signal processing methods attempt to extract the desired signal from corrupted signal plus noise. The optimum detector for the exactly known signal in white Gaussian noise is a matched filter followed by sampling and a decision device. The optimum processor for a signal of unknown phase consists of a matched filter followed by an envelope detector, sampling and a decision device. Now, if the uncertainties of the signal parameter increase or the signal is not deterministic but random, then the processor may be an energy detector. In short, conventional signals are processed coherently, semi-coherently, and incoherently depending upon the a priori knowledge of signal and noise.

In this concept, a unique approach is taken which is entirely different than that of traditional signal processing. The objective is to develop a signal processing tool based on information theory with emphasis on detection, localization and possible classification leading to a superior sonar system design. The fundamental quantities of information theory are entropy, relative entropy and mutual information which are functions of the probability distribution. In turn, they characterize the behavior of a long sequence of random variables and allow us to estimate the probability of rare events and to find the best error exponents in hypothesis testing.

The entropy is a measure of uncertainty of a random variable. In other words, entropy is the minimum descriptive complexity of a random variable X. The entropy, H(X) of a discrete random variable is defined by

$$H(X) = -\Sigma_i\, p_i(x) \log p_i(x) \text{ where } H(X) \geq 0 \text{ and } 0 \leq p(x) \leq 1.$$

The entropy is a measure of the amount of information required on the average to describe the random variable and does not depend on the actual values taken by the random variables, but only on the probabilities. The relative entropy is a measure of the distance between two distributions. In statistics, it arises as an expected logarithm of the likelihood ratio. The relative entropy

$$D(p||q) = \Sigma p(x) \log \frac{p(x)}{q(x)}$$

is a measure of the inefficiency of assuming that the distribution is q when the true distribution is p. The mutual information I(X;Y) is the relative entropy between joint distribution and the product distribution, i.e.,

$$I(X;Y) = \Sigma\Sigma_{xy}\, p(x,y) \log \frac{p(x,y)}{p(x)p(y)}$$

The mutual information I(X;Y) is a measure of the amount of information that one random variable contains about another random variable. It is the reduction in the uncertainty of one random variable due to knowledge of the other. It is symmetric and non-negative.[2]

Q. T. Zhang considered the problem of detecting stationary signals in the presence of additive white noise, both having unknown statistics; and a receiver was developed based on the concept of the entropy of a random process. In addition, he discussed the applications of the entropy-based receiver to compute simulation studies and to the moving target detection problem which involves the use of actual radar data.[3]

Under a constant energy constraint, the entropy of a received process having a finite number of sample points is maximum when hypothesis H_0 (that the received process consists of noise alone) is true and decreases when a correlated signal is present (H_1). Therefore, the following is the strategy:

(I) calculate entropy of the received process; then,
(II) compare this entropy with maximum values; and, finally,
(III) make a decision: H_1 is assumed if the difference is large compared to a pre-assigned threshold and H_0 otherwise.

The test statistics will be the difference between the entropies under H_0 and H_1. The test statistics will be used for detection and possible classification.

Mutual information between pairs of signals for a range of lag values, analogous to cross-correlation values, will provide time delays leading to localization of underwater sources.

SIMULATED AND EXPERIMENTAL RESULTS

In order to demonstrate the concept, the entropy H(X) can be used to detect active and passive sonar signals buried in noise. We shall present three examples.

The first example contains passive sonar that is composed of random noise and a non-stationary signal. Traditional signal processing techniques ignore this type of signal. Figure 1 shows a simulated passive sonar signal which is composed of random noise and a non-stationary signal. Experience indicates that underwater sources generate and radiate non-stationary signals which are often neglected in traditional signal processing methods, and non-stationarity is usually not exploited to enhance sonar system performance. Random white noise with zero mean and unit variance is generated using a Matlab program. A total of 30,000 digital samples of random noise with non-stationary signal superimposed to the random noise starting at 10,001 and ending at 20,000 digital samples is shown in figure 1. In other words, the first 10,000 points contain random noise, the second 10,000 points contain noise plus non-stationary signal, and the third 10,000 points contain noise only. The entropy H(X) is estimated on a window of 1,000 digital points. A sliding window of 1,000 points with 75% overlap is used to estimate the entropy. Figure

2 shows the estimated entropy as a function of window number. Note that the entropy goes down from window #40 to window #79 as expected because these windows contain signal plus noise, whereas other windows contain noise only, and, as a result, uncertainty increases.

The second example contains typical active sonar signals such as linear frequency modulated (LFM) signal. Figure 3 shows an active sonar signal superimposed on noise. Figure 3 contains 5,000 digital samples, the first 2,000 points are random noise with zero mean and unit variance, and the digital samples from 2,001 through 3,000 consist of random noise and a LFM signal (70-130 Hz) and the last 2,000 digital samples contain random noise only. Just like the first example, a sliding window of 1,000 digital points with 75% overlap is used to estimate the entropy. However, this time the entropy is estimated from cross-correlated signal of reference LFM signal and 1,000 digital points. The results showing entropy vs window are presented in figure 4. Notice that at window #9, the entropy is a minimum as expected because this window and the noise plus LFM signal are exactly synchronized.

The third example is the experiment conducted by Hughes who was interested in the application of the entropy concept to the analysis of echoes observed in an ultrasonic experiment. It was assumed that the RF signal acquired in an ultrasonic measurement should contain different amounts of information depending upon the details of the wave material interaction in the region of insonification. Of course, Shannon's theory does not provide a rigorous basis of such picturesque reasoning; however, this hypothesis has been tested experimentally and found to be operationally correct. In applying the concept, each ultrasonic signal is regarded as a string of symbols. By studying the distribution of these symbols, Hughes was able to make direct use of the Shannon entropy as a tool for analysis of raw RF data.[1] The objective of this effort was to investigate the Shannon entropy as a tool for analysis of ultrasonic waveforms. Hughes's results are shown next.

Ultrasonic echoes, which serve as messages for the entropy analysis, were acquired using a single Plexiglas bar immersed in water. As shown in figure 5, the specimen contained five holes and slots which served as artificial defects. The sides and bottom of the sample were covered with tape to permit immersion-mode measurements to be made without water flooding the voids. In addition, pennies were taped to the two ends of the specimen to keep it from floating; the penny on the left side of the specimen appears in subsequent c-scan images. The specimen was insonified from a height of 3.25 in. using a broad-band transducer having a 4-in. focal length, a 1-in. diameter, and a 10-MHz center frequency. This transducer was scanned over the specimen maintaining a constant water path. Stepsizes in the x and y direction of figure 5 were 0.025 in., and the region scanned measured $\Delta x X \Delta y$ = 5X1 in. At each of these positions, one hundred 2048 point waveforms were acquired, summed, and stored for later off-line analysis. The summation, equivalent to averaging without the division, was done to reduce the effects of electronic noise. The digital sampling oscilloscope used for this purpose was operated at 100 MHz. The gain settings on all amplifier stages in the experimental setup were adjusted to be as large as possible without distorting the internal echoes of interest. This was done to obtain maximum usage of the system's dynamic range. For this specimen, the internal echoes and front echoes were nearly the same size, permitting the reliable acquisition of all echoes without adjustments of gain during the scanning operation.

The Shannon entropy was extracted from each RF waveform in the scan. The results of the analysis are shown in figure 6. Figure 6 shows the value of the Shannon entropy as a function of transducer position. The three scans were performed with different amounts of signal summing using $N_A = 10^2$, 10^3, and 10^4 averages, respectively. The lowest curve corresponds to taking a slice along the long axis. Although somewhat obscured by noise, all five defects are evident. In the two-dimensional image, 40 of these curves are placed side by side making defect visual detection much easier. The five defect indicators are more apparent in the $N_A = 10^3$ curve, however, the height of the flat bottom slot indication above the background is seen to have decreased. The same effect is even more pronounced in the $N_A = 10^4$ curve.

Hughes concluded that a novel symbol oriented information theoretic approach to signal analysis based on the application or entropy can be a preferred alternative to conventional signal analysis. In particular, scanned images of a Plexiglas block containing artificial defects were constructed using attributes of back-scattered acoustic signal.

DISCUSSION AND CONCLUSION

We have described a novel signal processing technique based on the Shannon entropy to signal analysis. This is a unique and alternative approach to conventional signal processing. In particular, active and passive sonar signals are analyzed based on the application of the Shannon entropy. This analysis shows that active and passive sonar signals can be detected based on entropy. This complements the work of Hughes who has shown the ability of Shannon entropy to detect anomolies in a medium. A capability such as this, if extended to real sonar problems, has the potential to improve sonar system performance.

REFERENCES

1. Michael S. Hughes," Analysis of digitized waveforms using Shannon entropy", J. Acoust. Soc. Am., Vol. 92(2), February 1992, pgs 892-906.

2. Thomas M. Coner and Joy A. Thomas, "Elements of Information Theory", John Wiley & Sons, Inc., NY, 1991.

3. Q. T. Zhang, "An Entropy-Based Receiver for the Detection of Random Signals and Its Application to Radar", Signal Processing, Vol. 18, 1989, Elsvier Science Publishers, B.V., pgs 387-396.

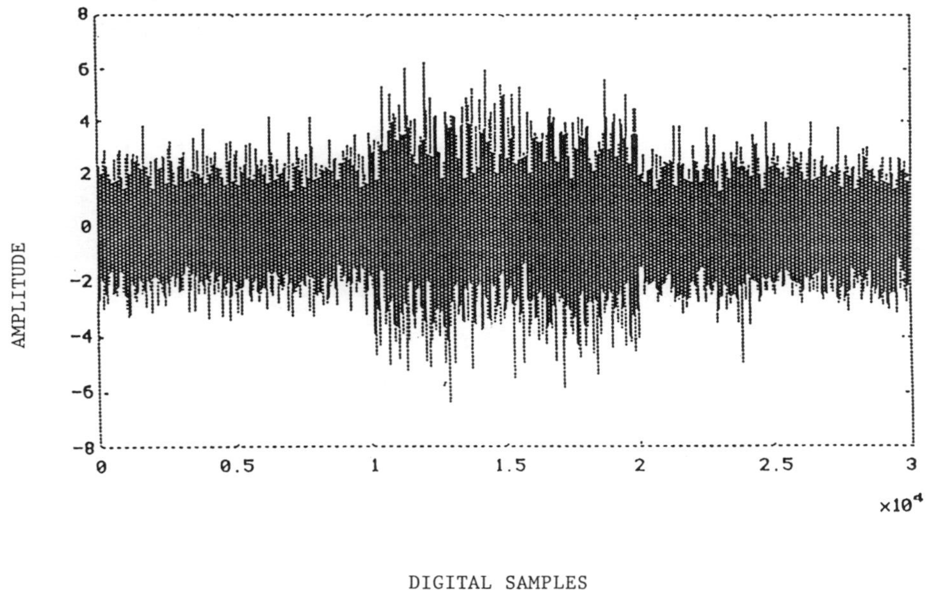

Figure 1. Random noise and non-stationary signal superimposed.

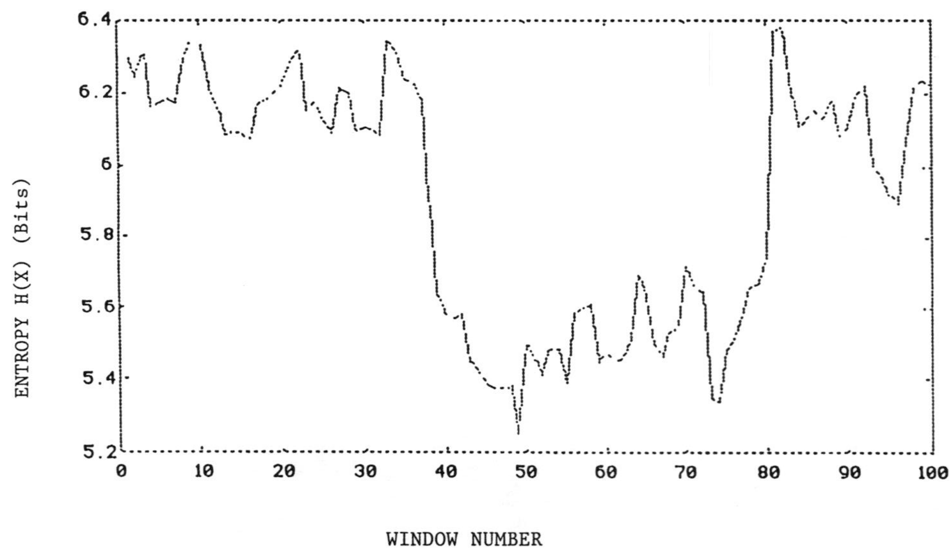

Figure 2. Entropy of random noise and non-stationary signal superimposed.

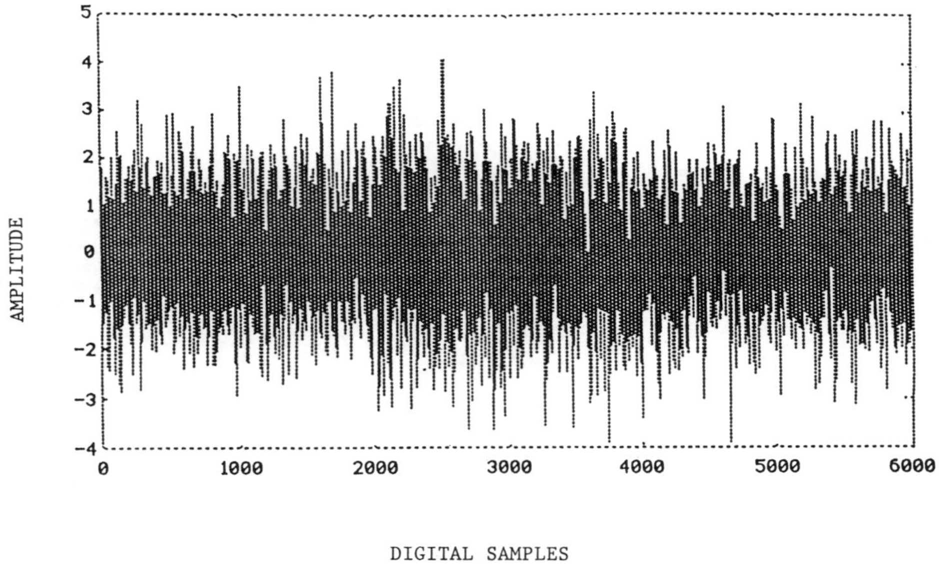

Figure 3. Random noise and linear frequency modulated signal superimposed.

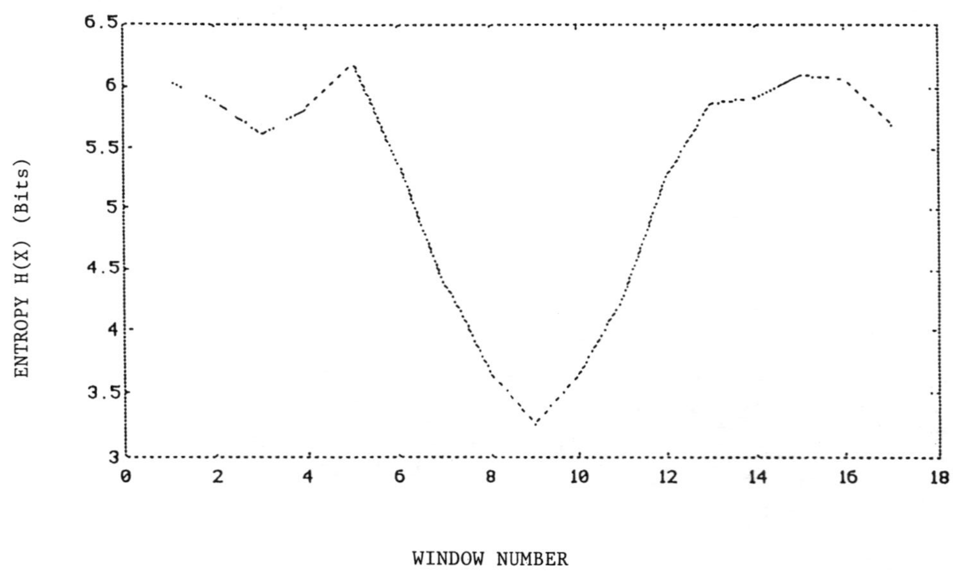

Figure 4. Entropy of random noise and linear frequency modulated signal superimposed.

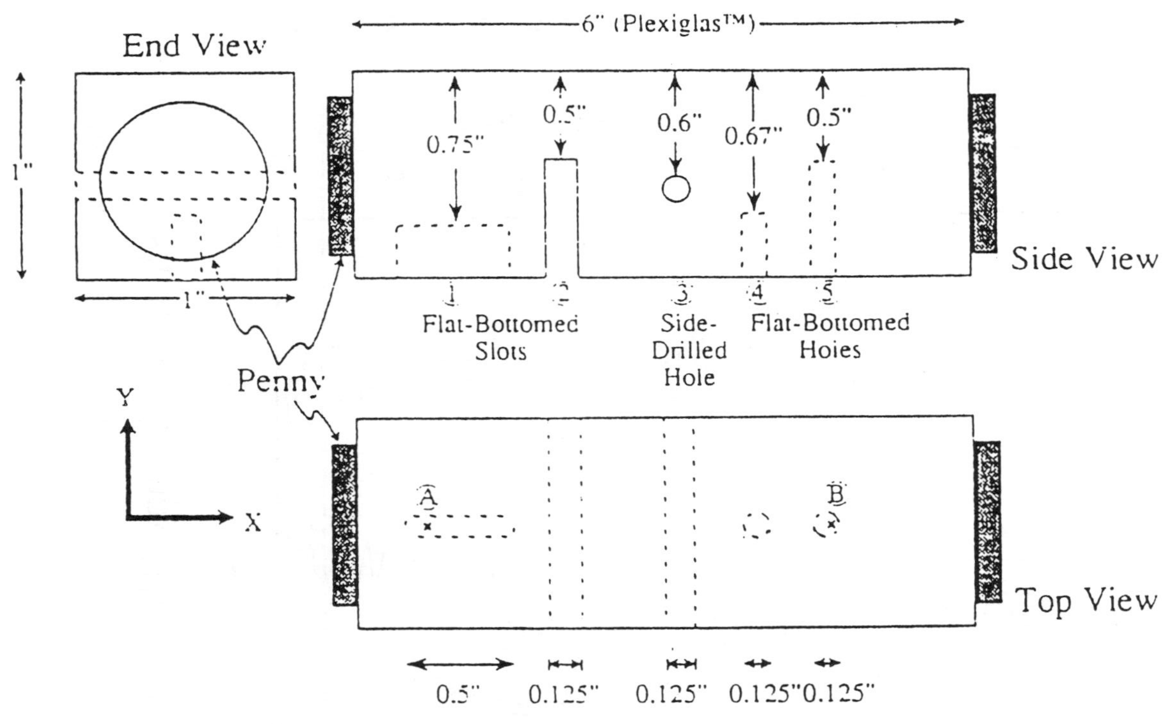

Figure 5. Side view, top view and end view of Plexiglas bar where the defects are numbered (1 through 5) (reference 1).

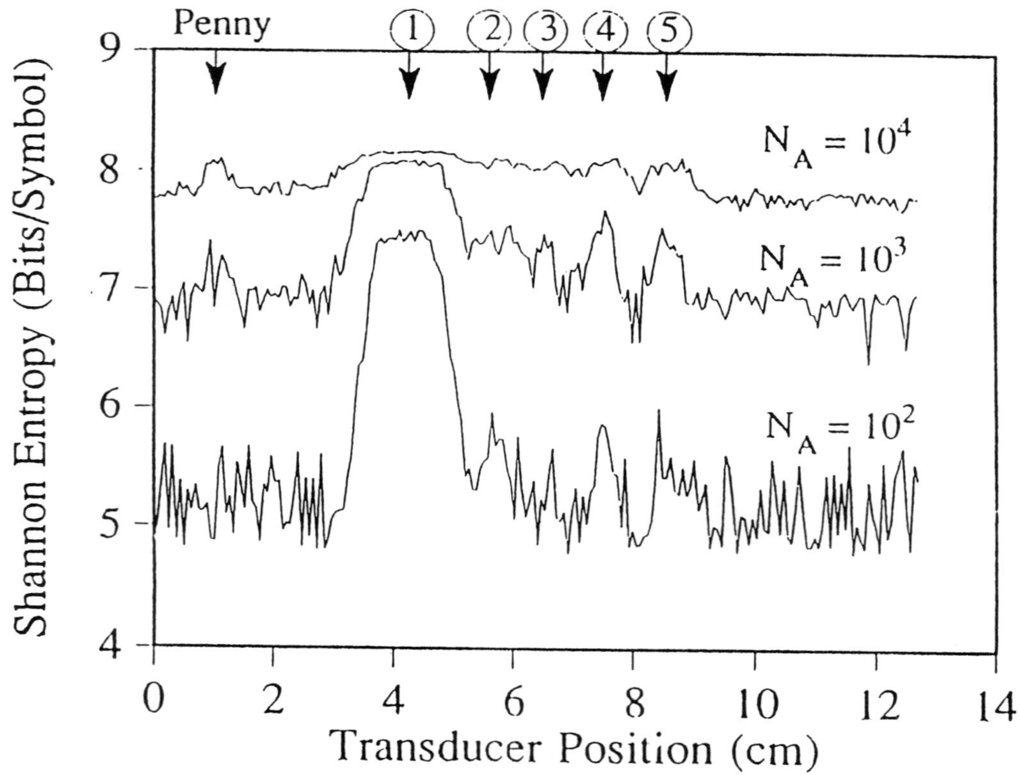

Figure 6. Entropy versus transducer position (reference 1)

Nonlinear Inverse Filtering in the Presence of Noise

D. S. Broomhead and J. P. Huke

DRA(Malvern), St. Andrew's Road, Great Malvern, Worcestershire, WR14 3PS, United Kingdom

Abstract. In an earlier paper we have considered the problem of separating signals from an additive mixture with nonlinear (chaotic) noise. Our approach has been based on an understanding of the effect of linear filters on nonlinear processes. In particular, we were able to demonstrate the usefulness, in this context, of nonlinear filters which act as the inverse of given linear filters. In this paper we extend this work by investigating the effects of stochastic noise on these methods. We discuss the robustness of the inverse filters in the presence of noise, and describe how nonlinear noise reduction methods developed for chaotic systems can be used in conjunction with inverse filters to improve the signal separation when stochastic noise is present. We illustrate this discussion with a synthetic example using speech data.

INTRODUCTION

Noise models in signal processing are usually linearly filtered, independent, identically distributed (i.i.d.) stochastic processes. However, a completely different kind of noise model has emerged from recent work on nonlinear dynamical systems. This new model regards a noise time series as a deterministic but nonlinear process. Conceptually this is to be thought of as the output of a chaotic, physical system, such as a machine or electronic circuit with nonlinear dynamics, or perhaps a turbulent fluid such as the sea. The signal processing required to treat these two types of noise appear to be completely different: linear noise models suggest linear methods such as Fourier spectral methods or linear filters, whereas chaotic noise models require nonlinear methods based on dynamical systems theory. Both of these models are idealisations; in practice we may be faced with noise which is an additive mixture of them, or even something more complicated. For example, in an electronic circuit there is thermal noise, which can be idealised as a stochastic process, and there may be chaotic noise due to nonlinear circuit elements. Furthermore, there may be interactions between the two; the stochastic variability of components may drive the nonlinear dynamics by modifying existing chaotic behaviour or by exciting long lived chaotic transients. Ultimately, we should like to develop signal processing strategies which are effective in the presence of these more general types of noise.

We have mentioned the possibility of mixtures of chaotic and stochastic noise as a step away from the idealisations of purely deterministic or purely random noise, but in fact the distinction between chaos and stochasticity is itself something of an idealisation. In practice, this distinction will depend on the methods adopted for the processing of signals, and the data to which they are applied. Indeed, much of the interest in chaotic noise models stems from the fact that they are indistinguishable from stochastic ones by linear (hence conventional) methods, but are distinguishable (at any rate potentially) by nonlinear methods. Any given nonlinear method, however, is likely to be able to deal with only a restricted range of chaotic systems: if the dimension of the chaotic system is too high, or the dynamics too complicated or the data too sparse or inaccurate, then the method will perform no better with the chaotic signal than with a corresponding stochastic one. We can imagine a continuum of chaotic systems, of increasing apparent stochasticity, with any given signal processing technique able to distinguish only a restricted subset from the stochastic limit.

Signal processing methods based on the assumption of chaotic noise are still in their infancy. In this paper we shall concentrate on one of them recently proposed by us [1]: a separation strategy based on the effects of linear FIR filters on chaotic time series. The theory underlying this method, and the examples that have been exhibited so far, assume that the noise in the signal has no stochastic component. We begin by examining how robust the method is when this assumption is not completely satisfied. We go on to describe and illustrate ways in which the method can be extended to account, at least partially, for the stochastic component.

DELAY RECONSTRUCTION AND SIGNAL SEPARATION

Dynamical systems theory can be applied directly to nonlinear signal processing because deterministic nonlinear systems are generally speaking *observable.* The concept of observability (which is central to control theory) means essentially that the state of system can be inferred by measuring its outputs. To be specific, suppose that $y(t_0)$, $y(t_0+\tau)$, $y(t_0+2\tau),\ldots$ is a series of measurements made on some time-evolving system at the uniformly spaced times t_0, $t_0+\tau$, $t_0+2\tau,\ldots$ and that at time t_0 the system is in some state $s(t_0)$. Then the system is said to be observable if there is a fixed integer n such that the measurements $y(t_0)$, $y(t_0+\tau),\ldots,y(t_0+n\tau)$ determine $s(t_0)$ uniquely. It has been shown [2, 3] that smooth dynamical systems are generically observable, if n is taken to be $2m+1$ or greater, where m is the dimension of the state space of the system. The upshot of this is that given a time series of measurements $\ldots,y(-\tau)$, $y(0)$, $y(\tau)$, $y(2\tau),\ldots$ we can construct a corresponding time series of vectors $\ldots,\mathbf{y}(-\tau)$, $\mathbf{y}(0)$, $\mathbf{y}(\tau),\ldots$ where $\mathbf{y}(t)$ is the *delay vector* $(y(t),y(t-\tau),\ldots,y(t-(n-1)\tau))$; and there is a one-to-one relationship (technically a diffeomorphism) between the trajectory in $\mathbb{R}^n$ described by the sequence of delay vectors and the trajectory followed by the system through its state space. While this relationship cannot usually be found explicitly, its mere existence allows us to infer a great deal about the system from the time series. The state space is said to be *reconstructed*, and the procedure is known as the *method of delays.*

Since the delay vectors correspond to states of the system, they do not fill out regions of the Euclidean space $\mathbb{R}^n$ in which they live: instead they lie on a submanifold of $\mathbb{R}^n$, say $\mathcal{M}$, whose dimension is m, (the state space dimension), and which is in fact the image of an *embedding* (i.e. a diffeomorphism) of the state space. $\mathbb{R}^n$ is often referred to as the *embedding space.*

Let us abbreviate $y(i\tau)$ as y_i. A linear finite impulse response (FIR) filter applied to the $\{y_i\}$ time series gives a filtered time series $\{u_i\}$ according to

$$u_j=\sum_{k=0}^{q-1}c_k y_{j-k} \tag{1}$$

where the coefficients c_k define the filter. It turns out that the filtered time series can also be used to reconstruct the state space using the method of delays [4, 5]. And since the reconstructions using the filtered and the unfiltered time series are both related to the original state space, there is a one-to-one (and in fact smooth) relationship between them. This relationship will be nonlinear, but since we have both reconstructions available to us we should be able to approximate it. We can think of this relationship as a kind of nonlinear inverse filter, because it can be used to determine the unfiltered time series from the filtered one. (But note that it is completely different from the usual linear inverse of an FIR filter, which is infinite impulse response (IIR), and not useful for the signal separation discussed here.)

To be more explicit, equation (1) shows that the length-n delay vector

$\mathbf{u}_k = (u_k, \ldots, u_{k-n+1})^T$ is related to the length-$(n+q-1)$ delay vector $\mathbf{y}_k = (y_k, \ldots, y_{k-n-q+2})^T$ by the relation $\mathbf{u}_k = \boldsymbol{F}_q \mathbf{y}_k$ where $\boldsymbol{F}_q$ is the $n \times (n+q-1)$ matrix

$$\boldsymbol{F}_q = \begin{pmatrix} c_0 & c_1 & c_2 & \ldots & c_{q-1} & 0 & 0 & \ldots & 0 \\ 0 & c_0 & c_1 & \ldots & c_{q-2} & c_{q-1} & 0 & \ldots & 0 \\ 0 & 0 & c_0 & \ldots & c_{q-3} & c_{q-2} & c_{q-1} & \ldots & 0 \\ \vdots & \vdots & \vdots & \ddots & \vdots & \vdots & \vdots & \ddots & \vdots \\ 0 & 0 & 0 & \ldots & c_0 & c_1 & c_2 & \ldots & c_{q-1} \end{pmatrix}$$

The observability property shows that, when restricted to $\mathcal{M}$, the mapping $\boldsymbol{F}_q$ is a diffeomorphism, and hence has an inverse, which we denote by $\boldsymbol{F}_q^{-1}$. Given a delay vector $\mathbf{u}_k$ we can find $\mathbf{y}_k = \boldsymbol{F}_q^{-1}\mathbf{u}_k$, which means we can find the unfiltered delay vectors from the filtered ones. Actually, this is more than we need. Given $\mathbf{u}_k$ it is sufficient to find y_k since these values constitute the unfiltered time series. Constructing the inverse filter thus amounts to finding the mapping from $\mathbf{u}_k$ to y_k.

This observation can form the basis for a signal separation technique [1], where one of the signals is a time series measured from a deterministic system. Suppose we have a time series which is the sum of a desired signal (which we sometimes refer to as the 'message') and a series of measurements of a deterministic, possibly chaotic, system. Also suppose that we can find an FIR filter which 'stops' the desired signal (gives zero output when the signal is fed into it). If the combined signal is passed through the filter, the output is just a (filtered) time series measured from the chaotic system. Now if we know the transformation $\boldsymbol{F}_q^{-1}$ between the reconstructions created using filtered and unfiltered chaotic time series, we can compute the latter series from the former. Once found, the unfiltered chaotic series can be subtracted from the original 'noisy' signal to yield the cleaned signal.

(Note that this discussion only applies to FIR filters: the output of an IIR filter cannot in general be used to reconstruct the state space of the dynamical system [6, 7]. It is possible to conceive of an analogous scheme for IIR filters [8] but the construction of the nonlinear inverse filter is more complicated.)

To use the method, the nonlinear inverse filter $\boldsymbol{F}_q^{-1}$ (or rather, a component of it) must be estimated from data. It is easiest to see how to go about this if it is assumed that the chaotic process can be observed in the absence of any other signal. Physically this corresponds for example to being able to observe a communications channel when there is no message being transmitted. A time series $\{y_k\}$ is collected, then filtered and delay vectors $\mathbf{u}_k$ constructed. Our task is now to fit the function relating $\mathbf{u}_k$ to y_k. We usually fit this function using the radial basis function technique [9, 10], although any nonlinear fitting method could be used.

So long as the FIR filter is effective at stopping the desired signal, it is not actually necessary to be able to observe the chaotic process on its own. The filtered delay vectors we construct will be the same whether the desired signal is present or not. If it is present, the problem becomes one of finding $\boldsymbol{F}_q^{-1}$ in the presence of 'noisy' y_k's. Given enough data, the contribution of the signal (assumed independent of the chaos) will be averaged away by the least squares fit.

In this paper, we are interested in the case where there is broadband, stochastic noise present as well as the chaos and the signal we are trying to separate from it. In this case it cannot be assumed that everything but the chaotic time series is removed by the FIR filter. The stochastic noise will affect the inverse filter signal separation method in two ways. Firstly, noise on the data that is used to fit the inverse filter $\boldsymbol{F}_q^{-1}$ will cause the approximation to deviate from that obtained in the absence of noise; and as discussed in the next section, this deviation is more than a random variability that disappears in the limit of large amounts of data. Secondly, even if a noiseless approximation to $\boldsymbol{F}_q^{-1}$ could be found, the stochastic noise on the unfiltered signal will still affect signal separation because some of it inevitably 'leaks through' the FIR filter that is supposed to pass only the chaotic time series. This means that delay vectors formed from the filtered time series are displaced from their proper positions, and application of $\boldsymbol{F}_q^{-1}$ to them fails to reconstitute precisely the unfiltered chaotic time series. Naturally, this leads to errors in the signal separation. It is worth noting that the first effect to some extent controls the deleterious effects of the second: uncontaminated chaotic data explores only a restricted subset of the embedding space, and an inverse filter fit using it can behave very poorly when presented with data outside this subset. The presence of noise on the fitting data prevents the inverse filter behaving in this unconstrained way. (A similar observation, though in a completely different context, has been made by Webb [11]). An example of this is presented later.

ESTIMATING NONLINEAR INVERSES IN THE PRESENCE OF NOISE

In this section we consider the complicating effect of broadband stochastic noise on the estimation of $\boldsymbol{F}_q^{-1}$. In the simple, noise free, case it is assumed that the FIR filter removes the desired signal completely and, therefore, that it is appropriate to use a least squares fitting procedure—even when the desired signal is present in the unfiltered time series. The effect of an additional broadband stochastic noise component is to confuse this convenient separation since some of the noise as well as the chaos is passed by the filter.

We need to estimate the nonlinear map $\boldsymbol{F}_q^{-1} : \boldsymbol{F}_q\mathcal{M} \to \mathcal{M}$. In practice this is done by finding a least squares estimate $\widehat{\boldsymbol{F}_q^{-1}}$ which minimises the following error

$$\varepsilon^2 = \int d\mu(\mathbf{x}) \int P(\chi) d\chi \, \|\mathbf{x} + \chi - \widehat{\boldsymbol{F}_q^{-1}} \circ \boldsymbol{F}_q(\mathbf{x}+\chi)\|^2 \tag{2}$$

Here $\mathbf{x}$ and χ are delay vectors of the chaotic and stochastic time series respectively, μ is the natural measure of the chaotic system, $P(\chi)$ is the density of the noise distribution and $\|\cdot\|$ denotes the Euclidean norm on $\mathbb{R}^{n+q-1}$. We must now choose $\widehat{\boldsymbol{F}_q^{-1}}$ such that the following first order variation in the error vanishes.

$$\delta\varepsilon^2 = 2 \int d\mu(\mathbf{x}) \int P(\chi) d\chi \, (\mathbf{x} + \chi - \widehat{\boldsymbol{F}_q^{-1}} \circ \boldsymbol{F}_q(\mathbf{x}+\chi)) \delta \widehat{\boldsymbol{F}_q^{-1}} \circ \boldsymbol{F}_q(\mathbf{x}+\chi) \tag{3}$$

Note that if $\boldsymbol{F}_q$ were invertible as a linear map onto $\mathbb{R}^n$ this could be achieved by setting $\widehat{\boldsymbol{F}_q^{-1}}$ to be the matrix inverse of $\boldsymbol{F}_q$. Since $\boldsymbol{F}_q$ is not invertible in this sense we need to specify the problem in more detail. Let us first make an orthogonal change of coordinates so that the null space of $\boldsymbol{F}_q$ becomes $\{\mathbf{0}_n\} \times \mathbb{R}^{q-1}$, (where $\mathbf{0}_n$ is the zero vector in $\mathbb{R}^n$) and for each $\mathbf{x} \in \mathbb{R}^{n+q-1}$ we write

$$\begin{aligned} \mathbf{x} &= (\mathbf{x}', \mathbf{0}_{q-1}) + (\mathbf{0}_n, \mathbf{x}'') && (4) \\ &= \mathbf{x}' \oplus \mathbf{x}'' && (5) \\ \boldsymbol{F}_q(\mathbf{0}_n, \mathbf{x}'') &= 0 && (6) \end{aligned}$$

where $\mathbf{x}'' \in \mathbb{R}^{q-1}$, and $\mathbf{x}' \in \mathbb{R}^n$.

Using this decomposition we can be more specific about the measures which appear in the variational equation. If we assume that the time series is corrupted by a gaussian, i.i.d. noise process then

$$P_{n+q-1}(\chi) d\chi = P_n(\chi') P_{q-1}(\chi'') d\chi' d\chi'' \tag{7}$$

where P_d is an isotropic multivariate gaussian distribution centred at the origin of $\mathbb{R}^d$. The invariant measure μ represents on the other hand a noise process which is far from isotropic, since the measure is concentrated on an m-dimensional submanifold, $\mathcal{M}$, of $\mathbb{R}^{n+q-1}$. In the new coordinates we write this as

$$d\mu(\mathbf{x}) = d\nu(\mathbf{x}') \delta(\mathbf{x}'' - g(\mathbf{x}')) d\mathbf{x}'' \tag{8}$$

where $g : \mathbb{R}^n \to \mathbb{R}^{q-1}$ is a smooth function, and ν is the projection of the natural measure onto the orthogonal complement of the null space of $\boldsymbol{F}_q$. This form for $d\mu$ is a consequence of the assumption that, generically, $\boldsymbol{F}_q\mathcal{M}$ is embedded in $\mathbb{R}^n$ (in particular that $n \geq 2m+1$). Therefore, we expect that the projection of $\mathcal{M}$ onto the n-dimensional linear subspace of $\mathbb{R}^{n+q-1}$ which is the orthogonal complement of the null space of $\boldsymbol{F}_q$ is an embedding.

It follows that $\mathcal{M}$ is a smooth graph over this subspace; we therefore write $\mathcal{M} \subset \{\mathbf{x} = \mathbf{x}' \oplus g(\mathbf{x}')\}$.

Equations (7) and (8) can be used to a simplify equation (3) to give:

$$\delta\varepsilon^2 = 2\int d\mathbf{y}' \int d\nu(\mathbf{x}')\, P_n(\mathbf{y}' - \mathbf{x}') \\ \times \{\mathbf{y}' \oplus g(\mathbf{x}') - \widehat{\boldsymbol{F}_q^{-1}} \circ \boldsymbol{F}_q(\mathbf{y}', \mathbf{0}_{q-1})\}\delta\widehat{\boldsymbol{F}_q^{-1}} \circ \boldsymbol{F}_q(\mathbf{y}', \mathbf{0}_{q-1}) \tag{9}$$

In this expression $\mathbf{y}' = \mathbf{x}' + \boldsymbol{\chi}'$ is the projection of the observed delay vector. The first order variation in the error, for an arbitary variation $\delta\widehat{\boldsymbol{F}_q^{-1}}$, vanishes if

$$\widehat{\boldsymbol{F}_q^{-1}} \circ \boldsymbol{F}_q(\mathbf{y}', \mathbf{0}_{q-1}) = \mathbf{y}' \oplus \frac{\int d\nu(\mathbf{x}')\, P_n(\mathbf{y}' - \mathbf{x}')g(\mathbf{x}')}{\int d\nu(\mathbf{x}')\, P_n(\mathbf{y}' - \mathbf{x}')} \tag{10}$$

This is the least squares estimate of the identity map, $\boldsymbol{F}_q^{-1} \circ \boldsymbol{F}_q$, on the manifold $\mathcal{M}$. Note that $\widehat{\boldsymbol{F}_q^{-1}} \circ \boldsymbol{F}_q$ cannot be the identity on $\mathbb{R}^{n+q-1}$ because

$$\widehat{\boldsymbol{F}_q^{-1}} \circ \boldsymbol{F}_q\mathbf{y} = \widehat{\boldsymbol{F}_q^{-1}} \circ \boldsymbol{F}_q(\mathbf{y}', \mathbf{0}_{q-1})$$

Equation (10) consists of two parts: a linear identity for the component of the argument which is orthogonal to the null space of $\boldsymbol{F}_q$; and a nonlinear part given by the second term on the right hand side. The nonlinear term is an estimate of the function g given the noisy data. In the limit of no noise it reduces to g. To prove this we imagine the density $P_{n+q-1}(\boldsymbol{\chi})$ is sharply localised at the origin. Then, since g is smooth, $P_n(\mathbf{y}' - \mathbf{x}')g(\mathbf{x}') \sim P_n(\mathbf{y}' - \mathbf{x}')g(\mathbf{y}')$. Using this in equation (10) gives the desired result

$$\begin{aligned} \widehat{\boldsymbol{F}_q^{-1}} \circ \boldsymbol{F}_q\mathbf{y} &= \mathbf{y}' \oplus g(\mathbf{y}') \\ &= \mathbf{y} \in \mathcal{M} \end{aligned}$$

It is also informative to think of the opposite limit of large amounts of noise. In this case we think of $P_{n+q-1}(\boldsymbol{\chi})$ being so broad as to appear uniform on a set which contains the support of ν in equation (10). In this case equation (10) reduces to

$$\widehat{\boldsymbol{F}_q^{-1}} \circ \boldsymbol{F}_q\mathbf{y} = \mathbf{y}' \oplus \frac{\int d\nu(\mathbf{x}')\, g(\mathbf{x}')}{\int d\nu(\mathbf{x}')} \tag{11}$$

That is, g is estimated by its mean with respect to the projected invariant measure ν. For intermediate amounts of noise the estimate of g will be a smoothed version of the true g, the degree of smoothing depending on the width of P_n. Thus the presence of the gaussian noise introduces a systematic error into the estimate of g. This error is not a sampling effect: it is present even in the limit of infinite amounts of data (which is the case considered above).

Equation (10) can be understood in terms of a probabilistic interpretation. $P_n(\mathbf{y}' - \mathbf{x}')$ is the conditional density associated with an observation $\mathbf{y}'$ given the system is in the state $\mathbf{x}'$. According to Bayes' Theorem, the nonlinear term in equation (10) is then the conditional expectation of the function g given an observation $\mathbf{y}'$, (and the denominator $\int d\nu(\mathbf{x}')\ P_n(\mathbf{y}' - \mathbf{x}')$ is the prior probability measure on the observations $\mathbf{y}'$).

We can use expression (10) to make a rough estimate of how large the errors in signal separation caused by the stochastic noise will be. Recalling that the nonlinear inverse filter is intended to produce unfiltered delay vectors of the chaotic time series from filtered ones, we write the error in the recovery of the delay vector $\mathbf{x}$ as

$$\epsilon = \mathbf{x} - \widehat{\boldsymbol{F}_q^{-1}} \circ \boldsymbol{F}_q(\mathbf{x} + \boldsymbol{\chi})$$

As described in the previous section, this error arises both because of the presence of the stochastic noise $\boldsymbol{\chi}$, and because of the systematic error in the fit $\widehat{\boldsymbol{F}_q^{-1}}$ to the inverse filter. The 'mean absolute error' in the recovery of the chaotic process is then

$$\epsilon = \int d\mu(\mathbf{x}) \int P_{n+q-1}(\boldsymbol{\chi}) d\boldsymbol{\chi}\ \|\mathbf{x} - \widehat{\boldsymbol{F}_q^{-1}} \circ \boldsymbol{F}_q(\mathbf{x} + \boldsymbol{\chi})\|$$

(We use the mean absolute error rather than the mean squared error because it leads to a compact expression for an upper bound.) Making the same changes of variables as above, this becomes

$$\epsilon = \int d\mathbf{y}' \int d\nu(\mathbf{x}')\ P_n(\mathbf{y}' - \mathbf{x}') \|(\mathbf{x}', g(\mathbf{x}')) - \widehat{\boldsymbol{F}_q^{-1}} \circ \boldsymbol{F}_q(\mathbf{y}', \mathbf{0}_{q-1})\|. \qquad (12)$$

From (10)

$$\begin{aligned} \|(\mathbf{x}', g(\mathbf{x}')) - \widehat{\boldsymbol{F}_q^{-1}} \circ \boldsymbol{F}_q(\mathbf{y}', \mathbf{0}_{q-1})\| &= \\ \|(\mathbf{x}' - \mathbf{y}', g(\mathbf{x}') - \frac{\int d\nu(\hat{\mathbf{x}}')\ P_n(\mathbf{y}' - \hat{\mathbf{x}}') g(\hat{\mathbf{x}}')}{\int d\nu(\hat{\mathbf{x}}')\ P_n(\mathbf{y}' - \hat{\mathbf{x}}')})\| & \\ \leq \ \|(\mathbf{x}' - \mathbf{y}')\| + \|g(\mathbf{x}') - \frac{\int d\nu(\hat{\mathbf{x}}')\ P_n(\mathbf{y}' - \hat{\mathbf{x}}') g(\hat{\mathbf{x}}')}{\int d\nu(\hat{\mathbf{x}}')\ P_n(\mathbf{y}' - \hat{\mathbf{x}}')}\| & \\ \leq \ \|(\mathbf{x}' - \mathbf{y}')\| + \frac{\int d\nu(\hat{\mathbf{x}}')\ P_n(\mathbf{y}' - \hat{\mathbf{x}}') \|g(\mathbf{x}') - g(\hat{\mathbf{x}}')\|}{\int d\nu(\hat{\mathbf{x}}')\ P_n(\mathbf{y}' - \hat{\mathbf{x}}')} & \end{aligned}$$

Using this inequality in (12) shows that ϵ is bounded by the sum of two terms, the first of which is $\int d\mathbf{y}' \int d\nu(\mathbf{x}')\ P_n(\mathbf{y}' - \mathbf{x}') \|(\mathbf{x}' - \mathbf{y}')\|$. By reverting to the $(\mathbf{x}', \boldsymbol{\chi}')$ variables this term is seen to be simply the mean length of an n-element delay vector of the gaussian noise series, and so is equal to $\sqrt{2}\,(\Gamma(\frac{n+1}{2})/\Gamma(\frac{n}{2}))\,\sigma$

where σ^2 is the variance of the gaussian distribution; for brevity we write this as $C\sigma$. The second term is

$$\int d\mathbf{y}' \int d\nu(\mathbf{x}')\, P_n(\mathbf{y}'-\mathbf{x}') \frac{\int d\nu(\hat{\mathbf{x}}')\, P_n(\mathbf{y}'-\hat{\mathbf{x}}')\|g(\mathbf{x}')-g(\hat{\mathbf{x}}')\|}{\int d\nu(\hat{\mathbf{x}}')\, P_n(\mathbf{y}'-\hat{\mathbf{x}}')}$$

which can be bounded by using the triangle inequality

$$\|g(\mathbf{x}')-g(\hat{\mathbf{x}}')\| \leq \|g(\mathbf{x}')-g(\mathbf{y}')\| + \|g(\mathbf{y}')-g(\hat{\mathbf{x}}')\| \tag{13}$$

giving

$$\begin{aligned} &\int d\mathbf{y}' \int d\nu(\mathbf{x}')\, P_n(\mathbf{y}'-\mathbf{x}') \frac{\int d\nu(\hat{\mathbf{x}}')\, P_n(\mathbf{y}'-\hat{\mathbf{x}}')\|g(\mathbf{x}')-g(\mathbf{y}')\|}{\int d\nu(\hat{\mathbf{x}}')\, P_n(\mathbf{y}'-\hat{\mathbf{x}}')} \\ &\quad + \int d\mathbf{y}' \int d\nu(\mathbf{x}')\, P_n(\mathbf{y}'-\mathbf{x}') \frac{\int d\nu(\hat{\mathbf{x}}')\, P_n(\mathbf{y}'-\hat{\mathbf{x}}')\|g(\mathbf{y}')-g(\hat{\mathbf{x}}')\|}{\int d\nu(\hat{\mathbf{x}}')\, P_n(\mathbf{y}'-\hat{\mathbf{x}}')} \\ &= 2\int d\mathbf{y}' \int d\nu(\mathbf{x}')\, P_n(\mathbf{y}'-\mathbf{x}')\|g(\mathbf{x}')-g(\mathbf{y}')\| \end{aligned}$$

Finally, let K be a Lipschitz constant for the function g so that

$$\|g(\mathbf{x}')-g(\mathbf{y}')\| \leq K\|\mathbf{x}'-\mathbf{y}'\|$$

then it is clear that the above term is bounded by $2KC\sigma$ so that

$$\epsilon \leq (2K+1)C\sigma. \tag{14}$$

It should be noted that there is a subtlety involved in the use of the triangle inequality (13). Recall that g is a function on the orthogonal complement of the null space of $\boldsymbol{F}_q$, defined so that its graph over the projection of the manifold $\mathcal{M}$ into this complement is $\mathcal{M}$ itself; its values at points outside the projection of $\mathcal{M}$ are not therefore well-defined. Since $\mathbf{y}'$ will in general lie outside this projection, in writing $g(\mathbf{y}')$ we are implicitly considering an extension of g to a neighbourhood of the projection. Many such extensions are possible.

If the variance of the gaussian noise is small the use of the Lipschitz constant in the error bound may not be appropriate, because we will only be interested in the magnitude of $\|g(\mathbf{x}')-g(\mathbf{y}')\|$ for points $\mathbf{x}'$ and $\mathbf{y}'$ such that $\|\mathbf{x}'-\mathbf{y}'\|$ is small. In that case it may be more reasonable to approximate $\|g(\mathbf{x}')-g(\mathbf{y}')\|$ by $\|Dg(\mathbf{x}')\|\|\mathbf{x}'-\mathbf{y}'\|$, in which case K in the expression (14) would be replaced by the average of $\|Dg(\mathbf{x}')\|$ over the invariant measure ν. In this case (14) implies that errors in signal separation may be anticipated to be large when the derivative of g is large. Geometrically it is clear that this derivative will be large if there is a tangent vector to $\mathcal{M}$ which lies close to the null space of $\boldsymbol{F}_q$, or if the chord joining two points in $\mathcal{M}$ lies close to this null space. In other words, the errors will be large if the projection of $\mathcal{M}$ induced by the filter nearly fails to be an embedding. It follows that in choosing the linear filter two criteria need to be considered: firstly, as discussed in the previous section, the filter needs to stop the signal that we wish to separate from the chaotic background; secondly, the function g corresponding to the filter should not have a large derivative. In some cases these may be conflicting requirements—the bandstop characteristics of the filter are related to the null space of $\boldsymbol{F}_q$ since each frequency stopped by the filter corresponds to a two-dimensional subspace lying in the null space. But though stopping the desired signal thus requires certain directions to lie in the null space of $\boldsymbol{F}_q$ there may be latitude in other directions: such extra dimensions to the null space will be permissible so long as tangents and chords of $\mathcal{M}$ do not lie close to them.

Example: Ikeda Time Series with a Sine Wave and Noise

The first example describes an application of the inverse filter method in a situation where stochastic noise is present, but where no attempt is made to account or correct for it. We shall compare the properties of two estimates of the inverse filter $\boldsymbol{F}_q^{-1}$: the first, $\overline{\boldsymbol{F}_q^{-1}}$ constructed when there is no stochastic noise present, and the second, $\widehat{\boldsymbol{F}_q^{-1}}$, constructed in the presence of gaussian random noise. In the previous paper we showed how a sine wave could be separated from a chaotic time series derived from the Ikeda map

$$\begin{aligned} z &\mapsto 1 + \mu z \exp i(\alpha - \frac{\beta}{1+|z|^2}) \\ z &\in \mathbb{C} \end{aligned}$$

The chaotic time series is of $x = \mathrm{Re}\{z\}$ with $\mu = 0.7$, $\alpha = 0.4$ and $\beta = 6.0$. The frequency of the sine wave is 9/32 cycles per unit time and the amplitude is 0.2—about the same as the standard deviation of the Ikeda data. The three coefficient filter (1.0, 0.3902, 1.0) is used to filter out the sine wave. The filtered time series is embedded using 4 delays, and an inverse filter $\overline{\boldsymbol{F}_q^{-1}}$ constructed using a $3 \times 3 \times 3 \times 3$ grid of radial basis functions (cubics) on a lattice surrounding the data. It was found that the accuracy of the recovered sine wave seemed to be limited only by the accuracy of the fit to $\boldsymbol{F}_3^{-1}$, which is not surprising because the linear filter stops the sine wave completely. Figure 1 shows a section of the combined Ikeda and sine wave time series, and figure 2 shows the original sine wave and its recovered version. These figures are taken from the previous paper [1]. In this case the fit to $\boldsymbol{F}_3^{-1}$ was performed using Ikeda data with no other signal present.

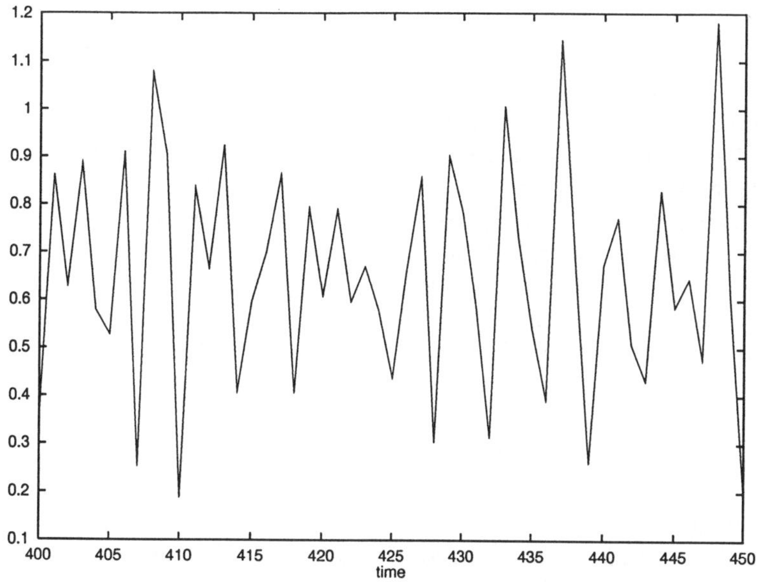

Figure 1: Sum of Ikeda and sine wave time series

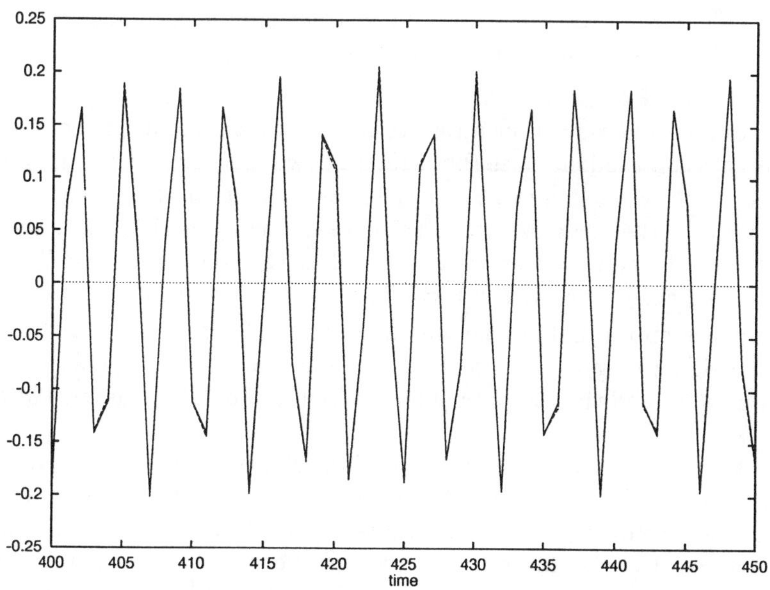

Figure 2: Coarsely sampled sine wave (solid) and its recovered version (dashed)

Now we perform a similar experiment, but with a 'stochastic' noise component present in the signal. This noise time series is of gaussian random numbers with a standard deviation of 0.1—a substantial fraction (half) of that of the chaotic series. To find the estimate $\widehat{F_3^{-1}}$ a 2000 sample time series of Ikeda data with gaussian noise added was filtered, and embedded using 4 delays. 1995 data pairs of the form $(\boldsymbol{F}_3(\mathbf{y}_k), y_k)$ were used to fit the inverse. The same arrangement of RBFs was used as in the previous version of this example. (As before, this calculation corresponds to the assumption that the noise process can be observed without the message; but here the noise process has both chaotic and stochastic components, and we do not assume that these can be observed separately.) The r.m.s. error in the fit was found to be 0.09, consistent with the standard deviation of the gaussian noise.

In this artificial example we can assess the accuracy of the fit by applying the estimate $\widehat{F_3^{-1}}$ to a section of filtered Ikeda time series with no random series present. The r.m.s. error found in this way was 0.062; the corresponding error in the fit to $\overline{F_3^{-1}}$ is 0.004, so the stochastic noise has led to a substantial increase. Partly, this increase in error is due to the systematic error in the fit that the random noise introduces, but it may also be due to the fact that more data is needed to obtain a given accuracy of fit because the effects of the random noise need to be averaged out. We can test this simply by repeating the fit using more data—doubling the number of data points turns out to make very little difference to the r.m.s. error, so we conclude that the value of 0.062 quantifies the systematic error due in the estimate $\widehat{F_3^{-1}}$.

How well does $\widehat{F_3^{-1}}$ perform at separating the sine wave from the 'noise'? To a 2000 sample sine wave time series we add corresponding lengths of Ikeda data, and gaussian random noise (both the last two are different to the series used in fitting the inverse filter). Figure 3 shows a section of the time series formed by adding the Ikeda data to the sine wave, and by adding the gaussian noise to the result, and gives an idea of the effect the stochastic noise has on the signal. The noisy sine wave is FIR filtered and the result passed through $\widehat{F_3^{-1}}$ to produce an estimate of the unfiltered noise. This is subtracted from the original signal to get an estimate of the sine wave. Figure 4 shows the original sine wave and the estimate obtained from the nonlinear inverse filter.

There are two ways that we might consider of assessing the effectiveness of the signal separation procedure. Since we are thinking of the sine wave as the signal that ultimately we wish to recover we could look at the difference between the result of the above operations and the sine wave; this is the difference between the two curves in figure 4. The r.m.s. difference is found to be 0.10, which is close to the standard deviation of the gaussian time series (indicating that on average there is little amplification of the error). However,

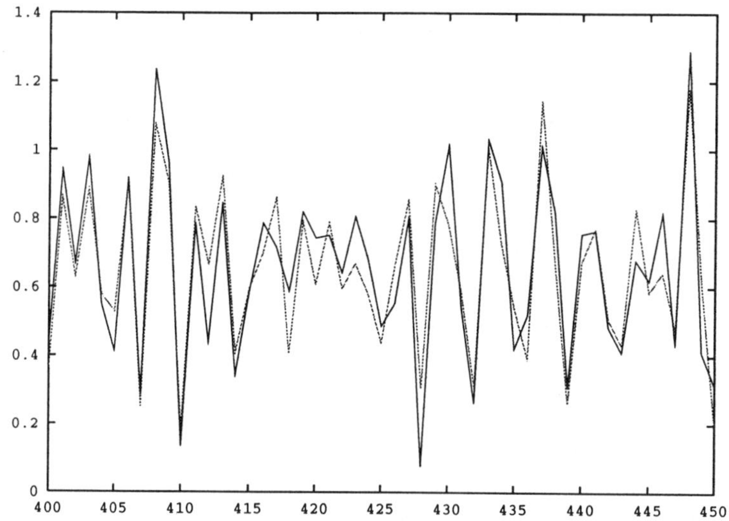

Figure 3: Sum of Ikeda and sine wave time series (dashed), and with gaussian random noise added (solid)

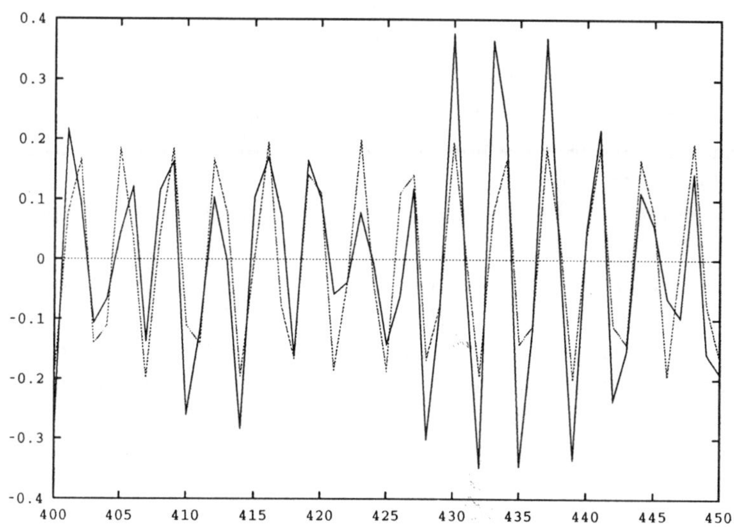

Figure 4: Sampled sine wave (dashed), and its recovered version (solid)

we could also take the view that the inverse filtering operation is only intended to remove the deterministic part of the noise, so the relevant comparison is between the output of the inverse filter and the sum of the sine wave and the random time series (this being what we would get if we were completely successful at removing the chaotic component). Here the r.m.s. difference is 0.11. This indicates that the presence of the random time series has compromised the ability of the inverse filter to remove the chaotic time series leaving the rest of the signal untouched. Much of the inaccuracy is due to the systematic error introduced into $\widehat{\boldsymbol{F}_3^{-1}}$ by the stochastic noise—recall that we concluded above that this error is of the order of 0.06; even were the random series to be removed from the original signal in this experiment we would still expect an r.m.s. difference of this size.

Considering the relatively high level of random noise, it is perhaps surprising that the method manages to recover the sine wave as well as illustrated in figure 4, where the difference between the true and estimated signals is of the same magnitude as the random noise. An interesting comparison here is with the results obtained using $\overline{\boldsymbol{F}_3^{-1}}$; we know from previous work that the RBF model we are using is capable of fitting $\overline{\boldsymbol{F}_3^{-1}}$ rather well, and the resulting estimate does not have the systematic error described above. Applying this estimate to the same mixture of sine wave, Ikeda data and random noise used above yields an r.m.s. difference between the recovered signal, and the sum of the sine wave and random series, of 0.66 (compared with 0.11 for $\widehat{\boldsymbol{F}_3^{-1}}$). Figure 5 shows a section of the recovered signal together with the corresponding result for $\widehat{\boldsymbol{F}_3^{-1}}$, and the original sine wave.

The poor performance of the estimate $\overline{\boldsymbol{F}_3^{-1}}$ results from the random noise causing the filtered data to explore regions of the embedding space not explored by the data used to estimate $\boldsymbol{F}_3^{-1}$—where the inverse filter is consequently not accurately modelled. In this case the errors introduced by this inaccurate modelling are considerably greater than those caused by the systematic error introduced when $\widehat{\boldsymbol{F}_3^{-1}}$ is computed in the presence of random noise. The lesson of this example seems to be that if the inverse filter is to be used in the presence of noise it needs to be estimated in the presence of noise. But the importance of this point will depend on the amount of noise: the spectacular difference in performance between the two filters considered here is no doubt due to the fact that the variance of the random time series is relatively high.

NOISE REDUCTION

The reduction of noise on chaotic time series has been one of the most active areas of applied nonlinear dynamical systems theory in the past few

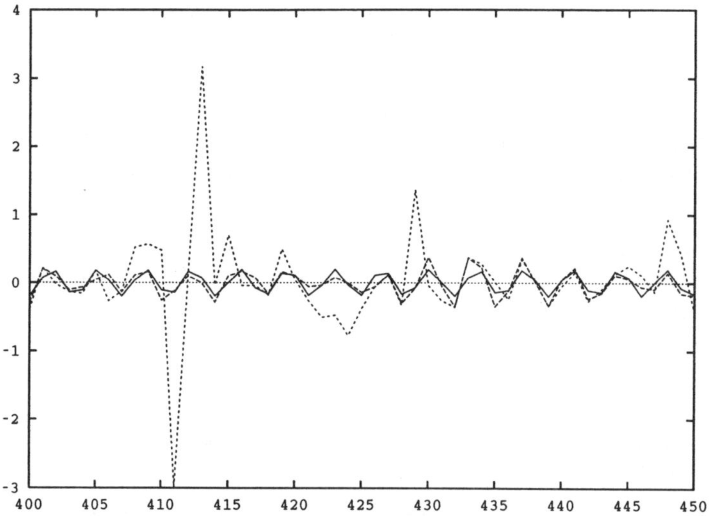

Figure 5: Sampled sine wave (solid), and its recovered versions using $\widehat{\boldsymbol{F}_3^{-1}}$ (long dashes) and $\overline{\boldsymbol{F}_3^{-1}}$ (short dashes).

years, (see [12] for a recent review). This problem is different from the signal separation problem because it assumes that the chaotic time series dominates any other signal present, and that the corrupting 'noise' is essentially i.i.d., (or more accurately, shows no structure—for example in the embedding space—that the noise reduction methods confuse with determinism). Nevertheless, the solution to both problems involves the production of two time series from the original, one of which is the uncontaminated chaotic series. The signal separation problem in the presence of both chaotic and stochastic noise is a combination of the problems addressed by the inverse filter method and by the noise reduction methods just mentioned, and calls for a combination of these approaches to tackle it.

The noise reduction method we shall use below is the one proposed by Davies [13]. We choose this because it is easy to implement using the RBF fitting method, and seems to be quite robust. Its central idea can be stated quite simply: suppose that we have a noisy time series measured from a dynamical system, and that (for the moment) we know what the dynamics of the system are—specifically, we know what the one step ahead predictor for the time series (in the absence of noise) should be. Given the noisy time series, applying the one step ahead predictor to it will yield a sequence of errors; by examining these errors we can decide how to change the time series so that the errors decrease, which means the new time series is more consistent with the predictor and hence (presumably) closer to a noiseless trajectory of the dynamical system.

In practice, the required predictor is not usually known, and must be estimated from the time series itself, using some technique such as the RBF method. The noise reduction procedure then becomes an alternation between the fitting of the predictor from the time series, and the adjustment of the time series to be consistent with the predictor. As with all such noise reduction algorithms, the convergence of this iteration is a delicate question, but in practice the method is found to work well, at least for the Hénon system that we use below. For more details, see [13] and [14].

It may be worth making a remark here about the processing of the received signal. Both the nonlinear inverse filter method, and the noise reduction method just outlined involve a least squares fit. Once this fit has been performed satisfactorily it should not be necessary to perform it again to process new data, unless we believe that the chaotic system has changed in some way, or the level of stochastic noise has changed. In the ideal case where this does not happen we would fix the inverse filter, or the one step predictor, and process any further incoming data using it. In the noise reduction case this would mean that we did not construct a new predictor at each iteration. Since least squares fitting can be performed in a recursive fashion there is scope for continually updating the inverse filter, and hence making the method at least partially adaptive. However the noise reduction, being iterative, seems to be inherently a batch procedure.

How can we use this kind of approach to deal with stochastic noise in the signal separation problem? Consider the situation where there is chaotic noise, a desired signal which is not especially small compared to the chaos, and stochastic noise which is small compared to the chaos (and the desired signal). In that case, the inverse filter method requires that we linearly filter the noisy signal to remove the desired signal. This filtering is not likely to remove the stochastic noise however, because this, like the chaos, is broadband. But the filtered chaotic signal still represents a time series of measurements that can be used for reconstruction of the chaotic noise generating system and so the noise reduction approaches should apply to it, (at any rate so long as the linear filtering does not destroy the properties of the stochastic noise on which these methods rely.) By using one of these methods we can remove the stochastic noise from the filtered signal, and hence remove the errors in the estimation of the inverse filter, and in the recovered signal. Note that the recovered signal will be the sum of the desired signal and the original stochastic noise: only the chaotic noise is removed. The noise reduction methods may even be useful for rectifying inadequacies of the linear filter: if some of the desired signal gets through this filter, noise reduction may be successful at eliminating this residuum—which is likely to be of small amplitude because most of the desired signal has been stopped.

This procedure has a simple interpretation in terms of the embedding spaces of the filtered and unfiltered signals. In the absence of contamination, the method of delays reconstruction of the unfiltered chaotic time series lies on a low-dimensional submanifold of the embedding space. The linear filter projects this embedding space onto a linear subspace—the embedding space of the filtered chaotic signal—and the reconstructed filtered time series again lies on a submanifold—the image of the original submanifold under the projection. The message time series causes the data in the unfiltered embedding space to move off its submanifold, but only in directions that are projected to zero by the filter, (this follows because the message is stopped by the filter). Stochastic noise pushes the data off its submanifold in all directions, including those not projected out, and so even in the filtered embedding space the data does not lie on its submanifold. The object of noise reduction is to move the data onto the submanifold so that the inverse filter can be applied to bring it back to the submanifold in the original embedding space. (It is clear that to do this some decision must be made for each data point, as to which point on the submanifold it is to be pushed back to. Different noise reduction methods make different decisions about this. But here the embedding space picture is perhaps slightly inadequate because the decision is not usually completely geometrical: it takes into account the relationship introduced between time consecutive data points by the method of delays.)

So now the attempt to isolate an uncontaminated version of the chaotic signal consists of two parts: an FIR filtering step to remove the message, and a subsequent noise reduction step to remove the stochastic noise. Neither step alone will generally be sufficient: the linear filtering cannot remove the broadband stochastic noise, and the noise reduction cannot remove the message, (because it is of too high amplitude and it may be insufficiently random). Once the contamination has been got rid of, an inverse filter can be built and applied, to remove the chaotic component from the original signal.

Example: Hénon Time Series with Speech

To illustrate the use of the noise reduction method with the inverse filter method we use a time series which is the sum of a chaotic series and a section of recorded speech. The chaotic series is generated from the Hénon map

$$x_{n+1} = 1 - ax_n^2 + bx_{n-1}$$

where $a = 1.4$ and $b = 0.3$. The resulting series (which is 20479 samples long) was normalized to have zero mean and unit variance.

The speech time series consisted of a 2.5 second recording of the utterance 'a b c', coarsely sampled at 8 kHz. This sampling rate ensures that the speech time series does not appear to be too narrowband compared to the chaos (power spectra are shown in figure 6) but on the other hand a good deal of the power occurs in the low frequency region. This means that a three coefficient

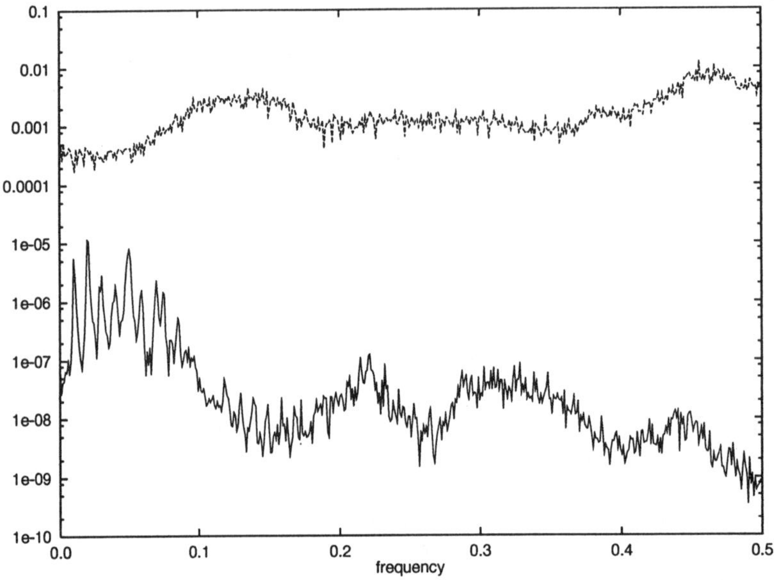

Figure 6: Power spectra for the Hénon time series (upper) and for the speech time series. (Frequency is in cycles per sampling interval: the sampling rate for the speech is 8 kHz)

FIR filter with a zero at 0.068 cycles per sampling interval (547 Hz) should stop a significant proportion of the power in the speech signal. The filter we use is $(1.0, -1.818, 1.0)$.

The speech signal is shown in figure 7. It is apparent that the signal has been severely clipped during the data acquisition (the signal was recorded directly onto a Sun Microsystems Sparcstation10 using the inbuilt recorder). The spoken sounds are separated by sections of 'silence'—which actually contain low levels of noise that we might reasonably assume to be stochastic. This time series was scaled to have zero mean and a standard deviation of 0.1, and added to the Hénon series. This standard deviation is quite small compared with the chaotic series, but includes the periods of silence; when present, the speech has a standard deviation of about 0.17, and an amplitude of about 0.4. A section of the Hénon time series, with and without the speech present, is shown in figure 8.

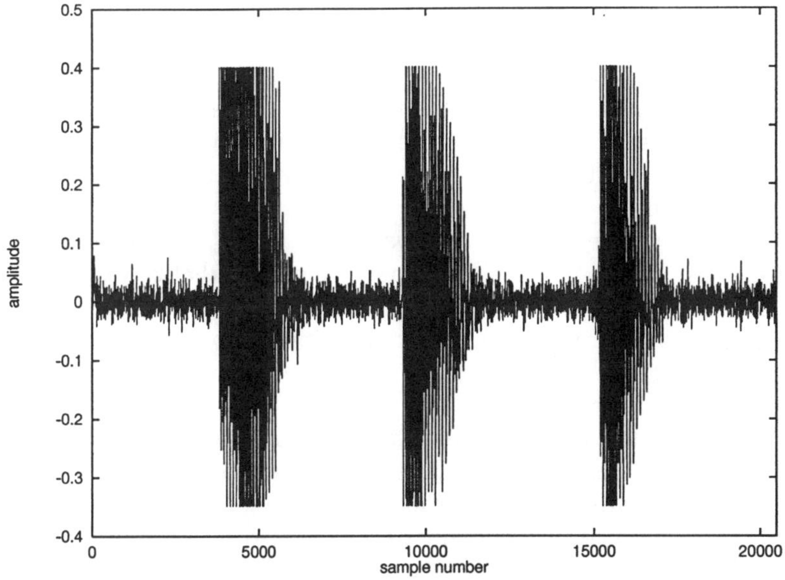

Figure 7: Recorded time series of the utterance ‘a b c’

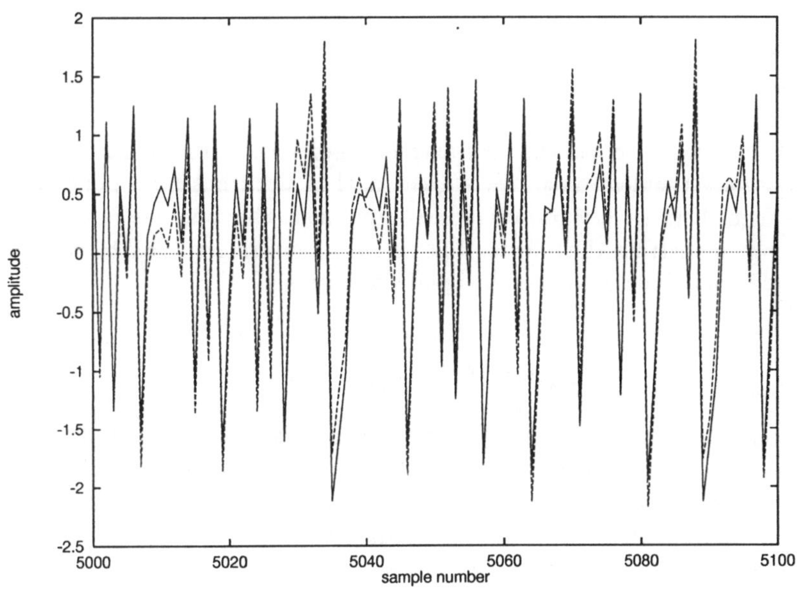

Figure 8: Section of the Hénon time series (solid), and with the speech signal added to it (dashed).

We estimate an inverse filter by reconstructing the filtered time series using 4 delays, and performing an RBF fit to the unfiltered time series using 300 cubics centred at points randomly chosen from the data. The data used here was the first 2000 samples of the sum of the Hénon and speech time series. Only background noise is present in the speech signal for this data, and this is very small compared with the chaotic series—its presence makes essentially no difference to the fit.

Signal separation is performed in two ways. In the first, the sum signal is linearly filtered and the inverse filter applied to the result—that is, we proceed as if the linear filter removed the speech signal completely. In the second, the sum signal is filtered, and then noise reduction is applied to the filtered signal to try to produce a clean, filtered Hénon time series; the inverse filter is then applied to this cleaned series.

When performing the noise reduction, we fit the one step ahead predictor using the first 2000 points of the combined (filtered) signal, but then use the predictor to make adjustments to the whole time series. As noted above, the first section of the time series is essentially Hénon data, with only a small, noisy contribution from the speech signal, so the fit should produce a good estimate for the predictor of the Hénon time series. This should in turn encourage the noise reduction iteration to converge to the clean Hénon signal.

The r.m.s. difference between the recovered and original speech time series in the first case (no noise reduction) is 0.083. In the second case it is 0.031. Sections of the recovered time series are shown in figures 9 and 10. It is clear from these figures that the noise reduction has been successful at improving the recovery of the speech signal.

(We should say, perhaps, that the extraction of speech from noise is a difficult and subtle problem on which a great deal of work has been done, and the experiment here is very primitive in comparison. Our purpose in this example is to demonstrate the effects of noise reduction, rather than to make a fundamental contribution to speech signal processing.)

CONCLUSIONS

The recent advances in nonlinear dynamical systems theory that have caused so much excitement in the physics and engineering communities also hold great promise for applications to signal processing. But to realise this promise the mathematical ideas and methods need to be converted to practical and robust algorithms. This is not necessarily a straightforward task, and attempts to do it, for example the recent interest in fractal dimension calcuation, have proved controversial. But the successful application of, for example, chaotic noise reduction methods to experimental data [15], have demonstrated what is possible.

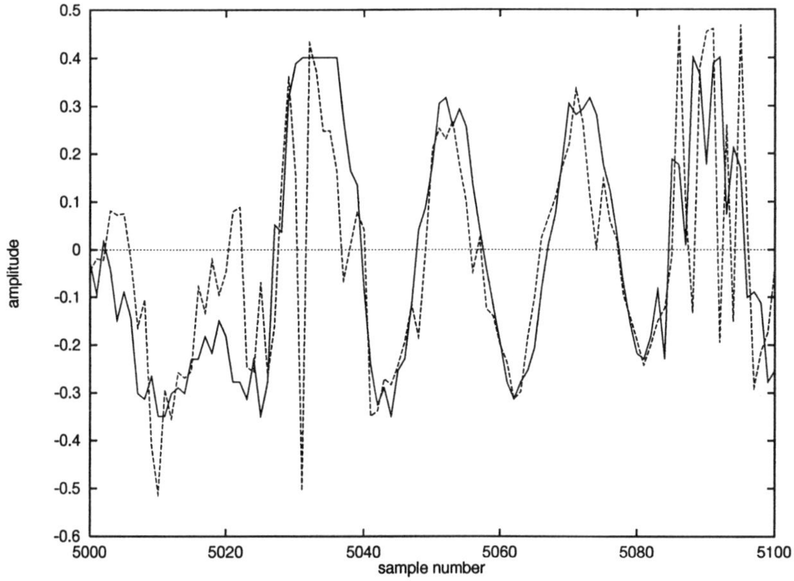

Figure 9: Section of the speech signal (solid) and its recovered version (dashed) with no noise reduction on the filtered time series.

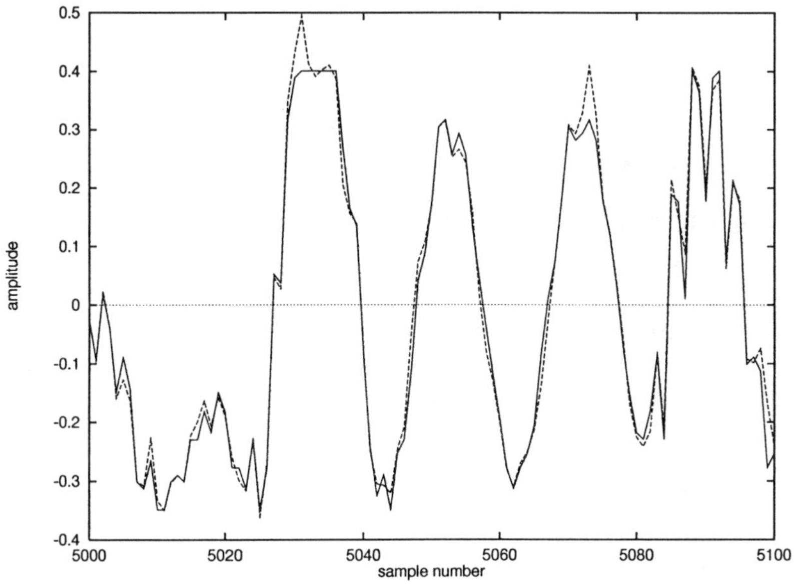

Figure 10: Section of the speech signal (solid) and its recovered version (dashed) with noise reduction performed on the filtered time series.

A major aspect of this transfer of theory to practice is the demonstration that proposed methods are robust when the assumptions on which they are predicated are relaxed. (Failures of this kind have caused much of the trouble with dimension estimation.) The results of this paper go some way to establishing this kind of robustness for our recently proposed signal separation algorithm. They also illustrate the general principle that practical signal processing applications may require a combination of several approaches to cope with the complexities of the real world. In particular, such applications will doubtless require the fusion of the new methods with conventional ones. In future we will need to develop noise reduction and signal separation strategies that will operate—adequately if not optimally—when confronted with a broad range of noise generating systems. Our intention is that the method discussed here will form part of such a strategy.

ACKNOWLEDGEMENTS

We would like to thank David Hughes for his help in the acquisition of the speech data.

REFERENCES

[1] D. S. Broomhead, J. P. Huke and M. A. S. Potts, *Physica D* to appear.

[2] D. Aeyels. *SIAM J. Control Opt.* **19** (1981) 595.

[3] F. Takens. Detecting strange attractors in turbulence, in *Dynamical Systems and Turbulence*, (Springer Lecture Notes in Mathematics, vol. 898) edited by D. A. Rand and L.-S. Young (Springer-Verlag, Berlin, 1981), 366.

[4] D. S. Broomhead, J. P. Huke and M. R. Muldoon. *J. R. Statist. Soc.* **54** (1992) 373.

[5] T. Sauer, J. A. Yorke and M. Casdagli. *J. Stat. Phys.* **65** (1991) 579.

[6] R. Badii and A. Politi, in *Dimensions and Entropies in Chaotic Systems*, edited by G. Mayer-Kress (Springer-Verlag, Berlin, 1986), 67.

[7] R. Badii, G. Broggi, B. Derighetti, M. Ravani, S. Ciliberto, A. Politi and M. A. Rubio. *Phys. Rev. Lett.* **60** (1988) 979.

[8] D. S. Broomhead, and J. P. Huke 'Filtering time series with topology', in *Advanced Signal Processing: Algorithms, Architectures and Implementations V*, Franklin T. Luk, editor, Proc. SPIE **2296** (1994) 305.

[9] M. J. D. Powell, Radial Basis Functions for Multivariate Interpolation: A Review, in *Algorithms for Approximation*, edited by J. C. Mason and M. G. Cox, IMA Conference Series (Oxford University Press, Oxford, 1987), 143.

[10] D. S. Broomhead and D. Lowe. *Complex Systems* **2** (1988) 321.

[11] A. R. Webb. *IEEE Trans. Neural Networks* **5** (1994) 363.

[12] E. J. Kostelich and T. Schreiber. *Phys. Rev. E* **48** (1993) 1752.

[13] M. Davies. *Int. J. Bif. Chaos* **3** (1992) 113.

[14] M. Davies. *Physica D* **79** (1994) 174.

[15] H. Kantz, T. Schreiber, I. Hoffmann, T. Buzug, G. Pfister, L. G. Flepp, J. Simonet, R. Badii and E. Brun *Phys. Rev. E* **48** (1993) 1529.

5. STOCHASTIC RESONANCE

Stochastic Resonance in Passive and Active Electronic Circuits.

Vadim S. Anishchenko, Igor A. Khovanov,
Boris V. Shulgin

*Laboratory of Nonlinear Dynamics,
Saratov State University, Saratov, Russia*

Abstract. The phenomenon of stochastic resonance in a bistable system modeling overdamped oscillator is studied by numerical simulations and experiments. Experimental data are compared with theoretical results. Stochastic resonance in Chua's circuit is investigated in detail for different regimes of its own dynamics. The main characteristics of stochastic resonance for different regimes under the adiabatic approximation are compared.

INTRODUCTION

One of the interesting phenomena which attract great attention of specialists on nonlinear dynamics is stochastic resonance (SR) in bistable systems (1, 2, 3, 4). The SR phenomenon manifests itself in the resonant amplification of a weak regular signal and in the increase of signal-to-noise ratio (SNR) due to the coherence between the process of switchings induced by external noise and the signal (2). Theoretical analysis of SR meets a series of mathematical difficulties, and for the present one is able to perform approximate analytical calculation only for bistable oscillators with the phase space dimension $N \leq 2$ (5, 6, 7, 8, 9, 10). Up to now, the most detailed theoretical and experimental investigation of the totality of stochastic resonance (SR) characteristics has been lead for bistable systems of Duffing oscillator type with two symmetrical interacting attractors being the stable equilibrium states (5, 8):

$$\ddot{x} + \gamma\dot{x} = -\frac{dU(x)}{dx} + F(t), \tag{1}$$

where $U(x)$ is a bistable potential, and $F(t)$ is an external signal which is the sum of a harmonic and noise components (2):

$$U(x) = -\frac{ax^2}{2} + \frac{bx^4}{4} \tag{2}$$

$$F(t) = A\sin[\omega_0 t + \varphi(t)] + \sqrt{2D}\xi(t), \tag{3}$$
$$< \xi(t+\tau)\xi(t) >= \delta(\tau),$$

A, ω_0, φ are the amplitude, frequency, and phase of the harmonic signal, respectively, $\varphi(t)$ is a random variable which is varied in the interval $0 \leq \varphi \leq 2\pi$, D is the intensity of external noise. Most of works are devoted to the study of a limit case, i.e. a so-called "overdamped" oscillator. In this limit the system's equations become one-dimensional what considerably simplifies the mathematical problems when using the methods of stochastic processes theory (8, 12, 13). In the case of large damping $\ddot{x} = 0$, and then in dimensionless variables (8) equation (1) will take the following form:

$$\dot{x} = -\frac{dU(x)}{dx} + F(t). \tag{4}$$

However, one can realize the bistable behavior in systems with dimensions $N > 2$ when the interacting attractors are symmetrical limit cycles, tori or chaotic attractors (14, 15, 16, 10). Moreover, the SR phenomenon arises in either multistable systems or in general systems (non-symmetrical). In the last case the SR may be realized when different kinds of attractors interact, for example, "equilibrium state - limit cycle", or "equilibrium state - chaotic attractor", etc. (14).

Theoretical analysis of SR in the systems pointed out above is, as a rule, impossible due to a number of reasons. The most principal of them are connected with the difficulties of solving the Fokker-Planck equations, the determining of potential, etc. Therefore, exploration of such systems is usually performed by numerical or experimental methods, as well as the method of analogous modeling (17, 18, 19).

In the SR regime, external noise of a certain intensity induces interaction of attractors (switchings between them) and with this the frequency of a weak external periodic signal becomes close to the average frequency of switchings (Kramers frequency) (20). With this, the main important for practice characteristics of SR phenomenon are the amplification and the signal-to-noise ratio (SNR) (7, 18, 21, 22, 23). Particularly, one of the most important (and up to now open (24)) questions is the question about the possibility of obtaining the SNR at the system output more then the one at the input. The answer is clear for the case of overdamped oscillator $SNR_{out} \leq SNR_{in}$. But for the more complicated systems there is no answer yet. Besides this, it is not quite clear how the amplification factor depends on the nature of motion on each of interacting attractors, i.e., on the type of intrawell dynamics of the system.

All mentioned above has determined the main aim of the present work. It seems to be interesting to consider a bistable system in which the SR phenomenon is realized when different types of attractors are interacting. This study allows one to answer a series of questions concerning the influence of intrawell dynamics on the SR characteristics. We shall study the SR by means of numerical and experimental methods which allow us to leave the framework of a number of restrictions used to a certain extent when constructing the theoretical models.

At first, we expound the results of computer simulation and experiments and compare them with theory for the classical overdamped oscillator; in next sections, we describe the active three-dimensional oscillator with bistable behavior (Chua's circuit), then, the results of numerical study of SR in Chua's circuit are given, after that, the results of experimental investigation of SR in Chua's circuit are presented as well as their comparison with the data of computer simulation; finally conclusions contain a short total of the study performed.

OVERDAMPED OSCILLATOR. TEST EXPERIMENTS.

The aim of the present section is to study the regularities in amplifying the signal by means of SR phenomenon in a real radiotechnical device which models a wide class of the simplest bistable systems and to find out the possibility to use in practice the bistable transformers in radioelectronics. The detailed comparison of the data of theoretical and experimental analysis will also allow us to determine the limits of using the theoretical results and the efficiency of a software which was created specially for numerically processing the data with the computer.

The electric circuit under study was worked out and built on the base of classical model of SR phenomenon, i.e., the bistable overdamped oscillator corresponding to the limit case of oscillator with large damping. The overdamped oscillator models the main features of bistable systems behavior irrespective of their physical nature. It is described by equation (4).

An experimental set was built modeling system (4) whose principal scheme is given in Fig.1.

The base of the built of set is the most simple realization of radiotechnical device with two stable and one unstable states. To further lead a series of comparative experimental and theoretical investigations, a nonlinear characteristics of the circuit was measured experimentally and fitted by a polynomial of the 7th order (Fig.2) as follows:

$$-\frac{dU(x)}{dx} = a_1 x - a_3 x^3 + a_5 x^5 - a_7 x^7, \tag{5}$$

$$a_1 = 0.802079,\ a_3 = 5.76467,\ a_5 = 10.1135,\ a_7 = 6.62694.$$

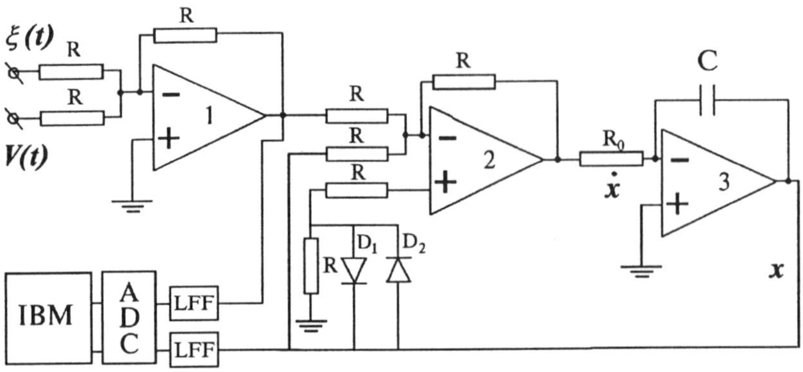

FIGURE 1. A simplified principal scheme of experimental set. 1,2 - the summators, 3 - integrator, $\xi(t)$ - white noise; $V(t)$ - regular signal $A\cos w_0 t$. D_1, D_2 - diodes with square - law nonlinearity; C and R_0 - the elements defining the integration constant.

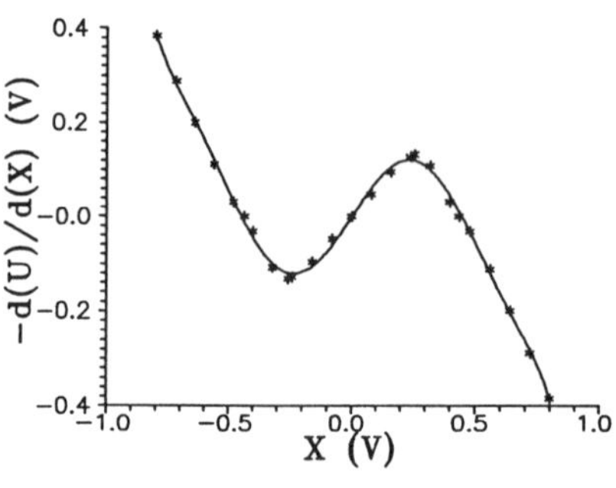

FIGURE 2. Nonlinear characteristics of bistable element of the system. The experimentally found values are shown with dots, and the solid curve is the fitting of characteristics by the 7th order polynomial (5).

When x are small ($|x(t)| \leq 1$), potential (5) approximately corresponds to the 4th order potential (2). The analytical fitting was further used to make theoretical and numerical calculations.

The potential $U(x)$ (5) has its minima at points $x_{min} \simeq \pm 0.447511$, its maximum at point $x_{max} = 0$, and the height of potential barrier $\triangle U \simeq 0.0347205$.

The values of potential curvature at the points of minima and maximum define the local time scale corresponding to the relaxation time when moving in the vicinity of minimum:

$$\tau_l^{-1} \sim \left.\frac{d^2U(x)}{dx^2}\right|_{x=x_{min}} = U''(min) = 1.00583, \tag{6}$$

and the global scale corresponding to the jumps from one well to another and defining the time of relaxing of probability density to the stationary one:

$$\tau_k^{-1} = \frac{\sqrt{|U''(max)|U''(min)}}{2\pi} e^{-\frac{\triangle U}{D}}. \tag{7}$$

When studying the SR, the signal was chosen to be weak: $A < \triangle U / x_{min}$. With this, it was possible to use the theory of linear response it was possible to use. The signal's period was chosen to be much smaller than the relaxation time of the system ($w_0 \ll U''(min)$), what corresponds to the adiabatic limit (5).

The signals corresponding to $x(t)$ and $F(t)$ from the scheme via ADC entered computer where their processing was performed with the help of specially designed software. Before the ADC (see Fig.1) the signals were low-passed to eliminate the aliasing effect (25).

The cutoff frequency of filters was equal to Nyquist frequency, i.e., the largest frequency of fast Fourier transform (FFT):

$$f_c = f_n = \frac{1}{2\triangle t} = \frac{f_{ADC sampling}}{2}, \quad f_c = 3kHz.$$

A number of approximate theories were worked out for stochastic resonance. The first proposed one was adiabatic theory of McNamara and Wiesenfeld (5). Hänggi and Jung suggested a theory in which theoretical solution of Fokker-Planck equation was found by numerical methods (7). A theory of linear response for SR was developed (8, 11). In the framework of our task it is advisable to use the linear response theory, because the experimental situation satisfy the conditions of applying this theory. To find the theoretical dependencies in which we are interested it is necessary to know only potential's curvature at points of minimum and maximum and the height of potential barrier. These are the parameters which the noise-directed global time scale depends on.

The signal amplification, i.e., the ratio of signal's spectrum power at the output to the one at the input is determined by the following expression (7, 8):

$$\eta = \frac{P_{in}}{P_{out}} = 4\frac{|k_1|^2}{A^2}, \tag{8}$$

where k_1 is the coefficient before the first harmonics of periodic asymptotic average value of stochastic process $x(t)$ at the output of system (4). It is defined by following formula:

$$|k_1| = \frac{A}{2}|\chi(w)|, \tag{9}$$

where $\chi(w)$ is the system's susceptibility which is the Fourier transform of the response function $R(\tau)$, which is, in its turn, may be found from fluctuation dissipation theorem. As a result, the amplification is determinated by the following formula (8):

$$\eta = \frac{1}{D} < x^2 >_{st}^2 \frac{1}{1 + w_0^2/\lambda_{min}^2}, \tag{10}$$

where λ_{min} is the minimum eigenvalue of Fokker-Planck operator for system (4):

$$L_{FP} = -\frac{\partial}{\partial x}(-\frac{\partial U}{\partial x}) + D\frac{\partial^2}{\partial x^2} ; \tag{11}$$

λ_{min} determines the time of relaxation of probability density to its stationary value. In the absence of periodic excitation λ_{min} equals the doubled Kramers frequency or is inversely proportional to the average first passage time:

$$\lambda_{min} = 2r_k = \frac{1}{T_1} = \frac{\sqrt{|U"(max)|U"(min)}}{\pi} e^{-\frac{\Delta U}{D}} ; \tag{12}$$

$< x^2 >_{st}$ in (10) is the average value of stochastic process in the absence of periodic force:

$$< x^2 >_{st} = \int x^2 P_s(x) dx, \quad P_s(x) = Nexp^{-U(x)/D}. \tag{13}$$

The amplification versus noise intensity for different frequencies are shown in Fig.3a by dashed lines.

Signal-to-noise ratio at the output is defined by the following formula (8):

$$SNR_{out} = \frac{\pi A^2}{2D^2} < x^2 >^2 \lambda_{min}. \tag{14}$$

Signal-to-noise ratio at the input is defined by:

$$SNR_{in} = \frac{A^2}{8D}. \tag{15}$$

The SNR at the input and at the output versus noise intensity is shown in Fig.3b by dashed lines. It follows from the results obtained that the bistable system increases the signal's power by a value of an order of 8-10 dB in the region of noise intensity where signal-to-noise ratio (SNR) at the output does not becomes worse in comparison with SNR at the input. The difference between the theory and experiment in the region of our interest do not exceed -1dB and seem to be caused by non-correspondence between the noise produced by a radiotechnical oscillator and the ideal white noise.

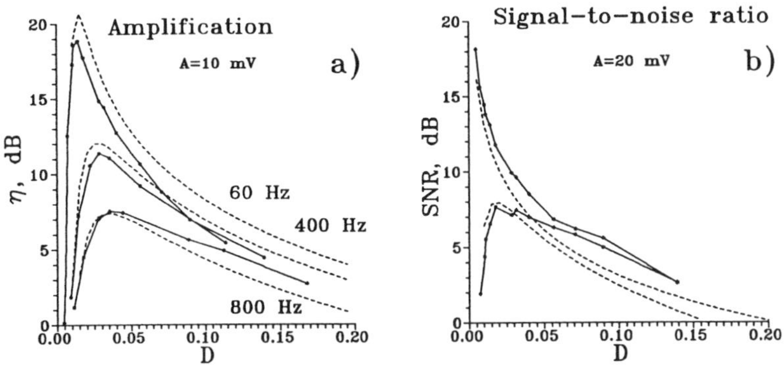

FIGURE 3. a) Amplification versus noise intensity for different frequencies; b) SNR versus noise intensity. Solid curves are experimental, and the dashed ones are theoretical.

System (4) was investigated by numerical methods. The realizations of stochastic processes have been being found by integrating the equation with potential (5) by Runge-Kutta method for stochastic differential equations (26). The ratio of signal's energy to the energy of noise in the band $\triangle f$, which is equal to the spectrum resolution of FFT was thought of as signal-to-noise ratio in computer simulation (as well as in experiment):

$$\triangle f = \frac{1}{N\triangle t}, \qquad (16)$$
$$SNR = \frac{P_s}{P_n(\triangle f)}.$$

Thus, to compare the theory with computer simulation and experiment it is necessary to add a factor $\triangle f = 1/(N\triangle t)$ into the denominator of (14). In the performed experiments, N was equal to 2048, $\triangle t$ was equal to $2\pi/(12w_0)$. The SNR at the input was determined by (15). Before calculating the output spectrum the signal from the output of bistable filter passed trough the Lower Frequency Filter to eliminate the aliasing effect. The cutoff frequency of the filter was equal to the Nyquist frequency of FFT:

$$f_c = \frac{1}{2\triangle t}. \qquad (17)$$

We used $f_c = 6f_0$. In Fig.4a., the numerically found amplification factor versus noise intensity is given for two different signal's frequencies. In Fig.4b., the SNR versus noise intensity is presented. Solid curves are numerical, and dashed ones are theoretical.

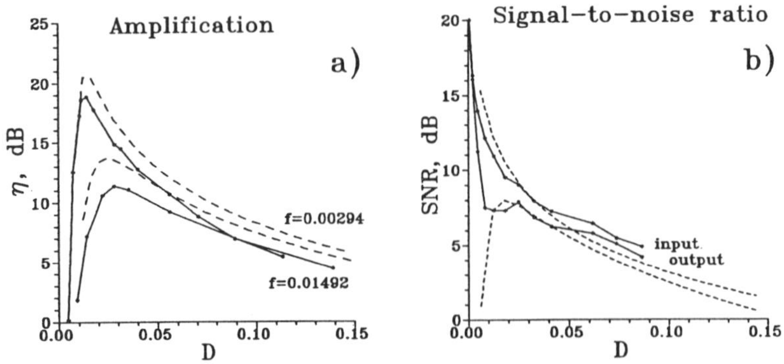

FIGURE 4. a) Amplification versus noise intensity for two different frequencies;b) SNR versus noise intensity. Solid curves are numerical, and the dashed ones are theoretical.

In Fig.5, the typical signals are shown obtained during experiments after having passed through ADC. The upper signal is the input one, $A\cos\omega_0 t + \xi(t)$, the middle one is the output signal $x(t)$, and the lower one is the signal of external modulation $A\cos\omega_0 t$.

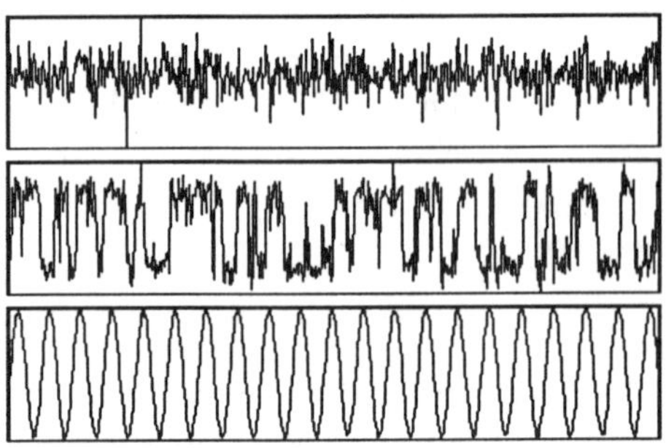

FIGURE 5. Typical experimental signals. The upper signal is the input one $A\cos\omega_0 t + \xi(t)$, the middle one is the output signal $x(t)$, and the lower one is the signal of external modulation $A\cos\omega_0 t$.

In Fig.6 the typical experimental input (to the left) and output (to the right) power spectra are given, the noise intensity growing from the upper figure to the lower one. In the first figure the noise intensity is small, the jumps between the states are rather seldom. That is why the lower-frequency components corresponding to a long residence of the system in the same well dominate in the spectrum. Signal-to-noise ratio (SNR) in the output spectrum is much less than in the input one.

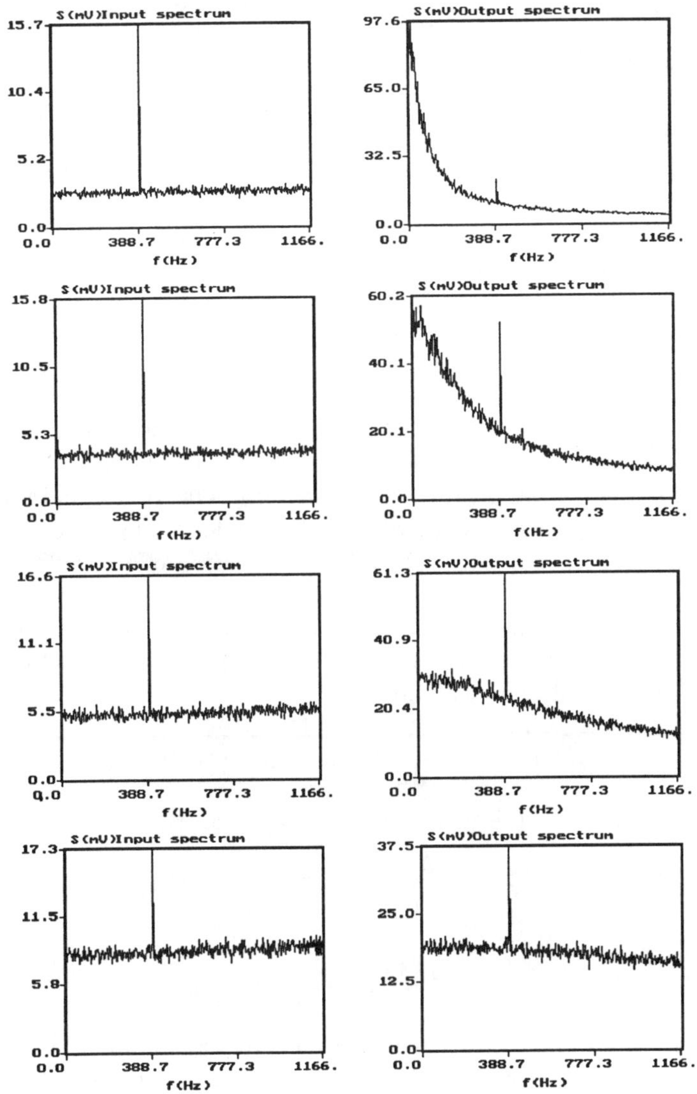

FIGURE 6. Typical experimental input (to the left) and output (to the right) power spectra; noise intensity is increased from the upper figure to the lower one.

The noise intensity in the second figure is somewhat more, the transitions become more frequent.

The third figure reflects the moment of stochastic resonance. The increase of noise at the input has lead to the increase of regular component at the output. The transitions of system from one state to another significantly correlate with periodic signal. The signal amplification is the largest, and $SNR_{in} \simeq SNR_{out}$.

In the fourth figure the noise intensity exceeds the optimal one. The signal amplification is less, $SNR_{in} \simeq SNR_{out}$ is as before.

The signals from the scheme before processing passed through the lower frequency filter because when sampling the signal the aliasing effect arises which consists in mirror-like reflection of spectrum from the band of frequencies higher than the maximum frequency of FFT, f_n to the lower frequency band (25). At the input of system the noise is white, and at the output it is rather colored, with the cutoff frequency $w_c \sim \tau_{kramers}^{-1}$. Therefore, the aliasing effect is maximum at the input where the noise energy is uniformly distributed along the spectrum, and is almost absent at the output where almost all energy of noise is concentrated in the band lower than the largest FFT frequency. Thus the noise at the input is artificially increased due to "reflection" of energy from the band of frequencies greater than f_n, SNR_{in} is decreased and the mistaken result $SNR_{out} > SNR_{in}$ is got which is the evidence of extracting the signal from noise by means of SR (21). When the aliasing effect is eliminated then $SNR_{out} \simeq SNR_{in}$.

To compare the theory and experiment it was necessary to pass from the physical quantities to the dimensionless ones (17). The scheme (Fig.1) is being modeled by the following equation:

$$RC\dot{x} = -\frac{dU}{dx} + A\cos w_0 t + \sqrt{2D}\xi(t). \tag{18}$$

Dividing the time by the time constant of integrator $\tau = RC$ we gain :

$$\frac{dx}{d\overline{t}} = -\frac{dU}{dx} + A\cos \overline{w}_0 \overline{t} + \sqrt{2D}\xi(\overline{t}), \tag{19}$$

$$\overline{t} = \frac{t}{\tau}, \ \overline{w}_0 = w_0 \tau. \tag{20}$$

The time constant of the circuit was measured and found to be $\tau = 49\ \mu s$. The input signal amplitude and nonlinear functions were divided by 1V. The noise intensity was determined by calculating the variance of digitized input noise signal.

The results of experiments are given in fig.3 with solid lines.

In Fig.7 the results of numerical simulation and experiment are compared with theory using as an example the signal amplification versus noise. The results are in good correspondence with each other, the latter confirming the correctness of the used methods and allowing one to use the created experimental set and software to investigate more complicated systems.

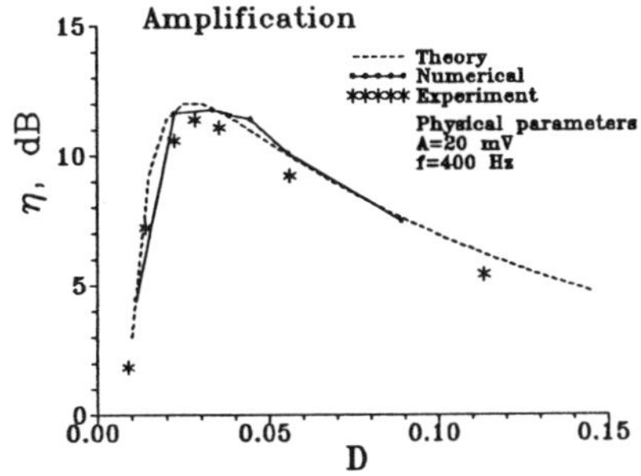

FIGURE 7. Amplification of signal versus noise. Comparing of theoretical, numerical and experimental data.

The theoretical and experimental results presented here give evidence when comparing them that in linear approximation the main characteristics of SR phenomenon obtained theoretically, by direct numerical simulation and experimentally consider within the error not exceeding $\pm 2\%$ (in numerical simulation) and $\pm 8\%$ in experiment. From this it follows that, 1 - the linear response theory is applicable to the analysis of dynamics of a system of overdamped oscillator with bistable potential type (2); 2 - the proposed technical scheme of a bistable system (Fig.1) admits description by equations (4) with potential (5); 3 - the software worked out for numerical and experimental investigation of SR phenomenon gives correct results and may be used in future as an independent method for analyzing with reference to the more complicated dynamical systems which realize the SR phenomenon.

The results of analyzing some certain characteristics of SR in the considered system are the evidence of the following: a bistable filter with noise-induced interaction of attractors being symmetrical equilibrium states provides amplifying the external harmonic signal by about 15-20 dB in power, with this, signal-to-noise ratio at the output of the filter comparing to its input value will at best not change, but can't be increased. From this it follows that such bistable transformers can be practically used as signal amplifiers working in the presence of strong external noise.

ACTIVE BISTABLE SYSTEM IN $\Re^3$ UNDER STUDY

To study the SR phenomenon in active systems we have chosen as an object the classic Chua's circuit described by the following equations (27):

$$\begin{aligned} \dot{x} &= \alpha[y - h(x)], \\ \dot{y} &= x - y + z, \\ \dot{z} &= -\beta y, \end{aligned} \tag{21}$$

where $h(x)$ - is a piecewise-linear characteristics of nonlinear Chua's diode (27):

$$h(x) = m_1 x + \frac{1}{2}(m_0 - m_1)(|x+1| - |x-1|). \tag{22}$$

System (21) is characterized by two control parameters α and β of Chua's circuit itself and two parameters m_0 and m_1 defining different slopes of Chua's diode characteristics. In our experiments we fixed the shape of nonlinear characteristics ($m_0 = -1/7$, $m_1 = 2/7$) and one of the parameters of Chua's circuit ($\beta = 14.286$). Thus, to control the oscillatory regimes in autonomous Chua's circuit a single control parameter was chosen, namely, parameter α.

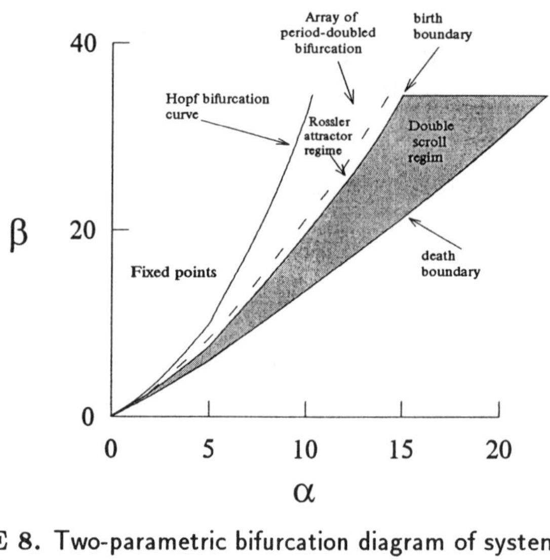

FIGURE 8. Two-parametric bifurcation diagram of system (21). ($m_0 = -1/7$, $m_1 = 2/7$).

The dynamical system (21) defines a vector field in $\Re^3$ which is invariant with respect to symmetry transformation

$$(x,\ y,\ z) \rightarrow (-x,\ -y,\ -z), \tag{23}$$

under the condition that m_0 and m_1 are not zeros. In Fig.8 bifurcation diagram of symmetrical oscillatory regimes in Chua's circuit is presented. It follows from the diagram that if we change parameter α in the interval $2 < \alpha < 10$ in section $\beta = 14.286$ then we will successively observe symmetrical regimes of two equilibrium states ($2 < \alpha < 6.8$), limit cycles T_1 ($6.8 < \alpha < 8.2$), cycles with doubled period T_2 ($8.2 < \alpha < 8.4$), chaotic attractors of Rössler type ($8.7 < \alpha < 8.9$) and, finally, "double-scroll" (DS) regimes ($8.9 < \alpha$) as it is shown in Fig.9.

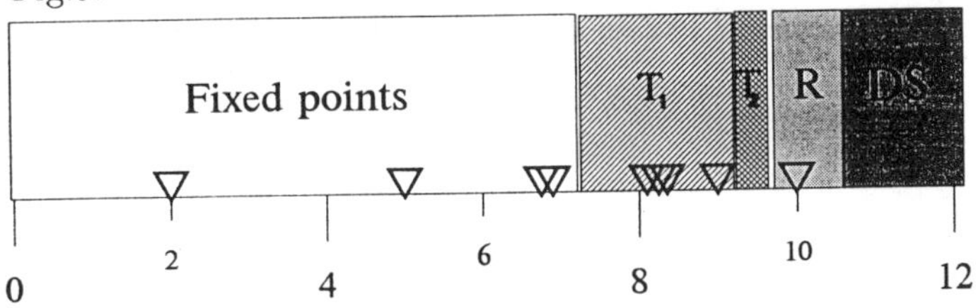

FIGURE 9. Sequence of dynamical regimes being the states of bistable system (21) when the parameter α is varied and $\beta = 14.286$. The value of α in the points under study are marked by triangles. T_n is the cycle with period n, R - Rössler attractor, DS is Double Scroll attractor.

To realize the SR phenomenon, signal $F(t)$ (3) was additively introduced into system (21). Two situations were analyzed, i.e., the signal was introduced into the last equation or into the first one of system (21). The process $x(t)$ whose characteristics were comprehensively analyzed was chosen as a signal at the input of system.

THE RESULTS OF NUMERICAL SIMULATION OF CHUA'S CIRCUIT UNDER THE ADIABATIC APPROXIMATION

When studying the influence of intrawell dynamics type on the SR characteristics we, on the first step, fixed the parameters of regular component of signal (3) which were optimized for the case of $\alpha = 5$. With this, SR was caused by switchings between the two equilibrium states in system (21) (see Fig.9). For the value of $\alpha = 5$ the optimal amplitude and frequency of harmonic signal's component were $A = 0.04$ and $f_0 = \frac{w_0}{2\pi} = 0.008$, respectively.

In Fig.10a the amplification factor, the input and output SNR versus noise intensity D are given, when the signal has been additively introduced into the right-hand part of the third equation of system (21). The similar data are presented in Fig.10b for the case when the signal $F(t)$ was introduced into the first system's equation. It is easy to see that Fig.10b. shows a considerable improvement of SR characteristics, i.e., amplification η is increased and the difference $\triangle SNR = |SNR_{out} - SNR_{in}|$ is decreased. Such a qualitative improvement of characteristics was noticed for all the other oscillatory regimes in Chua's circuit, and further we shall use the results corresponding to introduction of signal $F(t)$ into the first equation of system (21). In Fig.11 average frequency of switchings versus noise intensity is given. Kramers frequency corresponds to the doubled average "frequency of switchings" of system. "Frequency of switchings" is implied as the quantity which is inversely proportional to the average time of system's residence in one and then in another states. [1]

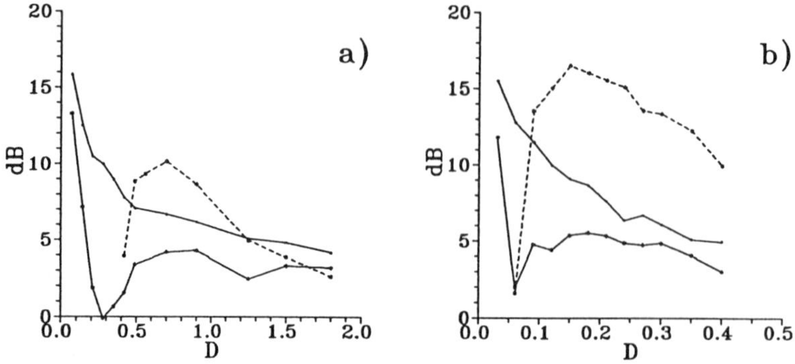

FIGURE 10. Amplification (dashed lines) and SNR (solid lines) versus noise intensity for $\alpha = 5$. **a)** Signal (3) is additively introduced into the third equation of system (21); **b)** Signal is additively introduced into the first equation.

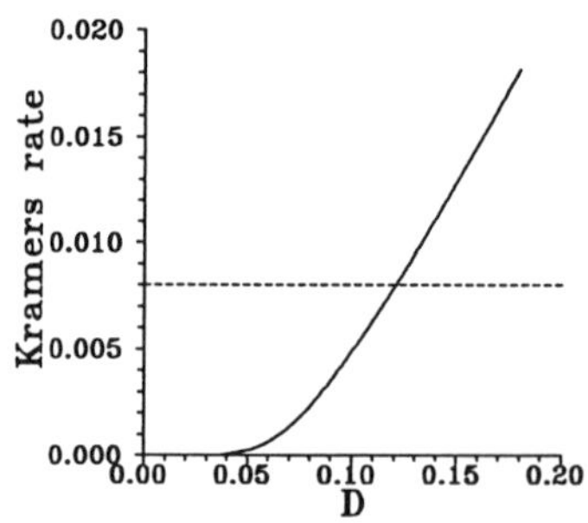

FIGURE 11. Average frequency of switchings versus noise intensity. The value of $f_{switching} = 2f_{kramers} = f_s$ is marked by dashed line.

As seen from the graph on Fig.11, as the noise intensity D is close to 0.12 the Kramers frequency fs is close to the frequency of signal $f_0 \simeq fs = 0.008$, and just for this noise intensity an abrupt increase of amplification is observed (see Fig.10b).

If one varies parameter α, a sequence of autooscillatory regimes becoming

[1]Kramers frequency was found by calculating the average residence time of a trajectory in one of interacting symmetric attractors $\tau_{av} = T_{full}/2$, $f_s = 2T_{full}^{-1} = f_{kram} = 2f_{switchings}$.

more and more complicated from limit cycles to chaos are realized in system (21) (see Fig.8, Fig.9). In Fig.12 the maximum value of amplification η in the region where $\triangle SNR$ is minimum and the average minimum difference $\triangle SNR$ versus parameter α for the fixed values of amplitude and frequency of external modulation are presented.

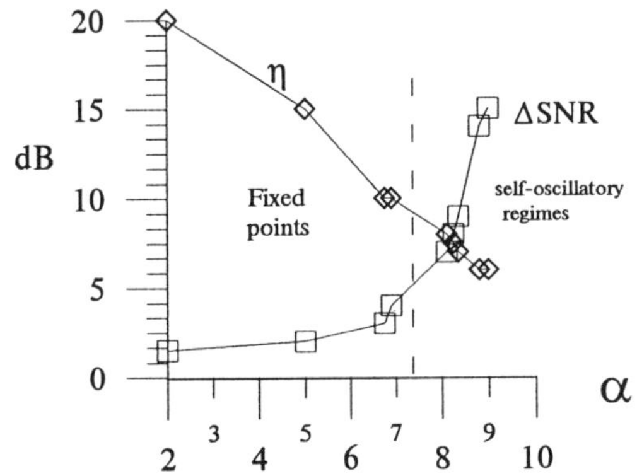

FIGURE 12. Maximum amplification in the region of minimum $\triangle SNR$ versus parameter α, and the average minimum $\triangle SNR$ versus parameter α. The regions of equilibrium states are divided from the regions with the own intrawell dynamics of local states by the dashed line.

The result visually show the worsening of SR characteristics as the type of interacting attractors is being complicated, i.e., amplification factor η almost linearly decreases and SNR abruptly increases with the growth of α.

When α is varied, the size and shape of potential barrier dividing the interacting symmetric attractors is also changed. The modulation signal, being optimal for $\alpha = 5$, naturally, will not satisfy these conditions in case when $\alpha > 5$. We performed the optimization of harmonic signal and noise parameters to increase the SR characteristic for $\alpha > 5$ to stay within the adiabatic approximation when the Kramers frequency is much less than relaxation frequency and signal's amplitude is much less than the height of potential barrier. Our investigation have shown that there are no qualitative changes comparing with data in Fig.12, though one can manage to obtain some improvement of quantitative SR characteristics.

EXPERIMENTAL RESULTS

Experiment on studying the SR characteristics in system (21) in our investigations has a significant advantage comparing to numerical simulation. Being practically not less accurate (c.f. the results in Fig.7) the experiment is lead in real time (ten times faster than the numerical simulation) and allows one to simply vary not only the system's parameters but also the statistical characteristics of external noise.

To process the output signal the computer is used with the same software as in numerical simulation. Thus, the "physics" of experiment is contained, in fact, only in that the signal at the output of bistable system (Chua's circuit) is a real electric signal for the analysis of whose characteristics a digital equipment is further used.

In the Chua's circuit under study corresponding to the mathematical model (21) parameters β; m_0 and m_1 are somewhat different from those used in numerical simulation. This is caused by the peculiarities of experiment. Due to this the values of α for which different dynamical regimes are observed are close to but not coincide with those given in Fig.9. The external force in the circuit corresponds to introduction of signal $F(t)$ into the last equation of system (21), and this is connected with experiment to be convenient.

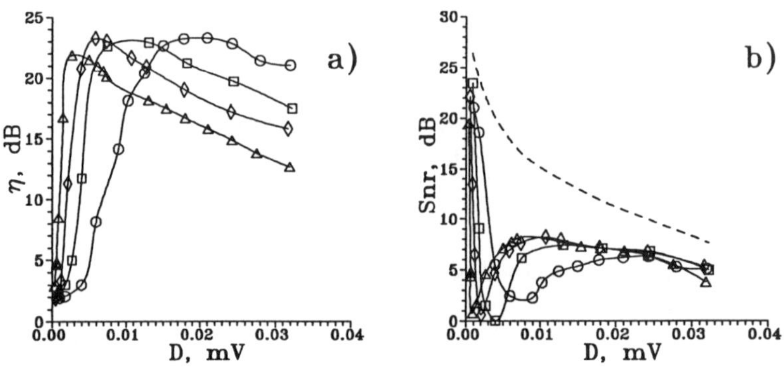

FIGURE 13. a) Experimental amplification of signal and b) SNR at the input (dashed line) and at the output (solid lines) versus noise intensity for different values of parameter α (stable equilibrium states: $\circ - \alpha \approx 4.62, \square - \alpha \approx 5.6, \diamond -$ $\alpha \approx 6.303$; limit cycle: $\triangle - \alpha \approx 7.022$).

In Fig.13a the amplification factor η versus the function of effective value of noise intensity D is presented for various regimes in Chua's circuit which are realized in the interval of values of α: $4.62 < \alpha < 7.022$. As seen from the results, the amplification factors under optimal conditions (optimal values of noise intensity) practically coincide ($\eta_{optim} \simeq 16.5 - 17.5dB$) but obviously decrease with the increase of α in the region $D > 0.015 - 0.02mV$. In Fig.13b the results of SNR_{in} and SNR_{out} versus intensity D are given for the same values of α as in Fig.13a. As seen from Fig.13b, the optimal difference $\Delta SNR = |SNR_{out} - SNR_{in}|$ is reached just in the region $D > 0.015$! Due to this, if we analyze the dependence of amplification factor measured in the region D which corresponds to the optimal difference of

ΔSNR (Fig.13b) we shall obtain qualitatively the same result as one presented in Fig.12. Fig.14 illustrates the results of measuring the amplification for optimal values of ΔSNR and the value of ΔSNR versus parameter α, including the regions of interaction of Rössler attractors ($8.2 < \alpha < 8.59$) and the DS regime ($8.59 < \alpha < 10$). Numerical (Fig.12) and experimental (Fig.14) data are in good qualitative correspondence. With the increase of parameter α, when stable equilibrium states in Chua's circuit are successively changed by the limit cycles, cycles with doubled periods, Rössler attractors and then by intermittency of "chaos-chaos" type (28) (DS regime), the characteristics of SR are worsened.

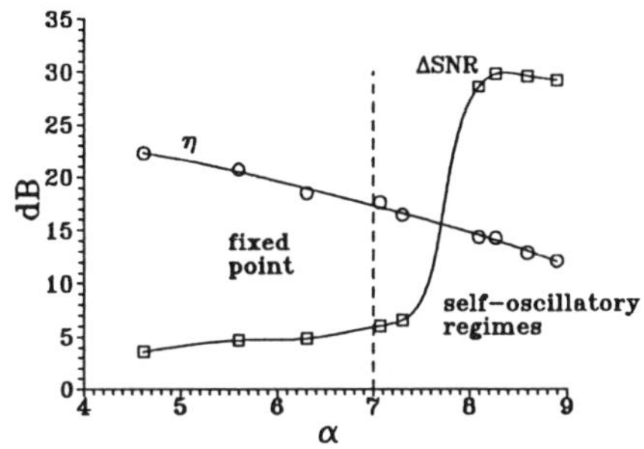

FIGURE 14. Experimental maximal amplification in the region of minimum ΔSNR versus parameter α and the average minimal ΔSNR versus parameterα. Regions of equilibrium states and self-oscillatory regimes of Chua's circuit are divided by a dashed line.

Notice, that in the results presented above the amplitude and frequency of harmonic component of signal $F(t)$ were fixed and corresponded to the condition of adiabatic approximation, i.e., frequency $\omega_0 = 60.06\ Hz$ and amplitude $A = 10mV$. Leaving the framework of these conditions (e.g., increasing the frequency of external regular force) can influence the quantitative characteristics of SR and lead to new phenomena of the so-called high-frequency stochastic resonance (10).

CONCLUSIONS

In the present work we have performed the calculation and measurement of SR characteristics (amplification factor and signal-to-noise ratios at the input and output) in Chua's circuit versus parameter α. With this, the SR phenomenon was realized in the presence of bistable interaction of two symmetric equilibrium states, limit cycles with period T_1, cycles with doubled period, Rössler attractor and DS-attractor. The investigations were held for the classical adiabatic approximation when the frequency of external harmonic force is much less than the frequency defined by relaxation time of the system (5) and amplitude is small comparing to the potential barrier. Under the condition of adiabatic approximation, when the structure of interacting attractors is being complicated (i.e., when the type of intrawell dynamics is being complicated) the SR characteristics are only worsened, i.e., the amplification is decreased and $\triangle SNR$ is increased!

However, it does not mean that there are no methods of optimizing the SR characteristics in this direction. The question is, that the complicated intrawell dynamics of system needs a more detailed account for new characteristic time scales and the search of new conditions of their coherent interaction with the external modulation signal. We are sure that to optimize the SR characteristics in more complicated cases it is necessary to leave the framework of adiabatic approximation, to consider the high-frequency SR (10) and to investigate in more detail the nonlinear aspects of SR phenomenon (29).

ACKNOWLEDGEMENTS

We express our sincere acknowledgements to A. Bulsara, K. Wiesenfeld and F. Marchesony for the discussion of the problem, F. Moss and L. Schimansky-Geier for the interest to this work, D.E. Postnov for his participance in creating the experimental set and N.B. Janson for the help in preparing the manuscript for publication.

We also thank International Scientific Program NATO (Grant HTECH.LC 930749), International Science Foundation of Soros (Grant NRO 000) and the State Committee of Russia on Science and Higher School (Grant #93-8.2-10).

REFERENCES

1. Benzi, R., Sutera, S., Vulpiani, A., *J. Phys. A* **14**, L453 (1981).
2. Moss, F., *Stochastic Resonance: From the Ice Ages to the Monkey Ear*, in Some problems in Statistical Physics ed. by G.Weiss, SIAM, Philadelphia (1992).
3. Proccedings of the NATO Advanced Research Workshop on Stochastic Resonance. *J. Stat. Phys.* **70** #1/2, special issue (1993).
4. Bulsara, A., Jacobs, E.W., Zhou, T., Moss, F., Kiss, L., *J. theor. Biol.* **152**, 531-555 (1991).
5. McNamara, B., Wiesenfeld, K., *Phys. Rev. A* **39**, 4854-4869 (1989).
6. Jung, P., Hänggi, P., *Europhys. Lett.* **8**, 505 (1989).
7. Jung, P., Hänggi, P., *Phys. Rev. A* **44**, 8032-8042 (1991).
8. Jung, P., *Phys. Rep.* **234**, 175 (1994).
9. Gammaitoni, L., Marchesoni, F., Menichaella-Saetta, E., and Santucci, S., *Phys. Rev. Let.* **62**, 349 (1989), *Phys. Rev. A* **40**, 2114 (1989).
10. Dykman, M., Luchinsky, D., Mannella, R., McClintock, P., Stein, N., Stocks, N., *J. Stat. Phys.* **70**, #1/2, 479-500 (1993).
11. Dykman, M., Haken, H., Hu, G., et.al., *Phys. lett. A* **180**, 332 (1993).

12. Gardiner, K., *Handbook of Stochastic methods*, Berlin: Springer, 1985.
13. Risken, H., *The Fokker-Planck Equations*, Berlin: Springer, 1988.
14. Anishchenko, V., Safonova, M., Chua, L., *Int. J. Bif. and Chaos* **2**, #2, 397 (1992).
15. Anishchenko, V., Neiman, A., Safonova, M., *J. Stat. Phys.* **70**, #1/2, 183-196 (1993).
16. Nicolis, G., Nicolis, C., McKernan, D., *J. Stat. Phys.* **70**, #1/2, 125-140 (1993).
17. Zhou, T., Moss, F., *Phys. Rev. A* **41**, #8, (1990).
18. Anishchenko, V., Postnov, D., Khovanov, I., Shulgin, B., *Radiotekhnika i Electronika* **39** #12, 2004 (1994).
19. Longtin, A., Bulsara, A., Pierson, D., Moss, F., *Biol. Cybern.* **70**, 569-578 (1994)
20. Kramers, H., *Physica* **7**, 284 (1940).
21. Gammaitoni, L., Menichella-Saetta, E., Marchesoni, F., *Phys. Lett. A* **142**, 59 (1989).
22. Anishchenko, V., Neiman, A., Safonova, M., Khovanov, I., *Radiorekhnika i elektronika* **39** #8/9, 1380-1392 (1994).
23. Neiman, A., Shimansky-Geier, L., *Phys. Rev. Lett* **72**, 2988-1991 (1994).
24. Moss, F., Pierson, D., O'Gorman, D., *Int. J. Bif. and Chaos* **4**, (1994).
25. Press, W., Flannery, B., Teukolsky, S., Vetterling, W., *Numerical Recipies in C*, Cambridge, Cambridge university press, 1988.
26. Nikitin, N., Razevig, V., *ZhVM* **18** #1, 106, (1978).
27. Chua, L., Komuro, M., Matsumoto, T., *IEEE Trans. in circuits and systems* **33**, 1073 (1986).
28. Anishchenko, V., Neiman, A., Chua, L., *Int. J. of Bifurcation and Chaos* **4**, #1, 99 (1994).
29. Dykman, M., Luchinsky, D., Mannella, R., McClintock, P., Stein, N., Stocks, N., Stochastic Resonance in Perspective, *special issue of Nuovo Cimento D, to be published.*

Possible Breakthrough: Significant Improvement of Signal to Noise Ratio by Stochastic Resonance

L.B. Kiss ∝

JATE University, Institute for Experimental Physics, Dóm tér 9, Szeged, H-6720 Hungary

Abstract. The *simplest stochastic resonator is used, a level crossing detector* (LCD), to investigate key properties of stochastic resonance (SR).

It is pointed out that successful signal processing and biological applications of SR require to work in the *large signal limit* (nonlinear transfer limit) which requires a completely new approach: *wide band input signal* and a *new, generalised definition of output noise*.

The new way of approach is illustrated by a new arrangement. The arrangement employs a special LCD, white input noise and a special, large, subthreshold wide band signal. *First time in the history of SR* (for a wide band input noise), the *signal to noise ratio becomes much higher at the output* of a stochastic resonator than *at its input*. In that way, SR is proven to have a potential to improve signal transfer. Note, that the new arrangement seems to have resemblance to *neurone models*, therefore, it has a potential also for biological applications.

I. INTRODUCTION

In the last decade's physics literature, stochastic resonance (SR) effect has been one of the most interesting phenomena taking place in noisy non-linear systems (see e.g. [1-25]. The input of *stochastic resonators* [12] (non-linear systems showing SR) has usually been fed by a Gaussian noise and a sinusoidal signal with frequency f_0 , that is, a random excitation and a periodic one are acting on the system. For the sake of convenience and without any restriction of generality, we shall call the measured input and output physical quantities "*voltage*" (when actual). There is an optimal strength of the input noise, where the system's output power density spectrum at the signal frequency f_0 has a maximal value. This effect is called SR. For a critical point of view about using sinusoidal signals, see Sections II and III.

From a practical point of view, the most important quantity is the "*signal to*

∝ Work carried out at Uppsala University, Institute of Technology, during a research leave (1992-1995) from JATE University, Inst. for Exper. Physics, Dóm tér 9, Szeged, Hungary.

noise ratio" (SNR) at the input (SNR_{inp}) and at the output (SNR_{out}) of the stochastic resonator. The SNR is defined as:

$$SNR = \frac{P_s}{S(f_0)}$$

(I.1)

where P_s is the mean squared value of the (background corrected) Fourier component of the input voltage at frequency f_0 , and $S(f_0)$ is the spectrum of background noise at f_0. In the case of SNR_{inp}, the background noise is itself the applied input noise voltage. In the case of SNR_{out} , the background noise is the total output voltage when no sinusoidal input signal is applied. In real SR systems, the SNR_{out} also shows a maximum, which is (in the small signal limit) located at the same strength of input noise as the maximum of P_s.

The "*old dream*" of scientists has been to find systems which can significantly increase the SNR at the output, that is,

$$SNR_{out} \gg SNR_{inp}$$

(I.2)

There have been certain hopes [e.g. 13,14] that the principle of SR might be widely applied by nature in biological systems in order to optimise the transfer of neural signals and that SR can be used at signal processing. In this paper, we shall show that this hope ***can be*** right, however, for a real breakthrough we have to give up some old beliefs, moreover, a vast majority of previous theories and experimental results is simply irrelevant.

After the introduction (where we describe our "tools"), in Section II, we outline the new approach to succeed. In Section III, we describe the arrangement, which can realise the "old dream": the significant improvement of the SNR (see Rel. I.2.).

I/1. Do We Need a Nonlinear *Dynamical* System to Observe Stochastic Resonance?

The answer is straightforward: ***no***. On the contrary of the fact that SR is a *nonlinear dynamical* phenomenon, we do *not* need a dynamical system to observe it. We need only a nonlinear system with *threshold-like nonlinearity*. The key point is that the *dynamical side* of SR is contained by the *level-crossing (threshold crossing) dynamics* of the Gaussian input noise.

Until last year, it has been a common belief that stochastic resonance (SR) phenomena occur only in (bistable, sometimes monostable [10] or multistable) *dynamical* systems [1-14]. Very recently, a new stochastic resonator has been described [15,16] (invention: Moss; theory: Kiss; experimental study: Gingl): the *level-crossing detector* (LCD), which is a very practical and simple non-dynamical linear system (widely used in signal processing as an FM demodulator). Both the theory of the SR in LCD and its extensive experimental study show that the *level crossing dynamics* of the Gaussian noise *inherently* contains the SR effect. The fact that the SR curve is *not sensitive* against changing the frequency of the sinusoidal signal and the shape of the spectrum ("colour") of input noise is a strong evidence that the LCD is the simplest possible stochastic resonator, where the SR curve is purely determined by the Gaussian noise and it is not influenced by the specific dynamics of the resonator. Due to its fundamental nature, simplicity and practical importance in signal processing, we are using the LCD both for theoretical and experimental investigations. As the detailed theory of Kiss [16] has not appeared yet in the open literature, for the benefit of the reader, we present the extended theory in the Appendix.

Note, that Bulsara et al. [17], Jung [18] and Wiesenfeld et al. [19] have also studied various other aspects of threshold-crossing SR phenomena.

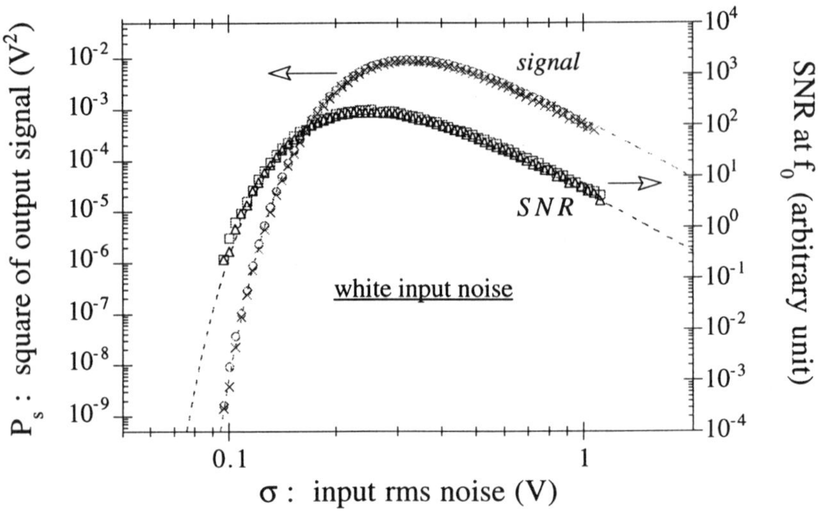

FIGURE I.1. Representative plot of the results of white noise experiments ***in the linear response limit*** (by Gingl [15,16]) on an asymmetric LCD system. Input sinusoidal signal: B=0.1V, f_0 = 38Hz and 305Hz; input noise: white, upper cut-off 12kHz; U_t=0.45V; output pulses: A=5V and τ_0=1μs. Data: P_S: square of the output Fourier amplitude at f_0; SNR: P_S divided by the output background noise spectrum at f_0 ; (x) and (*triangle*) 305Hz; (o) and (*square*) 38Hz; curve fits: dashed lines given by the *linear* theory of Kiss (Eqs. A.3.2 and A.3.4 in the Appendix).

I/2. Description of LCD Stochastic Resonators

Asymmetric LCD stochastic resonator. The asymmetric system consists of an LCD of the following kind: whenever the instantaneous strength of the input excitation (noise and small sinusoidal signal) crosses the positive threshold level

U_t *in increasing direction*, the LCD produces a positive, short pulse with amplitude A and duration τ_0 at its output. The resulting output response of the system is a random time-sequence u(t) of uniform, positive pulses.

Symmetric LCD stochastic resonator. The symmetric system consists of an LCD of the following kind: whenever the instantaneous voltage (noise and small sinusoidal signal) crosses the positive threshold level U_t *in increasing direction*, the LCD produces a positive, short pulse with amplitude A and duration τ_0 at its output; on the other hand, whenever the instantaneous strength of the input excitation (noise and sinusoidal signal) crosses the negative threshold level $-U_t$ *in decreasing direction*, the LCD produces a negative, short pulse with amplitude -A and duration τ_0 at its output. The resulting output response of the system is a random time-sequence u(t) of uniform, positive and negative pulses with zero time average.

It follows from the above definitions, that the role of a sinusoidal input signal of amplitude B and frequency f_0 can be viewed as a periodic modulation of the threshold level(s) which causes a periodic modulation of the mean frequency ν of the threshold crossings of the noise, that is the mean repetition frequency ν of pulses at the output of the system.

In Fig.1, a representative plot of the results of LCD experiments [15,16] with sinusoidal input signal and white input noise is presented and the excellent fit by the linear theory [15,16] (Eqs. A.3.2 and A.3.4 in the Appendix) is shown.

I/3. There is No Possibility to Improve the SNR in the Linear Response Limit

A corollary [20,25] given by Dykman et al. and DeWeese and Bialek, is a simple and exact proof that, in the *small signal limit* (*linear response limit, when the signal amplitude is small compared to the input noise rms amplitude*), the signal-to-noise-ratio (SNR) at the output of a stochastic resonator is *always* smaller than the SNR at the input of the resonator. The corollary is based on the fact that in the case of linear response, the input noise component belonging to an infinitesimally small bandwidth around the f_0 can also be considered as an additive input signal, due to the random-phase-oscillators representation [26] of Gaussian processes (see Eq. A.2.1. in the Appendix). Therefore, this noise component gets through the system by the same amplification as the signal does, thus, the output SNR at f_0 cannot be smaller than at the input. Moreover, if we cut out this component from the input noise, the stochastic resonator will still work and the output noise component around f_0 will be greater than zero. Due to linearity, this output noise component shows up as an excess output noise in the original case and it prohibits to reach *even* the input SNR at the output. The proof generally holds for any possible type of single stochastic resonators, for white or arbitrary coloured (Gaussian) noise. This result implies serious doubts about the applicability of those models of physical and biological systems, where the improvement of the SNR has been expected by (classical) stochastic resonance in the small signal limit.

Note that the theoretical analysis of the LCD resonator [15,16] (see Appendix) yields $SNR_{out} \leq 0.85\ SNR_{inp}$, which result is in good agreement with Gingl's simulations [15,16] and with the corollary described above.

In the scientific community of SR, the above fact empirically has been realised and this corollary explains the reason. Careful investigations on *coupled* stochastic resonator systems show, that while coupling can help to improve the SNR *compared to a single resonator*, the total gain cannot be greater than 1 , that is, the SNR at the output cannot be greater than at the input [21,22].

II. WHAT DO WE NEED FOR A BREAKTHROUGH?

We need a completely *new* way of approach. The fact that SR cannot improve the SNR in the small signal limit (the case of linear transfer) implies the need for investigations in the *strong signal limit* (the case of *nonlinear transfer*). As the research of physical, biological or other applications of SR are concerned, one has to keep in mind that the nonlinear response and its noise characteristics strongly depend on the input signal.

i) Therefore, at the experimental and theoretical study of systems with SR, the application of a *sinusoidal* input signal is *no more useful*, because a single stationary sinusoidal signal cannot transfer dynamical information. In contrast with this, an input signal with a *wide frequency spectrum* is desired, and its statistical properties are desired to *imitate* the properties of real physical, biological or other signals. Otherwise, the investigations do not yield a relevant conclusion for the applicability of SR.

ii) The output background noise can*not* be determined in the usual way, that is, by determining the output noise at no input signal. In the case of nonlinear transfer, the input signal yields extra cross-modulation products with the input noise, which is an extra output noise with a strong dependence on various characteristics of the input signal. Therefore, a new definition of the output noise is needed. To reach this new definition, first we define the "*generalised amplification*" K(f) of the generalised (non-sinusoidal) input signal:

$$K(f) = \frac{S_{inp,out}(f)}{S^{sig}_{inp}(f)} \quad , \qquad \text{(II.1)}$$

where $S_{inp,out}(f)$ is the cross-spectrum of the input noise voltage and the *total output voltage* (output signal + output noise), and $S^{sig}_{inp}(f)$ is the (power density) spectrum of input noise. Note that K(f) depends not only on the frequency, but also on the input signal and on the input noise. The "*generalised output signal*" can be defined by its spectrum $S^{sig}_{out}(f)$:

$$S^{sig}_{out}(f) = S^{sig}_{inp}(f)\,|K(f)|^2 = \frac{|S_{inp,out}(f)|^2}{S^{sig}_{inp}(f)} \qquad \text{(II.2)}$$

The definition of output noise spectrum $S_{out}^{noi}(f)$ is a straightforward consequence of the above definitions:

$$S_{out}^{noi}(f) = S_{out}^{tot}(f) - S_{out}^{sig}(f) \quad , \qquad (II.3)$$

where $S_{out}^{tot}(f)$ is the spectrum of the total output voltage. Note, that the above definitions restore the validity of the old definitions in the limit of small sinusoidal input signal (linear transfer and sinusoidal excitation). The new definitions work at arbitrary conditions and the only pre requirement is the stationarity of the input noise, input signal and stochastic resonator.

There the fundamental differences between the classical definition of SNR and the above described generalised quantities:

1. In the case of the *classical* definitions, the output signal is defined by the square of the frequency component of the total output voltage at the frequency of the input signal, so, that the output noise power at this frequency is subtracted.

2. In the case of the *classical* definitions, the output noise is the total output ac voltage in the case of no input signal.

3. In the case of the *generalised* quantities, the *output signal* is that part of the total output voltage which is *correlated with the input signal*.

4. In the case of the *generalised* quantities, the output noise is that part of the total output voltage which is *uncorrelated with the input signal*.

5. By comparing **2** and **4**, it is obvious that in the case of wide band signal and nonlinear response, the new definition (*correctly*) gives signal-dependent output noise and the *classical definition gives erroneously small output noise*, because of its definition with zero signal (missing cross-modulation products).

6. In the case of classical SR (linear response and sinusoidal excitation) both descriptions, the classical and the new work and *they give the same result.*

7. *In the case of nonlinear response and wide band signal, only the generalised terms have realistic physical meaning at signal processing.*

iii) Finally, a practical remark. Due to the nonlinearity of the transfer, the technical applications will probably have to use encoded signals (for example, pulse frequency modulation (PFM)) which are not sensitive against (non-dispersive) nonlinear transfer. Then, the necessary decoding can significantly influence the noise characteristics of the transfer, therefore this sort of investigation can also be crucial at real technical applications.

III. A POSSIBLE BREAKTHROUGH: IMPROVEMENT OF SIGNAL TO NOISE RATIO BY STOCHASTIC RESONANCE

In this section a new arrangement [23,24] is described which is able to improve the SNR by SR. In accordance with Section II, the new system is a stochastic resonator working in the strongly nonlinear transfer limit with wide band input signal. The arrangement has remarkable resemblance to neurones.

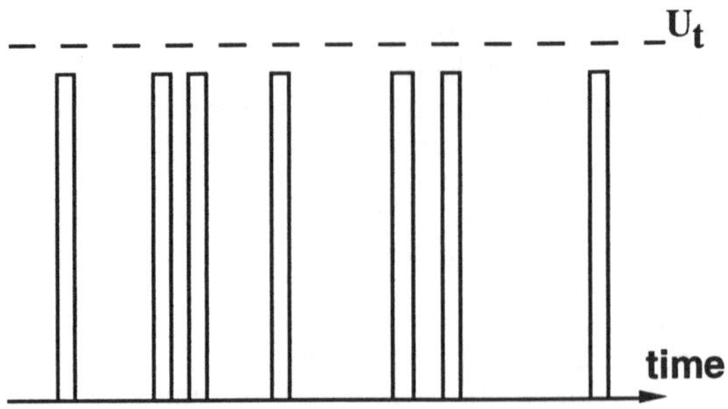

FIGURE III.1. Illustration of the wide band input signal of the new arrangement, which is similar to the input and output signal of neurones. The height, duration and mean repetition rate of pulses are A, τ_0 and ν_{sig}, respectively. The output voltage looks similar to this, except, some "jitter" of the initial times of pulses and some extra pulses at random times. The last effect corresponds to the classical output noise because it exists even without an input signal. The jitter effect yields a new sort of output noise ("jitter noise") which does not exist without input signal. It is a nice example of excess output noise described at the ii) point of Section II.

One of the simplest possible system is an asymmetric LCD, which is driven by the sum of a *Gaussian white input noise* and a *sequence of square input signal pulses*. To represent a *wide band input signal*, it is assumed that the initial times of pulses are generated by a Poissonian process with a rate ν_{sig}, however any stationary time sequence (even a periodical one) would give similar results as the improvement of the SNR is concerned. The present signal is a good representation of a real "pulse frequency modulated" (PFM) signal. As we want only to demonstrate that this system is able to *improve the SNR* significantly, for the sake of simplicity we choose the output characteristics that way, so they can make the calculations easy. Therefore, it is assumed that the square input signal pulses have the same duration τ_0 and size A *both* at the input and at the output of the LCD. Moreover, the rms amplitude σ of the input noise and A are assumed to satisfy the following relations with the threshold level U_t:

$$\sigma << U_t \quad , \qquad \text{(III.1)}$$

$$U_t - \sigma < A \qquad \text{(III.2)}$$

and the input white noise has a very high cut-off frequency f_c compared to the reciprocal length $1/\tau_0$ of pulses:

$$1/\tau_0 << f_c \qquad \text{(III.3)}$$

The quantity ν in Eq. A.2.8 (Appendix) gives the *unidirectional* threshold crossing frequencies $\nu(0)$ and $\nu(U_t\text{-}A)$ for the two possible cases: when the signal amplitude is zero and when the signal amplitude is A (that is, when the execution of a signal pulse is going on):

$$\nu(0) \approx \exp[-U_t^2/(2\sigma^2)]\ f_c/2 \qquad , \qquad \text{(III.4)}$$

$$\nu(U_t\text{-}A) \approx f_c/2 \qquad . \qquad \text{(III.5)}$$

The output voltage contains two different kinds of noises:

i) Noise due to unexpected pulses with mean repetition frequency $\nu(0)$. This noise corresponds the classical output noise in SR, because it exists even without input signal, as it is generated by the threshold crossings of input noise. In the practically interesting frequency limit ($f \ll \tau_0$), Campbell-theorem applies (see Eq. A.1.2) for the spectrum $S_{cla}^{noi}(f)$ of this classical noise:

$$S_{cla}^{noi}(f) = \nu(0)\, A^2\, \tau_0^2 \qquad . \qquad \text{(III.6)}$$

ii) Noise due to the random delay ("jitter") of an output pulse when following the input pulse. This noise is an excess noise compared to the output noise of classical SR and it exists clearly due to the *joint* existence of the signal and the input noise and it is clearly related to the new way of description outlined in Sec. II. To determine the strength of "jitter noise" spectrum, the new principle is used (described in Sec. II): the output noise is that part of the input noise which is uncorrelated with the input signal. Due to Rels. III.3 and III.5, the mean duration of the jitter is very short, $\approx 2/f_c$. As the small jitter effect does not affect the bandwidth $\Delta f \approx 1/\tau_0$ of output voltage, the noise spectrum $S_{jit}^{noi}(f)$ can be approximated by $P_{jit}/\Delta f$, where P_{jit} is the mean square of that fraction of output voltage pulses which do not overlap with the input pulse due to the jitter:

$$S_{jit}^{noi}(f) \approx P_{jit}/\Delta f \approx P_{ii}\, \tau_0 \approx \tau_0\, \nu_{sig}\, \frac{4\,A^2}{f_c\, \tau_0}\, \tau_0 = \frac{4\,A^2\, \tau_0\, \nu_{sig}}{f_c} \qquad . \qquad \text{(III.7)}$$

The total output noise $S_{out}^{noi}(f)$ is the sum of the classical and jitter noise spectra:

$$S_{out}^{noi}(f) = S_{cla}^{noi}(f) + S_{jit}^{noi}(f) \approx \nu(0)\, A^2\, \tau_0^2 + \frac{4\,A^2\, \tau_0\, \nu_{sig}}{f_c} \qquad . \qquad \text{(III.8)}$$

I our case, the very small jitter (see above) provides approximately the same signal spectrum at the output as at the input (see also Eq. A.1.2):

$$S_{out}^{sig}(f) \approx S_{in}^{sig}(f) \approx \nu_{sig}\, A^2\, \tau_0^2 \qquad . \qquad \text{(III.9)}$$

As the spectrum of the input white noise is $S_{inp}^{noi} = \sigma^2/f_c$, it is now easy to calculate the ratio of SNR at the output and input by Eqs. III.8 and III.9:

$$\frac{SNR_{out}}{SNR_{inp}} \approx \frac{2\,\sigma^2}{A^2\,\tau_0^2\,f_c^2\,\exp[-(U_t/\sigma)^2/2] + 8\,A^2\,\tau_0\,\nu_{sig}} \quad , \text{(III.10)}$$

where the first term in the denominator is due to the classical output noise, and the second term is due to the jitter noise. Eq. III.10 is our mean result [23,24] which shows that the SNR can significantly be improved by this system. In Fig. III.2, the effectiveness of the arrangement is illustrated with practical parameters (see figure caption). It can be seen that with the chosen parameters, the gain in the SNR is about 10 and 100 when the repetition rate of pulses is 10Hz and 1Hz, respectively.

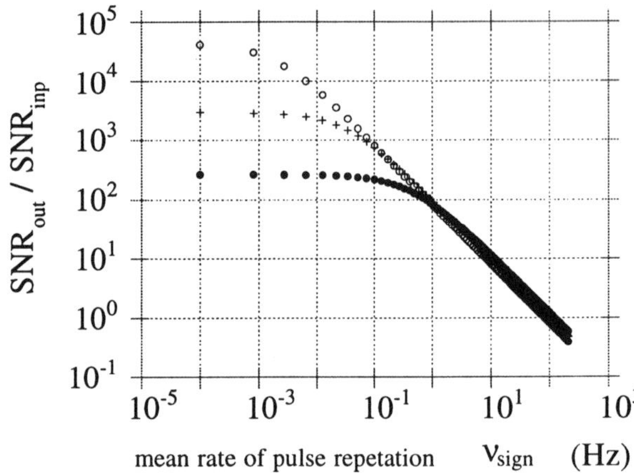

FIGURE III.2. Illustration of the effective SNR improvement of the new arrangement. Applied parameters: σ=1V; f_c=100kHz; τ_0=100μsec; A= 4.5V (filled), 5V (cross), and 5.5V (empty); U_t= 5V (filled), 5.5V (cross), and 6V (empty).

It is remarkable to note, that *close to the maximum* of possible repetition rate $\approx 1/\tau_0$, there is *no* gain, but a significant loss of SNR. As possible biological applications are concerned, the consequences of this fact has to be analysed in terms of information theory [25] and real neural signals and noise. The gain comes due to the *rare* repetition of pulses, which yields a strong reduction of the SNR at the input, but less strong SNR reduction at the output. It is due to the jitter noise component, which is proportional to the rate, similarly to the signal. At very small ν_{sig} (very rare pulses), the SNR gain is saturating due to the classical output noise which is independent of ν_{sig} . Note, the saturation limit of SNR gain can be *arbitrarily large* by choosing proper values of the A and threshold level.

Note, that Eq. III.10 is valid only in the limit given by the special conditions described by Eqs. III.1 - III.3. These conditions are chosen to prove the effectiveness of SNR improvement in a simple way. The system obviously shows SR phenomenon as a function of σ, however, the theory for arbitrarily chosen parameters is a more complicated issue.

APPENDIX: GENERALISATION OF RICE'S FORMULA FOR ZERO CROSSINGS AND THE LINEAR THEORY OF STOCHASTIC RESONANCE IN LCD SYSTEMS

At the time of completing this manuscript, the detailed Kiss-theory has not appeared yet in the open literature, therefore, for the benefit of the reader we present the theory here. The theory of asymmetric LCD system and symmetric LCD system gives the same result in the *linear limit* (sufficiently small input signal). In the non-linear case, the output signal of the symmetric system does not have even harmonics, otherwise the main features of its response is identical with the asymmetric system.

A.1 Spectra of (Weakly and Slowly) Frequency-Modulated Random Time-Sequence of Uniform Unipolar Pulses

Without frequency modulation, for sufficiently short pulses, the time-average U_{av} of the process is proportional to the mean repetition frequency ν of pulses, $U_{av} = \langle u(t) \rangle_t = \nu A \tau_0$. The small and slow modulation of the ν yields the linear modulation of the time-average U^*_{av} measured for short times (e.g. much shorter than the modulation frequency), so:

$$U^*_{av}(t) = \nu(t)\, A\tau_0 \qquad , \qquad (A.1.1)$$

which, in the case of a sinusoidal modulation, results in a sinusoidal component in the Fourier-spectrum of the pulse train. Moreover, this component has zero phase-shift and its amplitude is *independent* from the modulation frequency. The validity of this result requires to fulfil the following practical precondition:

(i) The modulation frequency $f_0 << \nu(t)$. This requirement is also necessary for the efficient transfer of modulating signal.

Under the condition i), Eq. A.1.1 describes the periodic response for the modulation in the pulse-train. However, it does not give information on the spectrum of the noise-component of the pulse-train. For pulses with completely random initial times, in the low frequency limit of $0 < f << \tau_0^{-1}$, the Campbell-theorem describes the spectrum $S_{no}(f)$ for the case of no frequency-modulation:

$$S_{no}(f) = \nu A^2 \tau_0^2 \qquad . \qquad (A.1.2)$$

Assuming, that the modulation of the mean repetition frequency of the pulses is symmetric, weak and small, we can expect Eq. A.1.2 correctly describes the background noise spectrum of the pulse-train. In Eq. A.1.2 , the values of ν for no frequency modulation applies.

It is obvious from Eqs. A.1.1 and A.2.2, that the quantity $\nu(t)$ has a crucial role in the determination of the Fourier-components of the pulse-train, so we devote the next section to the derivation of $\nu(t)$.

A.2. Generalisation of the Rice-Formula of Zero Crossings for the Case of Arbitrary Level Crossings

Rice [26] determined the mean frequency ν_0 of zero-crossings of the amplitude of Gaussian noises. In order to obtain the mean frequency $\nu(U_t)$ of crossings of arbitrary levels U_t we derive the Rice-formula in a different and simpler way which yields the required generalisation. We are concerned with physical noises which means that the amplitude y(t) of the Gaussian noise and its velocity dy/dt have finite root-mean-square (rms) values σ and Δ , respectively. Thus, their power density spectrum have cut-offs (at least for high frequencies).

First, it is shown that the amplitude y(t) of the Gaussian noise is statistically independent of its velocity dy/dt. According to the random-phase-oscillators representation [26] of Gaussian processes, the amplitude of a Gaussian noise can be written as:

$$y(t)=\sum_{n=1}^{\infty} a_n \sin(2\pi f_n t+\phi_n) \quad , \qquad \text{(A.2.1)}$$

where a_n , f_n and ϕ_n are the amplitude, the frequency and the random initial phase of the n-th oscillator, respectively. The oscillator frequency is given as $f_n= n\ \delta f$, where δf is infinitesimally small. The random set of ϕ_n (uniformly distributed in the range of $(0,2\pi]$) is different for each representation of the given noise process. From Eq. A.2.1, the velocity can be written, as:

$$\frac{dy(t)}{dt}=\sum_{n=1}^{\infty} 2\pi f_n a_n \cos(2\pi f_n t+\phi_n) \quad . \qquad \text{(A.2.2))}$$

It is easy to see that the cross-correlation function of Eqs. A.2.1 and A.2.2 is zero due to the orthogonality of sinus and cosine functions with the same frequency and due to the orthogonality of sinusoidal functions with different frequencies:

$$\langle y(t)\ dy(t)/dt \rangle_t = 0 \quad , \qquad \text{(A.2.3)}$$

where the index t represents time-averaging (as y(t) and dy/dt are stationary and ergodic processes). Using the well known fact that a zero cross-correlation of two different Gaussian processes implies their statistical independence, we can conclude that the *instantaneous amplitude and the instantaneous velocity of a Gaussian noise are statistically independent from each other.*

In the next step, we determine the functional form of the mean frequency $\nu_0(U_t)$ of the crossings of level U_t by the noise amplitude y(t). Let us consider the behaviour of the noise in an infinitesimally narrow amplitude interval $U_t-\partial U \leq y(t) \leq U_t-\partial U$ around U_t. The infinitesimal smallness of the interval $2\partial U$ has two important implications: i) the amplitude distribution function of the noise will be uniform in this range; ii) for a long but finite duration of the noise, the number of those amplitude trajectories which enter into the interval and, after changing their direction, leave the interval without crossing the level U_t , is zero. Note, the last implication pre-requires the frequency band limited property of the physical Gaussian noise (see above). On the other hand, the number of amplitude trajectories which enter into the interval and, without changing their direction, leave the interval via crossing the level U_t , is not zero and it is related to the mean frequency $\nu(U_t)$ of crossing this level. On the base of these considerations the following relations can be written for the probability of finding the instantaneous amplitude y(t) within the interval:

$$2\, g(U_t)\, \partial U = \nu(U_t)\, \partial t \quad , \qquad \text{(A.2.4)}$$

where g(U) is the amplitude distribution of the noise, that is, $g(U_t) = (2\pi)^{-1/2}\, \sigma^{-1} \exp[-(U_t/\sigma)^2/2]$ and ∂t is the mean passing time of trajectories via the interval. The right hand side of Eq. A.2.4 represents the fraction of time which the noise amplitude spends in the interval. From Eq. A.2.4, the level crossing frequency can be given as follows:

$$\nu(U_t) = 2\, \frac{\partial U}{\partial t}\, (2\pi)^{-1/2}\, \sigma^{-1} \exp[-(U_t/\sigma)^2/2] \qquad \text{(A.2.5)}$$

The quantity $2\partial U/\partial t$ is equal to the mean velocity of the noise in the interval. Due to the statistical independence of the velocity and the amplitude, $\partial U/\partial t$ will be independent from the location of the interval, so the value of U_t. On the other hand, $\partial U/\partial t$ is proportional to the rms velocity of the noise, consequently:

$$\nu(U_t) = c\, \Delta\, \sigma^{-1} \exp[-(U_t/\sigma)^2/2] \quad , \qquad \text{(A.2.6)}$$

where c is a constant. With the help of the power-density spectrum S(f) of the noise y(t) , the Δ term can be replaced by the integral of $(2\pi)^2 f^2 S(f)$:

$$\nu(U_t) = C\, \sigma^{-1} \exp\left[\frac{-U_t^2}{2\sigma^2}\right] \left[\int_0^{\infty} f^2\, S(f)\, df\right]^{1/2} \quad , \qquad \text{(A.2.7)}$$

where $C=4\pi^2 c$. Eqs. A.2.6 and A.2.7 describe the functional form of the dependence of ν on S(f) and U_t . The determination of the constants c and C is based on the following trick: we take $U_t=0$ as threshold (so the value of the exponent term becomes 1) and a narrow-band noise with the following spectrum: $S(f)=S_0$ for $f_0-\partial f < f < f_0+\partial f$ and otherwise it is zero. Then, in the limit $\partial f \rightarrow 0$, Eq. A.2.7 will approach the zero-crossing frequency of a sinusoidal signal with frequency f_0 , that is, $\nu = 2f_0$. In this way we have obtained our final formula:

$$\nu(U_t) = 2\,\sigma^{-1} \exp\left[\frac{-U_t^2}{2\sigma^2}\right] \left[\int_0^{\infty} f^2\, S(f)\, df\right]^{1/2} , \qquad (A.2.8)$$

which is the generalisation of the Rice-formula [26] (derived for the frequency of zero-crossings). Indeed, in the limit of U_t=0 , Eq. A.2.8 becomes identical with the formula derived by Rice [26].

A.3 Stochastic Resonance in Level Crossing Detectors: Theory for the Case of Small Sinusoidal Signal (Linear Signal Transfer)

When the pulse-sequence is generated in an LCD system (of threshold U_t) by the sum of a Gaussian noise and a sinusoidal signal $B\sin(\omega t)$, we can view the problem as the level-crossing problem of the pure noise with a modulation $-B\sin(\omega t)$ of the threshold level U_t . To describe the quantity $\nu(t)$ for this case, one can apply Eq. A.2.8 provided the following preconditions holds: condition i) and (ii). The modulation frequency $f_0 << \tau_{corr}$, (where τ_{corr} is the correlation time of the input noise). In the case of noises with a flat spectrum, the validity the condition i) (Section A.1.1) implies the validity of condition ii), however, for other noises (like band-limited 1/f noise) it does not necessarily imply that.

Then, condition i) and ii) provide that the pulse-sequence will represent its real statistical properties during the half-period $1/2f_0$ of the sinusoidal modulation, that is, Eq. A.2.8 does have a real meaning. Taking into the account that the ν at one-directional level crossing is half of the ν at bi-directional level crossing, the time-dependence $\nu(t)$ of the mean repetition frequency of output pulses in an asymmetric LCD system can be given as:

$$\nu(t) = \sigma^{-1} \exp\left\{\frac{-\,[U_t - B\sin(\omega t)]^2}{2\sigma^2}\right\} \left[\int_0^{\infty} f^2\, S(f)\, df\right]^{1/2} . \qquad (A.3.1)$$

From Eqs. A.1.1 and A.3.1, by Taylor-expanding the exponential function and keeping the linear term, we obtain the squared amplitude P_s of the output sinusoidal response at frequency f_0 :

$$P_s = (\nu_0\, B\, A\, \tau_0\, U_t)^2\, \sigma^{-4} \exp\,[-(U_t/\sigma)^2] , \qquad (A.3.2)$$

where the phase-shift of the first harmonic is zero compared to the input sinusoidal signal and ν_0 is the mean frequency of one-directional zero-crossings of the input noise and it can be given from the original Rice-formula [26] (see Eq. A.2.8):

$$\nu_0 = 2\,\sigma^{-1} \left[\int_0^{\infty} f^2\, S(f)\, df\right]^{1/2} . \qquad (A.3.3)$$

In order to satisfy the preconditions of the applicability of Campbell-theorem, that is, for Eq. A.1.2, we have to apply a new precondition:

iii) Let us assume that $\nu(t) << \tau_{corr}^{-1}$.

This condition provides random initial times of the pulses. Then, from Eq. A.1.2, Eq. A.3.1 with B=0 and Eq. A.3.2 we obtain the SNR at the output:

$$SNR_{out} = \nu_0 (B\, U_t)^2 \sigma^{-4} \exp[-(U_t/\sigma)^2/2] \quad , \qquad (A.3.4)$$

where $SNR_{out} = \frac{P_s}{S_{no}(f_0)}$ is the ratio of P_s and the background noise spectrum at f_0

An upper limit for the ratio of SNR at the output and at the input can be deduced:

$$\frac{SNR_{out}}{SNR_{inp}} \le \frac{4}{e\sqrt{3}} \approx 0.85 \qquad (A.3.5)$$

which is in agreement with experimental results and with Dykman's corollary [20] about the inapplicability of linear SR for the improvement of SNR.

A.4 Analysis of the Theoretical Results

Analysing Eqs. A.3.2 and A.3.4, it turns out that in the limit of our approximation, the *only important properties* of the input noise are its rms amplitude σ and its rms velocity, the particular structure of its spectrum S(f) does not have influence on the strength of output signal. Moreover, the frequency of the input signal does *not* play any role in Eqs. A.3.2 and A.3.4. It is important to note that these fundamental properties are present only in this threshold system, even the very recent threshold SR systems proposed by Jung [18] and Bulsara [19] do have a frequency dependent SR behaviour.

Note that, when the preconditions of the outlined theory were *not* fulfilled, we have naturally found deviations from Eqs. A.3.2 and A.3.4. For example, at very small input noise, the linear approach breaks down; and at very large input noise, correlations between level-crossing times can cause deviations from the linear SNR curve. The strongly nonlinear LCD arrangement described in Section III behaves in a completely different way.

ACKNOWLEDGEMENTS

I am very much indebted to the following colleagues for a direct and valuable impact on our new results presented here: A.R. Bulsara, M.I. Dykman and P.V.E. McClintock. I am grateful for the following colleagues for sharing relevant view and/or information: Z. Gingl, M. Inchiosa, P. Jung, F. Moss and A. Szilágyi. The work has been supported by OTKA grants (Hungary) and TFR (Sweden).

REFERENCES

1. Jung, P., Physics Report **234**, 176 (1993)
2. Bulsara, A.R. and Schmera, G., Phys.Rev.E **47**, 3734 (1993)
3. Benzi, R., Sutera, A. and Vulpiani, A., J.Phys.A **14**, L453 (1981)
4. Millonas, M.M., and Dykman, M.I., Phys.Lett.A **185**, 65 (1994)
5. Nicolis, C. and Nicolis, G., Tellus **33**, 225 (1981)
6. Benzi, R. , Parisi, G. , Sutera, A. and Vulpiani, A., Tellus **34**, 10 (1982)
7. Jung, P. and Hanggi, P., Phys. Rev. A, **44**, 8032 (1991)
8. Gammaitoni, L., Marchesoni, F., Menichella-Saetta, E., and Santucci, S., Phys. Rev. Lett. **62**, 349 (1989)
9. Dykman, M. I., Mannella, R., McClintock, P.V.E. and Stocks, N.G., Phys.Rev.Lett. **65**, 2606 (1990)
10. Stocks, N. G., Stein, N.D. and McClintock, P.V.E., J.Phys A: Math.Gen. **26**, L385 (1993)
11. McNamara, B., and Wiesenfeld, K., Phys.Rev.A **39**, 4854 (1989)
12. Kiss, L.B., Gingl, Z., Marton, Zs., Kertesz, J., Moss, F., Schmera, G. and Bulsara, A.R., J.Stat.Phys. **70**, 451 (1993)
13. Douglass, J.K., Wilkens ,L. , Pantazelou, E. and Moss, F., Nature **365**, 337 (1993)
14. Bulsara, A.R., Jacobs, E., Zhou, T., Moss, F. and L.B. Kiss, J.Theor.Biol. **152**, 531(1991); Longtin, A., Bulsara, A.R. and Moss, F., Phys.Rev.Lett. **67**, 656 (1991); Longtin, A., Bulsara, A.R, Pierson, D. and Moss, F., Biol.Cybern. **70**, 569 (1994)
15. Gingl, Z., Kiss, L.B. and Moss, F., "Non-Dynamical Stochastic Resonance", Europhys.Lett., **29**, 191 (1995)
16. Gingl, Z., Kiss, L.B. and Moss, F., "Non-Dynamical Stochastic Resonance: Theory and Experiments with White and Various Coloured Noises" presented at Intern. Workshop on Fluctuations in Physics and Biology, Elba-Island, Italy, June 1994, Nuov.Cim.D (Proc. editor R. Mannella) 1995, to be published
17. Bulsara, A. R., Lowen, S.B. and Rees, C.D., Phys.Rev.E **49**, 4989 (1994)
18. Jung, P., Phys.Rev.E **50**, 2513 (1994)
19. Wiesenfeld, K., Pierson, D., Pantazelou, E., Dames, C. and Moss, F., Phys.Rev.Lett. **72**, 2125 (1994)
20. Dykman, M.I., Luchinsky, D.G., Mannella, R., McClintock, P.V.E., Stein, N.D. and Stocks, N.G., "Stochastic resonance in Perspective", ibid 16.
21. Inchiosa, M.E. and Bulsara, A.R., "Periodically driven globally coupled nonlinear systems: response and its enhancement via stochastic resonance", to be published
22. Bulsara, A.R., private communication
23. Kiss, L.B., "Significant Improvement of Signal to Noise Ratio by Stochastic Resonance", *Proc. 13th Internat. Conf. on Noise in Physical Systems*, ed. V. Bareikis, London: World Scientific, May 1995
24. Kiss, L.B., "Stochastic Resonance: Significant Improvement of Signal to Noise Ratio", to be published
25. DeWeese, M. and Bialek, W., "Information Flow in Sensory Neurons", ibid 16.
26. Rice, S.O., Bell.Syst.Tech.J. **23** (1944) 282; 24 (1945) 46

Stochastic Resonance: A Residence Time Approach

L. Gammaitoni[1,2], F. Marchesoni[1,3], E. Menichella Saetta[2], S. Santucci[2]

[1] *Istituto Nazionale di Fisica Nucleare, Sezione di Perugia, I-06100 Perugia (Italy)*
[2] *Dipartimento di Fisica, Universita' di Perugia, I-06100 Perugia (Italy)*
[3] *Dipartimento di Fisica, Universita' di Camerino, I-62032 Camerino (Italy)*

Abstract. The Stochastic Resonance phenomenon is described as a synchronization process between periodic signals and the random response in bistable systems. The residence time approach as a useful tool in characterizing hidden periodicities is discussed.

Stochastic resonance (SR)(1) has become very popular in many fields of natural science as a paradigm which epitomizes noise-controlled onset of order in a complex system (2).

The archetype of SR models is represented by a simple symmetric bistable process $x(t)$ driven by both a random noise, for simplicity, white and gaussian, and an external sinusoidal bias. On keeping the amplitude and the forcing frequency fixed, the amplitude of the periodic component of the process, $< x >$, grows sharply with the noise intensity D until it reaches a maximum and, then, decreases slowly according to a certain power-law (3-8). It was initially suggested (1) that such a behavior results from the attuning of a deterministic with a stochastic time scale, that is, the forcing period T_0 and the switching time $T_K(D)$ of the un-biased bistable process $x(t)$, respectively. At variance with the notion of ordinary resonance $< x >$ is a monotonically decreasing function of the forcing frequency ω(5,6).

A deeper understanding of the SR mechanism can be gained by mapping the continuous stochastic process $x(t)$ into the stochastic point process t_i (4,9,11)of the relevant switching times (defined with respect to symmetric crossing levels (9)). The quantities $T(i) = t_i - t_{i-1}$ represent the residence times between two subsequent switching events. The normalized distribution of the residence times, $N(T)$, allows us to characterize SR as a syncronization effect of the switching mechanism driven by noise and periodic bias, simultaneously (4). Typically (see e.g. Fig. 1), $N(T)$ exhibits a structure of peaks centered at $T_n = (n - \frac{1}{2})T_0$, with exponentially decreasing amplitude (4) and finite size (4,10), superimposed on an exponentially decaying background. The background of $N(T)$ represents the switches activated at random by the noise alone; the peaks are a signature of the syncronization between switching mechanism and external bias: a switch tends to occur only when the bias helps noise activation, i.e. after a residence time close to an odd multiple of the half forcing-period.

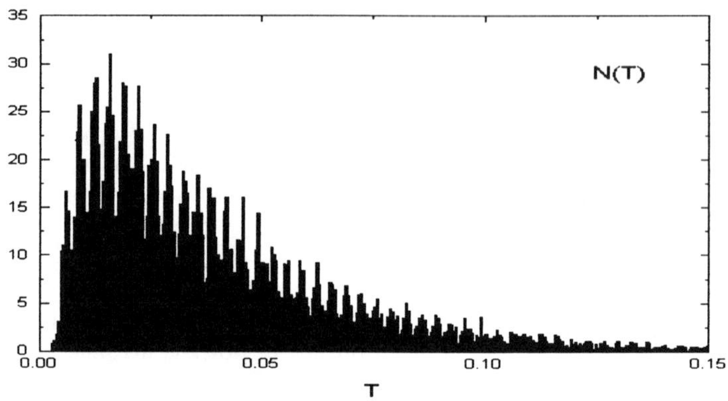

Fig.1 Residence times normalized distribution; T is in sec.

Of course, the probability that a switch occurs after an average residence time T_n decreases exponentially with n, whereas the width of the peaks is determined by the combined action of periodic and random drive (10). Syncronization is maximum when the first peak of $N(T)$ at T_1 dominates with respect to both the background and the remaining peaks. On increasing D from very small (multi-peaked $N(T)$) up to large values (exponentially decaying $N(T)$), one invariably reaches the condition of maximum syncronization. SR is the realization of such a condition.

Besides the SR, the residence time approach can be used to characterize also a different kind of resonance phenomenon which takes place in bistable systems driven by colored noise, in the presence of periodic forcing (9)(*resonant crossing*).

In real systems (12) subject to periodic bias the observable of interest $x(t)$ is coupled to one or more noise sources, the statistics of which is to be determined from the system response. An important conclusion of our investigation is that the triggered processes associated with $x(t)$ are extremely sensitive to the noise time-correlation, to the extent that the observed residence time distributions allow, in principle, a good characterization of both the periodic bias and the noise sources themselves. When the noise is colored the distribution $N(T)$ shows a rising branch for $T \ll \tau$, where τ is the noise correlation time (9), with a superimposed peak structure due to the presence of the periodic bias (see Fig. 1). On increasing ω well above the SR value, a qualitative difference in the peak structure of $N(T)$ for the white noise and the colored noise case becomes noticeable. In the white noise limit the peak height is maximum at SR and, then, decreases with increasing ω, that is with the number of detectable peaks. For very large values of ω, $N(T)$ approaches the corresponding unperturbed exponential distribution. In the strongly colored limit, instead, the peak height is much smaller than for the previous case; moreover, it grows through a maximum at $\omega \sim \sqrt{2a/\tau}$ (9), (here $2a$ is the curvature of the potential wells) thus, indicating a resonance effect in the colored *intra-well* dynamics.

The appearance of an enhanced peak structure in $N(T)$ for a certain value of ω can be viewed, indeed, as the signature of a higher order in the random point process t_n. However, the power spectrum of $x(t)$ has no sub-harmonics corresponding to the $N(T)$ peaks. Such a property shows a new remarkable resonance effect induced by colored noise that we called *resonant crossing*. Most importantly, resonant crossing occurs for frequencies much larger than the well-known SR frequencies; such an effect should not be mistaken for the SR induced by colored noise, either.

In conclusion we show that residence time distributions are very sensitive to the properties of the noise sources and to the presence of small periodic terms. In such distributions the spacing between adjacent peaks is controlled by the forcing frequency, whereas the position of the highest peaks is determined by the noise correlation-time and their sharpness is an apparent signature of the colored intra-well dynamics.

REFERENCES

1. R. Benzi, G. Parisi, A. Sutera and A. Vulpiani, Tellus **34** 10 (1982)
2. see, for example, *Proceedings of the NATO ARW Stochastic Resonance in Physics and Biology* edited by F. Moss, A. Bulsara and M. F. Shlesinger [J. Stat. Phys. **70** 1 (1993)]
3. B. McNamara and K. Wiesenfeld, Phys. Rev. **A39** 4854 (1989)
4. L. Gammaitoni, F. Marchesoni, E. Menichella-Saetta and S. Santucci, Phys. Rev. Lett. **62** 349 (1989)
5. C. Presilla, F. Marchesoni and L. Gammaitoni, Phys. Rev. **A40** 2105 (1989)

6. G. Hu, G. Nicolis and C. Nicolis, Phys. Rev. **A42** 2030 (1990)

7. L. Gammaitoni, F. Marchesoni, M. Martinelli, L. Pardi and S. Santucci, Phys. Lett. **A158** 489 (1991)

8. P. Jung and P. Hänggi, Europhys. Lett. **8** 505 (1989) and Phys. Rev. **A44** 8032 (1991)

9. L. Gammaitoni, F. Marchesoni, E. Menichella-Saetta and S. Santucci, Phys. Rev. Lett. **71** 3625 (1993), L. Gammaitoni, F. Marchesoni and S. Santucci, Phys. Rev. Lett. **74** 1052 (1995)

10. T. Zhou, F. Moss and P. Jung, Phys. Rev. **A42** 3161 (1990)

11. A. Simon and A. Libchaber, Phys. Rev. Lett. **68** 3375 (1992)

12. A. Longtin, A. Bulsara, D. Pierson, F. Moss, Biol. Cybern. 70, 569-578 (1994); A. Longtin, A. Bulsara and F. Moss, Phys. Rev. Lett. **67** 656 (1991)

Array Enhanced Stochastic Resonance: Implications for Signal Processing

Mario E. Inchiosa*, Adi R. Bulsara**

Naval Command, Control and Ocean Surveillance Center[¶]
Research, Development, Test and Evaluation Division
Materials Research Branch
San Diego, CA 92152-5000

John F. Lindner[†], Brian K. Meadows[††], William L. Ditto[†††]

Applied Chaos Laboratory[§]
School of Physics
Georgia Institute of Technology
Atlanta, GA 30332-0430

Abstract. In computer simulations, we enhance the response of a "stochastic resonator" by coupling it into a array of identical resonators. We relate this array enhanced stochastic resonance (AESR) to the global spatiotemporal dynamics of the array and show how noise and coupling cooperate to organize spatial order, temporal periodicity, and peak output signal-to-noise ratio. We consider the application of AESR to signal processing.

INTRODUCTION

A noisy nonlinear system exhibits *stochastic resonance* (SR) if its performance is optimized by a nonzero value of the noise. This surprising and robust phenomenon has generated considerable excitement over the last decade (1). Recently, we have demonstrated in computer simulations and theoretical calculations that the effect can be enhanced by coupling the nonlinear elements into arrays (2). Conventional SR enables the use of noise as a design parameter. *Array enhanced SR* (AESR) suggests an additional parameter: coupling strength. We have investigated AESR in a range of couplings, from global and nonlinear to local and linear. We believe that AESR is a novel use of noise which may be significant in biological systems and which may be important for engineers, especially in situations where the possibilities of linear systems have been exhausted, or where operating conditions are extreme.

In this report, we first illustrate AESR by studying an array of Duffing oscillators. We highlight the connection between the local response of an individual oscillator and the global spatiotemporal response of the array. We then consider more generally the implications of AESR to signal processing.

CASE STUDY: THE DUFFING ARRAY

We consider a one-dimensional array (or chain) of N overdamped Duffing oscillators, each coupled linearly to its nearest neighbor with identical springs of stiffness ε. We take the harmonic forcing to be $A\sin(\omega t)$. We assume the boundaries are free. In the presence of noise $N_n(t)$, the equation of motion of the n *th* oscillator is

$$\dot{x}_n = kx_n - k' x_n^3 + A\sin\omega t + N_n(t) + C_n(\varepsilon) \quad , \tag{1a}$$

where the coupling is given by

$$C_n(\varepsilon) = \begin{cases} +\varepsilon(x_2 - x_1), & n = 1 \\ +\varepsilon(x_{n+1} - x_n) - \varepsilon(x_n - x_{n-1}), & 1 < n < N \\ -\varepsilon(x_N - x_{N-1}), & n = N \end{cases} \quad . \tag{1b}$$

The (one-sided) spectrum of the Gaussian noise $N_n(t)$ is flat out to a cutoff frequency f_N, beyond which it is zero. In this article, we sometimes quantify the noise by its spectrum height $2D$ and sometimes by its mean squared amplitude *or noise power* $\sigma^2=2Df_N$. We consider both the case of local noise (uncorrelated from site-to-site) and global noise (same realization at each site).

We numerically integrate (Equation 1) using the Euler-Maruyama scheme (3) with a time step $dt=1/(2f_N)$. We use the time series of a single oscillator in the array to compute a power spectral density (PSD). We typically average 8 PSD segments of 32 forcing periods per segment and 8192 time steps per forcing period. This results in frequency bins of width $\Delta\omega=\omega/32$ and conveniently centers the forcing frequency on bin 32 of the PSD. This is defined as the signal bin. To reduce bin leakage we first convolve the time series with a Welch window.

We characterize SR by a signal-to-noise ratio, which we define as the ratio of the signal power divided by the noise power in the signal bin, expressed in decibels (dB). The noise power is estimated by performing a nonlinear fit to the PSD around, but not including, the signal bin. The signal power is estimated by subtracting this noise background from the total power in the signal bin, taking into account that Welch-windowing a long time series scales narrow-band peaks by a "processing gain" of about 0.83 (4). Thus,

$$\text{SNR} \equiv 10\log_{10}\left(\frac{(\text{total power - noise power}) / \text{gain}}{\text{noise power}}\right) \quad . \tag{2}$$

Variations of this definition do not qualitatively alter our results .

Since we are often interested in binary signals and quantized-output systems (e.g. some CCD arrays and neurons), we filter the time series to generate a binary output of ±1, reflecting which well the oscillator is in. To remove insignificant excursions across the potential barrier, we give the filter a small hysteresis; however this does not qualitatively alter our results. Furthermore, we emphasize that *both the SR peak and its enhancement in coupled arrays are clearly evident even in the unfiltered "analog" SNR curves.*

We choose our operating regime just below the deterministic switching threshold so that in the absence of noise the oscillator is confined to a single well of the (static part of the) Duffing potential, $V = -\frac{1}{2}kx^2 + \frac{1}{4}k'x^4$, but small noise can induce significant hopping between wells. The operating parameters k =2.1078, k'=1.4706, A=1.3039, ω=0.7301 (used throughout this section) are not otherwise special.

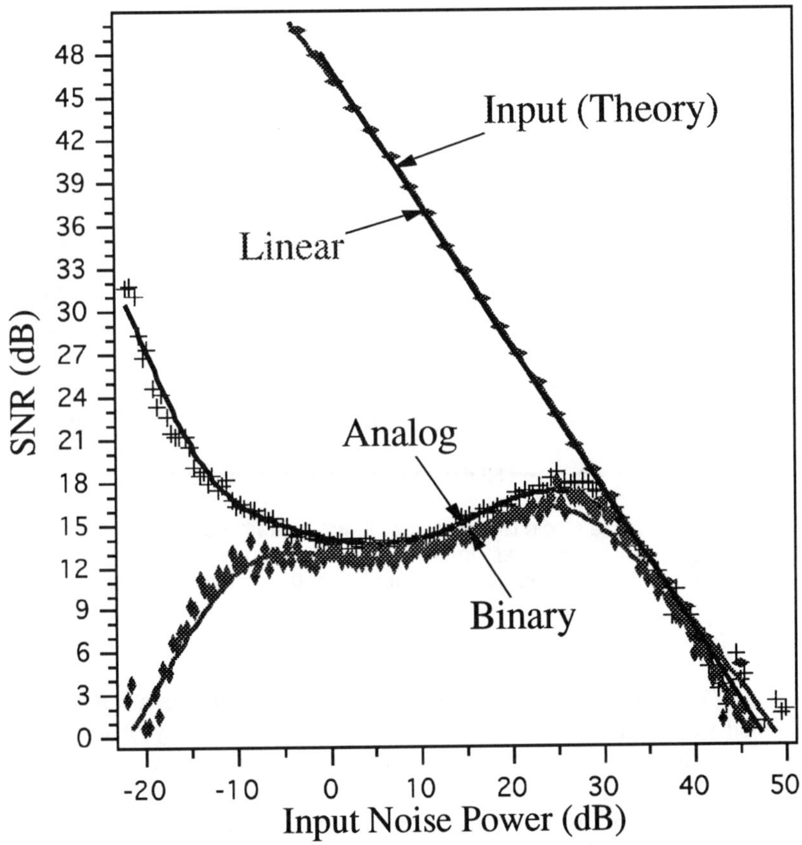

FIGURE 1. SNR of a single overdamped Duffing oscillator, with filtering (binary) and without filtering (analog). For reference, we also plot the response of the corresponding linear system (k'=0), and the input SNR $10 \log_{10} ((A^2/2) / (2D\, \Delta\omega /(2\pi)))$. We average 4 PSD segments of 32 periods per segment and 4096 time steps per period.

Figure (1) above illustrates how the SNR of a single Duffing oscillator varies with noise. Small noise induces no hopping, and consequently the binary curve drops below the analog curve at negative dB. The linear and input SNRs bound the nonlinear SNRs. Figure (2) below illustrates how the SNR curve changes as the amplitude of the forcing changes. We operate in a moderately nonlinear regime just below the deterministic switching threshold. As the amplitude increases and approaches the switching threshold from below, the SNR plateaus over many decibels of noise.

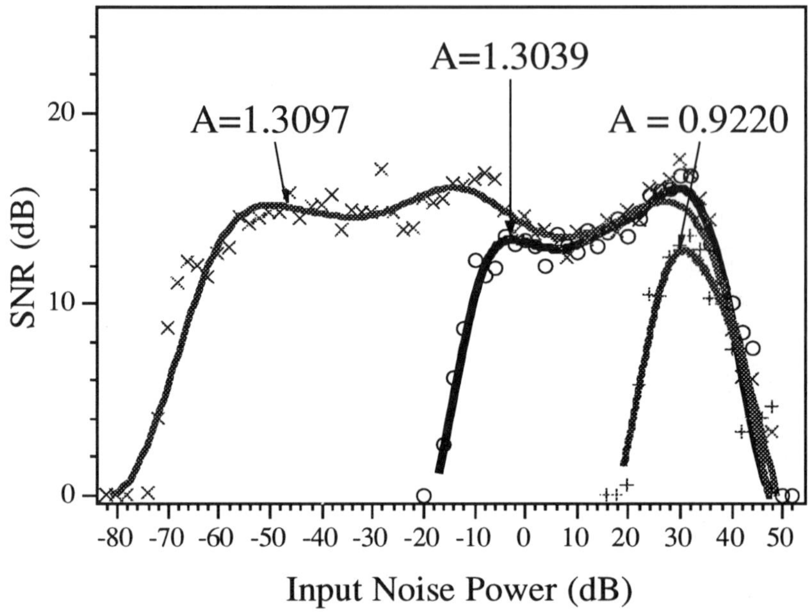

FIGURE 2. SNR of a single Duffing oscillator for 3 different amplitudes: strongly, moderately, and weakly nonlinear. In this case study, we use the moderate amplitude $A = 1.3039$. Cubic splines have been fit to the data to aid the eye.

Figure (3) below illustrates how the SNR curve changes as a single resonator is coupled more and more tightly into the middle of an array of nine identical resonators. As the noise increases, the SNR peak rises, shifts to higher noises, and then subsides. The binary curves shift to higher noise at larger coupling. With increasing coupling strength it becomes increasingly difficult for the noise to kick an oscillator, along with its coupled neighbors, over the barrier. The SR peak in the analog curves rises as the coupling strength increases from zero to two. Heuristically, coupling favors coherent motion over incoherent motion.

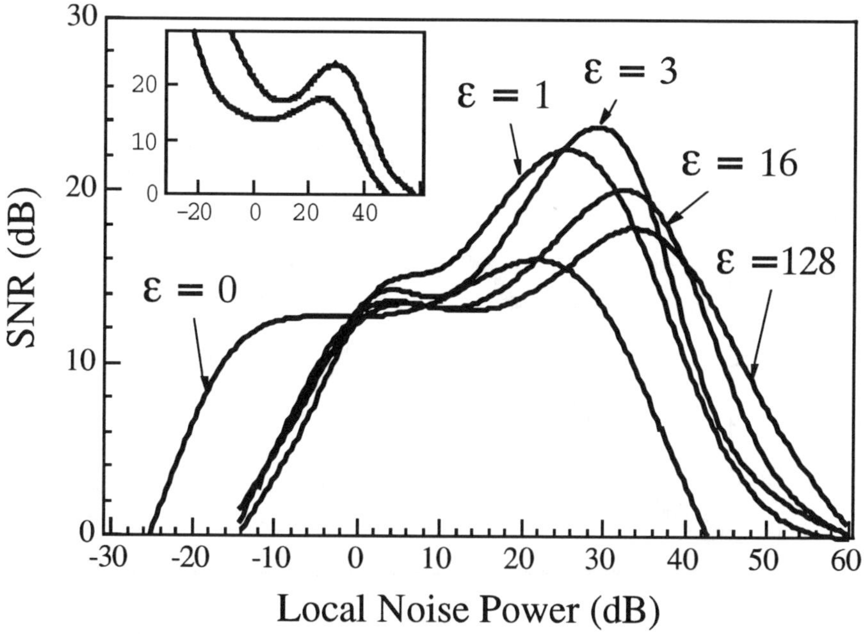

FIGURE 3. SNR of the middle of 9 oscillators, as a function of coupling. As the oscillator is coupled more and more tightly into the array, its binary SNR curves shift, rise, and subside. Inset: Corresponding unfiltered (analog) SNR increases at all noise power as the coupling varies from 0 to 2. Noise power is defined here as $10 \log_{10}(\sigma^2)$.

One also sees coupling-induced SNR enhancement in systems with global noise. However, in the case of linear coupling, if all the oscillators are identical and are started with identical initial conditions, they will evolve identically because the coupling terms $x_n - x_{n+1}$ vanish. In this case, only nonlinear coupling can realize an improvement in the SNR.

Figure (4) below illustrates how the SNR, maximized over noise, varies with coupling strength and chain length. In the limit of low coupling, the oscillators are independent and behave as if isolated. In the limit of high coupling, the oscillators are rigidly connected and behave as a single oscillator. In between, the coupling enhances the coherence of the oscillators. Even for a chain of length 9, the SNR enhancement is 6 dB. By summing the oscillator outputs in a local noise system, we can obtain even higher SNRs (e.g. 20 dB higher in a chain of 101 oscillators).

At low to moderate coupling, the SNR of the middle oscillator exceeds the SNR of the end oscillators. However, at large coupling the SNRs of the middle and end oscillators are indistinguishable, to within numerical uncertainty. Large coupling homogenizes the response of the chain.

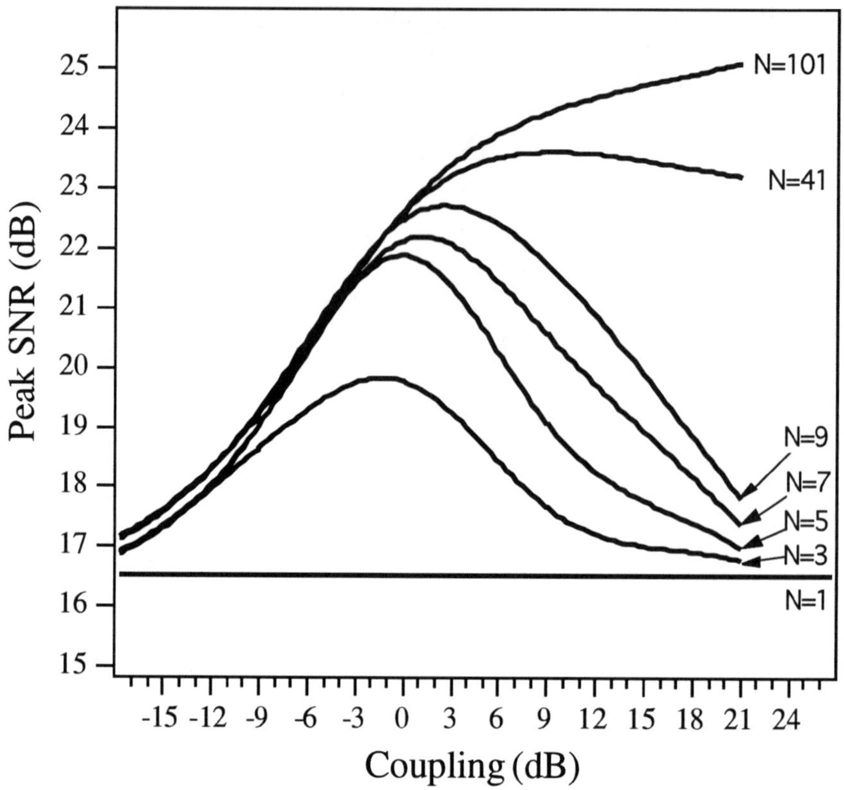

FIGURE 4. Peak SNR vs. coupling strength, for successively longer chains. For each chain length, there is an optimal local coupling strength that maximizes the SNR.

Figure (4) masks the fact that noise as well as coupling can enhance the SNR of the array. Figure (5) below illustrates in 3D the "design rules" for optimizing the SNR of a noisy coupled array of Duffing oscillators. The data have been smoothed to accentuate the SNR surface's shape.

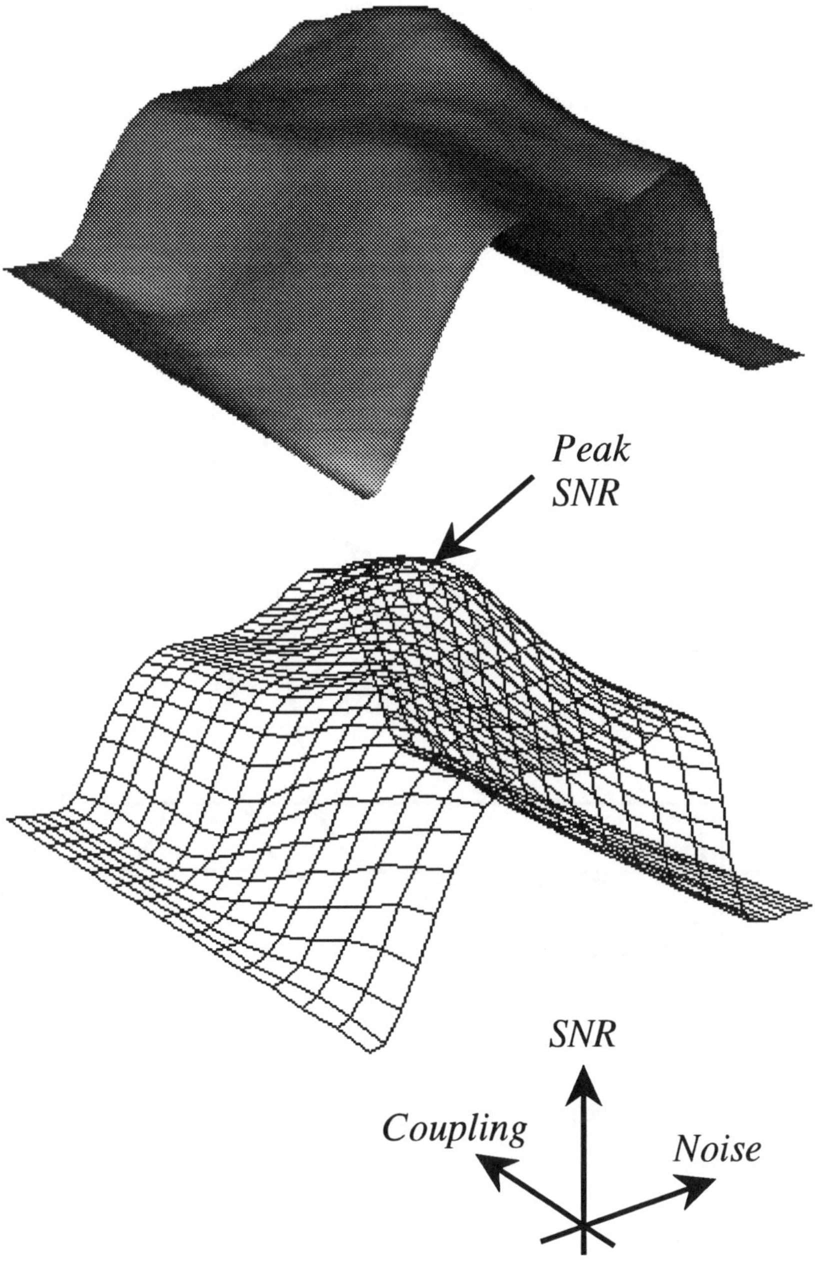

FIGURE 5. SNR of the middle of 11 oscillators as a function of (log) local noise power and (log) local coupling strength. A unique value of noise and coupling optimizes the performance of the array.

The oscillator hopping rate can further elucidate the shift in the SNR. Figure (6) below compares the hopping rate to the SNR, as a function of noise. The peak SNR corresponds (roughly) to a synchronous hopping rate of 2, namely once back and forth per forcing period.

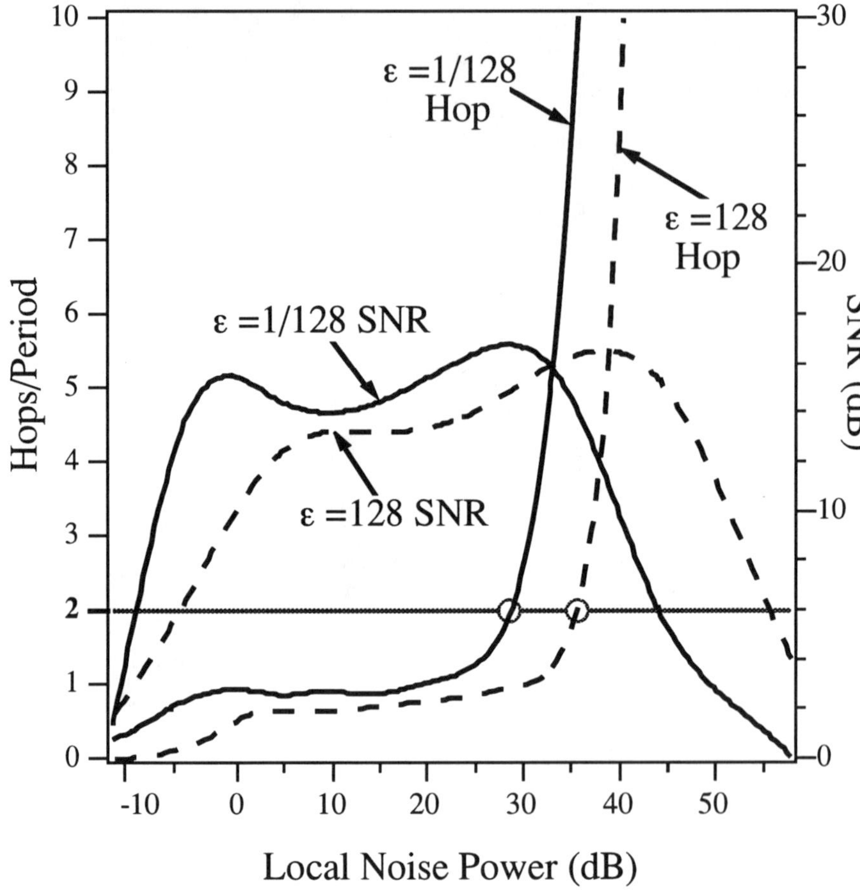

FIGURE 6. Hopping rate per period and SNR for the middle of 9 oscillators, for small and large local coupling. The shift in the peak SNR mirrors the shift in synchronization, as measured by a hopping rate of 2 (once back and forth). This synchronization naturally divides the dynamics into a potential-dominated regime and a noise-dominated regime.

We can relate the global dynamics of the array to the behavior of an individual oscillator of the array. The sequence in Figure (7ab) below reveals the spatiotemporal behavior of the chains at successively higher noises. Each sequence displays the evolution of a chain of 512 oscillators, time increasing upward. Black corresponds to one well and white to the other. The oscillators in the chain all begin in the white well, but with a random distribution of positions. As the chains evolve, distinct collective behaviors become apparent at different noise levels, and these correspond to features in the individual SNR curves. For the case of low noise, rarely does the noise force an oscillator in the chain into the opposite well. Consequently the chain as shown is mainly confined to the white well, except for small segments which make short-lived excursions into the black well. The binary SNR of a single oscillator is close to zero in this case. Higher noise begins to move the chain and large segments switch wells. The domains of black and white segments expand and contract as oscillators tug their coupled neighbors back and forth, and the binary SNR rises. At still higher noise (30 dB), the degree of entrainment between the noise and the signal is maximized. At this level of noise the *chain* is regularly forced between the two wells of the potential. It is at this point that the SNRs of individual oscillators are maximized. Greater noise overwhelms the coupling, forcing different parts of the chain to become uncorrelated. The SNR drops as oscillators occupy different wells. When the noise is adjusted to maximize the SNR of an individual oscillator, there is both spatial organization and temporal periodicity in the chain as a whole, as is clearly seen in Figure (7b). One might have expected disorganization in the presence of such a large noise. However the coupling, in cooperation with the noise and the bistable potential, organizes the chain in space and time.

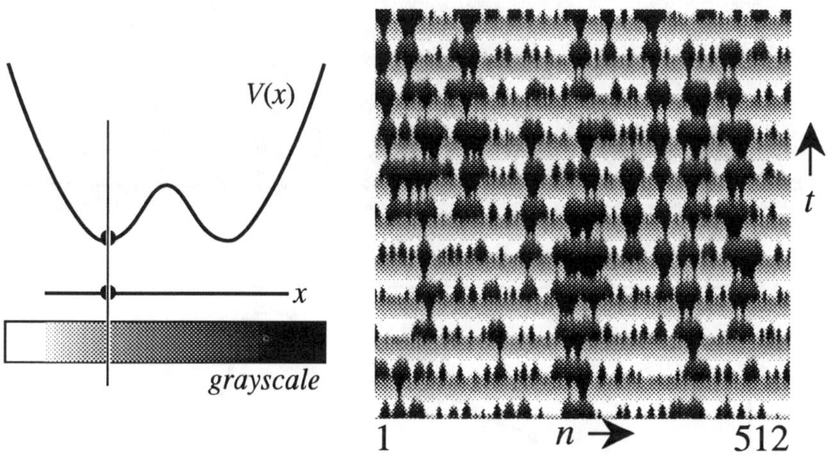

FIGURE 7a. Time evolution (upward) of a chain of locally coupled oscillators moving in a bistable potential, coded here as a black well and a white well.

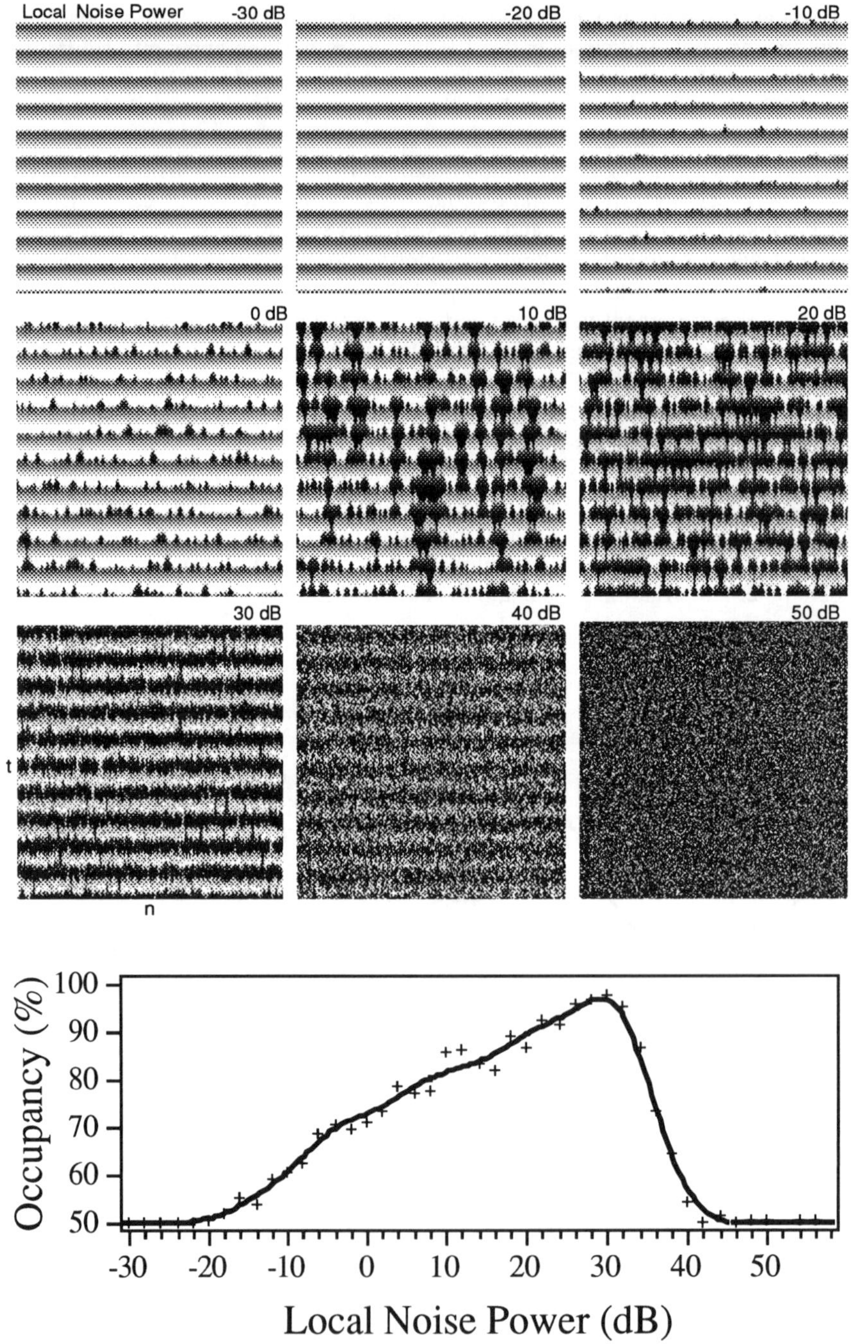

FIGURE 7b. In the lower left panel, the noise nearly synchronizes the chain, as quantified by the occupancy (defined below). This noise also maximizes the SNR.

This spatiotemporal organization of the array can be quantified by the *occupancy* function. Noting that the periodic forcing effectively rocks the bistable Duffing potential back and forth, we define the occupancy as the per cent of oscillators occupying the lower well at the extremes of the forcing cycle (namely, at $\omega t=\pi/2$ and $\omega t=3\pi/2$, if the forcing is $A\sin\omega t$):

occupancy = (% oscillators on left when left well is lower +
% oscillators on right when right well is lower)/2 (3)

At low noise the oscillators are confined to one well (say, the left) regardless of whether it is low or high and occupancy = (100% + 0%)/2 = 50%. At high noise an oscillator is equally likely to be in either well so that occupancy = (50% + 50%)/2 = 50%. At synchronization, the oscillators hop back and forth so as to remain always in the lower well and occupancy = (100% + 100%)/2 = 100%.

Figure (8) below illustrates how the occupancy of a chain (and hence its degree of synchronization) peaks at the same noise power as the SNR of a constituent oscillator. Since the occupancy is *much* faster to calculate numerically, it is a more convenient tuning parameter by which to optimize the response—both global (synchronization) and local (SNR)—of the chain

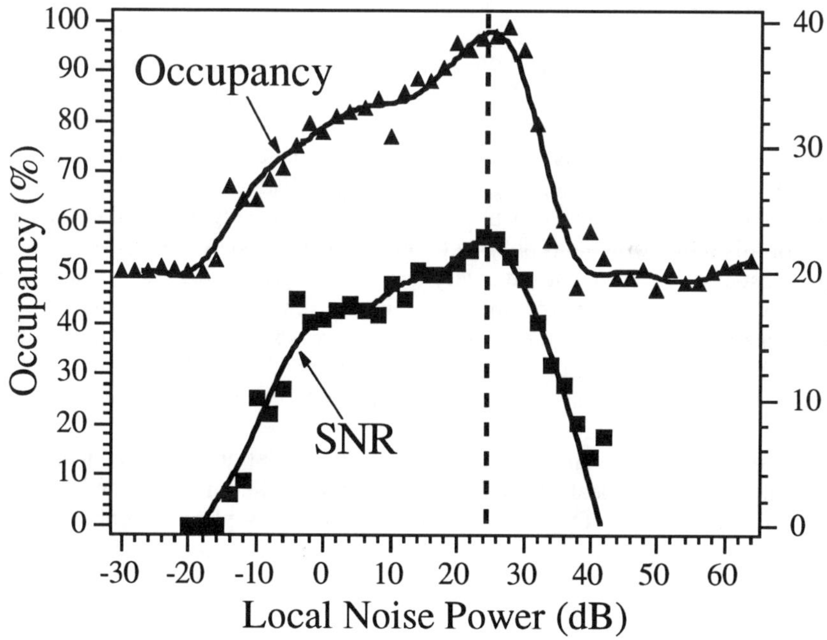

FIGURE 8. Occupancy of a chain of 101 oscillators superimposed on the SNR of the middle oscillator of the chain. The (global) average occupancy of the lower well peaks at the same noise power as the (local) SNR of the middle oscillator.

SIGNAL PROCESSING IMPLICATIONS OF STOCHASTIC RESONANCE

We know that the signal power component as well as the SNR of the response of a "stochastic resonator" goes through a maximum at a certain input noise level. Does the SR effect have signal processing applications? To address this question, we must define our signal processing task. Typical tasks are signal detection and signal estimation.

Important quantities affecting signal detection or signal estimation are the SNR of the input to and the SNR of the output from our signal processing elements. However, SNR alone is not a complete measure of the performance of a signal processing system. For example, a nonlinear signal processor may output a signal which has infinite SNR but is useless because it has no correlation with the input signal. What are good measures of performance? For signal detection, one must consider detection statistics: probability of detection and probability of false alarm. Such statistics may be summarized in plots of detection probability versus false alarm probability, known as the receiver operating characteristic (ROC). For signal estimation, relevant measures are mean square error or Bayesian tests (5).

Although SNR alone does not fully characterize the performance of a nonlinear signal processor, it is still a useful quantity. There are several alternative ways to define SNR. We can compare the signal power to the total noise power. This can cause problems when we make theories using ideal white noise, which is an infinite power signal. Alternatively, we can compare the signal power to the noise power over a limited range of frequencies. The latter approach has more relevance to practical signal processing. Real signals are confined to some frequency range. We can always use linear filtering to eliminate noise outside of this range which has nothing to do with the signal of interest.

As a test case, we have chosen a very basic signal processing task: detecting a sine wave of known frequency and phase in the presence of Gaussian white noise. One can prove that the best detector consists of an ideal linear filter followed by a "decision circuit" which compares the linear filter's output to a threshold (5). If the filter's output exceeds the threshold, then the system's decision is that the sine wave was present.

If we already know what is the best detector, why continue? In practice, there is no such thing as an ideal linear filter. All filters add some noise and deviate from perfect linearity. Our work relates to using coupling to enhance the performance of physically realizable (i.e. non-ideal) signal processing elements.

Global vs. Local Noise

An important factor affecting SNR enhancement in coupled arrays is whether the noise comes from external or internal sources. This affects whether the noise at different elements is correlated or uncorrelated. If the noise comes from sources internal to the system, then using many elements which are coupled and/or have summed outputs will enhance output SNR, compared to using only one element. In the case of summed outputs, and to some extent in the case of coupling, this is because signal components add coherently, while uncorrelated noise components add incoherently. In the external-noise case, one does not have this effect to exploit, but, somewhat surprisingly, coupling can still enhance output SNR (2).

The external/internal noise distinction also affects how we can measure SNR enhancement. If we are considering the external noise situation, then there is a well-defined input signal plus noise. We can compare the input and output SNR to determine if the system provides any gain in SNR. We can also look at output SNR as a function of N, the number of elements in our array. For internal noise, the input SNR is infinite, because the signal comes into the system with no noise added to it. However, we can still look at output SNR enhancement as a function N.

Measuring Detection Performance

A signal detection system's performance may be characterized by measuring its probability of detection and probability of false alarm. Probability of detection is the probability that the decision circuit will report that a signal is present when in fact a signal is present. Probability of false alarm is the probability that the decision circuit will report that a signal is present when in fact a signal is *not* present. Detection performance can be summarized by *receiver operating characteristics* (ROCs), which are plots of probability of detection versus probability of false alarm.

Our decision circuit operates by computing the power spectrum of the waveform of interest (e.g. the input signal, the response of a single isolated element, or the response of an element coupled to N-1 other elements) and comparing the power in a narrow frequency band of width $\Delta\omega$ centered at frequency ω to a threshold. If the power exceeds the threshold, the decision circuit concludes that a signal was present. A low threshold leads to high probability of detection and probability of false alarm, while a high threshold leads to low probability of detection and low probability of false alarm.

To produce a ROC, we repeatedly measure the power in the narrow frequency band and record the value. We do this both with and without the signal actually present. Figure (9) shows the power probability distributions for "signal"/"no signal". Using these statistics, the probability of detection for any given threshold power value may be computed by measuring the area under the "signal" curve to the right of that value. Similarly, the probability of false alarm equals the area under the "no signal" curve to the right of the threshold value.

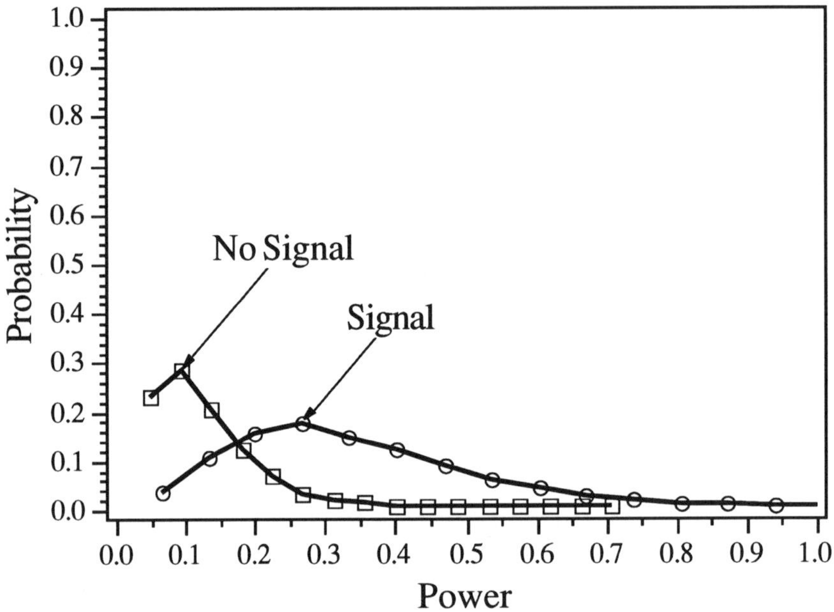

FIGURE 9. Typical probability distributions for the total power in a narrow band around the signal frequency.

Figure (10) shows signal detection (probability of detection vs. probability of false alarm) for the output of a single element and the output of the center element of an N=9 array. Each of the four panels shows detection statistics for a different noise strength D. The insets show SNR vs. D, with the vertical line indicating the value of D used for the ROCs in that panel. The figure illustrates a significant enhancement in signal detection for the N=9 system (broken line) compared to the single element (solid line).

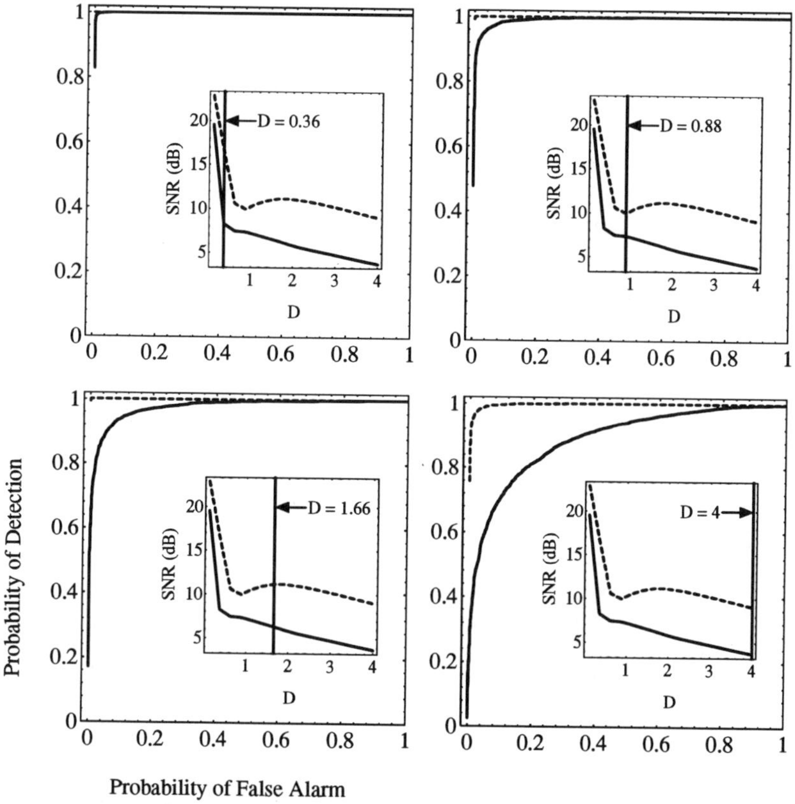

FIGURE 10. Duffing detection performance. Solid curve: output of a single nonlinear element driven by the input signal. Broken curve: output of the middle of 9 oscillators. Insets: SNR v. noise strength *D*, with the vertical line indicating the value of *D* used for the ROC's in that panel. The parameters used in this figure are: k=2.1078, k'=1.4706, ε=1, A=0.4, ω=0.7301, and $\Delta\omega = \omega/32$.

To check whether coupling-enhanced signal detection relies heavily on the specifics of our SR system, we also consider a second, very different system (2):

$$C_i \dot{x}_i = -\frac{x_i}{R_i} + \sum_{j=1}^{N} J_{ij} \tanh x_j + A \sin \omega t + \sqrt{D}\xi(t) \tag{4}$$

where $\xi(t)$ is Gaussian white noise with mean zero and autocorrelation $\langle \xi(t)\xi(t+\tau)\rangle = \delta(\tau)$. Systems of this form have been used to describe connectionist type electronic neural networks (6). In such networks, x_i denotes the ith neuron's activation function (membrane potential), and C_i, R_i denote the neuronal input capacitance and trans-membrane resistance, with the coupling coefficients (synaptic efficacies) J_{ij} usually determined via a "learning rule.'' Here, we shall choose the couplings to maximize the SNR of the response of oscillator i=1. The system illustrated by (Equation 4) differs from the previously considered system (1) in that (a) the noise is external (i.e. all oscillators are driven by an identical noise time series), (b) the coupling is nonlinear, (c) the coupling is global, and (d) a hyperbolic tangent function is used for the nonlinearity rather than the x^3 term used in the Duffing oscillator. However, the two systems share a wide range of qualitatively similar SR phenomena.

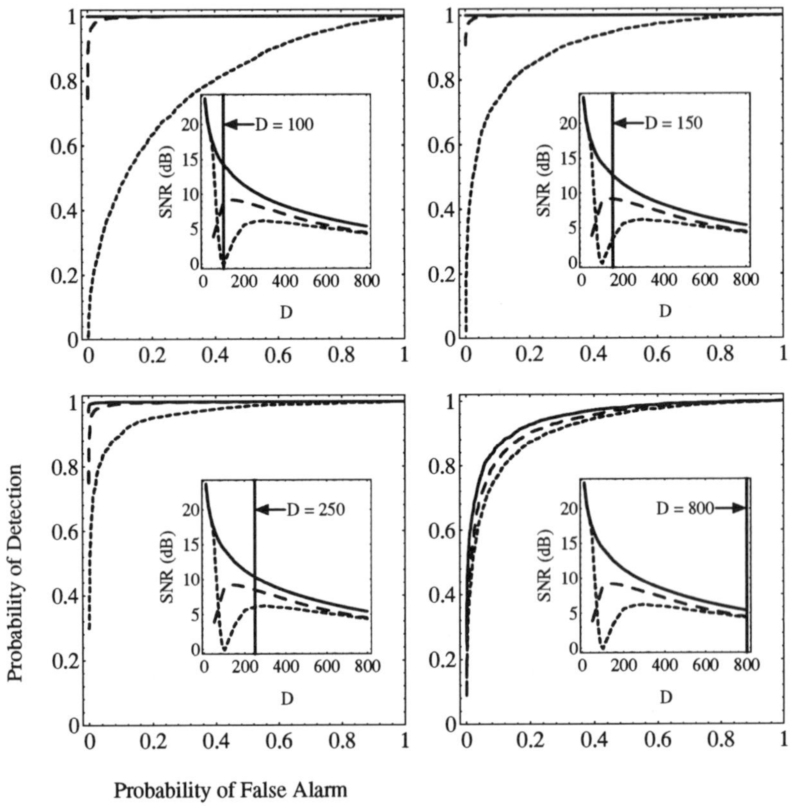

FIGURE 11. Neuron detection performance. Solid curve: input signal. Dotted curve: output of a single nonlinear element driven by the input signal. Dashed curve: output of coupled system (N=2). Insets: SNR vs. noise strength D, with the vertical line indicating the value of D used for the ROC's in that panel. The parameters used in this figure are: R_i=0.0186916, C_i=1, J_{ii}=216, J_{12}=50, J_{21}=-50, A=8, ω=1.22522, and $\Delta\omega$= ω/32.

Figure (11) shows signal detection statistics for (a) the input signal, (b) the output of a single element driven by the input signal, and (c) the output of an element coupled with a second element (N=2 array). Once again, we see a significant enhancement in signal detection for the array (dashed line) compared to the single element (dotted line). Signal detection is, however, highest for the input signal, so why bother with SR? The problem is to acquire the input signal perfectly. If the transducer which picks up the input signal has perfectly linear response, then it is ideal for our task of detecting a sine wave in Gaussian white noise. If the transducer has nonlinear, bistable response, then coupling it into an array can improve signal detection. Indeed, there exist special cases for which the nonlinearities (tanh x_i terms) in (Equation 4) identically cancel, giving the system a perfectly linear response to its input. For example, this occurs for a pair of identical oscillators ($N=2$, $C_1=C_2$, $R_1=R_2$, $J_{11}=J_{22}$) with couplings $J_{12}=J_{21}=-J_{11}$ and identical initial conditions.

Theoretical Limits to Performance

The optimal detector of a known signal in Gaussian white noise is the correlation receiver or matched filter, which is a linear filter whose output is compared to a threshold. The optimal estimator of a linearly modulated signal in Gaussian white noise is also a linear filter (5). This suggests why the output of a stochastic resonator circuit driven by Gaussian white noise has an output SNR bounded by the SNR obtained by a linear filter, or, equivalently, the input SNR. Indeed, for small signal and noise it is easy to prove that the output SNR of any nonlinear system is bounded by the output SNR of the linear filter (or input SNR) (7). This also follows from the threshold effect discussed below.

On the other hand, the output SNR of a nonlinear filter may exceed its input SNR. The bandpass limiter is a circuit which illustrates this effect (8). It consists of a bandpass filter followed by a threshold (or "infinite limiter") circuit. The bandpass filter removes all signals outside a narrow band of interest, and the threshold circuit converts the analog output of the bandpass filter to a two-level output ($+V$ if the bandpass filter output is positive, $-V$ if negative). This nonlinear device improves the SNR of a signal by tacitly assuming that the signal power exceeds the noise power. When the assumption holds, the SNR increases by several dB. However, when the assumption fails, the SNR decreases slightly. The fact that there is a positive SNR gain only above a certain threshold input SNR (the "threshold effect") is a universal phenomenon generic to all nonlinear processors (5).

When we drive one of our bistable nonlinear dynamic elements (Equation 4) with a sine wave plus bandpass-filtered noise which has most of its energy concentrated around the signal frequency, we should see behavior somewhat like that of the bandpass limiter. The limiting effect is due to the hyperbolic tangent function and does indeed result in a positive SNR gain, provided the input signal strength is large enough to overcome the threshold effect (2). Gang, et. al., have also observed a positive SNR gain for a different stochastic resonance circuit driven by band-limited noise (9). Similar gains have been predicted by Hänggi, et. al., for the driven Kramers equation at moderate to large friction (10).

CONCLUSION

Many variations are possible in the coupled systems we have studied. We have considered in detail locally, linearly coupled Duffing oscillators subject to local noise (Equation 1), and globally, nonlinearly coupled model neurons subject to global noise (Equation 4). We have not found great differences attributable to the choice of Duffing vs. neuron nonlinearity, nor global vs. local coupling. However, linear vs. nonlinear coupling and global vs. local noise can make a great deal of difference in certain cases. For example, linear coupling has little to no effect on identical oscillators subject to global noise; nor does summing of outputs. Also, nonlinear coupling can completely remove oscillator nonlinearites for some configurations of identical oscillators subject to global noise. Studies are underway to compare coupling with summing of outputs in systems subject to local noise.

In summary, we emphasize that *one can improve the performance of a stochastic resonator by coupling it into an array of identical resonators.* Significant enhancement can be obtained using even a small number of oscillators. Peak performance of a single oscillator corresponds to a global organization of the chain in space and time: noise, coupling, and bistable potential cooperate to create spatial order, temporal periodicity, and peak signal-to-noise ratio. We are currently studying the behavior of disordered arrays of nonidentical stochastic resonators. In the future, we plan to investigate experimentally the phenomenon of array enhanced SR in analog circuits and in Very Large Scale Integrated (VLSI) circuits. We are excited about possible applications to signal recovery in detector arrays and to focal plane detection devices such as CCD cameras.

ACKNOWLEDGMENTS

We acknowledge helpful discussions with K. Wiesenfeld (Atlanta), T. Albert (San Diego), L. Kiss (Uppsula), and C. Doering (Los Alamos). The assistance of B. J. Breen is warmly appreciated. A. R. B. acknowledges funding from the Office of Naval Research through an internal research grant at NCCOSC. This work was performed while M. E. I. held a National Research Council-NRaD Research Associateship. This work was supported in part by a grant of HPC time from the DoD Major Shared Resource Center at Wright-Patterson Air Force Base on the Intel PARAGON.

AUTHOR NOTES

*Electronic Address: inchiosa@nosc.mil
**Electronic Address: bulsara@nosc.mil
¶ http://www.nosc.mil/nrad
†Electronic Address: jlindner@acl1.physics.gatech.edu
††Electronic Address: bmeadows@acl1.physics.gatech.edu
†††Electronic Address: wditto@acl1.physics.gatech.edu
§ http://acl1.physics.gatech.edu/aclhome.html

John F. Lindner's permanent address is:
The College of Wooster, Wooster, OH 44691.

REFERENCES

1. F. Moss, A. Bulsara, M. Shlesinger, eds., *Proceedings of the NATO Advanced Resesarch Workshop on Stochastic Resonance in Physics and Biology*, J. Stat. Phys. **70**, no. 1/2, (1993); F. Moss in *Some Contemporary Problems in Statistical Physics;* G. Weiss, Ed.; (SIAM, Philadelphia, 1994); A. Bulsara, "Noise and Nonlinearity in Neuron Modeling," in *Proceedings of the Los Alamos meeting on "Fluctuations and Order: the New Synthesis"* M. Millonas, eds., (to be published); R. Mannella and P. V. E. McClintock; *Proceedings of the International Workshop on Stochastic Processes*, Elba, Italy, 1994 (to be published).
2. M. Inchiosa, A. Bulsara, "Coupling Enhances Stochastic Resonance in Nonlinear Dynamic Elements Driven by a Sinusoid Plus Noise", Phys. Lett. A **200**, 283 (1995); J. Lindner, B. Meadows, W. Ditto, M. Inchiosa, A. Bulsara, "Array Enhanced Stochastic Resonance", Phys. Rev. Lett. (to be published); M. Inchiosa and A. Bulsara, "Nonlinear dynamic elements with noisy sinusoidal forcing: enhancing response via nonlinear coupling," Phys. Rev. E **52**, 327 (1995); A. Neiman, L. Schimansky-Geier, Phys. Lett. A **197**, 397 (1995).
3. T. C. Gard *Introduction to Stochastic Differential Equations*; (Marcel Dekker Inc.; New York, 1988).
4. W. Press, S. Teukolsky, W. Vetterling, B. Flannery, *Numerical Recipes in C*; (Cambridge University Press; Cambridge, 1992).
5. See e.g. H. van Trees, *Detection, Estimation and Modulation Theory* (J. Wiley, New York, 1978).
6. See e.g. D. Amit, *Modeling Brain Function* (Cambridge University Press, 1989); G. Weisbuch, *Complex Systems Dynamics* (Addison Wesley, New York, 1991); J. Hertz, A. Krogh, and R. Palmer, *Introduction to the Theory of Neural Computation* (Addison Wesley, New York, 1991).
7. M. Dykman, R. Mannella, D. Luchinsky, P. McClintock, N. Stein, N. Stocks; F. DeWeese and W. Bialek; in *Proceedings of the International Workshop on Stochastic Processes,* Elba, Italy, 1994 (to be published).
8. W. B. Davenport, Jr., J. Appl. Phys. **24**, 720 (1953).
9. R. Li, G. Hu, C. Yang, X. Weng, G. Qing, and H. Zhu, "Stochastic Resonance in Bistable Systems subject to Signal and Monochromatic Noise," (to be published). We note in passing that the noise considered by Li, et. al., cannot truly be "monochromatic," for it would then be a sine wave.
10. P. Hänggi, P. Jung, C. Zerbe, and F. Moss, J. Stat. Phys. **70**, 25 (1993).

Chaotic Resonance: Birth of Double-Double Scroll Attractor

Vadim S. Anishchenko [1] and Tomasz Kapitaniak[2]

[1] *Physics Department, Saratov State University, Astrachanskaya 83, 410071 Saratov, Russia*

[2] *Division of Dynamics, Technical University of Lodz, Stefanowskiego 1/15, 90-924 Lodz, Poland*

Abstract. The dynamics of two nonautonomous Chua's circuits driven by small sinusoidal signal is investigated. The possibility to observe the phenomenon of stochastic resonance in purely deterministic system with four co-existing hyperchaotic attractors is discussed.

INTRODUCTION

The phenomenon of stochastic resonance is observed in multistable [1-9] and monostable [10] systems driven by a sum of a noise signal and a weak periodic signal. Under appropriate conditions a weak periodic signal can be amplified by noise signal i.e. the signal-to-noise ratio increases until it reaches a maximum at some noise intensity D. Stochastic resonance has been studied both theoretically and experimentally in a variety of contexts such as for example: meteorology [2], optical systems including lasers [3,4], electronic circuits [5,6,9], a magnetoelastic ribbon [7], sensory neurons [8], etc. Systems showing stochastic resonance are, in some sense, intermediate between regular and irregular ones, since they are described by a random process-the jumps do not follow a deterministic rule-which, nevertheless, exhibits a certain degree of regularity. In all these examples the theory describing stochastic resonance is statistical.

Recently it has been shown that stochastic resonance can be also observed when chaos, rather than noise is used as a nonperiodic component of driving force. In this case stochastic resonance can occur in a completely deterministic system as the presence of noise is not required, since deterministic chaos generates its own "effective" noise [11-15]. The presence of deterministic chaos plays the role of the random perturbation, so that it would be more correct to speak of *chaotic resonance* rather than of stochastic resonance. Chaotic resonance seems to be present in natural phenomena, such as the time evolution of weather, which is governed by a set of nonlinear equations (for example Lorenz equations) which surely exhibits deterministic chaos. As the effect of this resonance one observes some elements of regularity such as high predictability during summer and very poor predictability during winter. Generally we can expect chaotic resonance to be common behaviour in systems with different co-existing attractors.

In this paper we investigate the possibility of chaotic resonance phenomenon in a class of quasi-hyperbolic systems. We generalize the classical stochastic resenance phenomenon to the class of quasi-hyperbolic systems [17]. Many regular and strange attractors can coexist and interact in the phase space of such systems. The most general example of such interaction is a "chaos-chaos" intermittency [18]. Two chaotic attractors separated in the phase space by hypersurface separatrix can merge into one chaotic attractor. The phenomenon of this "chaos-chaos" intermittency is observed in a small range of some control parameters greater than some critical parameter value where the merging of the attractors takes place. Since the statistical properties of this kind of intermittency are identical to that observed from classical bistable systems, it can be considered as a generalized form of bistable behaviour.

HYPERCHAOS-HYPERCHAOS INTERMITTENCY

It is well-known that many regular (fixed points, limit cycles, tori), chaotic (with one positive Lyapunv exponent) and hyperchaotic (with at least two positive Lyapunov exponents) attractors can co-exist simultaneously in the phase space of such systems. These attractors can interact with each other giving rise to the phenomenon of intermittency. Intermittency is a type of chaotic behaviour commonly observed in deterministic systems [19]. It is characterized by long periods of regular motion interrupted by short chaotic bursts. When a burst starts at the end of a laminar phase this denotes an instability of the periodic motion due to the fact that the modulus of at least one floquet multiplier is larger than one. As it is well-known [19] it may occur in three different ways: a real Floquet multiplier crosses the unit circle at (+1) or at (-1) or two complex conjugate multipliers cross it simultaneously. In the pioneering work Pomeau and Manneville to all of these typical crossings associated one type of intermittency. Type-I is characterized by crossing at (+1), type-II by complex crossing and type-III by crossing at (-1). Beside these simple types, more complicated intermittent behaviour can take place in general case, such as the crisis-induced intermittency (chaos-chaos) intermittency [20]. This intermittency involves a collision in phase-space of two chaotic attractors as some parameter is varied, and it is again characterized by random switching between different aperiodic oscillations. In what follows particular attention is given to the interaction between hyperchaotic attractors, henceforth called a "hyperchaos-hyperchaos" intermittency. As an example of quasi-hyperbolic system we investigate a pair of coupled identical Chua's circuits which combined equations of motion are

$$\dot{x} = \alpha(y-x-f(x)) \quad (a)$$

$$\dot{y} = x-y + z + K(v-y) \quad (b)$$

$$\dot{z} = -\beta y + \epsilon \sin\omega t \quad (c)$$

$$\dot{u} = \alpha(v-u-f(u)) \quad (d) \qquad \textbf{(1)}$$

$$\dot{v} = u-v + w + K(y-v) \quad (e)$$

$$\dot{w} = -\beta v + \epsilon \sin\omega t \quad (f)$$

where $f(x)=bx+0.5(a-b)[|x+1|+|x-1|]$, $f(u)=bu+0.5(a-b)[|u+1|+|u-1|]$, α, β, ε, ω, a and b are constants. The first Chua's circuit (eqs 1(d-f)) is coupled to the second one (eqs 1(d-f) in such a way that the difference between the signals y and v: $K(y-v)$ or $K(v-y)$ is respectively introduced into the first or the second circuit as a negative feedback. $K>0$ is the stiffness of the perturbations.

In the absence of external perturbations ($\varepsilon=0$) eq.(1) has the following nine fixed points:

$P_1= (x=0,\ y=0,\ z=-D,\ u=D,\ v=0,\ w=-D)$,
$P_2=(x=D,\ y=0,\ z=D,\ u=D,\ v=0,\ w=D)$,
$P_3=(x=,\ y=0,\ z=D,\ u=-D,\ v=0,\ w=D)$,
$P_4=(x=-D,\ y=0,\ z=D,\ u=-D,\ v=0,\ w=D)$,
$P_5=(x=0,\ y=0,\ z=0,\ u=0,\ v=0,\ w=0)$,
$P_6=(x=0,\ y=0,\ z=0,\ u=-D,\ v=0,\ w=D)$,
$P_7=(x=0,\ y=0,\ z=-D,\ u=D,\ v=0,\ w=-D)$,
$P_8=(x=-D,\ y=0,\ z=D,\ u=0,\ v=0,\ w=0)$,
$P_9=(x=D,\ y=0,\ z=-D,\ z=-D,\ u=0,\ v=0,\ w=0)$,

where $D=(a-b)/(b+1)$.

In our numerical investigation we considered the following parameter values: $\beta=22.0$, $a=-1.1428571$, $b=-0.71428571$, and $K=M=0.0025$ and α was taken as a control parameter. Numerical computations have been performed using the fourth-order Runge-Kutta method. For $\alpha<8.78$, fixed points P_1, P_2, P_3 and P_4 are stable. When crossing the line $\alpha=8.78$ on parameter plane, the limit cycles with period T are born from these points due to Hopf bifurcation. The phase trajectory projections of these cycles on u-x plane are schematically shown in Figure 1.

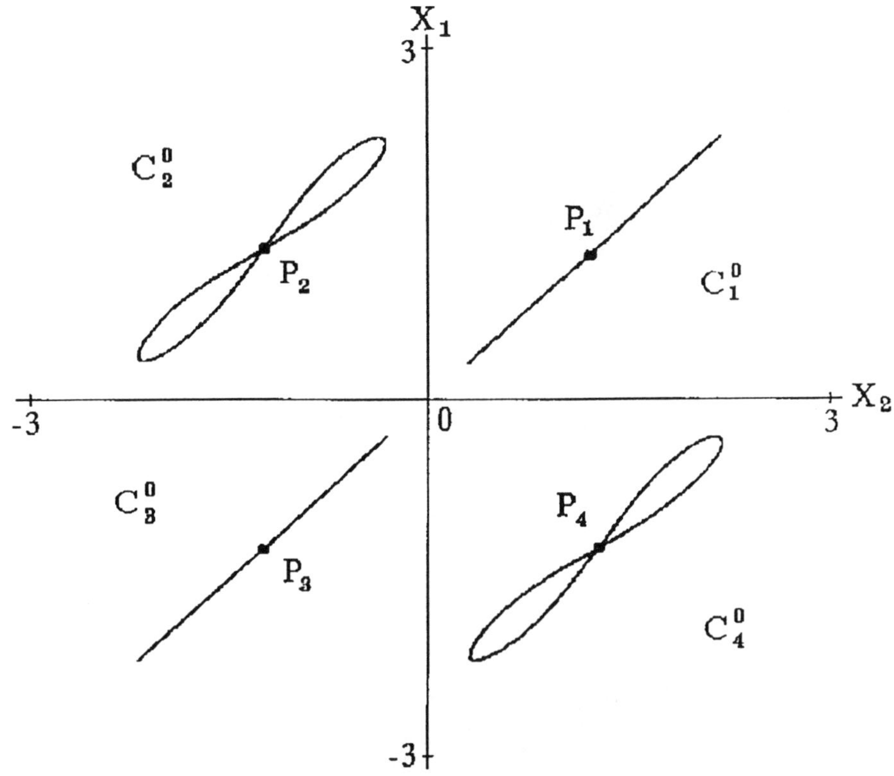

FIGURE 1. Location of fixed points in the phase space of Eqs (1).

Cycles C_1^0, C_3^0 as well as C_2^0, C_4^0 are symmetrical with respect to the transposition (5). Two of them (C_1^0, C_3^0) lie in symmetric manifold ($x=u$, $y=v$, $z=w$) of the complete phase space (they are located on the plane normal to the plane of Figure 1 and that is why they are visible as lines) and two of others are placed outside this subspace. It is necessary to note that the cycles C_2^0 and C_4^0 are selfsymmetrical with respect to substitution (5) in the sense that the point on the cycle transforms to the point on the same cycle under this substitution.

After a series of bifurcations [16] at $\alpha>11.8$ we observe four co-existing, symmetrical in respect to transposition (5) hyperchaotic attractors born from stable fixed points: P_1, P_2, P_3 and P_4. The example of such attractor is shown in Figure 2(a).

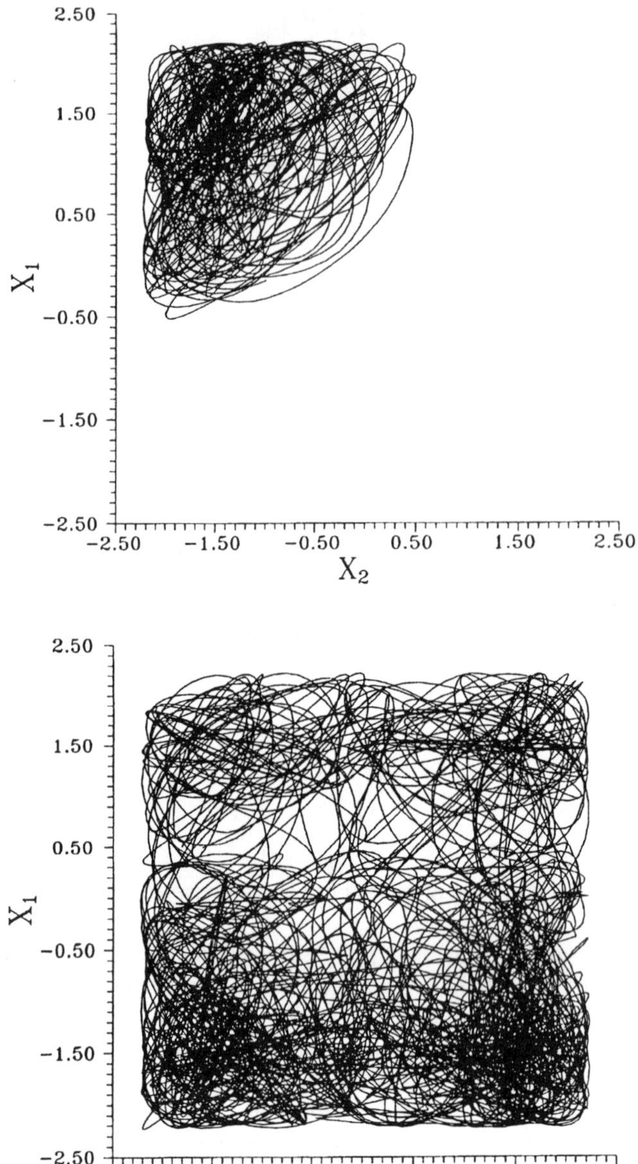

FIGURE 2. Hyperchaotic attractors of Eqs (1) shown in 2D *x-u* projection; **(a)** (top): α=11.8: Lyapunov exponents: λ_1=0.268, λ_2=0.236, λ_3=0, λ_4=-0.003, λ_5=-2.75, λ_6=-2.82; **(b)** (bottom): α=12.2; Lyapunov exponents: λ_1=0.416, λ_2=0.373, λ_3=0 λ_4=-0.004, λ_5=-3.151, λ_6=-3.577.

Phase space trajectories on this attractor are characterised by two positive Lyapunov exponents (λ_1=0.268, λ_2=0.236, λ_3=0, λ_4=0, λ_5=-2.75, λ_6=-2.82) and the system evolution takes place in the neighbourhood of one of the fixed points P_1, P_2, P_3 and P_4. Since the basins of these attractors are separated by five-dimensional invariant manifold (separatrics) they can not intersect each other. Any interaction between these two attractors can be induced only by external perturbations.

Let us consider the point α=12.0 for which four co-existing hyperchaotic attractors co-exist in the phase space as an initial state of the system (1) and investigate the possibility of the realization of the stochastic resonance phenomenon. The value of the amplitude and the frequency of the sinusoidal modulation input are fixed at ε=0.10 and ω=0.6 respectively. This external periodic modulation is too small to allow co-existing attractors to merge. With the increase of the parameter α four co-existing hyperchaotic attractors merge together and create one hyperchaotic attractor (the co-called double-double scroll) shown in Figure 2(b). To identify chaotic (stochastic) resonance we consider the signal-to-noise ratio defined by the relation

$$SNR = 10 \ln \frac{S_N(\omega) + S_M(\omega)}{S_N(\omega)}$$

where $S(\omega)$ is the power spectrum density of $x(t)$, $S_N(\omega_0)$ is the power spectrum density of the noise background at the frequency ω_0, and $[S_N(\omega_0)+S_M(\omega_0)]$ is superimposed power spectrum density of the modulation signal and the noise background at the same frequency. Hence, the value of SNR can be calculated from the computed data from the power spectrum $S(\omega)$ in the presence of the modulation signal.

The dependence of signal-to-noise ratio on different values of the parameter α is shown in Figure 3. These plots demonstrate a significant increase of signal-to-noise ratio characteristic for chaotic (stochastic) resonance phenomenon. A typical power spectra of signal $x(t)$ are shown in Figure 4(a,b). In Figure 4(a) we showed a power spectrum of $x(t)$ evolving on one of the co-existing attractors for α=11.8. Figure 4(b) presents the same power spectrum for α=12.2 after the birth of double-double scroll attractor. For lower values of ω we observe $1/f$ spectrum characteristic for intermittency. The power spectrum in the region near the onset of intermittency is described by the universal law $1/\omega^\delta$ with δ=1.2±0.1. These results clearly indicate that described example of chaotic resonance is caused by hyperchaos-hyperchaos intermittency.

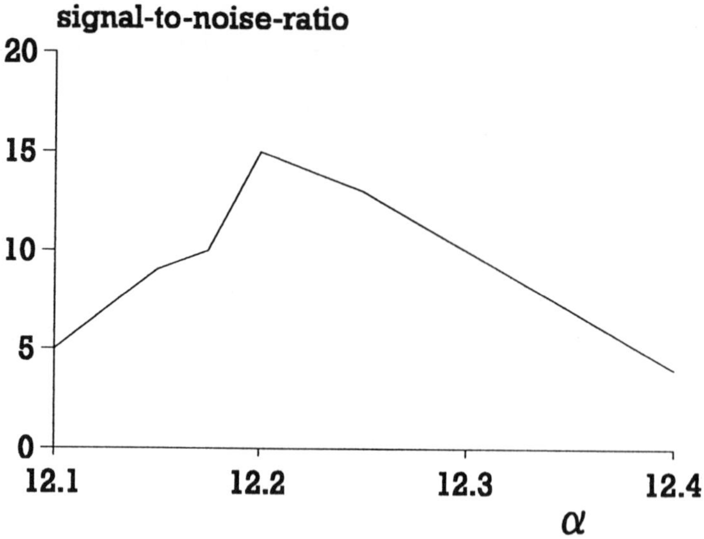

FIGURE 3. The signal-to-noise ratio versus the parameter α.

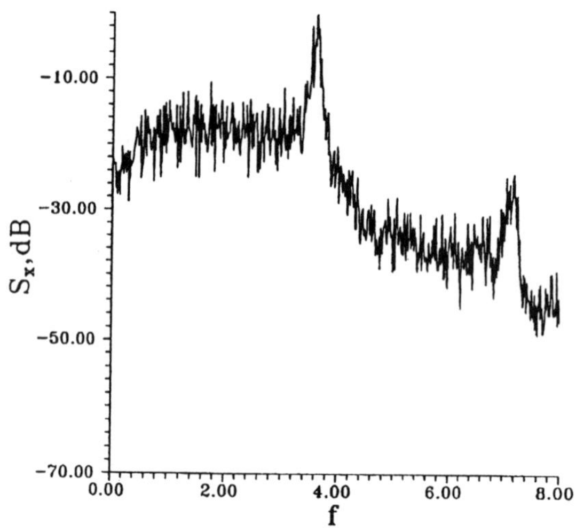

FIGURE 4(a). See caption on page 427.

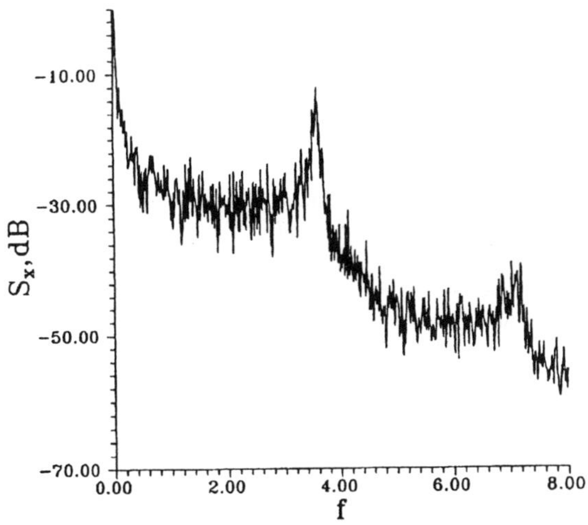

FIGURE 4(b)

FIGURE 4. Power spectra of signal $x(t)$; (a): α=11.8, (b): α=12.2.

CONCLUSIONS

To summarize the results of this paper we show that the stochastic resonance phenomenon is general in quasi-hyperbolic systems with strange attractors. In such systems unlike regular systems the random noise perturbation does not have a significant influence on the described phenomenon. The necessary condition for stochastic resonance to occur is an interaction of co-existing attractors. In chaotic or hyperchaotic systems this resonance can be induced either by an external noise (noise-induced intermittency) or strictly by the dynamical properties of nonlinear system (the dynamical intermittency). In the last case we propose to call this phenomenon as a chaotic resonance.

ACKNOWLEDGMENTS

This work has been supported by KBN (Poland) under the project no. PB 0266/P4/93/05.

REFERENCES

1. Moss, F.: in *Some Problems in Statistical Physics*, edited by G. Weiss (SIAM, Philadelphia, 1992), and references therein.
2. Benzi, R., Sutera, A. and Vulpiani, A.: *J. Phys. A: Math. Gen.* **14**, L453 (1981).
   C. Nicolis, *Tellus* **34**, 1 (1982).
   Benzi, R., Parisi, G., Sutera, A. and Vulpiani, A.: *Tellus* **34**, 10 (1982).
3. McNamara, B., Wiesenfeld, K. and Roy, R.: *Phys. Rev. Lett.* **60**, 2626 (1988).
4. Fox, R.F.: *Phys. Rev.* A39, 4148 (1988).
   Gang, H., Nicolis, G. and Nicolis, C.: *Phys. Rev. A* 42, 2030 (1990).
   Presilla, P., Marchesoni F. and Gammaitoni, L. *Phys. Rev.* 40, 2105 (1989).
5. Dykman, M.I., McClintock, P.V.E, Mannella, R. and Stocks, N.G.: *Phys Rev. Lett.* 65, 2606 (1990).
6. McNamara, B., Wiesenfeld, K.: *Phys. Rev.* **A39**, 4854 (1989).
   Gammaitoni, L., Marchesoni, F., Menichella-Saetta, E. and Santucci, S.: *Phys. Rev. Lett.* **62**, 349 (1989).
   Gammaitoni, L., Menichella-Saetta, E., Santucci, S., Marchesoni, F. and Presilla, C.: *Phys. Rev.* **A40**, 2114 (1989).
   Faure, S. and Heslot, F.: *Phys. Lett.* **97A**, 5 (1983).
7. Spano, M.L., wun-Fogle, M. and Ditto, W.L.: *Phys. Rev.* **46A**, 5353 (1992).
8. Longtin, A., Bulsara, A. and Moss, F.: *Phys. Rev. Lett.* **67**, 656 (1992).
9. Anishchenko, V.S., Safonova, M.A. and Chua, L.O., *Int. J. Bifurcations Chaos* **2**, 397 (1992).
10. Stocks, N.G., Stein, N.D. and McClintock, P.E. *J. Phys. A: Math. Gen.* **26**, L385 (1993).
11. Carroll, T.L. and Pecora, L.M.: *Phys. Rev. Lett.* **70**, 576 (1993), *Phys. Rev.* **47E**, 3365 (1993).
12. Kapitaniak, T.: *Phys. Rev.* **49E**, 4325 (1994).
13. Jung, P. and Hanggi, P.: *Phys. Rev.* **A44**, 8032 (1991).
14. Kapitaniak, T.: *Chaos, Solitons, and Fractals* **3**, 405 (1993).
15. Nicolis, G., Nocilis, C. and McKernan, D. *J. Stat. Phys.* **70**, 125 (1993).
16. Anishchenko, V.S., Kapitaniak, T., Safonova, M.A. and Sosnovzeva, O.V.: *Phys. Lett. A* **192**, 207 (1994).
17. Anishchenko, V.S., Neiman, A., and Safonova, M.A.: *J. Stat. Phys.* **70**, 185 (1993).
18. Anishchenko, V.S.: *Soviet Technical Physics Letters* **10**, 226(1984).
    Anishchenko, V.S. and Neiman, A.B.: *Soviet Technical Physics* **17**, 3-14 (1990).
19. Pomeau, Y., and Manneville, P.: *Communication in Mathematical Physics*, **74**, 189-197 (1980).
    Kohyama, T., and Aizawa, Y., *Progress in Theoretical Physics* **71**, 917 (1984).
20. Grebogi, C., Ott, E., Romeiras, F., and Yorke, J.A., *Phys. Rev.* **A36**, 5365 (1987).

6. MACHINERY DIAGNOSTICS

Chaos Concepts as Diagnostic Tools for Assessing Rotating Machinery Vibration Signatures

Maurice L. Adams* and Kenneth A. Loparo+

*Department of Mechanical and Aerospace Engineering and +Department of Systems Engineering
The Case School of Engineering, Case Western Reserve University, Cleveland, Ohio 44106

Abstract. Chaos content in measured vibration signals is of some practical importance in rotordynamical systems. Of particular interest is the relationship between the occurrence of determinsitc chaos and the diagnosis of mechanical failures in rotating machinery. Two nonlinear rotordynamical systems were studied using simulation and various forms of subharmonic, quasiperiodic and chaotic vibrations were observed. Different routes into and out of chaos show important signs for wear assessment and failure prediction. Experimental test facilities are currently under development and the next steps involve experimental verification of the simulation results and the development of signal processing techniques for extracting the dynamical features of the vibration signatures from meaured time series data.

INTRODUCTION

Common practice in the analysis of time-based vibration signatures is to transform the signals into the frequency domain using high-speed data acquisition systems and the Fast Fourier Transforms (FFT). Spectrum analysis is now a common diagnostic tool in assessing the vibration signatures of many types of rotating machinery (e.g., turbines, pumps, large fans, gear systems, etc.). However, there are numerous examples where spectrum analysis does not provide all the insight needed to quickly identify and correct a vibration problem. New techniques which can significantly improve the diagnostic capability in the processing of rotating machinery vibration signals would be of considerable benefit to power plant operating personnel. Chaos concepts and other model-based and advanced signal processing methods have the potential for advancing vibration analysis of rotating machinery. Such advancements are especially important for electric power generating plants, where continuing improvements in monitoring and diagnostic techniques can have substantial paybacks that far outweigh the monetary investments made in these systems.

CHAOTIC OSCILLATIONS IN ROTATING MACHINERY

Because special interpretations are necessary when using "frequency" based techniques in the analysis of non-stationary time series data., tracking the intricate dynamical behavior, such as the onset of chaos, will be beneficial in supplementing the currently available time and frequency based detection and diagnosis techniques. What makes chaos-tracking a potentially valuable monitoring and diagnostic tool for machinery vibration is the fact that excessive vibration and related root causes are quite often related to distinct types of nonlinear dynamic phenomena in the machinery, and the increasing presence of the nonlinear factors can be reliable indicators or predictors of a specific vibration or vibration related problem.

References [1] and [2] treat most of the basic rotating machinery vibration problems.

Simulation Results

Simulation results for two lower order models of rotor dynamical systems were used to evaluate the potential of using the detection of chaotic behavior in a time-series of machinery vibration data as an indicator of a failure or an impending failure. In each system, the rotor is modeled as a point mass with two-degrees of orbital motion. All of the systems analyzed are highly nonlinear and care was taken to ensure the accuracy of the numerical integration routines that were used to solve for the long term motions. Mathematical models and simulation details are given in [1] and [2].

Rotor Rub Impact

Rotor vibration orbits for the rotating system can exhibit steady state vibration signatures which include two-dimensional orbits with period-1, period-2, quasiperiodic and chaotic rotor vibrations. Bifurcation diagrams illustrate parameter regimes where alternating chaotic and periodic vibration orbits occur and provide a correlation between the qualitative properties of the orbit dynamics and the clearance values, friction coefficient values and damping ratios. This suggests that signal processing methods for detecting and diagnosing the chaos in monitored vibration orbits can be used to detect the wearing open of sealing clearances, loss of lubrication to the bearing or the entrapment of debris from wear in the bearing, as well as small changes in system damping. To emphasize this point, refer to Figure 1 where the Poincare' mapping of small chaotic components in the orbital motion is presented. This figure shows that a 3% difference of damping at off-resonance conditions, an event imperceptible from FFT processing of the same vibration signals is detectable from the Poincare' mapping of the data. The fractal dimension and the qualitative patterns of the orbit as viewed through the Poincare' map provide information so that differences between 8% and 11% damping are clearly discernible from the vibration orbit data. An important practical problem is whether or not such information can be extracted from contaminated vibration orbit data obtained from a rotating machine in a power plant. This is a focus of some of our current experimental and analytical work.

Cylindrical Journal Bearings

It is well known that the potential for instability does not arise until the rotational speed exceeds the lowest natural whirling frequency (critical speed) of the machine. Supercritical rotors are generally high energy systems with latent instabilities. Our analysis focused on cylindrical bearing rotor dynamics for supercritical operation and specifically the potential routes to and from chaos in these machines.

As with the rub-impact model, simulation results over a wide range of parameters produced steady state responses rich in subharmonic, quasiperiodic and chaotic motions. Similarly, routes to and from chaotic regions include Feigenbaum type, period doubling, quasiperiodic with period locking and sudden (crisis) bifurcation sequences. The observed chaotic phenomena for this rotor dynamical system, which are of considerable value as diagnostic tools, had not been previously investigated for fluid film journal bearings.

Figure 2 illustrates some familiar journal orbits for a large unbalance excitation and a zero static bearing load, at a dimensionless speed of nearly 3. The Poincare' map shows the emergence of a single closed loop, which indicates the presence of two incommensurate frequencies. Figure 3 shows the vibration orbits for a series of cases where the rotating speed of the machine is increasing. Period doubling and chaos occur at dimensionless speeds of about 7.7 and 8.5, respectively.

Figure 4 clearly illustrates the detectable changes in the fractal dimension as well as the pattern of the Poincare' map of the orbit that neither a direct measurement of the vibration orbit or the FFT clearly show. The two cases shown are comparable to a 1.2:2 change in viscosity.

Figure 5 shows the bifurcation diagram corresponding to a gradual increase in unbalance excitation and corresponding selected orbits, which illustrates a progression from low amplitude quasiperiodic motion, to period-3 motion, to fully developed chaotic motion. Again we observe that the intricate features of the chaotic motion, including both qualitative and quantitative information, are excellent indicators of the health of the rotor dynamic system.

The simulation studies indicate that for the two cases mentioned above, orbital vibration responses that exhibit various forms of dynamical behavior exist. These responses have rich dynamical behavior which includes subharmonic, quasiperiodic and chaotic motion, over a wide range of operating conditions and parameters. More importantly, discernible changes to the chaos structure, that are sensitive to parameter and operating conditions of the rotor, were observed. This information can be used to develop chaos tracking techniques for advanced detection and diagnosis of rotating machinery vibration problems.

Based on these preliminary results and the potential applications of chaos tracking as a diagnostic tool for rotating machinery, ongoing research is being conducted to experimentally verify the simulation results obtained to date and to extend the concepts to higher order (multi-degree-of-freedom) systems. Also, practical implementations of these ideas require developing methods to cope with the measurement uncertainties that result from transducers on real machines, as well as the development of automatic methods of detecting patterns of chaos in noisy measurements orbital vibrations.

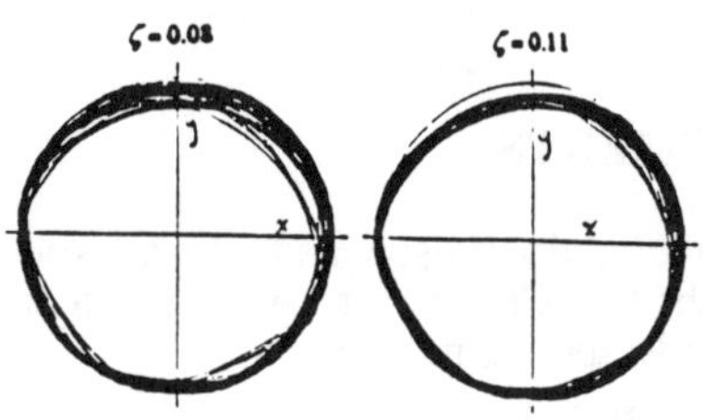

Figure 1: Poincare' map of a chaotic signal for small changes in damping (rub-impact).

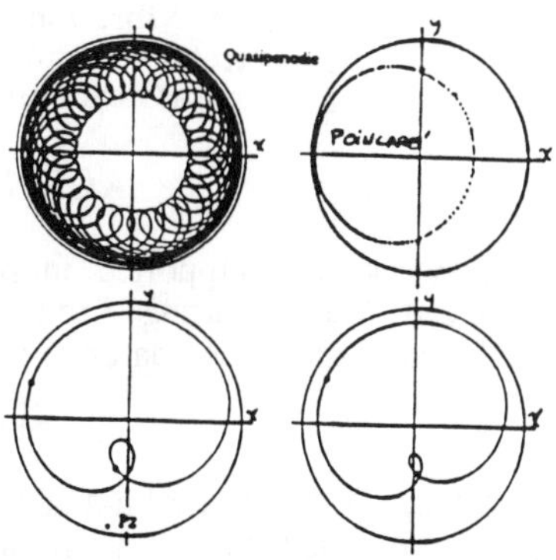

Figure 2: Vibration orbits in a cylindrical journal bearing.

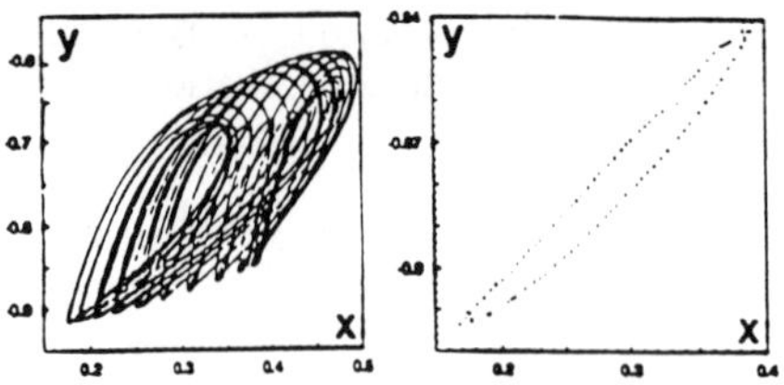

Figure 3: Bifurcation diagrams and vibration orbits for a cylindrical bearing.

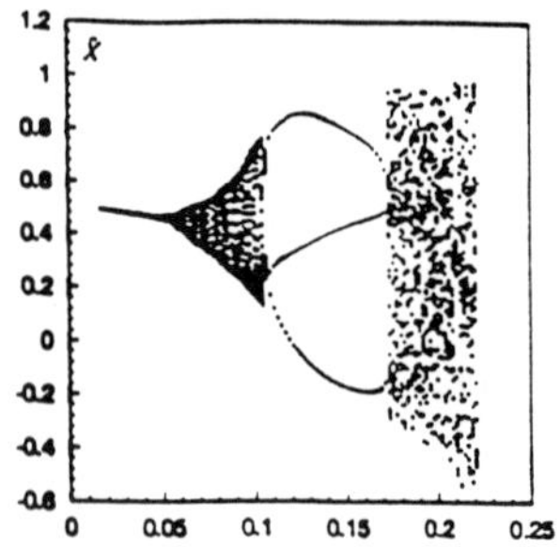

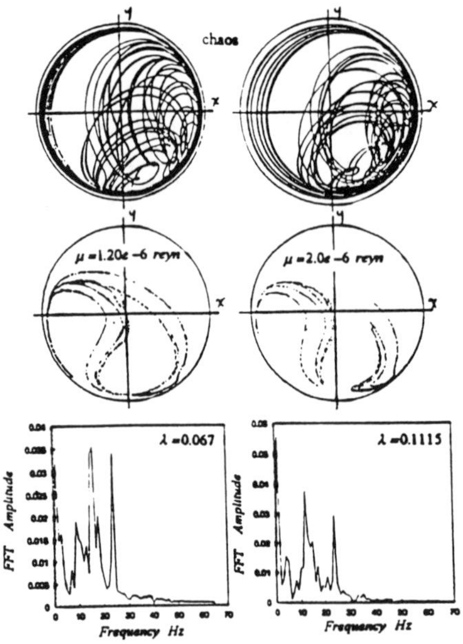

Figure 4: Vibration orbits, Poincare' maps and FFT spectra for a cylindrical journal bearing.

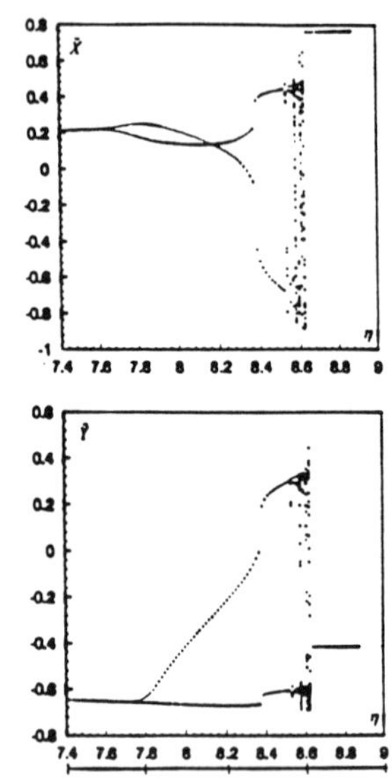

Figure 5: Bifurcation diagrams illustrating the onset of a chaotic orbit.

ACKNOWLEDGMENTS

This work was supported in part by EPRI, Ohio Edison and the Cleveland Advanced Manufacturing Programs through the Center for Automation and Intelligents Systems Research at Case Western Reserve University.

REFERENCES

1. Adams, M.L. and Abu-Mahfouz, I.A., "Exploratory research on chaos concepts as diagnostic tools for assessing rotating machinery vibration signatures," in *Proceedings, 4th International Conference on Rotor Dynamics, Internatinal Federation of Machines and Mechanisms*, September 7-9, 1994.

2. Abu-Mahfouz, I.A., "Routes to Chaos in Rotor Dynamics," Ph.D. Dissertation, Department of Mechanical Engineering, Case Western Reserve University, Cleveland, Ohio, May, 1993.

Dimension Determination of Precursive Stall Events in a Single Stage High Speed Compressor

Michelle M. Bright

NASA Lewis Research Center
Advanced Controls Technology and System Dynamics Branches
Cleveland, OH 44135

Helen K. Qammar and Tom T. Hartley

Departments of Chemical and Electrical Engineering
The University of Akron
Akron, Ohio 44325

Abstract. This paper presents a study of the dynamics for a single-stage, axial-flow, high speed compressor core, specifically, the NASA Lewis rotor stage 37. Due to the overall blading design for this advanced core compressor, each stage has considerable tip loading and higher speed than most compressor designs, thus, the compressor operates closer to the stall margin. The onset of rotating stall is explained as bifurcations in the dynamics of axial compressors. Data taken from the compressor during a rotating stall event is analyzed. Through the use of a box-assisted correlation dimension methodology, the attractor dimension is determined during the bifurcations leading to rotating stall. The intent of this study is to examine the behavior of precursive stall events so as to predict the entrance into rotating stall. This information may provide a better means to identify, avoid or control the undesireable event of rotating stall formation in high speed compressor cores.

MOTIVATION

In studying the dynamics of turbomachinery, specifically, axial-flow compressor behavior, several approaches have been developed to determine when the compressor is entering a rotating stall event. Rotating stall, as shown in Figure 1, is the development of a circumferentially non-uniform flow across the compressor face. Surge, on the other hand, is an axially oscillating flow event through the compressor. The rotating stall event is consequentially severe because stall can cause severe unsteady loading on the rotor and stator blades. Rotating stall causes a large drop in pressure rise delivered in the engine where an engine restart is often needed to clear the stall event. Rotating stall may also lead to surge or "flame out" of the engine. Obviously, this is a good behavior to avoid in axial-flow compressors.

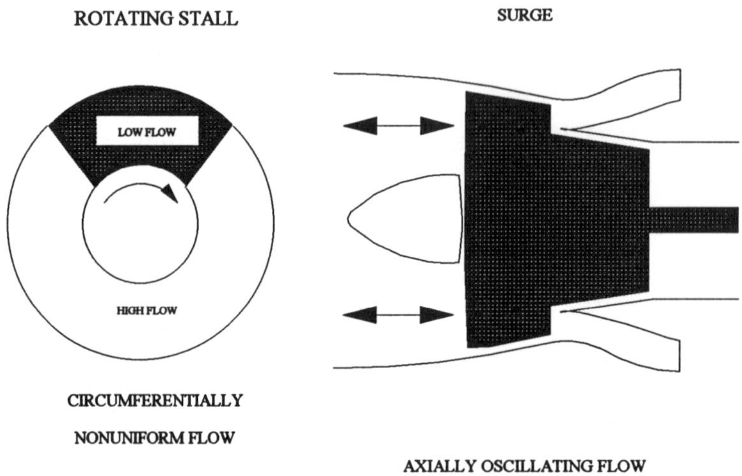

Figure 1. Rotating Stall and Surge Events in an Axial Flow Compressor

One approach for determining prestall behavior of the compressor exists based on the assumption that rotating stall is primarily a linear event and can be approximated by the first 3 spatial harmonics of the Fourier transform coefficients. "Measurement of prestall perturbations are made by acquiring time traces from a circular array of sensors. These signals are decomposed into spatial harmonics, a spectrum of spatial harmonics is used for identification of the evolution of 2-D rotating stall."[1]

Another approach uses the change in energy of the power spectral density traces of pressure data prior to stall. "In steady state behavior of the compressor, these power spectral density curves remain symmetric. However, during the evolution of stall cells these power spectral density curves change shape. This method identifies traveling wave energy in the compressor which may lead to the formation of stall cells."[2] This metamorphosis of the power spectral density curves is another precursive event for identifying rotating stall.

The approach used in this paper attempts to identify the stall precursive events as correlation dimension changes in the pressure data from a single sensor position. This paper will describe the compressor used for the experiment, equations of motion which capture some of the relevant 1-D compressor dynamics, the data acquisition and experimental setup, the correlation dimension technique used for analysis, and a summary of observations.

DESCRIPTION OF COMPRESSOR STAGE AND EXPERIMENT

The NASA rotor 37 axial-flow, high-speed compressor stage for advanced airbreathing engines is an inlet stage compressor designed as part of an eight-stage core compressor having a pressure ratio of 20:1. The compressor design has a constant meanline diameter, with an inlet hub-tip ratio of 0.7 and an inlet rotor-tip speed of 455 meters per second. The loading per stage is considerably high for a research test compressor. Additional details on the blading, geometry, and testing of this compressor can be found in references [3 and 4].

The rotor 37 compressor was instrumented with 8 circumferentially placed pressure transducers within 1 chord length upstream of the rotor. These wall static pressure transducers measured the pressure fluctuations during the transition into stall. The rotor was stalled by closing the throttle area in a linear manner. Data was acquired for 15 seconds prior to a full stall event at a sampling rate of 2 kilohertz. The rotor was operating at 100% speed, 17581 rpm, for the stall runs. The test facility for this compressor is shown in Figure 2 which details the flow path through the inlet, compressor, plenum and throttle. Figure 3 shows the 8 circumferentially placed transducers around the wall of the compressor.

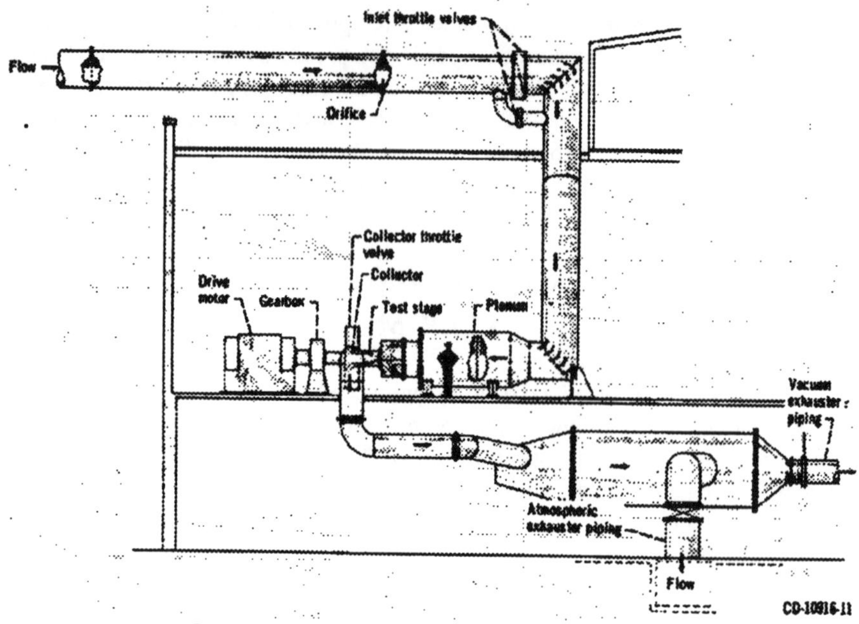

FIGURE 2. Compressor Facility Drawing of Flow Path

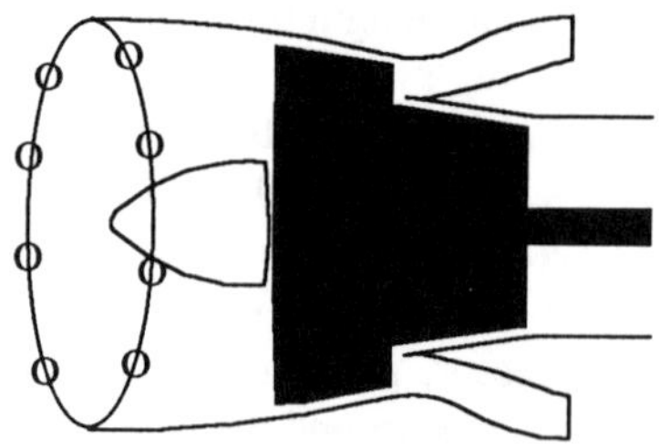

FIGURE 3. Circumferentially Placed Pressure Transducers

EQUATIONS OF MOTION

An approximate set of equations that describe surge and rotating stall transient formation are as follows from reference [5]:

1-D momentum: $$d\phi/dt = -\psi + \psi_c(\phi) - 3\phi R$$

1-D continuity: $$d\psi/dt = 1/\beta^2[\phi - \phi_T(\psi)]$$

Stall cell growth: $$dR/dt = \sigma R(1 - \phi^2 - R)$$

where ϕ = compressor mass flow, ψ = system pressure rise, R = (stall cell amplitude$)^2$.

From these equations one can estimate the rate of stall cell growth for a particular compressor geometry. With this estimated rate of stall cell growth, the data collection for the experiment was planned to capture the initial stall formation with the wall static pressure transducers. Additionally, initial estimates for the embedding dimension and time delay parameters for correlation dimension determination were based on the use of this physical model.

EXPERIMENTAL DATA

Data gathered from 1 of the circumferential pressure transducers is shown in Figure 4.

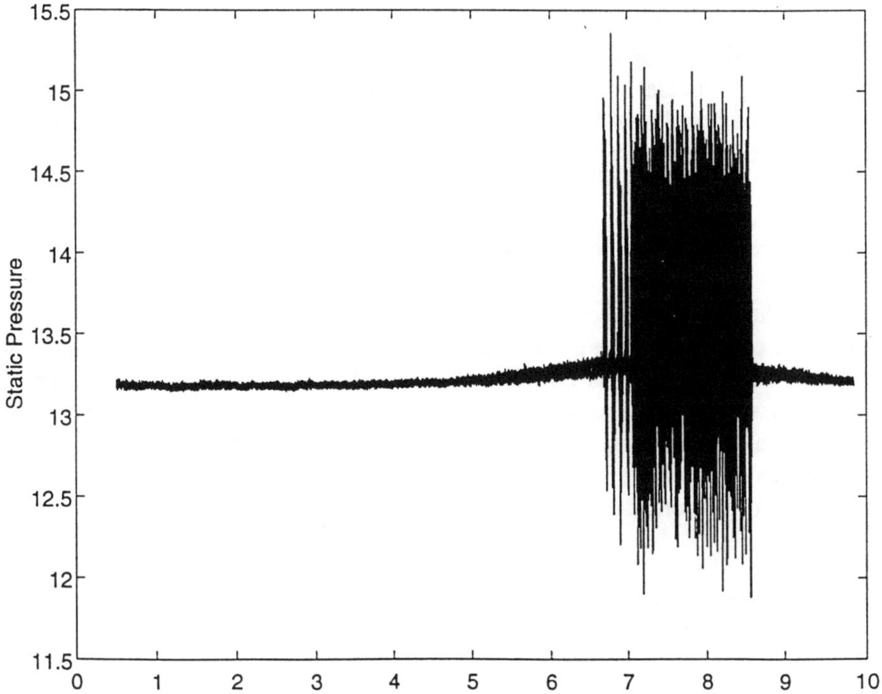

FIGURE 4. Pressure data

A phase plane portrait of the pressure data is reconstructed using 5000 data points. To reconstruct the attractor, an independent time series of data was generated using the following time delay embedding method[6]:

$$v(t) = (x(t), x(t+T), ..., x(t+(m-1)T).$$

where v(t) is the independent time series, m is the embedding dimension, and T is the reconstruction time delay. For this pressure data, the values of m=4 and T=240 were used for the reconstruction

DIMENSION DETERMINATION

Verification of the dimension variation of this compressor is provided through a chaos package software routine. A box-assisted correlation dimension algorithm [7,8] is used to determine the correlation dimension of the measured stall data.

For the correlation dimension method, 16000 pressure data samples are used prior to the onset of stall, of which the first 1000 points are thrown out as transient. Using a 5000 data point sliding window, the window is moved every 300 samples, thus creating 28 slices of 5000 points each of pre-stall data. Using 5000 data points allows for accurate prediction of correlation dimension up to approximately dimension 8, according to the following equation from Eckmann and Ruelle[9] which states:

Maximum Accurate Dimension < 2 log(number of data points).

With this limit in mind, the correlation dimension, (or correlation exponent), was calculated using a Grassberger-Procaccia correlation integral [10] for each of the 28 slices before stall. This correlation exponent is closely related to the fractal dimension and gives a lower bound to the fractal dimension. The Grassberger-Procaccia algorithm using Theiler's method[7,8] considers the spatial correlation between pairs of points on a reconstructed attractor, and is measured with the following correlation integral:

$$C(N,r) = \frac{2}{N(N-1)} \sum_{i} H(r - \| xi - xj \|)$$

where H(x) is the Heaviside step function. The summation counts the number of pairs (xi,xj) for which this distance is less than r, where r is the box size. Then

C(N,r) scales like a power of v so that:

$$C(N,r) = r^{v}$$

where v is the correlation dimension and is the slope of the log-log plot of C(N,r) vs. r.

Figure 5 shows the correlation dimension integral plots for the measured pressure data. It is shown that the correlation dimension for the first 20 slices of data remains constant. However, during the 21st slice of data, the correlation dimension begins to increase. This corresponds to a bifurcation occurring in the pressure data and subsequently to a stall precursive event as detected from the power spectral density curves shown in Figure 6. As the slices of data move further into the onset of stall the correlation dimension continues to increase. This is shown in Figure 7.

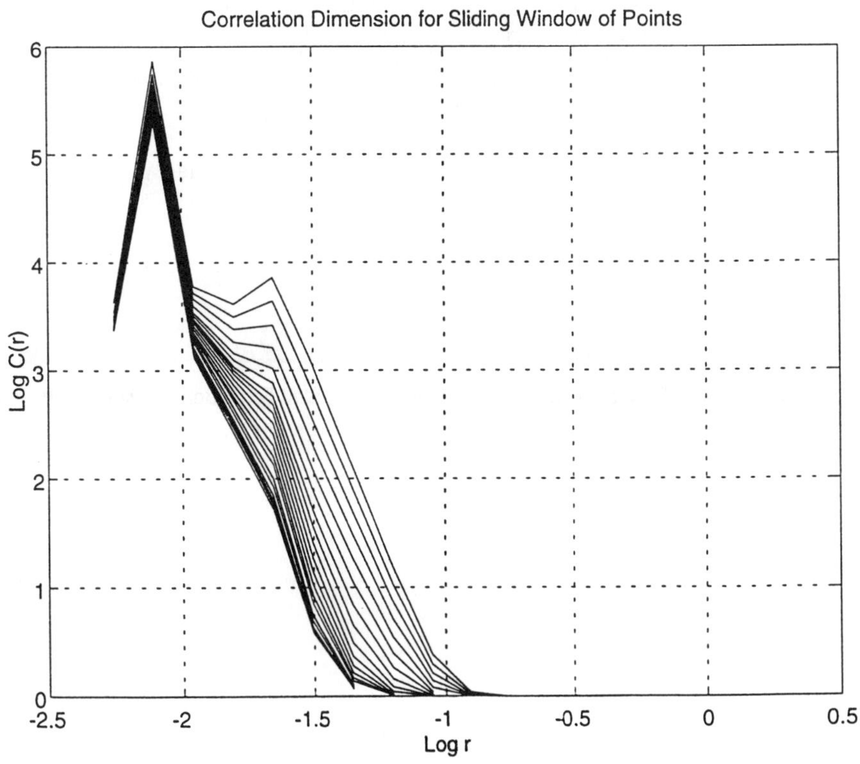

FIGURE 5. Correlation Dimension Changes at 20th Window and into Stall

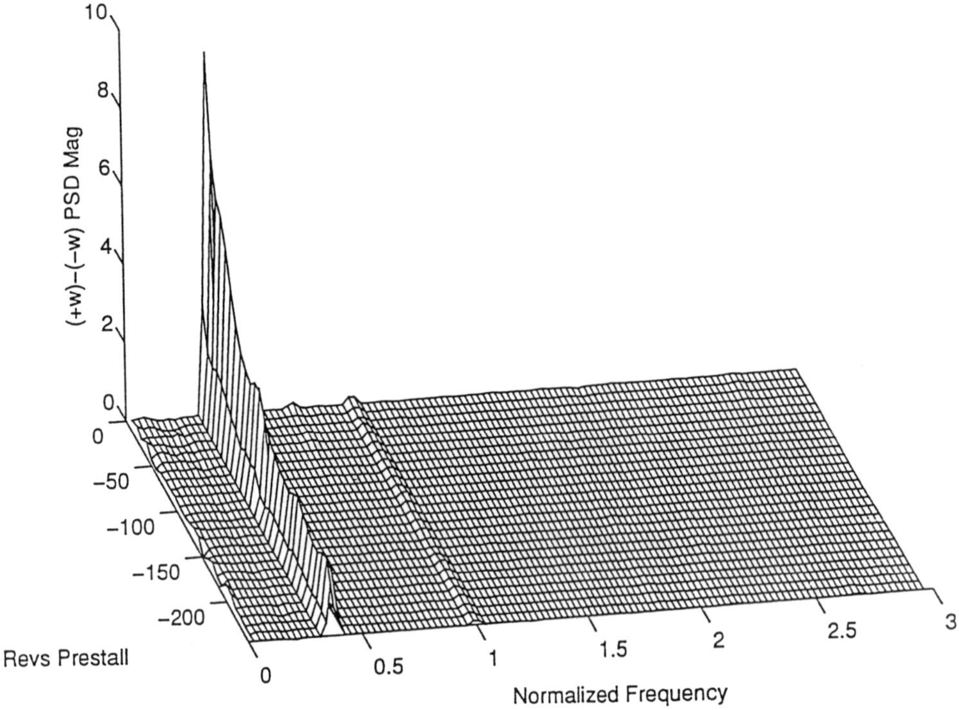

FIGURE 6. Power Spectral Density Plot Showing Stall Formation at 200 Revs Before Stall

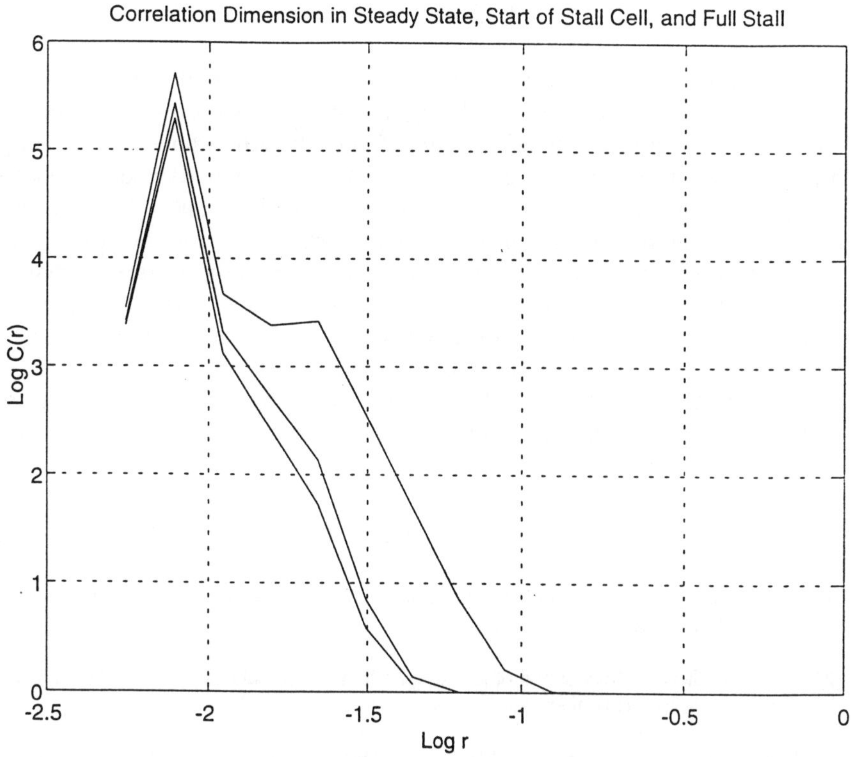

FIGURE 7. Correlation Dimension in Steady State, Prior to Stall, and In Rotating Stall

CONCLUSIONS

From this study it is shown that pressure data acquired from a single stage, high speed compressor bifurcates during the onset of rotating stall. During these bifurcations the power spectral density and the correlation dimension is analyzed from a single pressure sensor. It is shown that the correlation dimension increases during the stall precursive events leading to rotating stall. Changes in the correlation dimension integral match the results from analyzing the power spectral density curves for this pressure data. This indicates both the correlation dimension and the power spectral density curves begin to move several hundred rotor revolutions prior to a full blown rotating stall event. This shows that correlation dimension determination may help identify rotating stall precursors in high speed axial compressors. Further work is needed to compare these results with other existing methodologies for compressor stall identification and to accurately determine if rotating stall is a chaotic event.

REFERENCES

[1] J.D. Paduano, "Active Control of Rotating Stall in Axial Compressors," MIT PhD Dissertation, March 1992.

[2] Olivier Etchevers, "Evaluation of Rotating Stall Warning Schemes for Axial Compressors", MIT Masters Thesis, September 1992.

[3] Lonnie Reid and Royce Moore, "Design and Overall Performance of Four Highly Loaded, High-Speed Inlet Stages for an Advanced High-Pressure-Ratio Core Compressor", NASA Technical Paper 1337, 1978.

[4] Royce Moore and Lonnie Reid, "Performance of Single-Stage Axial-Flow Transonic Compressor with Rotor and Stator Aspect Ratios of 1.19 and 1.26, Respectively, and with Design Pressure Ratio of 2.05", NASA Technical Paper 1659, April 1980.

[5] E.M. Greitzer, "Overview of Flow Instabilities in Aeroengines and Methods for Stability Enhancement", Ohio Aerospace Institute Lectures, July 1992.

[6] N.H. Packard, et. al., "Geometry from a Time Series", Physical Review Letters, Volume 45, pp. 712-716, 1980.

[7] James Theiler, "Efficient Algorithm for Estimating the Correlation Dimension from a Set of Discrete Points", Physical Review A, Volume 36, No. 9, p. 4456-4462.

[8] James Theiler, "Spurious Dimension from Correlation Algorithms Applied to Limited Time-Series Data", Physical Review A, Volume 34, No. 3, p.2427-2432.

[9] J.-P. Eckmann and D. Ruelle, "Fundamental Limitations for Estimating Dimensions and Lyapunov Exponents in Dynamical Systems", Physica D, Volume 56, 1992, pp. 185-7.

[10] Peter Grassberger and Itamar Procaccia, "Characterization of Strange Attractors", Physical Review Letters, Volume 50, No. 5, p. 346-349.

Experimental Application of the Karhunen-Loève Decomposition to the Study of Modal Interactions in a Mechanical Oscillator

J. P. Cusumano*
Department of Engineering Science and Mechanics
The Pennsylvania State University
University Park, PA 16802

Abstract

The results of an experimental study of a magnetically-kicked flexible-beam oscillator are presented. The dynamics of the autonomous system are observed using nine strain gauges distributed along the length of the beam. The Karhunen-Loève decomposition is used to study the dimensionality of the system in terms of modal interactions. Two modes which interact to produce the observed dynamics over a wide range of the system's bifurcation parameter are found. It is concluded that a two degree of freedom model is adequate to describe the observed steady-state phenomena.

Introduction

In this paper, a statistical method known as the Karhunen-Loève decomposition [1] of spatial covariance measurements is used to study modal interactions in an magnetically-kicked flexible-beam mechanical oscillator. It is shown using this technique that two specific spatial modes interact to produce the observed dynamics over a wide range of the system's bifurcation parameter. It is concluded that a two degree of freedom model is adequate to describe the observed steady-state phenomena.

That the essential features of a dynamical system can be extracted from even a single scalar time series is a powerful fact that has been exploited in numerous numerical and physical experiments on nonlinear systems. Using delay reconstruction embedding [2,3], quantities such as generalized dimensions and Lyapunov exponents can be computed, and nonlinear predictive models can be constructed directly from data [4–6].

The power of these methods derives from the fact that they depend only on topologically invariant features of the underlying system, and hence require no knowledge of governing equations, or even of "physical" state variables. However in many applications, this strength becomes a weakness: viewed from a different perspective, current embedding methods *yield no information about a system that is not merely related to the topological structure of phase space*. In particular, potentially useful geometrical or spatial information is not readily accessible, and the mapping between the topologically invariant observables and relevant physical system parameters is almost never known.

In the study of mechanical and electromechanical systems of the sort usually encountered in engineering, a great deal of explicit knowledge is available in the form of governing "equations of motion". Typically, the general form of the equations is known up to unknown constitutive relations which must be supplied as hypotheses or obtained empirically (only in rare instances can constitutive relations be derived from first principles). The most basic advantage of starting with even approximate governing equations is that the role of material and geometrical system parameters (such as elastic moduli, damping constants, aspect ratios, etc.) can be clearly ascertained: this is crucial if the results of dynamical analyses are to be used for the redesign of system components. Furthermore, only with explicit governing equations can the physical parameters

*Associate Professor, email: jpc@coriolis.esm.psu.edu. Support of the Air Force Office of Scientific Research is gratefully acknowledged.

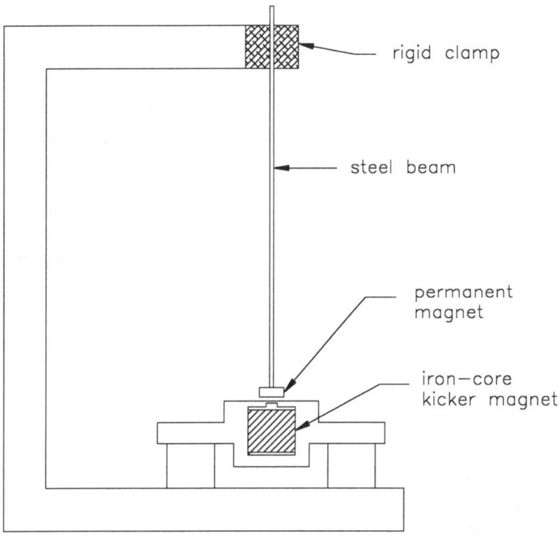

Figure 1: Schematic diagram of the kicked-beam oscillator experimental system.

be used to perform asymptotic, sensitivity and bifurcation analyses. However, first-principles formulations generally result in mathematical models having infinite-dimensional phase spaces (most typically coupled nonlinear partial differential equations). To facilitate the analysis and design of such high-dimensional systems, a procedure for generating reduced-order physical models from the first-principles equations is highly desirable. While scalar time series can be used in the model reduction process (see [7] for an example in which delay reconstruction and bifurcation data are used in an explicit reduction), a method which yields information about the spatial structure of solutions is to be preferred. The use of "shape functions" or "modes" is a mainstay of engineering analysis for linear systems. Modal analysis furnishes information with direct relevance for design (for example in suggesting structural modifications). Furthermore, a well-chosen set of active modes provides a physically-meaningful coordinatization of the reduced-order model's phase space.

The Karhunen-Loève (K-L) decomposition [1] is a method for obtaining shape functions or "K-L modes" from experimental data which is more generally applicable than conventional modal analysis, since it does not require the system under consideration to be linear. Applied originally in mechanics to the study of turbulent fluid flows [8,9], the K-L decomposition is widely used in a variety of fields, including, for example, pattern recognition [10], image compression [11] and system identification [12]. In this work, the K-L decomposition is used to study modal interactions by analyzing spatial covariance estimates obtained from an array of nine strain gauges uniformly spaced along the length of a steel beam. Space permits only a greatly abbreviated presentation of our experimental results: more experimental details, comparisons to dimensionality results obtained using other means, and application of the results to model reduction, will presented in a forthcoming paper.

Description of the Experimental System

The experimental system (see Figure 1) consists of a steel beam 6.5" × 0.5" × 0.009" (16.5 cm × 1.27 cm × 0.023 cm). At rest, the beam has its neutral axis oriented vertically, and the top of the beam is clamped. A 0.127" × 0.5"diameter (1.27 cm) rare-earth permanent magnet is epoxied onto the free end of the beam: the magnet modifies the free boundary condition of the beam through a combination of inertial and generalized

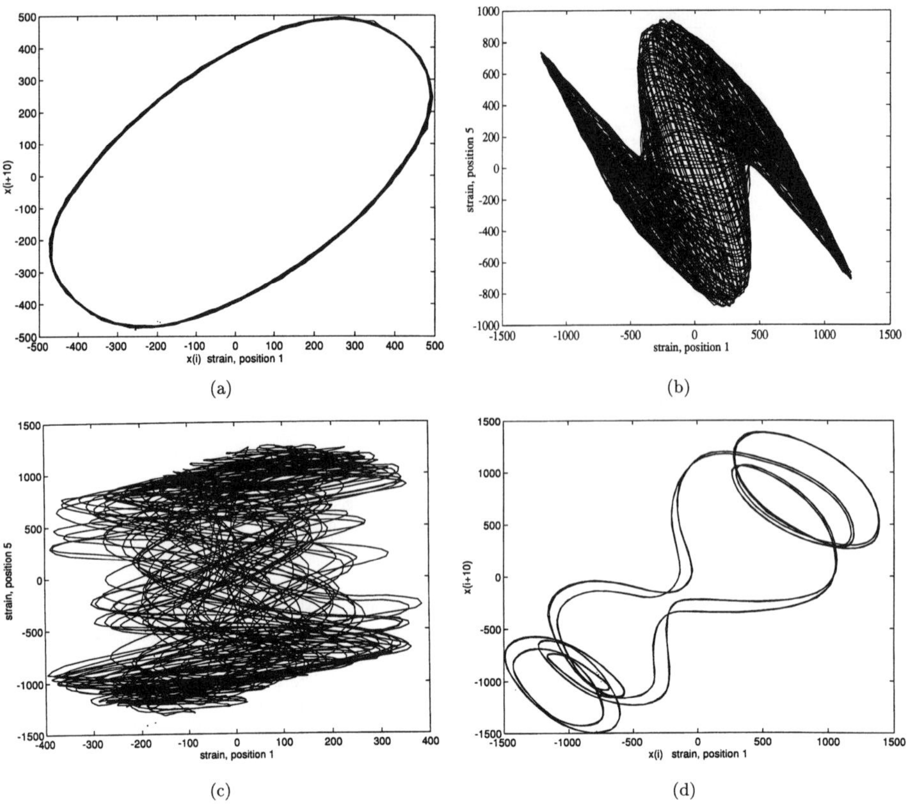

Figure 2: Four different dynamical regimes of the system: (a) small amplitude limit cycle; (b) quasiperiodic; (c) chaotic; and (d) large amplitude limit cycle.

restoring forces acting near the tip. The magnet interacts with the 0.225" (0.57 cm) diameter ferromagnetic core of a "kicker" electromagnet located nominally 0.06" (1.6 mm) from the permanent magnet: the result is an effective nonlinear restoring force at the beam tip which is very strong near the iron core, but falls rapidly to zero for tip oscillations exceeding the core radius. Thus, the linear regime of the system includes only those motions for which the tip displacements are appreciably smaller than half the iron core diameter.

The kicker electromagnet is spatially triggered, and hence the system is autonomous: when the beam tip passes with sufficient velocity $v^\star$ over the core, a sensor coil opens a transistor with supplies power to the electromagnet. This applies an approximately constant generalized force to the beam tip as it passes over the core. It is not hard to show that the change in generalized momentum provided by the kicker magnet is $\Delta p \propto (v_T)^{-1}$ where v_T is the beam tip velocity as it passes over the core (provided, of course, that $v_T > v^\star$). The strength of the kick (that is, the magnitude of Δp) is controlled by the DC voltage V applied to the kicker magnet coil, and hence V is suitable for use as a bifurcation parameter for the system.

The range of behaviors observed in the system is demonstrated by Figure 2. Data was collected from the system using nine strain gauges uniformly spaced along the length of the beam. For all values of the kicker magnet voltage V, motions starting with sufficiently small initial conditions asympotically approach the stable trivial solution of the system. However, above a critical voltage V_C, a bifurcation occurs giving rise to a stable limit cycle (in fact, since the trivial solution *always* exists, at least one *unstable* limit cycle must also be formed). Thereafter, four distinct dynamical regimes are observed: A small amplitude limit cycle regime (Figure 2(a)); a quasiperiodic regime (Figure 2(b)); a chaotic regime (Figure 2(c)); and a large amplitude limit cycle regime (Figure 2(d)).

Karhunen-Loève Analysis of Modal Interactions

The configuration space for our system can be taken to be the one-dimensional displacement field of the beam centerline, $u(x,t)$, where x is the position along the length of the beam and t is time. The spatial covariance $R(x,y)$ is then given by

$$R(x,y) = \langle\, u(x,t)\, u(y,t)\,\rangle\,, \tag{1}$$

where the angle brackets $\langle\cdot\rangle$ denote the time average, and we can assume that $\langle u\rangle = 0$ without loss of generality.

Consider the eigenvalue problem obtained by using $R(x,y)$ as the kernel of the Fredholm integral equation

$$\int_{\mathcal{D}} R(x,y)\, \phi(x)\, dy = \lambda\, \phi(x)\,. \tag{2}$$

The Karhunen-Loève decomposition theorem [1] states that the eigenfunctions $\{\phi_i(x)\}_{i=1}^{\infty}$ satisfying (2) form a complete set of orthonormal basis functions for the random process with trial functions $u(x,t)$, and hence one can write

$$u(x,t) = \sum_{i=1}^{\infty} q_i(t)\, \phi_i(x)\,. \tag{3}$$

We refer to the basis functions ϕ_i as K-L modes. The coordinates q_i with respect to the K-L modes can be shown to be related to the eigenvalues of equation (2) by

$$\lambda_i = \langle\, q_i^2(t)\,\rangle\,. \tag{4}$$

A key result of the K-L theorem is that the modal coordinates q_i are uncorrelated ($\langle q_i q_j\rangle = 0 \,\text{for}\, i \neq j$), and hence in a mean square sense the K-L modes are optimal: that is, they capture more of the mean square amplitude of u per mode than any other set of basis functions. Given this optimality, the K-L modes are ordered so that $\lambda_1 \geq \lambda_2 \geq \ldots$.

In the case of spatially discrete measurements $u(x_i,t) \equiv u_i(t)$ the covariance $R(x,y)$ becomes the Hermitian matrix

$$R_{ij} = \langle\, u_i\, u_j\,\rangle\,. \tag{5}$$

Thus given N measurement locations, R_{ij} is $N \times N$, and the K-L modes become N-dimensional vectors ϕ_i. In this discrete form, the K-L decomposition is often referred to as principle component analysis [13]. In this work we have elected to use this simple discrete approximation for the continuous K-L modes and then interpolate with splines as needed. We remark, however, that a method using Lagrange interpolants does exist for estimating the continuous modes directly from discrete data [14].

The results of a K-L analysis of strain gauge data from the attractors in Figure 2 are shown in Figures 3-4. The 9 gauges are uniformly spaced leaving 0.1 beam lengths at either end, and are numbered starting with the clamped end: the resulting discrete eigenproblem is therefore 9-dimensional with $\phi_i \in \mathbb{R}^9$. The K-L eigenvalues λ_i are plotted in Figure 3, normalized by $\sum \lambda_i$, which is the total mean square amplitude (or "power") of the data vector. As is clearly seen in 3(a), the first K-L mode captures almost all of the power of small amplitude limit cycle. Figures 3(b-d) show that only two K-L modes are needed to capture more

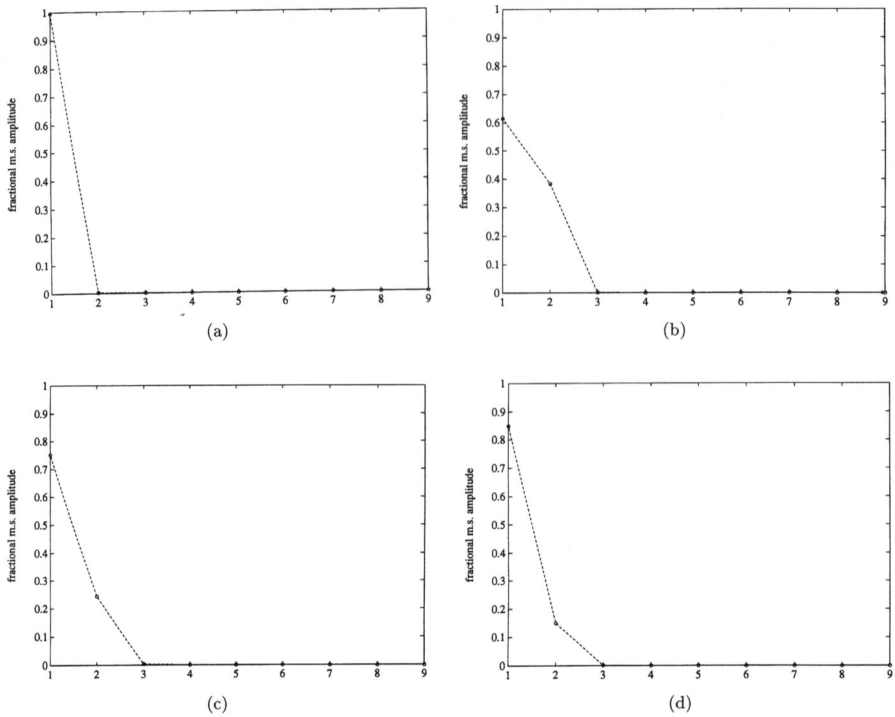

Figure 3: Approximate K-L spectra $\{\lambda_1, \dots, \lambda_9\}$ normalized by $\sum \lambda_i$ for the attractors shown in Figure 2: in all cases, ≤ 2 K-L modes are needed to capture most ($> 99.5\%$) of the mean square amplitude.

than 99.5% of the power in the other dynamical regimes. To rationally identify the total number of active modes, a simple procedure is to take the first integer N for which

$$\frac{\sum_{i=1}^{N} \lambda_i}{\sum_{i=1}^{9} \lambda_i} > C\,, \tag{6}$$

for some constant $C \in (0, 1)$. One might hope to find an N such that $\lambda_i = 0$ for $i > N$, in which case the cutoff criterion (6) would give $C = 1$, and would imply that the observed dynamics was invariant with respect to the N-dimensional subspace spanned by the first N K-L modes. In practice, however the maximum value of C is determined by the scale of noise present in the measurements. In our experiments, the measurement noise was on the order of 1 least significant bit in a 12 bit data acquisition system. Thus the maximum C was in excess of 0.999, and up to 4 K-L modes with amplitude above the noise floor were identified (in the quasiperiodic and chaotic cases). However, in this paper we are primarily concerned for the identification of the dominant modes and their physical interpretion: hence in the remainder of this section, we limit our discussion to the first two modes. It is important to keep in mind, however, that when applying such experimental results to the order reduction of a mathematical model, one needs to consider all empirically significant modes. The problem of obtaining a "closed" model from a reduced set if modes is far from being generally understood. Indeed, it has been shown [15] that modal selection based on amplitude (as implied by a cutoff criterion such as (6)) may not be the best way of identifying modes which interact to contain the

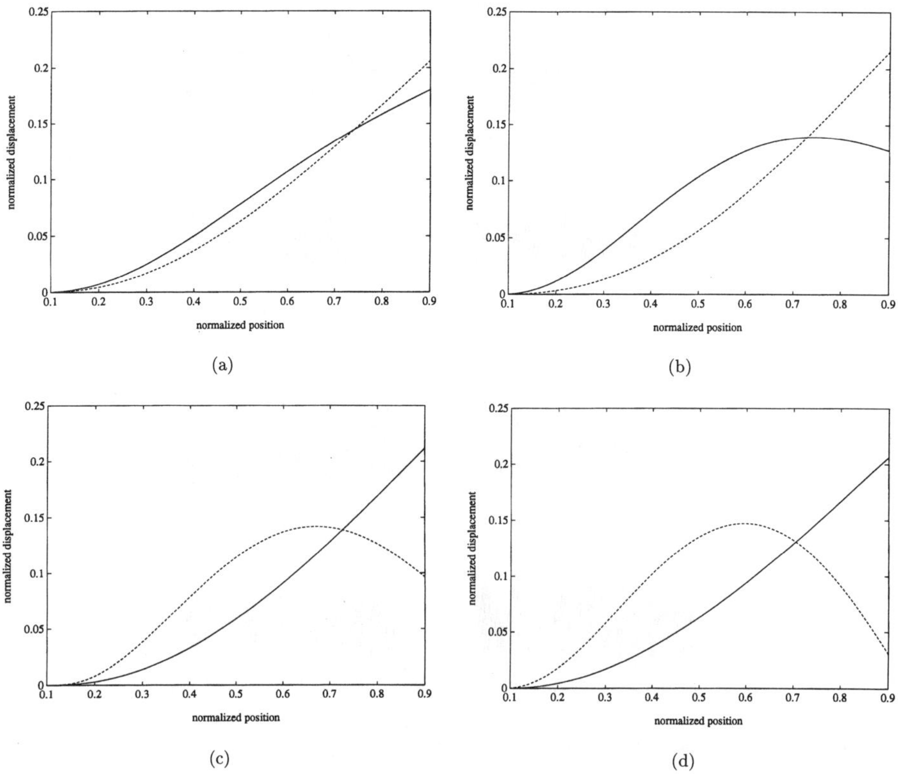

Figure 4: Approximate K-L modes for the attractors shown in Figure 2: (a) Φ_1 (—) compared to the theoretical first linear mode of clamped free beam (- - -); (b-d) Φ_1 (—) and Φ_2 (- - -). All K-L modes have been spline interpolated and spatially integrated twice to give displacements (see text for discussion).

attractor.

For bending vibrations of a beam, the strain at the surface is proportional to the curvature. Thus, for linear elasticity (entirely appropriate in this case), the sensor outputs at position x_i are not proportional to $u(x_i, t)$, but to $u''(x_i, t)$, and the actual K-L modes resulting from the data represent the curvature field along the beam. Rather than plot these "curvature modes", however, it is more physically meaningful to examine the modes in terms of displacements. Therefore, the K-L modes ϕ_i were fit with splines and integrated twice with respect to x using the boundary conditions at the clamped end of the beam ($u(0) = u'(0) = 0$). For convenience, we will denote the integrated K-L modes using upper case symbols, as Φ_i. Note that the Φ_i as presented here are no longer orthogonal, although it would be a simple matter to make them so: we leave the Φ_i in their as-integrated form to aid in the physical interpretation presented below.

In Figure 4(a), the dominant K-L mode for the small amplitude limit cycle is plotted together with the theoretical first mode for a linear clamped-free beam. The additional curvature near the free end of Φ_1 (for normalized x near about 0.6) when compared to the theoretical clamped free linear mode is precisely as expected due to the additional restoring force at the beam tip which results from the interaction of the permanent magnet with the kicker magnet core. The remaining cases are seen in Figures 4(b-d) to involve

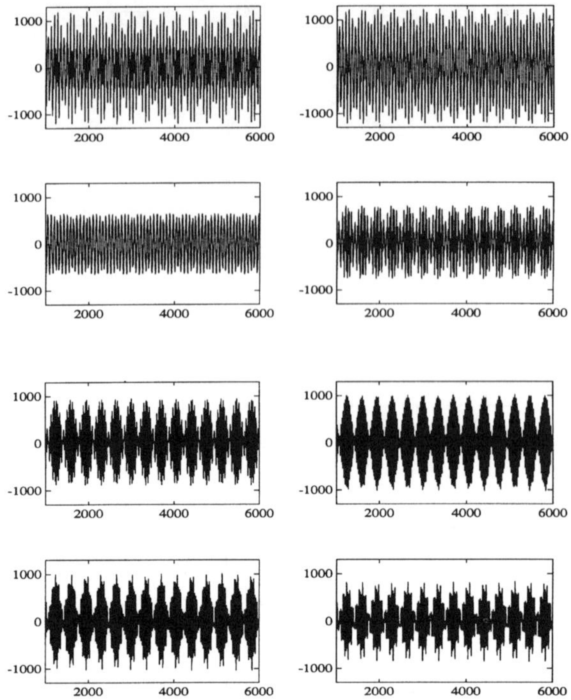

Figure 5: Time series data from first eight strain gauges for quasiperiodic orbit of Figure 2(b).

essentially the same two mode shapes with an interesting physical interpretation.

One shape, which we denote as Σ_S, has significant curvature near the free end (e.g., Φ_1 in Figure 4(b)), and results from relatively small oscillations in which the beam tip spends a significant amount of time near the core of the kicker magnet; the other, denoted as Σ_L (e.g. Φ_2 in Figure 4(b)), is identical to the first linear mode of a clamped free beam and arises from large oscillations in which the beam tip spends little time interacting with the kicker magnet core (and hence, on average, is "free"). It should be noted, however, that the latter shape function *could not* be found with standard linear modal analysis, since it does not satisfy the boundary conditions of oscillations in the small amplitude limit. Hence, despite its similarity to a linear mode of the clamped-free beam, it is a genuine nonlinear feature of the system studied experimentally. The slight differences between the empirical modes in the different cases, particularly Σ_L, are not significant, since any pair of them approximately span any other pair: the small additional error entailed by such projection could, if desired, be easily reduced by including one or more of the additional K-L modes available as desribed previously. The K-L modes obtained in this analysis are consistent with what one observes during chaotic oscillations of the experimental system: the system seems to intermittently hop between small amplitude motions with the tip remaining near the kicker magnet, and large amplitude motions which look like the first linear mode of the clamped-free beam. It is therefore satisfying that the K-L decomposition quantifies this qualitative observation. But K-L analysis does much more, since it indicates that the spatial structure of solutions visible to the eye during chaotic motions is in fact present in the other dynamical regimes. Thus the different regimes can be physically described as involving different partitions

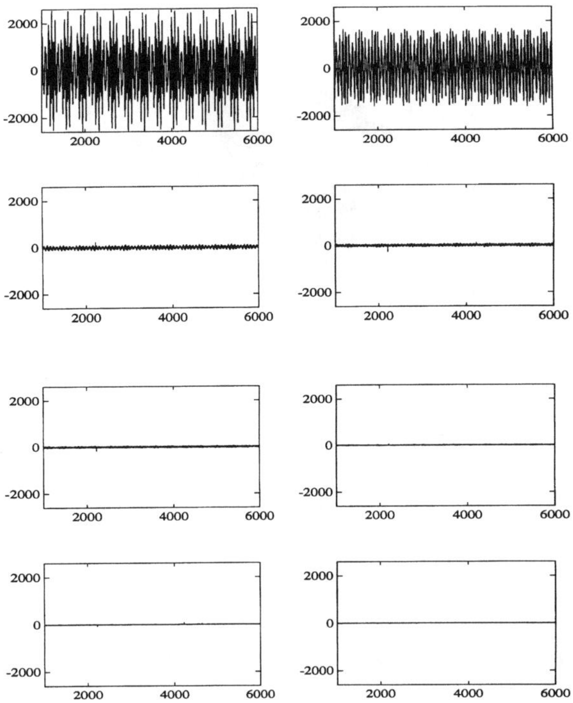

Figure 6: Same data as Figure 5 transformed into K-L coordinates.

of energy between the same two modes. In particular, we see that the transition from the quasiperiodic motion of Figures (2–4)(b) to the chaotic motion of Figures (2–4)(c) entails an exchange in which the the larger amplitude mode in the quasiperiodic case becomes the smaller amplitude mode of the chaotic case: that is, Φ_1 and Φ_2 exchange roles between Figures 4(b) and 4(c).

The K-L modes provide a coordinatization of the configuration space with reduced dimension. This can be seen by transforming the vector strain data $\mathbf{x}$ into K-L coordinates $\mathbf{q}$ with $\mathbf{x} = \Phi\mathbf{q}$, where the matrix Φ has as its ith column ϕ_i. An example of this is shown in Figures 5–6. Figure 5 displays output from the first eight strain gauges for the quasiperiodic signal of Figure 2(b), whereas Figure 6 shows the same data in K-L coordinates. Plotting just the first two of the K-L coordinates, q_1 and q_2, against each other as in Figure 7 shows that they recover the original data of Figure 2(b), as expected from the K-L spectrum. (That the attractor appears as the mirror image of Figure 2(b) is merely a result of the ordering of the K-L modes with respect to amplitude.)

Conclusions

We have applied the Karhunen-Loève (K-L) decomposition to the experimental study of a kicked oscillator. Over the entire range of behaviors observed, two dominant K-L modes were found to interact to generate the the motions of the oscillator, and thus it is reasonable to expect that a mathematical model with as few as two degrees of freedom (four phase space dimensions) would be required to capture the steady-state behavior

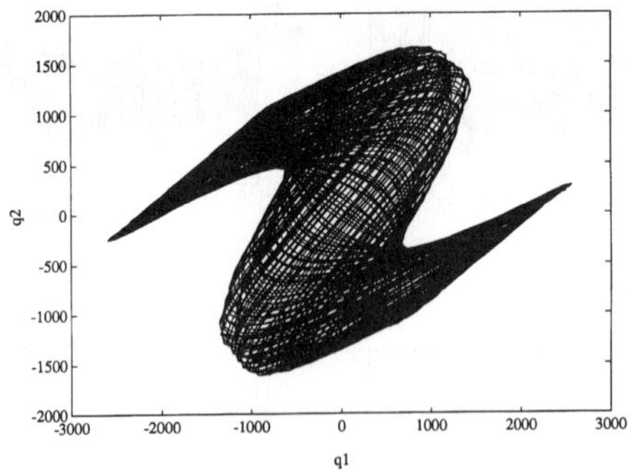

Figure 7: The quasiperiodic orbit of Figure 2(b) plotted in the (q_1, q_2) phase plane.

of the system. (Of course, higher modes, also found with the procedure, would also need to be considered in any general modeling effort for the system.) The analysis of the spatial data from the experiment provides information which is qualitatively superior to that available from delay embedding techniques alone: study of the K-L modes not only provides a dimensionality estimate, but also furnishes physical insight into the nature of the dynamics, and provides a set of coordinates which can be used for reduction of first principles equations.

References

[1] Loève, M. (1963) *Probability Theory*, D. Van Nostrand, Princeton, New Jersey.

[2] Packard, N. H., Crutchfield, J. P., Farmer, J. D., and Shaw, R. S., "Geometry from a Time Series." *Physical Review Letters*, **45**, 712-716.

[3] Takens, F. (1981) "Detecting Strange Attractors in Turbulence" in *Dynamical Systems and Turbulence*, D. A. Rand and L. S. Young, eds., Springer Lecture Notes in Mathematics, **898**, 266-281, Springer-Verlag: New York.

[4] Eckmann, J.-P., and Ruelle, D. (1985) "Ergodic Theory of Chaos and Strange Attractors", *Reviews of Modern Physics*, **57**, 617-656.

[5] Sauer, T., Yorke, J. A., and Casdagli, M. (1991) "Embedology", *Journal of Statistical Physics*, **65**(3,4), 579–616.

[6] Sauer, T. (1993) "Time Series Prediction Using Delay Coordinate Embedding", in *Time Series Prediction: Forcasting the Future and Understanding the Past*, A. S. Weigand and N. A. Gershenfeld, eds., Santa Fe Institute Studies in the Sciences of Complexity, Proceedings Volume XV, 175–193, Addison-Wesley: Reading, MA.

[7] Cusumano, J. P., and Sharkady, M. T. (1995) "An Experimental Study of Bifurcation, Chaos, and Dimensionality in a System Forced Through a Bifurcation Parameter", *Nonlinear Dynamics*, **8**, 467–489.

[8] Lumley, J. L. (1967) "The Structure of Inhomogeneous Turbulent Flows", in *Atmospheric Turbulence and Radio Wave Propagation*, A. M. Yaglom and V. I. Tatarski, eds., Nauka: Moscow, 166–178.

[9] Sirovich, L. (1987) "Turbulence and Dynamics of Coherent Structures, Part I: Coherent Structures", *Quarterly of Applied Math*, **45**, 561–571.

[10] Watanabe, S. (1985) *Pattern Recognition: Human and Mechanical*, Springer-Verlag: New York.

[11] Andrews, H. C., and Hunt, B. R. (1977) *Digital Image Restoration*, Prentice-Hall: Englewood Cliffs, New Jersey.

[12] Burl, J. B. (1993) "Reduced-Order System Identification using the Karhunen-Loeve Transform", *International Journal of Modelling & Simulation*,**13**(4), 183.

[13] Anderson, W. T. (1984) *An Introduction to Multivariate Statistical Analysis*, 2nd ed., Wiley: New York.

[14] Burl, J. B. (1989) "Estimating the Basis Functions of the Karhunen-Loève Transform", *IEEE Transactions on Acoustics, Speech and Signal Processing*, **37**(1), 99–105.

[15] Cusumano, J. P. and Lin, D.-C. (1995) "Bifurcation and Modal Interaction in a Simplified Model of Bending-Torsion Vibrations of the Thin Elastica", *ASME Journal of Vibrations and Acoustics*, **117**(1), 30–42.

7. TURBULENCE

A Model for the Two-Point Velocity Correlation Function in Turbulent Channel Flow

Anupam Sahay and K. R. Sreenivasan
Mason laboratory, Yale University
New Haven, CT 06520-8286

Abstract

A relatively simple analytical expression is presented to approximate the equal-time, two-point, double-velocity correlation function in turbulent channel flow. To assess the accuracy of the model, we perform the spectral decomposition of the integral operator having the model correlation function as its kernel. Comparisons of the empirical eigenvalues and eigenfunctions with those constructed from direct numerical simulations data show good agreement.

1 Introduction

Statistical description has been a major tool for studying turbulent flows. Predictive models are always validated against known statistics obtained from laboratory experiments or, these days, from direct numerical simulations (DNS). There is also the possibility of using statistics for extracting dynamical information of turbulent flows. For these and other reasons it is desirable to have easy access to various flow statistics. Presently, it is very difficult to extract multi-point statistics for full flow domains in laboratory experiments. Direct numerical simulations have started achieving this goal, although only for low Reynolds number flows and at great computational cost. Furthermore, they are not in a convenient analytical form. The purpose of this paper is to provide an example of an approximation of a statistic which can be used to obtain first-cut results. The statistic which we consider is the equal-time, two-point double-velocity correlation function in a turbulent channel flow.

The motivation for this exercise ensued from a very specific need for the correlation function. We have been investigating the possible use of the eigenfunctions of the linearized Navier-Stokes operator as a suitable basis for representing the Reynolds shear stress in turbulent channel flows. Computing the basis expansion of the shear stress requires the knowledge of the full two-point velocity correlation function. More specifically, the computations involve multi-dimensional integrals whose integrands are products of the correlation function and the adjoint eigenfunctions of the linearized Navier-Stokes operator. It seemed

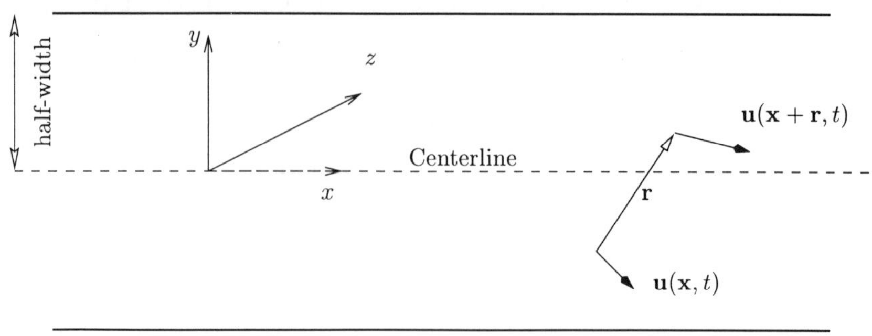

x	streamwise direction	$\mathbf{r} = (r_x, r_y, r_z)$	two-point separation vector
y	wall-normal direction	$U(y)$	mean velocity
z	spanwise direction	$\mathbf{u} = (u, v, w)$	instantaneous turbulent velocity fluctuation

Figure 1: Schematic of channel flow geometry and some preliminary notation.

prudent, in order to evaluate the legitimacy of a surrogate correlation function in this context, that we choose as a test application a similar and well-documented integral operation. Hence we choose to compute the empirical eigenvalues and eigenfunctions to validate our model.

The geometry of the channel flow and some preliminary notation are introduced in figure 1. All quantities are non-dimensionalized using the channel half-width as the length scale and the centerline mean velocity as the velocity scale. Since the flow is stationary in time and homogeneous in the x-z plane, the ensemble average will be taken as the average over all time and the whole x-z plane. The velocity correlation function $\mathsf{S}(y, \mathbf{r})$ is defined as

$$S_{ij}(y, r_x, r_y, r_z) = \langle u_i(\mathbf{x}, t) u_j(\mathbf{x} + \mathbf{r}, t) \rangle, \tag{1}$$

where $\langle \cdot \rangle$ denotes the ensemble average. The normalized correlation $\mathsf{R}(y, \mathbf{r})$ is defined as

$$R_{ij}(y, \mathbf{r}) = \frac{\langle u_i(\mathbf{x}, t) u_j(\mathbf{x} + \mathbf{r}, t) \rangle}{\langle u_i^2(\mathbf{x}, t) \rangle^{1/2} \langle u_j^2(\mathbf{x} + \mathbf{r}, t) \rangle^{1/2}}. \tag{2}$$

Construction of the velocity correlation function is presented in section 2, followed by the results of the empirical eigenfunction computations in section 3. All the major trends of the DNS-obtained eigenvalues and eigenfunctions are reproduced by the model. We conclude with a statement on the potential advantages of our model and on such modeling technique in general.

2 The model

$\mathsf{S}(y, \mathbf{r})$ has to satisfy the following constraints imposed by symmetry, kinematics and boundary conditions:

i) <u>x-z reflectional symmetry</u>

$$S_{ij}(y, r_x, r_y, r_z) = \begin{cases} -S_{ij}(-y, r_x, -r_y, r_z) & \text{if } i \text{ or } j = 2 \text{ and } i \neq j, \\ +S_{ij}(-y, r_x, -r_y, r_z) & \text{otherwise} \end{cases} \tag{3}$$

ii) <u>y-x reflectional symmetry</u>

$$S_{ij}(y, r_x, r_y, r_z) = \begin{cases} -S_{ij}(y, r_x, r_y, -r_z) & \text{if } i \text{ or } j = 3 \text{ and } i \neq j, \\ +S_{ij}(y, r_x, r_y, -r_z) & \text{otherwise} \end{cases} \tag{4}$$

iii) <u>Tensor index symmetry</u>

$$S_{ij}(y, r_x, r_y, r_z) = S_{ji}(y + r_y, -r_x, -r_y, -r_z) \tag{5}$$

iv) <u>Incompressibility</u>

$$\frac{\partial}{\partial r_j} S_{ij}(y, r_x, r_y, r_z) = 0 \tag{6}$$

v) <u>Boundary conditions</u>

$$\lim_{r_x \to \infty} S_{ij}(y, r_x, r_y, r_z) = 0, \tag{7}$$

$$\lim_{r_z \to \infty} S_{ij}(y, r_x, r_y, r_z) = 0, \tag{8}$$

$$S_{ij}(y, r_x, r_y, r_z) = 0 \quad \text{if } |y|{=}1 \text{ or } |y + r_y|{=}1, \ |\cdot| \text{ denoting absolute value.} \tag{9}$$

We have used these constraints along with data from experiments and direct numerical simulation as guidelines for constructing the model. All the modeled quantities will be indicated by an overset tilde ($\tilde{}$). The model construction entails the following steps.

1. $\tilde{S}_{11}$, $\tilde{S}_{22}$, $\tilde{S}_{33}$ and $\tilde{S}_{12}$ are chosen to be of the form

$$\langle u_i(\mathbf{x}, t) u_j(\mathbf{x}, t) \rangle \times \prod_{\alpha = r_x, r_y, r_z} \left[\left(1 - \frac{\alpha^2}{a_\alpha}\right) \exp\left(-\frac{\alpha^2}{s_\alpha (b_\alpha + |\alpha|)}\right) \right] \tag{10}$$

	α	b	so_α	ao_α	sw_α	aw_α	sp_α	ap_α
$\tilde{S}_{11}$	r_x	0.25	0.48	—[a]	0.80	—	-0.22	—
	r_y	0.12	0.18	—	0.50	—	0.90	—
	r_z	0.10	0.32	0.32	0.80	0.50	0.45	1.90
$\tilde{S}_{22}$	r_x	0.15	0.24	0.23	0.01	0.90	0.05	-2.00
	r_y	0.16	0.30	—	0.80	—	0.90	—
	r_z	0.10	0.21	0.35	0.80	0.80	0.56	1.40
$\tilde{S}_{33}$	r_x	0.12	0.30	0.32	0.01	0.90	0.05	-2.00
	r_y	0.12	0.18	0.14	0.50	0.10	0.90	15.0
	r_z	0.10	0.34	0.58	0.80	0.60	0.46	1.80
$\tilde{S}_{12}$	r_x	0.10	0.25	—	0.80	—	0.30	—
	r_y	0.14	0.30	—	0.99	—	0.30	—
	r_z	0.10	0.22	0.10	0.80	0.50	0.46	2.00

[a] A blank signifies that the factor containing the constant in the expression of the correlation function (10) is absent. For example, the blank in the row corresponding to the entries $\tilde{S}_{11}$ and r_x implies that the expression for $\tilde{S}_{11}$ does not contain the factor $(1 - r_x^2/a_{r_x})$.

Table 1: Values of constants in the definition of $\tilde{S}_{11}$, $\tilde{S}_{22}$, $\tilde{S}_{33}$ and $\tilde{S}_{12}$

where

$$s_\alpha = so_\alpha \left[(1 - sw_\alpha y^2)(1 - sw_\alpha (y + r_y)^2) \right]^{sp_\alpha}, \tag{11}$$

$$a_\alpha = ao_\alpha \left[(1 - aw_\alpha y^2)(1 - aw_\alpha (y + r_y)^2) \right]^{ap_\alpha}. \tag{12}$$

The constants ao_α, aw_α, ap_α, b_α, so_α, sw_α, and sp_α depend on the variable-valued α (taking on values r_x, r_y, and r_z) and the tensor index (i, j). The prefactors in (10) are the mean-square velocity fluctuations $\langle u^2 \rangle$, $\langle v^2 \rangle$, $\langle w^2 \rangle$ and the Reynolds shear stress $\langle uv \rangle$. The mean-square velocities have been obtained by fitting curves to the data from Laufer's experiments on turbulent channel flows [1]. The Reynolds shear stress $\langle uv \rangle$ was computed from the relation $Re^{-1}[dU/dy + y(dU/dy)_{y=-1}]$, where the mean velocity $U(y)$ is taken from experiment and Re is the Reynolds number of the flow based on channel half-width and centerline mean velocity. This method of obtaining $\langle uv \rangle$ was preferred to fitting a curve to the measured data because the mean velocity has been measured in much more detail, especially near the wall. For Laufer's flow Re=12,300. The $\langle uv \rangle$ curve and the fitted curves for the mean square velocities The determination of the constants for $\tilde{S}_{11}$, $\tilde{S}_{22}$ and $\tilde{S}_{33}$ has been guided by data on R_{11}, R_{22} and R_{33} from DNS of a turbulent channel flow ([2] for streamwise

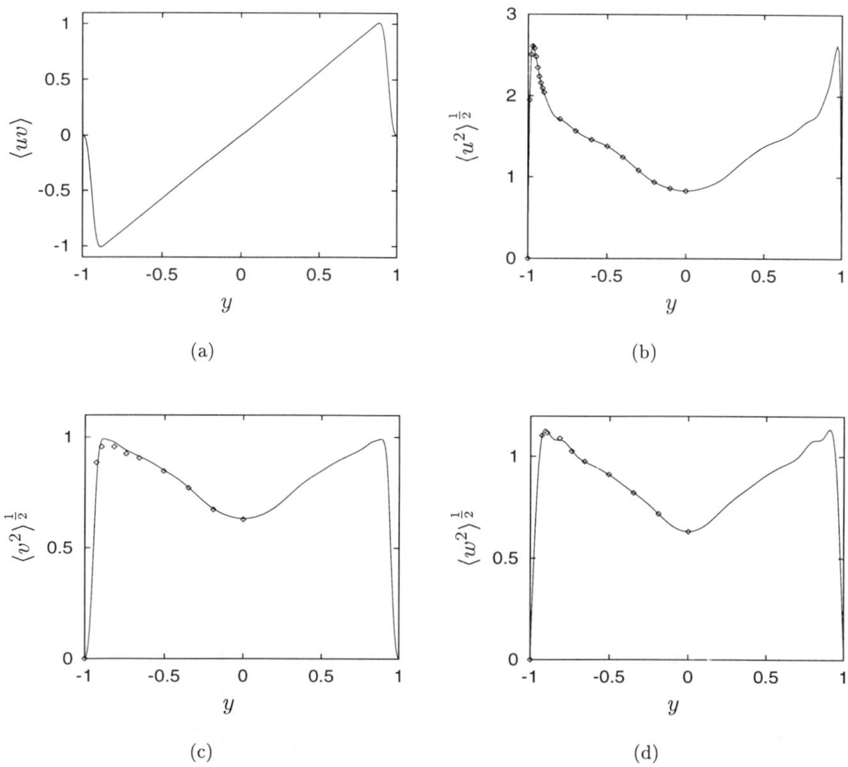

Figure 2: (a) shows the plot of the Reynolds shear stress $\langle uv \rangle$, calculated from the mean velocity distribution. (b), (c) and (d) are plots of the experimental data (points) and the curve fits to the mean square velocities $\langle u^2 \rangle$, $\langle v^2 \rangle$ and $\langle w^2 \rangle$ across the channel.

and spanwise variations and [3] for the wall-normal variation). The Reynolds number Re for the DNS data is 3,200. Due to lack of published data on R_{12} we have taken the corresponding constants to be averages of those of $\tilde{S}_{11}$ and $\tilde{S}_{22}$. are shown in figure 2. The values of the constants are shown in table 1. Figure 3 shows that the model correlation functions $\tilde{R}_{11}$, $\tilde{R}_{22}$ and $\tilde{R}_{33}$ are in good agreement with those obtained from DNS databases.

2. $\tilde{S}_{13}$ and $\tilde{S}_{23}$ are to be obtained, in principle, from $\tilde{S}_{11}$, $\tilde{S}_{22}$ and $\tilde{S}_{12}$ using the incompressibility condition. However the errors in the derivatives of the latter quantities yields $\tilde{S}_{13}$ and $\tilde{S}_{23}$ which violate the r_z boundary condition and the wall boundary

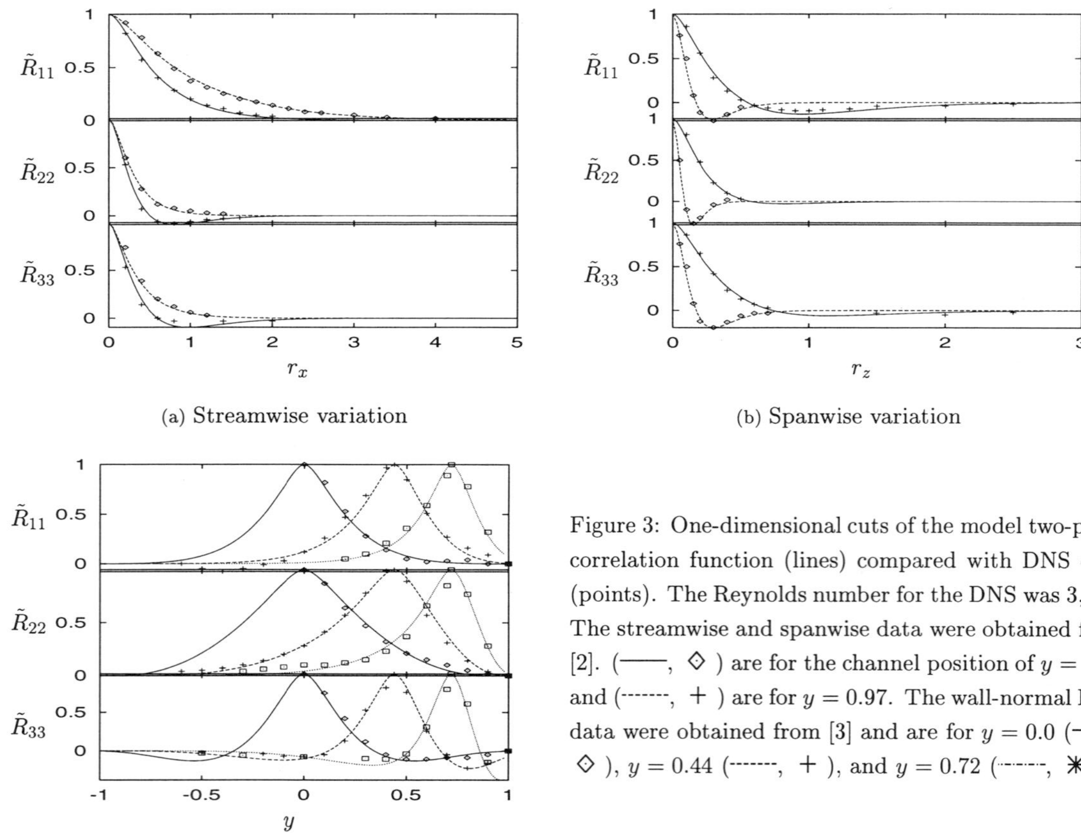

(a) Streamwise variation

(b) Spanwise variation

(c) Wall-normal variation

Figure 3: One-dimensional cuts of the model two-point correlation function (lines) compared with DNS data (points). The Reynolds number for the DNS was 3,200. The streamwise and spanwise data were obtained from [2]. (——, ◇) are for the channel position of $y = 0.17$ and (------, +) are for $y = 0.97$. The wall-normal DNS data were obtained from [3] and are for $y = 0.0$ (——, ◇), $y = 0.44$ (------, +), and $y = 0.72$ (-·-·-·-, ✳).

conditions. To remedy this aberration we multiply the correlation functions, obtained via incompressibility, with filters which enforce the boundary conditions in a smooth fashion. In doing so, we trade an exact relation between the derivatives of the components of the correlation tensor for boundary conditions on the functions themselves. The filters are

$$f_1(y, r_y) = \left(1 - (y + r_y)^8\right)^4 \tag{13}$$

$$f_2(r_z) = 1.0 - \mathrm{erf}\left(|r_z|, 1.0, 1.0\right), \tag{14}$$

where the error function $\mathrm{erf}(x, m, \mu)$ is defined as

$$\int_{-\infty}^{x} \mathrm{d}x' \ \frac{1}{\sqrt{2\pi}\mu} \exp\left(-\frac{(x' - m)^2}{2\mu^2}\right) \tag{15}$$

3. $\tilde{S}_{21}$, $\tilde{S}_{31}$ and $\tilde{S}_{32}$ are obtained from the tensor index symmetry.

The model satisfies all the symmetry and boundary constraints but does not exactly satisfy the incompressibility condition, for the reason just explained.

3 Model validation

Let the error in the approximation of a function be defined as

$$\epsilon(n) = \|\tilde{f}(x) - f(x)\| + \sum_{k=1}^{n} \|\tilde{f}^{(k)}(x) - f^{(k)}(x)\| \tag{16}$$

where $f^{(k)}(x)$ refers to the kth derivative of $f(x)$ and $\| \cdot \|$ represents a suitable norm, say the energy ($\mathbf{L}_2$) norm. The "quality" of approximation would be given by a series of numbers $\epsilon(1), \epsilon(2), \ldots, \epsilon(n)$ with n being determined by the application intended for a model of the function. For an application which involves explicitly only the function, and not its derivatives, a suitable choice would be $n = 1$. In the context of the two-point, double-velocity correlation $\mathsf{S}(y, \mathbf{r})$ such an application is the computation of the empirical eigenfunctions (also known as proper orthogonal decomposition functions or Karhunen-Loeve functions). We validate our model by comparing the empirical eigenfunctions and eigenvalues obtained from $\tilde{\mathsf{S}}(y, \mathbf{r})$ and those from DNS obtained $\mathsf{S}(y, \mathbf{r})$.

In turbulence literature empirical eigenfunctions [4, 5] refer to the eigenfunctions of the integral operator with kernel $\mathsf{K}(\mathbf{x}, \mathbf{x}') = \mathsf{S}(\mathbf{x}, \mathbf{x} - \mathbf{x}')$. If $\phi(\mathbf{x})$ represents the empirical eigenfunctions then

$$\mathsf{K} = \int \mathrm{d}\mathbf{x}' \ \mathsf{K}(\mathbf{x}, \mathbf{x}')\phi(\mathbf{x}') = \lambda\phi(\mathbf{x}), \tag{17}$$

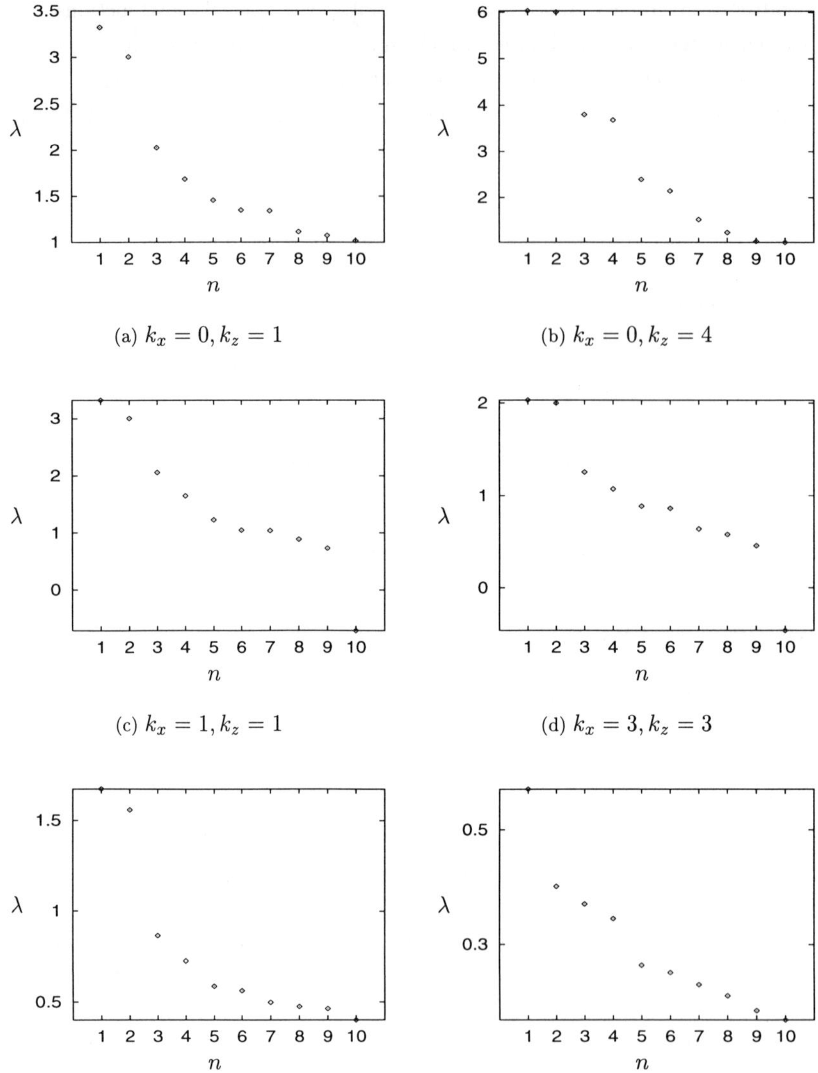

(a) $k_x = 0, k_z = 1$

(b) $k_x = 0, k_z = 4$

(c) $k_x = 1, k_z = 1$

(d) $k_x = 3, k_z = 3$

(e) $k_x = 1, k_z = 0$

(f) $k_x = 4, k_z = 0$

Figure 4: Empirical eigenvalues λ, for the model correlation function $\tilde{\mathrm{S}}(y, \mathbf{r})$, as a function of the index $[k_x, n, k_z]$.

where the integral is over the domain of the flow. The operator K is symmetric and positive. The real positive eigenvalues λ represent the energy associated with each eigenfunction $\boldsymbol{\phi}(\mathbf{x})$.

Making use of the translational symmetry in the x and z directions of the channel, equation (17) can be cast as the following equivalent eigenvalue problem

$$\int_{-1}^{1} \mathrm{d}y' \; \hat{K}_{ij}(y, y', \mathbf{k}) \hat{\phi}_j^*(y', \mathbf{k}) = \lambda(\mathbf{k}) \hat{\phi}_i(y, \mathbf{k}). \tag{18}$$

Here $\mathbf{k} = (k_x, 0, k_z)$ is the wavevector in the x-z plane, $\hat{\mathsf{K}}$ is the Fourier transform of K with respect to $r_x = x' - x$ and $r_z = z' - z$, $\hat{\boldsymbol{\phi}}$ is the Fourier transform of $\boldsymbol{\phi}$ with respect to x and z, and the superscript $(^*)$ represents complex conjugation. The operator $\hat{\mathsf{K}}$, apart from being symmetric and positive, is also square integrable and its spectral properties are given by the standard theory of Hilbert-Schmidt operators [6]. The eigenfunctions form a complete denumerable orthogonal set. In terms of the spectral empirical eigenfunctions (or spectral modes) $\hat{\boldsymbol{\phi}}(y, \mathbf{k}, n)$ a velocity fluctuation field in a channel can be represented as

$$\mathbf{u}(\mathbf{x}, t) = \int \mathrm{d}\mathbf{k} \left\{ \exp(i\mathbf{k} \cdot \mathbf{x}) \sum_{n=1}^{\infty} a(\mathbf{k}, n, t) \hat{\boldsymbol{\phi}}(y, \mathbf{k}, n) \right\} \tag{19}$$

where $\mathbf{k} \cdot \mathbf{x}$ represents the vector dot product between $\mathbf{k}$ and $\mathbf{x}$, n indexes the discrete set of solutions of (18) arranged in order of decreasing $\lambda(\mathbf{k}, n)$ and the coefficients $a(\mathbf{k}, n, t)$ are stochastic in time. We shall refer to a single spectral mode by the index $[k_x, n, k_z]$.

Now we present the results of computations of the spectral eigenvalue problem for the model correlation function $\tilde{\mathsf{S}}(y, \mathbf{r})$. Figure 4 shows the first ten eigenvalues for six different $\mathbf{k}$ values. The fact that all the larger eigenvalues are positive is non-trivial since the non-

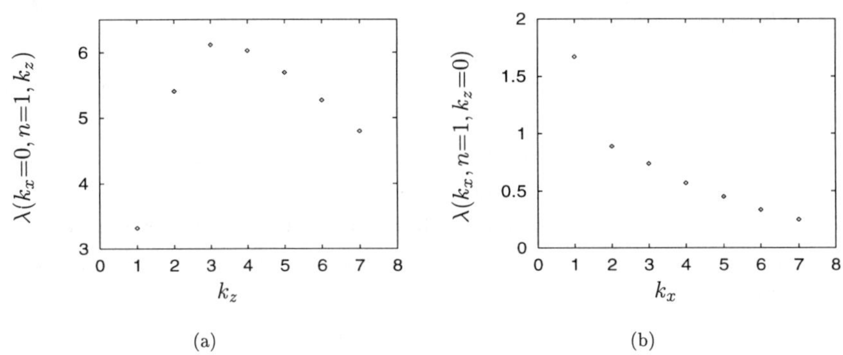

Figure 5: Variation of the largest eigenvalue (i.e. n=1) with the spanwise and streamwise wavenumbers.

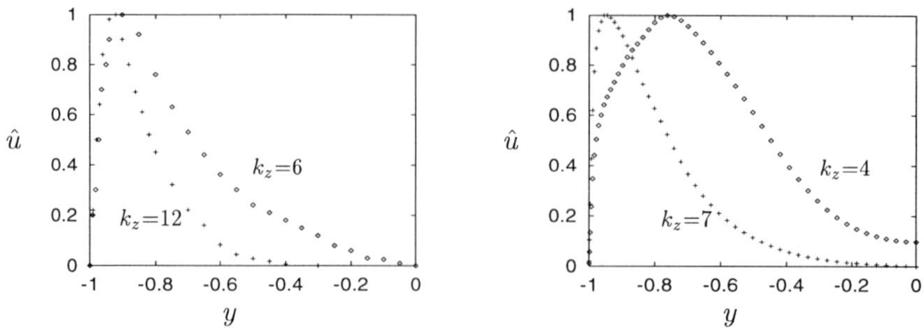

(a) (b)

Figure 6: Comparison of the spanwise u-eigenfunction $\hat{u}(y, k_x{=}0, n{=}1, k_z)$ obtained from (**a**) DNS databases, $k_z{=}6$ and $k_z{=}12$ and (**b**) model correlation function, $k_z{=}4$ and $k_z{=}7$. The spanwise modes are real and hence only one curve suffices to represent $\hat{u}$.

negativity of S is not built *a priori* into $\tilde{\mathsf{S}}$. The magnitude of λ as a function of the index $[k_x, n, k_z]$ follows the general trends observed in the DNS data [7, 8]. As shown in figure 5a the largest eigenvalue for the spanwise modes increases initially with k_z and then decreases. In the case of streamwise modes (figure 5b) the eigenvalues decrease monotonically with k_x. Finally, the shapes of the spectral u-eigenfunction $\hat{u}(y, k_x{=}0, n{=}1, k_z)$ are qualitatively similar to those observed in the DNS studies. Figure 6a is taken from [9] and shows $\hat{u}$ for the DNS data of reference [2]. The eigenfunctions from the model correlation function (figure 6b) reproduce the trends of their DNS counterparts, namely the shifting of the peak towards the wall and shrinking of the support of the eigenfunction as k_z is increased.

4 Conclusion

We have presented a model for the two-point double-velocity correlation function in a channel flow. The model was validated against the measured and DNS-computed components of the correlation function; it was also validated against the empirical eigenfunctions and eigenvalues computed from DNS databases.

At present experiments cannot obtain the full correlation function, which is a tensor and function of four space variables. In the last few years, direct numerical simulations have been able to fill the gap, although only for low Reynolds number flows and at a great computational cost. Even then, they do not provide the needed information in convenient analytical form. The usefulness of the model stems from the simplicity of its closed analytical form. The model can be improved further. Obvious improvements would have to focus on reducing the error in incompressibility condition and the addition of well-documented features like the shift in the maxima away from $\mathbf{r} = 0$ in some of the correlations and the upstream-downstream asymmetries. The effort needed to make such improvements would have to be weighed in the context of the intended application.

Finally, we mention a feature of the model which may be of some significance. It seems plausible that the same functional forms can provide an approximation to $\mathsf{S}(y, \mathbf{r})$ for a range of *Re*. The *Re* variation of $\mathsf{S}(y, \mathbf{r})$ can be viewed as having two components — one being the *Re* variation of the one-point correlations, namely $\langle u^2 \rangle$, $\langle v^2 \rangle$, $\langle w^2 \rangle$ and $\langle uv \rangle$, and the other being *Re* dependence of the normalized correlation $\mathsf{R}(y, \mathbf{r})$. It is possible that the latter is set by the physical mechanisms active in the flow, with the mean square velocities and the Reynolds shear stress determined by *Re*. If so, at least in the range of Reynolds number where the physical mechanisms do not change, the *Re* variation of $\mathsf{S}(y, \mathbf{r})$ can be subsumed exactly by the one-point correlations. These can be measured with relative ease in experiments.

References

[1] J. Laufer, Investigation of turbulent flow in a two-dimensional channel, Technical Report NACA-1053, 1950.

[2] J. Kim, P. Moin, and R. Moser, J. Fluid Mech. **177**, 133 (1987).

[3] J. Kim and A. K. M. F. Hussain, Phys. Fluids A **5**, 695 (1992).

[4] J. L. Lumley, *Stochastic Tools in Turbulence*, Academic Press, New York, 1970.

[5] L. Sirovich, Empirical eigenfunctions and low dimensional systems, in *New Perspectives in Turbulence*, edited by L. Sirovich, pages 139–163, Springer-Verlag, 1989.

[6] T. Kato, *Perturbation Theory for Linear Operators*, Springer-Verlag, Berlin, 1966.

[7] P. Moin and R. D. Moser, J. Fluid Mech. **200**, 471 (1989).

[8] L. Sirovich, K. S. Ball, and L. R. Keefe, Phys. Fluids A **2**, 2217 (1990).

[9] A. C. Poje and J. L. Lumley, J. Fluid Mech. **285**, 349 (1995).

Dynamics and Spectra of Quasi-LagrangianTracers

Oleg A. Druzhinin* and Lev. A. Ostrovsky†

Institute of Applied Physics, Russian Acad. Sci., Nizhni Novgorod 603600, Russia and † University of Colorado, CIRES/NOAA Environmental Technology Laboratory, Boulder, CO 80303, USA

Abstract

Dynamics of particles (tracers) with an effective density different from that of the ambient fluid is considered for the cases of small and large Reynolds numbers. Corrections to the particle energy spectra due to the density difference are found. These corrections may be responsible for the observed high-frequency cut-off in the buoy motion spectra in the ocean

1. Motion of particles in fluid flows is now a subject of a constantly growing amount of publications. One of the most interesting and important classes of such problems is referred to the role of inertial forces for particles with density different from that of the ambient fluid. Even if particle velocity changes only slightly due to these forces, the integral effect can be rather significant. Here we briefly discuss the motion of such a particle (quasi-Lagrangian tracer) which we represent here as a spherical particle with the radius a and density ρ_p slightly different from the fluid density ρ_f). in a fluid flow having a velocity $\mathbf{U}(\mathbf{x}, t)$. Such a motion can be described by the equation [1]

$$\rho_p \frac{d\mathbf{V}}{dt} = \rho_f \frac{D\mathbf{U}}{Dt} + (\rho_p - \rho_f)\,\mathbf{g} - \mathbf{F}_{dr}\,[\mathbf{V} - \mathbf{U}(\mathbf{r}(t), t)]\,, \tag{1}$$

where $\mathbf{r}(t)$ is the particle coordinate, V its velocity, $\mathbf{g}$ is the gravity acceleration and $\mathbf{F}_{dr}$ stands for effective drag force depending on the difference between the particle and fluid velocities. In the case of the low particle Reynolds numbers (when flow around the particle is a creeping Stokes flow), this force is given by

$$\mathbf{F}^{l}_{dr} = \frac{9\rho_f}{2a^2}\,\nu\,(\mathbf{V} - \mathbf{U}), \tag{2}$$

*On leave from the Institute of Applied Physics, Nizhni Novgorod 603600, Russia and in case of large Reynolds numbers, when the drag is caused mainly by the turbulent wake, we have

$$\mathbf{F}^t_{dr} = c\, \rho_f\, a^{-1}\, (\mathbf{V} - \mathbf{U})\, |V - U|, \tag{3}$$

where $c \simeq 0.2$ for a sphere.

Usually in advection problems the flow $\mathbf{U}$ is given and the problem is to find the particle trajectory $\mathbf{r}(t)$ by solving eq.(1) either numerically or, if possible, analytically, and then integrating equation $\dot{\mathbf{r}} = \mathbf{V}$ for a given initial position of the particle. However, in some cases the particle trajectory is given and one needs to retrieve the corresponding flow properties. Further, we shall consider the latter problem in the case of close densities ρ_p and ρ_f when inertia effects are weak.

2. Let us first consider the case of low particle Reynolds number when the drag force is described by (2). It is convenient to introduce dimensionless variables

$$U \to U/U_0, V \to V/U_0, t \to t/T, \tag{4}$$

where U_0 and T correspond to the characteristic velocity and time scale of the flow. The spatial scale of the flow $L = U_0\, T$ is assumed to be much larger than the particle radius a.

Then, equation of particle motion can be rewritten in a dimensionless form as [2]

$$\frac{\epsilon}{\delta}\frac{d\mathbf{V}}{dt} = \epsilon\frac{D\mathbf{U}}{Dt} + \mathbf{g}' - \frac{9}{2}(\mathbf{V} - \mathbf{U}), \tag{5}$$

where $\delta = \rho_f/\rho_p$, $\mathbf{g}' = (\delta^{-1} - 1)\epsilon\mathbf{g}T/U_0$ and parameter

$$\epsilon = \tau_s/T = \frac{U_0\, a^2}{L\, \nu} \tag{6}$$

is considered to be small. Also, notations d/dt and D/Dt are used for the derivatives along the particle and Lagrangian (fluid) trajectories, respectively.

Let us consider the case $\delta \sim 1$ when inertial effects are small compared to the drag. One can seek a solution for the flow velocity in the form of a series

$$\mathbf{U} = \mathbf{U}^{(0)} + \epsilon\, \mathbf{U}^{(1)} + \ldots \tag{7}$$

Substituting (7) into eq. (5) we derive expressions for the flow velocity in various orders of e. In the zeroth order the following solution is obtained:

$$\mathbf{U}^{(0)}(\mathbf{r}, t) = \mathbf{V} - \frac{2}{9}\,\mathbf{g}', \tag{8}$$

In the first order a basic correction caused by the inertial effects is derived in the form

$$\mathbf{U}^{(1)} = \frac{2}{9}\left((\delta^{(-1)} - 1)\frac{d\mathbf{V}}{dt} + \frac{2}{9}\,(\mathbf{g}'\boldsymbol{\nabla})\mathbf{V}\right) \tag{9}$$

It follows from (8) that in the leading order the gravity provides vertical motion of the particle with settling velocity proportional to $\mathbf{g}'$. Inertia corrections are modified by the gravity only if flow velocity $\mathbf{U}$ depends on the vertical coordinate. In many cases the flow may be regarded as two-dimensional, when vertical motion of the particle does not affect its horizontal dynamics which we shall consider below.

Thus, rewriting solutions (7-9) in original variables, we obtain

$$\mathbf{U}(\mathbf{r}(t), t) = \mathbf{V} - \gamma \frac{d\mathbf{V}}{dt}, \tag{10}$$

where the factor γ is given by

$$\gamma = \frac{2a^2}{9\,\nu}\left(1 - \frac{\rho_p}{\rho_f}\right). \tag{11}$$

According to the solution (10), (11), one can define the flow velocity $\mathbf{U}$ at a given point of the particle trajectory. Further we use this solution to relate the dispersion coefficient and energy spectrum for the moving particle with weak inertia to the parameters of a turbulent fluid flow.

2. Effect of inertia on the particle motion can be deduced from the particle velocity spectrum [3]

$$E_{pi}(\omega) = \frac{2}{\pi}\int_0^\infty d\tau\ \cos\omega\tau\ R_{pi}(\tau), \tag{12}$$

where $R_{pi}(\tau)$ is the autocorrelation function of i-th component of the velocity (we assume that $< V >= 0$),

$$R_{pi}(\tau) =< V_i(t)V_i(t+\tau) > . \tag{13}$$

Indeed, in the long-time limit one finds

$$\frac{d}{dt} < r_i^2 >= 2\int_0^t d\tau\ R_{pi}(\tau) \simeq 2\,T_p\ < V_i^2 > \tag{14}$$

for $t \gg T_p$, where

$$T_p = \int_0^\infty d\tau\ R_{pi}(\tau)/\ < V_i^2 > \tag{15}$$

is the particle time integral scale [3], and

$$< V_i^2 >= \int_0^\infty d\omega\ E_{pi}(\omega). \tag{16}$$

Thus, for the particle motion dispersion we have

$$< r_i^2 > \simeq 2\,D_p\,t, \quad D_p = T_p\ < V_i^2 > \tag{17}$$

As it follows from (14) and (15), dispersion coefficient D_p can be expressed via velocity spectrum as

$$D_p = E_p(0)\pi/2 \tag{18}$$

Let us assume that the (Lagrangian) spectrum of the flow motion is given by some function $E_{fi}(w)$. Then, taking the Fourier transform of the eq.(10), we derive the following relation between particle and flow spectra,

$$E_{fi}(w) = E_{pi}(\omega)\ \left(\omega^2\ \gamma^2 + 1\right) \tag{19}$$

From equations(17) and (18) we find that particle and fluid dispersion coefficients are equal, i.e.

$$D_p = D_f \tag{20}$$

This means that in the long-time limit the inertia effects do not affect the dispersion process, which is identical to the fluid particle dispersion. However, one should take into account that values T_p and T_f, as well as $< V_i^2 >$ and $< U_i^2 >$, differ. Indeed, from (16) and (19) we obtain

$$< V_i^2 >= \int_0^\infty d\omega\ \frac{E_{fi}(\omega)}{\omega^2\ \gamma^2 + 1} < \ < U_i^2 >= \int_0^\infty d\omega\ E_{fi}(\omega), \tag{21}$$

and, consequently, $T_p > T_f$. The asymptotics (16) are derived at different times for different density ratios δ which should be always taken into account in calculations of dispersion coefficients.

From eq. (19) we also find that the inertia correction to the particle spectrum does not depend on the sign of γ. Moreover, the inertia causes a cut-off in the particle spectrum $E_{pi}(\omega)$ at frequencies

$$\omega > \omega_c \simeq \gamma^{-1} \tag{22}$$

both for heavy ($\rho_p > \rho_f$) and light ($\rho_p < \rho_f$) particles.

3. Now let us consider the case of large particle Reynolds number, when the drag force which is given by eq. (3), is nonlinear. Given the terms related to the gravity omitted, in dimensionless variables (4) equation of the particle motion takes the form (cf. (5))

$$\frac{\tilde{\epsilon}}{\delta}\frac{d\mathbf{V}}{dt} = \tilde{\epsilon}\ \frac{d\mathbf{U}}{dt} - \ (\mathbf{V} - \mathbf{U})|\mathbf{V} - \mathbf{U}| \tag{23}$$

with parameter $\tilde{\epsilon} = a/cL \ll 1$. Solution of eq. (22) for the flow velocity can be found in the way analogous to the one used above. To the leading order in $\tilde{\epsilon}$ the following expression is obtained,

$$\mathbf{U}(\mathbf{r}(t), t) = \mathbf{V} - \ \tilde{\gamma}\ \frac{d\mathbf{V}}{dt}, \tag{24}$$

with the factor $\tilde{\gamma}$ given by

$$\tilde{\gamma} = \left(\frac{a}{cL}\right)^{1/2} \left(1 - \frac{\rho_p}{\rho_f}\right) \left|\frac{d\mathbf{V}}{dt} \left(1 - \frac{\rho_p}{\rho_f}\right)\right|^{-1/2} \tag{25}$$

Note, that expression (25) is nonlinear in V, so that there is no direct relation between the particle and Lagrangian motions spectra of type (18). Accordingly, corresponding dispersion asymptotics may be rather different from the result (19) found for the linear drag.

4. It is of interest to apply solutions obtained above to the case of a quasi-Lagrangian tracer with realistic parameters. As an example, we consider SOFAR floats which are widely used in geophysical oceanography. These floats are constructed to follow a horizontal flow along the isobaric surfaces and are believed to be nearly passive. However, they can cross isopycnal surfaces and thus ignore up- and downwelling of ambient water [4]. Obviously, vertical motions may cause an effective inertia due to the stratification. Observations of Lagrangian dynamics of such tracers [5] show that the high-frequency asymptotics of the kinetic energy spectrum defined from the float trajectory is usually well described as ω^{-2} (which corresponds to the Lagrangian spectrum of a fluid particle in a δ-correlated random fluid velocity field). However, there exists a cut-off at frequencies $\omega > \omega_c = O\left(10^{-4} sec^{-1}\right)$, the origin of which remained unclear.

According to eq. (22), such a cut-off can be caused by a difference between the tracer and fluid densities. Let us try to estimate this difference for the given value ω_c with the use of solutions discussed above.

Consider a tracer with a radius $a \sim 0.3\ m$ in a flow with characteristic velocity $U \sim 10\ cm/sec$ and spatial scale $L \sim 10\ km$. Let us assume also that the Reynolds number of the tracer

$$Re = |\mathbf{V} - \mathbf{U}|\ a/\ \nu \tag{26}$$

is not too large, so that the viscous drag force can be approximated by the linear Stokes formula (2). Then, the characteristic relative velocity of the float is estimated from solution (10) as

$$|\mathbf{V} - \mathbf{U}| \simeq \Delta V_l \sim \frac{\Delta\rho}{\rho_f}\ \epsilon\ U_0, \tag{27}$$

and, according to (6) with the viscosity $\nu \sim 10^{-2} cm\ sec$, the parameter ϵ is estimated as $\epsilon \sim 0.1$. Thus, for the parameters considered we can use the solution (10). Using the expression (11) for γ and the value $\omega_c \sim 10^{-4} sec^{-1}$, we obtain an effective density difference which could provide a corresponding spectrum cut-off

$$\Delta\rho/\rho \approx 5.10^{-2} \tag{28}$$

with the velocity difference $|\mathbf{V} - \mathbf{U}|$ of the order of 0.5 mm/s, and $Re \sim 10^2$ which seems a moderate enough value for such estimates.

Thus, we find that motion of floating buoys can be affected even by the weak inertia effects due to the density difference (26). Namely, the latter may result in the observed cut-off in the energy spectrum, according to (21). Note also, that the inertia of the float may especially strongly affect its trajectory in the case when the float is located at the flow stagnation point, near the separatrix. In this case the inertia results in the relative velocity of the float which generally may become of the order of the local fluid velocity [2]. This may result in trapping or detrapping of the float into (out off) the localized vortex structures. Thus, the corresponding dispersion characteristics of the float may differ significantly from that of the Lagrangian (fluid) parcels.

If the float's Reynolds number is large, one should use solution (24) to estimate the drift due to inertia. Then, for the corresponding relative velocity we have:

$$\Delta V_t \sim \left(\frac{\Delta\rho}{\rho}\frac{a}{L}\right)^{1/2} U_0. \tag{29}$$

Let us compare characteristic values of the relative velocity ΔV_l found from (27) for the moderate Reynolds number case with value ΔV_t given by (29). Thus,

$$\Delta V_l/\Delta V_t \sim \left(\frac{\Delta\rho}{\rho}\frac{L}{a}\right)^{1/2} \epsilon, \tag{30}$$

where ϵ is given by (6). Upon substituting characteristic values for the flow and tracer scales as $\Delta\rho/\rho \sim 10^{-2}$, $\epsilon \geq 1$, $L/a \sim 10^4$ we find:

$$\Delta V_l/\Delta V_t \gg 1. \tag{31}$$

Therefore, for the same density differences the drift caused by the tracer inertia is more pronounced in the case of moderate Reynolds number then in the case of large Reynolds number. The reason for this follows from the ratio of drag forces given by (2) and (3). For the characteristic scales considered

$$F^t_{dr}/F^l_{dr} \sim U_0 a/\nu \gg 1, \tag{32}$$

so that relaxation of particle velocity to the local fluid velocity is faster for a turbulent drag, which causes smaller values ΔV which is in agreement with (29).

These results are just rough estimates testifying that even weak corrections to the particle motions due to density (or effective density) differences, being cumulated, may provide noticeable deviations of the motion characteristics from those of passive (Lagrangian) tracers. In a more general perspective, such effects can play a significant role in heavy and light particle transport-a problem of an obvious environmental importance.

ACKNOWLEDGMENTS

This work was supported in part by the Russian Foundation of Fundamental Research, and in part by the joint NOAA/DOD Advanced Sensor Application Program.

REFERENCES

1. Soo, S. L., *Fluid Dynamics of Multiphase Systems.* Waltham, MA: Blaisdel, 1967.
2. Druzhinin, O. A., and Ostrovsky, L. A., *Physica D,* **76**, 34-43 (1994).
3. Tennekes, H., "Turbulence: diffusion, statistics, dynamics of the spectra". in *Handbook of Turbulence,* ed. Frost, W., and Moulden, T. H., vol.1, Plenum Press, 1977.
4. Shaw, P.-T., and .Rossby, H. T., *J. Phys. Oceanogr.*, **14**, 529-540 (1984).
5. Osborne, A. R., Kirwan, A. D., Provenzale, A., and L.Bergamasco, L., *Tellus,* 41a, 416-415, (1989).

Turbulent Wall Pressure and Wall Shear Fluctuations Calculated From the Orr-Sommerfeld Equation With Nonlinear Forcing Terms

Theodore R. Anderson

Submarine Electromagnetic Systems Department
Naval Undersea Warfare Center, Newport Division
New London Detachment, New London, Connecticut 06320

Abstract. The wavenumber-frequency spectral densities of turbulent wall pressure fluctuations are investigated over a rigid flat plate. Nonlinear Reynolds stress terms of the inhomogeneous Orr-Sommerfeld equation are regarded as a known forcing function. The forcing function is modeled after Bark's hydrodynamic bursting formulation. The inhomogeneous Orr-Sommerfeld equation is solved by the method of Eckhaus in terms of discrete homogeneous solutions. The method of Eckhaus is then extended and proved for the continuous Orr-Sommerfeld eigenfunctions.

Turbulent wall pressure fluctuations in terms of wavenumber-frequency spectral densities are numerically computed and compared to the experimental results of Martin as well as to his transformation of Blake's data fitted to a modified Corcos model. The wavenumber-frequency spectral densities numerically computed from the discrete eigenfunctions compared well with Martin's transformations on the convective ridge, but the continuous eigenfunctions made insignificant contributions there. However, it is shown that the continuous eigenfunction contributions compare well with the low-wavenumber, high-frequency wavenumber-frequency spectral density measurements of Martin.

INTRODUCTION

The Bark model (1) of hydrodynamic bursting is utilized as a forcing function for the Orr-Sommerfeld equation. It was shown by Landahl (2) that the lowest discrete Orr-Sommerfeld mode, corresponding to the least damped Tollmien-Schlichting wave, makes the largest contribution to turbulent wall pressure spectral densities. Bark, following Landahl, used only the lowest order Orr-Sommerfeld mode to determine the wavenumber-frequency spectral density of turbulent velocity fluctuations.

In this research, the lowest order mode was used to predict the contributions of turbulent wall pressure fluctuations from the discrete Orr-Sommerfeld spectrum. As in Bark's research, the method of Eckhaus (3) was employed to obtain the contribution of discrete Orr-Sommerfeld eigenfunctions to wavenumber-frequency spectral densities. The Eckhaus method was extended and proved for the inhomogeneous solution of the Orr-Sommerfeld equation in terms of continuous eigenfunctions. The inhomogeneous solution was then used to numerically predict the wavenumber-frequency spectral densities from continuous eigenfunctions.

The physical interpretation of bursting presented by Landahl (4), in which a closed cycle of large-scale motions creates conditions for the occurrence of bursting that, in turn, produces new large-scale motions, works well in this model, which uses bursting as a known forcing function.

The entire boundary layer with a mean velocity profile, as it appears explicitly in Landahl (2), was utilized to predict not only the discrete wave effects, but also the large-scale contributions of the continuous waves.

STATEMENT OF THE PROBLEM

The homogeneous solutions of the Orr-Sommerfeld equation were solved through a turbulent boundary layer. The computer code SUPORT by Scott and Watts (5) was used for integration and needed orthonormalization. The code was checked for discrete solutions against other numerical homogeneous solutions (Corner et al. (6), Jordinson (7)) in a Blasius boundary layer with good results. The homogeneous Orr-Sommerfeld discrete solutions were then computed through a turbulent boundary layer, showing close agreement with Landahl's (2) results for phase speeds. After various checks with known previous calculations, the continuous eigenfunctions were solved through a turbulent boundary layer for both spatial and temporal stability.

The inhomogeneous Orr-Sommerfeld equation, as given by Landahl (2), is

$$(U-c)(\phi-K^2\phi)-U''\phi+\frac{i}{\alpha R}\left(\phi^{IV}-2K^2\phi''+K^4\phi\right)=\frac{\hat{q}}{\alpha K}, \tag{1}$$

where

$$q=\nabla^2 T_2-\frac{\partial}{\partial y}(\nabla\bullet T) \tag{2}$$

and

$$T_i=\frac{\partial}{\partial x_j}\left(\overline{u_j u_i}-u_j u_i\right). \tag{3}$$

The terms $\overline{u_j u_i}$ and $u_j u_i$ are the time-averaged and instantaneous Reynolds stress values, respectively.

The turbulent mean velocity profile was assumed to be parallel and, hence, a function of y (distance from the wall) only. The turbulent mean velocity profile was taken from Landahl (2) and is given by

$$\frac{U}{u_\tau}=\frac{1}{K_v}\ln\left(1+K_v y^+\right)+\gamma_v\left\{1-\exp\left(-\frac{y^+}{\delta_v}\right)-\frac{y^+}{\delta_v}\left[\exp\left(-0.33y^+\right)\right]\right\}$$

$$+1.38\left\{1+\sin\left[(2y-1)\pi/2\right]\right\}\ , \tag{4}$$

where

$$y^+=\frac{yu_\tau}{\upsilon}\ . \tag{5}$$

The expression u is the wall friction velocity normalized with the freestream velocity. The constants K, α, and δ_v have the following values:

$$K_v=0.4,\quad \gamma_v=7.4,\quad \delta_v=11.0\,.$$

Landahl chose the friction velocity so as to give the same displacement thickness Reynolds number as in the experiments by Willmarth and Wooldridge (8). Reichart's (9) expression for the wall region and logarithmic region and Coles' (10) law of the wake (the last term) were utilized by Landahl in the construction of his semiempirical velocity profile.

Fourier transforming with the kernel $\exp[-i(\alpha x+\beta z-\omega t)]$, we obtain

$$\hat{q}=\nabla^2\hat{T}_2-\frac{\partial}{\partial y}\left(i\alpha\hat{T}_1+\frac{\partial}{\partial y}\hat{T}_2+i\beta\hat{T}_3\right).$$

To obtain values of $\hat{T}_1$, $\hat{T}_2$, and $\hat{T}_3$, the Bark (1) model of the burst is used. From the measurements by Willmarth and Lu (11) showing that the temporal behavior at a fixed point in space of the fluctuating transport of streamwise momentum perpendicular to the wall resembles a sequence of Dirac delta functions, Bark modeled the regions of high Reynolds stress in a burst as

$$uv=a_1\delta(t)\delta(x)\delta(z)y^3e^{-a_2y^2}\ . \tag{6}$$

Ignoring all other components of u_ju_i and u_ju_i, with the exception of uv and $\overline{uv}$ for lack of experimental data, Bark obtained values of a_1 and a_2 from a least squares fit to the measured data of Kim et al. (12). Bark indicates that instantaneous values of uv as large as 62 $\overline{uv}$ have been experimentally measured (Willmarth and Lu (11)) and that the $\overline{uv}$ term is ignored, leaving only the uv component.

The third power of y, where y is the distance from the wall, is a consequence of continuity and the fact that u varies linearly with y near the wall (Hinze (13) and Landau and Lifshitz (14)). This is seen from

$$\frac{\partial u}{\partial x}+\frac{\partial v}{\partial y}=0\quad (\textit{equation of continuity}).$$

Assuming $u = a(x)y$, the equation of continuity becomes

$$a'(x)y + \frac{\partial v}{\partial y} = 0 => v = -a'(x)y^2/2 + b(x)\ ;$$

hence, uv - y^3.

Bark chose the second degree of y in the exponent since it gave a better fit to experimental data than the first power of y.

Bark then considered the statistics of hydrodynamic bursting. Under the suggestions of Lahey and Kline (15), the statistical properties of the turbulent flow field may be modeled analogously to the "shot noise" current flowing in a resistor connected to a vacuum diode. This shot noise is due to randomly occurring electron emission from the heated cathode.

Assuming that the time between bursts was distributed according to a Poisson process, Bark was able to use the result of Rice (16) stating that the statistics for the response function are given by the response of an individual event. Hence, by Bark's model, the response to the statistics of bursting was reduced to the response to an individual burst.

Employing Bark's model of the burst in the inhomogeneous Orr-Sommerfeld equation with $u_1u_2 = uv$, where u_1u_2 is the only nonzero term in T_i, we obtain for T_i

$$T_1 = \frac{\partial(u_1u_2)}{\partial y} = \delta(t)\delta(x)\delta(z)a_1y^2e^{-a_2y^2}\left(2a_2y^2 - 3\right), \qquad (7a)$$

$$T_2 = -\frac{\partial(u_1u_2)}{\partial x} = -\delta(t)\delta'(x)\delta(z)a_1y^3e^{-a_2y^2}\ , \qquad (7b)$$

$$T_3 = 0. \qquad (7c)$$

Taking the Fourier transforms of the T_i's,

$$\hat{T}_1 = (2\pi)^{-3/2}\int\int\int_{-\infty}^{\infty} T_1 \exp[-i(\alpha x + \beta z - \omega t)]dxdzdt$$

$$= (2\pi)^{-3/2}a_1y^2\exp(-a_2y^2)(2a_2y^2 - 3), \qquad (8a)$$

$$\hat{T}_2 = -i\alpha(2\pi)^{-3/2}a_1y^3\exp(-a_2y^2)\ , \qquad (8b)$$

$$\hat{T}_3 = 0, \qquad (8c)$$

since $T_3 = 0$. Hence,

$$\hat{q} = \hat{\nabla}^2 \hat{T}_2 - \frac{\partial}{\partial y}\left(i\alpha \hat{T}_1 + \frac{\partial}{\partial y}\hat{T}_2 + i\beta \hat{T}_3 \right)$$

$$= i\alpha (2\pi)^{-3/2} a_1 e^{-a_2 y^2} \left[\left(6y - 14 a_2 y^3 + 4 a_2^2 y^5 \right) + K^2 y^3 \right] . \qquad (9)$$

The right-hand side of the inhomogeneous Orr-Sommerfeld equation becomes

$$Q \equiv \frac{\hat{q}}{\alpha K} = i(2\pi)^{-3/2} a_1 y e^{-a_2 y^2} \left[K y^2 + \frac{2}{K}\left(2 a_2^2 y^4 - 7 a_2 y^2 + 3 \right) \right] . \qquad (10)$$

The method of Eckhaus (3) is used to solve the inhomogeneous Orr-Sommerfeld equation by constructing a Green's function for non-self-adjoint operators. The inhomogeneous solution is expanded in terms of a series of homogeneous eigenfunctions. Bark (1) used this method to calculate the inhomogeneous solution for velocity fluctuations with a more restricted velocity profile applicable only to the sublayer and buffer layer.

The inhomogeneous solution expanded in discrete eigenfunctions of the Orr-Sommerfeld equation, as given by Eckhaus (3) for a bounded domain, is given here for a semibounded flow over a flat plate:

$$\phi = \sum_n A_n \phi_n , \qquad (11a)$$

where

$$A_n = \frac{1}{\left(c_n - c \right) N(n,n)} \int_0^\infty Q \tilde{\phi}_n dy , \qquad (11b)$$

with the normalization

$$\int_0^\infty S[\phi_n] \tilde{\phi}_m dy = N(n,m) \delta_{nm} . \qquad (12)$$

The inhomogeneous right-hand side, Q, is modeled in terms of hydrodynamic bursting.

The adjoint eigenfunctions ϕ_n satisfy the homogeneous adjoint Orr-Sommerfeld equation

$$(U-c)(\tilde{\phi}'' - K^2 \tilde{\phi}) + 2U' \tilde{\phi}' + \frac{i}{\alpha R}\left(\tilde{\phi}^{IV} - 2K^2 \tilde{\phi}'' + K^4 \tilde{\phi} \right) = 0 , \qquad (13a)$$

$$\tilde{\phi}(0) = \tilde{\phi}'(0) = \tilde{\phi}(\infty) = \tilde{\phi}'(\infty) = 0 . \qquad (13b)$$

The $\tilde{\phi}_n$'s are eigenfunctions of the homogeneous adjoint Orr-Sommerfeld equation $L[\phi_n] = c_n S[\phi_n]$ and the c_n's are the corresponding phase speed eigenvalues. The mathematical norm squared, $N(n,n)$, is given as

$$N(n,n) = \int_0^\infty S[\phi_n]\tilde{\phi}_n \, dy$$

$$= \int_0^\infty \left[\tilde{\phi}_n' \phi_n' + K^2 \tilde{\phi}_n \phi_n\right] dy \; , \tag{14}$$

as integration by parts shows.

The L and S operators in the inhomogeneous Orr-Sommerfeld equation appear as

$$L[\phi] = cS[\phi] + \frac{\hat{q}}{\alpha K} \, , \tag{15a}$$

where

$$L[\phi] = U(\phi'' - K^2\phi) - U''\phi + \frac{i}{\alpha R}\left(\phi^{IV} - 2K^2\phi'' + K^4\phi\right) \tag{15b}$$

and the self-adjoint S operator is defined by

$$S[\phi] = \phi'' - K^2\phi \, . \tag{15c}$$

The method of Eckhaus (3) was extended to calculate the continuous spectral response to hydrodynamic bursting. The summation over discrete eigenfunctions becomes an integral over continuous eigenfunctions, and the final form of the solution is

$$\phi = \int_0^\infty A_k \phi_k \, dk \, , \tag{16a}$$

where

$$A_k = \frac{1}{(c_k - c)N(k)} \int_0^\infty Q\tilde{\phi}_k \, dy \, . \tag{16b}$$

We get $N(k')$ from the property that the norm for the continuous eigenfunctions is defined by a Dirac delta function unlike a Kronecker delta function for the discrete case (Salwen and Grosch (17)):

$$\int_0^\infty S[\phi_k]\tilde{\phi}_k' \, dy = N(k,k')\delta(k-k'); \qquad (17a)$$

hence,

$$N(k') = \int_0^\infty \int_0^\infty S(\phi_k)\tilde{\phi}_k' \, dy dk. \qquad (17b)$$

More explicitly, *N(k')* can be rewritten as

$$N(k') = \int_0^\infty \int_0^\infty \left(\phi_k'' - K^2\phi_k\right)\tilde{\phi}_k' \, dy dk.$$

The *N(k')* was evaluated numerically in the boundary layer and analytically in the freestream. The numerical value of *N(k')* was negligibly small in the boundary layer due to the small values of ϕ_k, ϕ_k', ϕ_k'' and the corresponding adjoints in the boundary layer. Explicit values of the superposition coefficients of the freestream homogeneous solution

$$\phi = e^{-ik\alpha y} + Be^{ik\alpha y} + Ce^{-\alpha y}$$

(Grosch and Salwen (18)) were found by applying matching conditions at the top of the boundary layer (Drazin (19) and Drazin and Reid (20)) and by exploiting the small magnitude of ϕ_k and ϕ_k' inside the boundary layer.

Finally, the third derivative of the inhomogeneous solution, which enters in the solution to the pressure fluctuations resulting from the continuous spectrum, is obtained by numerically evaluating the integral

$$\phi''' = \int A_k \phi_k'''(y) \, dk.$$

The range of integration is determined by the asymptotic behavior of $A_k\phi_k'''(y)$. For large *k*,

$$(c_k - c) \to k^2\alpha,$$

$$N(k) \to \alpha k^2,$$

the burst integral

$$\int_0^\infty Q\phi_k \, dy \to (\alpha k)^{-1}$$

due to stationary phase effects, and ϕ_k''' goes approximately as $(\alpha k)^3$. Hence,

$$A_k\phi_k''' = \frac{1}{(c_k - c)N(k)}\left(\int_o^\infty Q\tilde{\phi}_k\, dy\right)\phi_k'''(y) \rightarrow \frac{1}{k_2}\,, \tag{18}$$

and the integral converges. However, we must go out far enough in the range of k so that the rapid oscillations in $\tilde{\phi}_k$ cause the integral of the product of the forcing function and the adjoint solution to vanish by stationary phase.

The Fourier-transformed pressure $\tilde{p}$ used in calculating the wavenumber-frequency spectral density can be expressed in terms of inhomogeneous Orr-Sommerfeld solutions, their derivatives, and Fourier-transformed nonlinear stress terms. Landahl (2) has derived this, with the following result:

$$\hat{p} = \frac{\alpha}{k}[(U-c)\phi' - U'\phi] + \frac{i}{kR}\left(\phi''' - k^2\phi\right) + \frac{i}{k^2}\left(\alpha\hat{T}_x + \beta T_z\right). \tag{19}$$

With Bark's model of the burst, $\hat{T}_z = 0$ and $\hat{T}_x = 0$ at $y = 0$. By application of the Orr-Sommerfeld boundary conditions at $y = 0$, $\hat{p}$ reduces to

$$\hat{p} = \frac{1}{kR}\phi'''(y=0). \tag{20}$$

Theoretically, the wavenumber-frequency spectral density for turbulent wall pressure fluctuations for a stationary, ergodic, and homogeneous turbulent field in the x- and z-directions is given as

$$\Phi(K,\omega) = \hat{p}\hat{p}^*\big|_{y=0}\,, \tag{21a}$$

where

$$\hat{p} = \hat{p}(\vec{K},\omega,y) \tag{21b}$$

(see Kinsman (21)).

In terms of the inhomogeneous solutions of the Orr-Sommerfeld equation, the spectral density becomes

$$\Phi(K,\omega) = \hat{p}\hat{p}^* = \frac{\phi'''\phi'''^*}{K^2R^2}\,, \tag{22}$$

where $\phi = \phi_D + \phi_C$, with ϕ_D and ϕ_C being the discrete and continuous inhomogeneous solution to the Orr-Sommerfeld equation

$$\phi_D = \sum_n A_n\phi_n$$

and

$$\phi_C = \int A_k \phi_k \, dk \, .$$

It can be shown that

$$\Phi(k,\omega) = \frac{1}{K^2 R^2}\Big[\left(\mathrm{Re}\,\phi_D'''\right)^2 + \left(\mathrm{Im}\,\phi_D'''\right)^2 + 2\,\mathrm{Re}\,\phi_D'''\,\mathrm{Re}\,\phi_C'''$$

$$+ 2\,\mathrm{Im}\,\phi_D'''\,\mathrm{Im}\,\phi_C''' + \left(\mathrm{Re}\,\phi_C'''\right)^2 + \left(\mathrm{Im}\,\phi_C'''\right)^2\Big] \, . \tag{23}$$

Finally, we can write the discrete, continuous, and cross-term contribution to the wavenumber-frequency spectral density as

$$\Phi(K,\omega) = \Phi_D + \Phi_X + \Phi_C \, , \tag{24}$$

where

$$\Phi_D = \frac{1}{K^2 R^2}\left(\mathrm{Re}\,\phi_D'''\right)^2 + \left(\mathrm{Im}\,\phi_D'''\right)^2 \, , \tag{25a}$$

$$\Phi_X = \frac{2}{K^2 R^2}\left(\mathrm{Re}\,\phi_D'''\right)\left(\mathrm{Re}\,\phi_C'''\right) + \left(\mathrm{Im}\,\phi_D'''\right)\left(\mathrm{Im}\,\phi_C'''\right) , \tag{25b}$$

$$\Phi_C = \frac{1}{K^2 R^2}\left(\mathrm{Re}\,\phi_C'''\right)^2 + \left(\mathrm{Im}\,\phi_C'''\right)^2 \, . \tag{25c}$$

The above terms are numerically evaluated in Tables 1, 2, and 3.

TURBULENT WALL SHEAR FLUCTUATIONS

The components of the fluctuating stress tensor that contribute to the wall shear fluctuations in cartesian coordinates are σ_{xy} and σ_{zy} evaluated at the surface $y = 0$.

The wavenumber-frequency spectral densities for wall shear fluctuations are given by

$$\pi_{xy} = \left.\frac{\hat{\sigma}_{xy}(k,\omega,y)\hat{\sigma}^*_{xy}(k',\omega',y)}{\delta(k-k',\omega-\omega')}\right|_{y=0} \tag{26}$$

and

TABLE 1. Theoretical Values of $\Phi_D(\alpha,\beta,\omega)$ in Decibels and a Comparison With Martin's (22,23) Transformation of Blake's (24) Data*

α	ω	**Theoretical Value of $\Phi_D(\alpha,0,\omega)$ in dB**	**Martin's Results in dB**
0.020	0.135	-6.8	
0.370	0.2505	-37.9	-40
1.00	0.620	-42.9	-50
2.20	1.690	-57.9	-60
6.60	3.790	-59.3	-70
11.49	7.790	-67.88	-80

* $\Phi_D(\alpha,\beta,\omega)$ is given in decibels by evaluating $10\log\{\Phi_D(\alpha_d,\beta_d,\omega_d)/q^2\delta^3/U_\infty\}$, where α_d, β_d, and ω_d are dimensional quantities, $q = \rho U_\infty^2/2$ is the freestream dynamic head, and δ is the scaling in units of the dimensional length. The term ρ is the fluid density, U_∞ is the freestream velocity with dimensional units.
R = 105,000 is scaled on boundary layer thickness. (This Reynolds number is approximately the average Reynolds number that Blake used, and since Martin transformed Blake's data, these results are compared to Martin's results.)
α and ω are scaled on displacement thickness.
$\beta = 0$.

TABLE 2. Comparison of the Wavenumber-Frequency Continuous Spectral Density in Decibels With Martin's (22,23) Low-Wavenumber, High-Frequency Measurements*

α	ω	**Theoretical Value of Φ_C in dB**	**Martin's Results in dB**
		(Scale With Displacement	***Thickness)***
0.185	01.58	-99.2	-94
0.37	1.75	-99.1	-99
0.45	2.21	-101.0	-99
0.47	3.4	-106.0	-105
0.54	1.8	-98.3	-97
0.58	4.82	-109.0	-115

* R = 50,000 is scaled on boundary layer thickness.
ω, α, and Φ_C are scaled on displacement thickness.

TABLE 3. Comparison of the Total Wavenumber-Frequency Spectral Density in Decibels With Martin's (22,23) Low-Wavenumber, High-Frequency Measurements*

α	ω	Theoretical Value of Φ_T in dB	Martin's Results in dB
		(Scale With Displacement	*Thickness)*
0.185	1.58	-68.6	-94
0.37	1.75	-67.8	-99
0.45	2.21	-70.9	-99
0.47	3.4	-75.1	-105
0.54	1.8	-68.4	-97
0.58	4.82	-78.2	-115

* R = 50,000 is scaled on boundary layer thickness.
ω, α, and Φ_T are scaled on displacement thickness.
$\beta = 0$.
$\Phi_T(\alpha,0,\omega) = \Phi_D(\alpha,0,\omega) + \Phi_X(\alpha,0,\omega) + \Phi_C(\alpha,0,\omega)$

$$\pi_{zy} = \left. \frac{\hat{\sigma}_{zy}(k,\omega,y)\hat{\sigma}^*_{zy}(k',\omega',y)}{\delta(k-k',\omega-\omega')} \right|_{y=0} , \tag{27}$$

where $\hat{\sigma}_{ik}$ is the Fourier transform of σ_{ik} using kernel $\exp[i(\alpha x + \beta z - \omega t)]$. The complex conjugate of σ_{ik} is σ^*_{ik}.

The Dirac delta function $\delta(k-k',\omega-\omega')$ appears both in the spectral density of the wall shear fluctuations (response) as well as in the wavenumber-frequency spectral density of the hydrodynamic bursting (source). Multiplying the source and response wavenumber-frequency spectral densities by the Dirac delta function and integrating over all numbers and frequencies yields

$$\tilde{\pi}_{xy} = \hat{\sigma}_{xy}(k,\omega,y)\hat{\sigma}^*_{xy}(k,\omega,y) \tag{28}$$

for the xy-component of the wall shear fluctuations wavenumber-frequency spectral densities. Similarly, we have

$$\tilde{\pi}_{zy} = \hat{\sigma}_{zy}(k,\omega,y)\hat{\sigma}^*_{zy}(k,\omega,y) \ . \tag{29}$$

The xy-component of the Fourier-transformed shear stress σ_{xy} is given as

$$\hat{\sigma}_{xy} = \frac{\partial \hat{v}_x}{\partial y} + \frac{\partial \hat{v}_y}{\partial x} = \frac{\partial \hat{u}}{\partial y} + \frac{\partial \hat{v}}{\partial x} = \frac{\partial \hat{u}}{\partial y}, \tag{30}$$

where $\hat{v}_x = \hat{u}$ and $\hat{v}_y = \hat{v}$.

The y-component of fluctuating velocity $v = v(y)$ is a function of y only and satisfies the inhomogeneous Orr-Sommerfeld equation. For $\hat{u}$, we combine Squire's transformation (25)

$$\alpha\tilde{u} = \alpha\hat{u} + \beta\hat{w} \tag{31a}$$

and

$$\alpha\tilde{w} = \beta\hat{u} - \alpha\hat{w} \tag{31b}$$

with the Fourier-transformed (using kernel $\exp[i(\alpha x + \beta z - \omega t)]$) continuity equation

$$i\alpha\hat{u} + D\hat{v} + i\beta\hat{w} = 0, \tag{32}$$

where $D = d/dy$, to obtain

$$\hat{u} = \frac{i\alpha}{k^2} D\hat{v} + \frac{\alpha\beta}{k^2}\tilde{w} \tag{33a}$$

and

$$\hat{w} = \frac{i\beta}{k^2} D\hat{v} - \frac{\alpha^2}{k^2}\tilde{w}\ . \tag{33b}$$

This technique was utilized by Bark (1) to obtain u. To calculate velocity spectral densities in the x-direction, Bark further showed the following:

$$\tilde{w} = L^{-1}\left[\left(\frac{i\beta}{\alpha}\right)(DU)\hat{v}\right] + \tilde{w}_H, \tag{34}$$

where U is the mean velocity profile given by Landahl (2). The operator L is defined by

$$L[\tilde{w}_H] = \left[(U - c) + \frac{i}{\alpha R}\left(D^2 - k^2\right)\right]\tilde{w}_H = 0. \tag{35}$$

When u is obtained from equations (33a), (34), and (36), the wavenumber-frequency spectral density $\tilde{\pi}_{xy}$ can be obtained from equations (28) and (30). The y-component of fluctuating velocity $\hat{v}$ satisfies the inhomogeneous Orr-Sommerfeld equation, with the solution given as an expansion in discrete and continuous Orr-

Sommerfeld eigenfunctions. The Orr-Sommerfeld solution ϕ is related to $\hat{v}$ by $\hat{v} = -ik\phi(y)$.

The zy-component of shear stress is given by

$$\hat{\sigma}_{zy} = \frac{\partial \hat{w}}{\partial y} + \frac{\partial \hat{v}}{\partial z} = \frac{\partial \hat{w}}{\partial y}, \tag{36}$$

where $\hat{v} = \hat{v}(y)$ and $\partial\hat{v} / \partial z = 0$.

The z-component of the fluctuating velocity $\hat{w}$ is obtained from equation (33b). The z-component of fluctuating velocity, as $\hat{u}$ and $\hat{v}$, will be expressed as an expansion in Orr-Sommerfeld discrete and continuous eigenfunctions. The wavenumber-frequency spectral density $\tilde{\pi}_{zy}$ is now obtained from equations (29) and (36).

The final forms of the wavenumber-frequency spectral density will be

$$\tilde{\pi}_{xy} = \tilde{\pi}_{xyD} + \tilde{\pi}_{xyX} + \tilde{\pi}_{xyC} \tag{37a}$$

and

$$\tilde{\pi}_{zy} = \tilde{\pi}_{zyD} + \tilde{\pi}_{yxY} + \tilde{\pi}_{zyC}\,, \tag{37b}$$

where the subscripts D, X, and C designate the contributions of the wavenumber-frequency wall shear fluctuations from discrete Orr-Sommerfeld eigenfunctions, crossproducts of discrete and continuous Orr-Sommerfeld eigenfunctions, and continuous Orr-Sommerfeld eigenfunctions, respectively.

Although no numerical results are given in this section, the techniques involved are similar to those already described.

CONCLUSIONS

The wavenumber-frequency spectral density predicted from continuous Orr-Sommerfeld eigenfunctions compared well with Martin's low-wavenumber, high-frequency measurements.

The discrete wavenumber-frequency spectral density (10 log $\Phi_D(\alpha, 0, \omega)$ along the convective ridge (where it dominates) predicts Martin's transformation of Blake's data fitted to a modified Corcos model (26) up to about the -70 dB contour on Martin's figure. Here it starts to deviate, with results about 10 dB higher than Martin's results. This could be due to the Bark's Dirac delta function model of the burst in space time, which gives a forcing function that is flatter than it should be in wavenumber-frequency space. Thus, as we get farther from the peak contour, the forcing function based on this model does not fall off fast enough, and the results begin to deviate from Martin's results.

The predicted results of 10 log $\Phi_D(\alpha, 0, \omega)$ are also higher on the convective ridge than Martin's transformation. The Bark model of the burst that was used in this research utilizes Poisson statistics in time. This results in a most probable time between bursts of zero with the consequence of too much bursting and, hence, too much momentum transferred to the wall. Thus, higher values of the theoretical discrete component of the values of the wavenumber-frequency spectral density are predicted than are measured.

The continuous component (10 log $\Phi_C(\alpha, 0, \omega)$) of the wavenumber-frequency spectral density agreed well with Martin's low-wavenumber measurements. However, the total wavenumber-frequency spectral density (10 log $\Phi_T(\alpha, 0, \omega)$) overpredicted Martin's low-wavenumber measurements. This is due to the large contribution of the discrete component of the spectral density into this region. Further work is needed to separate the continuous wave from the discrete wave contributions to the wavenumber-frequency spectral density.

ACKNOWLEDGMENTS

The author is indebted to Professor William Saric for his instruction on the use of the SUPORT code in solving the homogeneous Orr-Sommerfeld equation.

REFERENCES

1. Bark, F., *Journal of Fluid Mechanics* **70(2)**, 229–250 (1975).

2. Landahl, M. T., *Journal of Fluid Mechanics* **29(3)**, 441–459 (1967).

3. Eckhaus, W., *Studies in Non-linear Stability Theory*, Berlin: Springer-Verlag, 1965.

4. Landahl, M. T., *Proceedings of the Thirteenth International Congress of Theoretical and Applied Mechanics*, Moscow: Springer-Verlag, 1972, p. 177.

5. Scott, M. R., and Watts, H. A., "SUPORT, A Computer Code for Two-Point Boundary-Value Problems Via Orthonormalization," Applied Math Division 2646, Sandia Laboratories, Albuquerque, New Mexico, 1975.

6. Corner, D., Houston, D. J. R., and Ross, M. A., *Journal of Fluid Mechanics* **77(1)**, 81–103 (1976).

7. Jordinson, R., *The Physics of Fluids, Research Notes* **14(11)**, (1971).

8. Willmarth, W. W., and Wooldridge, C. E., *Journal of Fluid Mechanics* **14**, 187 (1962).

9. Reichart, H., *Z. Anger. Math. Mech.*, **31.208** (1951).

10. Coles, D., *Journal of Fluid Mechanics* **1**, 191 (1956).

11. Willmarth, W. W., and Lu, S. S., *Journal of Fluid Mechanics*, **65**, 55, (1972).

12. Kim, H. T., Kline, S. J., and Reynolds, W. C., *Journal of Fluid Mechanics* **50**, 133 (1971).

13. Hinze, J. O., *Turbulence*, New York: McGraw Hill, 1959.

14. Landau, L. D., and Lifshitz, E. M., *Fluid Mechanics*, Reading, Massachusetts: Pergamon Press, Inc., and Addison-Wesley Publishing Company, Inc., 1959, pp. 13, 121, 162.

15. Lahey, R. T., and Kline, S. J., *Report MD-26*, Themosc Division, Department of

Mechanical Engineering, Stanford University, 1971.

16. Rice, S. O., *Bell System Technical Journal* **23–24** (1944).

17. Salwen, H., and Grosch, C. E., *Journal of Fluid Mechanics* **104**, 445–465 (1981).

18. Grosch, C. E., and Salwen, H., *Journal of Fluid Mechanics* **87(1)**, 33–54 (1978).

19. Drazin, P. G., *Journal of Fluid Mechanics* **10**, 571–583 (1961).

20. Drazin, P. G., and Reid, W. H., *Hydrodynamic Stability,* New York: Cambridge University Press, 1984, p. 201.

21. Kinsman, B., *Wind Waves*, Englewood Cliffs, New Jersey: Prentice-Hall, Inc., 1967.

22. Martin, N. C., "Wavenumber Filtering by Mechanical Structures," Ph.D Thesis, Massachusetts Institute of Technology, Cambridge, Massachusetts, 1976.

23. Martin, N. C., and Leehey, P., *Journal of Sound and Vibration* **52(1)**, 95–120 (1977).

24. Blake, W., *Journal of Fluid Mechanics* **44(4)**, 637–660 (1967).

25. Squire, H. B., *Proceedings of the Royal Society A,* 1933, p. 621.

26. Corcos, G. M., *Journal of Sound and Vibration* **6**, 59–70 (1967).

Evolution of Frequency Content in Fluid Flow Noise

Patricia H. Carter

Nonlinear Dynamics and Wavelets Group
Naval Surface Warfare Center, White Oak
Silver Spring, Maryland 20903-5640

ABSTRACT

Time-frequency and time-scale techniques are appropriate to the analysis of signals which are broadband and highly non-stationary. These techniques are used to analyze unique features of noise data from a transient pipe flow experiment.

INTRODUCTION

The short time Fourier (or the Gabor) transform is a fundamental tool for the analysis of measurement data from processes in which frequency content is a natural characteristic. The mean velocity of the fluid in these experiments goes from zero to fully turbulent: therefore non-stationarity of the data is a fundamental attribute. The physics of fluid flow implies that the frequency content will vary with the mean velocity: as that velocity grows so does the total noise power, the relative power in high frequencies and the complexity of the power spectrum. The frequency content will also vary with the pipe geometry; a more complicated geometry producing more turbulent flow. In this experiment the fluid flow is coupled with the vibration of the pipe so it is useful to analyze flow noise together with pipe acceleration.

ANALYSIS TECHNIQUES

The frequency content of the signals can be characterized both globally, over the whole transient event, and locally by windowing. Suppose w is a window function of width d centered at time b, and s is the measured transient data time series. Then s(t)w(t-b) is that time series localized at b. Suppose **Q** is an operator, such as the variance, then

$$C = Q\,(s(t))$$

is a global characteristic of the time series, and

$$C(b) = Q\,(s(t)\,w(t-b))$$

is the corresponding local characteristic function.

Both the short time Fourier and the Gabor transform give a picture of the power as a function of time and frequency. The short time Fourier transform using window function w is

$$T(b,\gamma) = \int s(t)\,\overline{w(t-b)}\,e^{-2\pi i \gamma (t-b)}\,dt$$

where the overbar denotes complex conjugate. When γ is an integer and w is the Gaussian window this is the Gabor transform as given in [1]. In analogy to the Fourier spectrum, the time dependent power spectrum is

$$P(b,\gamma) = |T(b,\gamma)|^2$$

The particular transform data analyzed here has a large dynamic range so that it is reasonable to display the logarithm of the power, rather than the power itself. The log(P) is given as a *binned* function of (b, log(γ)). With log(γ) as an independent variable this is very much like the Littlewood-Paley wavelet [2], which is defined in the Fourier domain by

$$\phi(\gamma) = \begin{matrix} 1 & 2^i \le \gamma < 2^{i+1} \\ 0 & \textit{otherwise} \end{matrix}$$

If the operator **Q** in the paragraph above is the power in a given frequency range, then the resulting C is essentially a section of the binned log(P).

Fourier complexity is a measure of the concentration of the power spectrum - if all of the power is in a single frequency the complexity is 1, if it is equally spread out among n frequencies the complexity is n. Suppose ψ is the Fourier transform of a data series of length N then define the complexity

$$C = \sum_1^{N/2} |\psi(\gamma)|^e \,/\, \sum_1^{N/2} |\psi(\gamma)|^{2e}$$

where e is a positive parameter.

THE EXPERIMENT

The internal launchway flow experiment was conducted by Scott Hassan of the Naval Underwater Warfare Center, Newport. The facility was the Transient Flow Loop which is designed for the evaluation of transient hydrodynamic and hydroacoustic phenomena. This loop can be configured in two different flow geometries. In the baseline configuration a smooth 12" pipe section transitions to a 6" pipe section via conical collar; in the other the 6" pipe section is covered with a hemispherical cap, four symmetric slots are cut below the cap, and the 6" section is inserted into the 12" section, as the diagram in Figure 1 shows. Two different methods can be used to control the fluid flow mean velocity: motor driving and valve-controlled. These result in different fluid velocity time histories. The data was recorded under transient conditions, during which the mean fluid velocity was raised from 0 to a maximum of about 50 ft/sec over 4 seconds. The sampling rate was 2^{15} Hz. The data for each experimental run consists of three synchronous measurements: 1) the mean fluid velocity; 2) a hydroacoustic (pressure) measurement; and 3) an accelerometer measurement. The instrumentation for the accelerometer and pressure measurements is located at the downstream end of the transition section of the pipe. The data discussed below is from several runs of each of the four configurations of the experiment.

An intriguing aspect of this data is the appearance of simultaneous bursts in the hydroacoustic and the acceleration data during the last two seconds of the runs. It has been conjectured that these short, high frequency transients are due to vortex shedding events in the flow. These events have been observed in videos of steady state flows at comparable flow velocities, but a direct correspondence has not been established. The appropriate technique to detect and estimate the bursts is thresholding in the wavelet domain. An analysis using wavelet thresholding appears in [3].

DATA ANALYSIS

First we compute several scalar attributes based on frequency content. In almost all cases these, as individual features, suffice to distinguish between data from the four different experimental configurations. Here total power refers to the power in the de-meaned signals in arbitrary units. The low frequencies are .25 to 32 Hz and the high frequencies are 32 to 32768 Hz. The Fourier complexity is averaged over $e \in \{.75, 1., 1.25, 1.5, 1.75, 2.\}$ The total power, high frequency power and the complexity differences between the slot flow noise and the baseline flow noise are substantial enough to provide a robust classification, and they are consistent with the physics. The only feature that fails is the total power fails to distinguish between the valve driven and the baseline hydroacoustic data. Results are reported in the tables for three data series for every type except the motor-driven, slot flow accelerometer data for which there are only two.

Table 1. Global Observables of Baseline Hydroacoustic Data

power		power high freqs		high /low power		complexity	
valve	*motor*	*valve*	*motor*	*valve*	*motor*	*valve*	*motor*
1.67	1.65	.186	.210	.126	.145	1702.	1833.
1.82	1.73	.194	.214	.119	.142	1642.	1832.
1.80	1.55	.186	.219	.115	.164	1619.	1930.

Table 2. Global Observables of Slot Hydroacoustic Data

power		power high freqs		high /low power		complexity	
valve	*motor*	*valve*	*motor*	*valve*	*motor*	*valve*	*motor*
4.02	4.99	2.61	3.36	1.85	2.07	5064.	5249.
4.08	4.87	2.54	3.52	1.65	2.60	4813.	5565.
4.15	5.21	2.63	3.59	1.73	2.22	4868.	5243.

Table 3. Global Observables of Baseline Accelerometer Data

power		power high freqs		high /low power		complexity	
valve	*motor*	*valve*	*motor*	*valve*	*motor*	*valve*	*motor*
.014	.054	.008	.046	.710	.193	10920.	7211.
.018	.050	.011	.041	.555	.237	10005.	7854.
.019	---	.012	---	.514	---	9921.	---

Table 4. Global Observables of Slot Accelerometer Data

power		power high freqs		high /low power		complexity	
valve	*motor*	*valve*	*motor*	*valve*	*motor*	*valve*	*motor*
39.2	55.4	38.8	53.2	104.	24.6	36147.	27243.
36.7	56.7	36.3	54.5	88.6	25.4	35891.	27390.
38.1	56.4	37.7	54.1	86.1	23.2	35614.	27074.

Although the global attributes do suffice to discriminate between data from the different flow configurations, they are not sufficient to describe the data as dynamical processes, precisely because they ignore the intrinsic non-stationarity. Given that a fundamental aspect of these time series is their development as the mean velocity goes from zero to its maximum, the goal is to describe how the frequency content evolves as the mean velocity increases.

To obtain the localized attributes and the short time Fourier transform we use a Welsh (quadratic) window of .5 seconds and scan the data in 64 overlapping steps. The mean flow velocity is smoothed by using a spline interpolation of binned velocity measurements. The frequency content of the data time series is a function of this velocity and its derivative.

Figures 1 and 2 give plots of hydroacoustic and accelerometer measurements for each of the four experiment configurations. In Figures 3 and 4 are plotted the graphs of the windowed low and high frequency content of these data sets. The high frequency power tracks the mean velocity and the low frequency power appears to track the derivative of the velocity. From the contour plots of the normalized short time Fourier transforms in Figures 5 and 6 many differences between the data sets are apparent. For example, in the slot experiment the pipe accelerations have a sudden transition from not very much power, most of it in the low frequencies, to a lot of power, most of it in the high frequencies. In the baseline experiment pipe acceleration data the power is more spread out over time and frequency, and there are three frequency regimes of interest - there is substantial power in a mid-frequency band, more prominent in the valve controlled case. Figure 7 shows the Fourier complexity - in all cases it grows with time, and for the slot experiment it grows faster and to a higher level.

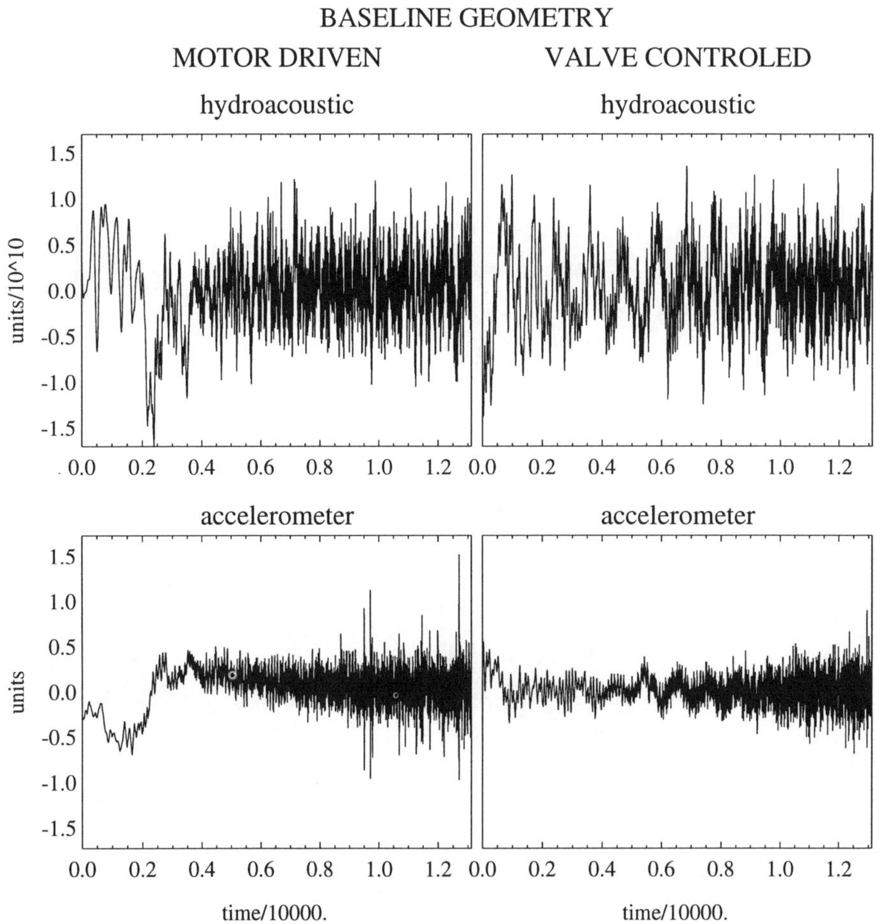

FIGURE 1. Examples of measured data from the baseline experiment configuration, in arbitrary units. The hydroacoustic data and the accelerometer data below it are from synchronous measurements taken during the same run.

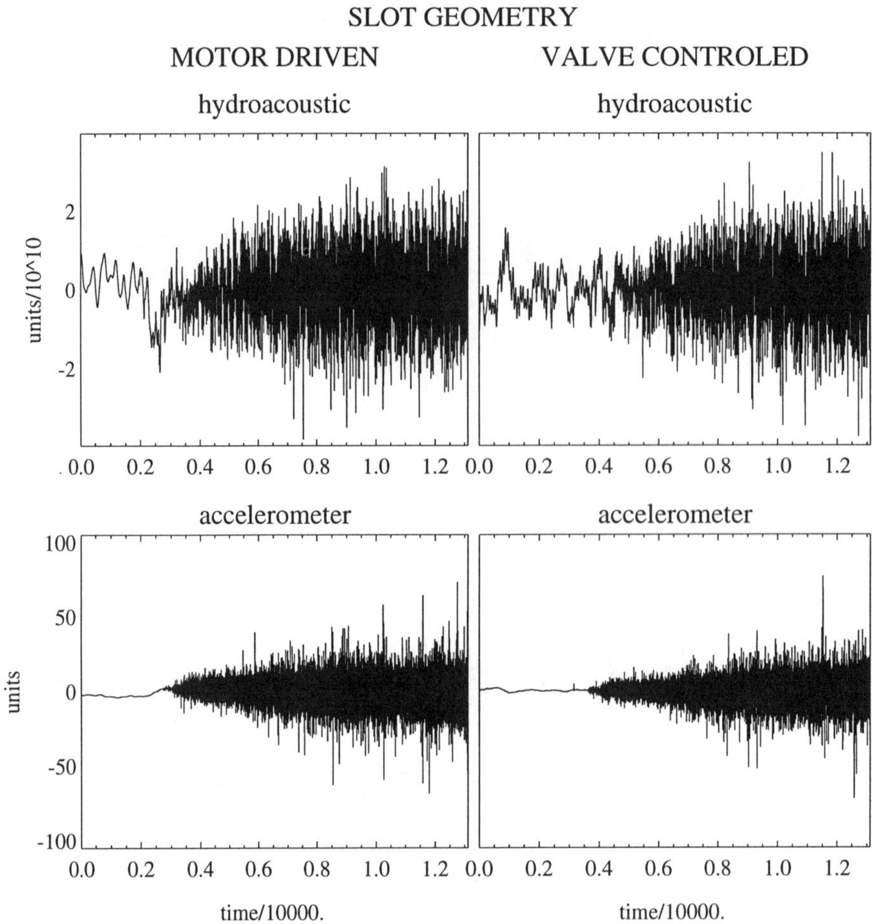

FIGURE 2. Examples of measured data from the slot experiment configuration, in arbitrary units. The hydroacoustic data and the accelerometer data below it are from synchronous measurements taken during the same run.

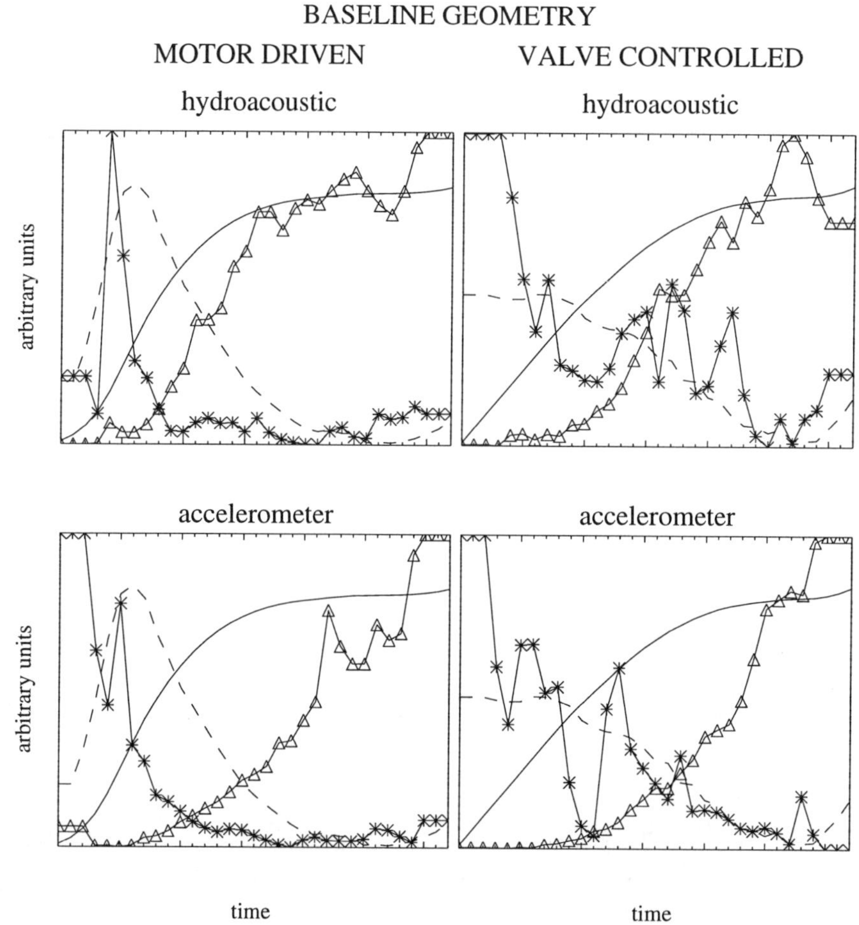

FIGURE 3. The localized power for the baseline data shown in Figure 1. In each plot there are graphs of the (smoothed) mean velocity, its derivative, the localized power in the high frequencies (128 to 32768 Hz) △△△△△ and the localized power in the low frequencies (2 to 127 Hz)**** . Each plot is scaled independently.

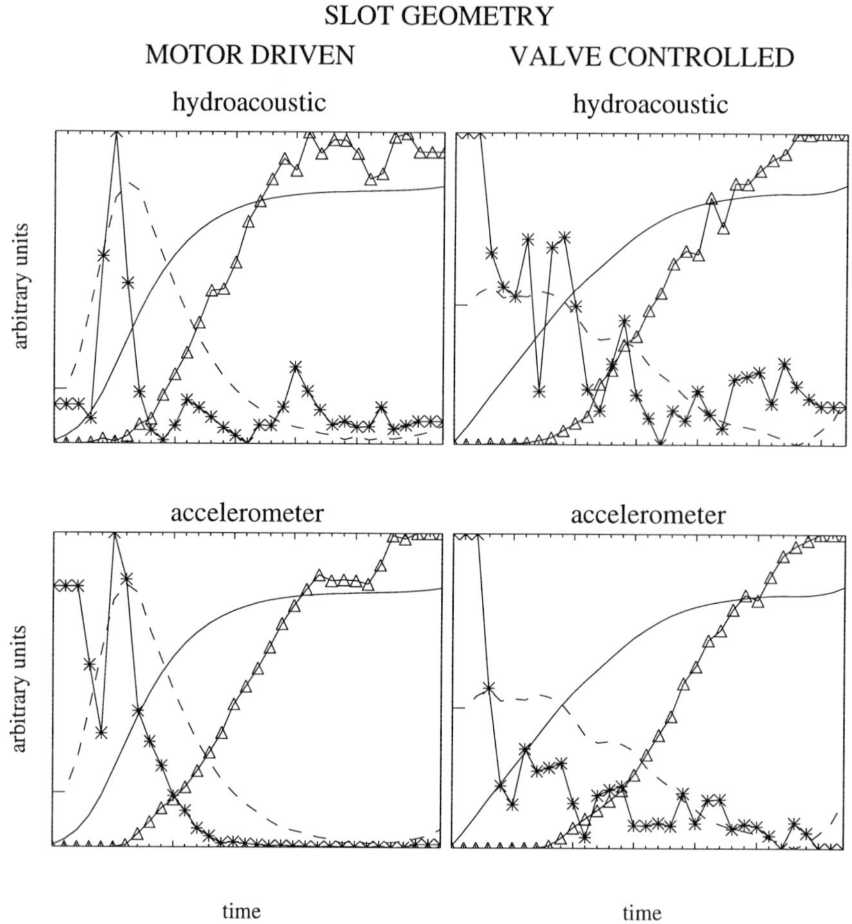

FIGURE 4. The localized power for the slot data shown in Figure 2. In each plot there are graphs of the (smoothed) mean velocity, its derivative, the localized power in the high frequencies (128 to 32768 Hz) ΔΔΔΔΔ and the localized power in the low frequencies (2 to 127 Hz) ****. Each plot is scaled independently.

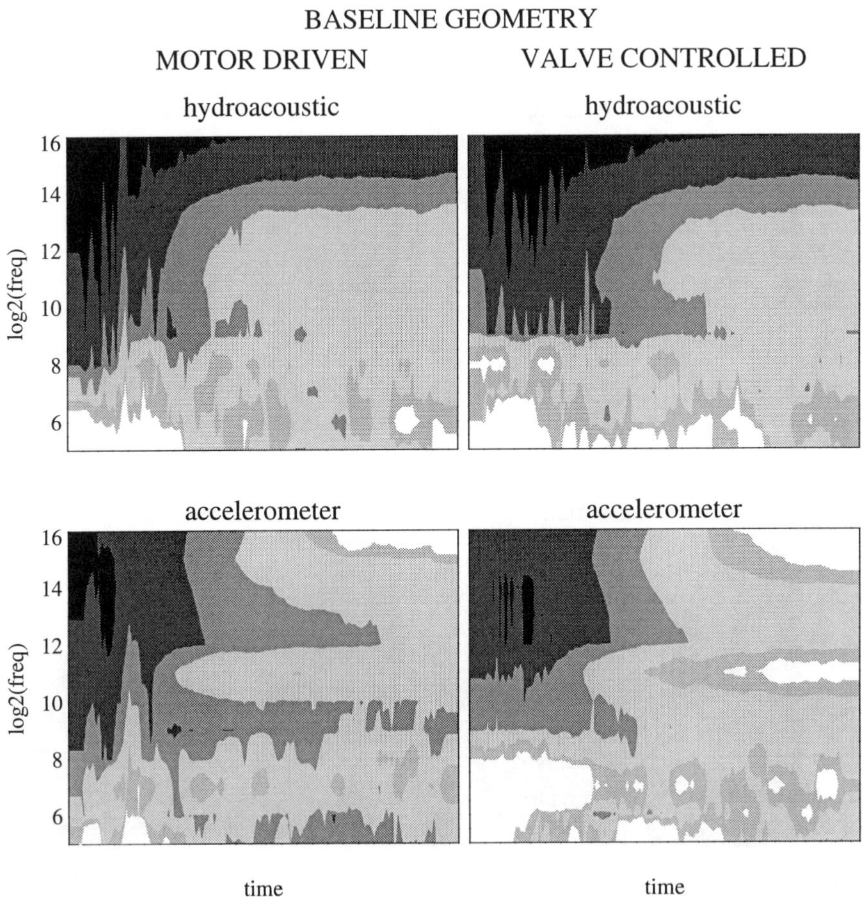

FIGURE 5. Contour maps of the logarithm of the time dependent power spectrum of each of the data sets in Figure 1. Each spectrum has been normalized by its maximum value. The frequency bins are equal logarithm (as for wavelets) and the levels are in equal increments. Black represents the minimum value and white the maximum value.

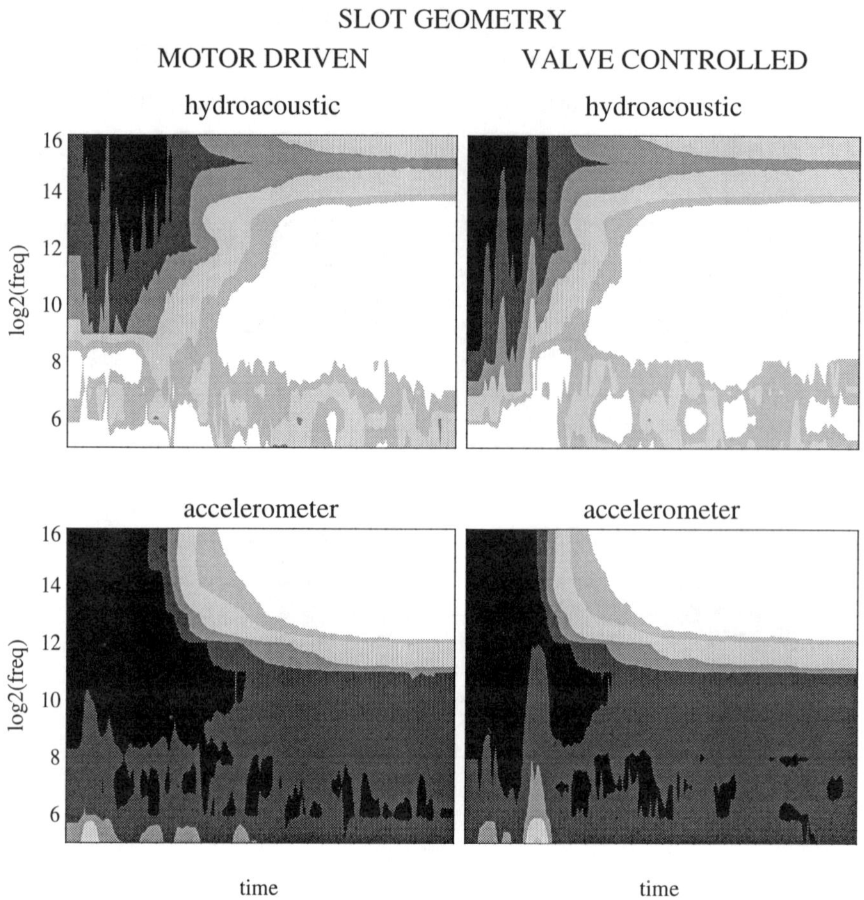

FIGURE 6. Contour maps of the logarithm of the time dependent power spectrum of each of the data sets in Figure 2. Each spectrum has been normalized by its maximum value. The frequency bins are equal logarithm (as for wavelets) and the levels are in equal increments. Black represents the minimum value and white the maximum value.

FOURIER COMPLEXITY

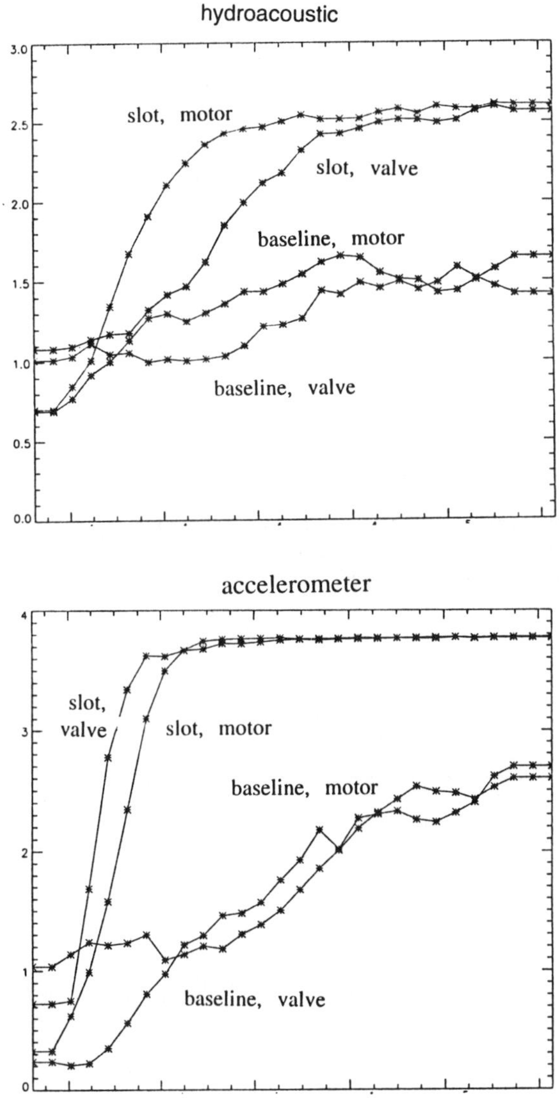

FIGURE 7. The logarithm of the localized Fourier complexity for hydroacoustic data from the four different experiment configurations is plotted in the upper graph and for the corresponding accelerometer data in the lower graph.

ACKNOWLEDGMENTS

The fluid flow noise data was provided by Scott Hassan, the experimentalist, and Jim Gutkowski, the Project Manager for Covert Launch Technology, both in the Launcher and Missile Systems Department, Naval Underwater Warfare Center, Newport. The author thanks Janet Hillenbrand who was especially helpful as the principal point of contact at NUWC. The author's research was supported by the Advanced Systems Technology Group (Carl Larson, Head) of the Systems Research and Development Department at the Naval Surface Warfare Center.

REFERENCES

1. C. Heil and D. Walnut, Continuous and discrete wavelet transforms, SIAM Review, pp. 628-666.

2. I. Daubechies, *Ten Lectures on Wavelets*, CBMS-NSF Regional Conference Series in Applied Mathematics Vol. 61, SIAM, Philadephia, 1992, p. 115.

3. P. H. Carter, Supertransients in fluid flow noise, presented at SPIE's International Symposium on Optical Science, Engineering, and Instrumentation, July 9-14, 1995, San Diego, Conference Proceedings Vol. 2569, to appear.

Spatio-Temporal Analysis of Rayleigh-Bénard Convection

CLAUDIA S. M. LAINSCSEK AND FERDINAND SCHÜRRER

Institut für Theoretische Physik
Technische Universität Graz
Petersgasse 16; A-8010 Graz; Österreich

Abstract. The analysis of Rayleigh-Bénard convection in a thin layer of an incompressible fluid caused by heating from below, is based on the Navier-Stokes equations. In planar geometry the Navier-Stokes equations in Bousinesq-approximation reduce to two nonlinear coupled partial differential equations for the velocity flux function ξ and the temperature deviation θ. These equations are analysed in form of spatial Fourier modes with time-dependent amplitudes. Only modes corresponding to free-free boundary conditions were selected. In this way, a set of ten coupled nonlinear ordinary differential equations for the mode amplitudes was obtained. These equations were solved numerically for different Rayleigh numbers. The temporal information in the ten dimensional phase space of the mode amplitudes is analysed with respect to the dimension of the attractor. In addition, a time series of flow patterns in real space is constructed. For this spatio-temporal patterns the empirical orthonormal functions are determined and used to find the temporal evolution from the projection onto the basic vectors. Finally the result of different types of analysis were compared. This should lead to a better understanding how to analyse real systems in terms of observational data, e.g. thermal convection on the surface of the sun.

CONVECTION EQUATIONS

Model

The analysis of convection in an incompressible fluid caused by heating from below is based on the equations for conservation of mass, momentum and energy, and on the equation of the state of the fluid.

The continuity equation $\frac{\partial \rho}{\partial t} + \operatorname{div}(\rho \mathbf{v}) = 0$ for an incompressible fluid ($\rho =$ *const.*) transforms into

$$\operatorname{div} \mathbf{v} = 0\,. \tag{1}$$

The Navier-Stokes equation reads as

$$\rho \left[\frac{\partial \mathbf{v}}{\partial t} + (\mathbf{v} \cdot \nabla)\mathbf{v} \right] = -\nabla p + \mu \Delta \mathbf{v} + \mathbf{g} \alpha \rho \Delta T\,, \tag{2}$$

and the equation of thermal conduction in the fluid is given by

$$\frac{\partial T}{\partial t} + \mathbf{v}\nabla T = \chi \Delta T + \frac{\nu}{2c_p}\left(\frac{\partial v_i}{\partial x_k} + \frac{\partial v_k}{\partial x_i}\right)^2 . \tag{3}$$

The state of the fluid obeys the equation

$$\rho = \rho_0(1 - \alpha T)\,. \tag{4}$$

The following abbreviations are used:

ρ	...	density of the fluid,
$\mathbf{g}$	...	acceleration of gravity,
μ	...	dynamic viscosity,
χ	...	thermal diffusivity of the fluid,
$\alpha = \frac{1}{\rho_0}\left(\frac{\partial \rho}{\partial T}\right)_p$	...	volume expansion coefficient.

The first step is to introduce the so-called Boussinesq approximation. This means that the variation with temperature of the material properties of the fluid such as the kinematic viscosity, the thermal diffusivity, and the volume expansion coefficient, are negligible provided that the temperature variation is small. The density of the fluid is also considered to be constant, except for the variation of the density in the vertical direction by the imposed vertical temperature difference. The vertical density variation provides the driving mechanism for convective motions.

The deviation of the temperature θ from the linear vertical temperature distribution is defined by

$$T = T_0 - \beta \boldsymbol{\lambda} \mathbf{x} + \theta \tag{5}$$

with

$$\beta = \frac{\Delta T}{H} \quad , \quad \boldsymbol{\lambda} = (0,0,1) \quad , \quad \theta = \theta(x,y,z,t)\,.$$

After some analysis, we get the equations for nonlinear convection in an incompressible fluid:

$$\frac{\partial \mathbf{v}}{\partial t} + (\mathbf{v}\cdot\nabla)\mathbf{v} = -\frac{1}{\rho_0}\nabla p + \nu\Delta\mathbf{v} + \mathbf{g}\alpha\Delta T\,, \tag{6}$$

$$\frac{\partial \theta}{\partial t} = \beta \mathbf{v}\boldsymbol{\lambda} - (\mathbf{v}\cdot\nabla)\,\theta + \chi\Delta\theta\,, \tag{7}$$

$$\nabla\cdot\mathbf{v} = 0\,, \tag{8}$$

where $\nu = \frac{\mu}{\rho}$ denotes the kinematic viscosity.

Taking the curl of equation (6) we obtain

$$\frac{\partial \boldsymbol{\omega}}{\partial t} - \operatorname{curl}(\mathbf{v}\times\boldsymbol{\omega}) = \nu\Delta\boldsymbol{\omega} + g\alpha\operatorname{curl}(\theta\boldsymbol{\lambda})\,, \tag{9}$$

where $\boldsymbol{\omega} = \operatorname{curl} \mathbf{v}$ is the vorticity.

To simplify the equations (7) and (9) we introduce the flux function ξ so that $\mathbf{v}$ is given by

$$\mathbf{v} = \left(0, -\frac{\partial \xi}{\partial z}, \frac{\partial \xi}{\partial y}\right) . \tag{10}$$

Boundary Conditions

The fluid is confined between the plates $z = 0$ and $z = H$. Regardless of the nature of these plates we must require

$$\begin{aligned} \theta &= 0 \\ v_z &= 0 \end{aligned} \quad \text{for} \quad z = 0, H . \tag{11}$$

Condition (11) must be introduced because the surfaces are maintained at constant temperatures. The further conditions depend on the nature of the bounding surfaces.

On a **rigid surface** no slip occurs, and therefore the horizontal components of the velocity vanish:

$$v_x = v_y = 0 . \tag{12}$$

These conditions are valid for all points on the surface. By taking into account the equation of continuity, we obtain

$$\frac{\partial v_z}{\partial z} = 0 . \tag{13}$$

On a **free surface** the components p_{xz} and p_{yz} of the viscous stress tensor are equal to zero:

$$p_{xz} = \mu \left(\frac{\partial v_x}{\partial z} + \frac{\partial v_z}{\partial x} \right) = 0 , \tag{14}$$

$$p_{yz} = \mu \left(\frac{\partial v_y}{\partial z} + \frac{\partial v_z}{\partial y} \right) = 0 , \tag{15}$$

and therefore

$$\frac{\partial v_x}{\partial z} = \frac{\partial v_y}{\partial z} = 0 , \tag{16}$$

because $v_z = 0$ for all x and y on the bounding surface. This leads with the equation of continuity to

$$\frac{\partial^2 v_z}{\partial z^2} = 0 . \tag{17}$$

In our notation the boundary conditions for **free surfaces at the bottom and at the top** are

$$v_z = 0 \to \frac{\partial \xi}{\partial y} = 0\,, \tag{18}$$

$$\frac{\partial^2 v_z}{\partial z^2} = 0 \to \frac{\partial^2}{\partial z^2}\frac{\partial \xi}{\partial y} = 0\,, \tag{19}$$

$$\theta = 0\,. \tag{20}$$

Fourier Modes Ansatz

From equations (7) and (9) we find that by taking into account the flux function ξ defined through equation (10)

$$\frac{\partial \Delta\xi}{\partial t} = \mathbf{g}\alpha\frac{\partial \theta}{\partial y} + \nu\Delta\left(\Delta\xi\right) - \left[\xi, \Delta\xi\right]\,, \tag{21}$$

and

$$\frac{\partial \theta}{\partial t} = \beta\frac{\partial \xi}{\partial y} + \left[\xi, \theta\right] + \chi\Delta\theta\,. \tag{22}$$

The Fourier series expansion of ξ and θ can be written as

$$\xi = \sum_{i=0}^{\infty}\sum_{j=0}^{\infty} \mathrm{A}_{2i+j+3}(t)\,\sqrt{2}^{i}\,\frac{(1+a^2)\chi}{a}\,\sin\left(iay\frac{\pi}{H}\right)\sin\left(jz\frac{\pi}{H}\right)\,, \tag{23}$$

$$\theta = \sum_{i=0}^{\infty}\sum_{j=0}^{\infty} \mathrm{A}_{2i+j-1}(t)\,\sqrt{2}^{i}\,\frac{\Delta T}{\pi R}\,\cos\left(iay\frac{\pi}{H}\right)\sin\left(jz\frac{\pi}{H}\right)\,, \tag{24}$$

with

with	the depth of the layer	$H = 1$;
	the dimensionless time	$\tau = \frac{(1+a^2)\chi\pi^2}{H^2}t$;
	the Prandtl number	$\sigma = \frac{\nu}{\chi}$;
	the Rayleigh number	$R_a = \frac{g\alpha H^3\Delta T}{\nu\chi}$;
	the critical Rayleigh number	$R_c = \frac{\pi^4(1+a^2)^3}{a^2}$;
	and the relations	$R = \frac{R_a}{R_c}$;
		$b = \frac{4}{1+a^2}$; $c_1 = 1 + 4a^2$;
		$c_2 = 4 + a^2$; $c_3 = a^2 - 1$.

Truncating this series at second order and inserting them into the equations (21) and (22) we obtain the following system of nonlinear ordinary differential equations:

$$\dot{\mathrm{A}}_0 = \frac{-b\,\mathrm{A}_0}{4} - \frac{\mathrm{A}_3\,\mathrm{A}_6}{2} - \frac{\mathrm{A}_2\,\mathrm{A}_7}{2} - 2\,\mathrm{A}_5\,\mathrm{A}_8 - 2\,\mathrm{A}_4\,\mathrm{A}_9\,,$$

$$
\begin{aligned}
\dot{A}_1 &= -b\,A_1 + A_2\,A_6 + 4\,A_4\,A_8\,,\\
\dot{A}_2 &= -A_2 + R\,A_6 - A_1\,A_6 + \frac{A_0\,A_7}{2} - \frac{3\,A_4\,A_7}{2} - \frac{3\,A_3\,A_8}{2}\,,\\
\dot{A}_3 &= \frac{-b\,c_2\,A_3}{4} + \frac{A_0\,A_6}{2} + \frac{3\,A_4\,A_6}{2} + R\,A_7 + \frac{3\,A_2\,A_8}{2}\,,\\
\dot{A}_4 &= \frac{-b\,c_1\,A_4}{4} - \frac{3\,A_3\,A_6}{4} + \frac{3\,A_2\,A_7}{4} + 2\,R\,A_8 - 2\,A_1\,A_8 + A_0\,A_9\,,\\
\dot{A}_5 &= -4\,A_5 + A_0\,A_8 + 2\,R\,A_9\,,\\
\dot{A}_6 &= \sigma\,A_2 - \sigma\,A_6 - \frac{9\,b\,c_3\,A_7\,A_8}{8}\,,\\
\dot{A}_7 &= \frac{4\,\sigma\,A_3}{b\,c_2} - \frac{b\,c_2\,\sigma\,A_7}{4} + \frac{9\,a^2\,A_6\,A_8}{2\,c_2}\,,\\
\dot{A}_8 &= \frac{8\,\sigma\,A_4}{b\,c_1} - \frac{9\,A_6\,A_7}{4\,c_1} - \frac{b\,c_1\,\sigma\,A_8}{4}\,,\\
\dot{A}_9 &= \frac{\sigma\,A_5}{2} - 4\,\sigma\,A_9\,.
\end{aligned}
\tag{25}
$$

If we take only three modes, our system will transform into the well known **Lorenz Equations**:

$$
\begin{aligned}
\dot{A}_6 &= -\sigma A_6 + \sigma A_2\,,\\
\dot{A}_2 &= -A_6A_1 - A_2 + RA_6\,,\\
\dot{A}_1 &= A_6A_2 - bA_1\,.
\end{aligned}
\tag{26}
$$

In this case only A_1, A_2, and A_6 in (25) are non-zero.

PHASE SPACE PLOTS AND BIFURCATION DIAGRAM

System (25) is integrated numerically by applying a Runge-Kutta method. We chose $\sigma = 10.0$ and $a = \frac{1}{\sqrt{2}}$, and use the Rayleigh number with respect to the critical Rayleigh number $R = \frac{R_a}{R_c}$ as Bifurcation parameter.

Fig. 2 shows for the initial data $(A_0, A_1, A_2, A_3, A_4A_5, A_6, A_7, A_8, A_9) = (1, 0, 0, 0, 0.001, 0, 0, 0, 0, 0)$ and different values of R the corresponding attractors in the A_0, A_4–plane. In order to distinguish between chaotic and periodic regions we computed the Bifurcation diagram. The 9–dimensional hyperspaces of the Poincare cut is the $(A_0 = 0)$–plane. In Fig. 1 the projection of this hyperspace on the A_4–axis is plotted.

The temporal evolution of all modes for $R = 35.0$ is plotted in Fig. 3. It is clearly seen, that the modes $\mathbf{A_{2,3,6,7}}$ are damped.

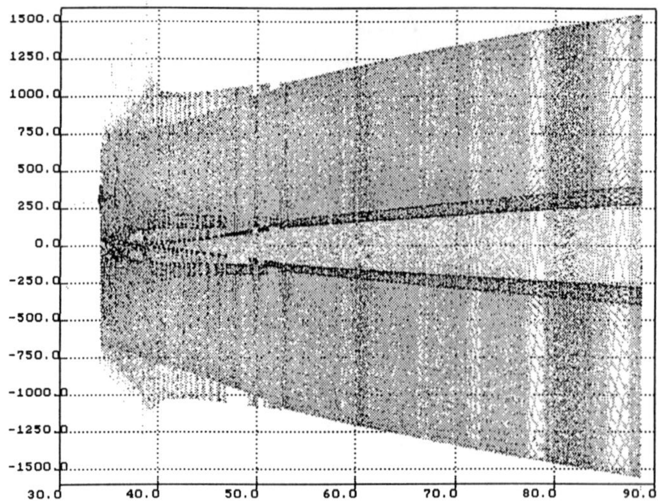

Figure 1 : Bifurcationdiagram

LINEAR ANALYSIS

System (25),

$$\frac{\mathrm{d}\mathbf{A}}{\mathrm{d}t} = \dot{\mathbf{A}} = \mathbf{F}(\mathbf{A}), \tag{27}$$

defines a vectorfield in the 10–dimensional phase space $\mathbb{R}^{10}$. The rate of change of an infinitesimal volume element with time is given by

$$\operatorname{div}\mathbf{F}(\mathbf{A}) = \sum_{i=0}^{9} \frac{\partial F_i(\mathbf{A})}{\partial \mathrm{A}_i} = -5 - \frac{5\,b}{4} - \frac{b\,c_1}{4} - \frac{b\,c_2}{4} - 5\,\sigma - \frac{b\,c_1\,\sigma}{4} - \frac{b\,c_2\,\sigma}{4}\,. \tag{28}$$

If $\mathcal{U}$ is a subspace of the $\mathbb{R}^{10}$ and $\mathbf{f}^t$ is the flux related to $\mathbf{F}$, $\mathcal{U}$ will be deformed under $\mathbf{f}^t$. Let $V(t)$ be the volume (the Lebesque-meassure) of $\mathbf{f}^t\mathcal{U}$. Hence the velocity of volume deformation at time t reads as

$$\frac{\mathrm{d}V(T)}{\mathrm{d}t} = \int_{\mathbf{f}^t\mathcal{U}} \operatorname{div}\mathbf{F}(\mathbf{A})\,\mathrm{d}\mathbf{A} = V(t)\operatorname{div}\mathbf{F}\,. \tag{29}$$

For system (25) $\operatorname{div}\mathbf{F} < 0$, which means that our system is dissipative.

CORRELATION DIMENSION

Grassberger and Procaccia [GP83] introduced a method for computing the correlation dimension $\mathcal{D}$ from a single variable time series.

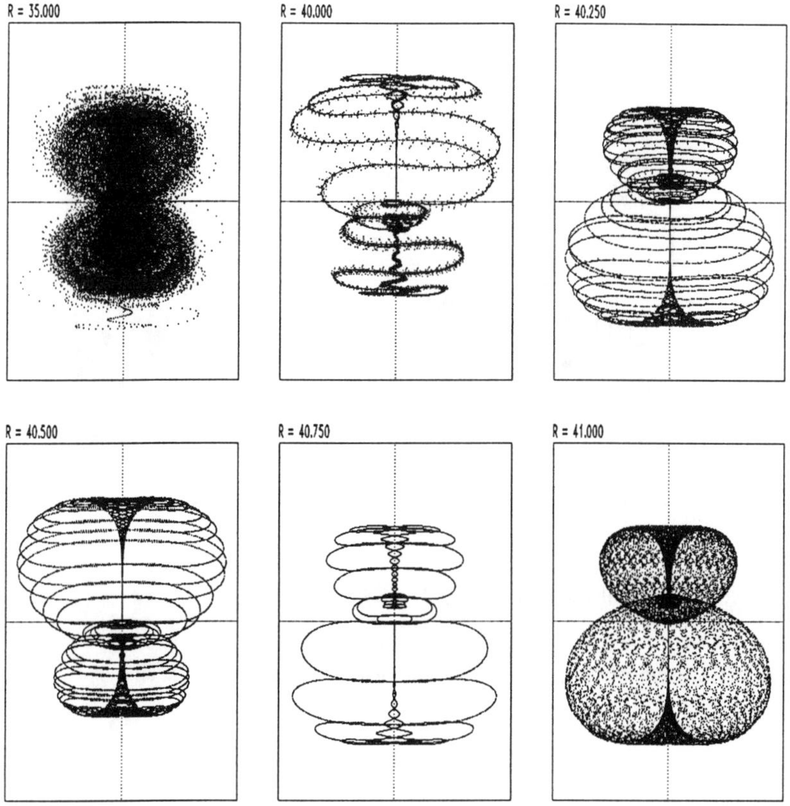

Figure 2 : Phase Space Plot in the A_0 , A_4–plane with $R = 35.0$, $R = 40.0$, $R = 40.25$, $R = 40.5$, $R = 40.75$, $R = 41.0$

The **correlation sum** is given by

$$C_d(\epsilon) = \frac{1}{N^2} \sum_{k=0}^{d-1} \sum_{i \neq j}^{N} \mathrm{H}\left(\epsilon - |\mathbf{x}_{i+k} - \mathbf{x}_{j+k}|\right) , \tag{30}$$

where H is the Heaviside step function.

If the number of points N is large enough, the correlation sum (30) obeys a power law:

$$C_d(\epsilon) \sim r^{\mathcal{D}} . \tag{31}$$

The exponent $\mathcal{D}$ is called **correlation dimension** and can be written as

$$\mathcal{D} = \lim_{\epsilon \to 0} \frac{\ln C_d(\epsilon)}{\ln \epsilon} . \tag{32}$$

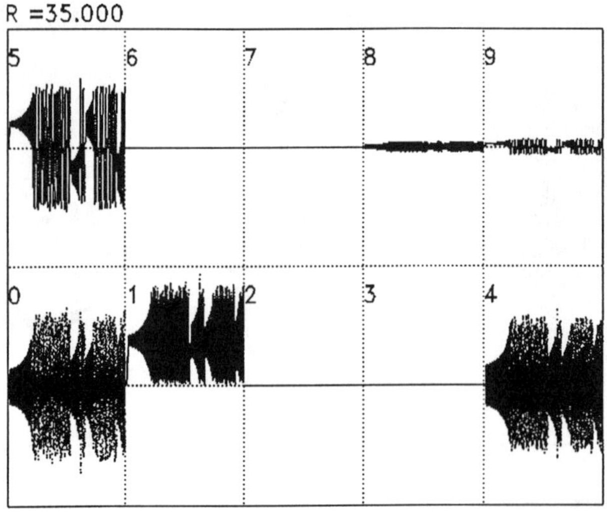

Figure 3 : Temporal Evolution of each mode at $R = 35.0$

For $R = 35.0$ we obtain $\mathcal{D} = 2.59$ and for $R = 40.0$ $\mathcal{D} = 1.12$.

TEMPERATURE FIELD

System (25) is integrated numerically for $\sigma = 10.0$, $a = \frac{1}{\sqrt{2}}$, $R = 35.0$ (chaotic region) and $R = 40.0$ (periodic orbit) and the initial conditions $\mathbf{A}(t = 0) = (1, 0, 0, 0, 0.001, 0, 0, 0, 0, 0)$. The results are time series $\mathbf{A}(t)$. The insertion of $\mathbf{A}(t)$ into equation (24) yields for $y = [0, 2\pi]$ and $z = [0, 1]$ sequences of 2–dimensional patterns (see Fig. 6 and 7), where the gray values represent the height above the y, z–plane.

Our sample pattern sequences contain 100 frames for $R = 40.0$ and 400 frames for $R = 35.0$ with 640x480 pixels each. Every vector can be treated as vector $\mathbf{w}_i$ of the $\mathbb{R}^{237150}$. From the correlation matrix $C_{ij} = \mathbf{w}_i \mathbf{w}_j$ we calculated the eigenvalues λ_i, shown in Fig. 4 and Fig. 5.

As expected for the periodic attractor at $R = 40.0$ only one dominating eigenvalue is seen (Fig. 5). In the chaotic case at $R = 35.0$ four modes are dominating. This result fits well to the dimension estimation and to Fig. 3, where only 6 modes remain.

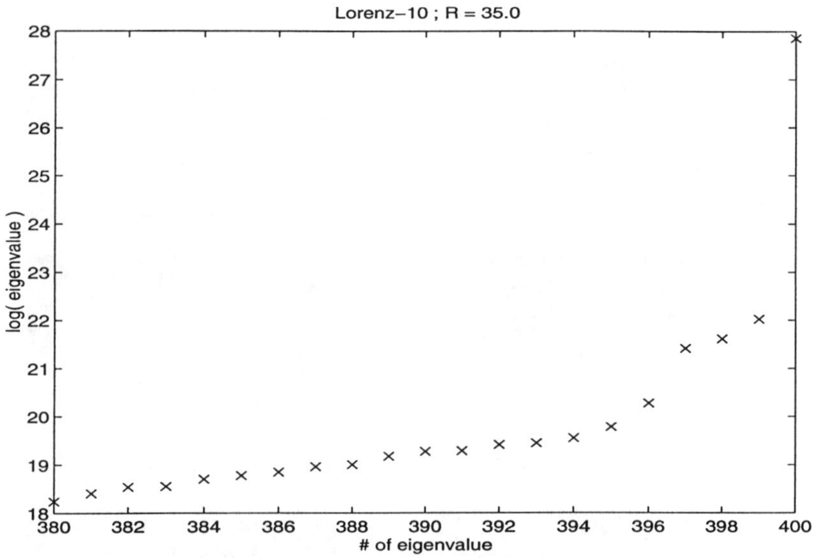

Figure 4 : Eigenvalues of the correlation matrix $\mathcal{C}$ for $R = 35.0$

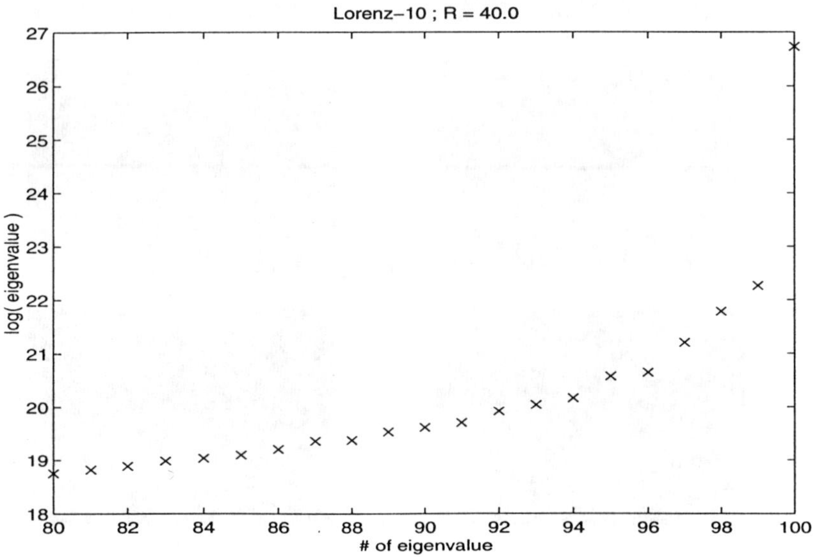

Figure 5 : Eigenvalues of the correlation matrix $\mathcal{C}$ for $R = 40.0$

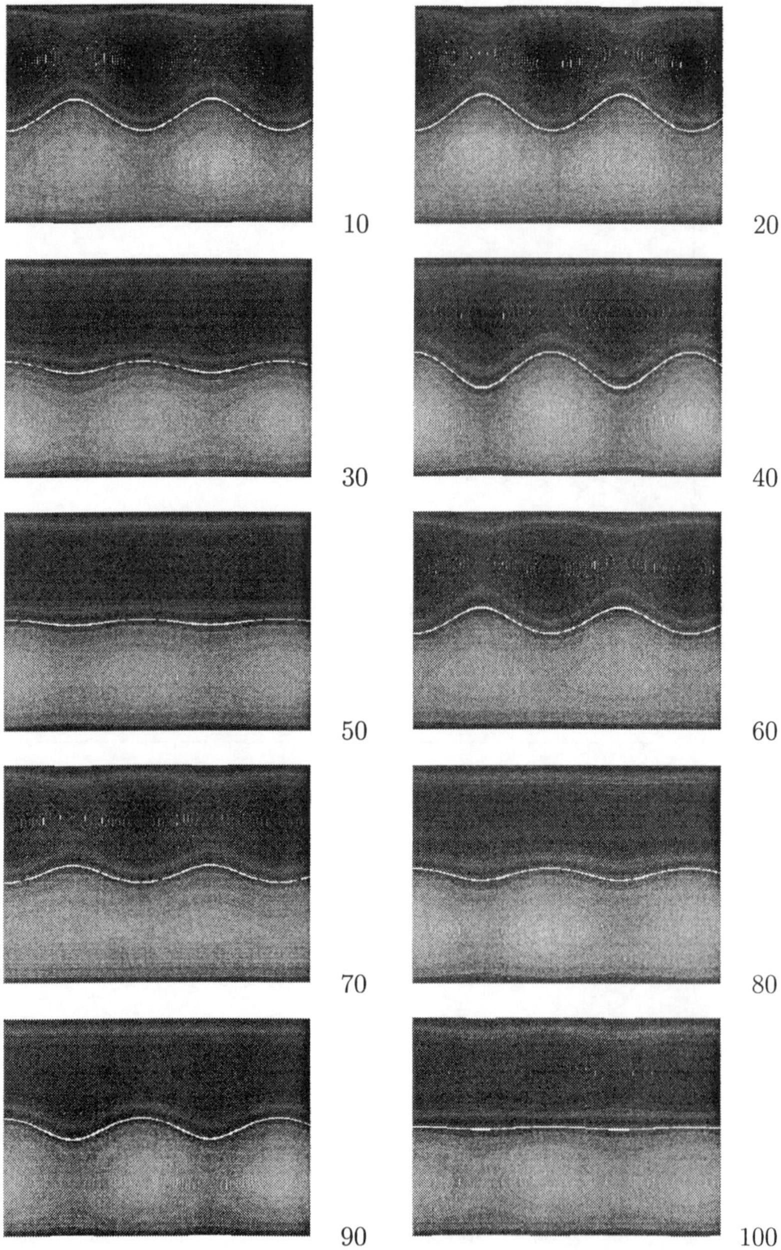

Figure 6 : Temporal evolution of the deviations from the linear temperature profile in the y, z–plane is represented by means of gray values. ($R = 35.0$)

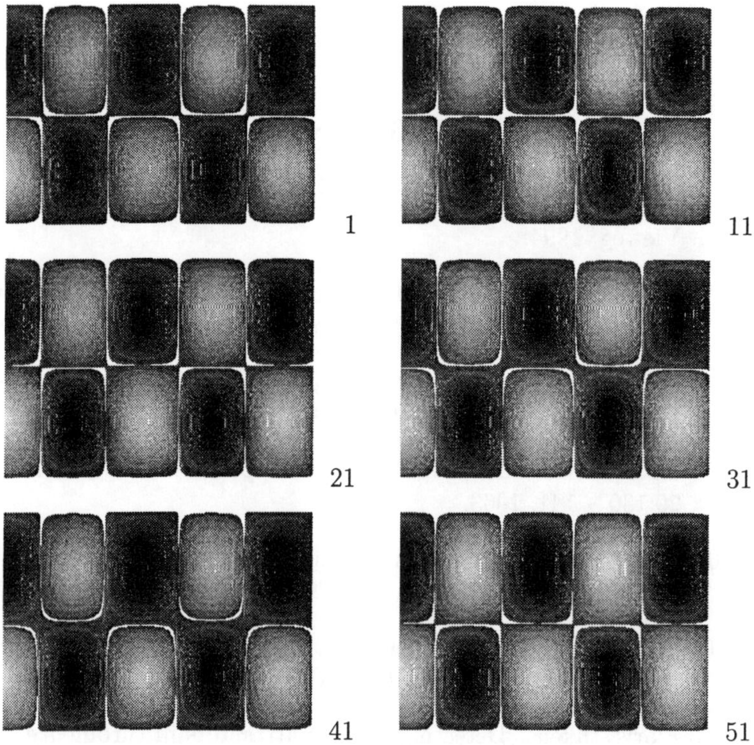

Figure 7 : Temporal evolution of the deviations from the linear temperature profile in the y, z–plane is represented by means of gray values. ($R = 40.0$)

CONCLUSION AND OUTLOOK

The methods presented in this paper were applied to simulated time series of an extended Lorenz system in a ten-dimensional phase space. This system was obtained from the Navier-Stokes equations by introducing a Fourier modes ansatz. We also used this ansatz to get a sequence of patterns of the temperature field. These frames were scanned in order to get the correlation matrix of the time evolution. From the number of dominant modes the number of necessary modes for a possible model could be assumed. In our case this number was lower than the dimension of the original phase space.

Therefore our next step will be to truncate the Fourier modes ansatz at third order and apply the same analysis. A system in a 64-dimensional phase space can be obtained. The interesting results of such an analysis will be a dimension estimate and the number of dominant modes due to the correlation matrix. If these results show a possible low dimensional (much less than 64) model, this model and methods could possibly be applied to real data sets of convection cells, e.g. the convection on the surface of the sun.

References

[AFH94] J. Argyris, G. Faust, and M. Haase. *Die Erforschung der Chaos.* Vieweg, 1994.

[BK86] D.S. Broomhead and G.P. King. Extracting qualitative dynamics from experimantal data. *Physica D*, 20:217, 1986.

[GP83] Peter Grassberger and Itamar Procaccia. Measuring the strangeness of strange attractors. *Physica D*, 9:189, 1983.

[Lor63] Edward N. Lorenz. Deterministic nonperiodic flow. *J. Atmos. Sci.*, 20:130 - 141, 1963.

[PCFS80] N. H. Packard, J. P. Crutchfield, J. D. Farmer, and R. S. Shaw. Geometry from a time series. *Phys. Rev. Lett.*, 45:712, 1980.

[PFTV90] W.H. Press, B.P. Flannery, S.A. Teukolsky, and W.T. Vetterling. *Numerical Recipes in C.* Cambridge University Press, 1990.

[Tak81] Floris Takens. Detecting strange attractors in turbulence. *Lecture Notes in Mathematics*, 898:366, 1981.

[Tso93] Anastasios A. Tsonis. *From Theory to Applications.* Plenum, 1993.

Spontaneous Bursting: From Temporal to Spatio-Temporal Intermittency

Nathan Platt and Stephen M. Hammel

Code B44, Nonlinear Dynamics and Wavelets Group,
Naval Surface Warfare Center, 10901 New Hampshire Ave,
Silver Spring, Maryland, 20903-5640

Abstract

A simple model for temporal bursting is introduced. This model invokes either dynamic or random forcing of a bifurcation parameter of some simple dynamical system in a way that makes the bifurcation parameter spend suitable amounts of time below and above the bifurcation threshold. This model is extended to coupled map lattices to produce spontaneous spatio-temporal burstings. It models physical systems which are embedded in a random background that is statistically homogeneous in space and time. An application of this model to optical turbulence is discussed.

1 Introduction

There is an abundance of physical phenomena exhibiting aperiodic oscillatory behavior. Such phenomena can be modeled using chaotic dynamical systems, and these models can be grouped into two classes according to the characteristics of their behavior. One class is characterized by aperiodic modulations of an underlying periodic signal while the other class is characterized by signals which exhibit apparently random switching between qualitatively different kinds of behavior. The latter behavior is called *intermittency* and this type of signal is the main concern of this work. The term intermittency most likely has its origins in fluid dynamics as in Prandtl's description of "intermittent occurrence of turbulence" [1]. In an investigation of the flow of a fluid in narrow tubes, Reynolds [2] observed that at the transition from laminar to turbulent flow, there is a region in parameter space, which includes fluid velocity as one of the parameters, where the flow could not be described as either laminar or turbulent. He described this state: "The disturbances would suddenly come on through a certain length of tube and pass away and then come on again, giving the appearance of flashes, and these flashes would often commence successively at one point in the flow". Couette [3] observed similar phenomena occuring in a jet of water flowing out of a large tank through the tubes. Examples of intermittency are abundant in nature. They include intermittent bursts of turbulence in otherwise laminar fluid flow, sunspot activity in astrophysics, and reversals of the geomagnetic field.

There is a large theoretical and experimental body of knowledge describing dynamical systems which exhibit weak aperiodic modulations of an established periodic signal, whereas intermittency is not studied as extensively. Dynamical systems models of intermittency as well as a partial classification of some types of intermittency were given by Pomeau and Manneville [4]. The three types of Pomeau and Manneville intermittency involve local bifurcations of maps corresponding to distinct ways for eigenvalues to pass through the unit circle. The fixed point in their model corresponds with a periodic orbit in the continuous time system that they implicitly describe. In general, signals produced by the Pomeau and Manneville scenario are periodic oscillations interrupted from time to time by some aperiodic bursts of activity. Another model of intermittency is called crisis-induced intermittency and it was introduced by Grebogi, Ott and Yorke [5]. This intermittency involves a collision in phase-space of two or more chaotic attractors as some parameter is varied, and it is again characterized by random switching between different aperiodic oscillations.

In both of these models of intermittency, the laminar phase consists of either periodic or aperiodic oscillations, while the intermittency is manifested by an apparently random switching between different kinds of behavior. But, for the case of a pipe flow or sunspot activity, the laminar phase is characterized by a *flat* signal (no cross flow in the pipe during a laminar phase, or almost no sunspot activity), while intermittency manifests itself by a short burst of activity. Thus, the laminar phase of such a phenomenon can be approximated by a point in a suitably chosen projection of the phase-space, instead of a limit cycle in Pomeau and Manneville scenario, or a chaotic attractor for crisis-induced intermittency.

A model of intermittency describing this On-Off switching of activity was introduced by Platt, Spiegel and Tresser [6]. Some features of this type of intermittency have been described by other authors [7]-[10]. In short, this model invokes either dynamic or random forcing of a bifurcation parameter of some simple dynamical system in a way that makes the bifurcation parameter spend suitable amounts of time below and above the bifurcation threshold. The simplest version of On-Off intermittency involves a master-slave formulation such that the dynamics of the slave system displaying intermittency does not affect or feedback to the dynamics of the master system responsible for the driving of the bifurcation parameter. A thorough treatment of the main qualitative aspects of On-Off intermittency could be found in [11] for the noise-free case and in [12] for the noisy case.

Models manifesting On-Off intermittency are temporal, and thus only applicable to averaged quantities of actual physical processes, such as sunspot number. But actual sunspots have both temporal and spatial extent, and their behavior should be captured by spatio-temporal models displaying spatio-temporal intermittency. The idea of using parametrically coupled systems extends naturally into the domain of coupled-map lattices to model space-time intermittency. We will discuss the extension of On-Off intermittency to dynamical systems which have spatial extent.

The organization of this work is as follows. Section 2 deals with On-Off intermittency in temporal systems, section 3 introduces the concept of parametrically coupled systems to extend On-Off intermittency to coupled-map lattices, and section 4 describes physical situations where these models could be applicable.

2 On-Off Intermittency in temporal domain

A simple model of On-Off intermittency which allows detailed numerical and theoretical study consists of a driven one-dimensional discrete map. On-Off intermittency was described in detail in [6] and will be discussed here only briefly. Consider a discrete map

$$y_{n+1} = g(y_n, \mu) \tag{1}$$

with $g(0, \mu) = 0$, and $\mu = \mu_c$ is a critical value of a bifurcation parameter μ for the fixed point $y_o = 0$. Let us assume that for $\mu < \mu_c$ the fixed point $y_o = 0$ is stable, while at $\mu > \mu_c$, its loses stability and the system bifurcates to a new, bounded, non-zero solution.

Now, replace μ by a bounded, suitably chosen time-dependent discrete function μ_n such that μ_n crosses μ_c-line irregularly. Then the resulting time series $\{y_n = g(y_{n-1}, \mu_{n-1})\}_{n=1}^{\infty}$ will display On-Off intermittent behavior. Here, μ_n can be either a random or a dynamic process.

For the actual calculations we use a logistic map. Let

$$x_{n+1} = f(x_n) \tag{2}$$

with $x_n \in (0, 1]$. Then set μ_n to

$$\mu_n = a x_n \tag{3}$$

and define the observable $\{y_n\}_{n=1}^{\infty}$ as

$$y_{n+1} = \mu_n y_n (1 - y_n) \tag{4}$$

Here, a is a bifurcation parameter allowed to vary. For $\mu_n < 1$, the fixed point $y_o = 0$ is stable, while for $\mu_n > 1$ it is unstable giving rise to bounded solutions.

In the results presented later f is taken to be either

$$\text{the dyadic map } f_1(x_n) = 2x_n \bmod 1 \tag{5}$$

$$\text{or the tent map, } f_2(x_n) = \begin{cases} 2x_n & \text{if } x_n \leq \frac{1}{2} \\ 1 - 2x_n & \text{otherwise} \end{cases} \tag{6}$$

or

$$f_3(x_n) \text{ — white noise uniformly distributed in } (0,1] \tag{7}$$

Since computers use finite precision arithmetic, some care must be exercised in the numerical implementation of the dyadic map and the tent maps. In our study, the dyadic map is implemented using the binary representation of x_n in a 51 bit array. After initialization, multiplication by 2 followed by the mod operation is performed by a left shift operation on the binary array. The least significant bit of the array is randomly set to either 0 or 1 using the first bit in the binary representation of a number returned by a "pseudo-random" number generator [13]. Similar care is taken in the numerical implementation of the tent map forcing.

The forcing functions f_i for this study were chosen with similar statistical properties so the resulting time series obtained from the map can be compared with each other. All of the forcing functions f_i have an identical distribution, uniform in $(0, 1]$, and all have the same power spectrum. In addition, both the tent map and white noise are decorrelated [14] while the correlation for the dyadic map is different from zero [14]. Also, both deterministic forcings have the same positive Lyapunov exponent of $\ln(2)$ [15].

In a previous work dealing with the statistics of the laminar phases for this system it was shown that the critical value of the parameter a_c to obtain On-Off intermittency depends only on the statistical distribution of the forcing and is independent of the particular forcing being used [11]. To calculate the bifurcation parameter value for the map under consideration, we note that expanding (2-4) in a Taylor series around $y = 0$, one obtains

$$y_{n+1} = ax_n(y_n + O(y_n^2)) \tag{8}$$

The solution of (8) keeping terms linear in y_n is

$$y_n = a^n(\prod_{j=0}^{n-1} x_j)y_0 \, , \tag{9}$$

and thus the long term behavior of y_n is determined from the asymptotic behavior of the product

$$P_n = a^n \prod_{j=0}^{n-1} x_j . \tag{10}$$

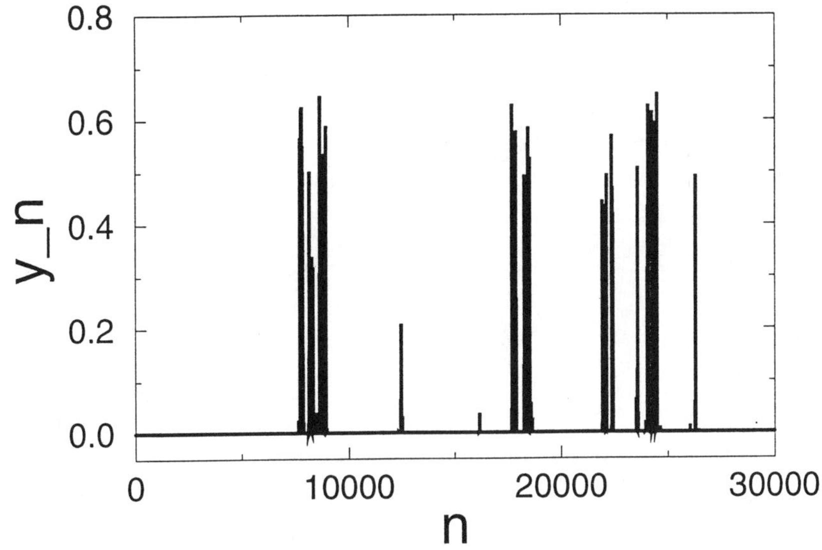

Figure 1: A typical signal y_n versus time n displaying intermittency at parameter $a = 2.74$.

By taking the logarithm of this product one obtains

$$\ln P_n = n\ln(a) + \ln(\prod_{j=0}^{n-1} x_j) = n\ln(a) + \sum_{j=0}^{n-1} \ln x_j \tag{11}$$

But $\sum_{j=0}^{n-1} \ln x_j \sim n < \ln x >$ from the law of large numbers [16]. Here $< \ln x >$ denotes the mean of the sequence $\{\ln x_i\}_{i=1}^{\infty}$. To compute this temporal average one uses the phase space average. Noting that the sequence $\{x_i\}_{i=1}^{\infty}$ obtained from (5-7) is uniformly distributed in the interval $(0, 1]$, one obtains

$$< \ln x >= \int_0^1 \ln x dx = -1 \; . \tag{12}$$

Hence,

$$\ln P_n \sim n\ln(a) - n \tag{13}$$

Finally, we arrive at

$$P_n \sim (\frac{a}{e})^n \; . \tag{14}$$

Thus, the critical value of the bifurcation parameter for the onset of intermittent behavior is $a_c = e = 2.71828...$ Figure 1 shows a typical signal obtained from intermittent system just above criticality.

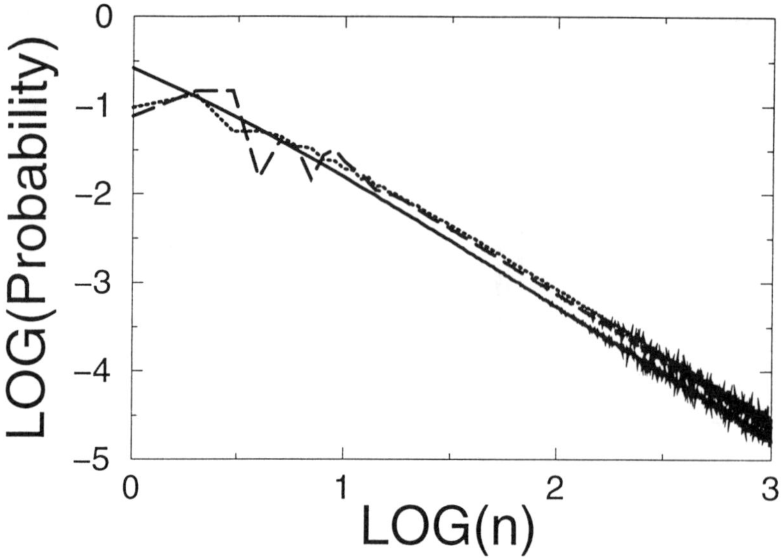

Figure 2: A numerically calculated log-log plot of the relative probability of a laminar phase of length n, versus n at $a = 2.72$. A solid line represents white noise forcing, a dashed line for tent map forcing, and a dotted line is for the dyadic map forcing. All of the lines asymptotically have a slope of $-3/2$.

The distinguishing characteristic of the intermittent signals such as those shown in Fig. 1 is the time between successive "on" or bursting events. The distribution of the inter-event times, or *laminar phases*, is easily measured (numerically or experimentally), and can serve as a potential classifier of the intermittent signals. Let Λ_n be the probability of obtaining a laminar phase of length exactly n. It is proved in [11] that for system (8) at the onset of intermittency, Λ_n is given by

$$\Lambda_n \sim n^{-\frac{3}{2}} \tag{15}$$

for uniformly distributed random driving. Thus, at the onset of intermittency, the distribution of laminar phases is a power law with exponent $-\frac{3}{2}$. In all cases we have studied so far, the asymptotic distribution of laminar phases at onset is well fit by a power law with exponent $-\frac{3}{2}$. This is true for both random *and* chaotic forcing. Moreover, it is proved in [11] that the $-\frac{3}{2}$ power law is a *universal* feature of On-Off intermittency at onset for a large class of *random* driving cases. Fig. 2 shows a log-log plot of numerically obtained distributions of laminar phases for various random and chaotic forcings x_n with uniform distribution in the interval $(0, 1]$.

3 Spontaneous Spatio-Temporal Bursting

Naturally occurring physical phenomena have both spatial and temporal extent. Moreover, there is a large number of physical situations where bursts are localized not only in time, but in space as well. For example, sunspots form at a particular latitude and then slowly drift towards the solar equator. Another example is the local roughening of the water surface, termed "cat's paws", which appear on a clear day with a light, gusting wind. To model such systems one needs to extend the model of temporal bursting to spatio-temporal systems. One of the simplest systems displaying spatio-temporal behavior is a coupled map lattice (CML). It is easily and efficiently implemented on computers, thus allowing a detailed study of the results. Moreover, when one solves partial differential equations numerically, after discretizing time and space one ends up with a CML. A large number of methods used for characterization of low-dimensional dynamical systems (ordinary differential equations and discrete maps) could be extended to CML's. Because of recent interest in spatio-temporal chaos, CML's are becoming increasingly popular [17]-[23].

In this section, we will extend ideas of On-Off intermittency to CML's to produce a model system for spontaneous spatio-temporal bursting. Again, for simplicity, we assume that the observed system could be decomposed into master/slave (background/observable) subsystems, and that the dynamics of the observable (slave) system *does not* feed back to the background (master). We will work with a one-dimensional diffusively coupled map lattice:

$$y_{t+1}^{(i)} = (1-\epsilon)F(y_t^{(i)}) + \frac{\epsilon}{2}\left[F(y_t^{(i-1)}) + F(y_t^{(i+1)})\right] \tag{16}$$

Here, t denotes the discrete time step, and i is the spatial lattice point. The function F prescribes the dynamics of each oscillator $y_t^{(i)}$ and ϵ denotes the strength of the nearest-neighbor coupling (diffusion). To complete this model for spatio-temporal bursting, we need to prescribe a form for the function $F(y_t^{(i)})$ to produce bursts. To that purpose, we use an On-Off intermittent process:

$$F(y_t^{(i)}) = ax_t^{(i)}(y_t^{(i)} + O\left[(y_t^{(i)})^2\right]) \tag{17}$$

Nonlinear terms in $y_t^{(i)}$ are needed to keep (17) bounded at all times, but they do not significantly effect the basic features of the bursting model. In actual calculations, a logistic map is used for computational efficiency:

$$F(y_t^{(i)}) = ax_t^{(i)}y_t^{(i)}(1 - y_t^{(i)}) \tag{18}$$

Now, to set this CML in motion, one needs to specify a driving function $x_t^{(i)}$. A large number of chaotic and random drivings $x_t^{(i)}$ would suffice, but again,

a)

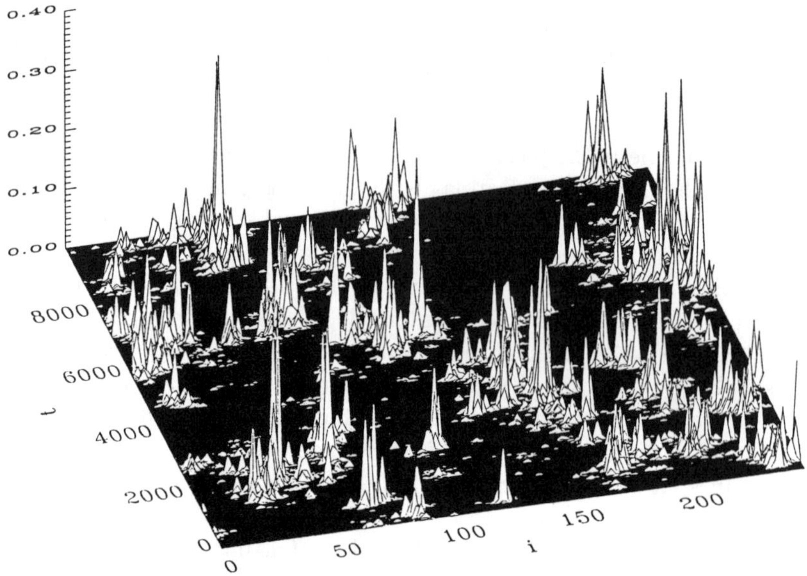
0.40
0.30
0.20
0.10
0.00
8000
6000
t
4000
2000
0
0
50
100
i
150
200

b)

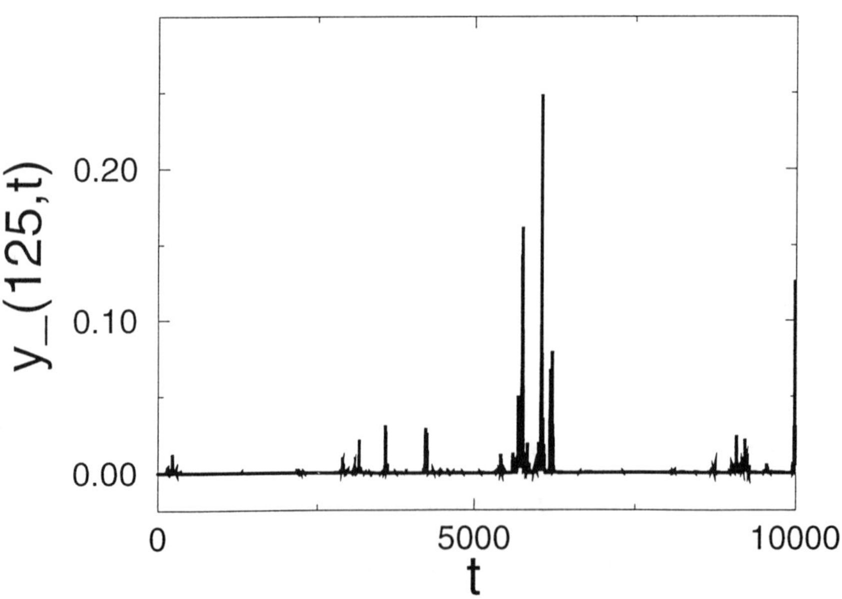
y_(125,t)
0.20
0.10
0.00
0
5000
10000
t

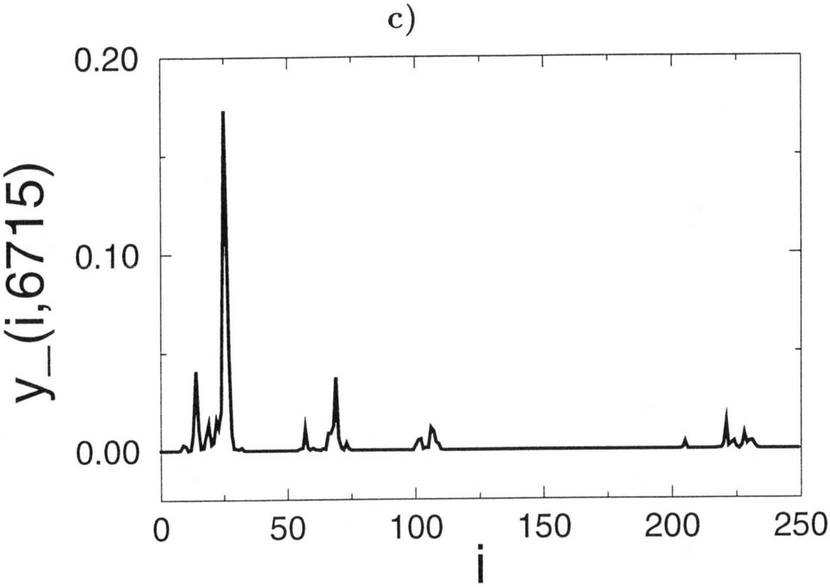

Figure 3: A typical signal $y_t^{(i)}$ displaying spatio-temporal bursting at $a = 2.15$, $\epsilon = 0.1$ and $L = 250$: a) Space-time surface plot of the CML, b) temporal evolution of a fixed probe in the middle of the lattice, c) instantaneous snapshot of the CML.

for simplicity, we set $x_t^{(i)}$ to white noise uniformly distributed in $(0, 1]$. In actual calculations, we use periodic boundary conditions: $y_t^{(1)} = y_t^{(L+1)}$ for a lattice of size L. A similar lattice model was derived in the study of symmetry breaking bifurcation in coupled CML's [24], [25]. Another model considered recently kept the driving term $x_t^{(i)}$ spatially homogeneous, i.e. $x_t^{(i)} = x_t$ for all lattice points [26]. We defer the discussion of possible physical situations which could be modeled by present CML's to the next section. Fig. 3a shows a typical signal produced by CML's at the criticality. Fig. 3b shows temporal evolution at a fixed spatial location, while Fig. 3c shows a spatial snapshot of the lattice at a fixed point in time.

There are both theoretical and numerical difficulties associated with the characterization of this CML model of spatio-temporal bursting. Numerical simulations become prohibitively expensive, since the computational effort to simulate a CML is directly proportional to the size of the lattice L. To illustrate this we compare theoretical and numerical estimation of the critical bifurcation parameter a_c. For a prescribed driving $x_t^{(i)}$, the critical value of the bifurcation parameter a_c is a function of the diffusion ϵ and the size of the lattice L. It was shown in the previous section that the critical bifurca-

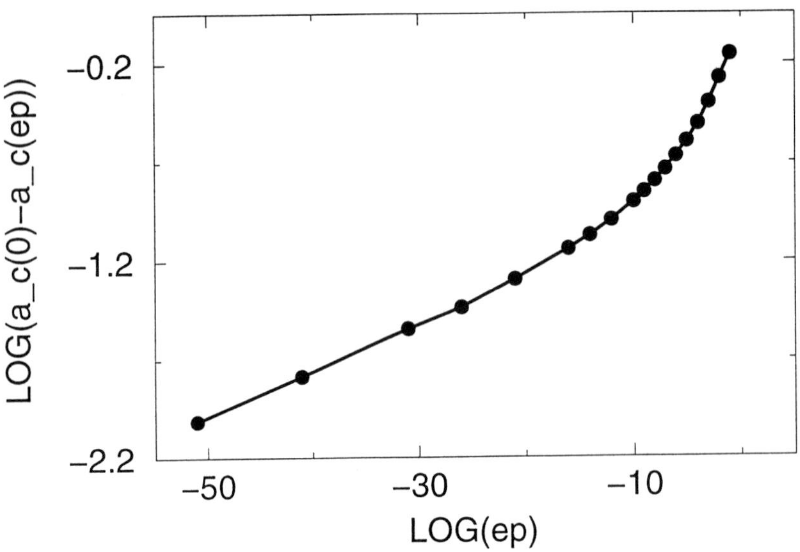

Figure 4: log-log plot of numerically calculated $a_c(0) - a_c(\epsilon)$ versus ϵ with lattice size $L = 100$.

tion parameter value for a single map is $a_c = e$ for x_t uniformly distributed in the interval $(0, 1]$ random driving. It follows that the theoretically predicted critical parameter value for the uncoupled map lattice at $\epsilon = 0$ is $a_c(0) = e = 2.71828...$ for any finite lattice size L. On the other hand, the numerically calculated value of a_c at $\epsilon = 0$ and lattice size $L = 100$ is 2.69886. In the calculation, the CML is settled for 10^6 iterations and then monitored for 500000 iterations to cross a prescribed threshold. These are not contradictory results, but different manifestations of the same phenomenon [27]. To obtain a better estimate for a_c, a longer run to settle transients would be needed.

One of the interesting and important consequences of a non-zero diffusion ϵ is a significant decrease in the critical value of the bifurcation parameter. Preliminary calculations of the critical parameter value a_c as a function of the diffusion strength ϵ are depicted in Fig. 4. It shows a log – log plot of numerically calculated $a_c(0) - a_c(\epsilon)$ versus ϵ suggesting a *power-law scaling* as diffusion approaches zero.

One of the most useful and frequently used classifiers of intermittent systems is the distribution of laminar phases (inter-spike statistics). For temporally intermittent systems there is an easy procedure to collect these statistics: a threshold is prescribed and then the signal is monitored for frequency and duration below threshold. Using that information, it is then possible to form

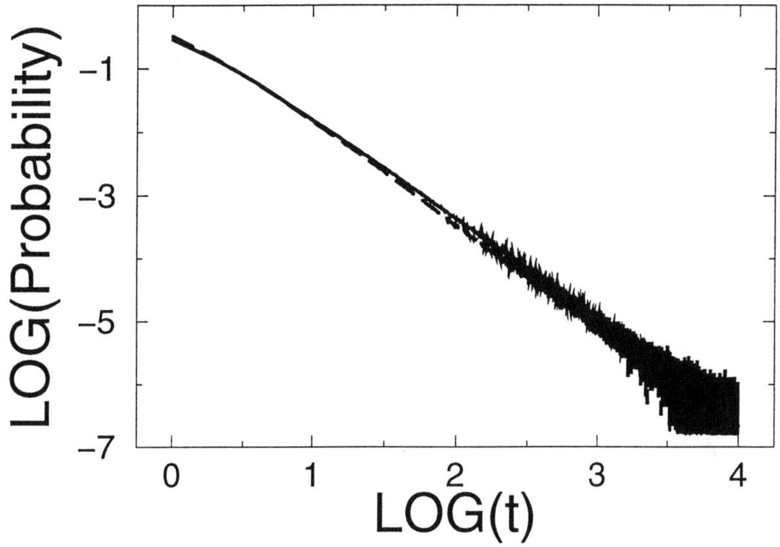

Figure 5: A numerically calculated log-log plot of relative probability of a laminar phase of length t versus t for max probe. A solid line is for $a = 2.346$, ϵ=0.001 and $L = 100$; a dashed line is for $a = 2.051$, $\epsilon = 0.25$ and $L = 20$. CML was settled for 10^7 iterations, and then laminar phase statistics was collected for 10^9 iterations.

a statistic for the distribution of laminar phases. This simplicity in obtaining the distribution of laminar phases makes it useful for identification of types of intermittency observed in experimental and real physical situations. One of the problems of using distribution of laminar phases for spatio-temporal bursting systems is the addition of *spatial* complexity to the problem. There is a number of ways one could assign a quantitative number to describe spatial complexity of the lattice at any given point in time. For example, as shown in Fig. 3b, one could use a point probe at a fixed location in space to classify the behavior of the lattice. Unfortunately, this completely ignores the spatial richness of the lattice, since bursts are intermittent in space as well as in time. A better possible classifier of the instantaneous state of the lattice is a maximum of the lattice at any given time, leading to use time series

$$p_t = \max_{1 \le i \le L} |y_t^{(i)}| \tag{19}$$

for the distribution of laminar phases. This probe still ignores most of the spatial structure of the bursts, but at least it captures total amount of time the CML stays laminar. Another possible classifier of the state of the lattice

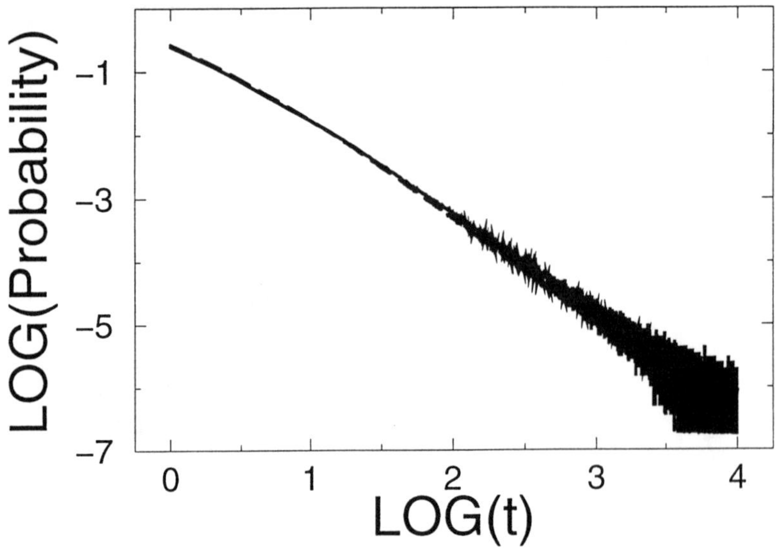

Figure 6: A numerically calculated log-log plot of relative probability of a laminar phase of length t versus t for averaged area probe. A solid line is for $a = 2.036$, ϵ=0.25 and $L = 100$; a dashed line is for $a = 2.245$, $\epsilon = 0.01$ and $L = 20$. CML was settled for 10^7 iterations, and then laminar phase statistics was collected for 10^9 iterations.

is the averaged area above laminar state:

$$A_t = \frac{1}{L}\sum_{i=1}^{L}|y_t^{(i)} - y_{laminar}| \qquad (20)$$

One of the possible benefits using this classifier is that in addition to the size of the burst, it also takes into account the number of bursts of various heights, and the spatial extent of the bursts. Fig. 5 and 6 show numerically calculated distributions of laminar phases for the maximum probe and the averaged area probe just above criticality. These calculations were performed for various diffusions $\epsilon < 1$ and lattice sizes L. Both figures strongly suggest that there is a *universal power law scaling* in the distribution of laminar phases at the onset of bursting for both types of probes.

4 Discussion

One of the most remarkable features of a bursting spatio-temporal CML model is the capability to organize apparently random behavior of a *micro-*

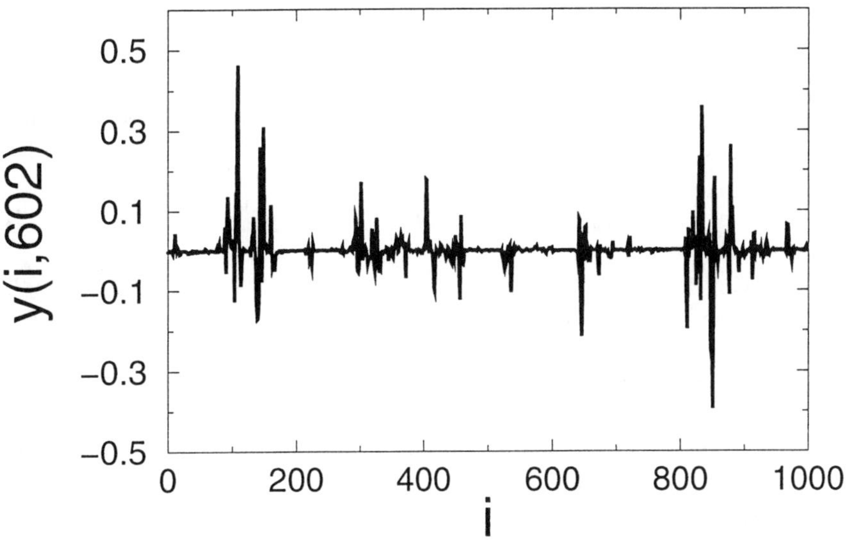

Figure 7: Instantaneous snapshot of CML to model phenomenologically atmospheric temperature fluctuations. Parameters used are $a = 2.3$, $\epsilon = 0.1$ and $L = 1000$.

scopic background into coherent, *macroscopic* observable bursts. The utility of this model lies in its genericity, and the simplicity of its ingredients, making it useful for phenomenological modeling of a variety of physical problems. For example, it models any scenario comprising a large number of identical coupled oscillators which are driven by some noise in the environment. If the noise is either uncorrelated or weakly correlated spatially, then in some parameter regime the system can produce spontaneous spatio-temporal bursting. Thus bursting should occur naturally in systems embedded in a *noisy* background that is *statistically homogeneous* both in space and time.

As a final note to illustrate the possible utility of phenomenological model such as CML, we will give a very brief description of a current attempts to inject this models into a refractive-effects simulator [28]. Ray bending in the atmosphere depends on the refractive index of the atmosphere, which is a function of temperature. There is strong evidence to suggest that atmospheric turbulence is highly spatio-temporally intermittent [29], [30], [31]. There are large portions of atmosphere where intense turbulence is confined to sparse active sections surrounded by broad quiescent regions. This is reflected in a figure shown in Tatarskii's paper [29] displaying spatial temperature fluctuations in the atmosphere at a height of $400m$ on a clear sunny day over the

Russian steppes (prairie).

A phenomenological CML model of atmospheric temperature fluctuations around a mean temperature must include negative as well as positive fluctuations. This is easily accomplished by replacing the quadratic logistic map in equation (18) with a cubic map:

$$F(y_t^{(i)}) = ax_t^{(i)}y_t^{(i)}(1 - y_t^{(i)})(1 + y_t^{(i)}) \tag{21}$$

and modifying the driving function $x_t^{(i)}$ to be uniformly distributed random numbers in the interval $[-1, 1]$. Fig. 7 shows resulting snapshot of the lattice.

In this paper we discuss models of spontaneous bursting. We started with a simple model for temporal bursting, called On-Off intermittency, and we extended it to spatio-temporal bursting. For a temporal On-Off intermittency there is a solid theoretical, numerical and experimental basis for the characterization of the master/slave formulation for a variety of chaotic and random drivings [11], [12], [9], [32], [33]-[34]. This type of intermittency manifests a variety of universal scaling relationships which hold at the onset of criticality. In the case of the CML model of spatio-temporal bursting, preliminary numerical experiments suggest that there are similar universal scaling relations that hold for a large class of random and chaotic drivings.

Acknowledgments: This work was supported in part by Naval Surface Warfare Center Independent Research Program, and Infrared Analysis, Measurements and Modeling Program (IRAMP) of the Office of Naval Research. N.P. is supported through NCEE postdoctoral fellowship.

References

[1] L. Prandtl and O.G. Titjens, *Applied Hydro- and Aeromechanics*, Dover Publications, (1957).

[2] O. Reynolds, *An experimental investigation of the circumstances which determine whether the motion of water shall be direct or sinuous, and of the law of resistance in parallel channels*, Phil. Trans. Roy. Soc. **174**, 935-82 (1883).

[3] M. Couette, *Investigation on the Friction of Fluids*, (French), Ann. Chim. Phys., vol 21, p. 433, (1890).

[4] Y. Pomeau, P. Manneville, Commun. Math. Phys., **74**, 1889 (1980).

[5] C. Grebogi, E. Ott, and J. A. Yorke, Phys. Rev. Lett. **50**, 93, (1983).

[6] N. Platt, E. A. Spiegel, and C. Tresser, Phys. Rev. Lett. **70**, 279 (1993).

[7] A. S. Pikovsky, Z. Phys. **B55**, 149, (1984).

[8] H. Fujisaka, and T. Yamada, Prog. Theor. Phys. **74**, 918, (1985).

[9] L. Yu, E. Ott, and Q. Chen, Phys. Rev. Lett. **65**, 2935 (1990).

[10] L. Yu, E. Ott, and Q. Chen, Physica D **53**, 102 (1991).

[11] J. F. Heagy, N. Platt, and S. M. Hammel, Phys. Rev. E. **49**, p. 1140-1150 (1994).

[12] N. Platt, S. M. Hammel and J. F. Heagy, Phys. Rev. Lett. **72**, p. 3498-3501 (1994).

[13] W.H. Press, S.A. Teukolsky, W.T. Vetterling and B. P. Flannery, *Numerical Recipes in Fortran*, Second Edition, Cambridge University Press (1992).

[14] The autocorrelation function for a discrete process $\{x_n\}_{n=1}^{\infty}$ is defined as $C(k) = \frac{E[x_n x_{n+k}]-E[x]^2}{E[x^2]-E[x]^2}$, where $E[x] = \lim_{N\to\infty} \frac{1}{N}\sum_{n=1}^{N} x_n$ is the expectation of the process. Note that $C(0) = 1$. For the forcing mappings under consideration the autocorrelation function is

$$C(k) = \begin{cases} 0 & \text{if } x_n \text{ comes from the uniformly distributed white noise} \\ 0 & \text{if } x_n \text{ comes from the tent map} \\ \frac{1}{2^k} & \text{if } x_n \text{ comes from the dyadic map} \end{cases}$$

for $k \geq 1$.

[15] Lyapunov exponent for the discrete map $x_{n+1} = f(x_n)$ is defined as $\lambda = \lim_{n\to\infty} \frac{1}{n} \ln |Df^{(n)}| = \lim_{n\to\infty} \frac{1}{n} \ln |Df(x_{n-1})Df(x_{n-2})...Df(x_0)|$, where $Df = \frac{df}{dx}$. It is a straight forward calculation to show that a discrete maps (5-6) have a Lyapunov exponent of $\ln 2$.

[16] W. Feller, *An Introduction to Probability Theory and Its Applications*, Vol. II, Third Edition, (Wiley and Sons, Inc., 1968)

[17] K. Kaneko, Physica D **34**, p. 1-41 (1989).

[18] T. Bohr and O.B. Christensen, Phys. Rev. Lett. **63**, 2161 (1989).

[19] F. Kaspar and H.G. Schuster, Phys. Lett **113A**, 451 (1986).

[20] G. Hu and Z.L. Qu, Phys. Rev. Lett. **72**, 68 (1994).

[21] J.D. Keeler and J.D. Farmer, Physica D **23**, 413, (1986)

[22] J.P. Crutchfield and K. Kaneko, Phys. Rev. Lett **60**, 2715 (198).

[23] P. Bak, Chao Tang and K. Wiesenfeld, Phys. Rev. A **38**, 364 (1988)

[24] A.S. Pikovsky and J. Kurths, Phys. Rev. E **49**, 898 (1994)

[25] J. Kurths and A.S. Pikovsky, *Symmetry breaking in distributed systems and modulational spatio-temporal intermittency*, to appear in Chaos, Solitons and Fractals.

[26] H.L. Yang and E.J. Ding, Phys. Rev. E **50**, R3295 (1994)

[27] For the definition of the burst used in the text, the numerical calculation of the bifurcation parameter for the uncoupled lattice map is equivalent to running L independent realizations for a single map and monitoring *any* of the trials crossing the threshold.

[28] S.M. Hammel and N. Platt, *Topological description of mirage effects*, will appear in SPIE proceedings (1995).

[29] V. I. Tatarskii, Radio Science **22**, p. 859-865 (1987).

[30] A.N. Kolmogorov, J. Fluid Mech. **13**, p. 82 (1962).

[31] D.K. Wilson and J.C. Wyngaard, in preparation

[32] M. Ding and W. Yang, *Distribution of First Return Time in Fractional Brownian Motion and Its Application to the Study of On-Off Intermittency*, to appear in Phys. Rev. E.

[33] P.W. Hammer, N. Platt, S.M. Hammel, J.F. Heagy and B.D. Lee, Phys. Rev. Lett. **73**, p. 1095-1098 (1994).

[34] Y.H. Yu, K. Kwak and T. K. Lim, Phys. Lett. A **198**, 34. (1995).

Time Periodic Spatial Disorder in a Complex Ginzburg–Landau Equation

M.Bazhenov, M.Rabinovich, and L.Rubchinsky

Institute of Applied Physics, Russian Academy of Science
46 Uljanov Str., 603600 Nizhny Novgorod, Russia

Abstract. The phenomenon of time-periodic evolution of spatial chaos (1) is investigated in the frame of a one and two-dimensional complex Ginzburg–Landau equation. It is found that there exists a region of the parameters at which a disordered spatial distribution of the field behaves periodically in time; the boundaries of this region are determined. A system of ordinary differential equations describing spatial disorder is derived. The effect of the size of the system on the shape and period of oscillations is investigated. It is established that in a two-dimensional case the regime of time periodic spatial disorder arises only in the narrow band and the critical width of the band is estimated. The phenomenon investigated in this paper indicates that a family of limit cycles with finite basins may exist in the functional phase space of complex Ginzburg–Landau equation in finite regions of the parameters.

INTRODUCTION

A complex Ginzburg–Landau equation (CGLE)

$$\partial_t a = a - (1 + i\beta)|a|^2 a + (1 + i\alpha)\partial^2_{xx} a \tag{1}$$

that describes the behavior of a space-extended system near the Hopf bifurcation point is one of the most popular models in the theory of nonequilibrium media. Equation (1) depends on two real parameters α and β and describes a great variety of phenomena even in a one-dimensional case. These include, first of all, the regimes of phase and amplitude turbulence (2–5), hysteresis (4, 5), and spatio-temporal intermittency (6). Our paper is concerned with investigation of a new phenomenon, namely, spatial chaos periodically oscillating in time (see also (1)). This regime, similarly to spatio-temporal intermittency, occurs in the domain $1 + \alpha\beta > 0$, where solutions of equation (1) in the form of plane waves

$$a(x,t) = \sqrt{1 - Q^2}\exp(-i\omega t + iQx) \tag{2}$$

with $\omega = Q^2(\beta - \alpha) + \beta$ are stable with respect to relatively small perturbations if $Q^2 \leq (1 + \alpha\beta)/(3 + \alpha\beta + 2\beta^2)$.

TEMPORAL SYNCHRONIZATION OF IRREGULAR PATTERNS

Consider equation (1) in a bounded domain $G_L = \{-L \leq x \leq L\}$, $L \gg 1$. We seek solutions to this equation among a set of functions of class $C^2(G_L)$ which meet periodic boundary conditions at the boundary $x = \pm L$ of domain G_L. We take as a basic space the Banach space $\mathcal{L}_4(G_L)$ of complex functions with the norm

$$\| a \|_4 = \left(\int_{G_L} |a|^4 dx \right)^{1/4} \tag{3}$$

Note that with such a choice of the norm, the boundedness of G_L is essential because equation (1) has a broad class of solutions that do not belong to $\mathcal{L}_4(G_\infty)$ in the region $G_\infty = \{-\infty < x < \infty\}$.

Equation (1) was investigated numerically by the pseudo-spectral method (7) based on FFT. We took as the initial conditions an arbitrary spatial chaotic regime with strong amplitude fluctuations (amplitude turbulence) and chaotic evolution of time series.

Results of computer experiment are presented in Fig. 1 for different values of the parameters. The symbol "o" labels the points at which the transition process is completed by the onset of the regime of spatial disorder with a complex amplitude $A(x, t)$ periodically oscillating in time. Of course, these points do not cover the entire multitude of the parameter values at which such a regime may be realized. This is explained by two facts. First, the periodic regime occurs not for all initial conditions and, second, the time needed for the periodic regime to set in is rather large ($T > 10^4$) and grows with decreasing distance to the upper boundary of the periodicity domain. The lower boundary of the domain evidently coincides with the boundary of the existence of the regime of amplitude turbulence (3, 4) (the dashed line in Fig. 1). The periodicity domain shown in Fig.1 is divided into two nonintersecting subsets. The latter may be attributed to the difficulty in finding the points corresponding to the periodic regime. However, we failed to obtain a periodic regime at intermediate points, in spite of a great variety of initial conditions we tried and long computation times ($T \sim 5 \cdot 10^4$). Note that the periodic regime at the points labeled by "o" in Fig. 1 may break at large times in the sense that the amplitude and phase of the field change periodically almost at all points of the domain G_L except the set of points of small (but non-zero) measure.

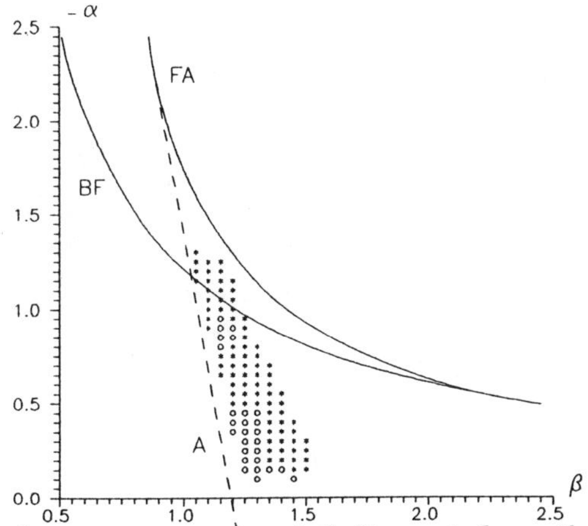

Figure 1: Plane of the parameters α and β. Presented are the Benjamin–Feir limit (solid curve, BF); the amplitude–phase turbulence boundary (solid curve, FA); the boundary of the region with non-zero density of defects (dotted curve, A). Marked are the values of the parameters corresponding to the regimes observed: o – time-periodic spatial disorder; * – spatio-temporal chaos.

The periodicity domain depicted in Fig. 1 is constructed for the system of length $2L = 600$ ($G_L = \{-L \leq x \leq L\}$). Besides, we considered a system of length $2L = 6000$. The oscillatory period was invariant relative to the length of the system for all values of the parameters.

Figure 2 shows the spatio-temporal plot of the real part of the field in the regime that sets in for the values of the parameters from the periodicity domain ($\beta = 1.15$; $\alpha = -0.9$). For comparison, Fig. 3 presents a spatio-temporal plot of the real part of the field corresponding to the chaotic regime ($\beta = 1.15$, $\alpha = -1.1$). It is clearly seen that developed amplitude turbulence having a large number of holes chaotically filling domain G_L occurs in both cases. Note that the spatial distribution of the holes depicted in Fig. 2 is chaotic on the whole but contains local subsets of the structures arranged quite regularly. The time evolutions of the spatial distributions shown in Figs. 2 and 3 are different in principle. In the first case, the motion of the holes is rather complicated, it includes processes of the birth and disappearance of the structures, but it is regular in time. Whereas in the second case, the motion is completely chaotic and no recursion of spatial distribution occurs. The time evolution of the field amplitude measured at the point $x = 0$ is shown in Fig. 4.

For a more detailed analysis of the oscillations of the complex amplitude $a(x,t)$ of the field we employ the metrics associated with the norm (3). We have

$$\rho(u,v) = \left(\int_{G_L} |u - v|^4 dx\right)^{1/4} \tag{4}$$

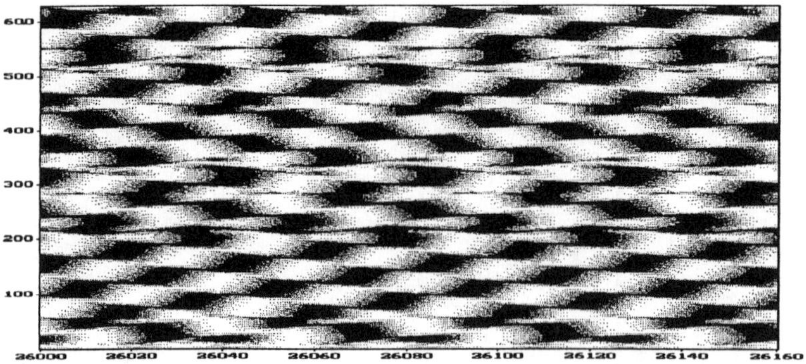

Figure 2: Spatio-temporal plot of the real part of the field for $\beta = 1.15$ and $\alpha = -0.9$. The phase of the field was multiplied by $e^{i\omega t}$ at each time step, which removed phase advance.

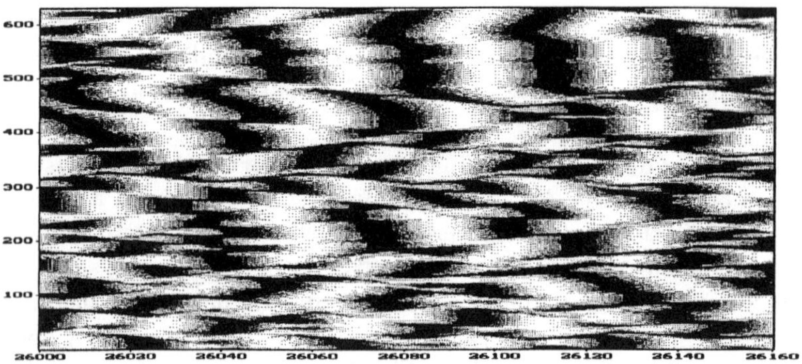

Figure 3: Spatio-temporal plot of the real part of the field for $\beta = 1.15$ and $\alpha = -1.1$. The periodic evolution of the field in time (Fig. 2) is completely replaced by chaotic behavior as the parameters are changed.

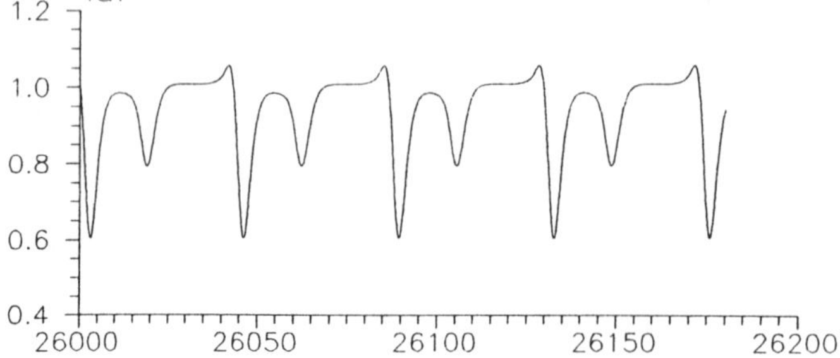

Figure 4: The evolution of the field amplitude in time for $\beta = 1.15$ and $\alpha = -0.9$ at the point $x = 0$.

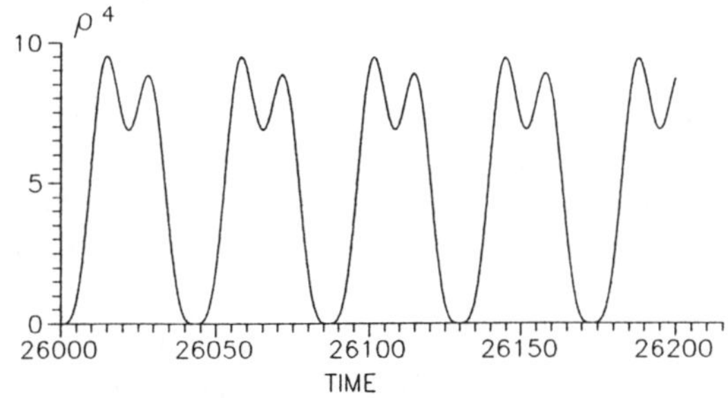

Figure 5: Distance $\rho^4(a(x,t_0), a(x,t_0+t))$ versus time for $\beta = 1.15$ and $\alpha = -0.9$. The function ρ^4 oscillates in time with period $T \approx 43$.

Figure 5 shows the time dependence of $\rho^4(a(x,t_0), a(x,t_0+t))$ for $\beta = 1.15$ and $\alpha = -0.9$. As was to be expected, it is an oscillatory dependence, the distance between the functions $a(x,t_0)$ and $a(x,t_0+t)$ periodically vanishes to zero, and the oscillatory period coincides with the period of the variations of $a(x,t)$ (see Fig. 4). It should be noted that we measured the distance between the complex amplitudes $a(x,t)$ at different instants of time. In order to eliminate the linear phase advance, the variable $a(x,t)$ was multiplied at each step by $\exp(i\omega t)$, where the frequency ω was determined in experiments.

The results obtained indicate that the regime of time periodic spatial disorder is stable not only in the linear approximation but also for a relatively broad class of initial perturbations. However, the problem of structural stability of this regime, i.e., of stability with respect to small perturbations of the right-hand side of equation (1), is unclear yet. In this connection we would like to recall the paper (8) in which the structural stability of hole solutions (9) was investigated and it was shown that perturbations of the form $\pm\delta|a|^4 a$, $\delta \ll 1$ may either reinforce or break stability of such solutions, depending on their sign. Following (8) we investigated the stability of the periodic regime of interest with respect to perturbations of the right-hand side of the equation of the form $\pm\delta|a|^4 a$ (where $\delta \sim 0.005$ and $\Delta t \sim 0.01$ is integration step). It was found that the temporally periodic regime is structurally stable in all the considered cases. The period of oscillations is a continuous function of the right-hand side of equation (1) although the shape of the oscillations varies significantly with the change of the sign of perturbation.

Results of numerical simulation of a 2D CGLE also confirm structural stability of the regime of time periodic spatial disorder considered above. We investigated a region in the form of a narrow stripe ($L_x \times L_y$) with periodic boundary conditions in both directions. The regime of spatio-temporal chaos that is inhomogeneous in both directions was chosen as initial conditions. It is found that, if the stripe width is smaller than the critical value ($L_y^{cr} \approx 29$), a regime of time periodic spatial disorder is established in the system for the values of the parameters from periodicity region (see Fig. 1). The field persists to be chaotic only along the x-axis and is homogeneous along the y-axis. If the width of the region exceeds L_y^{cr}, there is no time periodicity, and the regime of spatio-temporal chaos is retained in the system. Thus the regime of spatial disorder periodically oscillating in time is quasi-one-dimensional in the case of a 2D CGLE too, but its onset in such a system indicates that the regime is structurally stable.

The periodic regime described above may be born as a result of the Hopf bifurcation of a stable spatial distribution containing a family of holes. Such a scenario looks quite probable if we consider the system (1) as a chain of coupled oscillators. In this representation, such a regime may be interpreted as the result of complete synchronization of the oscillators having different intensities and, consequently, different oscillatory frequencies. A similar regime of synchronization for a 2D CGLE was investigated in (10).

DISCUSSION

The results of computer experiment presented above indicate that there exists a limit cycle with a finite basin of attraction in the phase space of system (1). In a general case, the phase space may evidently have a rather complicated organization and contain a family of periodic trajectories of equal periods. This is also confirmed by the fact that different oscillations may be self-excited in the system at different initial conditions. The period of oscillations is, actually, the only parameter invariant to the length of the system and to initial conditions.

Computer experiment shows that the transition from a periodic to a chaotic regime occurs via intermittency. The time needed for the periodic excitations to set in grows with decreasing distance to the boundary of periodicity domain and, starting from some values of the parameters, we fail to detect periodicity. The oscillations near the boundary are periodic at some points along the x-coordinate and aperiodic at the others. The aperiodicity domain changes its position in space with time and, thus, time-periodic oscillations are interrupted by chaotic outbursts.

The regime of time periodic spatial disorder obtained in numerical experiment may, in principle, be described analytically. Taking into consideration that any periodic function of period T may be expanded in the Fourier series in a complete system of orthogonal T-periodic functions, we can represent the solution $a(x,t)$ of equation (1) in the form

$$a(x,t) = \left(\sum_{k=0}^{\infty} A_k(x) \psi_k(t) \right) e^{-i\omega t}, \tag{5}$$

where $\{\psi_k(t)\}$ is a complete system of orthogonal functions of period T. One need not necessarily write the term $\exp(-i\omega t)$ in an explicit form, but such a representation is quite natural if one recalls that the spatially homogeneous solution to equation (1) has the form $a(t) = \exp(-i\omega t)$.

An approximate solution to equation (1) may be written as a finite series

$$a_m(x,t) = \left(\sum_{k=0}^{m} A_k(x)\psi_k(t)\right) e^{-i\omega t} \tag{6}$$

Since the solution (6) is not an exact solution of equation (1) the difference

$$F(a_m) = \frac{\partial a_m}{\partial t} - a_m + (1+i\beta)|a_m|^2 a_m - (1+i\alpha)\frac{\partial^2 a_m}{\partial x^2} \tag{7}$$

cannot be identical to zero. Hence there arises the problem of minimizing, in a certain sense, this difference by choosing an appropriate function $A_k(x)$ $(k = 0,\dots,m)$.

By using the Galerkin method we find a system of ordinary differential equations (ODE) for the variables $A_k(x)$:

$$\int_0^T F\left(\sum_{k=0}^{m} A_k(x)\psi_k(t)e^{-i\omega t}\right)\bar{\psi}_s(t)e^{i\omega t}dt = 0 \quad (s = 0,\dots,m), \tag{8}$$

where $\bar{\psi}$ is a complex conjugate of ψ.

The function $a_m(x,t)$ is not an exact solution of equation (1), but it can be represented as an exact solution of a reduced system $\partial a_m/\partial t = L_m(S(a_m))$ where S is the operator from the right-hand side of equation (1) and L_m is the operator of truncation of the Fourier series.

In passing from the series (5) to the finite series (6) the choice of a system of orthogonal T-periodic functions $\{\psi_k(t)\}$ becomes of fundamental significance. The system of functions of the form $\psi_k(t) = \exp(\pm ik\Omega t)$, $\Omega = 2\pi/T$ that is acceptable in many cases is not quite suitable here because the real oscillations observed in the system differ strongly from the harmonic ones (see Fig. 4) and a great number of harmonics must be taken into account for adequate description of the behavior of the system. Nevertheless we can assert that for any time-periodic solution $a(x,t)$ of equation (1) there exists a Galerkin approximation of a sufficiently high order which approximates this solution with preset accuracy. The approximate solution $a_m(x,t)$ at each fixed instant of time is described by a finite-dimensional ODE system. Consequently, at each time instant $t = t_0$, the spatial field distribution $a(x,t_0)$ observed in numerical experiment is approximated to arbitrary accuracy by the solution of a finite-dimensional dynamical system (along x) for $a_m(x,t_0)$. However, it is still to be found whether the exact solution $a(x,t_0)$ is finite-dimensional.

The considerations presented on the origin of the observed phenomenon are confirmed by the recent results obtained in (11) where an extended system in the form of a "one-way coupled logistic lattice" was investigated. It was revealed that spatial chaos with temporal periodicity may stably exist in a broad region of parameters, in particular, at sufficiently strong coupling. Such a regime appears to be quite natural in terms of the synchronization phenomenon since the medium is discrete and unidirectional. In this case, however, forced synchronization occurs instead of mutual synchronization of structures (or oscillators) because the coupling in the chain is unidirectional. Whereas the spatial disorder is determined by the difference of intensity pulsations in each element of the chain.

ACKNOWLEDGEMENTS

This research was made possible in part thanks to the Russian Foundation for Basic Research (project code N 94–02–03263-a) and the International Science Foundation grant (project N NOU 000).

REFERENCES

1. Bazhenov, M.V., Rabinovich, M.I., and Fabrikant, A.L., in *Proc.of the KIT Int. Workshop on the Physics of Pattern Formation in Complex Dissipative Systems*, ed. by S.Kai, Singapore: World Scientific, 1992, pp.361–373.
2. Sakaguchi, H., *Progr. Theor. Phys.* **84**, 792–800 (1990).
3. Gill, L., *Nonlinearity* **4**, 1213–1222 (1991).
4. Shraiman, B.I., Pumir, A., Saarloos, W., Hohenberg, P.C., Chate, H., and Holen, M., *Physica D* **57**, 241–248 (1992).
5. Bazhenov, M.V., Rabinovich, M.I., and Fabrikant, A.L., *Phys. Lett. A* **163**, 87–94 (1992).
6. Chate, H., in *Proc. of the KIT Int. Workshop on the Physics of Pattern Formation in Complex Dissipative Systems*, ed. by S.Kai, Singapore: World Scientific, 1992, pp.374–386.
7. Canuto, C., Hussaini, M.J., Quarteroni, A., and Zand, T.A., *Spectral Methods in Fluid Dynamics*, Berlin: Springer, 1988.
8. Aranson, I., Kramer, L., Popp, S., Stiller, O., and Weber, A., *Phys. Rev. Lett.* **70**, 3880–3883 (1993).
9. Nozaki, K., and Bekki, N., *J. Phys. Soc. Japan* **53**, 1581–1582 (1984).
10. Bazhenov, M., and Rabinovich, M., *Physica* **D 73**, 318–334 (1994).
11. Willeboordse, F.H., and Kaneko, K., *Phys. Rev. Lett.* **73**, 533–536 (1994).

8. GEOPHYSICS

An EOF study of the Earth's ozone field

D.Manin, R. Everson, and L. Sirovich
The Rockefeller University, New York City, 10021

Abstract

In a previous paper we discussed the spectral properties of the Earth's ozone layer, obtained using Empirical Orthogonal Function decomposition of the Total Ozone Mapping Spectrometer (TOMS) database. Here we present other aspects of the analysis, including the EOF method adapted for incomplete datasets, analysis of spatial structure and temporal variation of first several eigenfunctions, and an extended study of small-scale properties of ozone concentration fields.

Geophysical datasets collected by satellite-based devices currently provide a wealth of information which needs to be assimilated and interpreted. In this paper we analyze the global ozone concentration fields measured by Total Mapping Ozone Spectrometer [1]. The interest in atmospheric ozone is due to both its environmental role (ozone hole phenomenon [2, 3]), and the fact that it can be regarded as a passive tracer, thus providing an additional insight in the properties of atmospheric circulation. This is an extended version of a previous paper [4], where the principal attention was paid to the spectral properties of the ozone concentration fields.

The TOMS device was installed on a satellite, Nimbus, which completed sixteen rotations in a 24-hour period, on a south-north synchronous orbit. Thus, every global field (see example in Fig. 1) consists of 14 separate records taken roughly 1.5 hours apart. Every field is represented on a 288×180 grid, so the resolution is order 100 km. The measurement method is based on the absorption by ozone of light reflected from the surface, so it fails in the polar night regions (a black circle around North Pole in Fig. 1). Occasional technical failures also lead to missing pixels or whole records, as also can be seen in Fig. 1. The TOMS database contains daily ozone fields for the period of 13 years, which amounts to 270 Gigabytes of data. Due to computational requirements, we analyzed the subset consisting of 4 snapshots per month (approximately weekly).

The Karhunen-Loeve (KL) procedure [5, 6, 7] is ideally suited for analyzing massive spatiotemporal datasets, due to its ability to extract persistent features. In particular, the snapshot method [8] considerably reduces the needed computation effort. However, the incompleteness of data required modification of the methodology. We discuss the numerical procedure in Section 1. Section 2 is devoted to the large-scale temporal and spatial properties of the ozone field. In Section 3 we report the results on statistical properties on the smaller scale.

1 KL procedure for gappy data

The Karhunen–Loeve procedure, applied to the array of data $c(x,t)$, is aimed at finding the best sum-of-products representation

$$c(x,t) = \sum_k \phi_k(t)\psi_k(x) \tag{1}$$

The variational problem of minimizing the net error

$$\Delta = \left| c - \sum_k \phi_k(t)\psi_k(x) \right|^2 \to \min \tag{2}$$

reduces to the linear eigenproblem

$$\hat{K}_x \psi_k = \lambda_k \psi_k, \quad \phi_k = (\psi_k, c)_x \tag{3}$$

where round brackets denote averaging (inner product) in x: $(f,g)_x = \int fg\,dx$; and $\hat{K}_x$ is the spatial correlation operator:

$$K_x(x,y) = (c(x,t), c(y,t))_t, \quad \hat{K}_x\psi = (K_x(x,y), \psi(y))_y \tag{4}$$

The idea of the snapshot method [8] is that because of the symmetry between x and t in this formulation, one can instead solve the eigenproblem for ϕ_k

$$\hat{K}_t \phi_k = \lambda_k \phi_k, \quad \psi_k = (\phi_k, c)_t \tag{5}$$

where $\hat{K}_t$ is the temporal correlation operator. For practical applications, where both x and t are discrete variables, advantage of choosing the appropriate alternative can be essential. Typically with image data (when t enumerates images, and x indexes pixels), the number of images can be much smaller than the number of pixels in each image, and consequently a large gain in performance can be achieved by solving the eigenproblem for the $\hat{K}_t$ matrix, rather than for the much larger $\hat{K}_x$.

For discrete problems, the KL procedure produces the *complete basis* (ϕ_k, ψ_k), so that the net error (2) is zero. The total number of eigenfunctions required is the smaller of the number of snapshots and the number of pixels. Moreover, if the eigenfunctions are numbered in the order of decreasing energy, $\lambda_0^2 < \lambda_1^2 < \ldots < \lambda_k^2 < \ldots$, then every sub-basis comprising the first n eigenfunctions is the best of all n-dimensional bases.

Unfortunately, the problem is much more complicated, and the solution is much less elegant, when the data is partially missing. Denote by $D(x,t)$ the characteristic function (with values from $\{0,1\}$) describing the domain where the data is known. To generalize the optimization problem (2), we should consider integrating over the domain where the data is known:

$$\Delta = \int D(x,t)\Big(c(x,t) - \sum_k \phi_k(t)\psi_k(x)\Big)^2 dx\,dt \to \min \tag{6}$$

It can be easily seen that this problem has infinite number of solutions with $\Delta = 0$. Indeed, if we fill in the gaps in data with arbitrary values and apply the classical KL procedure, we will obtain the basis which represents the filled-in snapshots exactly everywhere, and in particular in regions with authentic data. This means that the KL procedure does not provide a preferable way to fill in the missing data points in the generic situation.

The approach we choose instead is based on the fact that KL eigenfunctions can be actually found independently of each other. Namely, the pair (ϕ_0, ψ_0) provides the best approximation $c(x,t) \approx \phi(t)\psi(x)$, the pair (ϕ_1, ψ_1) optimally approximates the *reduced* dataset $c_1(x,t) = c(x,t) - \phi_0(t)\psi_0(x)$, etc. In this way, we seek the k-th eigenfunction for gappy data by solving

$$\Delta = \int D(x,t)\Big(c_{k-1}(x,t) - \phi_k(t)\psi_k(x)\Big)^2 \, dx\, dt \to \min \tag{7}$$

with

$$c_k(x,t) = c(x,t) - \sum_0^{k-1} \phi_k(t)\psi_k(x) \tag{8}$$

Variational differentiation of (7) over ϕ and ψ gives a pair of nonlinear equations

$$\phi(t) = \frac{\int D(x',t)c(x',t)\psi(x')\, dx'}{\int D(x',t)\psi(x')^2\, dx'} \tag{9}$$

$$\psi(t) = \frac{\int D(x,t')c(x,t')\phi(t')\, dt'}{\int D(x,t')\phi(t')^2\, dt'} \tag{10}$$

Note that when all data is known, $D(x,t) \equiv 1$, and the denominators are constants (rather than functions of x or t), which reduces the task to solving the classical linear eigenvalue problem. Although the nonlinear problem (9,10) cannot be solved directly, it is well suited for solution by iterations (the next approximation being computed by substituting the previous one in the RHS). We cannot prove the convergence, but since the problem is in a sense close to the linear one, for which iterations do converge to the eigenvector with the greatest real part of eigenvalue, we can expect the similar behavior in (9,10). Indeed, the process converged well. However the amount of computation needed to iterate (9,10) for every eigenfunction was still very demanding (several hours per eigenfunction, on the average, for about 600 snapshots by 51840 pixels).

The resulting basis is not necessarily orthogonal. However, on practice the angle between different eigenfunctions,

$$\alpha_{nm} = \arccos \frac{(\psi_n, \psi_m)}{\sqrt{|\psi_n|^2\, |\psi_m|^2}}$$

did not exceed 1–3°. Thus, in what follows we consider ψ_n as a good approximation to the "true" eigenfunctions.

To evaluate the effect of this method, we compared the result with the naive technique of filling the missing data by time-average values and applying the classical KL procedure. The spectra of eigenvalues resulting from the two methods are very close, but there is essential difference in eigenfunctions in the polar regions. Fig. 2 shows the 6th eigenfunction for the southern hemisphere, obtained by both methods. The "naive" procedure creates a huge artifact in the polar night region, where the lack of data is most drastic. Beyond the polar circle, where gaps in the data are rare and random, the results agree well.

2 First eigenfunctions

The ozone content in the Earth's stratosphere is maintained due to the balance between complicated photochemical processes of its generation and destruction. The photochemical lifetime of ozone can range from days to months [9], depending on latitude and altitude, and thus the ozone layer exhibits seasonal variations. It turns out that the first three eigenfunctions capture the entire seasonal variation, and the entire temporal regularity, while the time courses of other eigenfunctions have an appearance of random functions.

The temporal course $\phi_0(t)$ of the zeroth eigenfunction (Fig. 3) is almost constant in time and, consequently, ψ_0 reflects the temporally averaged ozone concentration. The "Fourier coefficient" $\phi_0(t)$ demonstrates weak annual variation of total ozone content (with distinct annual minima in northern winter and secondary minima in southern winter), as well as a weak secular trend of 2.5% per decade (average for the period 1980–92), more pronounced between 1980 and 1986. (For exhaustive investigation of trends of atmospheric ozone, based on TOMS data, see [10]) No other eigenfunction possesses manifest secular trend.

The first and second eigenfunctions (Fig. 4) describe almost pure annual oscillation. The temporal coefficient ϕ_1 has extrema in Spring and Fall, while ϕ_2 is extreme in Winter and Summer. Especially interesting is the south-hemisphere part of ψ_1, with its negative extremum over Antarctica and positive extremum over moderate latitudes. Taking into account that the corresponding temporal coefficient reaches its annual negative extremum in September–October, we conclude that ψ_1 describes Antarctic ozone depletion in Fall, while sunlight generates ozone in mid-latitudes. In other words, ψ_1 is in a way related to the ozone hole phenomenon [11].

ψ_2 is probably related to two different mechanisms. In South Hemisphere, it is non-zero only over Antarctica, and positive in winter, corresponding perhaps to the late-Fall breakdown of Antarctic polar vortex, and consequent fast rise of ozone concentration. In Northern Hemisphere, ψ_1 is positive in Summer in tropics, when and where ozone is generated by sunlight. The peculiar pattern of ψ_2 over NH resembles generic mixing of a passive scalar blob (see, e.g. [12]), however it is not clear how significant this similarity is.

The general structure of following several eigenfunctions reflects the fact that the spatial spectrum of a typical ozone concentration field is a decreasing function of wavenumber (see below). EEFs from 3 to 8, shown in Fig. 5, appear as large-scale structures around the poles, where variation is much greater than in the equatorial region. Neither of them reveals any specific physical mechanism, though some qualitative observations can be made. Thus the temporal variation of EEFs 6 and 7 (Fig. 6) shows sharp peaks of activity in December–January, more pronounced in some years than in others. Spatially, both functions are mostly located around the North Pole, with little variation in the SH. Temporal courses of EEFs 4 and 8 suggest the presence of quasi-biennial oscillations, although the total observation time of 12 years is probably too short for a reliable spectral analysis.

Finally, note that processes in the two hemispheres seem to be largely decoupled, as demonstrated by hemispherical asymmetry of EEFs 3, 5, 6, and 7. Since the atmospheric transport across the equator is indeed small, and the patterns of atmospheric circulation are different in South and North Hemispheres (in part, because of greater influence of orography in the NH), it is natural to perform the EOF analysis for the two hemispheres separately.

This will be presented elsewhere.

3 Small-scale eigenfunctions and spectrum

The spectrum of eigenvalues λ_n, is very important because it shows for how much variance every eigenfunction is responsible [13, 4]. Indeed, the magnitude of projection of $c(x,t)$ onto $\psi_n(x)$ is

$$\begin{aligned} E_n &= \int \left(\int c(x,t)\psi_n(x)\,dx\right)^2 dt \\ &= \int \psi_n(x)\left(\int \psi_n(x')K(x,x')\,dx'\right)dx \\ &= \int \psi_n(x)\lambda_n\psi_n(x)\,dx \\ &= \lambda_n \end{aligned} \tag{11}$$

Hence eigenvalues are measures of average variance per eigenfunction, and the total variance is given by their sum total:

$$\left\langle \int c^2(\mathbf{x})d\mathbf{x} \right\rangle = \operatorname{Tr} K = \sum_n \lambda_n. \tag{12}$$

In Fig. 7 we display in doubly logarithmic form λ_n versus index n. As is seen the variance spectrum falls, to good approximation, on two different power laws

$$\lambda_n \propto \begin{cases} n^{\alpha_i};\ n < 35, & \alpha_i = -0.85 \pm .035 \\ n^{\alpha_o};\ n > 50, & \alpha_o = -1.56 \pm .022 \end{cases} \tag{13}$$

The error bounds appearing in (13) are based only on the least squares fit to the data, and not on the methods used in arriving at the spectrum which appears in Fig. 7.

It is possible to transform this spectrum to the traditional wavenumber spectrum [4]. The key observation here is that higher-order eigenfunctions (see samples in Fig. 5), i.e. those with smaller variance, consist of positive and negative patches of roughly the same length scale. This is natural, in view of the well-known fact that for processes possessing spatial homogeneity, empirical eigenfunctions are nothing else but sinusoids. Indeed, consider for simplicity a one-dimensional process $f(x,t)$, such that the spatial correlation function between points x and y depends only on the distance between them:

$$\langle f(x,t)f(y,t)\rangle_t = K(x-y) \tag{14}$$

Then Fourier transform (denoted by tilde) of the eigenproblem (5) reduces to the equation

$$\tilde{K}(k)\tilde{\psi}(k) = \lambda\tilde{\psi}(k) \tag{15}$$

which is in the generic case solved by sinusoids $\tilde{\psi}(k) = \delta(k-k_0)$, $\lambda = \tilde{K}(k_0)$.

On the global scale, the ozone distribution is nonhomogeneous, owing to latitudinal dependencies of the ozone source/sink and atmospheric circulation, influence of the underlying surface, etc. However, on smaller scales we can expect local homogeneity, and the appearance of higher-order eigenfunction supports this notion. To quantitively verify that small-scale eigenfunctions are characterized by a distinct length scale, and to determine it, we consider the correlation length of each eigenfunction as the position of the first minimum of the spatial correlation function

$$B_\psi(r) = \langle \psi_n(\mathbf{x})\psi_n(\mathbf{y})\rangle_{|\mathbf{x}-\mathbf{y}|)=r}$$

where $|\mathbf{x}-\mathbf{y}|$ stands for the distance between the two points on the sphere. In Fig. 9 a sample correlation function is presented for the eigenfunction ψ_{44}, shown in Fig. 8. These functions B_ψ decay with oscillations, and possess a distinct first minimum corresponding to the average distance L_n between adjacent maxima and minima of ψ_n. This distance can be interpreted as half wavelength.

Wavelength determined in this way can be related to the eigenfunction index by the following argument. Assume that the eigenfunctions are homogeneously distributed in the two-dimensional Fourier space. Then the number of those with wavenumber smaller than some k, is order

$$N \propto k^2, \text{ whence } k \propto N^{1/2}. \tag{16}$$

Assuming, in addition, that the spectral density is decreasing with wavenumber increasing, we can interpret N above as being proportional to the eigenfunction index. Finally, the correlation length, being in the inverse relation with wavenumber, should satisfy

$$L_n \propto n^{-1/2} \tag{17}$$

The correlation length is plotted in Fig. 10 versus the eigenfunction index in log-log coordinates. It is clear from this figure that (17) provides an excellent fit to the data in the two asymptotic regimes. The region of the *knee*, transitory between the two spectral asymptotics, is the only anomaly and it appears as a plateau in the figure and corresponds to just one scale.

$$2\pi/k_* = L_* \approx 4000 \text{ km}. \tag{18}$$

We refer the reader to ref. [4] for the discussion of the knee region.

Now that the relation (17) between eigenfunction index and its wavenumber is established, we can translate the power laws for λ_n, (13) to the wavenumber form. In keeping with customary practice we consider the variance per wavenumber $E_c(k) = k\mathcal{E}_c(k)$. It follows from (13) and (16) that

$$E_c(k) \propto \begin{cases} k^{-2/3}, & k < k_* \\ k^{-2}, & k > k_* \end{cases} \tag{19}$$

The more precise exponents are entered for suggestive reasons. Surprisingly enough, neither value has been predicted by existing theories of passive scalar advection, which propose either $k^{-5/3}$ (Obukhov and Corrsin [14, 15, 16]), or k^{-1} (Batchelor [17], see also Chertkov *et al.* [18]). Leaving the theoretical discussion for elsewhere, here we will present an additional empirical analysis in order to establish the exponent -2 in (19) in a more solid manner.

4 Confirmation of the k^{-2} spectrum

The k^{-2} spectrum of a scalar field can be interpreted as an evidence that the scalar variance is dominated by (curvi)linear discontinuities. In fact, it is this reasoning that lead Saffman [19] to propose the k^{-4} energy spectrum for the two-dimensional turbulence, by arguing that the vorticity field should develop jumps, and concluding that k^{-2} vorticity spectrum means k^{-4} spectrum of kinetic energy.

Fig. 11 shows the field of squared gradient of the concentration (for the arbitrarily picked snapshot of March 1, 1995). It is clearly seen that large gradient values are mostly concentrated in nearly one-dimensional filaments. Consider the structure function of such a field, $D_c(r) = \langle(c(\mathbf{x}) - c(\mathbf{y}))^2\rangle_{|\mathbf{x},\mathbf{y}|=r}$. For the distances $l < r < L$, where l is the characteristic thickness of a jump, and L is the characteristic distance between jumps, the average difference between two points is $\langle(c(\mathbf{x}) - c(\mathbf{y}))^2\rangle = P\delta c^2$, where P is the probability to have a jump between the points $\mathbf{x}$ and $\mathbf{y}$, and δc is the typical amplitude of a jump. Since $P \propto r$, the structure function, should be proportional to r, and correlation should behave as $B_c(r) = \langle c^2\rangle - D_c(r)/2 \sim \langle c^2\rangle - O(r)$. This also implies the power spectrum $E_c(k) = O(k^{-2})$. The actual structure function is shown in Fig. 12. It is practically linear for distances between about 300 to 1500 km.

The correlation function of a scalar field determines the correlation function of its gradient [20] by

$$B_{ll}(r) = \langle g_l(\mathbf{x})g_l(\mathbf{y})\rangle_{|\mathbf{x},\mathbf{y}|=r} = -B_c''(r) \qquad (20)$$

$$B_{nn}(r) = \langle g_n(\mathbf{x})g_n(\mathbf{y})\rangle_{|\mathbf{x},\mathbf{y}|=r} = -\frac{B_c'(r)}{r} \qquad (21)$$

where primes denote derivatives over r, and subscripts l and n refer to longitudinal and normal components of the gradient, that is, its projections onto the straight line joining the two points, and onto its normal, respectively. For the linear $B_c(r)$, this gives $B_{ll} = 0$, which is easily understandable, since the difference in concentration between two points can be written as

$$c(\mathbf{y}) - c(\mathbf{y}) = \int_0^r g_l(s)\, ds \qquad (22)$$

where s is the coordinate along the line joining the two points. That this difference grows on the average as $r^{1/2}$ requires the integrand be delta-correlated on the distances order r. In Fig. 13, B_{ll} and B_{nn} are shown for a particular snapshot. Although B_{ll} is non-zero, it is essentially smaller than B_{nn} at small scales, which is a reasonable agreement, making the allowance for the amplification of errors in numerical differentiation of experimental data. Actually, it is easy to see from (21), (20) that $B_{ll} \ll B_{nn}$ implies that $B_c(r)$ is linear in r.

Thus, the scalar power spectrum $E_c(k) = O(k^{-2})$ seems to pass several tests. We will attempt to provide a theoretical explanation and discuss the failure of available theories to provide such.

5 Conclusions

Global atmospheric ozone concentration fields from TOMS device were analyzed in terms of Karhunen–Loeve method (Empirical Orthogonal Functions). This analysis reveals characteristic spatio-temporal patterns, including the Antarctic ozone hole phenomenon. A new variant of the KL procedure was developed to deal with data containing gaps resulting from both measurement method limitations (polar nights), and occasional device malfunction. By analyzing spatial structure of eigenfunctions, it is shown that the eigenvalue spectrum can be reliably related to the wavenumber spectrum. This wavenumber spectrum contains two power-law domains with a knee between them. An explanation is proposed for the locations and extent of the knee region. Neither of the exponents in the power-law domains is predicted by available theories of passive scalar advection by 2D fluid flow. For the small-scale power-law domain, an additional empirical evidence is obtained in favor of the hypothesis that the corresponding k^{-2} asymptotic is due to the spatial distribution of ozone, typically consisting of regions with almost-constant concentration, separated by one-dimensional near-discontinuities.

The authors are grateful to P. K. Bhartia, B. W. Knight, V. Yakhot, and G. Falkovich for helpful conversations. This work was supported under a grant from NASA-Goddard (NAG 5-2336).

References

[1] P. Grimares and R. McPeters, editors. *Total Ozone Mapping Spectrometer (TOMS) Data, 1978–1993.* NASA Goddard Space Flight Center, 1993.

[2] M.R.Schoeberl, R.S.Stolarski, and A.Kreuger. *International Journal of Supercomputer Applications*, 16(377), 1989.

[3] M.H. Proffitt et al. *Nature*, 347:31, 1990.

[4] L. Sirovich, R. Everson, and D. Manin. Turbulent spectrum of the earth's ozone field. *Phys. Rev. Letters*, 74(13):2611–2614, 1995.

[5] L. Sirovich and R. Everson. *International Journal of Supercomputer Applications*, 6(1):50–68, 1992.

[6] G. W. Stewart. *SIAM Rev.*, 35, 1993.

[7] G. Berkooz et al. *Ann. Rev. Flu. Mech.*, 25, 1993.

[8] L. Sirovich. Turbulence and dynamics of coherent structures. *Quarterly of Applied Mathematics*, 155(3):561–590, 1987.

[9] R. Garcia and S. Solomon. *J. Geophys. Res.*, 90(D2):3850, 1985.

[10] J.R.Herman and R.McPeters. *J. Geophys. Res.*, 98(D7):12783–12793, 1993.

[11] R.P. Wayne. *Chemistry of Atmospheres.* Oxford, 1991.

[12] M. R. Schoeberl and J. T. Bacmeister. *NATO ASI Series*, 18:135, 1992.

[13] B. Knight and L. Sirovich. *Phys. Rev. Lett.*, 65:1356, 1990.

[14] A.M. Obukhov. *Izv. Acad. Nauk SSSR Geogr. Geofiz*, 13:58–69, 1949.

[15] S. Corrsin. *Appl. Phys.*, 22:469, 1954.

[16] A.S. Monin and A. Yaglom. *Statistical Fluid Mechanics*, volume 2. MIT, 1971.

[17] G.K. Batchelor. *J. Flu. Mech.*, 5:113, 1958.

[18] M. Chertkov, G. Falkovich, I. Kolokolov, and V. Lebedev. Statistics of a passive scalar advected by a large-scale velocity field: analytic solution. (preprint wis/94/t/jan-ph). *Phys. Rev. E (to appear)*.

[19] P.G. Saffman. *Stud.Appl. Math.*, 50:377–383, 1971.

[20] S. Panchev. *Random Functions and Turbulence*. Pergamon Press, 1971.

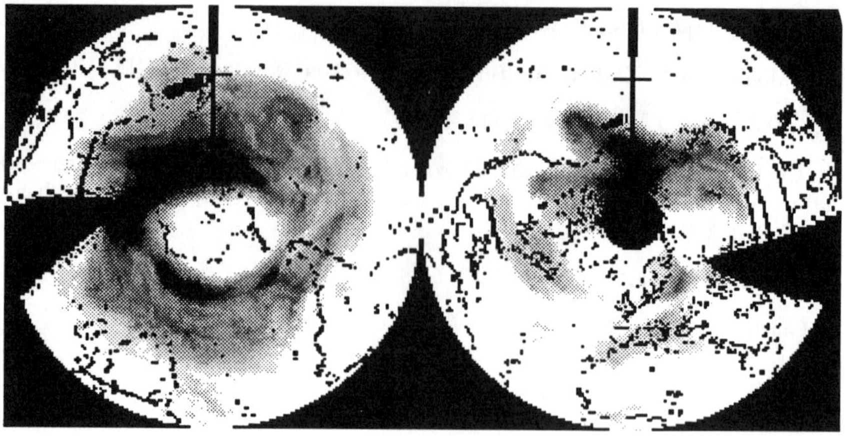

Data of 10/22/80

Figure 1: A typical snapshot of the global ozone field. Outlines of continents are shown.

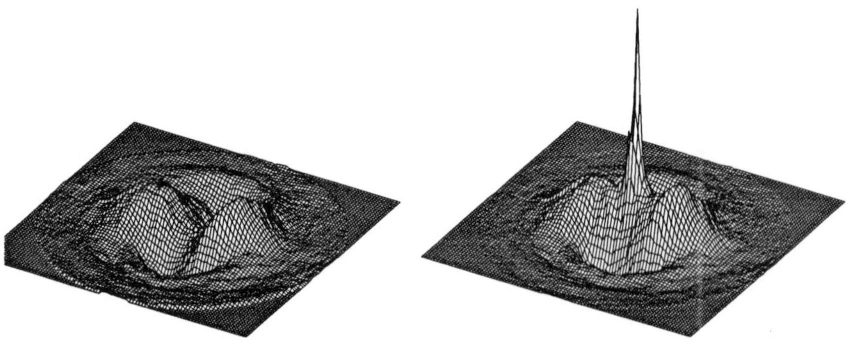

a b

Figure 2: Southern-hemisphere topographic view of ψ_6, obtained by iterations (a) and by the naive method (b).

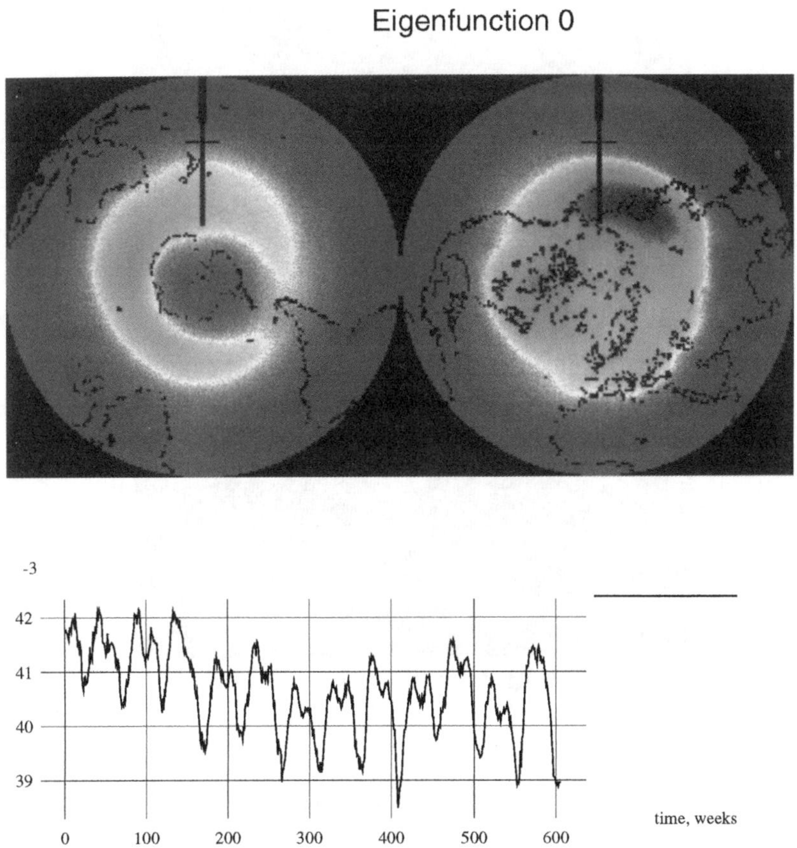

Figure 3: Eigenfunction 0, ψ_0, and its time history.
(See color Plate 3.)

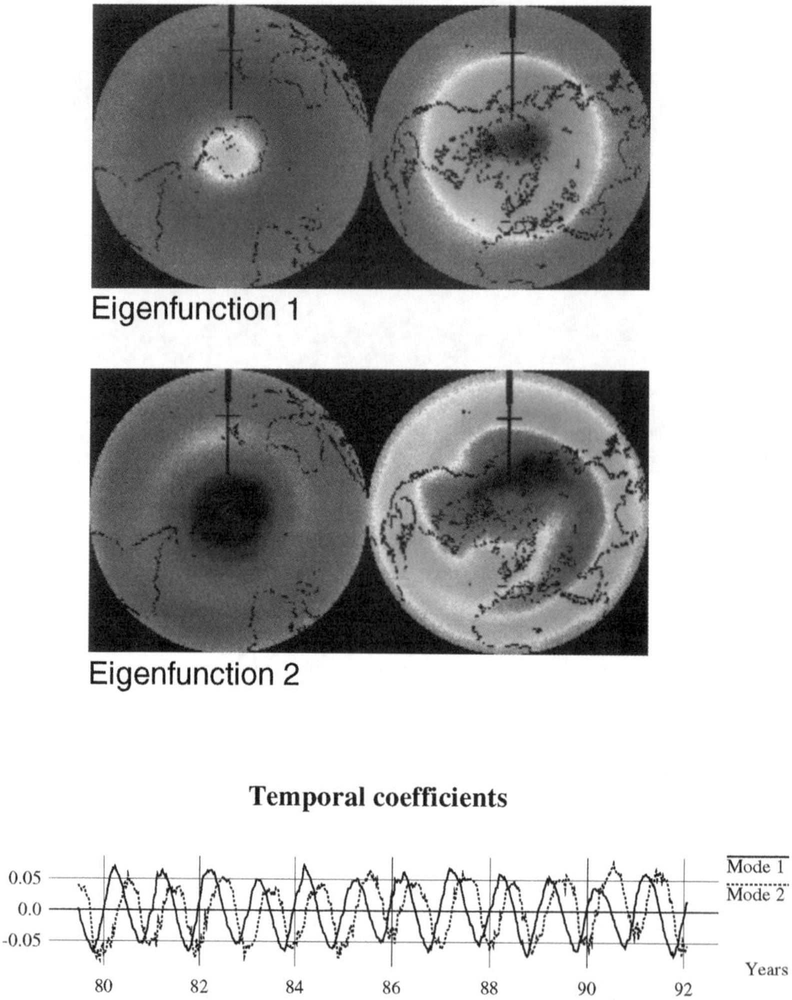

Figure 4: Eigenfunctions 1 and 2, and their time histories.
(See Color Plate 4.)

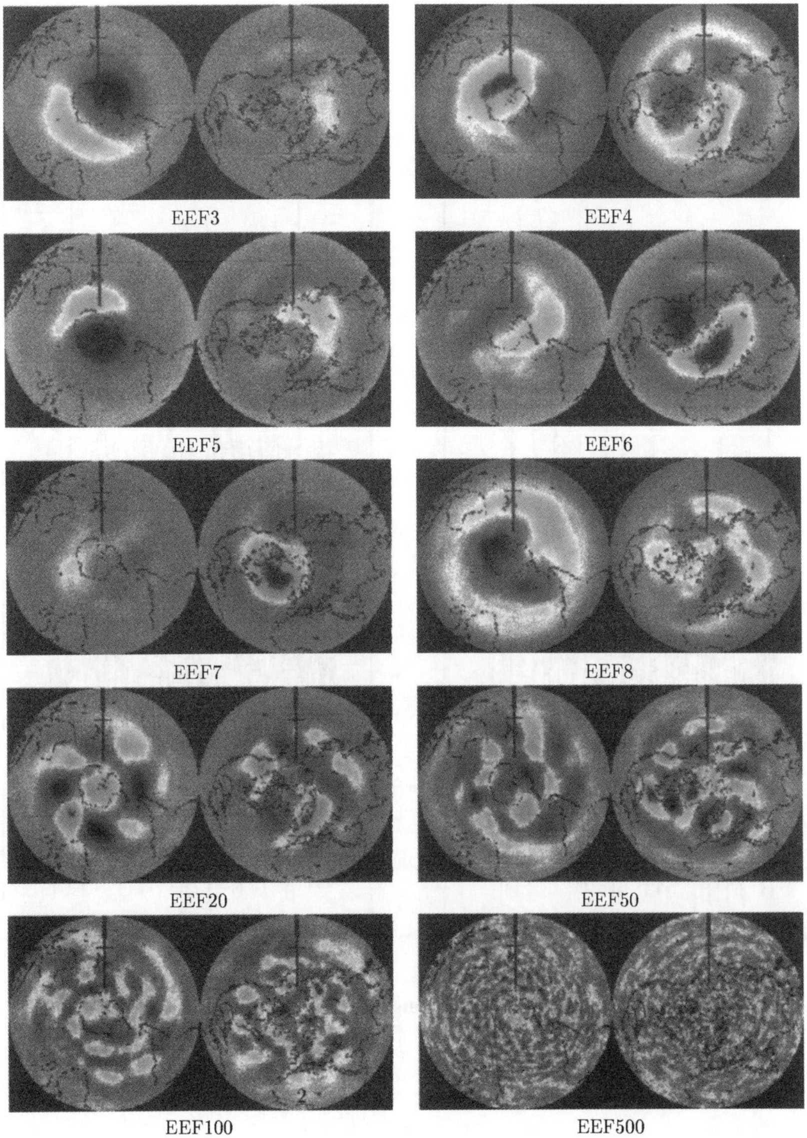

Figure 5: A selection of eigenfunctions.
(See Color Plate 5.)

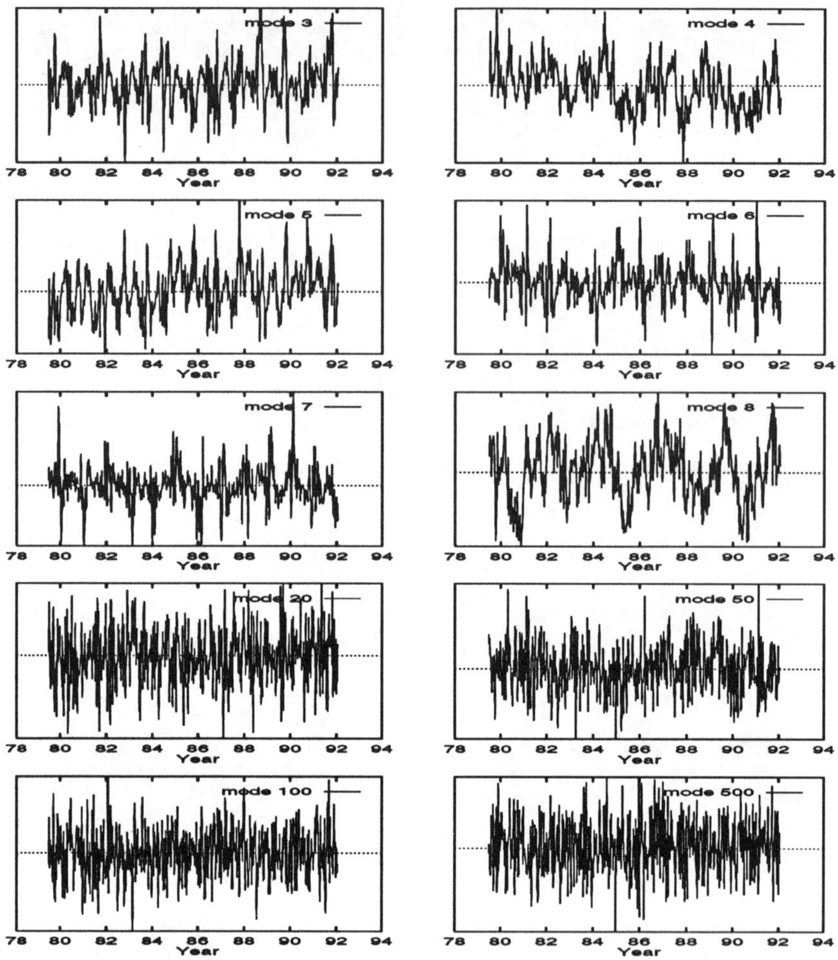

Figure 6: Time histories for eigenfunctions shown in Fig. 5

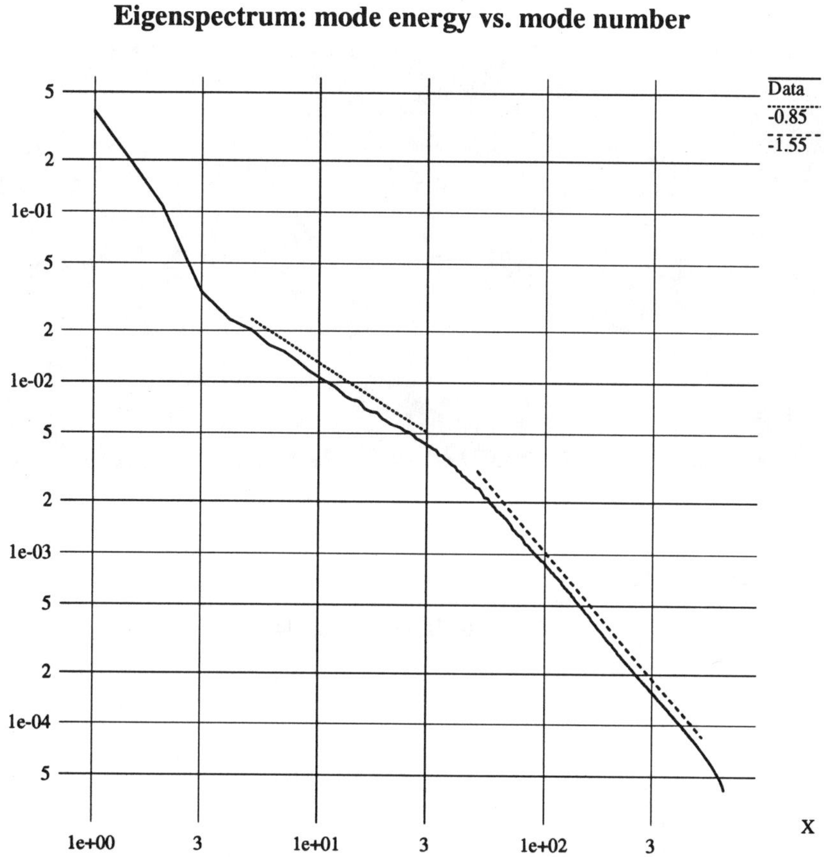

Figure 7: Log–log plot of the global eigenvalue spectrum.

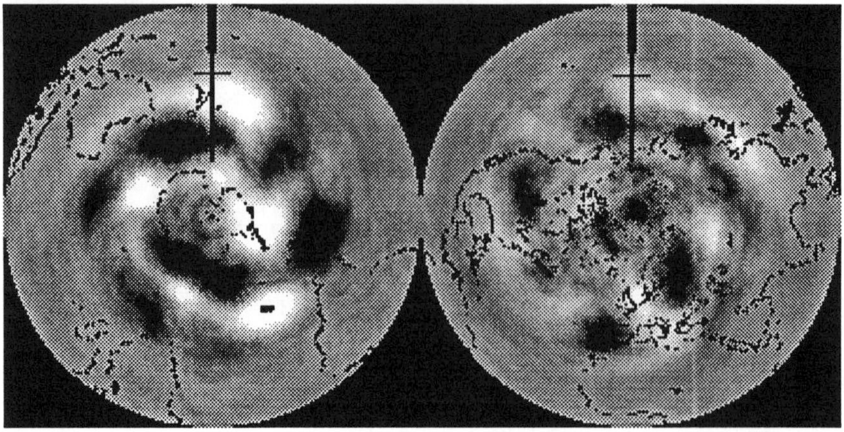

Eigenfunction 44

Figure 8: Eigenfunction 44.

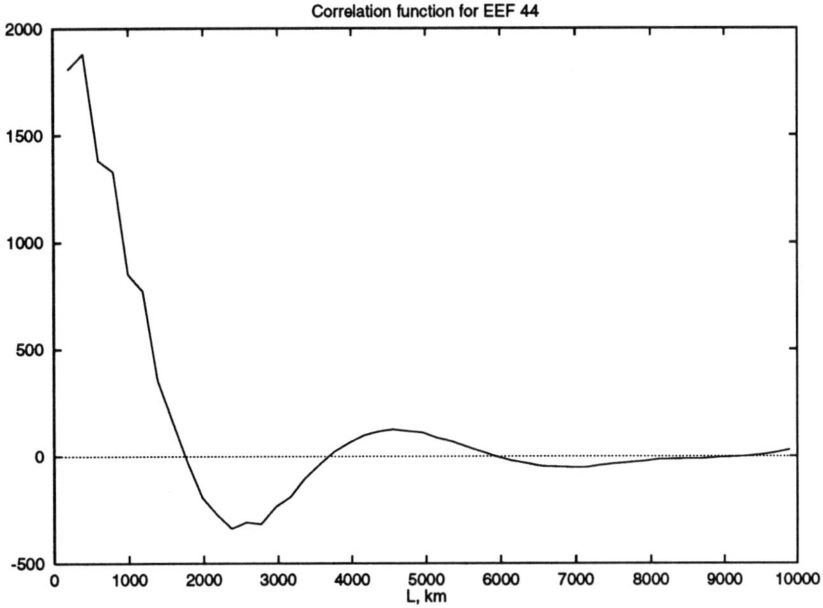

Figure 9: Spatial correlation function of the eigenfunction ψ_{44}, shown in Fig. 8

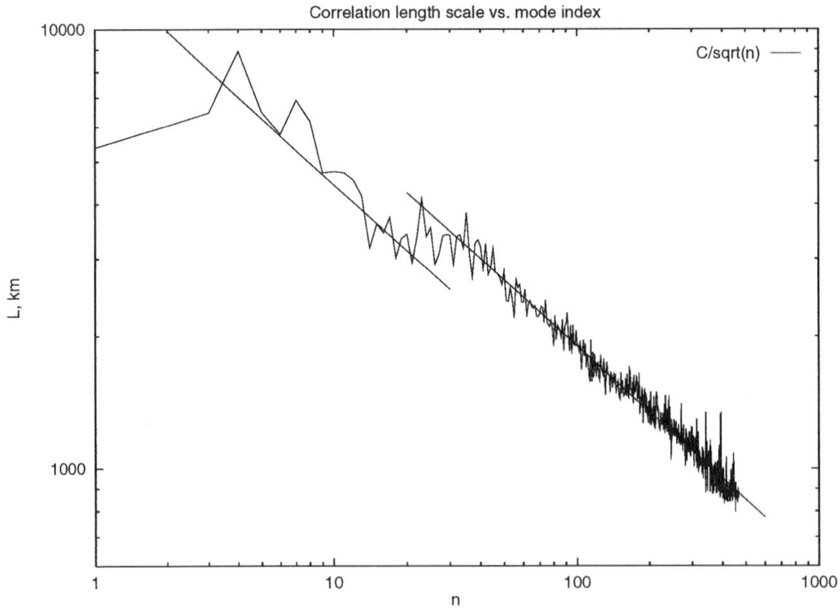

Figure 10: Correlation length scale of eigenfunction ψ_n vs. its index n.

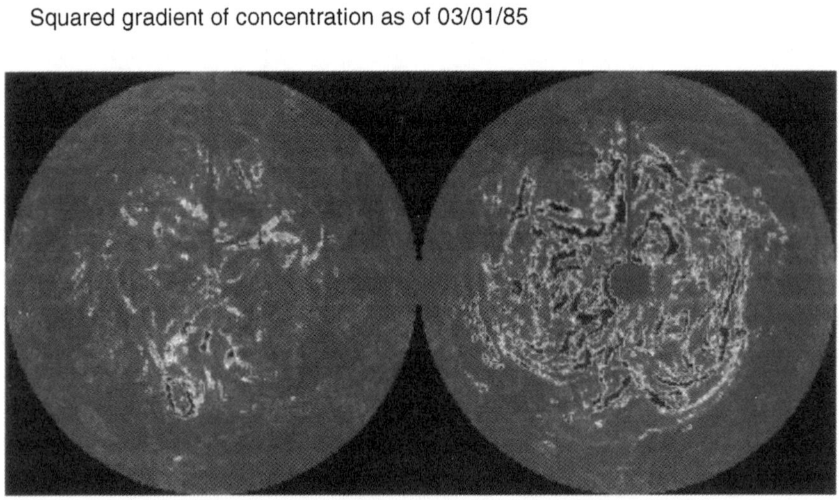

Figure 11: Squared gradient of the ozone concentration field.

(See Color Plate 6.)

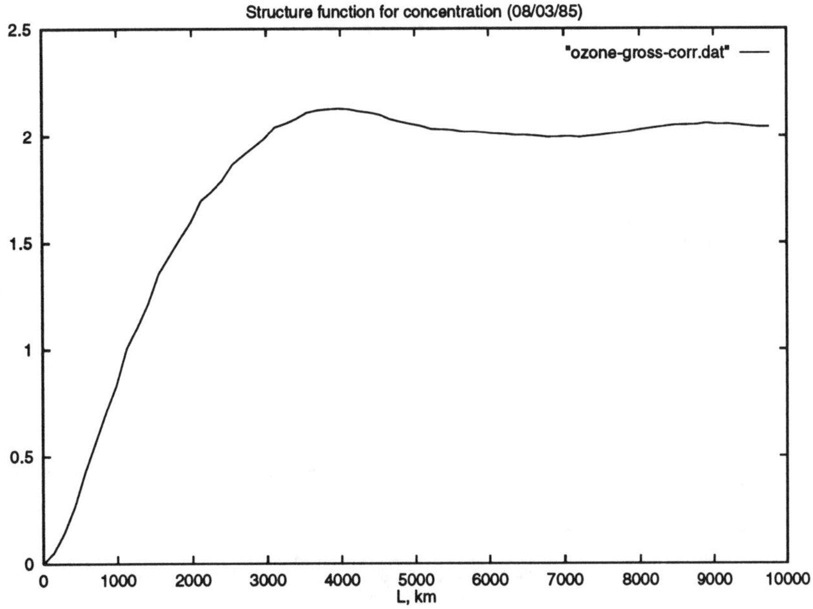

Figure 12: A concentration structure function.

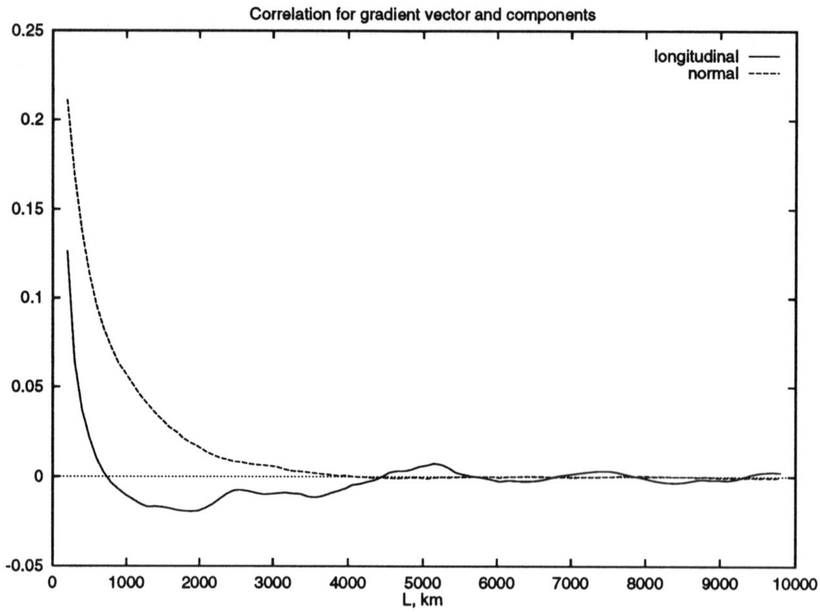

Figure 13: Correlation functions for the gradient components.

Empirical Eigenfunction Analysis of Sea Surface Temperatures in the Western North Atlantic

R. Everson,* P. Cornillon,‡ L. Sirovich,*† A. Webber†

** The Rockefeller University, New York City, 10021-6399*

† Center for Fluid Mechanics, Brown University, Providence, RI 02912-1966

‡ Graduate School of Oceanography, University of Rhode Island, Narragansett Bay Campus, Narragansett, RI 02882-1197

Abstract

The Karhunen-Loève decomposition is used to analyse time records of AVHRR sea surface temperature observations of the Western North Atlantic. A manually declouded dataset covering the spring of 1985 is analyzed. The majority (80%) of the variance about the mean is accounted for by an empirical eigenfunction which is identified with seasonal warming. The empirical eigenfunction describes the rates of warming in different regions of the ocean. Additional eigenfunctions associated with the meandering of the Gulf Stream are found. Techniques to deal with missing and partially cloudy data are advanced and applied to relatively cloud-free data selected from the period 1985 to 1991. Again, the dominant effect is identified as seasonal warming.

Introduction

This paper presents a study of Advanced Very High Resolution Radiometer (AVHRR) sea surface temperature observations in the Western North Atlantic using the Karhunen-Loève expansion. Assimilation and interpretation of the wealth of satellite observations is a major task facing the scientist. The Karhunen-Loève expansion gives an optimal modal expansion of the data and allows the identification of particular modes with relevant physical processes.

This study examines a 1200 $\times$ 1200 km region of the Western North Atlantic from the eastern seaboard of North America to the Grand Banks and from the Sargasso Sea to Nova Scotia. The predominant contribution to sea surface temperature variability is found to be the overall seasonal warming and cooling of the ocean. A mode describing this effect are found from both a manually declouded dataset covering the Spring of 1985 and also from data covering 1986 – 1991. The spatial form of the mode shows distinct warming rates in the Sargasso Sea, over the Grand Bank and in the Shelf and Slope waters. Two modes describing the meandering of the Gulf Stream are used to the obtain average phase velocity and wavelength of the meander during the spring of 1985.

The Karhunen-Loève procedure is closely related to factor analysis [1], principal components analysis [2], and singular value decomposition [3, 4]. The technique has a long history [5, 6] and in recent times has found wide application in turbulence [7] (see [8] for a survey). Within the oceanographic community the empirical eigenfunctions associated with the Karhunen-Loève procedure are usually referred to as empirical orthogonal functions (EOF, [9]). The technique has been applied to the analysis of sea surface temperature (SST) fields obtained from both in situ observations (primarily eXpendable Bathy Thermographs, XBT's) and satellite-derived fields. The source of the data has in turn determined the scales, both temporal and spatial, of the analysis. Given the sparse coverage of in situ observations, EOF analyses of SST fields derived from these data have focussed on large spatial and long temporal scales. Tanimoto *et al.* [10] analyzed 37 years of SST anomalies in the North Pacific; Naiyou *et al.* [11] considered 35 years of data for the western North Pacific, and Panitz & Speth [12] performed an EOF analysis on data for 1979 in the Equatorial Atlantic. The primary focus of such studies has been large scale atmospheric forcing and the response of the SST field as a measure of the response of the upper ocean in general.

EOF analyses of satellite-derived SST fields have tended to focus on much smaller spatial scales ($\approx$ 100km) and temporal scales ($\approx$ 100 days): Kelly investigated the relationship between sea surface temperatures, winds and topography over the north California Slope with EOFs [13]; Lagerloef and Bernstein have used EOFs to investigate currents in the coastal waters off California [14, 15]; Paden *et al.* have used the technique to investigate tidal and atmospheric forcing in the Gulf of California [16]; Gallaudet and Simpson examined large scale and mesoscale processes in the open ocean on the opposite side of the Baja Peninsula from the region considered by Paden *et al.* [17]; Fang and Hsieh analyzed the variability of the summer SST field off Vancouver Island with 8 years of AVHRR data [18]; Garzoli *et al.* compared EOFs derived from satellite SST fields with those derived from numerical model runs for both the Brazil/Malvinas region and the Kuroshio/Oyashio region [19], and Chiswell investigated the waters off New Zealand [20]. It is interesting to note that except for the work of Garzoli *et al.* [19] and Chiswell [20], EOF analyses of AVHRR data have been for the eastern North Pacific. To date there have been no published EOF analyses of AVHRR-derived fields for the North Atlantic despite Lagerloef and Bernstein's [15] suggestion that "...EOF analyses of remote sensing data could be directed to such strong frontal features as the Gulf Stream front ...". The work presented herein attempts to redress this omission.

In the next section we describe the data used in this analysis and preliminary processing of these data. This is followed by a brief outline of the Karhunen-Loève procedure. A 120 day spline interpolated set of SST fields obtained from the AVHRR sensor is then analyzed. The next two sections examine the application of the procedure to gappy data covering a much longer period in time. The manuscript ends with a discussion of the use of the procedure for compressing satellite SST fields and a summary of the general findings of this study.

Data Preparation

The measurements which we analyze consist of AVHRR images of the North Atlantic collected by the NOAA polar orbiting satellites. These images stretch from the eastern seaboard of North America to the Grand Banks (62.7° to 76.3°W) and from the Sargasso Sea in the south to Nova Scotia in the north (32.9° to 43.6°N). This region covers the development of the Gulf Stream and its meandering into the mid-Atlantic. Each image is comprised of 512 $\times$ 512 pixels on a rectangular grid with a pixel spacing of approximately 2.3 km. The sea surface temperature is derived from multi-channel infrared measurements. See [21] for details of the processing.

Cloud cover and missing data are major hindrances to the analysis of sea surface temperatures.

Data may be missing from a particular image because the area of interest does not lie completely in the swath of the observing satellite. A rough estimate of the cloud cover and missing regions (from the NOAA polar orbiting satellites) over the North Atlantic for the months March to June 1985 suggests that on average 12% of the sea-surface is missed by the satellites and 28% is obscured by clouds. In order to extract useful spatial and temporal information we have analyzed a declouded and spline interpolated dataset produced at the University of Rhode Island. This data set was obtained by first manually flagging all pixels in each image thought to be contaminated by cloud cover. The remaining "clear" pixels define an irregularly spaced time series at each pixel location in the study area. For each of these locations (approximately 170,000 excluding land), a continuous function of sea surface temperature in time was estimated by performing a least squares approximation to the temperature time series using cubic splines with fixed knots [22]. A total of seventeen cubic segments were used to fit the data at each location and the knots connecting these segments were fixed at the same times for all locations. A new set of images was generated by sampling each of the splines every 12 hours and then combining all interpolated values corresponding to the same times. Finally, the images were median filtered with a 7×7 pixel box. This last step tended to replace extraneous sea surface temperature values resulting from splines fit with insufficient data. The dataset covers the spring of 1985 from March 1 to June 30 and consists of 244 images; a typical image is shown in Figure 1. The cold continental Shelf and Gulf of Maine Waters show up as shades of blue in the upper half of the image. The Gulf Stream, extending from the lower left hand corner of the image across its middle, is defined by the warmest (reddest) water, with a very sharp sea surface temperature gradient on its northern edge. To the south the slightly cooler (reddish-yellow) water defines the Sargasso Sea. Between the Gulf Stream and the Shelf Water is the Slope Water (greenish-yellow in the figure). The warmer approximately circular features in the Shelf Water are warm-core rings. Two cold-core rings are evident to the south of the stream. One, immediately to the east of the stream in the western part of the image, is defined by the semi-circular band of warm water drawn off of the stream by the ring. The other is south of the stream on the eastern edge of the image and shows up as a relatively cooler near circular region surrounded by a slightly warmer band.

We first analyze this *cloudless* dataset and then make comparisons with relatively cloudy data drawn from a much longer period. In [23] we present a technique based on the Karhunen-Loève expansion which may be used to detect and remove cloudy regions, and in [24] a method for determining eigenfunctions from flawed images is given.

The Karhunen-Loève procedure

We denote by $\Phi(\mathbf{x}, t)$ the sea surface temperature at location $\mathbf{x}$ and time t. The dataset thus consists of the ensemble of images $\{\Phi(\mathbf{x}, t_n)\}$ measured at M discrete times t_n, though it is useful to consider time as a continuous variable. The mean temperature is given by

$$\overline{\Phi}(\mathbf{x}) \;=\; <\Phi(\mathbf{x})> \;\doteq\; \frac{1}{T}\int_{-T/2}^{T/2} \Phi(\mathbf{x}, t)dt \quad , \tag{1}$$

where T represents the duration of a sufficiently long record. We consider fluctuations about $\overline{\Phi}$:

$$\phi(\mathbf{x},t) \;=\; \Phi(\mathbf{x},t) \;-\; \overline{\Phi} \;. \tag{2}$$

We seek a set of orthonormal functions $\{\psi^{(n)}(\mathbf{x})\}$ to serve as a basis for the ensemble of images; that is, we write an N term expansion for $\phi(\mathbf{x},t)$ as

$$\phi(\mathbf{x},t) \;\approx\; \sum_{n=1}^{N} a_n(t)\psi^{(n)}(\mathbf{x}) \;, \tag{3}$$

where the modal coefficients (the coordinates in the new basis), $a_n(t)$, are given by the inner product

$$a_n(t) \;=\; (\phi,\psi^{(n)}) \;=\; \int \phi(\mathbf{x},t)\psi^{(n)}(\mathbf{x})d\mathbf{x} \;. \tag{4}$$

The basis which best represents the ensemble is found by demanding that, for any N, the approximation error, when averaged over the ensemble is minimized.That is, the criterion average error,

$$<\epsilon_N> \;=\; <\int d\mathbf{x} \mid \phi(\mathbf{x},t) - \sum_{n=1}^{N} a_n(t)\psi^{(n)}(\mathbf{x}) \mid^2 > \tag{5}$$

is required to be a minimum.

The variational problem posed here is solved if the basis functions satisfy

$$\int K(\mathbf{x},\mathbf{y})\psi^{(n)}(\mathbf{y})d\mathbf{y} \;=\; \lambda_n\psi^{(n)}(\mathbf{x}) \tag{6}$$

where the kernel, K, is the correlation function:

$$K(\mathbf{x},\mathbf{y}) \;=\, <\phi(\mathbf{x})\phi(\mathbf{y})> \;. \tag{7}$$

The eigenfunctions which solve (6) therefore form the best basis in which to represent the ensemble. The basis is *intrinsic* to the particular dataset, because it is wholly determined by the dataset itself. This technique is generally known as the Karhunen-Loève procedure [25, 26] or Principal Components Analysis [1, 2] and the eigenfunctions are termed *empirical eigenfunctions* or *empirical orthogonal functions* (EOFs [9]).

The modal coefficients, too, are orthogonal:

$$<a_n(t)a_m(t)> \;=\; \lambda_n\delta_{nm} \;, \tag{8}$$

and the eigenvalues are given by

$$\lambda_n \;=\, <a_n^2(t)> \,=\, <(\phi,\psi^{(n)})^2> \;. \tag{9}$$

The eigenvalue, λ_n, therefore measures the mean squared projection of the data onto the nth empirical eigenfunction. The empirical eigenfunctions are ordered in descending order of the corresponding eigenvalues.

By choosing to represent the dataset in terms of the first N empirical eigenfunctions one therefore captures, on average, a fraction $\sum_{n=1}^{N} \lambda_n / \sum_{n=1}^{\infty} \lambda_n$ of the variance or energy, the maximum that can be captured using any N basis functions.

Solution of the eigenfunction equation (6) for the sea surface temperature dataset by a direct method is computationally impossible on current machines. In fact, since each image is composed of $512^2 \approx 2.5 \times 10^5$ pixels finding the empirical eigenfunctions via (6) amounts to finding the eigenstructure of a 2.5×10^5 square matrix. An alternative approach, called the *snapshot method* [27], is to observe that the eigenfunctions may be written as a superposition of the original images, or snapshots. This leads to a formulation that requires finding the eigen-decomposition of a matrix whose side is the number of snapshots, M. Here $M = 244$, making the problem computationally tractable.

Spline Interpolated Data

We have computed the empirical eigenfunctions for the 244 manually declouded and spline interpolated sea surface temperature images. The spectrum of eigenvalues is shown in Figure 2. The mean temperature distribution, $\overline{\Phi}(\mathbf{x})$, about which the eigenfunctions are fluctuations, is shown in Figure 3 and the first four eigenfunctions are depicted in Figure 4.

It is clear from the spectrum of eigenvalues that essentially all the energy in the ensemble is captured by only 17 empirical eigenfunctions. This degree of compression is an artifact, however; a result of the low-pass filtering and smoothing implicit in the spline interpolation. In fact, prior to median filtering, there are only 17 degrees of freedom (corresponding to the 17 spline functions used in the interpolation) in the spline interpolated image set and one would expect to find only 17 non-zero eigenvalues. The median filtering of the spline interpolated images distributes some of the energy from the first 17 eigenfunctions to the remaining ones. We shall concentrate on the information provided in the first four eigenfunctions, noting that over 99% of the energy is captured by these modes.

In Table 1, the characteristics of the primary EOFs obtained in this study are compared with those of the AVHRR studies cited in the Introduction. Also included in this table are the number of input images, the area covered by each image as well as the number of pixels, the period covered and the general geographic location of the study. It is clear from the table that the fraction of variance explained by the first eigenfunction varies significantly from study to study, with the value obtained in this study near the middle of the range. Except for the Gallaudet and Simpson [17] results, the fraction of the variance explained tends to decrease with the increase in the number of temporal degrees of freedom as one would expect. Also as one might expect, the fraction of variance explained is much higher for semi-enclosed regions (the Gulf of California [16] and the Santa Barbara Channel [15] compared with open ocean areas. Our results are consistent with these trends.

We now examine each of the first four eigenfunctions in detail. As indicated in Table 1, on average 80% of the energy in the fluctuations is accounted for by the first eigenfunction. As can be seen in Figure 4, $\psi^{(1)}(\mathbf{x}) > 0$ everywhere, and the eigenfunction therefore represents a warming or cooling of the entire region. The modal coefficients $a_k(t)$ are plotted in Figure 5. The coefficient a_1 increases almost linearly with time, indicating a general warming with the advancing season (recall that the time series of images covers the period from March through June), and we deduce that this eigenfunction represents the seasonal warming of the sea surface. The first EOF in all of the EOF analyses performed to date on temporally demeaned AVHRR SST fields have been everywhere positive and have been associated with seasonal warming/cooling.

Temperature time series (Figure 6, obtained from the 244 manually declouded and spline interpolated images) for the Sargasso Sea (the location marked with an "a" in Figure 4.1), for the Gulf Stream (point "b"), for the coastal water off New Jersey (point "c"), and for the Gulf of Maine (point "d") all show a steady increase in temperature with season. The uppermost curve in this figure corresponds to the Gulf Stream (point "b" in Figure 4.1) and the least square fit straight line to this curve has the smallest slope; i.e., the increase in temperature for the Gulf Stream was smaller than that for the other three regions considered. This is consistent with the first mode eigenfunction (Figure 4.1) in which the Gulf Stream has the smallest values. This is not surprising in that Gulf Stream water is advected north from the Gulf of Mexico with little cooling and the seasonal change in sea surface temperature of water in the Gulf of Mexico is quite small.

The curve immediately beneath that of the Gulf Stream is representative of the Sargasso Sea and corresponds to location "a" in Figure 4.1. As with the Gulf Stream there is very little change in temperature with season at this point. This point is quite representative of the entire Sargasso Sea as can be seen from Figure 4.1.

The steepest curve in Figure 6 corresponds to a point (point "c" in Figure 4.1) on the continental shelf east of the entrance to the Delaware Bay. The temperature time series shows rapid warming consistent with the fact that the water on the shelf is relatively shallow; hence there is a smaller vertical range in comparison with the deeper Slope and Sargasso Sea Waters over which the absorbed heat can be mixed. From Figure 4.1 it is clear that this is true for all shallow water except for that on Georges Bank. The difference in the water on Georges Bank results from the strong tidal mixing which tends to bring cold deep water from the Gulf of Maine to the north and from the Slope Water region to the south up onto the banks.

Rapid warming of shelf waters in contrast to adjacent deep water has been observed by others [13, 14, 17, 18] (although in the second mode); and [16]. An eigenfunction performing a similar function has also been found in an EOF analysis of both surface and deeper water temperatures in the Levantine Basin [28].

The coldest curve in Figure 6 corresponds to point "d" of Figure 4.1 located in the Gulf of Maine off Cape Cod. It warms slightly more quickly with season than the other open ocean points, but substantially more slowly than water on the continental shelf.

The first eigenfunction shows significantly more spatial variability in the Slope Water region than in the water masses represented by the four points discussed above. This is because of the presence of warm core rings which drag warm water off the Gulf Stream and cold water from the shelf as they propagate to the west.

When the sequence of images is viewed as a movie loop, the motion of the Gulf Stream is readily apparent in March and April because of the large contrast with adjacent water. However, it is later masked as the cooler water to the north warms relatively more rapidly. It can be unmasked by subtracting out the effect of seasonal warming; that is, we may project the data onto all the modes *except* $\psi^{(1)}$:

$$\hat{\Phi}(\mathbf{x},t) = \overline{\Phi}(\mathbf{x}) + \sum_{k\neq 1} a_k(t)\psi^{(k)}(\mathbf{x}) \tag{10}$$

Figure 7 shows the sea surface temperature for June 21, and the projection without the contribution from the seasonal warming. The Gulf Stream and other oceanographic features are apparent in the projection. Removal of seasonal trends using the empirical eigenfunctions is generally superior to *ad hoc* de-trending, because the empirical eigenfunctions compactly capture uncorrelated trends in the data (recall that the modal coefficients are orthogonal).

The second eigenfunction, the most subtle of those contributing significantly, is a correction to the first eigenfunction; i.e., as discussed below this eigenfunction provides for a slight change in the warming trend defined by the first eigenfunction. This correction appears to result from several different processes. These are identified by region with the help of Figure 8 which shows only those areas of the second eigenfunction for which the absolute value of the eigenfunction exceeds 0.025; i.e., only those areas contributing significantly to the correction. Values from -0.025 to 0.025 are colored grey in the figure. Superimposed on the second eigenfunction in this figure are the continental outline in red, the 200 m isobath in white and, crosshatched in black, regions of persistently high gradient. The persistent gradient regions were obtained by applying a Sobel gradient operator to each of the images in the time series, averaging the resulting gradient images and then marking locations in the figure where this average exceeded a given threshold.

The different areas contributing significantly to the correction term are: (1) a narrow band (yellow) hugging the coast which extends south from the mouth of the Delaware Bay ($\approx 39°N$) to Cape Hatteras ($\approx 35°N$); (2) a moderately large region (yellow) seaward of the southern edge of the Gulf Stream off Cape Hatteras, (3) a band (yellow) that follows the northern edge of the Gulf Stream, (4) a band (blue/green) that follows the 200 m isobath from approximately the entrance of Chesapeake Bay ($\approx 37°N$) to the middle of Georges Bank ($\approx 67°W$) and along most of the extent of this isobath in the Gulf of Maine, and (5) alternating bands (yellow and blue/green/violet) between regions 3 and 4. The processes involved in the correction differ from one area to another. The correction due to region 1 clearly results from coastal runoff from the Delaware and the Chesapeake bays. The region becomes wider to the south both because the Chesapeake adds to the runoff and because the plume is mixed to greater distances off shore as it moves to the south. The modification here results in more rapid warming early in the period. This occurs because the shallow water in the Delaware Bay and the Chesapeake Bay is warmed more rapidly than the shelf water. Figure 9a shows the contribution to the temperature at ($75.8°W$, $36.9 \circ N$), the mouth of the Chesapeake Bay, as a function of time for the first EOF (thin) and for the sum of the first two EOFs (bold).

Regions 2 and 5 are related to rapid changes in SST caused by warm water being advected off the Gulf Stream by a cold core ring for region 2 and by two warm core rings for region 5. Early in the time series the cold core ring/stream interaction is clearly visible (not shown here) as a thin band of Gulf Stream water being drawn off of the stream as the ring begins to interact with the it. Toward the middle of the interval the ring/stream interaction is much more vigorous and large quantities of water are drawn off of the stream into the adjacent Sargasso Sea. The interaction between the stream and the warm core rings occurs as large eastward propagating meanders touch the rings again drawing off warm Gulf Stream water. Both the warm and cold core ring/stream interactions contribute to the second EOF because they occur near the middle of the interval. Had they occurred near the beginning or near the end of the time series, we believe that they would not have played a role in this eigenfunction; their impact would either have been incorporated in the first eigenfunction or would have given rise to a new eigenfunction.

The remaining regions, 3 and 4, are, we believe, associated with vertical motion in the water column which tends to modify the warming rate when compared to regions with relatively little vertical motion. An alternate hypothesis is that regions 3 and 4 in this EOF result from the simple horizontal displacement of strong fronts over the study period. We discount this hypothesis however because the bands in regions 3 and 4 are offset relative to the bands containing the persistent fronts (Figure 8) and because motion of the fronts would require another EOF in quadrature with this one and none is apparent. Large vertical motion on the other hand can result in a temporal change in the SST that may be explained by one EOF. Such motion is often related to strong fronts associated with steep gradients in bathymetry (the shelf break near the 200 m isobath) and steep gradients in dynamic topography (the northern edge of the Gulf Stream). Figure 8 shows the relationship between persistent fronts, bathymetry, the northern edge of the Gulf Stream and regions in which the magnitude of second EOF is large.

Eigenfunctions 3 and 4 (Figure 4.3 and 4.4), are large only in the Gulf Stream and Slope Water regions. In the Gulf Stream region they have the approximate form of a pair of modes that are $\pi/2$ out of (spatial) phase with each other. The corresponding modal coefficients (Figure 5) are roughly sinusoidal oscillations one quarter period out of phase with each other. It can be seen that the period is approximately 40 days. The superposition

$$a_3(t)\psi^{(3)}(\mathbf{x}) + a_4(t)\psi^{(4)}(\mathbf{x}) \tag{11}$$

is thus approximately a traveling wave, representing the meandering of the Gulf Stream. The 40-day period agrees well with that of the most energetic meanders (46 days) obtained by [29], from an analysis of eight years of satellite-derived Gulf Stream paths. The wavelength of these meanders, measured from either the third or fourth eigenfunction, $\approx$ 375km, is comparable to that obtained by Lee and Cornillon, 427 km, and can be regarded as the representative scale of a traveling wave. From the period and wavelength we estimate the phase velocity as 9.4cm s^{-1}, a little smaller than the mean phase speed of $\approx$ 12cm s^{-1}, observed by [30].

Partial Data

We have also applied the Karhunen-Loève procedure to an ensemble of 73 fairly cloudless and complete images. This ensemble covers approximately the same region as the previous ensemble, but was obtained by searching several thousands of images from the years 1986 to 1991 to find those with putatively less than 10% cloud cover or missing regions.

Nonetheless, these images are flawed and the Karhunen-Loève procedure must be modified to cope with missing data. When an observation is missing or the observed temperature is very cold, (below 0.5° C) indicating the unambiguous presence of cloud cover, we must insert a sensible numerical value in place of the missing data at $\mathbf{x}'$ when calculating the covariance. The value that assumes the least about the data is just the mean at that point; that is, the fluctuation is set to zero in the demeaned images: $\phi(\mathbf{x}') = 0$.

With complete data the ensemble average temperature is calculated as:

$$\overline{\Phi}(\mathbf{x}) = \frac{1}{T}\int_{-T/2}^{T/2} \phi(\mathbf{x})d\mathbf{x} \approx \frac{1}{M}\sum_{k=1}^{M} \Phi(\mathbf{x}, t_k) , \tag{12}$$

where the t_k are the M times at which the images were collected. With incomplete data the ensemble average is calculated from

$$\overline{\Phi}(\mathbf{x}) \approx \frac{1}{M(\mathbf{x})}\sum_{k=1}^{M(\mathbf{x})} \Phi(\mathbf{x}, t_k) , \tag{13}$$

where $M(\mathbf{x})$ is the number of images which have *good data* at position $\mathbf{x}$. More sophisticated schemes, based on the Karhunen-Loève expansion, for the detection and replacement of cloudy regions are possible; this issue is discussed in [23].

The spectrum of eigenvalues for the relatively cloud free ensemble is shown in Figure 11. The fact that the number of eigenfunctions that contribute significantly is greater than for the declouded, smoothed and interpolated ensemble, may be attributed to two facts: (1) the data now records sea surface temperatures in four seasons over a number of years, rather than a single season, and (2) as noted previously, the process of fitting a spline to the declouded data set constrains the number of degrees of freedom. Even so, for the multi-year data set, only 32 modes are required to capture 90% of the energy and we estimate that modes with indices greater than about 20 are dominated by noise.

As before, the first empirical eigenfunction (Figure 12) carries the vast bulk of the variance (62%). The huge change in variance between eigenvalues 1 and 2 is a feature in common with other studies (see Table 1) and indicates that the sea surface temperature is dominated by seasonal warming and cooling. The empirical eigenfunctions has approximately the same form as the first eigenfunction derived from the spline-interpolated data, and apparently represents the gross seasonal warming and cooling of the ocean. Empirical eigenfunctions corresponding to meandering of the Gulf Stream are now not evident. This is not surprising both because the average separation in time between images is approximately 30 days, close to the period of the dominant meanders, and because the stream moves laterally from year-to-year [31].

Modal Coefficients

The modal coefficients, $a_k(t)$, are calculated from the inner product

$$a_k(t) = (\phi(\mathbf{x}, t), \psi^{(k)}(\mathbf{x})) . \tag{14}$$

When much of the data is missing or obscured by clouds application of equation (14) can not be accomplished without some assumption on the missing data points. We advance the following alternative scheme for estimating the modal coefficients from partial data. Suppose that the data, $\phi(\mathbf{x}, t)$, is present in the (usually disconnected) region $\mathbf{x} \in \mathcal{D}(t)$. The modal coefficients are estimated by demanding that the error of the approximation, over the region $\mathcal{D}(t)$, is minimized. That is, we seek, at time t, the coefficients, $\hat{a}_k(t)$, that minimize

$$\epsilon_N(\mathcal{D}(t)) = \int_{\mathcal{D}} \|\phi(\mathbf{x},t) - \sum_{k=1}^{N} \hat{a}_k(t)\psi^{(k)}(\mathbf{x})\|^2 d\mathbf{x} . \tag{15}$$

This is a standard least-squares problem and is readily solved via the normal equations found by differentiating $\epsilon_N(\mathcal{D})$ with respect to $\hat{a}_k$. If this procedure is carried out at each instant the time record, $\hat{a}_k(t)$, is obtained.

This procedure allows estimation of the modal coefficients with good accuracy from even rather small amounts of data [24]. It should be noted that this method relies on the fact that the Karhunen-Loève procedure provides an optimal basis for the ensemble.

The coefficient $\hat{a}_1$, which gives the magnitude of the seasonal warming, has been estimated from the cloudy data using this procedure. Figure 13 shows the coefficient as estimated from the ensemble of 73 snapshots spanning 1986 – 1991 and used to find the eigenfunctions. Excluded from the calculation were missing pixels, $\Phi(\mathbf{x},t)$, that were colder than the mean, $\overline{\Phi}(\mathbf{x})$, by $5°C$ or pixels colder than $0.5°C$. Choosing these relatively unobscured snapshots means that a large portion of data comes from the months of July and August, when the skies are clear; the clustering of the data is indicated in the figure by the locations of the data points. Although the data is irregularly spaced, it is clear that the interpretation of the first eigenfunction as a mode describing the seasonal warming is compelling.

In Figure 14 the coefficient of seasonal warming, $\hat{a}_1$, for a single year (54 snapshots at roughly weekly intervals throughout 1990) is shown. Each data point is derived from a single snapshot, and the eigenfunctions from the relatively cloud-free ensemble. The annual warming and cooling cycle, peaking in August, is apparent. These data were arbitrarily chosen, rather than being selected for their quality and clearly illustrate the possibility of estimating oceanographic parameters from partial data. In this case no useful information can be obtained from the higher modes. This is due to the fact that the data were collected over a wide area and included several years, during which time the position of the Gulf Stream was very variable.

Compression

The potential for lossy compression of satellite images should be mentioned here. If it is sufficient to approximate the data with a 10% mean squared error, then only 32 numbers (the modal coefficients, a_n) are required to specify each image. This makes the rapid transmission of images between sites that possess the empirical eigenfunctions extremely rapid. In fact, the Karhunen-Loève formulation provides the assurance that the empirical eigenfunctions are the optimal basis with which to describe the ensemble; using any other orthonormal basis of 32 functions will incur at least as much error. It is anticipated that the number of eigenfunctions required to achieve a given accuracy will decrease as more cloud free images are included in the ensemble.

Summary

In the preceding we analyzed using the Karhunen-Loève procedure (also known as Empirical Orthogonal Function (EOF) analysis), a series of sea surface temperature images covering the Gulf

Stream and surrounding ocean from the point of separation (off Cape Hatteras, North Carolina, $\approx 75°W$) to the New England Seamounts, $\approx 65°W$. These images were spatially complete, clouds having been replaced by an interpolated value obtained from cubic splines fit at each pixel location. The first mode, explaining 80% of the variance, was shown to correspond to seasonal heating/cooling. This is typical of the results of other EOF analyses as is the fraction of the variance explained by this mode. Also consistent with previous analyses, waters on the continental shelf were shown to heat up more quickly than those in the open ocean as Spring progresses into summer. Less typical are the different rates of warming seen in the open ocean. Unlike other results obtained from EOF analyses, the higher order EOFs were shown to correspond to propagating meander. The average phase velocity and wavelength of the meanders was obtained from the EOFs.

The technique was also applied to a number of images covering several years. These images were not spatially complete so the Karhunen-Loève procedure had to be modified to handle gappy data. In this case the first mode explained only 60% of the variance and the higher order modes could not associated with propagating meanders. These results are not surprising in that in the longer time series the images used were too far apart in time for a given meander to appear in more than one or two, and the irregular temporal spacing of the images did not allow for propagating meanders to be followed.

Acknowledgment

This paper was prepared with support from the Office of Naval Research, the Naval Research Laboratory, Stennis Space Center, from the National Aeronautics and Space Administration, and from the State of Rhode Island and Providence Plantations through salary support to P. Cornillon. The satellite data processing software used to process the imagery was developed by R. Evans, O. Brown, J. Brown and A. Li of the University of Miami. Their continued support is greatly appreciated.

References

[1] H. Harman. *Modern Factor Analysis.* Univ. Chicago Press, Chicago, 1960.

[2] H. Hotelling. Analysis of complex statistical variables in principal components. *J. Educ Psy.*, 24:417–441, 498–520, 1933.

[3] G.H. Golub and C.F. Van Loan. *Matrix Computations.* North Oxford Academic, Oxford, 1983.

[4] C. Moler and D. Morrison. Singular value analysis of cryptograms. —*AMEMM*—, 90(2):78–87, 1983.

[5] L. Sirovich and R.M. Everson. Analysis and management of large scientific databases. *International Journal of Supercomputing Applications*, 6(1):50–68, 1992.

[6] G.W. Stewart. On the early history of the singular value decomposition. *SIAM Review*, 35(4):551–566, 1993.

[7] L. Sirovich. Empirical eigenfunctions and low dimensional systems. In L. Sirovich, editor, *New Perspectives in Turbulence.* Springer, 1991.

[8] G. Berkooz, P. Holmes, and J.L. Lumley. The proper orthogonal decomposition in the analysis of turbulent flows. *Annual Reviews Fluid Mechanics*, 25, 1993.

[9] E.N. Lorenz. Empirical orthogonal functions and statistical weather prediction. Technical Report 1, Statistical Forecasting Project, Department of Metrology, MIT, Cambridge, 1956.

[10] Y. Tanimoto, N. Iwasaka, K. Hanawa, and Y. Toba. Characteristics of sea surface temperature with multiple time scales in the north pacific. *J. Climate*, 6:1153–1160, 1993.

[11] X. Naiyou, X. Qichun, and M. Tiansong. Relationship between the subtropical high in summer and the thermal regime anomaly in the northwest Pacific. *J. Ocean Univ. Qingdao Qingdao Haiyang Daxue Xuebao*, 22:29–38, 1992.

[12] H.J. Panitz and P. Speth. The influence of the surface wind stress over the equatorial atlantic on oceanic upwelling processes during FGGE 1979. *Oceanogr. Trop.*, 21:185–203, 1986.

[13] K.A. Kelly. The influence of winds and topography on sea surface temperature patterns over the Northern California slope. *Journal of Geophysical Research*, 90(C6):11783–11798, 1985.

[14] G.S.E. Lagerloef. The Point Arena Eddy: a recurring summer anticyclone in the California current. *Journal of Geophysical Research*, 97(C8):12557–12568, 1992.

[15] G.S.E. Lagerloef and R.L. Bernstein. Empirical orthogonal function analysis of advanced very high resolution radiometer surface temperature patterns in the santa barbara channel. *Journal of Geophysical Research*, 93(C6):6863–6873, 1988.

[16] C.A. Paden, M.R. Abbott, and C.D. Winant. Tidal and atmospheric forcing of the upper ocean in the Gulf of California. 1. Sea surface temperature variability. *Journal of Geophysical Research*, 96(C10):18337–18359, 1991.

[17] T.C. Gallaudet and J.J. Simpson. An empirical orthogonal function analysis of remotely sensed sea surface temperature variability and its relation to interior oceanic processes off Baja California. *Remote Sens. Environ.*, 47:375–389, 1994.

[18] W. Fang and W.W. Hsieh. Summer sea surface temperature variability off Vancouver Island from satellite data. *J. Geophys. Res.*, 98:14391–14400, 1999.

[19] S.L. Garzoli, Z. Garraffo, G. Podesta, and O. Brown. Analysis of a general circulation model product 1. Frontal systems in the Brazil/Malvinas and Kurshio/Oyashio regions. *J. Geophys. Res.*, 97:29117–20138, 92.

[20] S. Chiswell. Variability in sea surface temperature around New Zealand from AVHRR images. *N.Z. J. Mar. Freshwat. Res.*, 28:179–192,, 1994.

[21] P. Cornillon, C. Gilman, L. Stramma, O. Brown, R. Evans, and J. Brown. Processing and analysis of large volumes of satellite derived data. *J. Geophys. Research*, 92(12):993–13002, 1987.

[22] C. de Boor and J.R. Rice. Cubic spline approximation I-fixed knots. Technical Report TR20, Computer Science Department, Purdue University, 1968.

[23] R.M. Everson and L. Sirovich. Cloud detection using the Karhuen-Loève transform. Technical report, The Rockefeller University, 1995. *(in preparation)*.

[24] R.M. Everson and L. Sirovich. The Karhunen-Loève transform for incomplete data. *JOSA A*, 12(8), 1995.

[25] K. Karhunen. Zur Spektraltheorie Stochasticher. *Prozesse Ann. Acad. Sci. Fennicae*, 37, 1946.

[26] M.M. Loève. *Probability Theory*. Van Nostrand, Princeton, N.J., 1955.

[27] L. Sirovich. Turbulence and the dynamics of coherent structures, pt i: Coherent structures, pt ii: Symmetries and transformations, pt. iii: Dynamics and scaling. *Quarterly of Applied Mathematics*, XLV(3):561–590, 1987.

[28] R.M. Everson, L. Sirovich, and A. Robinson. Empirical eigenfunction analysis of ocean currents in the Levantine Basin. *(unpublished)*.

[29] T. Lee and P. Cornillon. Propagation of gulf stream meanders between 74° and 70°W. *J. Phys. Oceanogr. (submitted)*, 1994.

[30] T. Lee and P. Cornillon. Propagation and growth of gulf stream meanders between 75° and 45°W. *J. Phys. Oceanogr.*, 1994. *(in press)*.

[31] T. Lee and P. Cornillon. Temporal variation of meandering intensity and domain-wide lateral oscillation of the gulf stream. *Journal Geophysical Research*, 1994. *(in press)*.

Reference	Galludet '94	Paden '91	Kelly '85	Fang '93	Lagerloef '88	Garzoli '92	Garzoli '92	This Study	This Study
Location	Southern California	Northern Gulf of California	Northern California	Vancouver Island	Santa Barbara Channel	Brazil/Malvinas Current	Kuroshio/Oyashio Current	Gulf Stream	Gulf Stream
Size (km)	500 × 500	150 × 750	512 × 512	105 × 253	50 × 100	512 × 512	512 × 512	1200 × 1200	1200 × 1200
# pixels	512 × 512	75 × 350	100 × 100	150 × 350	50 × 100	15 × 15	20 × 30	512 × 512	512 × 512
Period	'84-'89	3/84-2/86	4/81-7/81		1/84-1/85	7/84-7/88	11/81-12/89	3/85-7/85	86-91
# of Images	35	20	56	133	36	263	434	244	73
First Mode	38.7%	98%	74%	55.8%	91.6%	85.7%	84.6%	80%	62%
Second Mode	6.8%	1.1%	11%	6.9%	2.5%	2.7%	5.5%	4.4%	3.4%
Third Mode	3.9%	0.3%	4%	6.6%		1.5%	1.5%	3.1%	2.7%
Fourth Mode	3.5%		2%					2.3%	1.9%

Table 1: Comparison of AVHRR SST EOF Studies

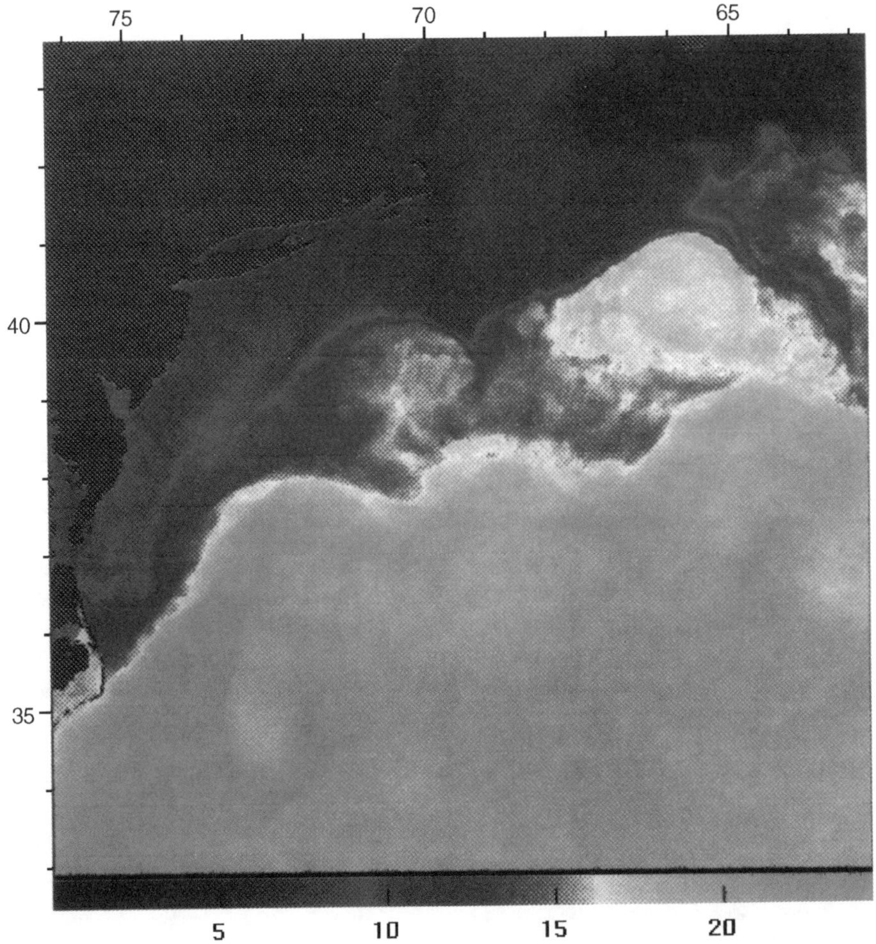

Figure 1: AVHRR image (9 March 1985) of the Western North Atlantic after spline interpolation and manual declouding. The color-bar shows the temperature in Celsius and the scales at the top and left show the latitude and longitude. *(See Color Plate 7.)*

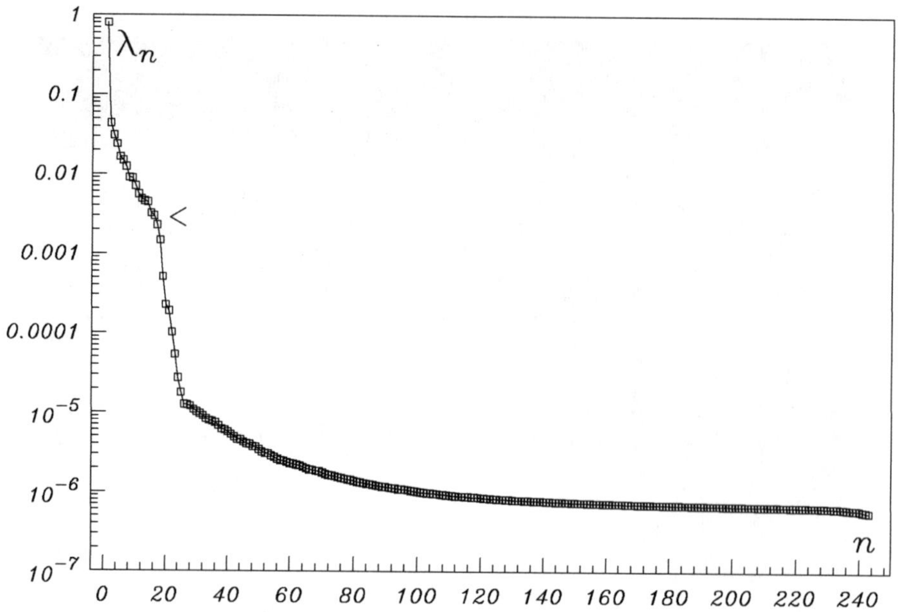

Figure 2: Eigenvalues derived from the spline-interpolated dataset. The 17th eigenvalue is indicated with an arrow.

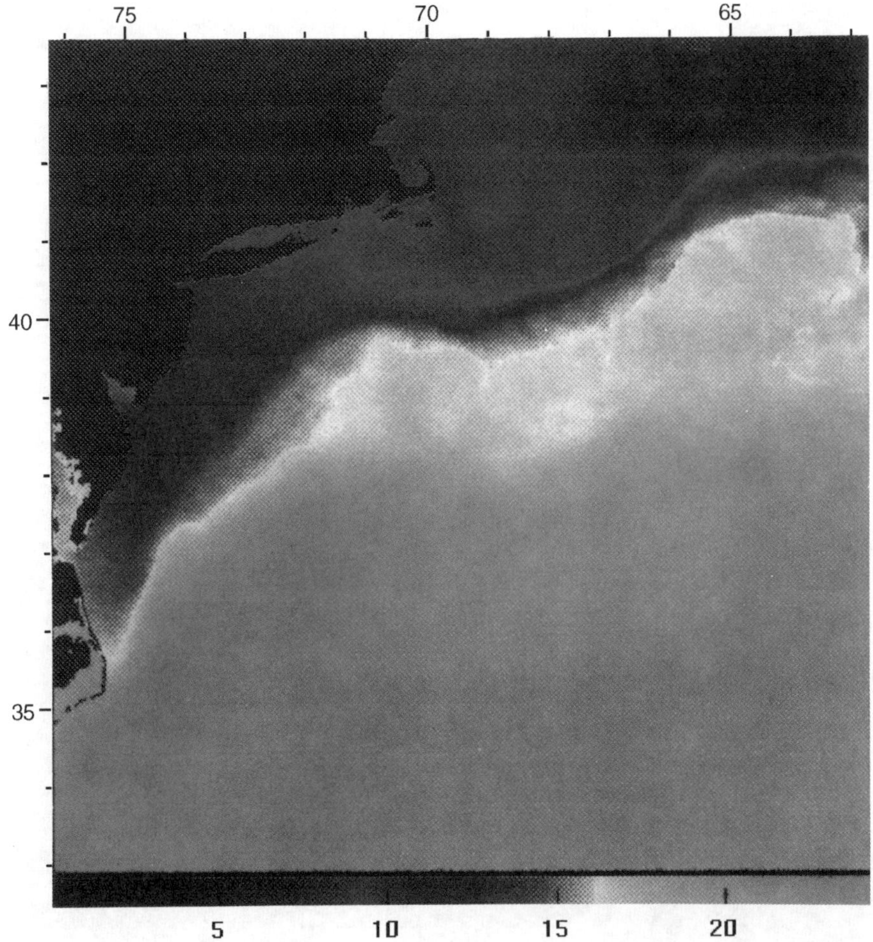

Figure 3: Ensemble average sea-surface temperature, $\overline{\Psi}(\mathbf{x})$, derived from the spline-interpolated ensemble. *(See Color Plate 8.)*

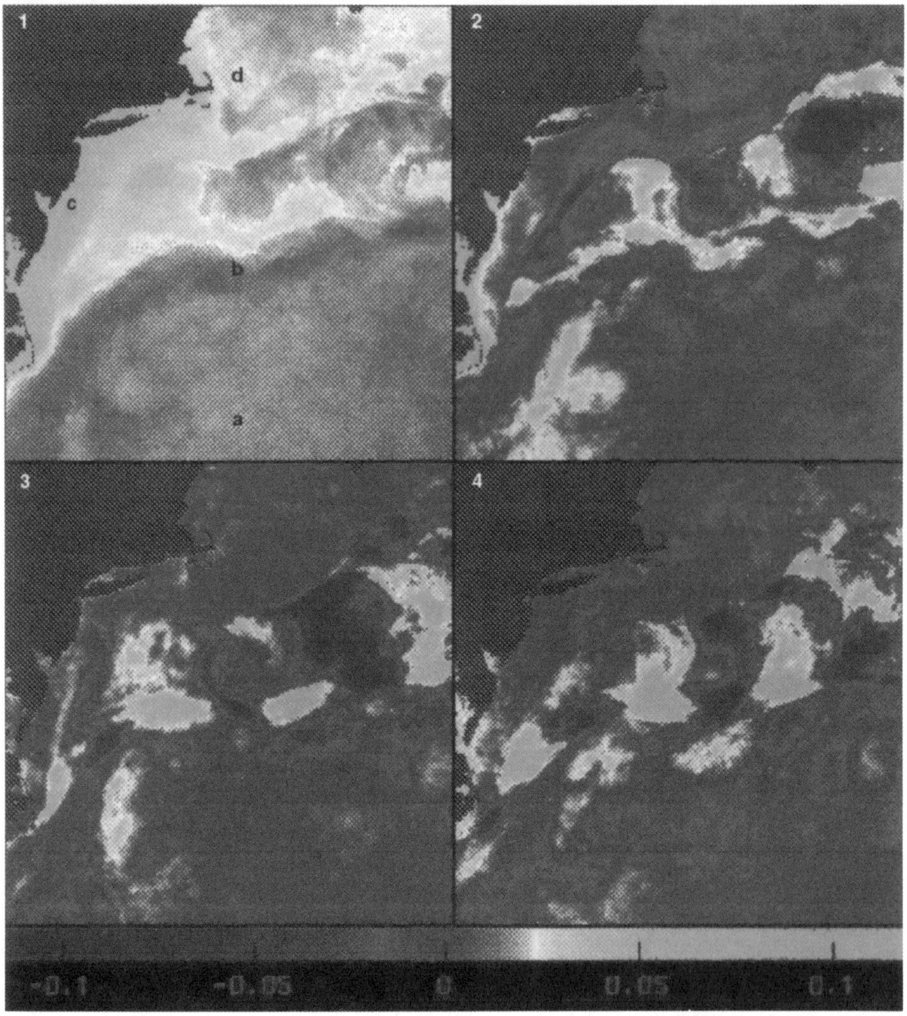

Figure 4: First four empirical eigenfunctions derived from the spline-interpolated ensemble. Projections onto these four eigenfunctions capture 99% of the variance on average. *(See Color Plate 9.)*

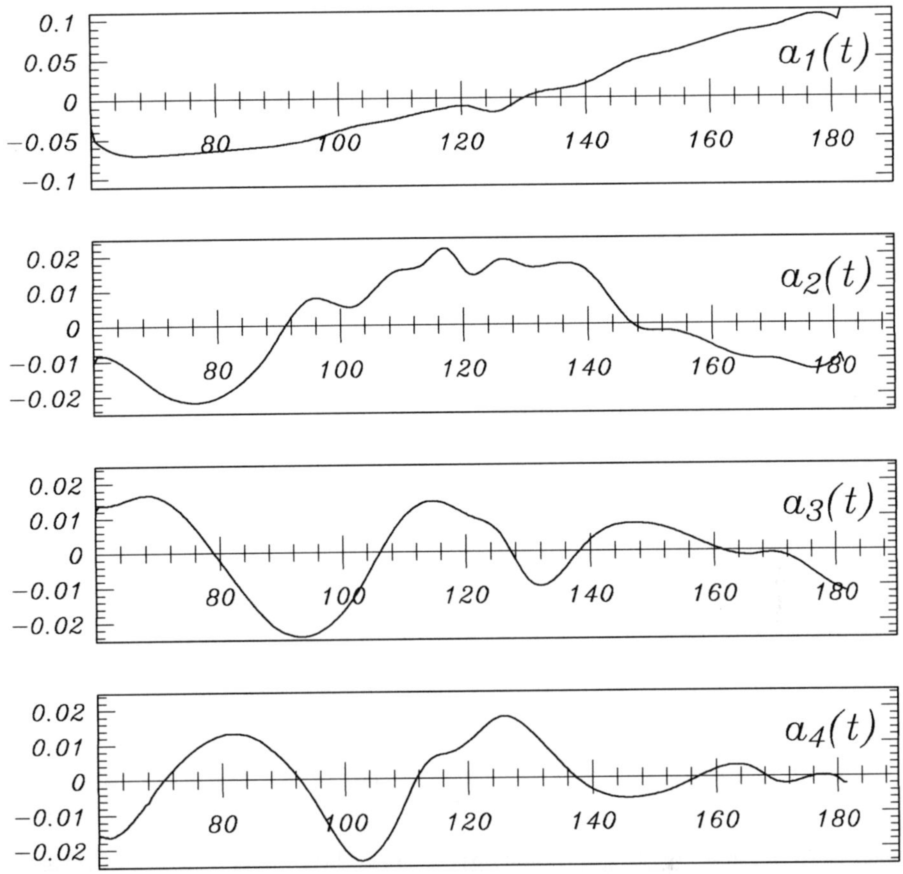

Figure 5: Modal coefficients, $a_k(t)$, for the first four empirical eigenfunctions. The scale on the abscissa shows the day of the year 1985. The steady increase in a_1 indicates the seasonal warming of the sea surface. The coefficients, a_3 and a_4 are a quarter period out of phase which, when combined with the spatial phase shift in eigenfunctions $\psi^{(3)}$ and $\psi^{(4)}$ produces a traveling wave.

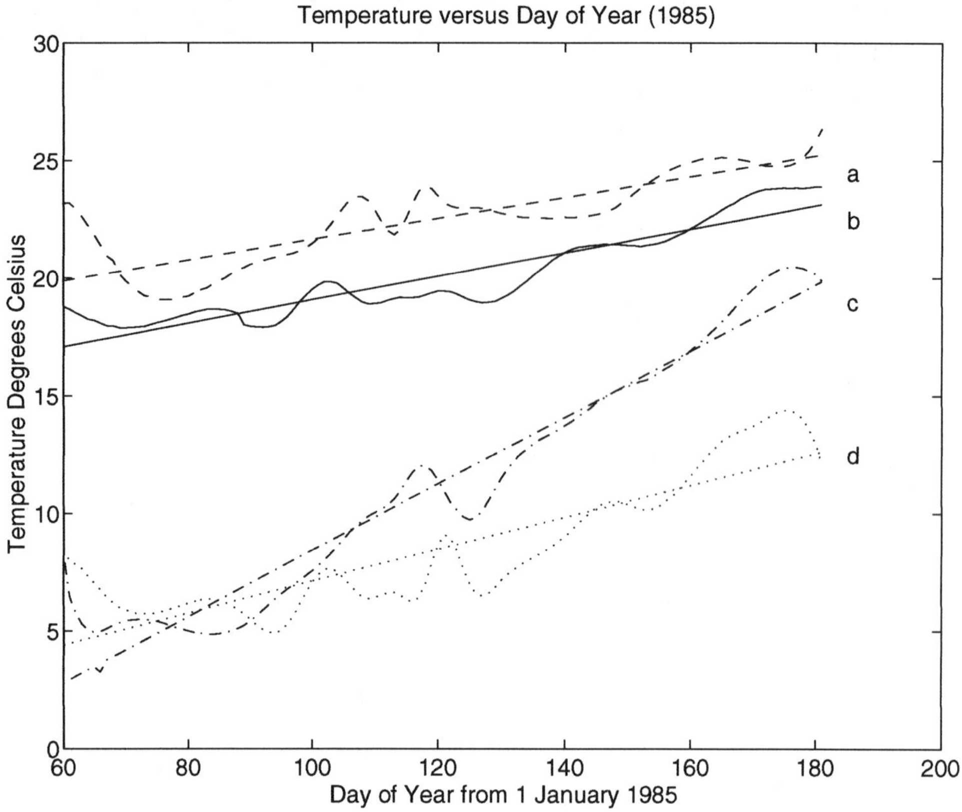

Figure 6: Temperature as a function of time averaged over 11.5 by 11.5 km^2 areas (5x5 pixels) centered on ($69°W$, $34°N$) (dashed); ($69°W$, $37.5°N$) (solid); ($69°W$, $42°N$) (dotted); ($74°W$, $39°N$) (dash-dotted). Also indicated are the best fit, in a least square sense, straight lines.

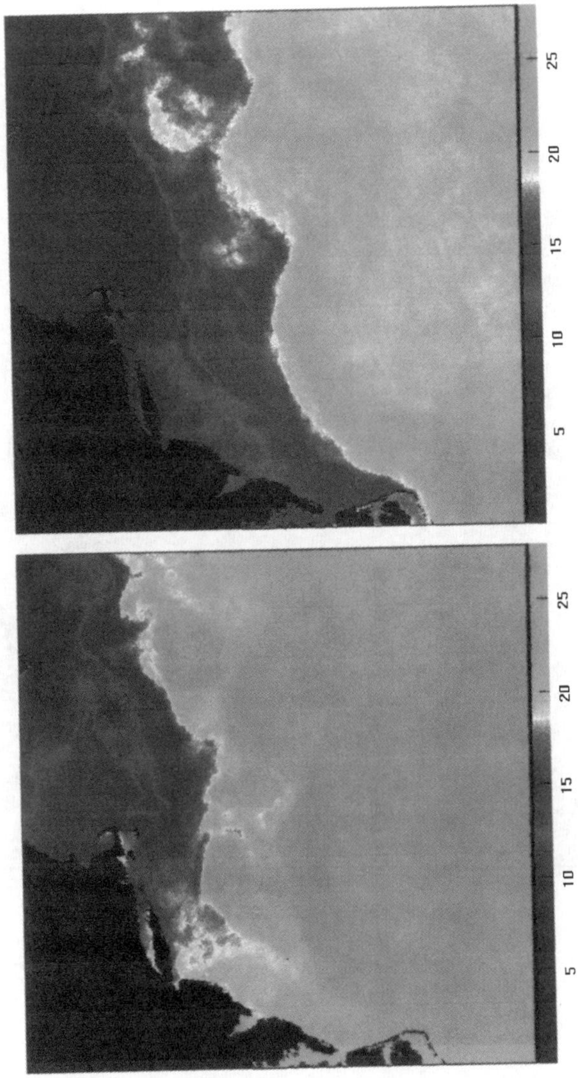

Figure 7: Sea surface temperature for 21 June 1985, and the same image with the seasonal warming contribution removed. The Gulf Stream, which was obscured by the general seasonal warming, is revealed in the projected image. *(See Color Plate 10.)*

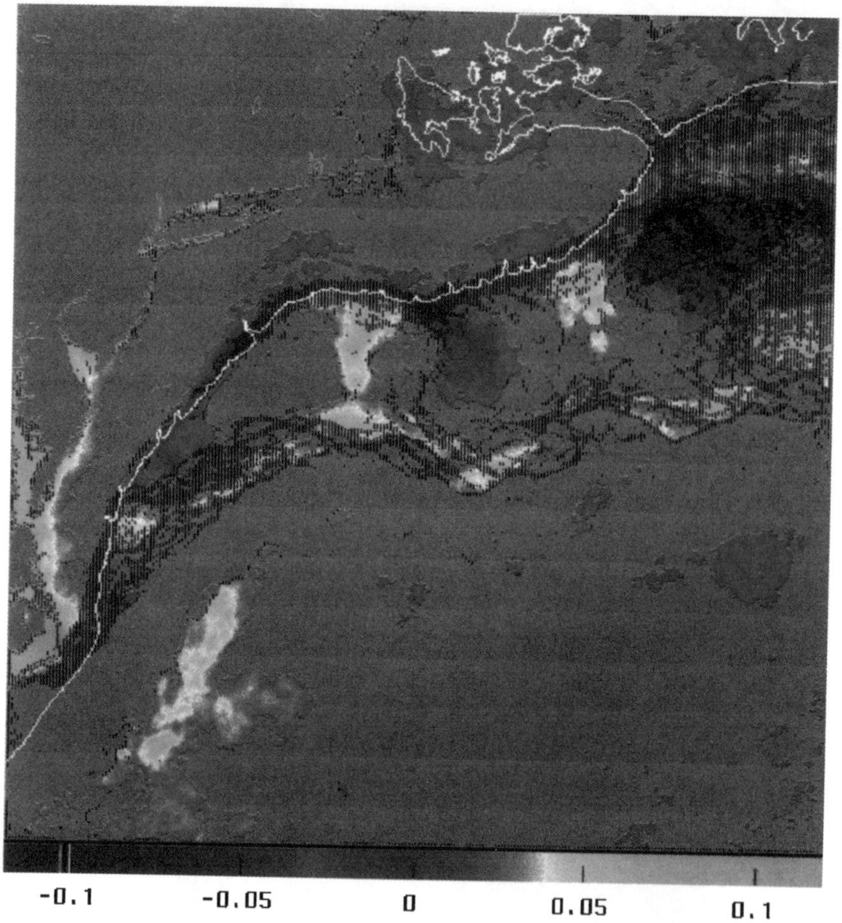

Figure 8: Second EOF with regions of magnitude less that 0.025 masked (grey), 200 m isobath (white), continental outline (red), persistent fronts (crosshatched in black). *(See Color Plate 11.)*

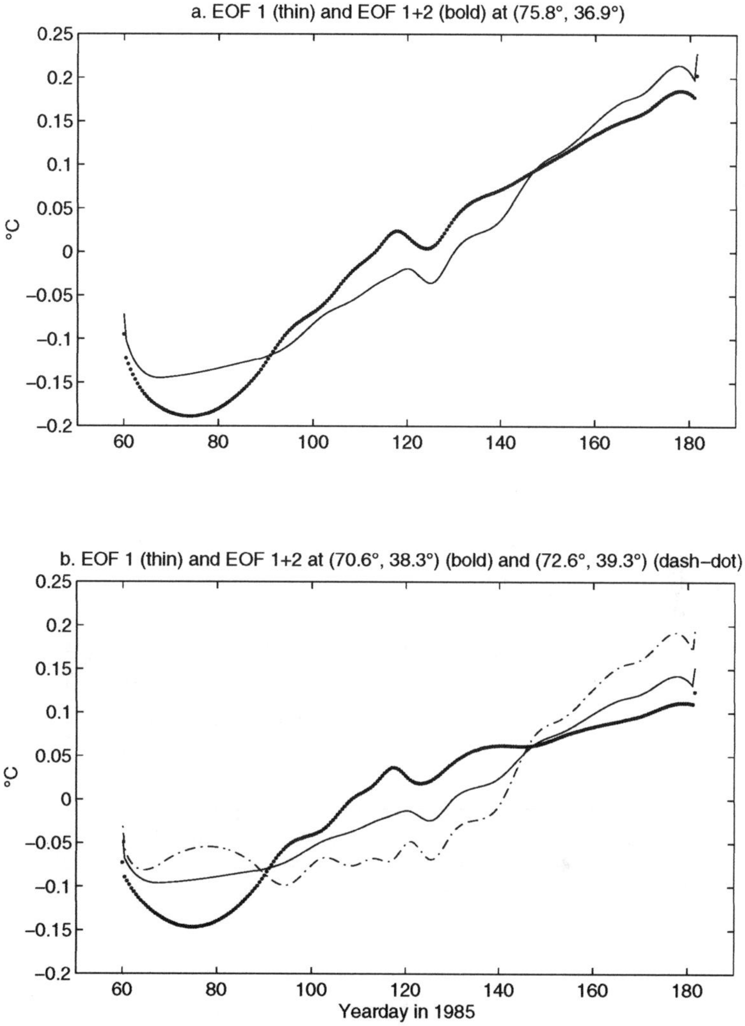

Figure 9: (a) Time series of the first EOF (thin solid) and the sum of the first two EOFs (bold) at the entrance to the Chesapeake Bay ($75.8°W$, $36.9°N$). (b) Time series of the first EOF (thin solid) and of the sum of the first two EOFs (bold) along the northern edge of the Gulf Stream ($70.6°W$, $38.3°N$) and of the sum of the first two EOFs (dash-dot) in the shelf/slope frontal band ($72.6°W$, $39.3°N$).

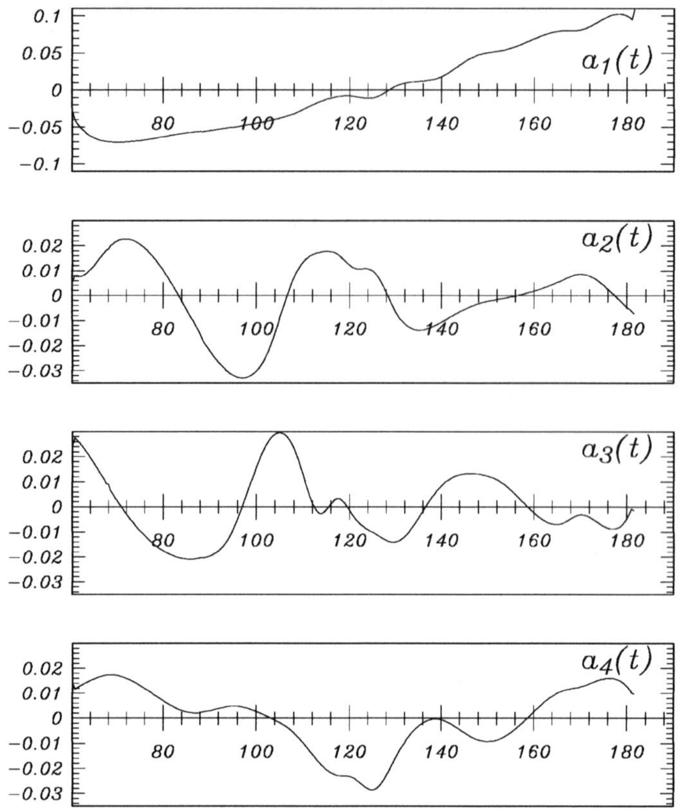

Figure 10: Modal coefficients, $a_n(t)$ for EOFs in the regions associated with persistent fronts.

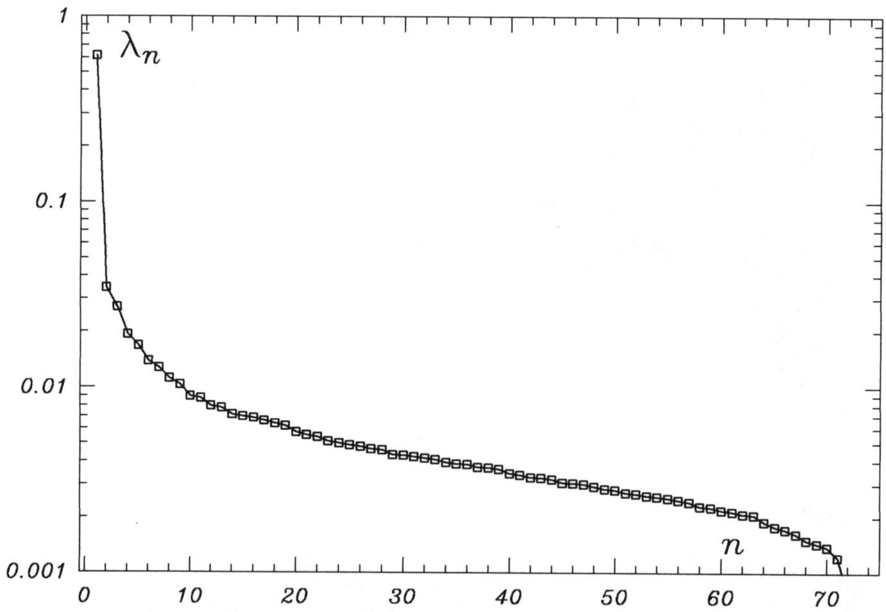

Figure 11: Eigenvalues derived from the relatively cloud-free ensemble of 73 images. 90% of the variance is captured by the first 32 eigenfunctions.

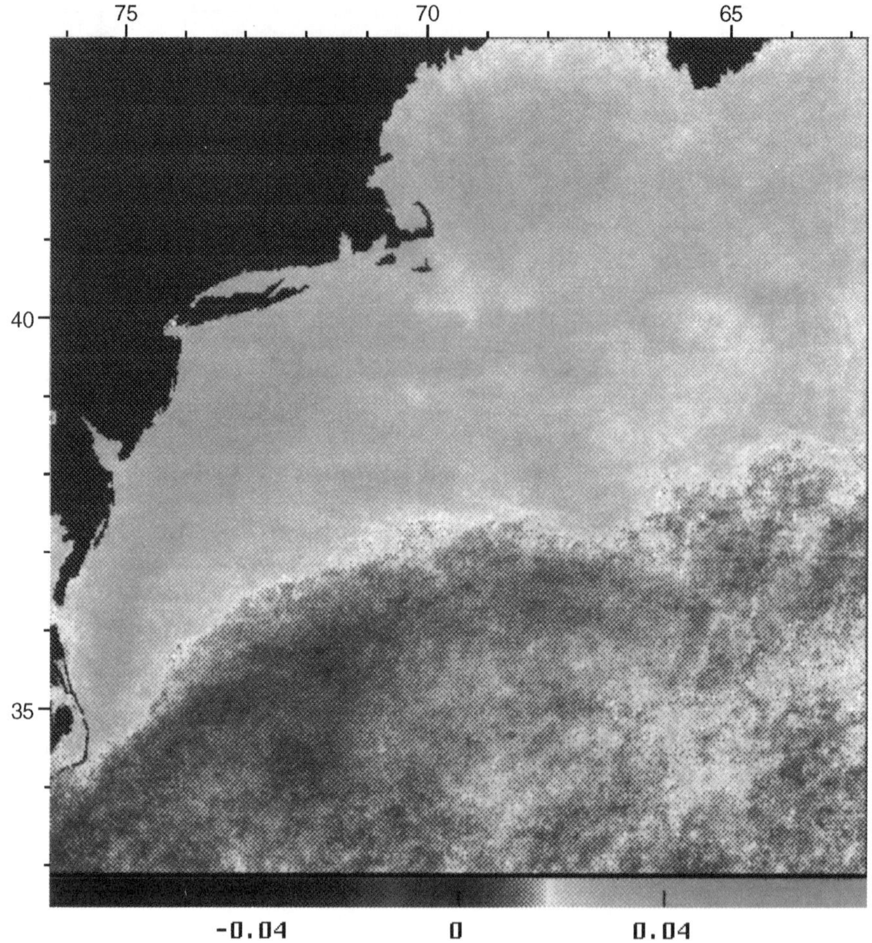

Figure 12: The first empirical eigenfunction derived from the relatively cloud-free ensemble.
(See Color Plate 12.)

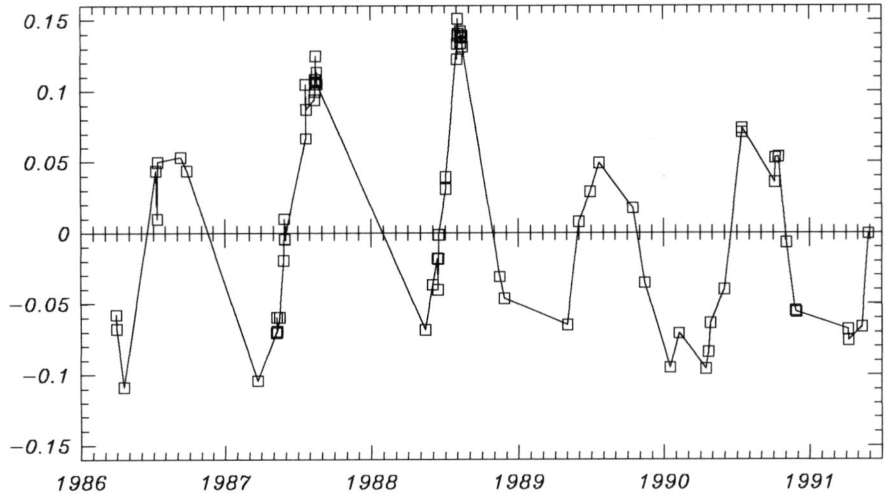

Figure 13: The coefficient of the first empirical eigenfunction estimated from the warm regions of snapshots drawn from the years 1986 – 1991. Though the data is irregularly spaced (and missing in the cloudy spring months) the annual cycle of seasonal warming is apparent.

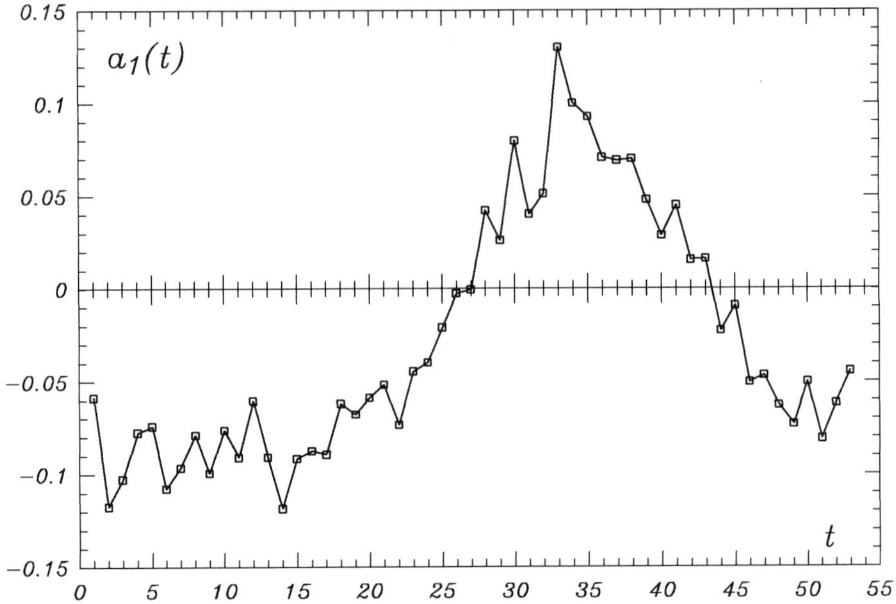

Figure 14: The coefficient of the first empirical eigenfunction estimated from weekly snapshots in 1990. The abscissa is labeled by week of the year.

Chaotic Mixing Across Oceanic Jets

P. Miller, C.K.R.T.Jones, G.Haller and L. Pratt[†]

Division of Applied Mathematics
Brown University, Providence, RI 02912

[†]Woods Hole Oceanographic Institution
Woods Hole, MA 02543

Abstract. The perspective of geometric dynamical systems is used to study the transport of fluid across oceanic jets. We study the mixing associated with the simplest analytical models for jets, namely, neutral modes superimposed on a base mean flow, where the base flow and the neutral modes are approximately potential vorticity conserving. The base jet plus a single neutral mode is an integrable flow in the appropriate moving frame, and heteroclinic orbits act as impenetrable boundaries separating different regions of phase space. Superimposing more than one neutral mode results in the breakup of these heteroclinic orbits and associated chaotic mixing. Using a cusped jet model we study the case where the perturbation is periodic in time. We present numerical simulations of the Poincaré map along with calculations of the Melnikov integral which characterizes the exchange rate across such boundaries. The analytical and numerical results show that these models explain mixing along the edges of the jet, but do not appear to explain mixing across the body of the jet.

Introduction

The mechanism for the transport of fluid particles across such oceanic jets as the Gulf Stream is not well understood. Near the surface it is expected that high potential vorticity gradients at the center of the jet should act as barriers to this transport and indeed, this has been experimentally corroborated by Behringer, et al. [1]. Nevertheless, transport across jets appears to occur either by ring detachment or straightforward advection. Various models have been formulated based on the meander structure of the Gulf Stream as evident, for example, from the tracking of isopycnal floats (see the float data presented by Bower and Rossby [3]). Bower [2] and Samelson [14] have considered models which are purely kinematic in studying the transport associated with these meander structures. More recently, Pratt, Lozier, and Beliakova [10], and del-Castillo-Negrete and Morrison [5] have proposed analytical models of velocity fields which are approximately potential vorticity conserving.

The analytical models considered in [10] and [5] consist of a base shear flow perturbed by neutral modes where the perturbing modes are constrained to satisfy the linearized potential vorticity equation. The formulation of such models as base flow plus perturbing modes can provide insight into characteristic flows that may truly exist as the nonlinear equilibration of such "neutral" or "unstable" directions from the base flow. Explicit analytic expressions for such fully nonlinear Eulerian flow fields are comparatively rare, and these "linear" approximations offer, with the existing technology, the best way to make meaningful, analytic predictions about the nature of Lagrangian trajectory motion in the full flow field.

In this work we use the perspective of geometric dynamical systems to study the transport and mixing associated with these models. The flow consisting of the base jet plus a single neutral mode is integrable when viewed in a reference frame translating with the same speed as the perturbing wave. For this integrable flow there exist two rows of heteroclinic orbits which define the boundaries of the meandering jet and serve as barriers between regions of qualitatively different fluid motion. Superimposing additional neutral modes creates a time-dependent perturbation which breaks the integrability and generally results in some transport of fluid between these different regions of the phase space. When the perturbation is time-periodic this transport is viewed as the transverse intersection of stable and unstable manifolds of the associated Poincaré map. These transverse intersections and transport rates can be characterized via the Melnikov integral evaluated along the heteroclinic orbit of the unperturbed flow.

We present a general description of these models and the derivation of the Melnikov integral for the case where there are exactly two perturbing neutral modes. We use numerical simulations to illustrate the mixing associated with the cusped jet model proposed by Pratt, et al. [10], where the amplitude of the second mode is taken to be much smaller than the first. For modest values of the perturbation parameters this model shows significant mixing along the edges of the jet but does not demonstrate any transport through the body of the jet. Moreover, the mixing depends strongly on the relative wavespeeds of the two perturbing modes. The most significant mixing occurs between regions of circulating fluid and the body of the jet in the case when the weaker neutral mode is moving faster than the dominant neutral mode. We include numerical calculations of the Melnikov integrals for a range of parameters associated with the cusped jet model.

Modeling the Jet

In this section we present a general description of the dynamically consistent models proposed by Pratt, Lozier and Beliakova [10], and del-Castillo-Negrete and Morrison [5]. In these models the flow is considered to be two-dimensional and incompressible and therefore determined by a streamfunction $\psi(x, y, t)$.

$$\begin{aligned} \frac{dx}{dt} &= -\frac{\partial \psi}{\partial y} \\ \frac{dy}{dt} &= \frac{\partial \psi}{\partial x} \end{aligned} \tag{1}$$

Here x is taken to be the easterly direction and positive y points northward. If q denotes the potential vorticity, the dynamics of the flow are governed by the conservation of potential vorticity along particle paths,

$$\frac{\partial q}{\partial t} - \frac{\partial \psi}{\partial y}\frac{\partial q}{\partial x} + \frac{\partial \psi}{\partial x}\frac{\partial q}{\partial y} = 0. \tag{2}$$

These "dynamically consistent" models assume that $\psi(x, y, t)$ is a small perturbation of an equilibrium streamfunction $\psi_0(y)$ defining a base shear flow (zonal flow),

$$U(y) = -\frac{\partial \psi_0}{\partial y}.$$

The perturbations are neutral modes of the form

$$\psi(x, y, t) = \mathrm{Re}\left[a\phi(y)e^{ik(x-ct)}\right],$$

and should satisfy conservation of potential vorticity to order $\mathcal{O}(a^2)$.

In the work by del-Castillo-Negrete and Morrison [5] the beta-plane approximation is used, so that $q = \nabla^2\psi + \beta y$. Substituting the streamfunction into (2) and retaining just the leading order terms leads to the Rayleigh-Kuo equation for $\phi(y)$,

$$(U - c)\left(\frac{d^2\phi}{dy^2} - k^2\phi\right) + \phi\frac{dq_0}{dy} = 0. \tag{3}$$

Here $q_0(y)$ denotes the potential vorticity for the base shear flow. It is possible to have neutral modes (c real-valued) if the potential vorticity gradient dq_0/dy vanishes whenever $U(y) = c$. For the case where the base flow is the Bickley jet, $U(y) = \mathrm{sech}^2 y$, there exist two distinct neutral modes for each value of β in the interval $0 < \beta < 2/3$.

Pratt et al. [10] use a $1\frac{1}{2}$-layer model where the potential vorticity is given by $q = \nabla^2\psi - \psi$. For this model the linearization of (2) about the base flow gives the following equation for $\phi(y)$,

$$(U - c)\left(\frac{d^2\phi}{dy^2} - (k^2 + 1)\phi\right) + \phi\frac{dq_0}{dy} = 0. \tag{4}$$

The base flow is a cusped jet profile, $U(y) = e^{-|y|}$, derived from a piecewise constant potential vorticity model (see Pratt and Stern [11]). The center of the jet is characterized by a discontinuous front separating two regions of constant potential vorticity. For this model the condition $dq_0/dy = 0$ is trivially satisfied for all $y \neq 0$. Since these cusped jets are linearly stable, it is reasonable to model perturbations in the full flow field by superimposing on the base flow selected neutral modes, and this model will be the focus of our later analysis.

For simplicity, we assume that there are exactly two neutral modes superimposed on the base shear flow,

$$\psi(x, y, t) = \psi_0(y) + \psi_1(x - c_1t, y) + \psi_2(x - c_2t, y).$$

The perturbations are of the form prescribed earlier,

$$\psi_1(x - c_1t, y) = a\phi_1(y)\cos k_1(x - c_1t),$$

$$\psi_2(x - c_2t, y) = \varepsilon\phi_2(y)\cos k_2(x - c_2t).$$

Here we only consider a and ε to be small parameters, $0 \leq \varepsilon \ll a \ll 1$. The first neutral mode introduces a characteristic speed c_1 and by switching to the moving frame, $x \to x - c_1t$, the explicit time-dependency is removed from ψ_1. With respect to this moving frame the Lagrangian trajectories are governed by a time-dependent vector field where the order ε time dependency is T-periodic with angular frequency $\omega = 2\pi/T = k_2(c_2 - c_1)$.

$$\begin{aligned} \frac{dx}{dt} &= U(y) - c_1 - a\phi_1'(y)\cos k_1x - \varepsilon\phi_2'(y)\cos(k_2x - \omega t), \\ \frac{dy}{dt} &= -ak_1\phi_1(y)\sin k_1x - \varepsilon k_2\phi_2(y)\sin(k_2x - \omega t). \end{aligned} \tag{5}$$

First we describe how the system (5) contains the general structure of a meandering jet by considering the unperturbed flow with $\varepsilon = 0$. The flow at $\varepsilon = 0$ is a one-degree-of-freedom Hamiltonian system with streamfunction,

$$\psi_m(x, y) = \psi_0(y) + \psi_1(x, y) + c_1y. \tag{6}$$

The steering lines, denoted by $y = y_s$, are defined by the condition $U(y_s) = c_1$. Assume $U(y) > 0$ with a single maximum $U = U_0$ occurring at $y = 0$, and

$U(y)$ decaying monotonically to zero as $y \to \pm\infty$. Then for $0 < c_1 < U_0$ there will be exactly two steering lines, the northern (upper) line being denoted by $y_s = y_s^+ > 0$ and the southern (lower) steering line denoted by $y_s = y_s^- < 0$. For every integer n, there exists a pair of rest points located at $x = 2n\pi/k_1$ with $y = y_s + \mathcal{O}(a)$ determined by the equation,

$$U(y) - c_1 - a\phi'(y) = 0.$$

Similarly, at $x = (2n+1)\pi/k_1$ there are two rest points $y = y_s + \mathcal{O}(a)$ determined by,

$$U(y) - c_1 + a\phi'(y) = 0.$$

The eigenvalues for the linearization at the rest points are determined by,

$$\mu^2 = -ak_1^2\phi(y)(U'(y) - a\phi''(y)), \qquad \text{for } k_1 x = 2n\pi,$$

$$\mu^2 = ak_1^2\phi(y)(U'(y) + a\phi''(y)), \qquad \text{for } k_1 x = (2n+1)\pi.$$

With the assumption that $\phi(y) > 0$ for all y, and a being a small parameter, the rest points alternate between hyperbolic ($\mu^2 > 0$) and elliptic ($\mu^2 < 0$). In the vicinity of the northern steering line the hyperbolic rest points occur at $x = 2n\pi/k_1$, and along the southern steering line the hyperbolic points are at $x = (2n+1)\pi/k_1$.

For any integer n, the unperturbed flow is invariant under a change of variables which reverses time and reflects x about the axis $x = n\pi/k_1$.

$$t \to -t, \qquad x \to -x + \frac{2n\pi}{k_1}, \qquad y \to y.$$

Therefore any solution parametrized by $x(0) = n\pi/k_1$ satisfies

$$x(-t) = -x(t) + \frac{2n\pi}{k_1}, \qquad y(-t) = y(t). \tag{7}$$

From this symmetry it follows that there are two rows of heteroclinic orbits, one along each of the steering lines (the cat's eyes structure). We are also assuming here that the steering lines are not so close to each other as to create homoclinic orbits in the phase space, as in the separatrix reconnection discussed in [5]. Figure 1 shows the streamlines for such an integrable flow using the cusped jet model from [10]. The parameters have been chosen to satisfy $a \ll 1$ and still show the pattern of a meandering jet. Increasing the speed c_1 draws the steering lines closer together, reducing the width of the jet and accentuating the meander.

Let q denote the solution vector $q = (x, y)^t$, and use q^h to denote a heteroclinic orbit connecting adjacent hyperbolic stagnation points of the integrable flow. We refer to the connecting orbits along the exterior of the jet as the

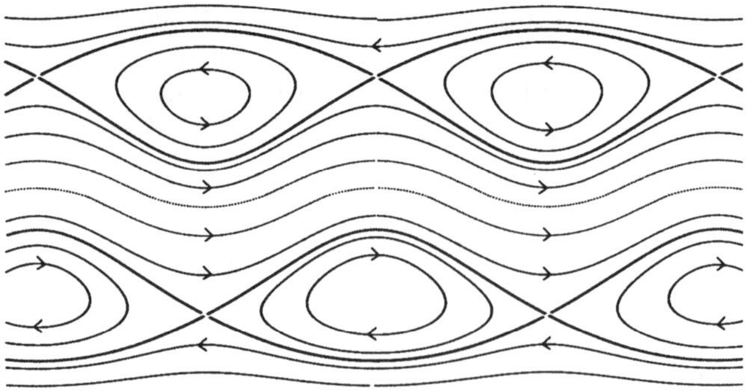

Figure 1: Flow associated with the single cusp jet model with a single neutral mode superimposed on the base shear flow ($c_1 = 0.80$, $a = 0.01$). When viewed in a reference frame moving with the speed of the neutral wave, the flow is time-independent and admits two rows of critical points at the steering lines. The flow is integrable and the heteroclinic trajectories act as barriers separating the phase space into three distinct regions.

outer heteroclinics, and the connecting orbits along the interior of the jet as the *inner* heteroclinics. As seen in Figure 1, these invariant manifolds separate the unperturbed flow into three distinct regions, the *interior* or body of the jet where the flow is left to right, the *exterior* region where the flow is right to left (retrograde), and the *circulation* regions where fluid is trapped by a pair of heteroclinic orbits. With $\varepsilon = 0$ the integrability of the flow ensures that these regions remain isolated from one another, that is, fluid can never pass from one region to another.

Chaotic Mixing in the Jet

Setting $\varepsilon > 0$ in (5) results in a time-periodic perturbation of an integrable flow with period $T = 2\pi/\omega$. Here it is most useful to consider the associated Poincaré map generated by following the flow of each point in phase space through exactly one time period. The T-periodic solutions of the time-dependent flow become fixed points of the map and for ε sufficiently small the hyperbolic rest points of the unperturbed flow perturb to hyperbolic fixed points for the Poincaré map. Generically, one expects the heteroclinic orbits to break into an infinite intersection of invariant manifolds. This *heteroclinic tangle* is associated with chaotic trajectories and transport of fluid between the different regions previously separated with respect to the integrable flow (see Rom-Kedar and Wiggins [13] and [15]).

Fix a parametrization for the unperturbed heteroclinic orbit by specifying $q^h(0)$. Then the vector $\nabla\psi_m(q^h(-t_0))$ is normal to the heteroclinic trajectory at the point $q^h(-t_0)$ and can be used to measure the signed distance between the stable and unstable manifolds of the Poincaré map, denoted $d(t_0, \varepsilon)$. This gap between the stable and unstable manifolds provides a mechanism for fluid to be transported from one region of phase space to another, and the size of $d(t_0, \varepsilon)$ characterizes the amount of mixing which occurs. This distance function is $\mathcal{O}(\varepsilon)$ with the leading term expressed in terms of the Melnikov function $M(t_0)$,

$$d(t_0, \varepsilon) = \varepsilon \frac{M(t_0)}{\|\nabla \psi_m(q^h(-t_0))\|} + \mathcal{O}(\varepsilon^2).$$

With the Hamiltonian structure in (5) the Melnikov function can be expressed in terms of the streamfunctions,

$$M(t_0) = \int_{-\infty}^{\infty} \left\{ \psi_m(q^h(t)), \psi_2(q^h(t), t + t_0) \right\} dt \, .$$

Here $q^h(t)$ indicates that the integrand is being evaluated along the unperturbed heteroclinic orbit connecting the rest points for $\varepsilon = 0$, and $\{\cdot\,,\cdot\}$ denotes the Poisson bracket in the x and y variables,

$$\{F, G\} \;=\; \frac{\partial F}{\partial x}\frac{\partial G}{\partial y} - \frac{\partial F}{\partial y}\frac{\partial G}{\partial x}.$$

Note that the distance function also depends on the phase coordinate θ_0 used to define the Poincaré map, where $\theta(t) = \omega t \pmod{2\pi}$. The distance function is periodic in t_0 with period $T = 2\pi/\omega$ and a different choice of θ_0 just phase shifts the result. There is no loss in generality in choosing θ_0 for our convenience. For more details on these methods see Guckenheimer and Holmes [6], and Rom-Kedar and Wiggins [13], [15].

Showing that $M(t_0)$ has a simple zero implies the transverse intersection of the stable and unstable manifolds and the associated heteroclinic tangle and chaotic trajectories, for sufficiently small values of the perturbation parameter ε. For the system in (5) the Melnikov integral has the form,

$$\begin{aligned} M(t_0) \;&=\; \int_{-\infty}^{\infty} \;\{-\dot{x}^h(t) k_2 \phi_2(y^h(t)) \sin(k_2 x^h(t) - \omega(t + t_0)) \;+ \\ &\qquad \dot{y}^h(t) \phi_2'(y^h(t)) \cos(k_2 x^h(t) - \omega(t + t_0))\} \, dt \, . \end{aligned}$$

This can be expressed as

$$\begin{aligned} M(t_0) \;&=\; \int_{-\infty}^{\infty} [-\dot{x}^h k_2 \phi_2 \sin(k_2 x^h - \omega t) + \dot{y}^h \phi_2' \cos(k_2 x^h - \omega t)\,]\, dt\; \cos(\omega t_0) \;+ \\ &\quad \int_{-\infty}^{\infty} [\;\dot{x}^h k_2 \phi_2 \cos(k_2 x^h - \omega t) + \dot{y}^h \phi_2' \sin(k_2 x^h - \omega t)\,]\, dt\; \sin(\omega t_0)\, . \end{aligned}$$

Parametrizing the heteroclinic orbit such that $x(0) = n\pi/k_1$ allows us to apply the symmetry in (7) and conclude that the first integrand is an odd function about $t = 0$ and integrates to zero. It follows that the Melnikov function can be expressed as

$$M(t_0) = A\sin(\omega t_0),$$

where the amplitude A is given by the integral,

$$A = \int_{-\infty}^{\infty} [\, \dot{x}^h k_2 \phi_2 \cos(k_2 x^h - \omega t) + \dot{y}^h \phi_2' \sin(k_2 x^h - \omega t) \,]\, dt\,. \tag{8}$$

From this expression it is clear that $M(t_0) = 0$ at $t_0 = n\pi/\omega$, for any integer n. Since the derivative at $t_0 = 0$ is given by $M'(0) = \omega A$, proving the existence of transverse intersections and the heteroclinic tangle requires showing that the integral in (8) is nonzero.

The amount of flux crossing from one region to another in the course of one time period is characterized by the area in phase space enclosed by the invariant manifolds between two adjacent intersection points. The leading order term for this lobe area $\mu(L)$ can be found by integrating $M(t_0)$ ([13], [15]). The time averaged rate of exchange across the boundary is given by dividing the total amount of flux crossing the boundary by the length of the time period T.

$$R \;=\; \frac{\omega}{2\pi}\int_0^{\frac{2\pi}{\omega}} |M(t_0)|\; dt_0 + \mathcal{O}(\varepsilon^2) \;=\; \varepsilon\frac{2|A|}{\pi} + \mathcal{O}(\varepsilon^2). \tag{9}$$

Chaotic Mixing with Cusped Jet Model

We have calculated solutions to the flow in (5) using the cusped jet model from [10]. For this model the base velocity profile is $U(y) = e^{-|y|}$, and the solutions to the linearized potential vorticity equation are given by $\phi(y) = e^{-\kappa|y|}$. The wavenumber k, the wave speed c, and the cross-stream decay rate κ are related by

$$k^2 = \frac{1}{(1-c)^2} - 1\,, \qquad \kappa = \sqrt{1+k^2}\,. \tag{10}$$

An example with a single neutral mode ($\varepsilon = 0$) is shown in Figure 1 with $a = 0.01$ and $c_1 = 0.80$. Here the system is integrable and we see two rows of heteroclinic orbits bracketing the meander structure of the jet. The streamfunction in (6) is a first integral for the flow and the heteroclinic trajectories act as barriers to the flow inhibiting transport across the boundaries of the jet.

To break the integrability a second neutral mode is added to the streamfunction with amplitude $\varepsilon = 0.001$ and wavespeed $c_2 = 0.90$. In the reference

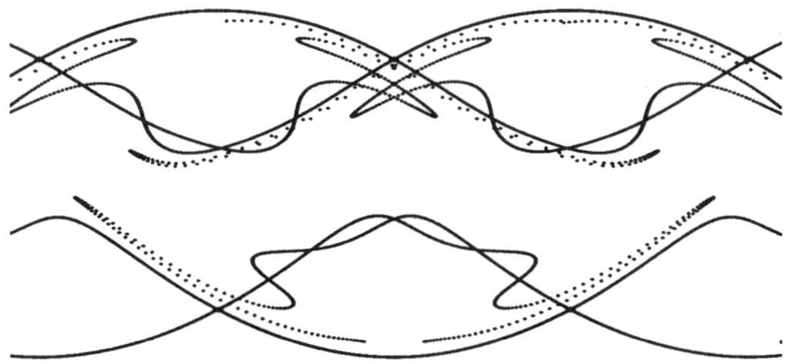

Figure 2: After adding a second neutral mode to the jet model ($c_2 = 0.90$, $\varepsilon = 0.001$) the flow is time-periodic with period $T = 6.31$. Shown here are the stable and unstable manifolds for hyperbolic fixed points of the Poincaré map. For the case $c_2 > c_1$ the transverse intersection of these manifolds are easily observed along the interior boundary of the jet while the outer boundaries appear unbroken.

frame moving with speed c_1 this second mode appears as a time-periodic perturbation with frequency $\omega = k_2(c_2 - c_1)$. The Poincaré map is determined by integrating the flow over exactly one time period. For small ε it is known that the hyperbolic rest points of the integrable flow perturb to hyperbolic fixed points of the map. Figure 2 shows the stable and unstable manifolds for some of these hyperbolic fixed points of the map. These simulations clearly illustrate the distinctive looping and heteroclinic tangle associated with transverse intersections of the stable and unstable manifolds.

An interesting and consistent observation in all of these simulations is that when the speed of the second perturbation is greater than the speed of the moving frame, it is significantly harder to detect the presence of transverse intersections along the outside edge of the jet. This is similar to a feature noted by Samelson [14] for the kinematic meander model. The intersections on the outer boundary seem to occur with a higher order of transversality, and it appears that the outside heteroclinics do not break at all. This asymmetry in the breakup of the heteroclinic orbits is due to the perturbation being in the form of a traveling wave. More specifically, the evaluation of the Melnikov integral in (8) depends strongly on the sign of ω. When the wave speed of the second perturbation in (5) is slower than the speed of the moving frame ($c_2 < c_1$), the sign of ω is reversed and the more observable breakup occurs at the outer boundary, along the retrograde region. An example of this situation is shown in Figure 3 where the wavespeed for the second perturbation is set to $c_2 = 0.65$, with amplitude $\varepsilon = 0.001$. Even though the break up along the outer boundary is now more significant than in Figure 2, it does not appear to be as dramatic as the mixing along the inner boundary with $c_2 > c_1$.

These transverse intersections of stable and unstable manifolds provide

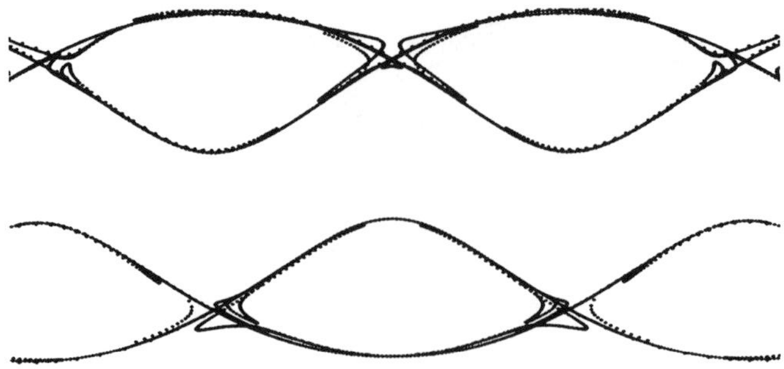

Figure 3: Here the second neutral mode is moving slower than the characteristic speed c_1 ($c_2 = 0.65$, $\varepsilon = 0.001$), and the time period is $T = 15.65$. For $c_2 < c_1$ the transverse intersections of stable and unstable manifolds for the Poincaré map are more easily observed along the outer boundaries of the jet.

a dynamical mechanism for fluid to pass from one region of phase space to another where the motions in the different regions may be qualitatively quite different. This transport occurs via the dynamics of the *turnstiles* defined by the intersecting stable and unstable manifolds (see [13] and [15]). In this cusped jet model with $c_2 > c_1$, the breakup of the boundaries indicates that there is some transport of fluid between the body of the jet and the regions of circulating motion. This transport of fluid across boundaries of the stable and unstable manifolds is illustrated in Figure 4 where a small region of fluid particles are followed for six iterations of the Poincaré map.

In Figure 4 the particles begin in what is considered the interior of the jet where the motion is essentially left to right. After two time periods they have entered a region where the fluid is circulating. After several more iterations of the map some of these particles may reenter the turnstile and pass back into the body of the jet, resuming a left to right motion. These numerical observations agree qualitatively with the observations of float data discussed by Bower and Rossby in [2] and [3] where it was observed that floats exiting the Gulf Stream to the south generally did so as they approached a meander trough.

Our final calculations involve computing the integral in (8) for the amplitude of the Melnikov function. The calculations are performed for the heteroclinic loop at the southern steering line connecting $x = -\pi/k_1$ with $x = \pi/k_1$. The solutions for the heteroclinic orbits at $\varepsilon = 0$ are parametrized so that $x^h(0) = 0$, making the integrand in (8) an even function about $t = 0$. The initial value $y^h(0)$ is determined from the value of the streamfunction $\psi_m(x, y)$ at the rest points. The heteroclinic solution $(x^h(t), y^h(t))$ is determined by integrating the flow in (5) with $\varepsilon = 0$. The resulting solution is then used in approximating the integral for the amplitude of the Melnikov function.

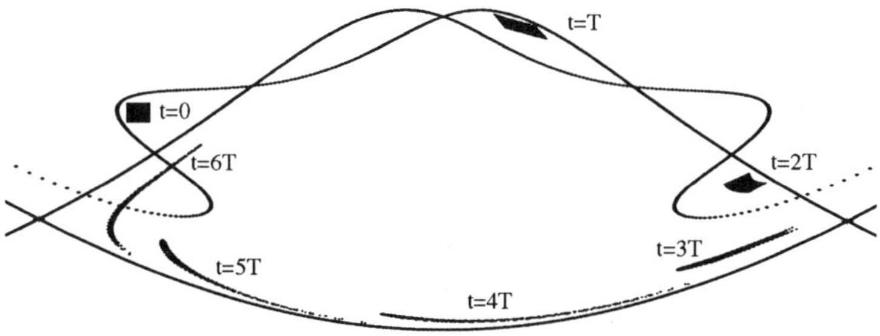

Figure 4: Poincaré map for the time-periodic flow in Figure 2 with $c_2 > c_1$. The dark block at the left represents 400 initial particles and their flow is mapped through six time periods. This illustrates how the fluid trajectories pass from the body of the jet into a region of circulating motion. To an observer the floats appear to exit the body of the stream as they approach a trough of the jet meander.

In the cusped jet model there are three parameters, the amplitude a, and the wavespeeds c_1 and c_2. This is a perturbation result for ε sufficiently small and ε does not appear explicitly in the calculation of A. As stated in (9) the calculation of the integral A gives the leading order term for the average rate of fluid exchange across the heteroclinic boundary, where the leading order term is $\mathcal{O}(\varepsilon)$. In the following calculations we have fixed the amplitude of the first neutral mode at $a = 0.01$.

Figure 5 shows the calculation of A as a function of the second wavespeed c_2, with the first wavespeed fixed at $c_1 = 0.80$. This value of c_1 coincides with the value used in the simulations presented in the previous figures. The graph shows that no fluid is exchanged at $c_2 = 0.80$ where the speed of the second mode matches the speed of the first and the system is still integrable. As observed in the earlier simulations the exchange across the inner boundary (solid line) is significant only for a range of values with $c_2 > c_1$. Exchange across the outer boundary (dashed line) is significant only for values satisfying $c_2 < c_1$.

The final two figures present the calculations of the integral A over the two-dimensional parameter space (c_1, c_2). In the cusped jet model admissible values of the wavespeed are $0 < c_{1,2} < 1$. Again the amplitude of the first neutral mode is fixed at $a = 0.01$. Figure 6 shows the exchange rate across the inner boundary and Figure 7 shows the same calculation across the outer boundary. These results verify that the most significant exchange of fluid takes place along the interior boundaries over a narrow range of parameter values satisfying $c_2 > c_1$. Exchange across the exterior boundaries is significant only when $c_2 < c_1$, consistent with our earlier simulations.

For the purpose of this study we have chosen to keep the parameter a on the order of 0.01 and ε is kept small with respect to a. This is in keeping

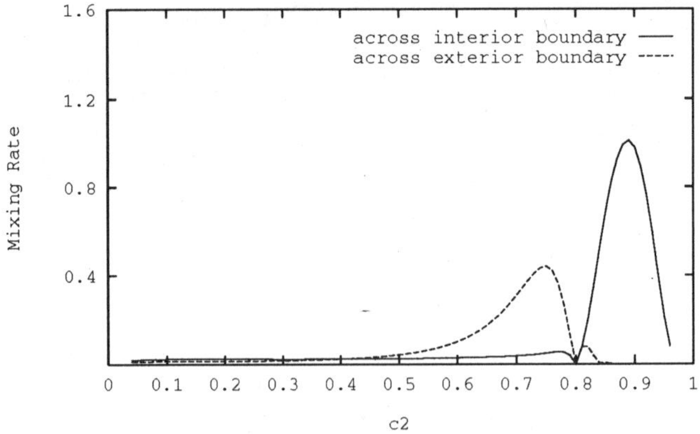

Figure 5: Amplitude of the Melnikov function in (8), as computed for the cusped jet model with $a = 0.01$ and $c_1 = 0.80$. The solid line represents the exchange of fluid along the interior boundary and shows that significant mixing occurs only for $c_2 > c_1$. The dashed line verifies that exchange across the outer boundary occurs only for $c_2 < c_1$.

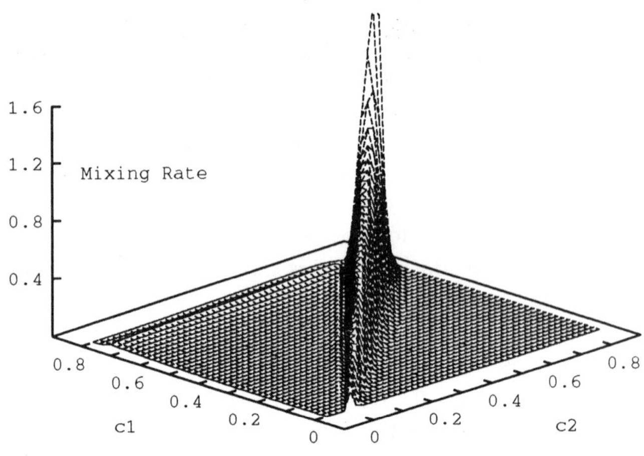

Figure 6: Calculation of the exchange rate as computed along the inner heteroclinic orbit for the cusped jet model with $a = 0.01$. Significant exchange occurs along the interior boundary of the jet only for values of $c_2 > c_1$.

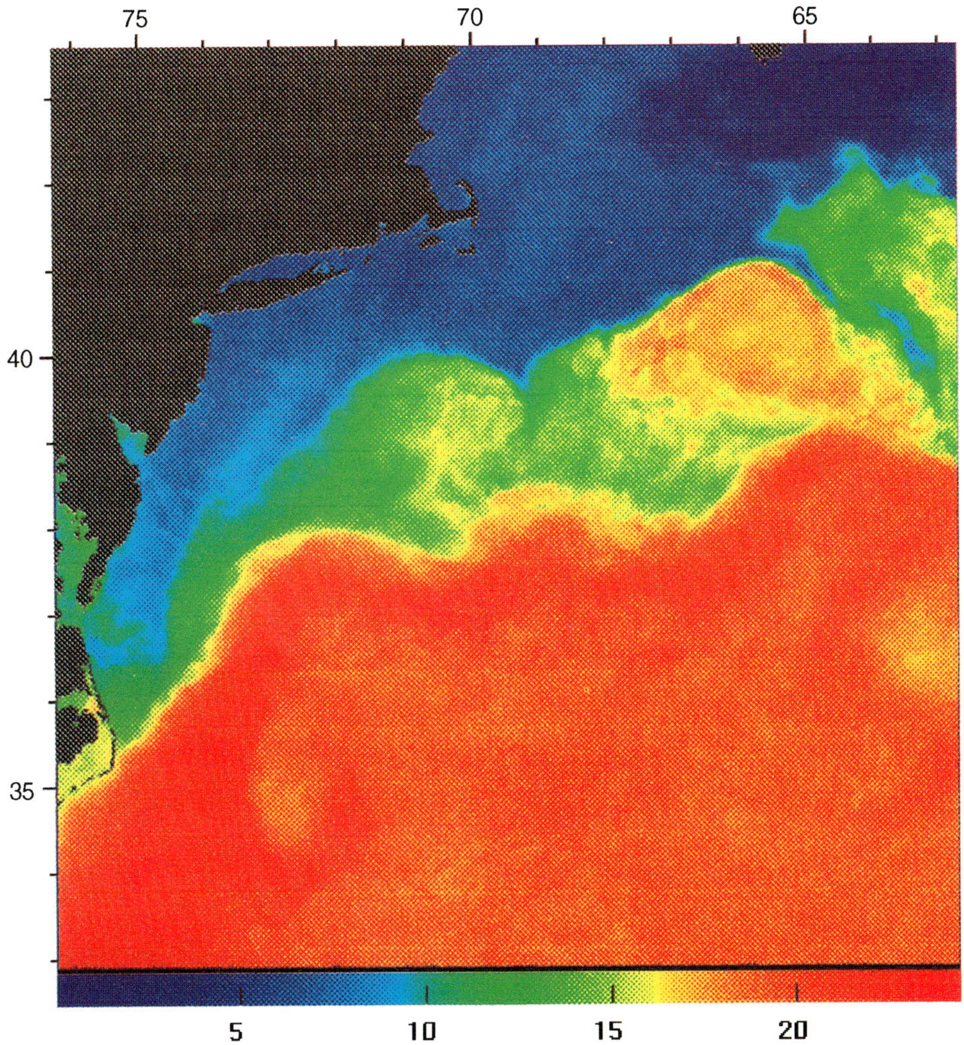

Color Plate 7. AVHRR image (9 March 1985) of the Western North Atlantic after spline interpolation and manual declouding. The color-bar shows the temperature in Celsius and the scales at the top and left show the latitude and longitude. (Grey-tone version of figure available on p. 577.)

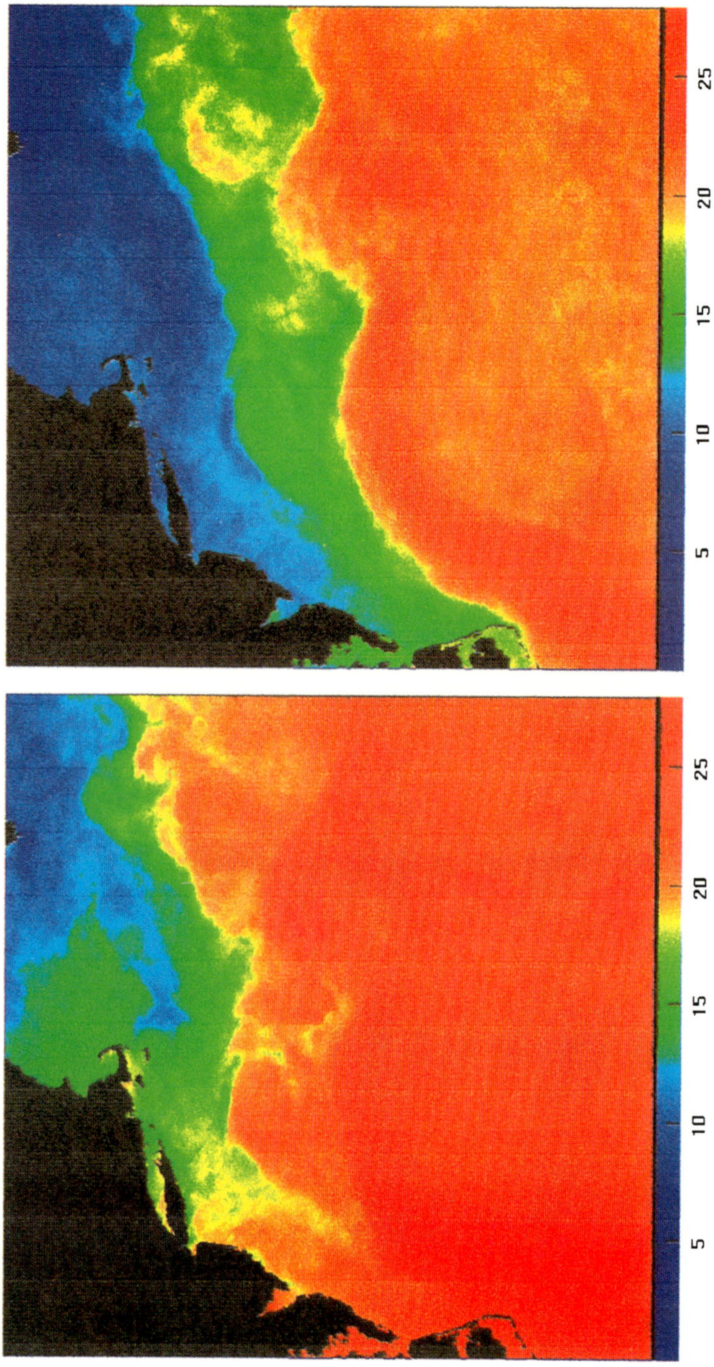

Color Plate 10. Sea surface temperature for 21 June 1985, and the same image with the seasonal warming contribution removed. The Gulf Stream, which was obscured by the general seasonal warming, is revealed in the projected image. (Grey-tone version of figure available on p. 583.)

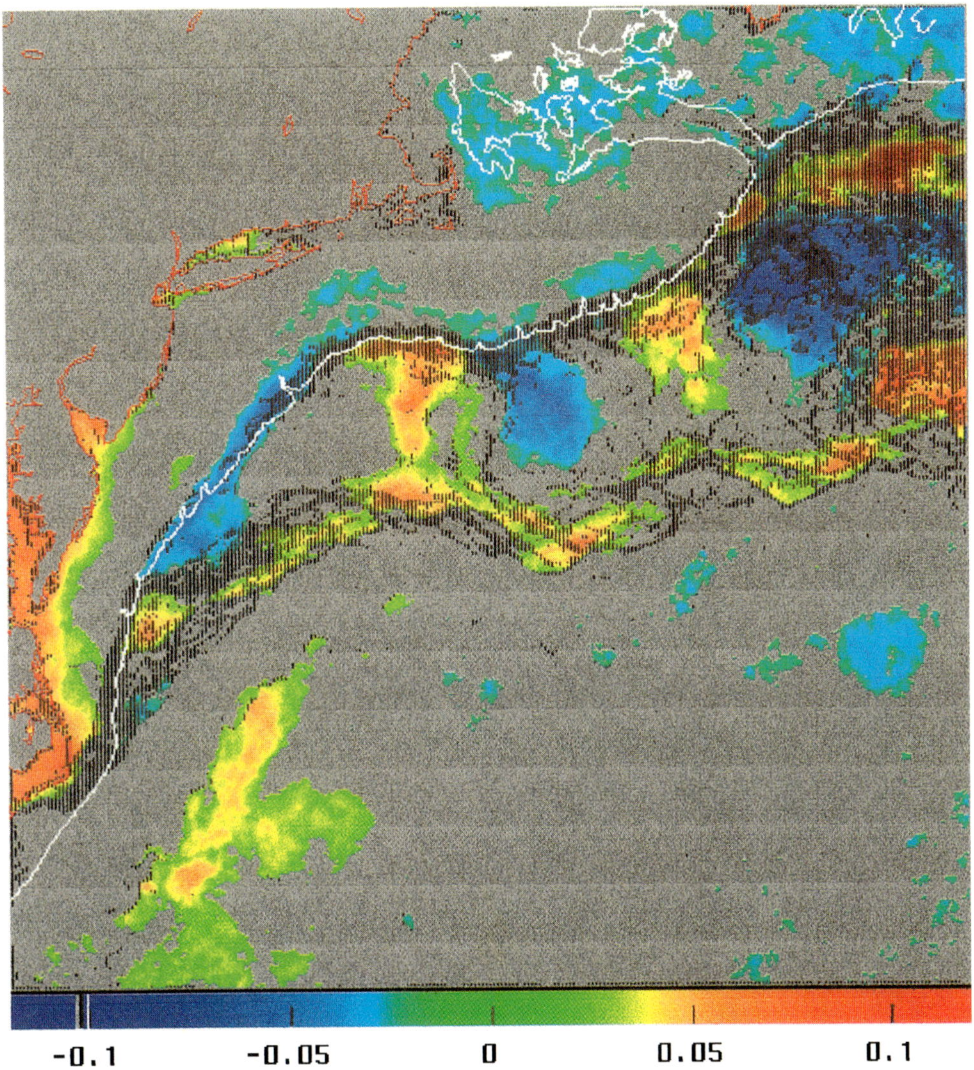

Color Plate 11. Second EOF with regions of magnitude less that 0.025 masked (grey), 200 m isobath (white), continental outline (red), persistent fronts (cross-hatched in black). (Grey-tone version of figure available on p. 584.)

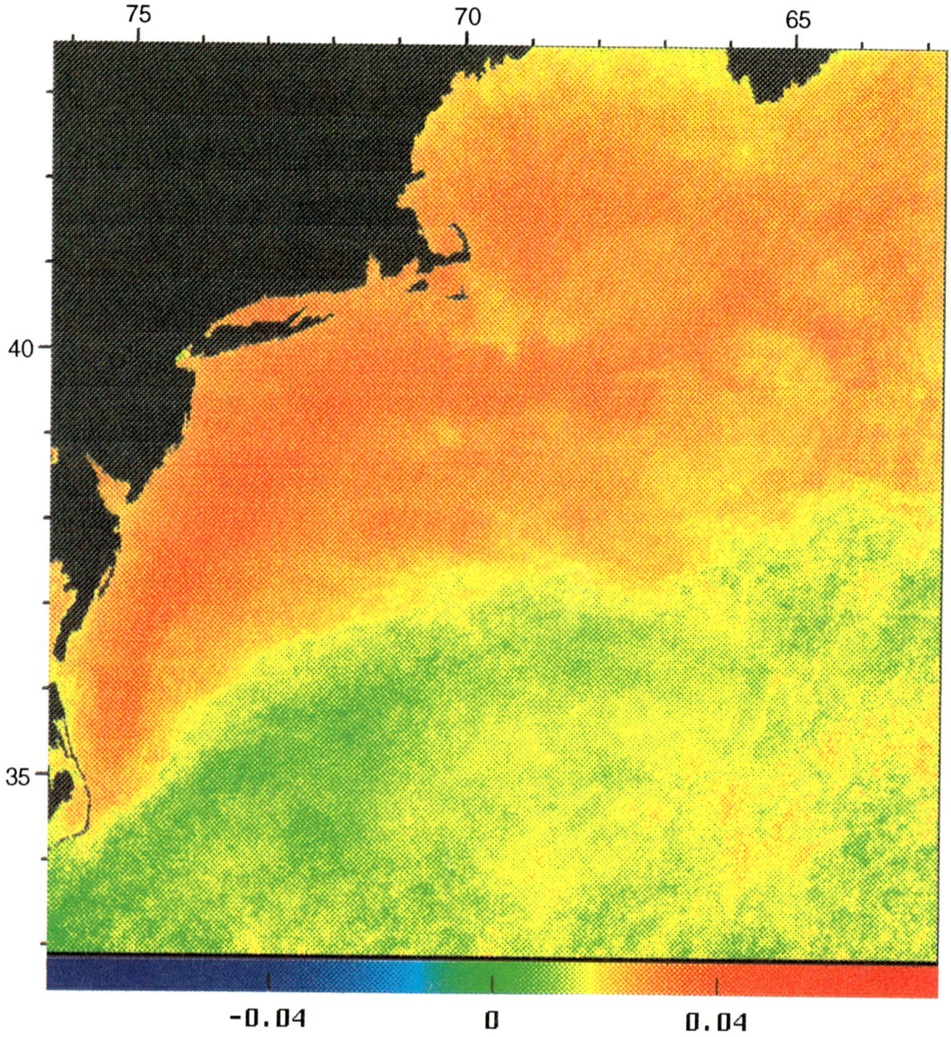

Color Plate 12. The first empirical eigenfunction derived from the relatively cloud-free ensemble. (Grey-tone version of figure available on p. 588.)

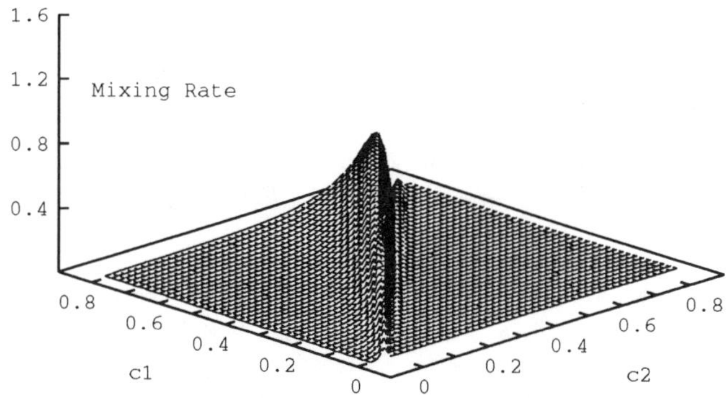

Figure 7: Calculation of the exchange rate as computed along the outer heteroclinic orbit for the cusped jet model with $a = 0.01$. Significant exchange occurs along the exterior boundary of the jet only for values of $c_2 < c_1$.

with the linearization used in deriving the analytical model and the fact that the Melnikov theory is a perturbation result with respect to $\varepsilon > 0$. Therefore changing the geometry of the jet requires changing the characteristic speed of the first neutral perturbation. Increasing the speed c_1 while keeping $a = 0.01$ draws the steering lines closer together, narrowing the width of the jet and making the meander more pronounced. However, for small values of ε our simulations show that the inner body of the jet still acts as a barrier to transport through the core of the original jet. Though there is considerable mixing at the boundaries, there remains a core region to the jet where the fluid particles never mix with the fluid in the circulating regions.

Acknowledgments

The research described in this article has been made possible by support from the Office of Naval Research under grant number N00014-92-J-1481. The phase space plots and Poincare maps were generated using *DsTool.*

References

[1] R.P.Behringer, S.D.Meyers & H.L.Swinney, Chaos and mixing in geostrophic flows, *Physics of Fluids A* 3 (1991) 1243-1249.

[2] A.S.Bower, A simple kinematic mechanism for mixing fluid parcels across a meandering jet, *Journal of Physical Oceanography* 19 (1991) 174-180.

[3] A.S.Bower & T.Rossby, Evidence of cross-frontal exchange processes in the Gulf Stream based on isopycnal RAFOS float data, *Journal of Physical Oceanography* 19 (1989) 1177-1190.

[4] M.G.Brown & R.M.Samelson, Particle motion in vorticity-conserving, two-dimensional incompressible flows, *Physics of Fluids* 6 (1994) 2875-2876.

[5] D.del-Castillo-Negrete & P.Morrison, Chaotic transport by Rossby waves in shear flow, *Physics of Fluids A* 5 (1993) 948-965.

[6] J.Guckenheimer & P.Holmes, *Nonlinear Oscillations, Dynamical Systems, and Bifurcations of Vector Fields*, Springer-Verlag, NY, 193.

[7] M.S.Lozier & S.C.Riser, Potential vorticity sources and sinks in a quasi-geostrophic ocean: beyond western boundary currents, *Journal of Physical Oceanography* 20 (1990) 1608-1627.

[8] J.Pedlosky, *Geophysical Fluid Dynamics*, Springer-Verlag, New York, 1979.

[9] L.J.Pratt, J.Earles, P.Cornillon & J.-F.Cayula, The nonlinear behavior of varicose disturbances in a simple model of the Gulf Stream, *Deep Sea Research* 32 (1991) 591-622.

[10] L.J.Pratt, M.S.Lozier & N.Beliakova, Parcel trajectories in quasi-geostrophic jets part 1: neutral modes, to appear in *Journal of Physical Oceanography* (1994)

[11] L.J.Pratt & M.E.Stern, Dynamics of potential vorticity fronts and eddy detachment, *Journal of Physical Oceanography* 16 (1986) 1101-1120.

[12] H.Ridderinkhof & J.T.F.Zimmerman, Chaotic stirring in a tidal system, *Science* 258 (1992) 1107-1111.

[13] V.Rom-Kedar & S.Wiggins, Transport in two-dimensional maps, *Arch. Rat. Mech. Anal.* 109 (1990) 239-298.

[14] R.M.Samelson, Fluid exchange across a meandering jet, *Journal of Physical Oceanography* 22 (1992) 431-440.

[15] S.Wiggins, *Chaotic Transport in Dynamical Systems*, Springer-Verlag, NY, 1992.

Enhanced Predictability in Chaotic Geophysical Systems

John Brindley[1] and Tomasz Kapitaniak[2]

[1] *Department of Applied Mathematical Studies, University of Leeds, Leeds LS2 9JT, UK*

[2] *Division of Dynamics, Technical University of Lodz Stefanowskiego 1/15, 90-924 Lodz, Poland*

Abstract. Using the Lorenz equations as an example we show that one chaotic system can be controlled by synchronizing its behaviour with the chaotic behaviour of another system. We particularly discuss the implications of this phenomenon in geophysical systems.

The Lorenz equations have for many years been the subject of much attention as one of the simplest systems of nonlinear ODEs to contain within their asymptotic solution space a strange attractor [1]. Their origins as a model for atmospheric dynamics (or more accurately thermal convection in a fluid [2] have perhaps rightly been played down as presenting far too simplistic a view of such a complex dynamical problem. There persists, however, a vigorous debate in the geophysical fluid dynamics community over the relevance and applicability to at least some features of atmospheric and/or ocean dynamics of low-order models in the form of simple nonlinear ODEs. Such models might of course, at least at certain phases of their cycles, offer the prospect of much enhanced predictability [3]. Prime examples are the certain coherent features, localised in either or both of space and time, which occur sporadically and unpredictably but with a certain statistical regularity, including atmospheric blocking patterns, and spectacularly, the El Nino phenomenon of the Pacific Ocean [4]. Though time series of real data are scarcely long enough to provide a conclusive argument, synthetic time series from typically plausible models [5,6] are highly suggestive of chaotic behaviour, and at the same time strongly reminiscent of real data. Certain randomly forced systems, on the other hand, eq [7] though yielding results similar to real data, do not suggest a low-dimensional attractor.

These models in general consist of coupled systems of equations, one or both of which themselves can exhibit chaotic behaviour, and their popularity motivates a more general examination of properties of coupled chaotic systems. Thus, in this paper, we focus on an example of coupled chaotic systems in which the effect of the coupling is to enhance predictability in the coupled system, relative to its separate components. We follow the broad approach of continuous chaos control scheme described in [8-11].

Thus we consider two chaotic systems, which we call A and B respectively,

$$\dot{x} = f(x) \qquad \dot{y} = g(y) \tag{1}$$

where $x,y \in \mathbf{R}^n$, and we use the controlling strategy which is schematically illustrated in Figure 1; the two systems are coupled through the operators λ, μ, which have a very simple linear form. We assume that some or all state variables of both systems A and B can be measured, so that we can measure signal $x(t)$ from the system A and signal $y(t)$ from B, and that the systems are coupled in such a way that the differences $D_{1,2}(t)$ between the signals $x(t)$ and $y(t)$ are

$$F_1(t) = \lambda[x(t) - y(t)] = \lambda D_1(t) \qquad F_2(t) = \mu[y(t) - x(t)] = \mu D_1(t) \tag{2}$$

used as control signals introduced respectively into each of the chaotic systems A and B as a negative feedback. We take λ, $\mu>0$ to be experimentally adjustable weightings of the perturbation.

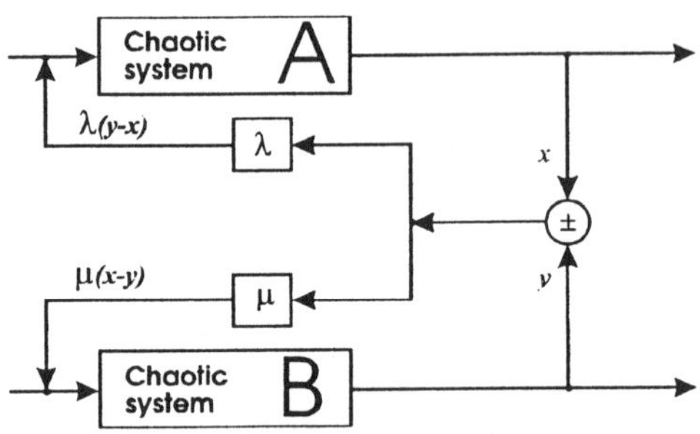

FIGURE 1. Scheme of controlling procedure.

Using the coupling schematically shown in Figure 1, we have shown that one chaotic system coupled with the other one can significantly change the behaviour of one of them (unidirectional coupling, i.e. λ or μ=0) or of both systems (mutual coupling, i.e. $\lambda,\mu\neq 0$). This property allows us to describe the above procedure as the controlling chaos by chaos method. In [12] we gave rigorous conditions under which chaotic attractors of systems A and B are equivalent, or the evolution of one of them is forced to take place on the attractor of the other one. Here we describe some applications of controlling chaos by chaos in geophysical systems.

In our numerical examples we first consider two Lorenz models [2,13] mutually coupled in the following way

$$\begin{aligned}
\dot{X}_1 &= -\sigma X_1 + \sigma Y_1 + \lambda(X_2 - X_1) \\
\dot{Y}_1 &= -X_1 Z_1 + r_1 X_1 - Y_1 + \lambda(Y_2 - Y_1) \\
\dot{Z}_1 &= X_1 Y_1 - b Z_1 + \lambda(Z_2 - Z_1) \\
\dot{X}_2 &= -\sigma X_2 + \sigma Y_2 + \mu(X_1 - X_2) \\
\dot{Y}_2 &= -X_2 Z_2 + r_2 X_2 + \mu(Y_1 - Y_2) \\
\dot{Z}_2 &= X_2 Y_2 - b Z_2 + \mu(Z_1 - Z_2)
\end{aligned} \tag{3}$$

where σ, $r_{1,2}$ and b are constants. The Lorenz model has often been proposed as a paradigm for the "chaotic" extratropical atmospheric circulation [14]. The variables X, Y and Z then represent in some broad sense Rossby wave components of the extratropical general circulation. Coupling between two Lorenz models introduced in eq.(3) might then be interpreted as mutual interdependence of extratropical circulations in two regions characterized by different r parameter value, say an intensive storm track and a relatively stable anticyclonic region. The concept of teleconnections of this kind, achieved through the mechanism of quasi-linear Rossby trains, has both theoretical and observational support [15-17].

Numerical computations have been carried out using software INSITE [18]. In Figure 2 (a-b) we show the chaotic attractors of single Lorenz models ($\lambda,\mu=0$) for $\sigma=10.0$, $r_1=197.4$, $b=8/3$ (Figure 2(a)) and $r_2=211.0$ (Figure 2(b)). These attractors are characterized by the following spectra of Lyapunov exponents $\lambda_1=1.87$, $\lambda_2=0$, $\lambda_3=-15.54$ (Figure 2(a)) and $\lambda_1=0.78$, $\lambda_2=0$, $\lambda_3=-14.44$. In Figure 2(c) we show the behaviour when the above Lorenz systems are mutually coupled with $\lambda=100$ and $\mu=1$, while in Figure 2(d) Lorenz systems are uniderectionally coupled with $\lambda=100$ and $\mu=0$. Althouth both attractors are still chaotic ($\lambda_1=0.79$, $\lambda_2=0$, $\lambda_3=-14.34$), trajectory behaviour on them is more predictable as both

Lyapunov dimensions, d_L=2.053 and d_L=2.044 are smaller than this dimension the of original attractor (d_L=2.121).

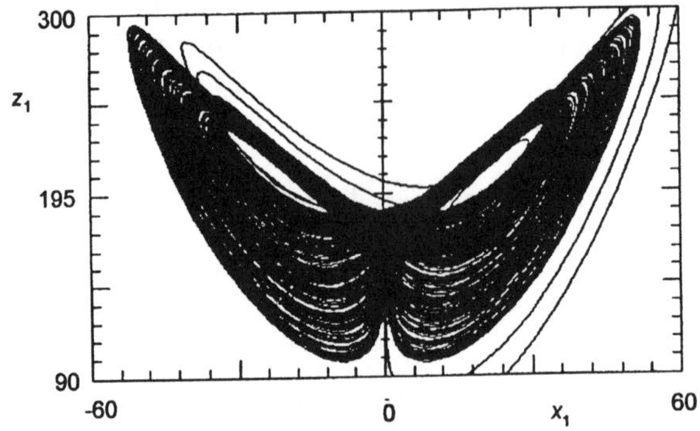

(a)

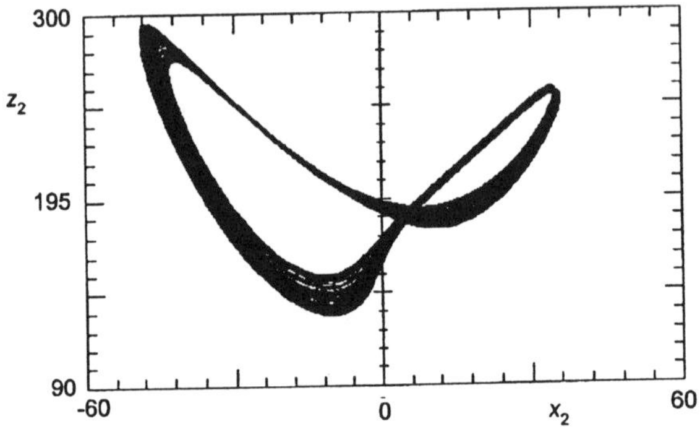

(b)

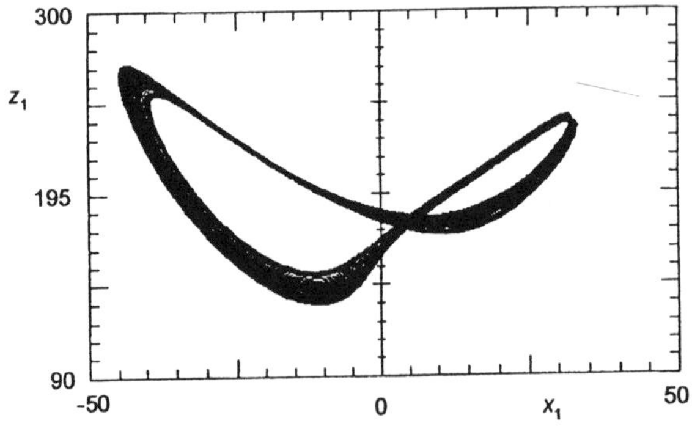

(c)

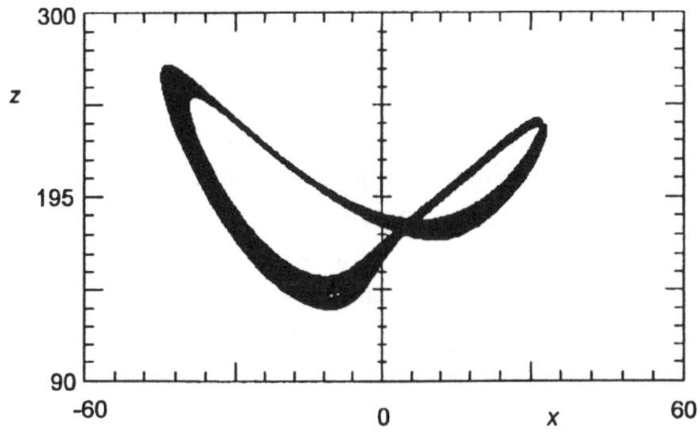

(d)

FIGURE 2. Evolution of Lorenz system: **(a)** unmodified first system: σ=10.0, *r*=197.4, *b*=8/3; **(b)** unmodified second system: σ=10.0, *r*=212.0, *b*=8/3; **(c)** attractor of the first system controlled by the secon system: λ=100.0, μ=1.0; **(d)** attractor of the first system controlled by the second one λ=100, μ=0.

This dimension increase is produced by a significant decrease of positive Lyapunov exponent ($\lambda_1=0.79$ in comparison with $\lambda_1=1.87$ of the original attractor).

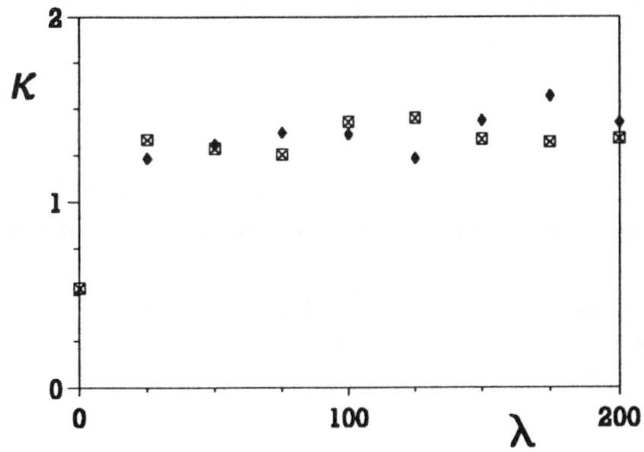

FIGURE 3. Measure predictability κ versus coupling coefficient λ_j (♦ - μ=1, ⊠ - μ=0).

In Figure 3 we show the plots of the measure of predictability $\kappa=1/\lambda_{max}$, where λ_{max} is the largest Lyapunov exponent, versus coupling coefficient λ for both mutual and unidirectional coupling. The analysis of Figure 3 shows that for sufficiently large λ the predictability is significantly increased both in the case of mutual coupling (diamonds) and uniderectional coupling (squares).

We would like to note that, despite the fact that in neither of our numerical examples have the conditions of Propositions 1 and 2 of [12] been precisely fulfilled, nevertheless the application of our controlling procedure has allowed us to convert one type of chaotic behaviour to chaotic behaviour which is more predictable as we choose parameters λ and μ close to those that fulfil the above-mentioned conditions. For λ and μ which are far away from the conditions stated in [12] our controlling procedure does not work.

In the cases considered in this paper the application of a continuous control scheme did not result in obtaining periodic behaviour because the external periodic perturbation was not taken to coincide with an unstable periodic orbit of the original chaotic system. If that is done, a similar controlling scheme allows us to convert the original chaotic behaviour into an appropriate periodic one [8-11]. It should be mentioned here that, if $f=g$ and $\mu=0$ in eqs (1), our controlling procedure simplifies itself to the method of synchronization chaos using continuous control [10].

In summary we propose that this procedure of controlling chaos by chaos might be regarded as a possible mechanism for the so-called extended-range atmospheric predictability sometimes observed in geophysical systems. The results of this simple coupling show the great potential influence of the behaviour of one chaotic system on that of another. It is also clear that, in a stationary state, when the variables $\boldsymbol{x}$ and $\boldsymbol{y}$ are close together, the control signals (2) are small (under conditions of Proposition 2 they converge to each other). This probably means that such a coupling might not allow easy experimental verification in real geophysical systems. It is, however, a mechanism which may account for the existence of many unexpected short and relatively predictable features in the situations when strongly chaotic behaviour might be expected. We hope that the results of this letter could be of direct benefit concerning the problem of "predicting predictability", which is currently a research topic of great interest.

Finally, we remark that a similar chaos by chaos controlling mechanism, having a different type of coupling, was recently proposed to be responsible for stabilization of the Earth's obliquity by the Moon [19] .

ACKNOWLEDGEMENT

This work has been supported by the British Council-KBN (Warsaw) grant No. WAR/992/033.

REFERENCES

1. Sparrow, C. *The Lorenz Equations; Bifurcations, Chaos and Strange Attractors*, New York: Springer, 1982.
2. Lorenz, E.H., *J. Atmos. Sci.* **20**, 130-146 (1963)
3. Palmer, T.H., Buizza, R., Molteri, F., Chen, Y.-Q. and Corti, S., *Phil. Trans. Roy. Soc. A*. (in press).
4. Philander, S.G.H. *El Nino, La Nina, and the Southern Oscillations*, London: Academic Press, 1990.
5. Jin, F.F., Neelin, J.D. and Ghil, M., *Science* **264**, 70-72 (1994)
6. Tziperman, E., Stone, L., Cane, M.A. and Jaroth, H.., *Science* **264**, 72-74 (1994).
7. Barcilon, A., *Phys. Lett.* **167A**, 179-184 (1992).
8. Pyragas, K., *Phys. Lett.* **170A**, 421-428 (1992).
9. Pyragas, K., *Phys. Lett.* **181A**, 203-210 (1993).
10. Kapitaniak, T., *Phys. Rev. E.*, **50**, 1237-1239 (1994).
11. Kapitaniak, T., Kocarev, L. and Chua, L.O., *Int. J. Bifurcation and Chaos*, **3**, 459-468 (1993).
12. Brindley, J., Kapitaniak, T. and Kocarev, L., *Geophys. Res. Lett.* (in press).
13. Lorenz, E.N., *Tellus* **17**, 231-338 (1965).
14. Palmer, T.N., *Bull. Amer. Meteorological Soc.* **74**, 49J-65 (1993).
15. Madden, R., and P.R. Julian, *J. Atmos. Sci.* **28**, 702-708 (1971).
16. Mc Creary, J.P., and D.L. Andeson, D.L., *J. Geophys. Res.* **96**, 3125-3150 (1991).
17. Weickmann, J., *Geophys. Res.* **96**, 3187-3195 (1991).
18. Parker, T., and Chua, *Practical Numerical Algorithms for Chaotic Systems*, New York: Springer 1989.
19. Laskar, J., F. Joutel, and P. Robutel, *Nature* **361**, 615-617 (1993).

9. MEDICINE

Fractal Scaling Properties in Nonstationary Heartbeat Time Series

C.-K. Peng,[1,2] S. Havlin,[2,4] H. E. Stanley[2] and A. L. Goldberger[1,3]

[1] Cardiovascular Division, Harvard Medical School, Beth Israel Hospital, Boston, MA 02215
[2] Center for Polymer Studies and Department of Physics, Boston University, Boston, MA 02215
[3] Department of Biomedical Engineering, Boston University, Boston, MA 02215
[4] Department of Physics, Bar Ilan University, Ramat Gan, ISRAEL

Abstract

Under healthy conditions, the normal cardiac (sinus) interbeat interval fluctuates in a complex manner. Quantitative analysis using techniques adapted from statistical physics reveals the presence of long-range power-law correlations extending over thousands of heartbeats. This scale-invariant (fractal) behavior suggests that the regulatory system generating these fluctuations is operating far from equilibrium. In contrast, we find that for subjects at high risk of sudden death (e.g. congestive heart failure patients) these long-range correlations break down. Application of fractal scaling analysis and related techniques provides new approaches to assessing cardiac risk and forecasting sudden cardiac death, as well as motivating development of novel physiological models of systems that appear to be "hetero-dynamic" rather than "homeo-static."

1 Introduction

Scale-invariant properties in biological systems have received much attention recently [1,2]. The absence of characteristic length (or time) scales may confer important biological advantages, related to adaptability of response [2,3]. Here we present some recent progress in applying scale-invariant (fractal) analysis to interbeat interval time series. We will concentrate on the concept of long-range (power-law) correlations which is a consequence of the scaling properties. We will also emphasize the difficulties of analyzing heartbeat time series that arise mainly from their nonstationarity and sometimes short data length.

A system is said to exhibit long-range correlations when some physical properties of the system are correlated at different times (or positions) and the corresponding correlation function decays much slower than exponentially as a function of time or distance. In physics and mathematics, long-range correlations typically refer to a power-law decay of the correlation function [4]. The mechanism for generating such long-range correlations is not always obvious. Usually, long-range correlations are a result of the collective behavior of a complex system (under unique conditions) with the multiple components interacting through "local" (short-range) interactions. A well-studied example in statistical physics is a system at or near its *critical point*—such as the critical point of liquid-gas transition [5], for which the interactions among the molecules are quite short range, being essentially nearest-neighbor only. In this case, the origin for generating long-range correlations is the balance between two competitions—"order" (interaction between molecules) and "disorder" (thermal fluctuations)—over all scales.

2 Human Heartbeat Dynamics

Clinicians often describe the normal activity of the heart as "regular sinus rhythm." But in fact cardiac interbeat intervals normally fluctuate in a complex, apparently erratic manner [2, 6] (Fig. 1). This highly irregular behavior has recently motivated researchers [7, 8] to apply time series analyses that derive from statistical physics, especially methods for the study of critical phenomena where fluctuations at all length (time) scales occur. These studies show that under healthy conditions, interbeat interval time series exhibit long-range power-law correlations reminiscent of physical systems near a critical point [5, 9]. Furthermore, certain disease states may be accompanied by alterations in this scale-invariant (fractal) correlation property. Here we explore the potential utility of such scaling alterations in the detection of pathological states.

Our analysis in this paper is based on the beat-to-beat heart rate fluctuations of digitized electrocardiograms recorded with an ambulatory (Holter) monitor. The time series obtained by plotting the sequential intervals between beat i and beat $i+1$, denoted by $B(i)$, typically reveals a complex type of variability (Fig. 1). The mechanism underlying such fluctuations appears to be related primarily to countervailing neuroautonomic inputs. Parasympathetic stimulation decreases the firing rate of pacemaker cells in the heart's sinus node. Sympathetic stimulation has the opposite effect. The nonlinear interaction (competition) between the two branches of the autonomic nervous system is the postulated mechanism for the type of erratic heart rate variability recorded in healthy subjects [6, 10].

An immediate problem facing researchers applying time series analysis to interbeat interval data is that the heartbeat time series is often highly non-stationary. Several approaches can be taken to reduce these nonstationary effects. We will discuss one particular method and its physiological implications here.

3 Detrended Fluctuation Analysis

To describe quantitatively the properties of normal interbeat fluctuations, we can first examine the *moments* of their distribution—such as mean value, variance, etc. However, these measurements contain no information about the dynamics, i.e., the sequential ordering of data points. The correlation function is a natural statistical measurement that does reveal some important properties of the dynamics. We will introduce a useful analysis to quantify the correlation properties of the interbeat time series. The method is derived from the concept of self-similar processes.

The concept of self-similar processes was first proposed by Kolmogorov [11] in theoretical physics and later introduced into mathematics through the influential work of Mandelbrot on fractals [12]. An object is self-similar if its subset can be rescaled to resemble (statistically) the original object itself. A scaling exponent (also called the *self-similarity parameter*) can be defined by this rescaling process. A stationary time series with long-range correlations can be integrated, i.e., form an accumulated sum, to generate a self-similar process. Therefore, measurement of the self-similarity scaling exponent of the integrated series can tell us the long-range correlation properties of the original time series. Hurst analysis [4] and root mean square analysis of random walks [13] are both based on this concept.

As discussed above, an immediate problem facing researchers applying time series analysis to

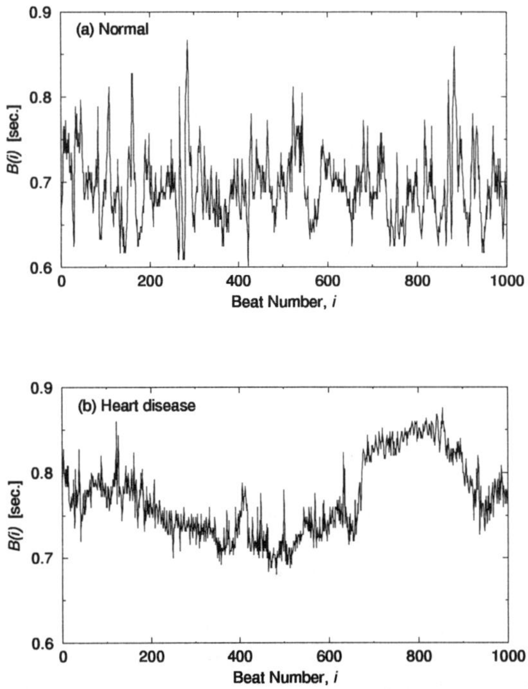

Figure 1: The interbeat interval time series $B(i)$ of 1000 beats for (a) a healthy subject and (b) a patient with severe cardiac disease (dilated cardiomyopathy). The healthy heartbeat time series shows more complex fluctuations compared to the diseased heart rate fluctuation pattern.

heartbeat data is that interbeat time series are often highly non-stationary. An important question is whether this heterogeneous structure arises trivially from external and intrinsic perturbations which push the system away from a homeostatic "set point". An important alternative hypothesis is that such fluctuations are, at least in part, due to the underlying dynamics of the system. If this hypothesis is correct, i.e., neurophysiological control systems behave like dynamical systems far from equilibrium, such long-range correlations are consequences of the underlying control mechanisms and these long-range correlation properties in the signals should be very robust with respect to other varying trends existing in the data. The key problem is how to decompose subtle fluctuations with long-range correlations from other nonstationary trends.

To this end, we introduced a modified root mean square analysis of a random walk—termed *detrended fluctuation analysis* (DFA)[1] [14]—to the analysis of heartbeat data. The advantages of DFA over conventional methods are that it permits the detection of long-range correlations embedded in a seemingly nonstationary time series and also avoids the spurious detection of apparent

[1]Computer software of DFA algorithm is available upon request; contact C.-K. Peng (e-mail: peng@chaos.bih.harvard.edu).

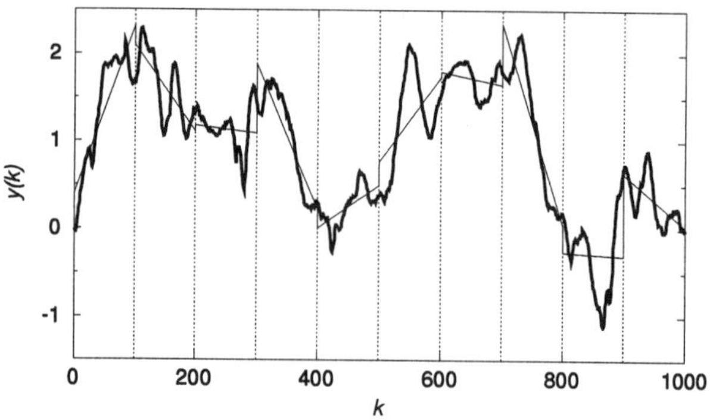

Figure 2: The integrated time series: $y(k) = \sum_{i=1}^{k}[B(i) - B_{\text{ave}}]$, where $B(i)$ is the interbeat interval shown in Fig. 1(a). The vertical dotted lines indicate box of size $n = 100$, the solid straight line segments represent the "trend" estimated in each box by a linear least-squares fit.

long-range correlations that are an artifact of non-stationarity. This method has been tested on control time series that consist of long-range correlations with superposition of a non-stationary external trend. It has also been successfully applied to detect long-range correlations in highly heterogeneous DNA sequences [14,16,17].

To illustrate the algorithm of DFA, we use the interbeat time series shown in Fig. 1a as an example. Briefly, the interbeat interval time series (of total length N) is first integrated, $y(k) = \sum_{i=1}^{k}[B(i) - B_{\text{ave}}]$, where $B(i)$ is the i-th interbeat interval and B_{ave} is the average interbeat interval. Next the integrated time series is divided into boxes of equal length, n. In each box of length n, a least squares line was fit to the data (representing the *trend* in that box) (Fig. 2). The y coordinate of the straight line segments is denoted by $y_n(k)$. Next we detrend the integrated time series, $y(k)$, by subtracting the local trend, $y_n(k)$, in each box. For a given box size n, the characteristic size of fluctuation for this integrated and detrended time series is calculated by

$$F(n) = \sqrt{\frac{1}{N}\sum_{k=1}^{N}[y(k) - y_n(k)]^2}. \tag{1}$$

This computation is repeated over all time scales (box sizes) to provide a relationship between $F(n)$ and the box size n (i.e. the number of beats in a box). Typically, $F(n)$ will increase with box size n. A linear relationship on a double log graph indicates the presence of scaling, i.e., the integrated and detrended time series is a self-similar process. In other words, fluctuations in small boxes are similar to the fluctuations in bigger boxes in a power-law fashion. The slope of the line relating log $F(n)$ to log n determines the scaling exponent (self-similarity parameter), α. For a process where the value at one interbeat interval is completely uncorrelated from any previous values, e.g. white noise, the integrated value, $y(k)$, corresponds to a random walk and therefore $\alpha = 0.5$ [13]. If there are only short-term correlations, the initial slope may be different from 0.5, but α will approach 0.5 for large window sizes. An α greater than 0.5 and less than or equal to 1.0 indicates persistent long-range power-law correlations, while $0 < \alpha < 0.5$ indicates anti-persistent power-law correlations [15]; $\alpha = 1$ corresponds to $1/f$ noise. For α greater than 1, correlations exist

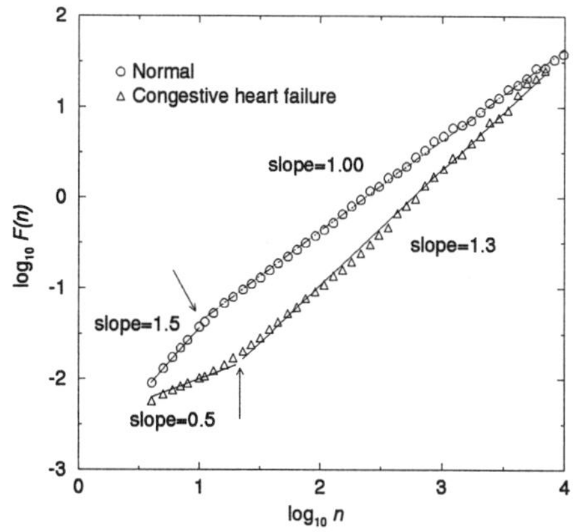

Figure 3: Plot of $\log F(n)$ vs $\log n$ (see description of DFA computation in text) for two very long interbeat interval time series ($\sim$ 24 hours). The circles are from a healthy subject while the triangles are from a subject with congestive heart failure. Arrows indicate "crossover" points in scaling. After [34].

but cease to be of a power-law form; $\alpha = 1.5$ indicates brown noise, the integration of white noise. The α exponent can also be viewed as an indicator that describes the "roughness" of the original time series: the larger the value of α, the smoother the time series. In this context $1/f$ noise can be interpreted as a "compromise" or balance between the complete unpredictability of white noise (very rough "landscape") and the much smoother "landscape" of Brownian noise [18].

Figure 3 compares the DFA analysis of representative 24 hour interbeat interval time series of a healthy subject (○) and a patient with congestive heart failure (△). Notice that for large time scales (asymptotic behavior), the healthy subject interbeat interval time series shows almost perfect power-law scaling over two decades ($20 \leq n \leq 10000$) with $\alpha = 1$ (i.e., $1/f$ noise) while for the pathologic data set $\alpha \approx 1.3$ (closer to Brownian noise). This result is consistent with our previous finding that there is a significant difference in the long-range scaling behavior between healthy and diseased states [7,8].

4 Normal vs. Pathologic Time Series

To test for statistical significance using the DFA method, we analyzed cardiac interbeat data from two different groups of subjects reported in our previous work [7]: 12 healthy adults without clinical evidence of heart disease (age range: 29–64 years, mean 44) and 15 adults with severe heart failure[2] (age range: 22–71 years; mean 56). Data from each subject consist of approximately

[2]ECG recordings of Holter monitor tapes were processed both manually and in a fully automated manner using our computerized beat recognition algorithm (Aristotle). Abnormal beats were deleted from each data set. The deletion has practically no effect on the DFA analysis since less than 1% of total beats were removed. Patients in the heart failure group were receiving conventional medical therapy prior to receiving an investigational cardiotonic drug; see Ref. [19].

24 hours of ECG recording. Data from patients with heart failure due to severe left ventricular dysfunction are likely to be particularly informative in analyzing correlations under pathologic conditions since these individuals have abnormalities in both the sympathetic and parasympathetic control mechanisms [20] that regulate beat-to-beat variability. Previous studies have demonstrated marked changes in short-range heart rate dynamics in heart failure compared to healthy function, including the emergence of intermittent relatively low frequency ($\sim$ 1 cycle/minute) heart rate oscillations associated with the well-recognized syndrome of periodic (Cheyne-Stokes) respiration, an abnormal breathing pattern often associated with low cardiac output [20].

We observe the following scaling exponents[3] (for time scales $10^2 \sim 10^4$ beats) for the group of healthy cardiac interbeat interval time series (mean value $\pm$ S.D.): $\alpha = 1.00 \pm 0.11$. This result is consistent with previous reports of $1/f$ fluctuations in healthy heart rate (by spectral analysis) [3, 21]. The pathologic group shows a significant ($p < 0.01$ by Student's t-test) deviation of the long-range correlation exponent from normal. For the group of heart failure subjects, we find that $\alpha = 1.24 \pm 0.22$. Of interest, some of the heart failure subjects show an α exponent very close to 1.5 (Brownian noise), indicating random walk-like fluctuations, also consistent with our previous findings in this group. The group-averaged exponent α is less than 1.5 for the heart failure patients, suggesting that pathologic dynamics may only transiently operate in the random walk regime or may only approach this extreme state as a limiting case. We obtained similar results when we divided the time series into three consecutive subsets (of $\sim$ 8 hours each) and repeated the above analysis. Therefore our findings are not simply attributable to different levels of daily activities.

5 Crossover Phenomena

Although this asymptotic scaling exponent may serve as a useful index for selected diagnostic purposes, a drawback is that very long data sets are required (at least 24 hours) for statistically robust results. For practical purposes, clinical investigators are often interested in the possibility of using substantially shorter time series. In this regard, we note that for short time scales, there is an apparent *crossover* exhibited for the scaling behavior of both data sets (arrows in Fig. 3). For the healthy subject, the α exponent estimated from very small n (< 10 beats) is larger than that calculated from large n (> 10 beats). This is probably due to the fact that on very short time scales (a few beats to ten beats), the physiologic interbeat interval fluctuation is dominated by the relatively smooth heartbeat oscillation associated with respiration, thus giving rise to a large α value. For longer scales, the interbeat fluctuation, reflecting the intrinsic dynamics of a complex system, approaches that of $1/f$ behavior as previously noted. *In contrast, the pathologic data set shows a very different crossover pattern* (Fig. 3). For very short time scales, the fluctuation is quite random (close to white noise, $\alpha \approx 0.5$). As the time scale becomes larger, the fluctuation becomes smoother (asymptotically approaching Brownian noise, $\alpha \approx 1.5$). These findings are consistent with our previous report of altered correlation properties under pathologic conditions [7, 8].

[3]Typical regression fit shows excellent linearity of double log graph (indicated by correlation coefficient $r > 0.97$) for both groups. However, usually data from healthy subjects show even better linearity on log-log plots than data from subjects with heart disease.

5.1 Stochastic Model for Pathologic Data

The physiologic mechanism for the long-range correlations represented by the $1/f$ spectrum of normal interbeat intervals remains to be established. Both stochastic and deterministic models have been proposed to account for such scale-invariant behavior in physical systems. We introduce a simple three-parameter stochastic model (without relating to actual neuroautonomic control mechanisms) that can quantitatively describe the crossover scaling behavior *under extreme pathologic conditions*. The simple stochastic model is based on two assumptions:

(i) For short time scales (less than $10-20$ beats) the cardiac interbeat intervals with congestive heart failure can be described as white noise. Consider that the sinus node tends to maintain a constant ("homeostatic") firing rate. However, the actual beat-to-beat time intervals will deviate from a perfectly regular oscillation due to random fluctuations described by a distribution with a zero mean value and a well-defined variance (Δ^2). The typical period (the characteristic time) that the sinus node keeps its firing rate constant is denoted by τ.

(ii) The system responds to other driving forces (environmental influences or intrinsic factors) by increasing or decreasing the firing rate. The typical change of the firing rate is characterized by a parameter δ. Once the sinus node adjusts its firing rate, it will tend to maintain it for a period of τ beats as described in (i).

Figs. 4 shows the comparison of the actual interbeat data for one of the heart failure subjects to that generated by the model. The effect of assumption (i) is to generate white noise for time-scale less than τ, i.e., if we let $\tau \to \infty$ then the DFA plot of this model will be a straight line with slope 0.5 for all ranges of n. On the other hand, the effect of (ii) is to create the kind of noise associated with Brownian motion ("brown noise"), i.e., if we set τ very small ($\sim$ 1 beat) then the DFA plot will be a straight line with slope 1.5. In order to simulate the observed *crossover* between these two regimes, we need to set τ in the model to be of the same order as the crossover time observed in Fig. 4c. Beside the parameter τ, we only need to select the other two parameters, Δ^2 and δ, to fit the observed data.

The simple model described above is useful because it shows how two apparently different pathologic scaling mechanisms are in fact connected by the emergence of a characteristic time, corresponding to the observed "crossover" behavior of the real data. However, this model is limited in its scope because it (i) does not account for the $1/f$ ($\alpha = 1$) scaling behavior of the healthy heart rate dynamics, and (ii) it does not relate the parameters to specific neuroautonomic control mechanisms. We note that this model implies that under extreme pathologic conditions, the system attempts to maintain a constant interbeat interval for short time scales while responding to other factors over longer time scales by a smooth variation of the interbeat interval. This behavior is dramatically different from the observed dynamics of interbeat interval under normal (healthy) condition which shows a more complex pattern of sinus rhythm fluctuations than can be accounted for with traditional homeostasis models.[22]

5.2 Clinical Application: Scaling Behavior in Health and Disease

The above observation of a differential crossover pattern for healthy versus pathologic data motivated us to extract two parameters from each data set by fitting the scaling exponent α over two different time scales: one short, the other long. To be more precise, for each data set we calculated

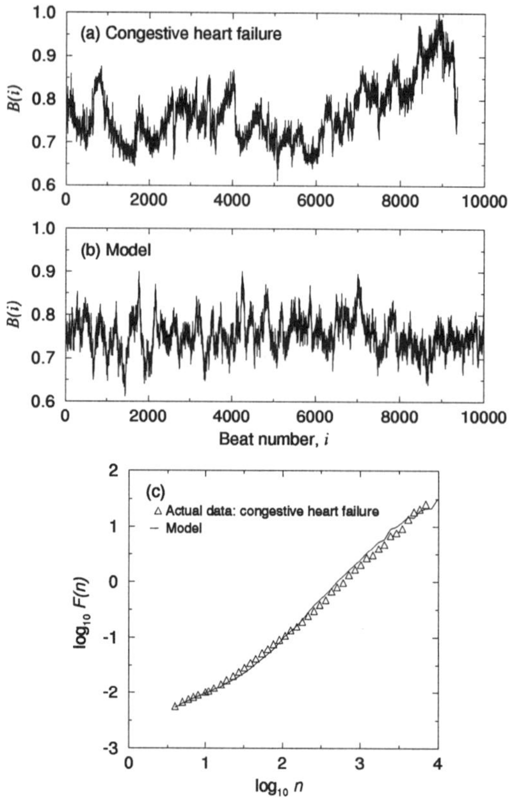

Figure 4: (a) Interbeat interval time series from a patient with heart failure (the same subject described in Fig. 1b). (b) Time series generated by model. Only part of the total time series is shown here. (c) The DFA of the interbeat interval function $B(i)$ for data in (a) and (b). The triangles represent actual data from the subject and the solid lines are generated by the model. The actual simulation of the model is carried out as follows: (i) Choose a firing rate, R_1, such that $B(1)$ equal to the average value, $\overline{B}$, of the real interbeat data set. (ii) The subsequent interbeat values will be $B(1)$ plus a random fluctuation (described by a distribution with zero mean and a finite variance, Δ^2). (iii) With a probability $1/\tau$, the firing rate R_1 will change to a new value $R_2 = R_1 + \xi$. The magnitude of this drift, ξ, is random and described by a distribution with zero mean and standard deviation δ. For this simulation, the value of the parameters are: $\tau = 20$ beats, $\Delta = 0.011$ sec. and $\delta = 0.008$ sec. The parameter τ is chosen to fit the crossover time, where Δ and δ are chosen for fitting the data over short and long time scales, respectively. To account for the physiological constraint that the firing rate cannot become arbitrarily large or small, we also add a instantaneous restoring force that is proportional to the difference between the current firing rate and the mean (average) firing rate (measured from the actual data). This restoring force only affects very long time scales ($\sim 10^4$) fluctuations.

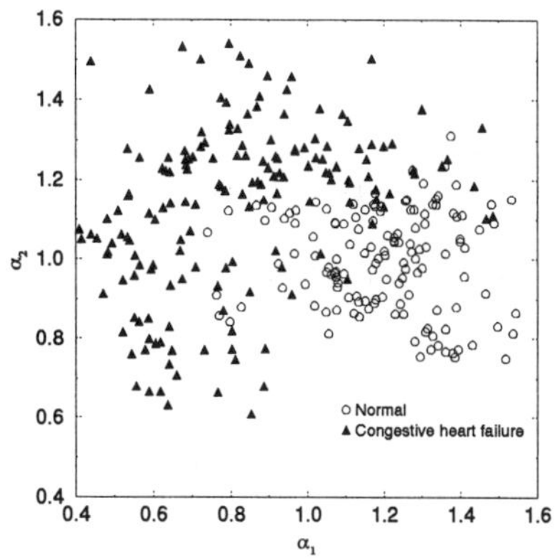

Figure 5: Scatter plot of scaling exponents α_1 vs α_2 for the healthy subjects (○) and subjects with congestive heart failure (△). The α's were calculated from interbeat interval data sets of length 8192 beats. Longer data set records were divided into multiple data sets (each with 8192 beats). Note good separation between healthy and heart disease subjects, with clustering of points in two distinct "clouds." After [34].

an exponent α_1 by making a least squares fit of $\log F(n)$ vs $\log n$ for $4 \le n \le 16$. Similarly, an exponent α_2 was obtained from $16 \le n \le 64$. Since these two exponents are not extracted from the asymptotic region, relatively short data sets are sufficient, thereby making this technique applicable to "real world" clinical data.

We applied this quantitative fluctuation analysis to the two different groups of subjects mentioned above to measure the two scaling exponents α_1 and α_2. All data set records were divided into multiple sub-sets (each with $N = 8192$ beats $\sim$ 2 hours) and the two exponents were calculated for each subset. For healthy subjects, we find the following exponents (mean value $\pm$ S.D.) for the cardiac interbeat interval time series: $\alpha_1 = 1.201 \pm 0.178$ and $\alpha_2 = 0.998 \pm 0.124$. For the group of congestive heart failure subjects, we find that $\alpha_1 = 0.803 \pm 0.259$ and $\alpha_2 = 1.125 \pm 0.216$, both significantly ($p < 0.0001$ for both α_1 and α_2) different from normal. Furthermore, we show in Fig. 5 that fairly good discrimination between these two groups can be achieved by using these two scaling exponents. We note that not all subjects in our preliminary study show an obvious crossover in their scaling behavior. Only 8 out of 12 healthy subjects exhibited this crossover, while 11 out of 15 pathologic subjects exhibited a "reverse" crossover. However, the two scaling exponents (α_1 and α_2) measured from relatively short data sets can still be potentially useful indicators to distinguish normal from pathologic time series.

To test the effect of data length on these calculations, we repeated the same DFA measurements for longer data sets ($N = 16384$) and also for shorter data sets ($N = 4096$). As expected, the results for shorter data sets are less reliable (more overlap between two groups) due to anticipated statistical error related to finite sample size [23]. On the other hand, longer data sets result in little improvement for the distinction between groups. Therefore, the data length of 8192 seems to be a statistically reasonable choice[4].

Furthermore, we note that data from normal interbeat interval time series are tightly clustered suggesting that there may exist a "universal" scaling behavior for physiologic interbeat time series. In contrast, the pathologic data show more variation, a finding which may be related to different clinical conditions and varying severity of the pathologic states.

5.3 Forecasting Clinical Outcomes: Survival Rate Assessment

Based on the hypothesis that there is a region of scaling behavior (in Fig. 5) over which the normal (healthy) cardiac control operates, we have recently found another promising application of DFA in analyzing data sets from the Framingham Heart Study—a prospective, population based study [24]. The primary group of interest was individuals with congestive heart failure (CHF); 28 CHF cases and 41 sex and age-matched healthy control cases were analyzed by our scaling analysis. Briefly, using Holter monitor data (approximately 2 hours) from each subject of the Framingham Study, we assigned an index (range from 0 to 1) to each individual by estimating the probability that this particular heartbeat time series was operating in the appropriate region in Fig. 5 (normal vs. pathologic). Note that the data studied in Fig. 5 are treated as an representative "training set." Of course, none of the Framingham Heart Study subjects is included in this training set. The question we asked was: What is the utility of this new fractal index? In particular, does this measure add independent information to conventional measures? In comparison with other 10 time and frequency measures [24], we found that the DFA index carries prognostic information about mortality not extractable from these traditional methods of heart rate variability analysis [24]. Most recently the DFA algorithm was successfully applied to the study of alteration of interbeat correlations due to physiological aging [25].

6 Conclusions

Our finding of non-trivial long-range correlations in healthy heart rate dynamics is consistent with the observation of long-range correlations in other biological systems that do not have a characteristic scale of time or length [2, 26–32]. Such behavior may be adaptive for at least two reasons. (i) The long-range correlations serve as an organizing principle for highly complex, non-linear processes that generate fluctuations on a wide range of time scales. (ii) The lack of a characteristic scale helps prevent excessive *mode-locking* that would restrict the functional responsiveness of the organism. Support for these related conjectures is provided by observations from severe diseased states such as heart failure where the breakdown of long-range correlations is often accompanied by the emergence of a dominant frequency mode (e.g., the Cheyne-Stokes frequency). Analogous transitions to highly periodic regimes have been observed in a wide range of other disease states including certain malignancies, sudden cardiac death, epilepsy and fetal distress syndromes [3].

[4] We also tested these calculations by varying the fitting range for α_2. We find that the results are very similar when we measure α_2 from 16 beats to 128 beats. However, when we move the upper fitting range for α_2 from 128 beats to 256 beats or more, the pathologic data sets show larger variation of α_2 leading to less obvious separation from normal subjects. This is partly due to the fact that, for finite length data sets, the calculation error of $F(n)$ increases with n [23]. Therefore, scaling exponents obtained over larger values of n will have greater uncertainty.

The complete breakdown of normal long-range (fractal) correlations in cardiovascular system could theoretically lead to three possible diseased states: (i) a random walk (brown noise), (ii) highly periodic behavior, or (iii) completely uncorrelated behavior (white noise). Cases (i) and (ii) both indicate only "trivial" long-range correlations of the types observed in severe heart failure. Case (iii) may correspond to certain cardiac arrhythmias such as fibrillation. More subtle or intermittent degradation of long-range correlation properties may provide an early warning of incipient pathology. In addition to the correlation analysis discussed here, a recent study of heartbeat variation [33] demonstrated that nonlinear Fourier phase correlations are also important for the understanding of heartbeat dynamics. Finally, we note that the long-range correlations present in the healthy heartbeat suggests that the neuroautonomic control mechanisms actually drives the system away from a single steady state. Therefore, the classical theory of homeostasis, according to which stable physiological processes seek to maintain "constancy" [22], should be extended to account for this dynamical, far from equilibrium, behavior of these "hetero-dynamic" systems.

Acknowledgments.

We are grateful to S. V. Buldyrev, J. M. Hausdorff, K. K. L. Ho, L. Glass, J. E. Mietus, G. B. Moody, L. A. Lipsitz, P. Purdon and D. R. Rigney for valuable discussions. Partial support was provided to CKP by NIH/NIMH, to HES and SH by NSF and to ALG by the G. Harold and Leila Y. Mathers Charitable Foundation, NIDA, and NASA. This review is an updated version of Ref. [34].

References

[1] A. Bunde, S. Havlin, eds.: *Fractals in Science* (Springer-Verlag, Berlin 1994).

[2] A. L. Goldberger, D. R. Rigney and B. J. West, "Chaos and fractals in human physiology," *Sci. Am.* **262**, 42–49 (1990).

[3] A. L. Goldberger and B. J. West, "Fractals in physiology and medicine," *Yale J. Biol. Med.* **60**, 421–423 (1987).

[4] J. Beran, *Statistics for Long-Memory Processes,* (Chapman & Hall, New York, 1994).

[5] H. E. Stanley, *Introduction to Phase Transitions and Critical Phenomena,* (Oxford University Press, London, 1971).

[6] R. I. Kitney and O. Rompelman, *The Study of Heart-Rate Variability*, Oxford U. P., Oxford, 1980.

[7] C.-K. Peng, J. Mietus, J. M. Hausdorff, S. Havlin, H. E. Stanley, and A. L. Goldberger, "Long-range anti-correlations and non-Gaussian behavior of the heartbeat," *Phys. Rev. Lett.* **70**, 1343–1346 (1993).

[8] C.-K. Peng, S. V. Buldyrev, J. M. Hausdorff, S. Havlin, J. E. Mietus, M. Simons, H. E. Stanley, and A. L. Goldberger, "Fractal landscapes in physiology and medicine: Long-range correlations in DNA sequences and heart rate intervals," in *Fractals in Biology and Medicine*, T. F. Nonnenmacher, G. A. Losa and E. R. Weibel, eds. (Birkhaüser Verlag, Basel, 1994), pp. 55–65.

[9] P. Bak, C. Tang & K. Wiesenfeld, "Self-organized criticality: an explanation of 1/f noise," *Phys. Rev. Lett.* **59**, 381–384 (1987).

[10] M. N. Levy, "Sympathetic-parasympathetic interactions in the heart," *Circ. Res.,* **29**, 437-445 (1971).

[11] A. N. Kolmogorov, "Local structure of turbulence in fluid for very large Reynolds numbers," *Transl. in Turbulence.* S. K. Friedlander and L. Topper, eds. (Interscience Publishers, New York, 1961), pp. 151–155.

[12] B. B. Mandelbrot, *The Fractal Geometry of Nature,* (Freeman, San Francisco, 1982).

[13] E. W. Montroll and M. F. Shlesinger, "The wonderful world of random walks," in *Nonequilibrium Phenomena II. From Stochastics to Hydrodynamics*, eds. J. L. Lebowitz and E. W. Montroll, pp. 1-121 (North-Holland, Amsterdam, 1984).

[14] C.-K. Peng, S. V. Buldyrev, S. Havlin, M. Simons, H. E. Stanley and A. L. Goldberger, "On the mosaic organization of DNA sequences," *Phys. Rev. E* **49**, 1685–1689 (1994).

[15] S. Havlin, R. B. Selinger, M. Schwartz, H. E. Stanley and A. Bunde, "Random multiplicative processes and transport in structures with correlated spatial disorder," *Phys. Rev. Lett.* **61**, 1438–1441 (1988).

[16] S. V. Buldyrev, A. L. Goldberger, S. Havlin, C.-K. Peng, H. E. Stanley and M. Simons, "Fractal landscapes and molecular evolution: Modeling the myosin heavy chain gene family," *Biophysical Journal* **65**, 2673–2679 (1993).

[17] S. M. Ossadnik, S. V. Buldyrev, A. L. Goldberger, S. Havlin, R. N. Mantegna, C.-K. Peng, M. Simons, and H. E. Stanley, "Correlation approach to identify coding regions in DNA sequences," *Biophysical Journal* **67**, 64–70 (1994).

[18] W. H. Press, "Flicker noise in astronomy and elsewhere," *Comments Astrophys.*, **7**, 103–119 (1978).

[19] D. S. Baim et al., "Survival of patients with severe congestive heart failure treated with oral milrinone," *J. Am. Coll. Cardiol.* **7**, 661–670 (1986).

[20] A. L. Goldberger, D. R. Rigney, J. Mietus, E. M. Antman & S. Greenwald, "Nonlinear dynamics in sudden cardiac death syndrome: Heart rate oscilations and bifurcations," *Experientia* **44**, 983–987 (1988).

[21] M. Kobayashi and T. Musha, "1/f fluctuation of heartbeat period," *IEEE Trans. Biomed. Eng.* **29**, 456 (1982).

[22] W. B. Cannon, "Organization for physiological homeostasis," *Physiol. Rev.* **9**, 399 (1929).

[23] C.-K. Peng, S. V. Buldyrev, A. L. Goldberger, S. Havlin, M. Simons, and H. E. Stanley, "Finite size effects on long-range correlations: Implications for analyzing DNA sequences," *Phys. Rev. E* **47**, 3730–3733 (1993).

[24] K. K. L. Ho, G. B. Moody, C.-K. Peng, J. E. Mietus, M. G. Larson, A. L. Goldberger, and D. Levy, "Assessing survival in heart failure cases and controls using nonlinear and conventional indices of heart rate dynamics: The Framingham heart study," *Circulation* **90**, I-330 (1994).

[25] N. Iyengar, C.-K. Peng, R. Morin, A. L. Goldberger, and L. A. Lipsitz, "Age-related alterations in the fractal scaling of cardiac interbeat interval dynamics," (preprint).

[26] E. R. Weibel, "Fractal geometry: A design principle for living organisms," *Am. J. Physiol.* **261**, L361 (1991).

[27] J. B. Bassingthwaighte and R. P. Beyer, "Fractal correlation in heterogeneous systems," *Physica D* **53**, 71–84 (1991).

[28] C.-K. Peng, S. Buldyrev, A. L. Goldberger, S. Havlin, F. Sciortino, M. Simons and H. E. Stanley, "Long-range correlations in nucleotide sequences," *Nature* **356**, 168-170 (1992).

[29] S. V. Buldyrev, A. L. Goldberger, S. Havlin, C.-K. Peng and H. E. Stanley, "Fractals in biology and medicine: From DNA to the heartbeat," in *Fractals in Science*, A. Bunde and S. Havlin, eds (Springer-Verlag, Berlin, 1994), 49–83.

[30] C.-K. Peng, S. Buldyrev, A. L. Goldberger, S. Havlin, M. Simons and H. E. Stanley, "Statistical properties of DNA sequences," *Physica A* **221**, 180-192 (1995).

[31] J. M. Hausdorff, C.-K. Peng, Z. Ladin, J. Y. Wei and A. L. Goldberger, "Is walking a random walk? Evidence for long-range correlations in the stride interval of human gait," *J. Appl. Physiol.* **78**, 349–358 (1995).

[32] J. M. Hausdorff, P. Purdon, C.-K. Peng, Z. Ladin, J. Y. Wei and A. L. Goldberger, "Fractal scaling of human gait: Stability of long-range correlations in stride interval fluctuations," *J. Appl. Physiol.* (in press).

[33] P. Ch. Ivanov, M. G. Rosenblum, C.-K. Peng, J. Mietus, S. Havlin, H. E. Stanley and A. L. Goldberger, "Scaling behavior of heartbeat intervals," (preprint).

[34] C.-K. Peng, S. Havlin, H. E. Stanley, and A. L. Goldberger, "Quantification of scaling exponents and crossover phenomena in nonstationary heartbeat time series," *Chaos* **5**, 82–87 (1995).

Nonlinear Analysis and Prediction of Pulsatile Hormone Secretion

Klaus Prank[*†], Mirko Kloppstech[*], Steven J. Nowlan[†], Heio M. Harms[*], Georg Brabant[*], Rolf-Dieter Hesch[*], and Terrence J. Sejnowski[†]

[*]*Abteilung Klinische Endokrinologie, Medizinische Hochschule Hannover, D-30623 Hannover, Germany;* [†]*Howard Hughes Medical Institute and Computational Neurobiology Laboratory, The Salk Institute, San Diego, California 92186-5800*

Abstract. Pulsatile hormone secretion is observed in almost every hormonal system. The frequency of episodic hormone release ranges from approximately 10 to 100 pulses in 24 hours. This temporal mode of secretion is an important feature of intercellular information transfer in addition to a dose-response dependent regulation. It has been demonstrated in a number of experiments that changes in the temporal pattern of pulsatile hormone secretion specifically regulate cellular and organ function and structure. Recent evidence links osteoporosis, a disease characterized by loss of bone mass and structure, to changes in the dynamics of pulsatile parathyroid hormone (PTH) secretion. In our study we applied nonlinear and linear time series prediction to characterize the secretory dynamics of PTH in both healthy human subjects and patients with osteoporosis. Osteoporotic patients appear to lack periods of high predictability found in normal humans. In contrast to patients with osteoporosis patients with hyperparathyroidism, a condition which despite sometimes reduced bone mass has a preserved bone architecture, show periods of high predictability of PTH secretion. Using stochastic surrogate data sets which match certain statistical properties of the original time series significant nonlinear determinism could be found for the PTH time series of a group of healthy subjects. Using classical nonlinear analytical techniques we could demonstrate that the irregular pattern of pulsatile PTH secretion in healthy men exhibits characteristics of deterministic chaos. Pulsatile secretion of PTH in healthy subjects seems to be a first example of nonlinear determinism in an apparently irregular hormonal rhythm in human physiology.

Present address: Dr. Steven J. Nowlan, Synaptics Inc., San Jose, CA 95134; Prof. Rolf-Dieter Hesch, University of Konstanz, D-78464 Konstanz, Germany.

INTRODUCTION

Hormonal systems are regulated dynamically (1). There has been growing evidence in recent years that oscillations of the plasma concentration of parathyroid hormone (PTH) on short time scales play an important role in the physiological bone remodeling process which ensures a balance of bone resorption and bone formation (2-4). PTH in healthy humans is secreted in an episodic manner in distinct pulses with a frequency ranging from 1 pulse/hour for large pulses and 1 pulse/10 minutes for smaller pulses (2). In animal experiments intermittent administration of PTH increases bone mass and may enhance normal bone connectivity, whereas continuous infusion of the same dose leads to a dramatic loss of bone mass and structure (5-7). As these changes occur in osteoporotic patients, it is conceivable that this disease may also be caused by a disruption of the normal temporal dynamics of episodic PTH secretion. On the other hand patients with hyperparathyroidism have a well preserved trabecular bone structure (8). These results suggest that the temporal pattern of PTH secretion in osteoporotic patients, patients with hyperparathyroidism, and healthy controls would be different. And in fact these differences were found in recent studies in healthy subjects, patients with osteoporosis, and hyperparathyroidism by drawing blood every two minutes over several hours (2, 9).

In recent work the technique of time series prediction has been used to effectively characterize irregular complex time series and separate nonlinear deterministic (chaotic) from different forms of stochastic behavior (10-12). To reliably discriminate between nonlinear determinism and stochastic dynamics in a time series the technique of surrogate random data has been used (13, 14) . Predictive models have shown to be particularly effective when applied to short time series containing on the order of a couple of hundred data points (11, 12). Differences in the system dynamics are reflected in different degrees of predictability. This approach has been applied to the classification of various biological time series (15) as for example EEG (electroencephalogram) data (16), ECG (electrocardiogram) data (17) and experimental data from simple neuronal circuits (18, 19).

Here we apply the technique of time series prediction in conjunction with the technique of generating random surrogate data for classifying and characterizing the dynamics of PTH secretion in nine healthy subjects, six patients with osteoporosis, and four patients with hyperparathyroidism. In a recent study using standard nonlinear analytical techniques such as the calculation of the correlation dimension (20), the Lyapunov exponents (21) and Kolmogorov entropy (22) compared with surrogate random data we could demonstrate nonlinear determinism (chaos) in three 21h viz. 24h PTH time series from healthy subjects (23). These PTH time series (Fig. 1) exhibit an irregular temporal pattern of secretion corresponding to broadband power spectra (Fig. 2).

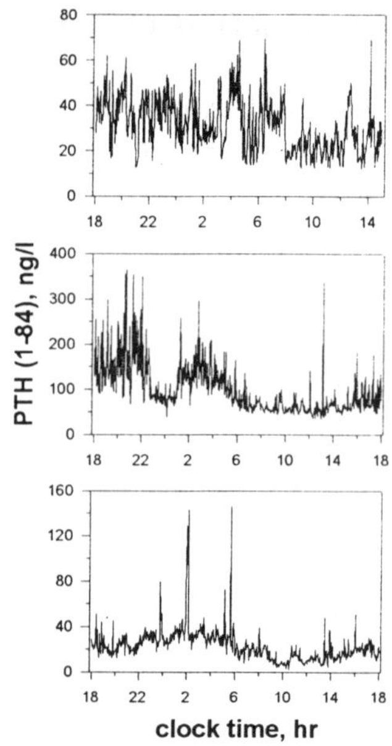

FIGURE 1. Time series of parathyroid hormone (PTH)-(1-84) plasma concentration in 3 healthy male subjects.

METHODS

Subjects

Twelve healthy men (aged 24-42 years), three women with postmenopausal osteoporosis (aged 55-62 years), three men with idiopathic osteoporosis (aged 31-42 years), one male (56 years), and three female patients (aged 60-65 years) with primary hyperparathyroidism took part in this study. The studies reported here were approved by the local Committee on Medical Ethics, and all subjects gave their informed written consent. Some of the subjects in this study as well as the details on design and measurements can be found in detail earlier (2, 9, 23).

Hormone analysis

In three of the healthy subjects, blood samples were taken every two minutes over an extended period (21 hours for one subject and 24 hours for the other two) for

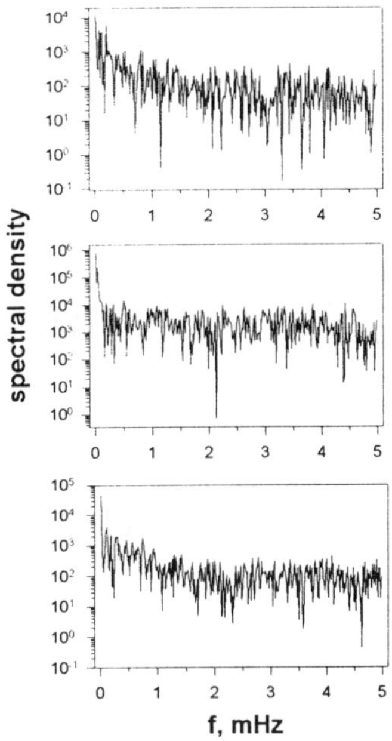

FIGURE 2. Fourier power spectrum of PTH concentration time series from Fig. 1.

determining the PTH plasma concentration. In the remaining 9 healthy subjects, 6 osteoporotic patients, and 4 patients with hyperparathyroidism PTH serum concentrations were also measured at two minute intervals, but over much shorter periods of time (3.5 to 9 hours) as shown in Figure 3. All PTH concentrations were measured in duplicate using either a two-site chemiluminometric (sandwich) immunoassay (Magic Lite Intact PTH, Ciba-Corning Diagnostics Corporation, Medfield, MA; intra-assay coefficient of variation (CV) 3.4%, inter-assay CV 4.3%) or an intact PTH immunoradiometric assay (Nicols, San Juan Capistrano, CA; intra-assay CV 5.1%; inter-assay CV 7.8%). Blood samples (1 ml) were drawn via a central venous catheter. Throughout the study, the subjects rested in bed.

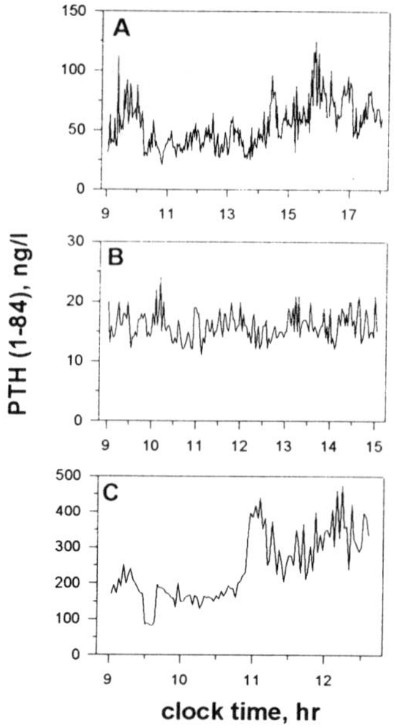

FIGURE 3. Three representative PTH plasma concentration time series in a **A**) healthy subject, **B**) osteoporotic patient, and **C**) patient with hyperparathyroidism.

Time series prediction

Neural network approaches have been demonstrated to be powerful tools for time series prediction (24). Such a neural network model predicts the next value $x(t_i)$ of a time series from a number of previous values:

$$x(t_i) = f(x(t_{i-1}), x(t_{i-2}), .., x(t_{i-m})) + \epsilon_i \qquad (1)$$

where $x(t_{i-1})$ is the value at time t_{i-1}, m is the number of previous values used for prediction, and ϵ_i represents noise or fitting error. The model f is selected by minimizing some measure of misfit (such as mean square error) in a set of training examples selected from the samples of the time series.

In our study we used feedforward networks with nonlinear (sigmoidal) activation functions as well as linear activation functions (equivalent to autoregressive models) to predict future values of the time series of PTH serum concentrations. This form of a preditor was chosen because it is relatively easy to control over-fitting using regularization functions and cross-validation. Such connectionistic models have shown to be effective to predict short (noisy) time series (24-27).

The neural networks were trained to predict one time step into the future. To predict multiple time steps ahead the value predicted one time step ahead was iterated by feeding it back into the input of the network. If the predicted value one step into the future is:

$$\hat{x}(t_i) = f(x(t_{i-1}), x(t_{i-2}), \ldots, x(t_{i-m})) \tag{2}$$

then the predicted value two time steps ahead is defined as:

$$\hat{x}(t_{i+1}) = f(\hat{x}(t_i), x(t_{i-1}), \ldots, x(t_{i-m+1})). \tag{3}$$

To predict any desired number of time steps into the future this procedure can be repeated. The prediction error was estimated using the *average relative variance (arv)*:

$$arv = \frac{\langle (x_i - \hat{x}_i)^2 \rangle}{\sigma^2(x_i)}, \tag{4}$$

where the angle brackets denote the over all mean and $\sigma^2(x_i)$ denotes the variance of the indexed variable.

To improve the signal to noise ratio in our experimental data the raw PTH time series were filtered prior to the connectionist approach using an acausal filter (28) which has been shown not to affect certain nonlinear measures as for example the correlation dimension.

Training and testing of predictive models

A large variety of different network architectures was explored, applying feedforward networks with linear as well as nonlinear (sigmoidal) activation functions with 5 to 15 input units, 0 to 5 hidden units, and 1 output unit. A regularization technique (26) was used to control overfitting to the training data, with the weight assigned to the regularization term chosen by cross validation. The value for the regularization term varied from 10^{-5} for the nonlinear networks to 10^{-10} for the linear networks. The weights were updated by a conjugate gradient descent method. (For a linear model, this fitting technique is formally equivalent to a least-squares linear fit.)

Networks for the healthy group were trained using pooled time series data from the three healthy male subjects whose PTH serum concentrations were measured every 2 minutes over extended periods of 21 viz. 24 hours. Predictive networks for the patients with osteoporosis viz. hyperparathyroidism were trained using a "leave one out" technique where all data from the patient group besides the time series to be predicted were used for training. The performance of a large variety of network architectures as described above was compared using the *arv* as measure for the prediction error.

Because of their predictive performance a nonlinear 15 input units, 5 hidden units and 1 output unit network with sigmoidal activation functions and a 15 input units and 1 output unit linear network were used for the simulations discussed in the remainder of this paper After training with pooled data within each subject viz. patient group the network was used for predicting each of the time series from the three groups.

To measure the degree of predictability the prediction error *arv* was computed for each of the nineteen test subjects as a function of the number of steps predicted into the future as well as for ten surrogate time series generated by two different procedures as explained below. The number of time steps for each time series where the *arv* reached a value of 0.5 was recorded. This cut-off criterion ensures that the predictive model is still useful for predicting the system dynamics whereas an *arv* of 1.0 may be achieved by always guessing the mean of the time series as the predicted value.

Surrogate data

We generated surrogate data as statistical controls for our experimental data to test the null hypothesis that the predictive results can be explained by a stochastic process. Surrogate data are randomized versions of the original data sets preserving certain statistical features of their originals. For each original time series we generated ten surrogate data time series by two different procedures as proposed by Theiler *et al.* (13).

Phase randomized surrogate data

In this procedure the original time series is Fourier transformed to obtain phase and amplitude spectra. The phases are randomized using i.i.d. (independent identically distributed) random numbers from an $[0, 2\pi]$ interval and after transforming back to the time domain one gets a randomized version of the original time series which preserves the power spectrum, the mean, the standard deviation, and the autocorrelation. This sort of surrogate data is used to test the null hypothesis that the results can be explained by using data from a Gaussian linear stochastic process. Other types of surrogate data are necessary since this procedure does not preserve the histogram of the signal amplitudes.

Gaussian scaled surrogate data

Gaussian scaled surrogate data are generated by shuffling the original data in such a way that the amplitude distribution of the original is preserved and that the power spectrum remains very similar but not identical as in the case of the phase randomized surrogates. Using this sort of surrogates we can test the null hypothesis that our results can be explained by testing data from a nonlinear static transformation of a linear stochastic process.

Statistical significance

To measure the amount of deterministic structure that might have been destroyed by either of the randomization techniques we used the significance level S, defined as

$$S = \frac{\langle T_{surrogate} \rangle - T_{original}}{\sigma} \qquad (5)$$

where T is the prediction time step where the *arv* exceeds a value of 0.5. The angle brackets denote the over all mean of ten surrogate time series generated for each original by either of the two methods and σ is the standard deviation of the results obtained from the surrogates.

RESULTS

Using a large variety of linear as well as nonlinear neural network predictors we found a linear increase of the prediction error *arv* within the osteoporotic patients whereas the healthy subjects and the patients with hyperparathyroidism showed a saturation of the *arv* on a much lower level than the osteoporotic group. Figure 4 displays the predictive results of a 15 input units, 5 hidden units and 1 output unit network with sigmoidal activation functions and a 15 input unit and 1 output unit network with linear activation functions. However, the performance between linear and nonlinear predictors in general was not significantly different within each group (Fig. 5).

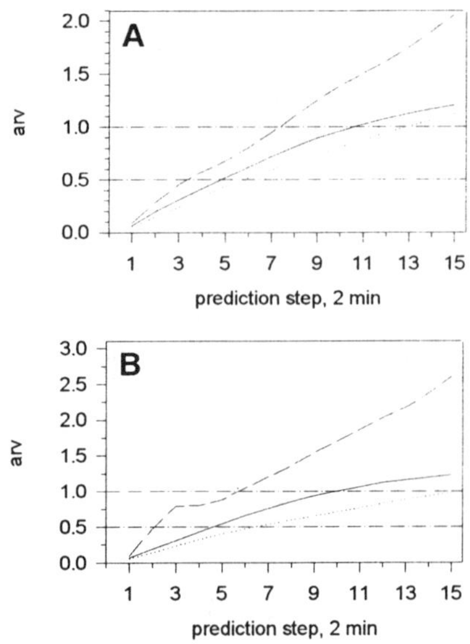

FIGURE 4. Prediction error *arv* versus prediction step (2 min) for normal subjects (n=9, solid line), patients with osteoporosis (n=6, dashed line), and patients with hyperparathyroidism (n=4, dotted line). **A**) 15 input units, 5 hidden units, and 1 output unit neural network with sigmoidal activation functions. **B**) 15 input units, 1 output unit neural network with linear activation functions.

To test for nonlinear determinism we evaluated the predictive ability of our neural network on 10 phase randomized and 10 Gaussian scaled surrogate data sets. We could demonstrate a significance level S between 1.9 and 2.3 for the predictabilty of time series from all three groups compared to their corresponding Gaussian scaled surrogate data using a nonlinear network (Table 1). Only the normal subjects showed a relatively low significance level S of 1.0 for the difference between the predictability of the original time series and the corresponding phase randomized data sets whereas the osteoporotic patients and hyperparathyroid patients did not show a significant difference. Using a linear predictor for comparing its predictive ability for the original data with that for the surrogates we found similar results (Table 2). A significant difference from both types of surrogate data could be found only for the normal subjects with significance levels S above 2.0. Although the predictability in some of the osteoporotic patients and

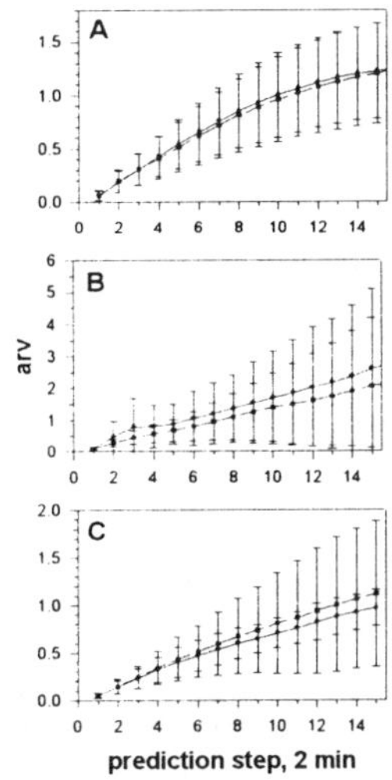

FIGURE 5. Predictability of a nonlinear 15 input unit, 5 hidden unit, and 1 output unit neural network(■) versus linear 15 input unit and 1 output unit neural network(●). Mean arv ± standard deviation. **A**) normal subjects, **B**) patients with osteoporosis **C**) patients with hyperparathyroidism.

patients with hyperparathyroidism was above a significance level of 2 (more than two σ different from the surrogates), no significant difference could be found for both patient groups.

Whereas some time series from the osteoporotic and the normal group had a significantly higher predictability than their phase randomized counterparts such behavior could not be demonstrated for any of the hyperparathyroid patients. Figure 6 displays the prediction error *arv* versus the prediction step for the original time series and both types of surrogate data from one healthy subject viz. patient of each group using the 15 input units sigmoidal neural network. The prediction error for the normal subject lies below the scatter of both types of surrogates for most of the time steps whereas the prediction error for the osteoporotic patient keeps within the scatter of both types of surrogate data sets for all prediction time steps. In the hyperparathyroid patient we find an almost undistinguishable prediction error *arv* for the original time series from that of the phase ranomized surrogate data. However, the *arv* of the original data lies outside the scatter of the Gaussian scaled surrogates for all prediction steps.

Using a linear network with the same length of the input window (Fig. 7) the difference of the prediction error *arv* between the surrogate scatter and the original

TABLE 1. Prediction step (in 2min) where the *arv* exceeds 0.5 (15-5-1 nonlinear network). Statistics using the significance level S as described above.

group	original (I)	phase surrogates (II)	Gaussian surrogates (III)	I vs. II	I vs. III
normals	6.6±3.2	6.0±2.7	4.7±1.7	1.0	2.3
osteoporosis	4.6±2.6	5.1±2.6	3.1±0.6	0.1	1.9
hyperparathyroidism	8.0±4.1	6.5±2.5	2.2±0.5	0.5	2.2

TABLE 2. Prediction step (in 2min) where *the arv* exceeds 0.5 (15-1 linear network). Statistics using the significance level S as described above.

group	original (I)	phase surrogates (II)	Gaussian surrogates (III)	I vs. II	I vs. III
normals	6.2±2.7	4.9±3.0	4.0±2.2	2.1	2.9
osteoporosis	3.6±1.5	4.2±2.5	2.7±0.9	0.6	1.6
hyperparathyroidism	7.3±2.6	6.9±2.7	3.9±1.7	0.2	1.8

in the normal subject decreases and the ostoporotic subject displays a clearly reduced predictability than using a nonlinear predictor. The prediction error of the hyperparathyroid subject is not distinguishable from both types of surrogate data.

DISCUSSION

We found no evidence of a highly predictable saturating component in any of the osteoporotic time series using our predictive approach. Although the hyperparathyroid patients demonstrated high predictability similar to that of the normal subjects it was only significantly different from their corresponding Gaussian surrogate data sets using a nonlinear predictor. The osteoporotic group did not show a significant difference between the predictability of the original data and both types of surrogates using linear as well as nonlinear predictors.

Confirming results from a former study on nonlinear determinism in extended time series of PTH concentration in healthy subjects we could observe significant nonlinear determinism in nine shorter time series of healthy subjects using linear and nonlinear time series prediction and statistical tests against phase randomized as well as Gaussian scaled surrogate data sets. This suggests that osteoporotic patients and patients with hyperparathyroidism have lost at least in part nonlinear

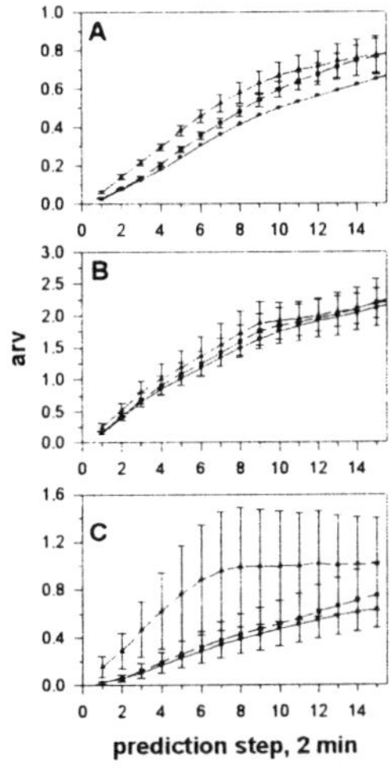

FIGURE 6. Predictability of a 15 input units, 5 hidden units, and 1 output unit nonlinear neural network for original time series(●) and mean ± standard deviation of 10 phase randomized surrogate data sets (■) viz. 10 Gaussian scaled surrogate data sets (▲). **A)** normal subject, **B)** patient with osteoporosis, **C)** patient with hyperparathyroidism.

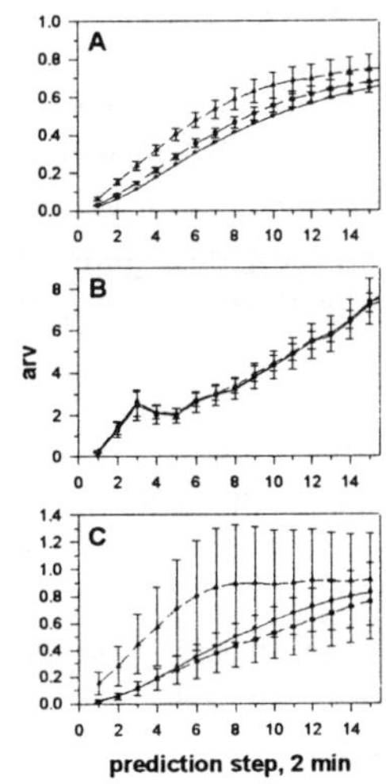

FIGURE 7. Predictability of a 15 input units and 1 output unit linear neural network for original time series (●) and mean ± standard deviation of 10 phase randomized surrogate data sets (■) viz. 10 Gaussian scaled surrogate data sets (▲). **A)** normal subject, **B)** patient with osteoporosis, **C)** patient with hyperparathyroidism.

deterministic structure in their secretory dynamics of PTH.

The major dynamic regulator for the normal coupling of bone resorption and formation in the bone remopdeling process appears to be found in the predictability of the secretory dynamics of PTH.

However, a detailed understanding of the secretory dynamics of PTH and bone anabolic properties is still lacking. Our data provide first evidence that states of high predictability are a regular finding in subjects with normal bone mass and metabolism as well as patients with hyperparathyroidism both of which have a preserved normal bone structure.

Consequently, if a high rate of change of PTH concentration is the significant factor in maintaining normal bone remodeling, frequent application of PTH with large doses may be even more effective than daily injections as are used in current treatment protocols of osteoporosis. Application of large pulses of PTH on an hourly scale might be feasable with the use of a hormone pump, an important field for future investigation.

We found no significant difference in the predictive ability of linear and nonlinear networks for the PTH concentration time series from all groups. However, using classical nonlinear analysis, we have seen evidence for nonlinear determinism and conceivably low-dimensional deterministic chaos in the irregular pattern of PTH secretion in healthy human subjects (23). These results are confirmed by comparing time series prediction of the original data with phase randomized as well as Gaussian scaled surrogate data. Our results are also in accordance with the findings of Blinowska and Malinowski (22) who demonstrated that the irregular nonlinear EEG signal could be predicted equally well or even better by a linear autoregressive model than by the nonlinear prediction model proposed by Sugihara and May (20).

These results differ from others who have compared linear and nonlinear predictive models. This may be due to the relatively small amount of data available in our study (11, 24, 25).

One limitation in our studies is the time period over which PTH concentrations can be measured. Longer periods may only be performed by an online biosensor for PTH to avoid the blood loss due to frequent blood sampling. Such a biosensor for PTH is not yet available although biosensors containing biological receptors such as the nicotinic acetylcholine receptor (29-31) and the *L*-glutamate receptor (32) have been developed.

Time series prediction has been established as a valid tool for the characterization and classification of the dynamics in a variety of biological systems. Using the technique of surrogate data null hypothesis can be constructed to statistically distinguish between stochastic and deterministic behavior in the dynamics and gain new insight into the physiological regulation of episodic hormone secretion in cases where simple linear methods such as computing the mean value of a time series or the power spectrum might fail to distinguish normal from disturbed patterns. Our results suggest that the PTH secretory pattern in healthy subjects is an example of nonlinear determinism which seems to be lost in part in two bone diseases, osteoporosis and hyperparathyroidism.

ACKNOWLEDGMENTS

S. J. N. and T. J. S. were supported by Howard Hughes Medical Institute. K. P. and G. B. were supported by the *Deutsche Forschungsgemeinschaft* under grant Pr 333/1-2 and grant Br 915/4-1.

REFERENCES

1. Brabant, G., Prank, K., and Schöfl, C., *Trends Endocrinol. Metab.* **3**, 183-190 (1992).

2. Harms, H.M., Kaptaina, U., Külpmann, W.R., Brabant, G., and Hesch, R.D., *J. Clin. Endocrinol. Metab.* **69**, 843-851 (1989).

3. Kitamura, N., Shigeno, C., Shiomi, K., Lee, K., Ohta, S., Sone, T., Katsushima, S., Tadamura, E., Kousaka, T., and Yamamoto, I., *J. Clin. Endocrinol. Metab.* **70**, 252-263 (1990).

4. Kripke, D.F., Lavie, P., Parker, D., Huey, L., and Deftos, L.F., *J. Clin. Endocrinol. Metab.* **47**, 1021-1027 (1978).

5. Liu, C.C., Kalu, D.N., Salerno, E., Echon, R., Hollis, B.W., and Ray, M., *J. Bone Miner. Res.* **6**, 1071-1080 (1991).

6. Podbesek, R., Edouard C., Meunier, P.J., Parsons, J.A., Reeve, J., Stevenson R.W., and Zanelli, J.M., *Endocrinology* **112**, 1000-1006 (1983).

7. Tam, C.S., Heersche, J.N., Murray, T.M., and Parsons, J.A., *Endocrinology* **110**, 506-512 (1982).

8. Parisien, M., Silverberg, S.J ., Shane, E., de la Cruz, L., Lindsay, R., Bilezikian, J.P., and Dempster, D.W., *J. Clin. Endocrinol. Metab.* **70**, 930-938 (1990).

9. Harms, H.M., Schlinke, E., Neubauer, O., Kayser, C., Wüstermann P.R., Horn, R., Külpmann, W.R., von zur Mühlen, A., and Hesch R.D. *J. Clin. Endocrinol. Metab.* **78**, 53-57 (1994).

10. Casdagli, M., Des Jardins, D., Eubank, S., Farmer, J.D., Gibson, J., Hunter, N., and Theiler, J., "Nonlinear modeling of chaotic time series: theory and applications." *Technical Report No.* LA-UR-91-1637, Los Alamos National Laboratory (1991).

11. Sugihara, G., and May, R.M., *Nature (Lond.)* **344**, 734-741 (1990).

12. Tsonis, A.A., and Elsner, J.B., *Nature (Lond.)* **358**, 217-220 (1992).

13. Theiler, J., Eubank, S., Longtin, A., Galdrikian, B., and Farmer, J.D., *Physica D* **58**, 77-94 (1992).

14. Schiff, S.J., Sauer, T., and Chang, T., *Integrative Physiol. Behav. Sci.* **29**, 246-261 (1994).

15. Sugihara, G., *Phil. Trans. R. Soc. Lond. A* **348**, 477-495 (1994).

16. Blinowska, K.J., and Malinowski, M. *Biol. Cybern.* **66**, 159-165 (1991).

17. Lefebvre, J.H., Goodings, D.A., Kamath, M.V., and Fallen, E.L., *Chaos* **3**, 267-276 (1993).

18. Chang, T., Schiff, S.J., Sauer, T., Gossaed, J.P., and Burke, R.E., *Biophys. J.* **67**, 671-683 (1994).

19. Schiff, S.J., Jerger, K., Chang, T., Sauer, T., and Aitken, P.G., *Biophys. J.* **67**, 684-691 (1994).

20. Grassberger, P., and Procaccia, I., *Phys. Rev. Lett.* **50**, 346-349 (1983).

21. Wolf, A., Swift, J.B., Swinney, H.L., and Vastano, J.A., *Physica D* **16**, 285-317 (1985).

22. Grassberger, P., and Procaccia, I., *Phys. Rev. A* **28**, 2591-2593 (1983).

23. Prank, K., Harms, K., Dämmig, M., Brabant, G., Mitschke, F., and Hesch, R.D., *Am. J. Physiol.* **266**, E653-E658 (1994).

24. Weigend, A.S., and Gershenfeld, N.A. (eds.), *Time series prediction: Forecasting the future and understanding the past.* SFI Studies in the Sciences of Complexity, Proc. Vol. XV, Reading, MA, Addison-Wesley, (1993).

25. Lapedes, A.S., and Farber, R.M., "Nonlinear signal processing using neural networks: prediction and system modeling." *Technical Report No.* LA-UR-87-2662, Los Alamos National Laboratory (1987).

26. Nowlan, S.J., and Hinton,G.E., *Neural Comput.* **4**, 473-493 (1992).

27. Weigend, A.S., Huberman, B.A., and Rumelhart, D.E., *International Journal of Neural Systems* **1**, 193-209 (1990).

28. Mitschke, F. *Phys. Rev. A* **41**, 1169-1171 (1990).

29. Eldefrawi, M. E., Sherby, S.M., Andreou, A.G., Mansour, N.A., Annau, Z., Blum, N.A., and Valdes, J.J.*Anal. Lett.* **21**, 1665-1680 (1988).

30. Gotoh, M., Tamiya, E., Momoi, M., Kagawa, Y., and Karube, I. *Anal. Lett.* **20**, 857-870 (1987).

31. Gotoh, M., Tamiya, E., and Karube, I. *J. Membrane Sci.* **41**, 291-303 (1989).

32. Uto, M., Michaelis, E.K., Hu, I.F., Umezawa, Y., and T. Kuwana. *Anal. Sci.* **6**, 221-225 (1990)..

Biphonation in Voice Signals

Hanspeter Herzel* and Robert Reuter†

*Institute of Theoretical Physics, Technical University
Hardenbergstr. 36, Sekr. PN 7-1, D-10623 Berlin, Germany

†Institute of Electronics, Technical University
Einsteinufer 17, D-10587 Berlin, Germany

Abstract

Irregularities in voiced speech are often observed as a consequence of vocal fold lesions, paralyses, and other pathological conditions. Many of these instabilities are related to the intrinsic nonlinearities in the vibrations of the vocal folds. In this paper, a specific nonlinear phenomenon is discussed: The appearance of two independent fundamental frequencies termed biphonation. Several narrow-band spectrograms are presented showing biphonation in signals from voice patients, a newborn cry, a singer, and excised larynx experiments. Finally, possible physiological mechanisms of instabilities of the voice source are discussed.

INTRODUCTION

The vocal folds, together with glottal airflow, constitute a highly nonlinear self-oscillating system. According to the accepted myoelastic theory of voice production, the vocal folds are set into vibration by the combined effect of subglottal pressure, the visco-elastic properties of the folds, and the Bernoulli effect [1, 2]. The effective length, mass and tension of the vocal folds are determined by muscle action, and in this way the fundamental frequency ("pitch") and the waveform of the glottal pulses can be controlled. The vocal tract acts as a filter transforming the primary signals into meaningful voiced speech [3].

Normal sustained voiced sound appears to be nearly periodic, although small perturbations (in the order of a percent) are important for the naturalness of speech. Under certain circumstances, however, much larger irregularities are observed in vocalizations which are often associated with the term "roughness". In the phonetic literature phenomena such as "octave jumps",

*E-mail: herzel@itp1.physik.tu-berlin.de
†E-mail: rreuter@tubife1.ee.tu-berlin.de

"biphonation"(two independent pitches), and "noise concentrations" have been reported for decades [4, 5]. However, only within the past few years it has been pointed out that these observations might be interpreted as period-doubling bifurcations, tori, and chaos, respectively [6]-[10]. It has been shown using Poincaré sections and estimations of attractor dimensions and Lyapunov exponents that newborn cries [6] and vocal disorders [7]-[10] are intimately related to bifurcations and low-dimensional attractors.

On the one hand, subharmonics and irregularities appear occasionally in normal vocalizations (newborn cries [6], "vocal fry" phonation [9], and speech [11]). On the other hand, vocal instabilities due to bifurcations are often symptomatic in voice pathology [7]-[10]. Laryngeal stroboscopy reveals that various voice disorders lead to irregular vibratory patterns of the vocal folds resulting in a rough voice quality [12]. The corresponding acoustic signals often show sudden jumps to subharmonic regimes (period-doubling or -tripling) and low-frequency modulations (tori) [7, 8, 10]. It is the aim of current research to understand the underlying physiological mechanisms of these bifurcations with the aid of high-speed films, excised larynx experiments [13, 14], and biomechanical modeling [15, 16].

Subharmonics and low-frequency modulations have been studied most extensively in previous studies. In this paper we focus on another phenomenon - the appearance of two independent audible fundamental frequencies termed "biphonation". Biphonation corresponds to a torus of the underlying dynamical system. Of course, also low-frequency modulations mentioned above represent a torus. However, modulations below, say, 70 Hz are usually not perceived as independent pitch but as tremor (1-15 Hz) or roughness (20-70 Hz). Contrarily, two independent frequencies in the range of normal pitch are perceived as a highly dissonant sound and reflect, consequently, a serious deviation from normal phonation.

A powerful tool for the analysis of independent pitches are narrow-band spectrograms. Spectrograms allow the detection of bifurcations in systems with slowly varying parameters [4, 5, 6, 17, 18]. They are based on many subsequent short-time power spectra of overlapping segments. The spectral amplitude is encoded as a grey scale and, in this way, the (dis)appearance of spectral peaks owing to bifurcations of the underlying dynamical system can be monitored. For example, period-doubling is reflected in additional peaks in the middle of the harmonics of the original pitch. Low-frequency modulations lead to side-bands of the main spectral peaks. Biphonation is characterized by nonparallel dark lines which represent independently varying spectral peaks.

In the following sections it will be shown that biphonation is a normally occuring event in a variety of voice signals including voice disorders, newborn cries, an artistic voice, and excised larynx experiments. We discuss possible physiological mechanisms leading to biphonation and review results from computer simulations.

BIPHONATION IN VOICE DISORDERS

The characteristic periods of vocal fold vibrations are a few milliseconds. Consequently, stationary segments in sustained vowels contain many cycles and the underlying attractor can be characterized by narrow-band spectrograms. A Hamming window of 4096 points (about 200 ms) and a shift of 512 points are used to generate the overlapping spectral segments.

Normal (almost periodic) phonation leads to a series of parallel lines (see, e.g., the segment around 1.5 s in Fig. 1). Owing to slowly changing parameters such as subglottal pressure or muscle tension, transitions to other regimes can be induced. Such sudden changes reflect bifurcations of the underlying dynamical system. Fig. 1 displays, for example, the appearance of a second series of spectral components around 2 s. This biphonation is audible as a rather dissonant sound.

Two other examples of spectrograms with independent pitches are shown in Figs. 2 and 3. Note that the dominant spectral components in these graphs are not harmonically related. A part of the time-series of the vowel displayed in Fig. 3 has been analyzed in Ref. [8]. For a relatively stationary episode

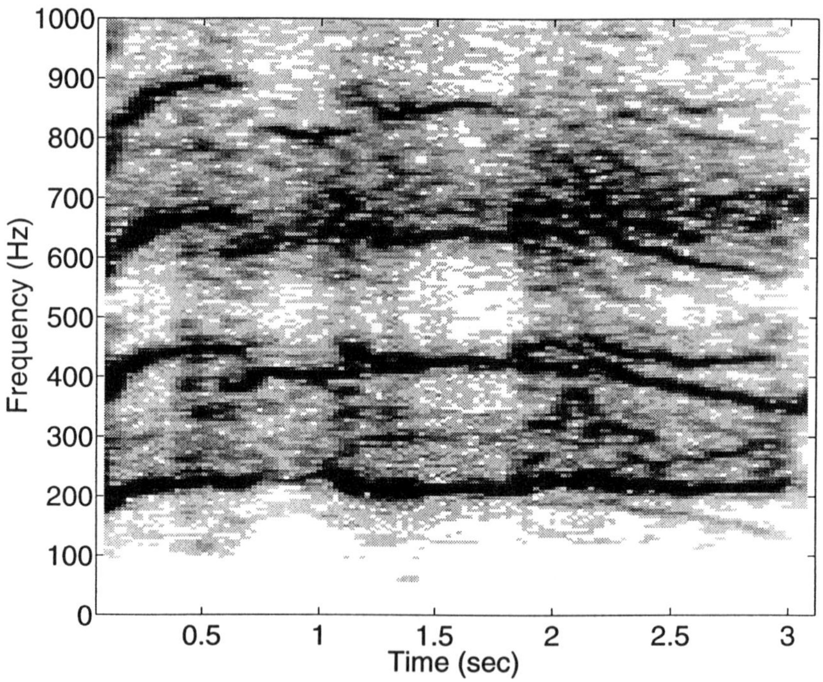

Figure 1: Spectrogram showing biphonation in a sustained vowel "u" of a male patient with edema. Note the nonparallel lines in the final second. The signals in Figs. 1-3 have been recorded at the university hospital Charité [7, 19].

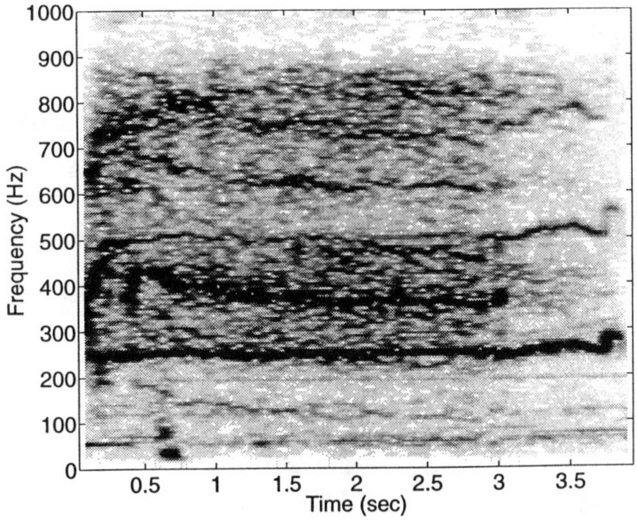

Figure 2: Biphonation in a sustained vowel "u" of a male patient with papilloma.

around 0.5 s an attractor dimension of 2.6 has been estimated. The examples presented in the Figures 1-3 are chosen from a data base of 21 patients with very hoarse voices. Obviously, biphonation seems to be a rather common

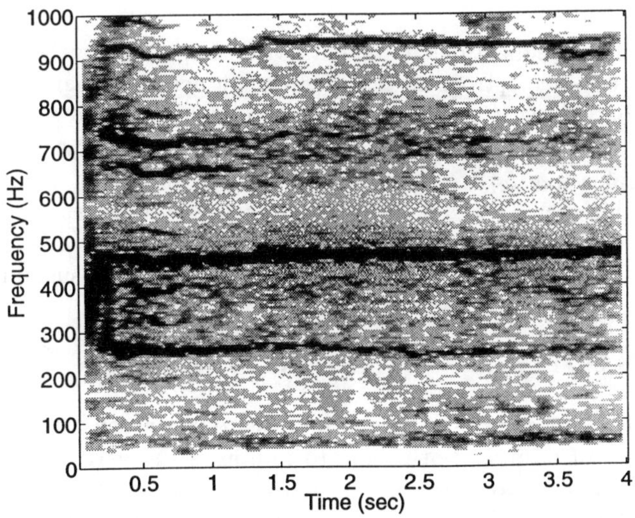

Figure 3: Biphonation in a sustained vowel "u" of a female patient with papilloma.

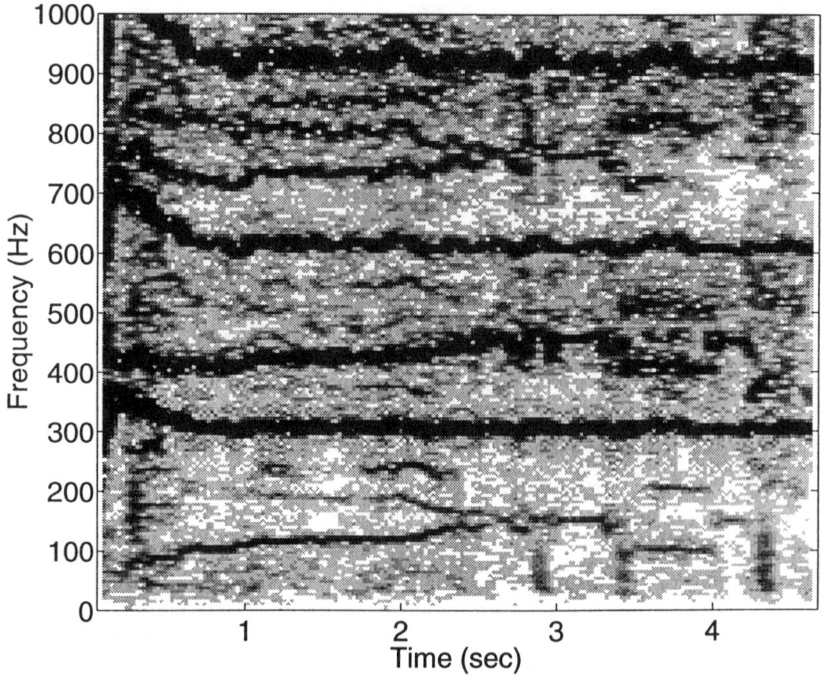

Figure 4: Biphonation and subharmonic regimes in a sustained vowel "a" due to acute laryngitis.

phenomenon in vocalizations of patients with strongly perturbed voices.

Fig. 4 shows a spectrogram recorded during acute laryngitis of one of the authors (RR). There is spectral energy at $f_1 \approx 300$ Hz and at its harmonics and another frequency $f_2 \approx 100$ Hz. Up to 2.5 s several linear combinations are discernible ($f_1 + f_2, 2f_1 + f_2, 3f_1 - f_2, ...$). An episode of 2:1 entrainement can be seen between 3 s and 3.3 s . In addition, close to the end (3.5 s - 4 s) 3:1 entrainement can be detected. This example illustrates that different dynamic regimes can be visualized by spectrograms even in a single sustained vowel.

BIPHONATION IN NEWBORN CRIES

The acoustic properties of newborn cries are in a sense comparable to sustained vowels of adults. Both serve to minimize the variation in vocal fold state and, consequently, one can find relatively stationary segments of a few hundred milliseconds. Due to the smaller size of the vocal apparatus the characteristic frequencies are shifted to higher values. The pitch is usually in the range of 400-600 Hz and the main resonances of the vocal tract are between 2 kHz and 3 kHz.

The sounds of newborn infant cries are used by clinicians for the evaluation of neurological and respiratory function. Already 30 years ago nonlinear phenomena such as subharmonics, two independent pitches, frequency jumps, and "turbulence" have been described in phonations of 30 normal infants [4]. However, only recently these observations have been related to bifurcations and chaos [6].

The authors of Ref. [20] claim that markedly complicated phonation is perhaps as common as the simple basic cry. They also display two spectrograms which clearly indicate two fundamental frequencies (biphonation).

Robb and Saxman [21] analyzed 1200 non-cry vocalizations of 14 healthy infants with an age of 11-25 months. They found subharmonics, frequency shifts, and biphonation to be normally occurring phonatory events. Subharmonics appeared much more often (54 instances in all 14 children) than biphonation (5 instances in 4 children).

It is worth discussing, why these nonlinear phenomena are more frequent in infants vocalizations than in the speech of adults. A possible explanation is that due to the relatively short vocal folds and the strong airflow the role of

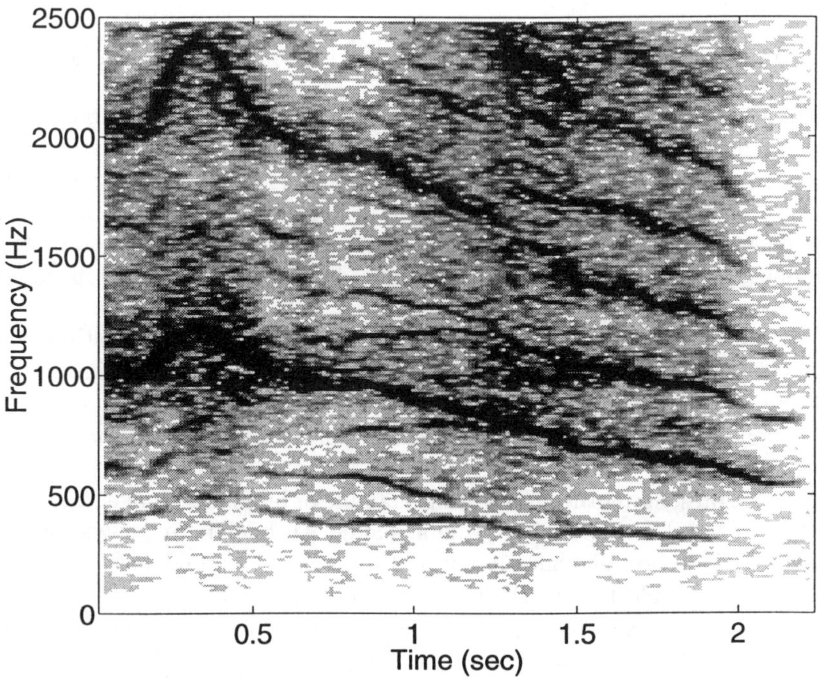

Figure 5: Spectrogram of an infant (4 years old) with papilloma. Biphonation is indicated by a series of lines at multiples of about 400 Hz and another spectral component with descending frequency. In this figure a window of 2048 points was used.

nonlinearities is more pronounced. Moreover, the instabilities may reflect laryngeal neuromotor immaturity. In other words, older speaker learn to control those phenomena.

There is an intense research to exploit narrow-band cry spectrograms in the neonatal period for assessing the diagnosis and prognosis [22, 23, 24]. Chromosomal aberrations such as Down's syndrome or the cat-cry syndrome perturb newborn cries very markedly. Diseases affecting the central nervous system change cry characteristics and, therefore, cry features can indicate disturbances of brain function due to oxygen deficiency, menengitis, or hydrocephalus [22]. It was found that especially biphonation is more common in infants with cerebral lesions [23].

Hirschberg and Szende compiled in a monograph [24] more than 100 representative sound signals from infants with various diseases. Biphonation was detected in cries of children with *Soor laryngis* and with *Papilloma laryngis*. Fig. 5 is obtained from the record which accompanies the monograph [24].

In summary, biphonation is found occasionally in vocalizations of normal healthy infants. However, it is also symptomatic for brain damages and can be exploited, therefore, as an early diagnostic tool.

BIPHONATION IN NORMAL VOICES

Sustained vowels of normal healthy adults are predominantly periodic. Only occasionally [11] or under extreme conditions (pressed voice, vocal fry, simulated creaky voice, inspiratory phonation) subharmonics and irregularities occur. Biphonation seems to be very rare in normal subjects. The impressive Mongolian overtone singing cannot be regarded as biphonation, since these singers use only cultivated boosting of higher harmonics in order to mimic two independent frequencies.

An example of reproducible subharmonics, amplitude modulations, and inspiratory biphonation was reported by Mazo et al. [25] for Russian lament. In these emotionally charged expressions extreme parameter regions of the human voice source are explored and, therefore, these phenomena are likely to occur.

Now we present another interesting example of biphonation in a voice without known vocal pathology. A young female Australian singer was able to produce biphonation at high pitches and intensities. Fig. 6 shows a convincing example. In addition to the intended constant pitch $f_1 \approx 600$ Hz and its harmonics another series of spectral lines is clearly visible. If we denote the frequency slightly below 200 Hz as f_2, one can easily identify various linear combinations of the two frequencies: $f_1-f_2, f_1+f_2, f_1+2f_2, 2f_1-f_2, 2f_1+f_2, \ldots$. Unfortunately, no stroboscopic or high-speed observations were possible to analyze the generation of this remarkable sound in detail.

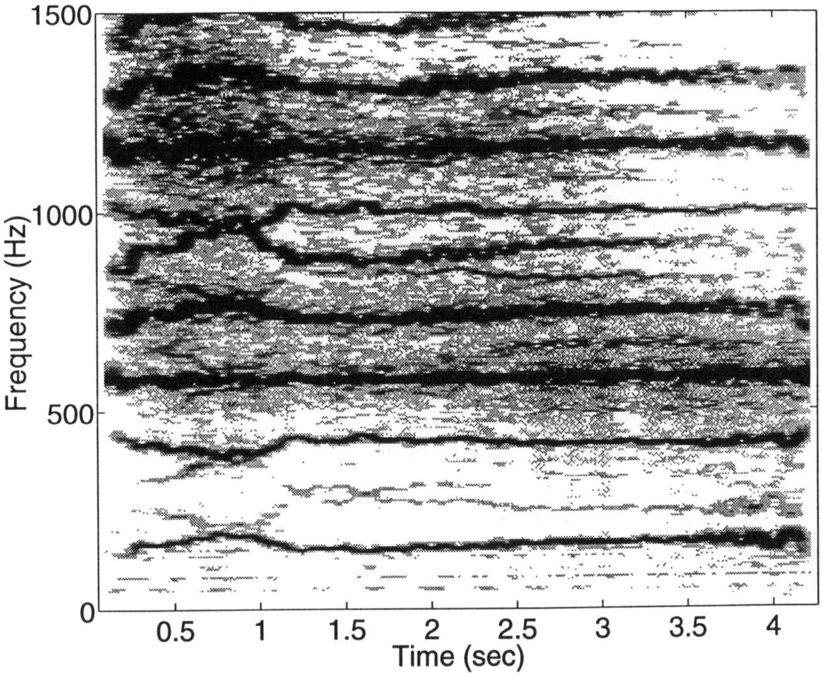

Figure 6: Voluntarily produced biphonation recorded by Pam Davis and Richard Troughear at St. Vincent's Hospital Sydney.

EXCISED LARYNX EXPERIMENTS

Experiments with human or animal larynges serve as a link between the human voice source in vivo and computer models [1, 14, 26]. They allow controlled and systematic parameter variations and easy observation of vibratory patterns.

We have examined five larynges from large (about 25 kg) mongrel dogs coming from coronary research units at The University of Iowa. The dissected larynges were mounted on an apparatus described in detail elsewhere [14]. Heated and humified air was supplied from below as the driving force of the oscillations. The device was attached to several micrometers to control the adduction and the elongation of the vocal folds. To facilitate observation of vocal fold movement, a strobe light with adjustable frequency was placed above the glottis. The data were recorded on a color video system and afterwards digitized with 16 bit resolution and a sampling rate of 20 kHz.

In our experiments instabilities have been studied for varying subglottal pressure and for asymmetric adduction and elongation of the vocal folds. 2-parameter bifurcation diagrams can be found in Ref. [14]. Here we only summarize briefly the various dynamic regimes which have been observed for

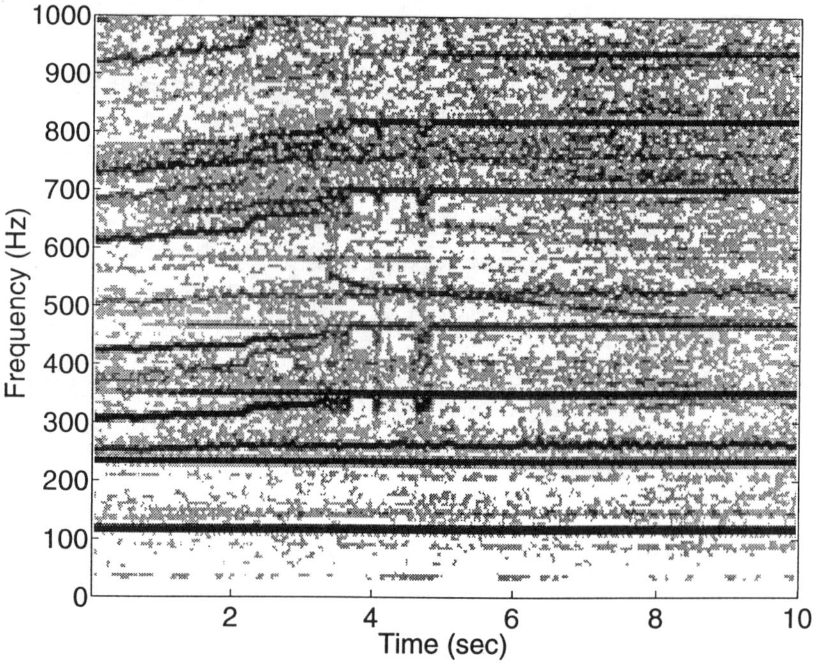

Figure 7: Biphonation in an excised larynx experiment. Details about the experimental setup and the parameter values are given in [14].

overcritical asymmetry and pressure:

- symmetric periodic phonation in head-like and chest-like registers
- periodic vibrations of the lax fold only
- subharmonics, modulations, and irregular vibrations with both folds involved
- whistle-like sound
- aphonia, i.e. vibrations ceased for very strong asymmetric tension

Typically, the parameter ranges of these regimes overlap, i.e. hysteresis is observed.

In a parameter region of medium asymmetry and pressure we have found three different phonatory modes ("registers"): A chest-like phonation revealed large amplitudes of vibration, complete closure, a pronounced phase lag between the lower and the upper edge ("mucosal wave"), and very intense harmonics. A falsetto-like phonation was characterized by small amplitude vibrations, incomplete closure, a weak mucosal wave, and less intense harmonics. A whistle-like phonation had a higher pitch, small vibration without any visible mucosal wave, and very weak harmonics. A preliminary explanation of the latter regime would be an interpretation as "vortex-induced vibrations" [14]. Sometimes spontaneous transitions between different registers appeared without external parameter changes. In particular, Fig. 7 shows the coexistence of falsetto register (note the lines at multiples of about 115 Hz) and whistle register (series of nonparallel lines).

In this case the origin of biphonation can be traced back to the coexisting registers. There are two different mechanisms at work to transfer energy from the vertical airflow to vocal fold vibrations: The phase lag between the lower and the upper edge (see [16, 26, 27] for details) and vortex shedding.

MECHANISMS OF INSTABILITIES

Biphonation is a specific kind of vocal instability. Other phenomena such as subharmonics and chaos are often found in nearby parameter regions. In the following we compile possible mechanisms leading to aperiodicities.

Without making claims about their relative importance, possible candidates are listed:

- Neurological sources: Unsteadiness of muscle contraction due to variations in motor unit firing is a possible source of physiological jitter and shimmer [28]. Moreover, vibrato and vocal tremor in the frequency range of 3-12 Hz seem to be related to a neural feedback loop [29].

  Modulations of higher frequencies (say, e.g., 40 Hz) and subharmonic vocalization have predominantly biomechanical origin, i.e., the muscle tension can be regarded as approximately constant. This point of view is supported by computer simulations and excised larynx experiments which exhibit these phenomena without any variations in neuromuscular control.

- Aerodynamic instabilities: Even if the flow is not predominantly turbulent (as for a breathy voice) vortex shedding and instabilities of the jet emerging from the glottis may cause perturbations [30]. The interaction of vortex shedding with acoustic resonances may generate whistle-like sound. Moreover, by means of transverse "lift forces" periodically generated vortices can induce self-sustained oscillations of vibrating structures ("vortex-induced vibrations"). Such a vibratory regime was described above for the excised larynx experiments and might be associated with the so-called whistle-register of the soprano voice [14]. Moreover, the interaction of vortices with vocal resonances might induce secondary frequencies. Biphonation as discussed above is typically associated with a large airflow generating vorticity and, therefore, it seems likely that vortex shedding is involved.

- Coupling between vocal folds and sub- and supraglottal resonances: This coupling might play a role in extreme situations such as pressed voice where supraglottal pressure changes play an essential part in driving the vocal folds.

- Mucus: It is a frequent observation in clinical practice that inappropriate management of mucus riding on vocal fold tissue leads to a perturbed voice.
- Asymmetry of left and right fold: For certain disorders such as unilateral vocal fold paralyses or highly asymmetric vocal fold lesions the desynchronization of both fold might be the dominant mechanism. Such asymmetry is explored extensively in computer simulations [16] and excised larynx experiments [14].
- Anterior-posterior modes: Such horizontal motion can be seen frequently in stroboscopy and high-speed movies [31]. However, in many cases these modes are entrained by the main vibratory modes, and hence, they do not induce much irregularity.
- Desynchronization of horizontal motion ("10-mode" related to opening and closing) and vertical modes ("11-mode" related to the well-known phase shift of upper and lower edge of the folds): These two vibratory modes are dominating for normal chest voice. Their desynchronization might play a prominent role in voice disorders. For example, vocal fry, which is often symptomatic in voice pathology, seems to be related to this mechanism. Computer simulations have shown [15] that a lax cover which slows down the 11-mode gives rise to vibratory pattern known from high-speed films of vocal fry [32].

At the current state of the art, the presented list of sources of irregularities has a somewhat speculative character. It is the aim of intense research at the National Center for Voice and Speech and in other groups to substantiate the above conjectures.

DISCUSSION

Computer models of speech production are valuable to understand the basic mechanisms of normal and pathological phonation. There are conceptionally simple two-mass models [9, 16, 33, 34] and sophisticated continuum models simulating the visco-elastic equations [2, 8, 15, 35, 36].

Biphonation was found in the intensively studied Ishizaka-Flanagan model [33]. In that model the bifurcations to toroidal oscillations can be induced by detuning of the two normal modes (in-phase and out-of-phase motion of both masses) [37, 38]. It is worth mentioning that in a simplified version of the two-mass without vocal tract only very small instability regions were found

[9, 39]. This result indicates an important role of the acoustic loading of the vocal tract for the generation of biphonation.

In summary, with respect to these biomechanical models and for the excised larynx data we have some ideas about the mechanisms of biphonation. Unfortunately, the physiological mechanisms of biphonation in newborn cries and voice pathologies are still unclear.

The following tools seem to be essential for further progress in understanding the underlying mechanisms of instabilities:

- Two-parameter-bifurcation diagrams: This means that the instabilities are located in planes of parameters of physiological interest (subglottal pressure, muscle tension, ...). For example, voice range profiles (phonetograms) can be regarded as such diagrams where regions of sustained oscillations are separated from regions of no phonation. Bifurcation diagrams have been calculated for the asymmetric 2-mass model [16] and measured in excised larynx studies [14].

- Mode concept: The characterization of instabilities as desynchronization of vibratory modes provides a framework for the classification of biomechanically induced irregularities in models and observations.

- Excised larynx studies: Experiments with human or animal larynges serve as a link between the human voice source in vivo and computer models [26]. They allow controlled and systematic parameter variations and easy observation of vibratory patterns.

- High speed digital imaging: A visualization of irregular patterns is of central importance for detecting the physiological mechanisms of instabilities.

These techniques and broad interdisziplinary cooperation will hopefully lead to a better understanding and treatment of voice disorders in the future.

ACKNOWLEDGMENTS

Partial funding for this research was provided by the Deutsche Forschungsgemeinschaft and Grant No. P60 DC00976 from the National Institutes on Deafness and Other Communication Disorders. We thank J. Wendler and M. Cebulla for providing data from patients with voice disorders and acknowledge gratefully P. Davis and R. Troughear for providing the voice signals of the singer. One of us (HH) thanks for the warm hospitality at the National Center for Voice and Speech in Iowa-City and the fruitful cooperation with I. R. Titze, D. Berry, and B. Story.

References

[1] van den Berg, J., *J. Speech Hearing Res.* **1**, 227-244 (1958).

[2] Titze, I. R. and Alipour-Haghighi, F., *Myoelastic Aerodynamic Theory of Phonation,* (forthcoming book).

[3] Fant, G., *Acoustic theory of speech production,* Mouton: The Hague, 1960.

[4] Lind (Ed.), J.,*Newborn infant cry,* Uppsala: Almquist and Wiksells Boktrycken, 1965.

[5] Kelman, A. W., *Folia phoniat.* **33**, 73-99 (1981).

[6] Mende, W., Herzel, H., and Wermke, K., *Phys. Lett.* **A 145**, 418-424 (1990).

[7] Herzel, H. and Wendler, J., "Evidence of chaos in phonatory samples," in *EUROSPEECH,* ESCA, Genova, 1991, pp.263-266.

[8] Titze, I. R., Baken, R., and Herzel, H., "Evidence of Chaos in Vocal Fold Vibration," in *Vocal Fold Physiology: Frontiers in Basic Science* edited by I. R. Titze, Singular Publ. Group, San Diego 1993, pp.143-188.

[9] Herzel, H., *Appl. Mech. Rev.* **46**, 399-413 (1993).

[10] Herzel, H., Berry, D. A., Titze, I. R., and Saleh, M., *J. Speech Hearing Res.* **37**, 1008-1019 (1994).

[11] Dolansky, L. and Tjernlund, P., *IEEE Trans.* **AU-16**, 51-56 (1968).

[12] Hirano, M., *Folia phoniat.* **41**, 89-144 (1989).

[13] Herzel, H., Berry, D., Titze, I. R., and Steinecke I., *CHAOS* **5**, 30-34 (1995).

[14] Berry, D. A., Herzel, H., Titze, I. R., and Story, B. H., *J. of Voice*, in press.

[15] Berry, D. A., Herzel, H., Titze, I. R., and Krischer, K., *J. Acoust. Soc. Am.* **95**, 3595-3604 (1994).

[16] Steinecke, I. and Herzel, H., *J. Acoust. Soc. Am.* **97**, 1875-1884 (1995).

[17] Lauterborn, W. and Cramer, E., *Phys. Rev. Lett.* **47**, 1445-1448 (1981).

[18] Liauw, M. A., Koblitz, K., Jaeger, N. I., and Plath, P., *J. Phys. Chem.* **97**, 11724-11730 (1993).

[19] Wendler, J., Rauhut, A., and Krüger, *J. Phonetics* **14**, 485-488 (1986).

[20] Truby, H. M. and Lind, *J. Acta Paediat. Scand. Suppl.* **163**, 7-92 (1965).

[21] Robb, M. P. and Saxman, J., *J. Acoust. Soc. Am.* **83**, 1876-1882 (1988).

[22] Sirviö, P., and Michelsson, K., *Folia phoniat.* **28**, 161-173 (1976).

[23] Michelsson, K., Raes, J., and Rinne, A., *Folia phoniat.* **36**, 219-224 (1984).

[24] Hirschberg, J., and Szende, T., *Pathologische Schreistimme, Stridor und Hustenton im Säuglingsalter,* Stuttgart: Gustav Fischer Verlag, 1985.

[25] Mazo, M., Erickson, D., and Harvey, T., "Emotion and Expression: Temporal Data on Voice Quality in Russian Lament" in *Vocal Fold Physiology* edited by O. Fujimura and M. Hirano, Singular Publ. Group, San Diego 1995, pp.173-178.

[26] Baer, T., *ASHA Reports* **11**, 38-47 (1981).

[27] Titze, I. R. and Strong, W., *J. Acoust. Soc. Am.* **57**, 736-744 (1975).

[28] Titze, I. R., *J. Speech Hearing Res.* **34**, 460-472 (1991).

[29] Smith, M. and Ramig, M. O., "Neurological disorders and the voice," National Center for Voice and Speech Progress and Status Report 7, 1995, pp. 207-227.

[30] Pelorson, X., Hirschberg, A., van Hassel, R. R., and Wijnands, A. P. J., *J. Acoust. Soc. Am.* **96**, 3416-3431 (1994).

[31] Hess, M. M., Gross, M., and Herzel, H., *Otorhinolaryngologia NOVA* **4**, 307-312 (1994).

[32] Moore, G. and von Leden, H., *Folia phoniat.* **10**, 205-238 (1958).

[33] Ishizaka, K. and Flanagan, J. L., *Bell Syst. Techn. J.* **51**, 1233-1268 (1972).

[34] Smith, M. E, Berke, G. S., Gerrat, B. R., and Kreimann, J., *J. Speech Hearing Res.* **35**, 545-554 (1992).

[35] Titze, I. R. and Talkin, D. T., *J. Acoust. Soc. Am.* **66**, 60-74 (1979).

[36] Alipour-Haghighi, F. and Titze, I. R., *J. Acoust. Soc. Am.* **90**, 1326-1331 (1991).

[37] Herzel, H., Steinecke, I., Mende, W., and Wermke, K. "Chaos and bifurcations during voiced speech," in *Complexity, Chaos and Biological Evolution* edited by E. Mosekilde and L. Mosekilde (Plenum Press, New York, 1991), pp. 41-50.

[38] Herzel, H. "Non-Linear Dynamics of Voiced Speech" in *Nonlinear Dynamics: New Theoretical and Applied Results* edited by J. Awrejcewicz, Akademie Verlag, Berlin 1995, pp. 256-274.

[39] Herzel, H. and Knudsen, C., *Nonlinear Dynamics* **7**, 53-64 (1995).

CHAOTIC DATA ANALYSIS OF HEART R-R INTERVAL EKG DATA

Ted W. Frison
Randle, Inc.
P.O. Box 1010
Great Falls, VA 22066
ted@chaotic.com

C. K. Peng
Ary Goldberger
Cardiovascular Division
Harvard Medical School
Beth Israel Hospital
Boston, MA 02215
peng@astro.bih.harvard.edu
ary@astro.bih.harvard.edu

Richard A. Katz

Abstract

Cardiac beat-to-beat interval data is analyzed with a chaotic data analysis toolkit. The embedding dimension of ten data sets from healthy subjects is 7 or at most 8. Ten of the eleven pathological data sets have an embedding dimension of 9 or greater. Statistically, the first local minimum of average mutual information for healthy hearts is larger than the pathological cases. But, there is a large standard deviation for this metric that blurs the distinction between the healthy and pathological data.

Introduction

Using the tools of chaotic data analysis[1], we characterize ten data sets of healthy heart R-R intervals and 11 data sets from pathological subjects. These data are derived from thresholded EKG data collected over long periods of time with ambulatory monitors.

The general procedure is to use a single time series (in this case, the R-R interval data) to form a reconstructed attractor[2,3,4]. The attractor is a geometric time evolving object that is formed by taking a time series,

$$x(n),\ x(n+1),\ x(n+2)\ \ldots,$$

and forming vectors,

$$y(n) = [x(n),\ x(n+T),\ \ldots,\ X(n+(d_E-1)T],$$

where T is a time delay and d_E (the embedding dimension) is the number of geometric coordinates needed to represent the system. The geometric properties of the attractor and information loss rates characterize the underlying system.

Average mutual information provides a recommendation for the time delay and shows how information is lost over time.

False nearest neighbor calculations determine the global embedding dimension and give some sense of the amount of "noise" on the signals. These calculations are confirmed with *local embedding dimension* estimates that may indicate that for finite horizons, the dynamics may be represented by a local dimension, $d_L <= d_E$.

Finally, the *Lyapunov exponents* determine how fast trajectories on the attractor are separating, providing a measure of the forecasting horizon. If one or more exponents are positive (there are d_E exponents), the system is chaotic[5].

One problem that occurred in all the data is quantization rounding. The recorded R-R intervals are rounded to only a few significant digits, which causes many of the values to be identical. When the vectors are formed, the attractor occupies discrete points (a sort of hyperspace Bucky-ball), making it difficult to separate true close neighbors from neighbors that are close because of the rounding.

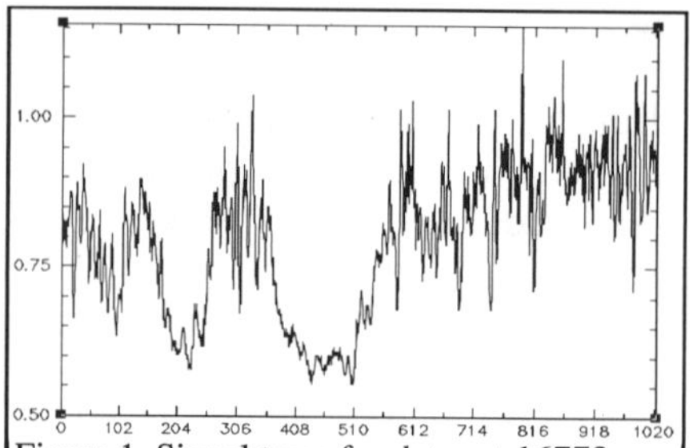

Figure 1. Signal trace for data set 16773.

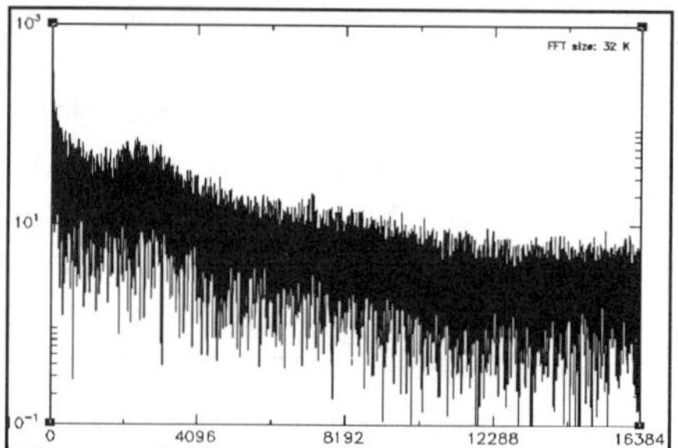

Figure 2. Power spectrum for data set 16773.

Data Analysis

We present the details of the results for only one representative data set within each category. With one exception, all the data within each category is similar, but have varying amounts of contamination. The exception is that one of the pathological cases has characteristics that are typical of the healthy cases.

Healthy Subjects

Figures 1 and 2 show a typical signal trace and fourier power spectrum. The power spectrum is unrevealing beyond the observation that it has a 1/f-like broad-band spectrum.

Figure 3 shows the average mutual information for these data and the first minimum at T = 6. The recommendation for attractor reconstruction is to use the first minimum of the average mutual information to select the time delay because this is the point where each component of the vector adds new information, but is not yet so independent that the vector components appear to be independent with respect to each other [6].

Average mutual information determines how much information (in bits) measurements at a time have relative to measurements at some other time. For N observations it is defined as

$$I(T) = \sum_{n=1}^{N} P\left(x(n),x(n+T)\right) \log_2\left[\frac{P\left(x(n),x(n+T)\right)}{P\left(x(n)\right)P\left(x(n+T)\right)}\right]$$

where P(x(n)) and P(x(n+T)) are the distributions of x(n) and x(n+T) and

P(x(n),x(n+T)) is the joint distribution. If the measurements are independent, each term vanishes because P(a,b) = P(a)P(b). In the fully independent case, I(T) = 0 for all T.

The first minimum in figure 3 is reasonably well defined, so it is probably correct. That is, using more data would not significantly change the estimate of T_{opt}. The first local minimum for the healthy subjects ranged from 2 to 9 with the predominate values being 6 or 7. The average value is 6.0 samples.

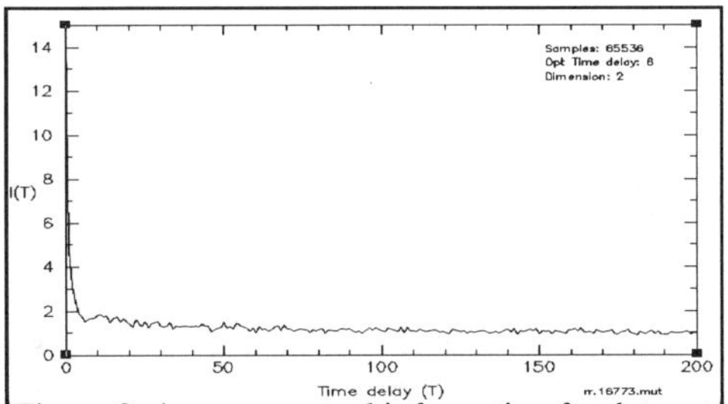

Figure 3. Average mutual information for data set 16773 showing the first minimum at T = 6.

Figure 3 also shows that there are no predominant periodic events. If there were, there would be pronounced peaks in the plots. These peaks would correspond to peaks in the power spectrum, which are also absent.

Using the prescription for T_{opt} provided by the average mutual information, the global embedding calculations come next. The method of false nearest neighbors[7] uses the geometric basis for the reconstruction technique. If the dimension is too low, points are projected down into the lower dimension causing ambiguities in the trajectories because they cross. As the dimension is increased, attractors "unfold." Trajectories that are nearby in dimension *d* may move apart in dimension *d+1*. These are "false" neighbors in *d* and the method measures the percentage of false neighbors as *d* increases. Points that are close in *d* are tallied, and the number of these points that become widely separated in *d+1* are calculated.

The nearest neighbor to y(n) is

$$y^{NN}(n) = (x^{NN}(n), x^{NN}(N+T), \ldots, x^{NN}(n+(d-1)T)).$$

The neighbor is false if

$$\left[\frac{R^2_{d+1}(n,r)}{R^2_d(n,r)}\right] = \frac{\left|x(n+Td) - x^{(r)}(n+Td)\right|}{R_d(n,r)} > R_{toll}$$

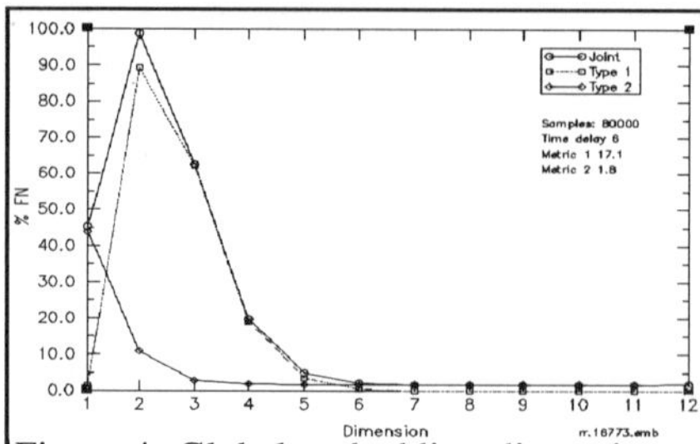

Figure 4. Global embedding dimension for 16773.

where R_d is the distance between a vector and its nearest neighbor and R_{tol} is a separation criteria.

But, a nearest neighbor may not necessarily be "close," especially in higher dimensions. In higher dimensions, volume occupied by the signal will decrease and the distance to neighbors will increase. If the nearest neighbor to a point is false, then the distance in d+1 will be ~ $2R_a$. So, a second requirement is

$$\frac{R_{d+1}(n)}{R_A} > R_{tol2}$$

where

$$R_A^2 = \frac{1}{N} \sum_{n-1}^{N} \left[x(n) - \overline{x}\right]^2$$

A nearest neighbor is false if either test fails.

For a noiseless signal, the number of false neighbors becomes zero when the minimum embedding dimension d_E is reached. A noisy signal has some kind of minimum, but does not become zero.

Figure 4 shows the results for data set 16420. Usually, a criteria of 1 or 1/2 percent joint false nearest neighbors is used as the threshold for a decision on the dimension. If a system is noisy, then the dimension with the lowest joint false nearest neighbor count is taken as the global embedding dimension.

Several features are notable. First, these are not "clean" data. The fact that the false neighbor counts have a floor rather than descend to zero is an indication of very high dimension "stuff." What this contaminant is, we cannot say at this time. It may be electrical noise in the data collection equipment or even quasi-random events within the biological system itself. Finally, the obviously low counts in dimension 1 are caused by the quantization. The toolkit module that does these calculations compensates for quantization error and they do not affect the results in dimensions above one or two.

A consistent estimate of d_E for all the healthy cases is 7 or 8. The extremely cautious might select 8 for some data sets and risk-takers might select 6, but the average value is 7.3.

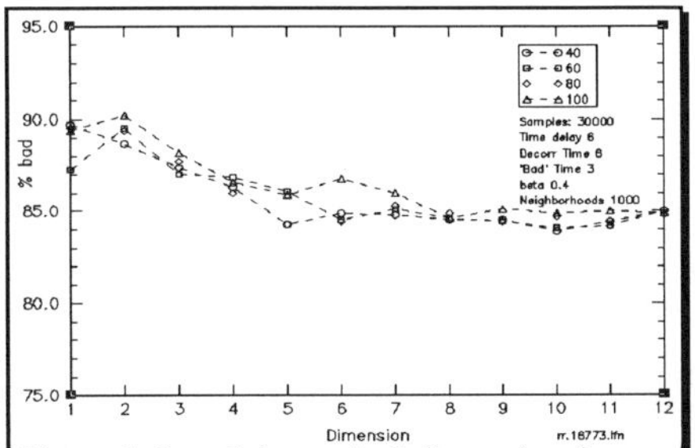

Figure 5. Local dynamical dimension for 16773.

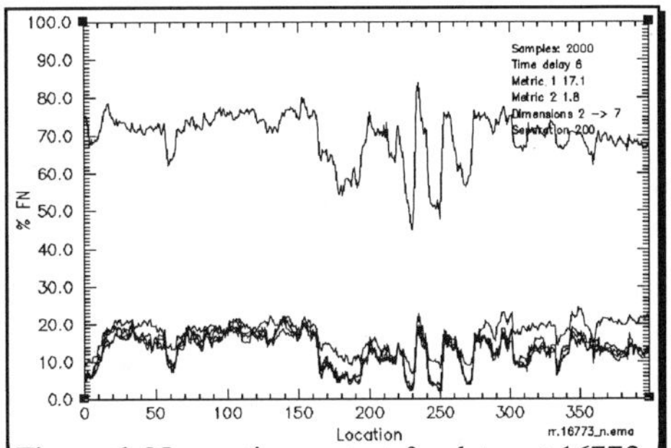

Figure 6. Nonstationary test for data set 16773.

However, short time periods may be described in dimensions less than d_E[8]. The local dimension, d_L, describes the aggregate behavior for all finite regions on the attractor. If $d_L < d_E$, then models of the system can be simplified.

If trajectories that are close at some point separate faster than is expected, it is possible that the paths cross because the dimension is too low. To test whether trajectories remain close for reasonable periods, neighborhoods are selected and, for all pairs of neighbors in a neighborhood, the percentage for which the trajectories remain "close" is computed. "Close" is arbitrarily defined as some size relative to the overall size of the attractor. At some embedding dimension, the percentage of "bad" predictions becomes independent of the dimension and N_b. This is d_L.

Figure 5 shows the results of the local dynamical dimension calculations for the representative data set. Certainly by dimension 8, the failures are independent of the number of neighbors and the trial dimension. So, in this case, $d_L = d_E$.

Computation of valid Lyapunov exponents is difficult. The sum of all of the exponents must be negative in order to pull the trajectories back to the attractor. If not, the implication is that the trajectories do not remain within the finite attractor. Often, the computations produce a positive sum, which cannot be.

There are probably several reasons for this. First, there are indications that the signals are not stationary. The data collections ran for several hours (typically more than 10 hours). A moving average of the embedding dimension calculations shows that the false neighbor counts vary significantly across the data -- figure 6. Further, the minimum of the counts changes dimension over time. The interpretation could be that

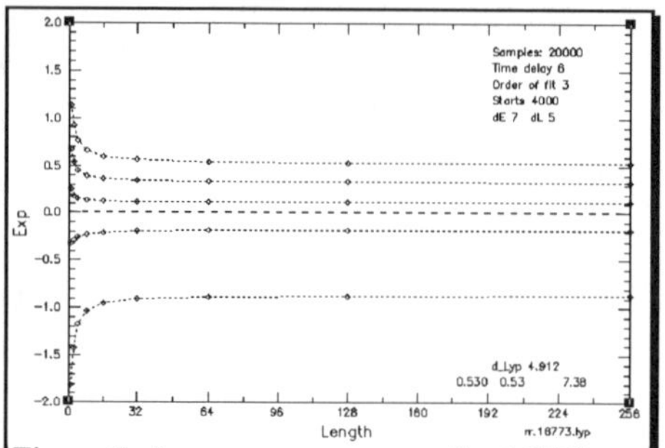

Figure 7. Lyapunov exponents for 16773.

there is a parameter change that is masked in the synoptic calculations presented in figure 4. Additionally, the quanta nature of the data makes it difficult, if not impossible, to form a map of the trajectories.

Figure 7 shows the Lyapunov exponents for the region between locations 50 to 160 in figure 6 (samples 10,000 to 32,000). This region appears to be relatively stationary.

The largest exponent has a value of .536. So, the growth rate between nearby trajectories is $e^{\tau\lambda_1} = e^{.536}$. Because the largest exponent is positive, the system can be categorized as chaotic. For trajectories that are initially separated by 1% of the mean attractor size, the magnitude of the largest exponent means that, on average, by 2.3 samples (beats) the trajectories will have separated to 50% of the mean attractor size. This is one metric for the inherent predictability of chaotic signals.

The results for the healthy subjects are summarized in Table 1.

DSN	T_{opt}	d_E Daring	d_E Best	d_E Cautious	%FNN @ Floor
16265	7	7	8	8	1.3%
16273	2	7	8	8	2.3%
16420	2	7	7	9	.5%
16483	9	6	7	8	2.0%
16539	7	6	7	7	8.4%
16773	6	6	7	7	1.6%
16786	2	7	8	9	0.4%
17052	7	7	7	8	2.6%
17453	6	6	7	7	6.0%
C4	12	7	7	8	0.96

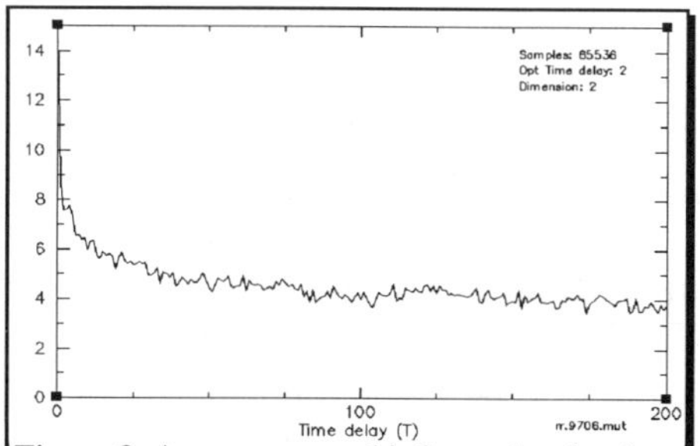

Figure 8. Average mutual information for data set 9706 showing the first minimum at T = 2.

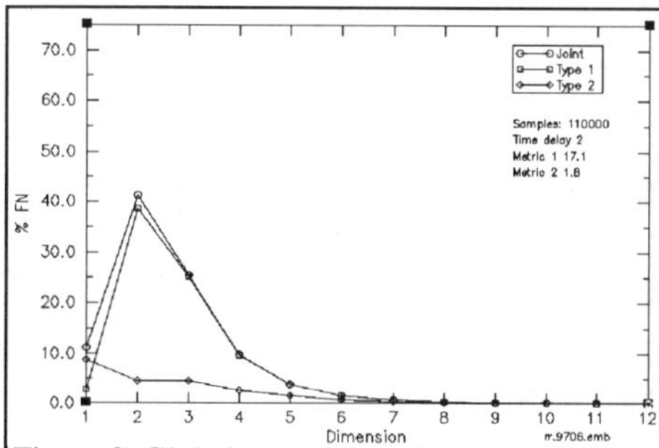

Figure 9. Global embedding dimension for 9706.

Pathological Cases

Figures 8, 9, and 10 show the average mutual information, false nearest neighbor calculations, and the local embedding dimension for one of the representative pathological cases.

As with the healthy cases, the first minimum of average mutual information at T = 2 is comfortably deep and there is no evidence of periodic behavior.

The severe quantization effects are noticeable in the global embedding dimension calculations. Even so, the counts have reached a near zero plateau by dimension 9.

The discernable differences from the healthy heart EKG data are a shorter average T_{opt}, which is 3.3 for the unhealthy cases and 5.3 for the healthy cases. While the average is shorter for the unhealthy cases, this is not, by itself, a reliable discriminator for a specific case.

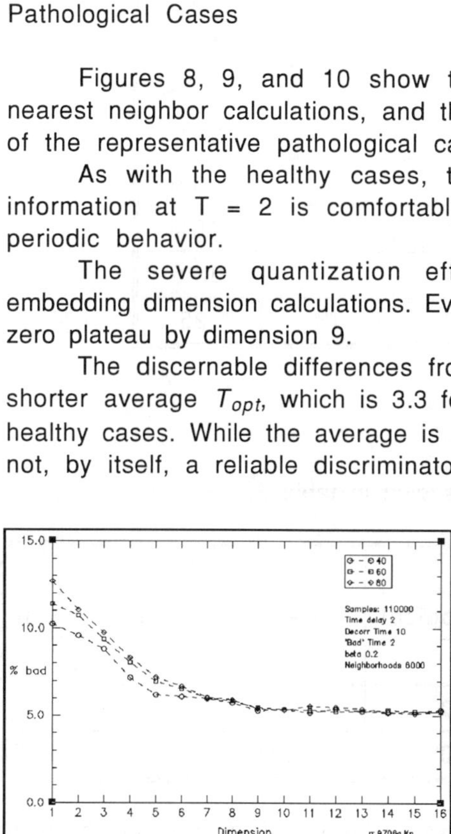

Figure 10. Local dynamical dimension for 9706.

However, the global embedding dimensions for the two conditions are noticeably different. The average best estimate for the unhealthy cases is 10.3 while the average best estimate for the healthy cases is 7.3. Table 3 compares the statistics for the two sets.

The one anomaly is data set 9377. It has a long time delay, and an embedding dimension of 8. These characteristics are typical of the healthy cases and are the extreme values for both these parameters within the unhealthy data.

Table 2 summarizes the results for the unhealthy EKG data.

DSN	T_{opt}	d_E Daring	d_E Best	d_E Cautious	%FNN @ Floor
8519	2	11	12	12	5.8%
8679	3	10	10	10	0.35%
9049	2	9	10	10	0.57%
9377	9	7	8	8	1.5%
9435	2	10	10	11	4.1%
9643	3	> 13	> 14	> 15	10.3%
9674	3	8	9	10	0.6%
9706	2	9	9	10	.14%
9723	3	11	12	12	6.0%
9778	2	8	9	10	0.25
9837	5	9	10	11	.25

Summary

A comparison of the two data sets is shown in table 3:

	T_{opt}		d_E	
	Healthy	Pathological	Healthy	Pathological
avg	6.0	3.27	7.3	10.3
std dev	3.1	2.0	.46	1.7
var	9.6	4.0	.21	2.7

The immediate result is that analysis of the EKG data should be done in the appropriate dimension using the time delays provided by average mutual information. That is, return maps that normally plot x(t) vs. x(t+1), ought to be constructed using x(t), $x(t+T_{opt})$, $x(t+2^*T_{opt})$ ... $x(t+(d_E-1)^*T_{opt})$. One might try a simple two-dimension return map of x(t) vs $x(t+T_{opt})$, but given the evidence of a higher dynamic (which leads to ambiguities in the lower dimension space), the results should be viewed with caution.

Future Studies

The absolute value of the residual false neighbors at the embedding dimension is, on average, not large. The healthy cases had about 33% higher residuals than the pathological cases (after excluding the pathological case that is suspected to be anomalous). The causes for this bear investigation.

More interesting, however, is the test for nonstationary behavior. We plan to investigate the use of this algorithm to assist in detrending data and identify transient behavior.

References

[1] H. D. I. Abarbanel, R. Brown, J. J. Sidorowich, and L. S. Tsimring, "The Analysis of Observed Chaotic Data in Physical Systems, *Reviews of Modern Physics*, **65** 4 pp. 1331 - 1392, October 1993.

[2] R. Mañè, "On the Dimension of the Compact Invariant Sets of Certain Nonlinear Maps," in *Dynamical Systems and Turbulence, Warwick 1980,* eds. D. Rand and L.S. Young, *Lecture Notes in Mathematics* **898**, 230, (Springer, Berlin, 1981).

[3] F. Takens, "Detecting Strange Attractors in Turbulence," in *Dynamical Systems and Turbulence, Warwick 1980,* eds. D. Rand and L.S. Young, *Lecture Notes in Mathematics* **898**, 366, (Springer, Berlin, 1981).

[4] Packard, N. H., J. P. Crutchfield, J. D. Farmer, and R. S. Shaw, "Geometry from a time series," *Physical Review Letters*, **45**(9), pp. 712-716, Sept. 1, 1980.

[5] Ruelle, D., Proc R. Soc. A **427** (1990) 241.

[6] A. M. Fraser and H. L. Swinney. "Independent coordinates for strange attractors from mutual information," *Phys. Rev. A*, 33, pp. 1134-1140, Feb. 1986.

[7] M. B. Kennel, R. Brown, and H. D. I. Abarbanel, "Determining embedding dimension for phase-space reconstruction using a geometrical construction," *Phy. Rev. A* **45** pp. 3403-3411, 15 March 1992.

[8] H. D. I. Abarbanel, and M.B. Kennel, "Local False Nearest Neighbors and Dynamical Dimensions from Observed Chaotic Data", Phys. Rev. E., **47**, pp. 3057-3068 (1993).

[9] H. D. I. Abarbanel, R. Brown, and Matthew Kennel, "Local Lyapunov Exponents Computed from Observed Data",*Journal of Nonlinear Science*, **2** pp. 343-365 (Sept. 1992).

Correlations in Heart Beat Data as Quantitative Characterization of Heart Pathology

J. Ulbikas[*], A. Čenys[*], D. Žemaitytė[⊕] and G. Varoneckas[⊕]

[*]*Semiconductor Physics Institute,*
Goštauto 11, 2600 Vilnius, Lithuania
[⊕]*Institute of Psychophysiology and Rehabilitation,*
Vydūno 4, 5720 Palanga, Lithuania

Abstract. Correlation between heart pathology and statistical properties of heart beat data has been studied. It is shown that heart beat data has different scaling behavior for healthy and disease cases. Possibilities to develop new monitoring technique based on the permanent control of the correlations in heart beat data are discussed.

INTRODUCTION

Although everyday experience tells us that the heart electrical activity is periodic, deeper, quantitative analysis of the time series reveals a number of irregularities in it. Constructed by Babloyantz and Destexhe (1) power spectrum from electrocardiogram (ECG) data shows a broad band structure typical for chaotic as well as for stochastic processes. Reconstructions of the ECG or heart rate attractor (2) and calculations of the correlation dimension using the Grassberger - Procaccia correlation function (3) give some evidence of the nonlinear processes involved in heart functioning. However obtained values for correlation dimension from ECG (1) (ν in the range 3.5 - 5.2) and heart rate data (1) (ν about 5.9) and huge error bars (up to 0.4) do not allow reliably identify deterministic chaotic dynamics in heart beat data. These results strongly suggest that investigated system is essentially influenced by random biological and environmental noise. Therefore many of chaotic parameters can serve only as system complexity indicators, but not as a quantitative parameters for system state description.

Problems related with data requirements for reliable estimation of chaotic parameters are discussed in many papers devoted to the investigations of the biological systems (1,2,4,5). Moreover, recently it was shown that the standard chaotic dynamics algorithms can give spurious results; i.e. they can indicate the presence of chaos in the stochastic systems. Surrogate data techniques have been developed to distinguish nonlinear chaotic systems from linearly correlated noise (6). It is obvious that now every experimental data should be checked for nonlinearity before estimating chaotic parameters.

Due to the above problems up to now there is little or no use of the chaotic parameters in practical physiology. It is even not clear if sophisticated chaotic parameters such as dimensions or Lyapunov exponents are more useful for practical purposes than simple Poincare maps visualizing short range correlations. Recently it was shown that also long range correlations in the heart beat are different for healthy and disease heart (7). Our study is devoted to comparison of the results obtained applying different methods. We have analyzed sensitivity of different parameters to detect differences in four groups of adult patients data: 1) healthy, 2) well trained sportsman, 3) patients with clinical evidence of ischemic disease and 4) patients with ischemic disease after miocardia arrest.

ALGORITHMS

In our analysis we used three different methods to estimate correlations in heart beat data. We have calculated correlation dimension using modified Grassberger-Proccacia algorithm (3) suggested in (8) to eliminate random noise influence. In the standard algorithm saturating value of the correlation integral scaling exponent on the dimension of the reconstructed phase space is defined as a correlation dimension. In our experiments as in many cases of the biological systems non-saturating behavior of the scaling exponent was observed. The modified algorithm (3), based on the analytical results, suggests following simple graphical procedure to estimate correlation dimension of the noise free system in the nonsaturating case. The linear part of the plot of correlation exponent $\nu(M)$ vs embedding dimension M in the range of sufficiently large values of M should be detected. This line is extended until it intersects the line $\nu(M)=M$. The needed correlation dimension ν of the underlying attractor is determined by this intersection point.

To estimate short range correlations we have used standard Poincare maps. For the quantitative characterization of the long range correlations in heart beat data the mean fluctuation function $F(N) = \langle (x_n - x_{n-N})^2 \rangle$ introduced in (7) have been calculated. Function $F(n)$ by its definition quantifies the magnitude of the fluctuations over different time scales.

PHYSIOLOGICAL IMPORTANCE OF THE HEART RATE VARIABILITY

An opinion that reduced heart rate (HR) autonomic control, otherwise, reduced HR variability (HRV), is the main independent indicator of negative prognoses in ischemic heart disease (IHD) patients is widely accepted by scientific community (9). Reduced HR autonomic control is seen as predisposing to development of HR disturbances, particularly dangerous ventricular premature beats (VPB's) leading to ventricular fibrillation (VF) and sudden death. There are some evidence, that loss of physiologic 'sinus HRV', appears to be a common feature of pathologic dynamics, e.g., cardiac arrest follows a progression, often involving abrupt changes, from apparently chaotic dynamics of the normal 'sinus' heart rate to the pathologic preconditions and loss of HRV of the dying heart.

The classification of rhythmogram (RG), e.g., recording of successive RR intervals, according to periodical structure of HRV have shown two main ways towards reduced HRV: a) reduced HRV on the basis of increased HR frequency and b) reduced HRV on the bases of bradycardia, particularly severe bradycardia. The first situation develops through an increase of sympathetic control, followed by an increase of medium level HR periodicity related to blood pressure variations, to reduction of HRV due to prevalence of humoral HR control. The second situation of HRV reduction with severe bradycardia, might be related to maximal parasympathetic control paralleled by reduced sympathetic influence, or reduced HRV might be the result of reduced autonomic HR control of both, sympathetic and parasympathetic control together, leading to maximal influence of humoral control. Thus, it might be the result of reduction of either, sympathetic or parasympathetic influence alone, or both together, e.g., while HRV reduction appears on the basis of expressed bradycardia, tachycardia or normal HR frequency, correspondingly.

RESULTS

The time series of the 2^{11} up to 2^{15} RR intervals have been used in our analysis. First of all we have checked experimental data for nonlinearity, using surrogate data technique developed in (6). An ensemble of so called "surrogate" data sets was created from the original data. For this purpose a Fourier transform of the original data was performed, and after randomization of the phases, invert Fourier transformation was applied. Estimation of the correlation dimension using the same modified algorithm for both original and surrogate data sets has shown no significant differences. Due to this result calculations of the correlation dimension in our experimental data can not be related with the nonlinear dynamics. Moreover estimated values of the

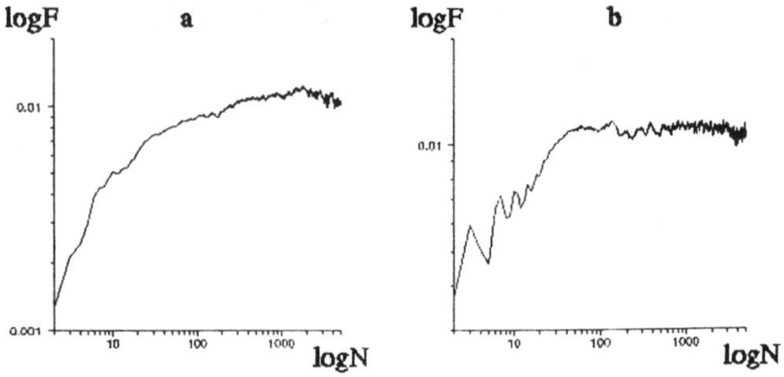

FIGURE 1. log-log plot of F(n) for the data obtained from healthy adult subjects: a - normal heart activity, b - well trained sportsman heart activity

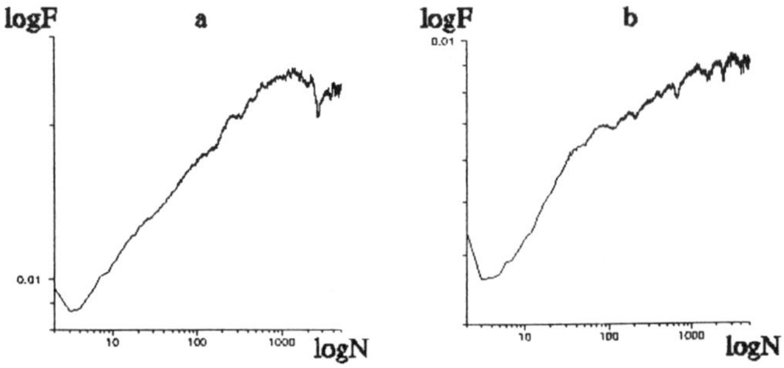

FIGURE 2. log-log plot of *F(n)* for the data obtained from patients with heart disease: a) - ischemic heart disease, b) - ischemic heart disease after miocardia arrest.

correlation dimension was in the range ν = 5.6 - 5.9 which is close to mentioned in (1), but without significant differences between different groups of patients.

On the contrary analysis of the long range correlations using fluctuations function *F(n)* has shown differences over physiologically relevant time scales (200 - 3000 heart beats) in different investigated groups (typical results for all four groups are presented in Fig. 1 and Fig. 2). First of all in the case of ischemic disease linear part in the log-log plot corresponding to the scale invariant power law behavior has a same exponent in the range from very small to very large time scales. For all other cases change of behavior at about 100 heart beats is clearly seen. Scaling exponent α estimated from the liner fit in the range of 200-3000 heart beats also are different for the different groups ranging from 0 for well trained patients to 0.2 for IHD patients.

An analysis of short-range correlation of successive RR intervals and stroke volume (SV) values was performed using Poincare maps in IHD

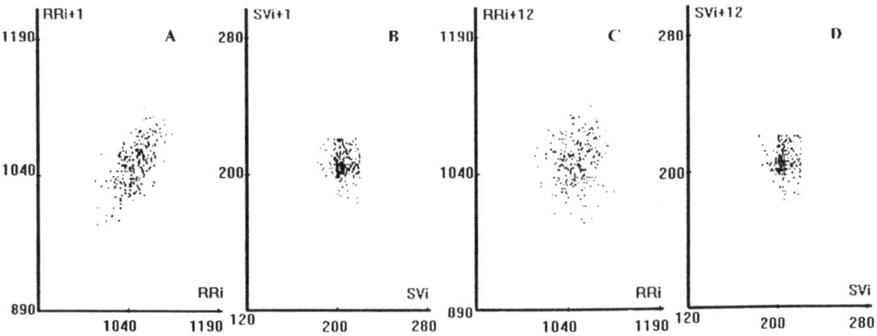

FIGURE3. Poincare maps of successive RR intervals and stroke volume records in IHD patient with normal 'sinus' rhythm: correlation of RR intervals (A and C) and stroke volume (B, D); without (RRi/RRi+1) delay (A, B) and with (RRi/RRi+n) delay (C, D) of successive values.

patients having normal sinus rhythm or some dysrhythmias: atrial fibrillation (AF), premature beats (PB's), sino-atrial (SA) or atrioventricular (AV) blocks. Fig. 3 demonstrates phase-space diagrams of successive RR intervals (A, C) and stroke volume values (B, D) of IHD patient with normal sinus rhythm, demonstrating prevalence of parasympathetic control (respiratory arrhythmia). Correlation between of successive RR intervals (RRi/RRi+1) shown ellipse-form cloud of points, which is dependent on the HR variability level and the period of the dominating HR waves. After inclusion of some delay (n=12) this cloud becomes a random distribution of the points. SV distribution did not change its pattern, while changing delay. Reduction of HR variability is followed by the changes in the shape of diagrams tending to the pattern of point. On the other hand, some HR disturbances shows particular specific patterns of diagrams (Fig. 4).

Atrial fibrillation (AF), as it is seen on the Fig. 4A, is followed by random distribution (RR intervals) of the points, tending to concentrate at the level of highest HR frequency, while SV values are more concentrated than RR intervals. Complicated recording of sinus rhythm with AV blocks and separate ventricular PB's (Fig.4, panels C,D) demonstrated quadrangular distribution of RR intervals with the point at center, corresponding to reduced HRV of sinus rhythm, related to PB's and postextrasystolic cycles, and the cycles with AV blocks concentrating on the upper part of the Fig.4C. SV values (D) distribution shows random feature.

Very specific pattern demonstrates a map of RR intervals, while ventricular PB's (parasystolia) with episodes of allorhythmias is present (Fig.4, panel E). It has three separate clouds of elliptic form, related to correlation between of normal 'sinus' rhythm, extrasystolic and post-extrasystolic intervals, while at the graphic with delay (RRi/RRi+n) pattern becomes crucifix-like (Fig.4G). Stroke volume values distribution (Fig.4F and 4H) demonstrates

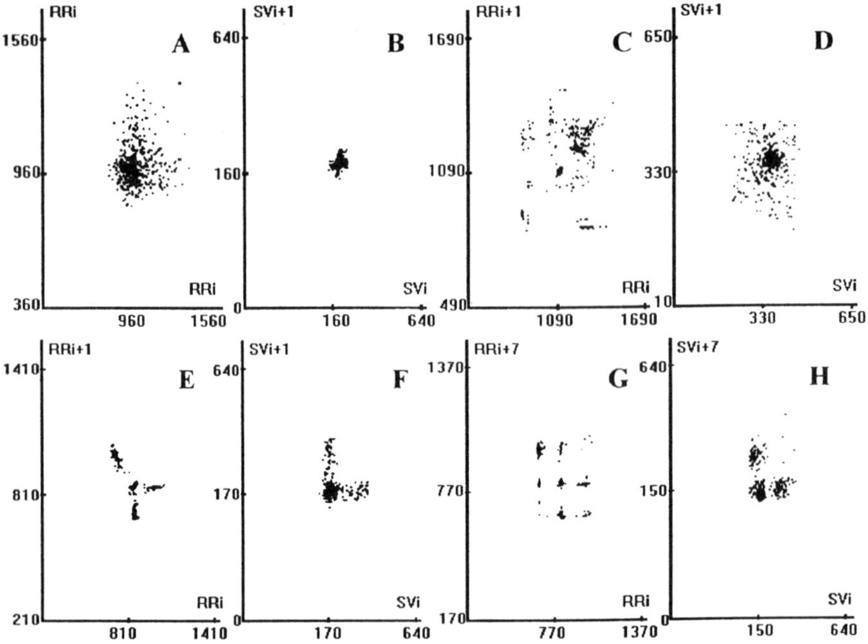

FIGURE 4. Poincare maps of successive RR intervals and stroke volume record in IHD patients with atrial fibrillation (A,B), combined sinus rhythm, atrioventricular blocks and premature beats record (C,D), sinus rhythm with episodes of parasystolia (E,F): correlation without delay (RRi/RRi+1) of RR intervals (A, C, E) and stroke volume values (B,D,F) and with delay (G, H).

three levels of hemodynamics, corresponding to 'sinus' rhythm, extrasystolic and postextrasystolic intervals.

The results presented on Fig.3 and Fig.4 demonstrate big variety of the different patterns in RR intervals and SV Poincare maps depending on the heart pathology. These differences can be used as very simple and convenient measure for the preliminary indication of the heart diseases.

CONCLUSIONS

Our investigations on heart rate data has shown, that for physiologists practical purposes correlations analysis can be very useful. First of all long-range correlations in successive RR intervals can serve as characteristic for heart pathology detection. Scaling exponent in plot log-log F(n) was obtained markedly different for the healthy and diseased states, i.e. shows increase in it's absolute value in relation to development of cardiac pathology.

On the other hand simple Poincare mapings also shows sufficient sensitivity to the heart pathology development and can serve as basic idea for developing new monitoring techniques in medicine practice. Short-range correlation pattern of 'sinus rhythm' depends on HR variability level and tends to the point with reduction of HRV. Some HR disturbances, particularly PB's, demonstrated specific patterns on two-dimensional phase-space trajectories (or Poincare maps), which might be effective for evaluating of hemodynamic efficiency of PB's.

ACKNOWLEDGMENTS

This work was supported in part by EEC under a PECO associated contract of the Universite Libre de Bruxelles (ULB) research project, Contract No. ERBCIPDCT940011 within the STD-2 EC Programme.

REFERENCES

1.Babloyantz, A., Destexhe, A., *Biol. Cybern.* **58**, 203-211 (1988).

2. Bruce J.West, *Fractal physiology and chaos in medicine*, Singapore: World Scientific, 1990, pp. 208-222.

3.Grassberger, P., Procaccia, I., *Phys.Rev.Lett.* **50**, 346-352 (1983).

4. Layne, S.P., Mayer-Kress, G., Holzfuss, *Dimensions and entropies in chaotic Systems*, Berlin: Springer-Verlag, (1986) pp. 246-256.

5.Glass, L., Hunter, P., *Physica D* **43**, 1-16 (1990).

6.Theiler, J., Eubank, S., Longtin, A., Galdrikian, B., Farmer, J.D., "Testing for nonlinearity in time series: the method of surrogate data," in *Proc. of the IUTAM Symposium and NATO Advanced Research Workshop on interpretation of Time Series for Nonlinear Mechanical Systems* University of Warwick, Coventry, England, 26-30 August 1991, pp. 1-27.

7.Peng, C.K., Mietus, J., Hausdorff, J.M., Havlin, S., Stanley, H.E., Goldberger, A.L., *Phys.Rev.Lett.* **70**, 1343-1346 (1993).

8.Čenys, A., Lasienė, G., Pyragas, K., *Lith.Phys.Coll.* **30** 297- (1990).

9. Schwartz P.J., La Rovere M.T, Vanoli E., *Circulation* **85** 177-191 (1993).

Applying Non-Linear Dynamics to Atrial Appendage Flow Data to Understand and Characterize Atrial Arrhythmia

Shalabh Chandra*, Richard A Grimm*, Richard Katz$^{\Psi}$, James D. Thomas*

*Cleveland Clinic Foundation, Cleveland, OH 44195

$^{\Psi}$Naval Undersea Warfare Center, New London, CT 06320

ABSTRACT

The aim of this study was to better understand and characterize left atrial appendage flow in atrial fibrillation. Atrial fibrillation and flutter are the most common cardiac arrhythmias affecting 15% of the older population. The pulsed Doppler velocity profile data was recorded from the left atrial appendage of patients using transesophageal echocardiography. The data was analyzed using Fourier analysis and nonlinear dynamical tools. Fourier analysis showed that appendage mechanical frequency (f_f) for patients in sinus rhythm was always lower (around 1 Hz) than that in atrial fibrillation (5 to 8 Hz). Among patients with atrial fibrillation spectral power below f_f was significantly different suggesting variability within this group of patients. Results that suggested the presence of nonlinear dynamics were: a) the existence of two arbitrary peak frequencies f_1, f_2, and other peak frequencies as linear combinations thereof ($mf_1 \pm nf_2$), and b) the similarity between the spectrum of patient data and that obtained using the Lorenz equation. Nonlinear analysis tools, including Phase plots and differential radial plots, were also generated from the velocity data using a delay of 10. In the phase plots, some patients displayed a torus-like structure, while others had a more random-like pattern. In the differential radial plots, the first set of patients (with torus-like phase plots) showed fewer values crossing an arbitrary threshold of 10 than did the second set (8 vs. 27 in one typical example). Furthermore, the outcome of cardioversion was different for these two set of patients. In conclusion, Fourier analysis helped to a) differentiate between sinus rhythm and atrial fibrillation, b) understand the characteristics of the wide range of atrial fibrillation patients, and c) provide hints that atrial fibrillation could be a nonlinear process. Nonlinear dynamical tools helped to further characterize and sub-classify atrial fibrillation. These insights into atrial fibrillation could, eventually, lead to improved risk stratification of patients with regard to stroke risk and long term outcome.

INTRODUCTION

Atrial fibrillation and flutter are among the most common of all cardiac arrhythmias, affecting 15% or more of the population above 65 years of age. In patients with atrial arrhythmias, the mechanical function of the left atrial appendage appears to play a pivotal role in thrombogenesis and systemic embolization. The left atrial appendage, an accessory chamber that is an actively contractile structure, empties into the left atrium and has a propensity to house thrombi within its cone-shaped cavity. Figure 1 illustrates the left atrial appendage with a thrombus.

Although attempts have been made by several investigators[1,2,3] to understand and characterize the mechanical function of the left atrial appendage in atrial fibrillation using Doppler echocardiographic techniques, these attempts have merely been qualitative. A more comprehensive characterization of left atrial appendage function has been difficult largely due to the complexity and temporal variability of these flow profiles.

In this paper we present more sophisticated analyses of appendage velocity patterns. Fourier transformation is used to differentiate between atrial

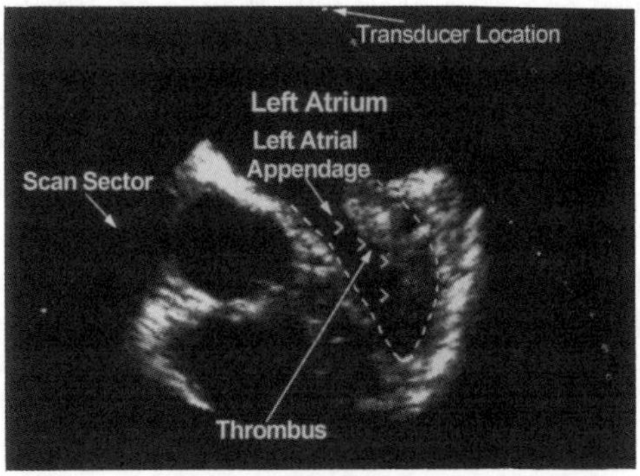

Figure 1. Left atrial appendage with a thrombus.

flutter, fibrillation and sinus rhythm. Additionally, the apparently "irregularly irregular" flow pattern of the left atrial appendage in atrial fibrillation suggests that the application of non-linear dynamical tools may be useful in characterizing the signal. Theories of deterministic chaos have previously been applied to mammalian systems, specifically to arrhythmias. This is because cardiac rhythms undergo complex frequency dependent patterns of electrical excitation. Atrial fibrillation has classically been described as a “chaotic” rhythm, however, this assumption has not been previously confirmed or rejected. We addressed this problem by applying Fourier transformation and non-linear dynamical tools to the left atrial appendage flow[1] obtained using Doppler echocardiography.

METHODS

Data Acquisition

The left atrial appendage flow data was recorded using transesophageal Doppler echocardiography. In order to collect the flow data using this technique, the endoscope is inserted into the esophagus of the patient with the ultrasonic transducer immediately behind the left atrium, and the imaging plane is adjusted to visualize the left atrial appendage. The pulsed Doppler echocardiographic technique used to record the velocity uses short bursts of sound energy transmitted towards the target, and the returning echoes are recorded within the appropriate time window, determined by the range of the target from the transducer. In this study, data from the mouth of the left atrial appendage was recorded on a strip chart with the y-axis as velocity and the x-axis as time. At every instant of time there is more than one velocity recorded from the sample volume, the envelope of the recorded data being the maximum velocity with time. It is the maximum velocity that we used for our analysis. These data were recorded from patients in atrial fibrillation, atrial flutter and sinus rhythm undergoing transesophageal echocardiography. Figure 2, 3a and 3b show a representative segment of the velocity vs. time patterns recorded on a strip chart recorder for a patient in sinus rhythm and two patients in atrial fibrillation respectively (for ease of reference, we will call the atrial fibrillation patients P1 and P2).

Three to five seconds of the envelope of the temporal velocity data were manually traced and digitized on a Summagraphics Bitpad II interfaced to a PC (Gateway 2000, 486/66MHz). The software used was LabVIEW 3.1 (National Instruments). The sampling interval of the digitized data was 0.5ms. Figure 3c and 3d show the manually traced and digitized velocity envelopes corresponding to the flow shown in Figure 3a and 3b. The digitized data was analyzed using Fourier transform and non-linear techniques.

[1] The actual strip chart recording used for the analysis is velocity vs. time. Since flow is related to the velocity, in the medical field it is also referred to as the flow data.

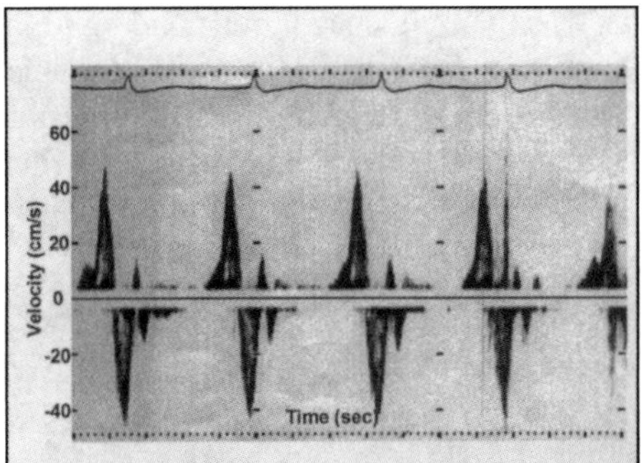

Figure 2. Atrial Appendage Flow for Sinus Rhythm (normal)

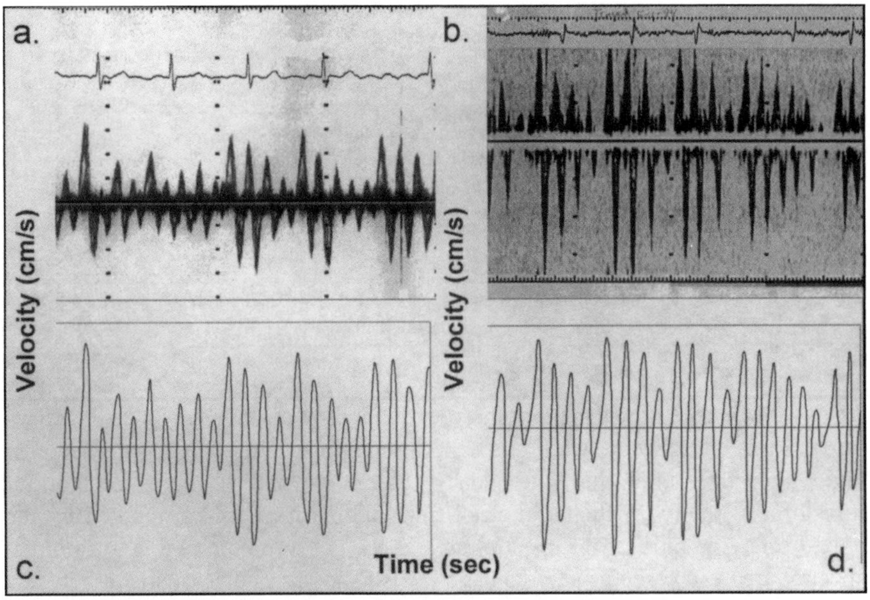

Figure 3. Atrial Appendage Flow for two patients (P1 and P2) in Atrial Fibrillation, and the corresponding digitized velocity vs. time graph. a and c for P1, and b and d for P2.

Analysis

Digitized data for atrial fibrillation, atrial flutter and sinus rhythm were transformed to the frequency domain. For this analysis, we defined the *dominant frequency* f_d as the frequency with maximal spectral power and the *fundamental frequency* f_f as the frequency corresponding to the atrial mechanical cycle length. Subharmonic frequencies were frequencies below f_f while superharmonic frequencies were those above f_f. In atrial fibrillation and flutter, f_d and f_f are the same frequency, but in the case of sinus rhythm these could be different frequencies. Figure 4a shows the spectrum of sinus rhythm with f_f and f_d marked.

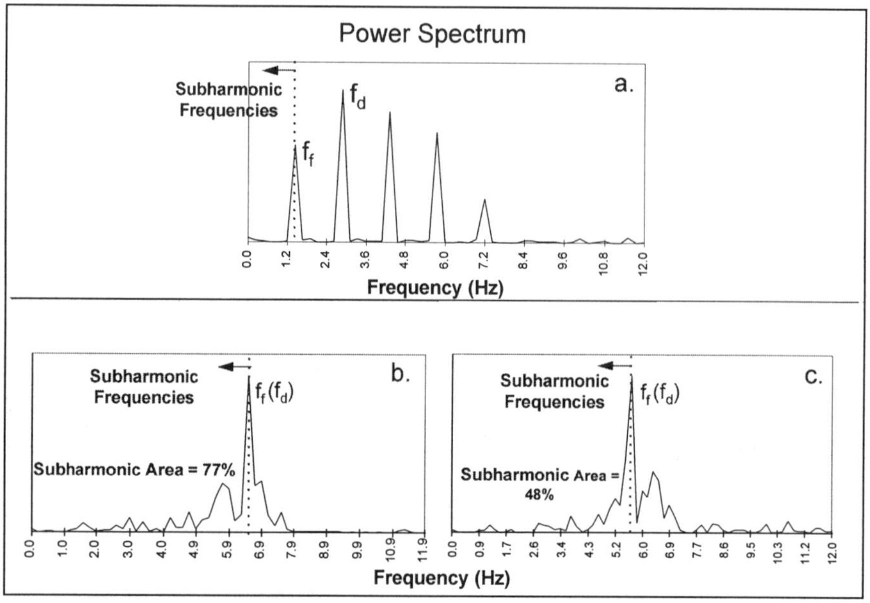

Figure 4. Power Spectrum a) Sinus Rhythm, b) P1, c) P2.

Also, we calculated the subharmonic spectral energy as the area under the curve prior to the f_f frequency, while the superharmonic spectral energy was the area under the curve after the f_f frequency. Figure 4b and 4c show the Fourier transform of the two patients in atrial fibrillation with f_f and f_d marked by the dotted line. The ratio of the spectral area in the subharmonic region to the total spectral area expressed as a percentage is also shown. The comparison between sinus rhythm, atrial fibrillation and atrial flutter was performed with respect to the general shape of the spectrum, the frequency value of f_f, and the percent subharmonic power.

In an attempt to determine if the flow data for atrial fibrillation obeys the laws of dynamical chaos we performed the following analyses: 1) the Fourier spectrums of atrial fibrillation patients were analyzed to determine if there were two basic peak frequencies (f_1 and f_2) present together with other peaks at frequencies that are harmonics or integer combinations ($mf_1 \pm nf_2$). Literature on chaos[4] in fluid dynamics suggests that if such basic frequencies f_1 and f_2 exist, then this state could be a possible route to chaos. 2) The output of the Lorenz equation was generated with parameters carefully chosen so that the output is chaotic. This output and its Fourier transform were compared to the left atrial appendage flow and its Fourier transform.

Based on the results obtained and the apparent non-linear nature of atrial fibrillation, non-linear dynamical tools were used to further analyze the data. Two measures used were: a) the Poincare phase plots, b) the Differential Radial plots. In order to generate the phase plots, the value of the first zero-crossing of the autocorrelation function of the data set was calculated, expressed as the number of data points. The values were 41 and 42 in the two patients (corresponding to about 20 msec). A fraction of this value (zero crossing value minus 30) was used to decimate the data set. Therefore, for the autocorrelation value of 41, the data set was sub-sampled by every 11th point to yield a velocity time series v_n. The main reason for decimation was to reduce the data set and make it manageable by the software used for analysis purposes. The decimated data sets were used to generate Poincare phase plots by plotting v_n vs $v_{n+\tau}$, where τ is the delay time.

Qualitative comparison was made based on the shape of the phase plots, for example whether the plot showed a random pattern or displayed a torus-like pattern with a data void in the center. For the Differential Radial plot, the original data set was used with a delay ranging between 10 to 30 sample intervals. These data were used to create the phase plots as described above. The radial distance (r_n) for each instant of time was then calculated from the phase plot as the distance of the points from the origin.

$$r_n = \sqrt{\left((v_n)^2 + (v_{n+\tau})^2 \right)}$$

The difference between consecutive radii (r_d) was then plotted with time to generate the differential radial plot: $r_d = r_{n+1} - r_n$.

The differential radius is one of the many quantitative measures of "randomness" in a phase plot. If the differential radial plot has large values, it implies the points jump around from one time instant to another, in contrast, if the values are small, it means the points are closer together, and, hence, less random. In order to quantify this measure of randomness, we arbitrarily defined a threshold of 10 for this study, and counted the number of peaks that cross it. Figure 6a and 6b show the differential radial plots for the two patients, with a horizontal line drawn at ±10. The value at the right bottom of these figures are the number of differential radial peaks that cross this threshold.

RESULTS

Fourier Analysis

Fourier analysis of atrial flutter, fibrillation and sinus rhythms yielded patterns that were characteristic to each flow and distinct from each other. The Fourier transform of the flow data of patients in atrial flutter showed a dominant peak at 4 Hz (f_f and f_d), corresponding to the flutter emptying wave cycle length of 250 ms. A prominent subharmonic peak at 2 Hz corresponds to the mechanical alternans, which can also be seen in the tracing.

Patients with atrial fibrillation also revealed a characteristic pattern. The Fourier spectrum of these patients revealed a dominant frequency at 8 Hz (f_f and f_d), corresponding to an atrial mechanical cycle length of approximately 125 msec. Also, subharmonic frequencies below 2 Hz could be detected, which appears to be low frequency modulation of the velocity amplitude over time.

For comparison purposes the Fourier spectrum of patients in sinus rhythm were also calculated. In contrast to the patients with atrial fibrillation and flutter, the Fourier spectrum of the left atrial appendage flow in sinus rhythm tends to be broad-banded, with a series of distinct peaks at and above the fundamental mechanical frequency. In this example, f_f is 1.5 Hz (representing a heart rate of 90beats/min) but f_d is at 3 Hz. The majority of the spectral power is concentrated in the superharmonics.

For 71% of the atrial fibrillation data analyzed, two basic peak frequencies (f_1 and f_2) existed such that the other peaks could be described as a linear combination of the two frequencies ($mf_1 \pm nf_2$). Moreover, the output of the Lorenz equation and its Fourier transform correlated very well with flow data and the Fourier transform.

Non-Linear Dynamical Analysis

Phase plot data showed that the two patients in atrial fibrillation with apparently similar temporal flow patterns actually have quite different phase plots, as shown in Figures 5a and 5b. While one figure shows a torus-like pattern without data points near the origin, the other one is more random. The difference was even more revealing when the differential radial plot of the data was generated for the two sets (Figures 6a and 6b).

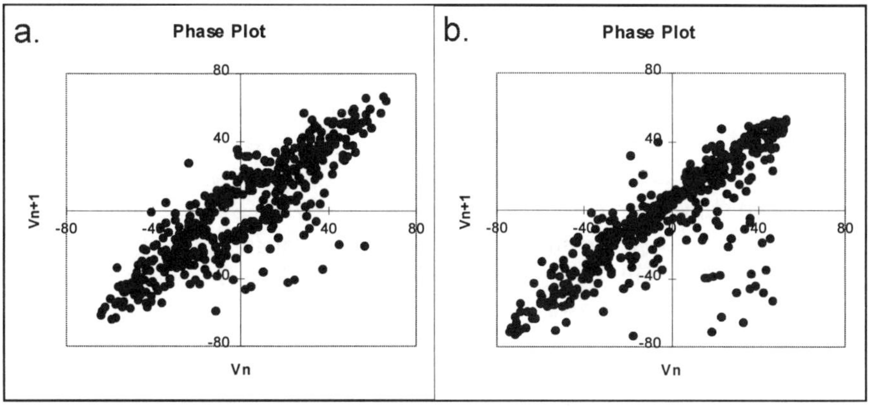

Figure 5. Phase plots a) P1, b) P2.

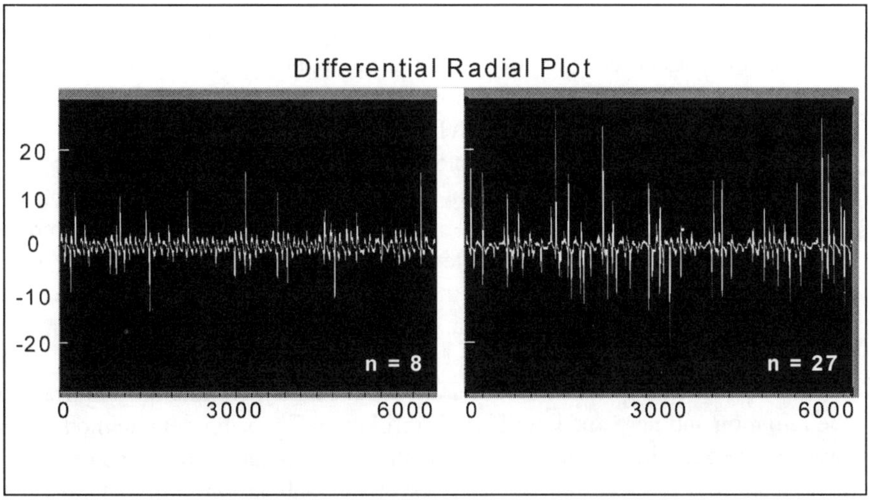

Figure 6. Differential Radial Plots. a) P1, b) P2.

By comparing these plots one notices that the patient with the torus-like phase plot has a less random differential plot as compared to the patient with the more homogeneous phase plot. The number of values that exceed a threshold of ±10 for figure 6a is 8, which is considerably less than 27 observed for figure 6b.

DISCUSSION

The left atrial appendage has become the focus of recent investigation, particularly in atrial fibrillation, for two principal reasons: a) the appendage has a propensity to form thrombus and a) the widespread use of transesophageal echocardiography makes the appendage easy to image. Several investigators have studied flow patterns in the atrial appendage[5,6] for patients in atrial fibrillation, which have demonstrated that blood flow emptying velocities in the appendage can be correlated with the presence of spontaneous echo contrast and thrombus formation. Additionally, lower emptying velocities have been associated with recent episodes of systemic embolization.

Despite its clear importance, the mechanical function of the left atrial appendage is still not well understood, nor has the electrophysiologic implications of these flows been elucidated. Preliminary results from our lab suggest that these Doppler flows represent the mechanical correlate of atrial electrical activity. The rapid, repetitive and irregular emptying and filling Doppler flow patterns, seen primarily in atrial fibrillation, make characterization and interpretation difficult. Previous studies have been limited to measuring peak flow velocities and velocity time integrals (which are likely largely dependent upon left atrial cavity pressure changes). Fourier analysis and non-linear mathematical tools, applied to left atrial appendage flow profiles, provide techniques which can analyze the flow data and provide us with some insight into the mechanical function of the left atrial appendage.

Analysis of the pulsed Doppler patterns of the left atrial appendage by Fourier transformation in the present study revealed that atrial fibrillation and atrial flutter have characteristic patterns of frequency and regularity. Moreover, both of these patterns are very different from the Fourier transform of sinus rhythm. Atrial fibrillation typically has f_f in the 6 to 8Hz range, while atrial flutter has f_f around 4Hz. The percentage spectral power before the f_f frequency in atrial fibrillation is more than that in atrial flutter. Previous studies[7,8] have suggested that multiple continuous intra-atrial reentry circuits are required for the initiation and perpetuation of atrial fibrillation. The subharmonic modulation may represent these multiple wavelets of depolarization, and/or tissue participation. Therefore, further analysis of this subharmonic modulation with simultaneous atrial electrical mapping, may add insights into mechanisms of atrial fibrillation.

Fourier transform not only seems to be helpful in differentiating various types of arrhythmias and characterizing them more comprehensively, it also gives an insight into whether atrial fibrillation follows the laws of deterministic chaos and whether chaotic activity is a mechanism of arrhythmogenesis. There are arguments both for and against fibrillation being a chaotic phenomenon,[9,10,11] but recent studies have suggested that fibrillation is not a chaotic phenomenon. In similar studies, Fourier analysis of *ventricular* fibrillation showed that the frequency spectrum occurred across fewer and fewer frequencies. Because chaotic systems typically have a "broad-band" spectrum, this led to the conclusion that ventricular fibrillation is non-chaotic. The narrow-band Fourier transform suggests a very regular or highly periodic rhythm versus the broad-banded spectral pattern associated with temporal variability.

In contrast, the broad band nature of the sinus rhythm signal is because of the sharp nature of the flow signal. Higher frequencies are required to represent this sharp signal in sinus rhythm, and thus a broad spectrum does not necessarily reflect chaotic phenomenon. Furthermore, on closer inspection of the atrial fibrillation spectrum, there are a number of subharmonic (below f_f) frequencies exhibited by the atrial appendage Doppler profiles (Figure 4b and 4c), which suggest that the frequency spectrum for the atrial appendage is actually broader than initially thought. We, therefore, analyzed the signal further in order to test if the signal was chaotic in nature.

There seem to be various routes to chaos, the period doubling route, intermittancy route, etc. One of such routes is a spectrum that show presence of two basic peak frequencies (f1 and f2) together with other peak frequencies as their harmonics and integer combinations ($mf1 \pm nf2$). This was observed in most of the atrial fibrillation Fourier transform data sets. This could suggest that atrial fibrillation is at the verge of chaos. To further strengthen our hypothesis, we compared the left atrial appendage flow signal and its Fourier transform with that of the output of the Lorenz equation with constants selected to model chaos. There was striking similarity between the two.

Based on these initial results, non-linear dynamic techniques were applied to characterize the atrial fibrillation better. The Poincare phase plot (v_n vs $v_{n+\tau}$) generated from the decimated velocity data set demonstrated different patterns between the patients with apparently similar flow in the temporal domain. Differential radial plots calculated for the same set of patients showed the results similar to those obtained through Poincare phase plot, demonstrating that atrial fibrillation lends itself to analysis using non-linear dynamical tools.

This preliminary work appears to have several potential clinical implications. An important goal would be to risk stratify patients in atrial fibrillation. Currently, all patients with atrial fibrillation are classified into one homogenous group, and usually undergo electric shock (electric cardioversion) to try to bring the atrium back to normal sinus rhythm. Unfortunately, not all patients convert to normal sinus rhythm and some patients who are converted successfully return to atrial fibrillation over days to months following cardioversion. At present there is no technique to differentiate reliably between patients who successfully cardiovert and those who do not. Perhaps patients who cardiovert successfully could have a different non-linear dynamical signature than those who cannot be cardioverted. Moreover, understanding and characterizing atrial fibrillation may prove useful in analyzing the mechanism of thrombogenesis and thromboembolism and predicting which patients are likely to have these complications.

One of the limitations of this study is related to converting data to a form that can be analyzed digitally. Currently, the velocity envelopes have to be manually digitized before Fourier transformation and nonlinear analysis. This is cumbersome and limits the number of seconds of data that can be digitized and, hence, limits the number of points available for analysis. Ideally, several minutes of data acquisition and analysis should be used to tease out more subtle relationships than those observed with these short data sets. Secondly, the pulsed Doppler flows recorded from the left atrial appendage occasionally demonstrates entirely passive or akinetic flow. The analysis in the current study was limited to a selected population as no patients with passive flow profiles were included in the study.

CONCLUSIONS

Fourier analysis has been used to define the frequency and regularity of mechanical contraction within the left atrial appendage especially during atrial fibrillation and flutter. Atrial fibrillation and flutter have different fundamental frequencies, and the percentage of spectral power below the fundamental frequency is higher in atrial fibrillation than flutter. Analysis of the fibrillation spectrum suggested that the appearance of two frequencies and their linear combination could be one of the routes to chaos. Similarity with the Lorenz equation gave more credibility to the hypothesis that atrial fibrillation follows the laws of non-linear dynamics. The non-linear dynamical tools like phase plots and differential radial plots were successfully used to demonstrate differences between patients. These tools could possibly be used to differentiate between patients who successfully cardiovert and those who do not.

Fourier analysis and non-linear dynamical tools seem to be powerful tools for characterizing left atrial appendage flow and risk stratifying patients with atrial fibrillation. Characteristic dominant frequency and subharmonic modulation along with the phase plots could yield important insights into the mechanisms of atrial flutter and fibrillation.

In the future we hope to overcome the small data point limitation of this study by digitizing several minutes of velocity data, along with more automated signal processing. If successful, this would make the nonlinear analysis reported in this paper more applicable to the general clinical population with atrial fibrillation and flutter.

REFERENCES

1 . Pollick C., Taylor D. Assessment of left atrial appendage function by transesophageal echocardiography. Implications for the development of thrombus. *Circulation.* 1991;84:223-31.

2 . Pozzoli M., Febo O., Torbicki A., Tramarin R., Calsamiglia G., Cobelli F., Specchia G., Roelandt J.R. Left atrial appendage dysfunction: a cause of thrombosis? Evidence by transesophageal echocardiography-Doppler studies. *J Am Soc Echo.* 1991;4:435-41.

3 . Grimm RA, Stewart WJ, Maloney JD, Cohen GC, Pearce GL, Salcedo EE, Klein AL. Impact of electrical cardioversion for atrial fibrillation on left atrial appendage function and spontaneous echo contrast: characterization by simultaneous transesophageal echocardiography. *J Am Coll Cardiol.* 1993;22:1359-1366.

4 . Tritton DJ. *Dynamical Chaos in Physical Fluid Dynamics.* Oxford Science Publications, Great Britain. 1988;2nd Ed:393-416.

5 . Garcia-Fernandez MA, Torrecilla EG, San Roman D, Azevedo J, Bueno H, Moreno MM, Delcan JL. Left atrial appendage Doppler flow patterns: implications on thrombus formation. *Am Heart J.* 1992;124:955-961.

6 . Verhorst PM, Kamp O, Visser CA, Verheugt FW. Left atrial appendage flow velocity assessment using transesophageal echocardiography in nonrheumatic atrial fibrillation and systemic embolism. *Am J of Cardiol* 1993;71:192-196.

7 . Moe GK, Abildskov JA. Atrial fibrillation as a self sustaining arryhthmia independent of focal discharge. *Am Heart J* 1959;58:59-70.

8 . Moe GK. On the multiple wavelet hypothesis of atrial fibrillation. *Arch Int Pharmacodyn Ther* 1982;140:183-8.

9 . Golderberger A, Rigney D, Meitus J, Antman E, Greenwald S. Nonlinear dynamics in sudden cardiac death syndrome: Heart oscillations and bifurcations. *Experentia.* 1988;44:984-987.

10 . Evans SJ, Khan SS, Garfinkel A, Kass RM, Albano A, Diamond GA. Is ventricular fibrillation random or chaotic? (abstr) *Circulation* 1989;80II-134.

11 . Jalife J. Chaos theory and the study of arrhythmogenesis: PartII. *ACC Current Journal Review* July/August 1993:13-16.

Modeling the Dynamical Systems on Experimental Data

Natalie B. Janson and Vadim S. Anishchenko

Laboratory of Nonlinear Dynamics,
Saratov State University, Saratov, Russia

Abstract. An attempt is made in the work to create qualitative models of some real biological systems, i.e., isolated frog's heart, a human's heart and a blood circulation system of a white rat. Sampled one-dimensional realizations of these systems were taken as the initial data. Correlation dimensions were calculated to evaluate the embedding dimensions of the systems' attractors. The result of the work are the systems of ordinary differential equations which approximately discribe the dynamics of the systems under investigation.

INTRODUCTION AND STATEMENT OF THE PROBLEM

The problem of reconstructing the dynamical systems (DS) on experimental data has a rather old history. One may extract two parts of this problem. The first one is the restoration of the phase portrait of the system using the data of measuring the time dependence of only one system's variable $a(t)$, which we will further call realization. The result of solving this problem is the N-dimensional restored trajectory $\vec{x}(t)$ which with the given accuracy reproduces the phase portrait of initial system. This task also includes the forecasting of the dependence $a(t)$ for the time $t > t_0$, t_0 is the duration of the experimental realization. A great number of works is devoted to this problem and constructive results are achieved (1-3). The second and the more general look on this problem includes solving a non-unique task of mathematical modeling of the system in the form of corresponding maps or differential evolution equations. If the model is found then the solution of evolution equations must with the given accuracy reproduce the experimental dependence $a(t)$ and give the forecast for the times $t > t_0$. Besides (and this is the main result), the presence of reconstructed equations gives the opportunity to describe not only the process $a(t)$, but also its dependence on the control parameters of the model system. A rather great number of papers was also devoted to this problem using some famous model systems as the examples (see, for example, (4-5) and references in them). One of the problems arising with this is that the resulting equations appear to be non-rough, i.e., the small variation of their parameters leads to the trajectories tending to infinity. Note, that solving the problem of reconstruction of mathematical model, certainly, includes restoring the N-dimensional phase trajectory.

The aim of our work is an attempt to reconstruct the biological systems under study. We shall interpret the reconstruction as the finding of explicit form of equations qualitatively modeling the systems' behavior.

DESCRIPTION OF METHOD

The first step of reconstruction is to restore the phase portrait on the known initial realization. There are a few ways of solving this problem (4-7) but here we shall point out the two of them. The first one uses the Takens's dealy method

$$\vec{x}(t) = \{a(t), a(t+\Delta t), \ldots, a(t+p(N-1)\Delta t)\}, \tag{1}$$

where $\vec{x}(t)$ is the phase vector in N-dimensional phase space, N is the embedding dimension, p is the delay, t is the current time, Δt is the sampling step of the initial realization $a(t)$. The second way uses the successive differentiating of coodinates

$$x_1 = a, \; x_{i+1} = \dot{x}_i, \; i = 2, N. \tag{2}$$

Note, that usually the derivatives are fitted using the following formula

$$\dot{x}_i = \frac{x_i(t+\Delta t) - x_i(t)}{\Delta t}, \tag{3}$$

where Δt is the sampling step. But the used realizations were measured with the sampling step which was too big for such fitting. Therefore, to more precisely compute the derivatives we had to interpolate the phase coordinate between the two known points with the help of polynomials of the m-th power to obtain the point $y(t + \Delta t^*)$ where $\Delta t^* \ll \Delta t$. In this case, the derivatives were computed using the formula (3), in which Δt was replaced by Δt^*.

The next step of solving our problem is to construct the mathematical model of a system under study, for example, to gain a system of N first order ordinary differential equations (ODE)

$$
\begin{aligned}
\dot{x}_1 &= f_1(x_1, x_2, \ldots, x_N), \\
\dot{x}_2 &= f_2(x_1, x_2, \ldots, x_N), \\
&\cdots\cdots\cdots \\
\dot{x}_N &= f_N(x_1, x_2, \ldots, x_N)
\end{aligned}
\tag{4}
$$

or a map in R^N

$$
\begin{aligned}
x_{k+1,1} &= f_1(x_{k,1}, x_{k,2}, \ldots, x_{k,N}), \\
x_{k+1,2} &= f_2(x_{k,1}, x_{k,2}, \ldots, x_{k,N}), \\
&\cdots\cdots\cdots \\
x_{k+1,N} &= f_N(x_{k,1}, x_{k,2}, \ldots, x_{k,N}).
\end{aligned}
\tag{5}
$$

There is a great number of ways how to fit the functions f_j and the most simple of them is to represent the sought functions as the polynomials of the ν-th power

$$
f_j(x) = \sum_{l_1, l_2, \ldots, l_N = 0}^{\nu} C_{j, l_1, l_2, \ldots, l_N} \prod_{k=1}^{N} x_k^{l_k}, \qquad \sum_{k=1}^{N} l_k \leq \nu,
\tag{6}
$$

where x_k is the k th coordinate of vector $\vec{x}$, and to find the coefficients of these polynomials[1] by means of least squares method. Just this way was chosen in the current work. The use of the least squares method needs a great amount of computer's memory, that is why the attractor must be considered not in all the available points, but only in a relatively small number of them i_{max} with the chosen step s along the trajectory.

SYSTEMS UNDER STUDY

The algorithms described were tested on well-studied models, particularly, the equations of Anishchenko–Astakhov oscillator (modified oscillator with inertial nonlinearity, or MOIN) which were considered in the chaotic regime with $m = 1.5, g = 0.2$ (8), and also the Lorenz system with the following standard parameter values: ($\sigma = 10, r = 28, \gamma = 8/3$) and the Rössler system with the values of parameters as follows: $c = 0.38, b = 0.3, a = 4.82$.

[1]Legendre polynomials were used.

After the reconstruction method had been tested on the models we began to check it on the more complicated systems of medical and biological origin.

The first such system was an isolated frog's heart (IFH). The experiment on measuring the initial realization was lead as follows: an isolated frog's heart was put into the special physiological solution where it began to contract. A metal hook was stuck into it which moved together with the heart's surface. A digitized time dependence of a current coordinate of the hook gave the initial time series. The sampling frequency was 60 Hz and the amount of points in the time series was 8192. For the experiments with reconstruction of the described system a most typical realization of a point's oscillation on the frog's heart surface was chosen[2].

The second system under investigation was a human's heart (HH). The initial realization was a typical electrocardiogramme (ECG) of a healthy man which was measured with the sampling frequency 200 Hz and consisted of 10240 points.

And the last system with which we tried to work was a system of blood circulation of a white rat. As the initial data a time dependence of arterial blood pressure (ABP) of rat was taken which was measured with the help of a sensor injected directly into the animal's artery. The sampling frequency and the number of points were the same as for the ECG of a human's heart.

The time series obtained in the course of experiments consisted of integer numbers varying in the interval [-256; 256] (for IFH) or [-512; 512] (for HH and ABP). To make the work with them more comfortable a renormolization of each of them was performed to the interval [0; 1] (this does not refer to the model systems). Moreover, to obtain more smooth and "filled" phase portraits for systems HH and ABP we, by means of interpolation, determined in the space between each two neighbouring points 4 additional points. Thus, the number of points in the initial phase portrait was increased by a factor of 5 and, respectively, the sampling time was decreased by the same factor. Further, we shall regard this new time as the sampling step Δt.

RESTORATION OF PHASE PORTRAITS

When restoring a phase portrait on realization using the method of derivatives we have only one kind of result because all the coordinates are uniquely connected with each other according to the expression (2). Certainly, one may vary the suitable parameters to achieve the best accuracy of this relation. In particular, when interpolating the data by polynomilas, one may vary their

[2]The experimental data were given to us by the Central Scientific Laboratory of Medical University (supervisor is professor G. Ye. Brill).

order and, therefore, the number of used neighbouring points which can influence the result. One can also vary Δt^*. But these variations do not cause any qualitative changes of a phase portrait.

Another situation arises when restoring the phase portrait by delay method which is more simple to realize. In this case, one must properly choose delay time $\tau = p\Delta t$. In ref. (9) it is stated that, in principle, it may have any value for the noiseless data of infinite length. However, if it will be too small, the i-th and the $(i+1)$-th coordinates of a phase point will be practically equal to each other. In this case, the reconstructed attractor will be situated near the main diagonal of the embedding space, the latter complicating its diagnostics. When choosing a too large τ the coordinates become non-correlated, and the structure of reconstructed attractor is lost (10).

In a series of works it is suggested to use as a time delay the first zero of auto-correlation function (ACF) τ_0, the first minimum of function of mutual information (11-12) or the first minimum of logarythm of generalized correlation integral (13).

In Fig. 1, the two-dimensional projections of the reconstructed by delay method phase portraits of systems IFH, HH and ABP are presented for different τ. To the right $\tau = \tau_0$ and to the left $\tau \ll \tau_0$. For systems IFH and ABP the choice of delay time seems to be of no principal importance, but for system HH the increase of τ highly copmplicates the phase portrait, though the attractor's topology is maintained. Therefore, in the present work we have chosen the phase portraits shown in Fig. 1 to the left which correspond to the condition that $\tau \ll \tau_0$[3].

ESTIMATING THE EMBEDDING DIMENSION OF SYSTEMS UNDER STUDY

To estimate the minimum embedding dimension for further reconstruction of equations we used correlation dimension D_c due to the advantages of its calculation which were pointed out in (14-19). A standard method for calculating the correlation dimension of DS was used, i.e., by defining the slope of a linear segment of a $\log_{10}(C)$ plot versus $\log_{10}\epsilon$, where C is the correlation integral determined by the following formula:

$$C(\epsilon) = \log_{10} \sum_{i=1}^{M(\epsilon)} p_i^2. \tag{7}$$

Here, ϵ is the size of a box covering the phase space, p_i is the probability for a phase point to fall within the box having number i, $M(\epsilon)$ is the amount

[3]Further, we shall give τ_0 in the units of Δt.

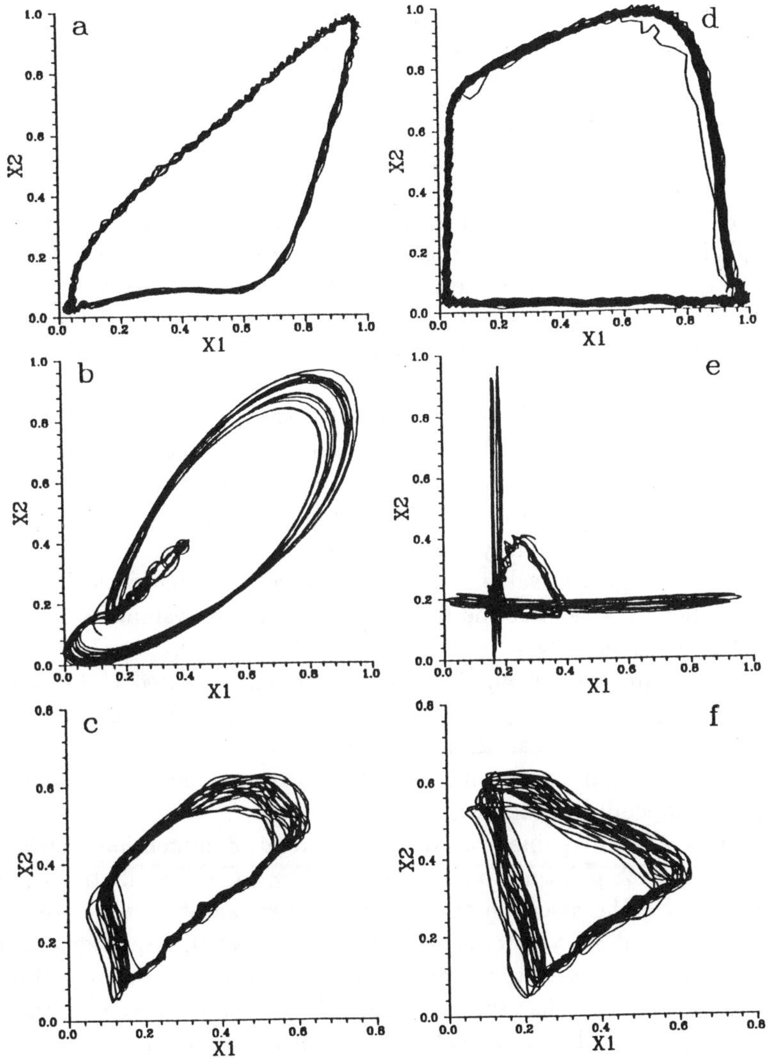

Fig.1. Phase portraits of systems under study restored by delay method with different values of time delay: a) IFH with $p = 5$, d) IFH with $p = 54 = \tau_0$, b) IIH with $p = 5$, e) HH with $p = 70 = \tau_0$, c) ABP with $p = 5$, f) ABP with $p = 45 = \tau_0$.

Table 1: The results of caclulating the correlation dimension

System	Correlation dimension D_c
IFH	1.19
HH	2.79
ABP	2.14

of non-empty boxes. Note, that due to the performed renormalization to the interval [0; 1] the maximum size of a box is 1. Dimension D_c was calculated for the range of embedding dimensions N from 1 to 10. As a "true" D_c a dimension was chosen which had been obtained for $N = 10$, though the convergence to this value began for significantly less values of N. The results are tabulated in Table 1.

Correlation dimension D_c of attractor of system IFH turned out to be smaller than 2 and close to 1. We suppose that the attractor of system IFH is a limit cycle. This is proved by the shape of realization, slowly decreasing ACF, the spectrum consisted of peaks with the noisy background and the one-dimensional probability density which is typical for the limit cycle (20). A somewhat greater value of D_c seems to be connected with the use of noisy experimental data. For the reconstruction of this system we used the embedding dimension $N \leq 3$.

Correlation dimension computed for system HH appeared to be significantly non-integer number. This is evidence that we deal with chaotic attractor. Note, that when calculating D_c for $N = 3$ we obtained a value of $D_c = 2.83$[4]. We shall try to limit ourselves to the value of $N = 3$ when choosing the embedding dimension.

For system ABP we gained the value of correlation dimension which is a little greater than 2, pointing to a possible chaotic type of oscillations. This result agrees with the assertion that in this system an attractor is realized which is close to a destroyed torus (22). To reconstruct the equations of this system we confine ourselves to the embedding dimension which is equal to 3.

As a whole, the fact that the correlation dimension of these systems are small gives hope to us, because it follows that the dynamics of systems under study is not more complicated than the dynamics of systems with a single positive Lyapunov exponent.

[4]This value agrees with the result obtained in (21).

RECONSTRUCTION OF EQUATIONS

Testing the Method on Model Systems

Consider the Anishchenko–Astakhov oscillator, whose equations are given below

$$\begin{aligned} \dot{x} &= mx + y - xz \\ \dot{y} &= -x \\ \dot{z} &= g[I(x)x^2 - z] \end{aligned} \tag{8}$$

where $I(x) = 1$ *as* $x > 0$ *and* $I(x) = 0$ *as* $x \leq 0$, in the chaotic regime ($m = 1.5,\ g = 0.2$). We numerically solve this system of equations by Runge-Kutta method of the first order storing the phase trajectory points with the time step $\Delta t = 0.25$.

First, we take as the initial data all the three coordinates $x(t), y(t), z(t)$ and try to obtain the original system (8) by reconstruction method. We fit the right-hand parts of equations by the 3-rd order polynomials. For the step along the trajectory $s^5 = 3$ and $i_{max} = 2000$ we obtained a system of three equations. When comparing it with system (8), one can find out that i) a substitution of variables $x \to z,\ y \to y,\ z \to x$ took place, ii) the coefficients in the first two equations were reconstructed with the accuracy 10^{-3} (if one takes into account the substitution), iii) the right-hand part of the third equation is 3-rd order polynomial which can't be reduced. The third observation has a natural explanation, namely, due to non-smoothness of function $I(x)$ the right-hand part of the third equation cannot be fitted by the second-order polynomial. The result of solving the reconstructed system is shown in Fig. 2(a) (the phase portrait) and Fig. 2(b) (the time dependence of initial coordinate $x(t)$ and the first of reconstructed coordinates).

Second, we use coordinate $y(t)$ as the initial realization and restored the phase portrait by its differentiating (Fig.2(c)). The reconstruction of the corresponding equations was also successful.

Third, the attractor was reconstructed on $y(t)$ by delay method with $p = 8$ (Fig. 2(d)). For the following values of reconstruction parameters: $s = 3, \nu = 3, m = 10, \Delta t^* = 10^{-10}$ we obtained a system of three equations with all nonzero coefficients. After solving it with the sampling step $\Delta t = 0.25$ the phase portrait was obtained shown in Fig. 2(e), as well as the time dependence of the first variable together with the initial variable $y(t)$ (Fig. 2(f)).

The same procedure was applied to Lorenz and Rössler systems and the similar results were obtained. The material of this section gives evidence that our software works well and one may try to apply the method of reconstruction to the more complicated systems.

---

[5] s is given in the units of Δt.

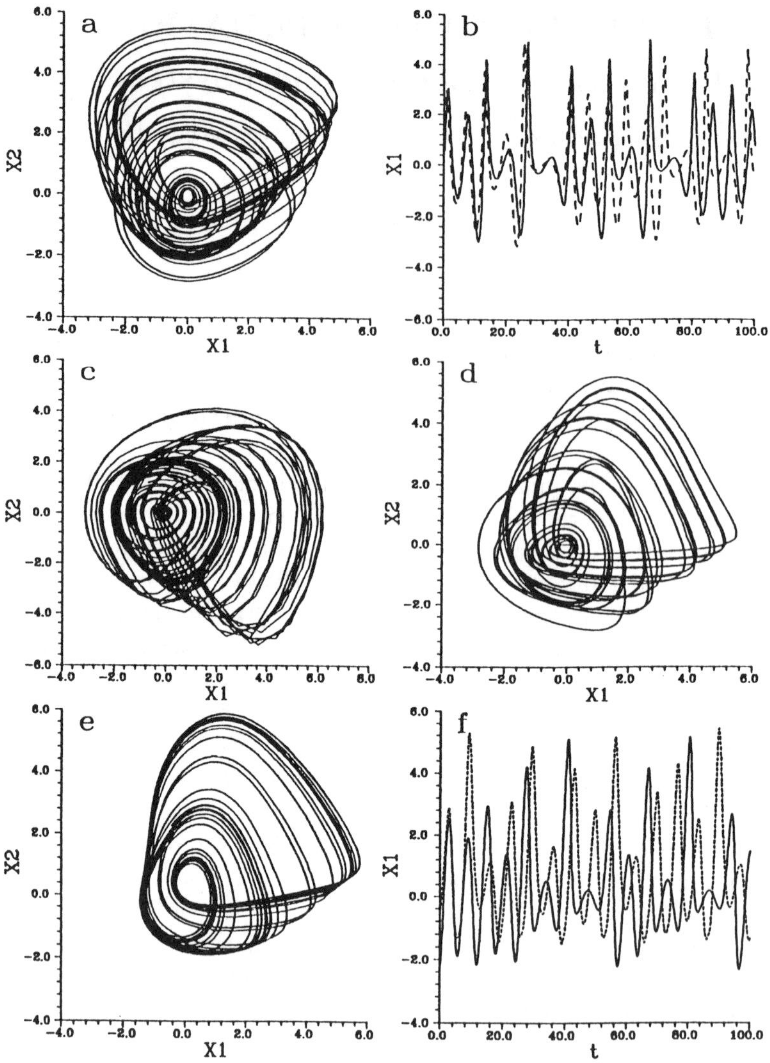

Fig. 2. Solution of ODEs reconstructed on three coordinates of MOIN: a) phase portrait, b) $x(t)$ (solid line) together with the first reconstructed coordinate (dashed line); c) phase portrait restored by differentiating $y(t)$ of MOIN; d) phase portrait restored on $y(t)$ of MOIN by delay with $p = 8$; solution of ODEs reconstructed on phase portrait in d): e) phase portrait, f) initial (solid line) and reconstructed (dashed line) realizations

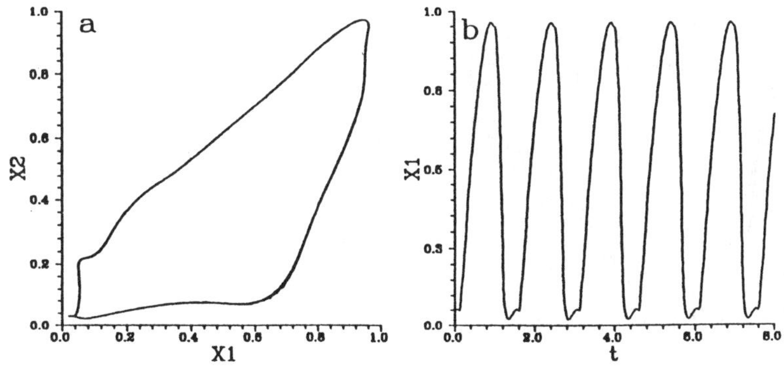

Fig. 3. Solution of three-dimensional ODEs obtained for Fig. 1 (a): a) phase portrait b) realization.

Reconstruction of Real Medical and Biological Systems

The first our attempt to create a model of IFH was (20), in which we managed to obtain only the corresponding two-dimensional map. In the present work a system of two differential equations was gained approximately modeling this system. For this purpose the phase portrait, as in (20), was restored by delay method with $N = 3$, $p = \frac{\tau}{\Delta t} = 5$ for $\Delta t = 0.016666...sec$ (see Fig. 1(a)). The projection of the obtained solution of ODE's is shown in Fig. 3(a), and the time dependence of $x_1(t)$ is given in Fig. 3(b) (compare Fig. 5(a)).

The next step of our work with IFH was restoration of the phase portrait by finding successive derivatives of the initial realization using the embedding dimension $N = 3$. Due to the presence of rather intensive noise in the initial realization the phase portrait obtained in such a way appeared to be much more chaotic than the one in Fig. 1(a) and did not remind the limit cycle at all (see Fig. 4(a,b,c)). But, nevertheless, we made an attempt to reconstruct the equations on these data. One could consider this attempt to be successful only after very thorough selection of reconstruction parameters ($s = 54(\tau_0 = 26), \Delta t^* = 10^{-10}, m = 10, \nu = 3$). For other values of these parameters no reasonable results were obtained. The solution of reconstructed equation was a limit cycle reminding of the initial phase portrait in its shape (see Fig. 4(d,e,f)). The time dependence of the corresponding coordinates also looks quite similar (Fig. 5 (a,b,c) and (d,e,f)). But their power spectra have shown that the accuracy of reconstruction of this system is not enough.

Now, consider system HH. Its phase portrait restored by delay method

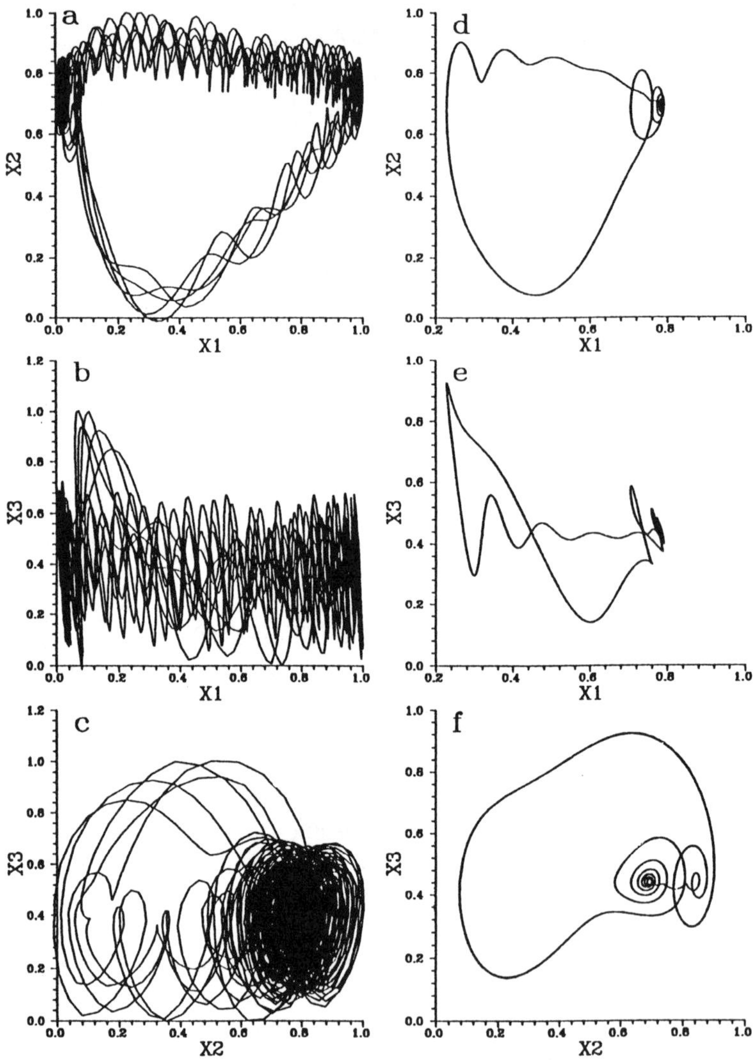

Fig. 4. a), b), c) projections of attractor reconstructed by differentiating the realization of IFH; d), e), f) projections of solution of ODEs reconstructed on a), b) and c).

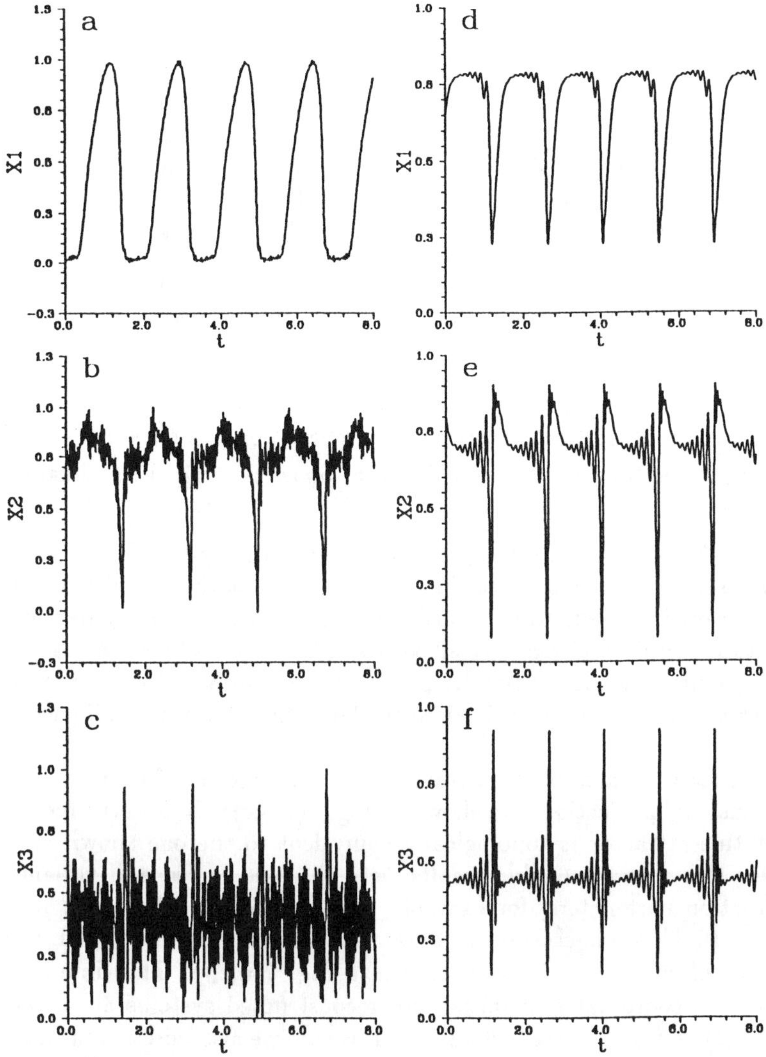

Fig. 5. a), b), c) time dependences of a, $\dot{a}$, $\ddot{a}$ of IFH; d), e), f) time dependences of coordinates of solution of corresponding reconstructed ODEs.

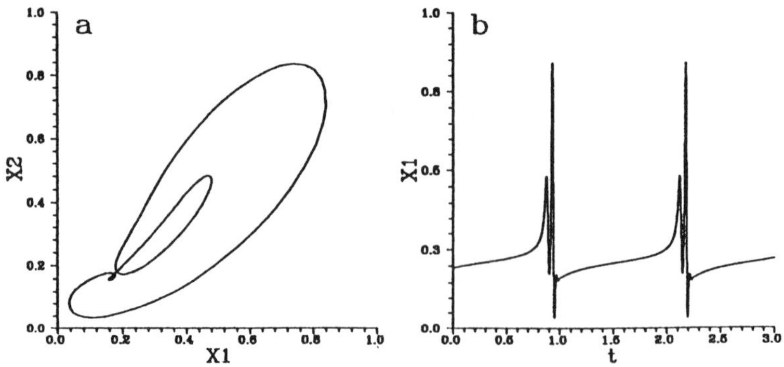

Fig. 6. Solution of ODEs reconstructed on phase portrait obtained by delay embedding of the first derivative of ECG: a) phase portrait, b) realization.

is shown in Fig. 1(b). An attempt to reconstruct the system HH using this phase portrait was not a success. Further, we chose as the initial realization the first derivative of experimental time series. We applied delay method to these data to obtain a phase portrait very similar to that shown in Fig. 1(b). With $s = 1, m = 10, \Delta t^* = 10^{-10}, i_{max} = 3500, \nu = 4 (N = 3)$ the equations were obtained with a limit cycle close to the initial phase portrait (see Fig. 6(a,b)).

After that we restored the attractor of HH by means of derivatives (its two-dimensional projections are shown in Fig. 7(a,b,c)). It is worth mentioning that this attractor is topologically equivalent to the one shown in Fig. 1(b). For this case reasonable results were obtained for several variants of reconstruction parameters; for example, for $s = 16, (\tau_0 = 70), \nu = 3, m = 10, \Delta t^* = 10^{-10}$, we succeeded in obtaining a system of three ODE's, the projections of whose solutions are shown in Fig. 7(d,e,f). The time dependences of the coordinates of initial and reconstructed systems are given in Fig. 8(a,b,c) and (d,e,f), respectively. In this case we also posess a limit cycle as a result of reconstruction, though the initial attractor looked like chaotic one (this is ensured by the ACF besides the dimension). The effect when instead of chaotic regime a regular one is reconstructed may be explained by the averaging action of the least-squares method[6]. The phase trajectory of system HH is not periodic, but a phase point moves in the neighbourhood of some periodic regime which is reconstructed by our method. The shape

[6]We do not exclude that the attractor of HH is in fact regular and the results of Fig. 1(b) and Fig 7(a,b,c) are the consequence of fluctuation influence.

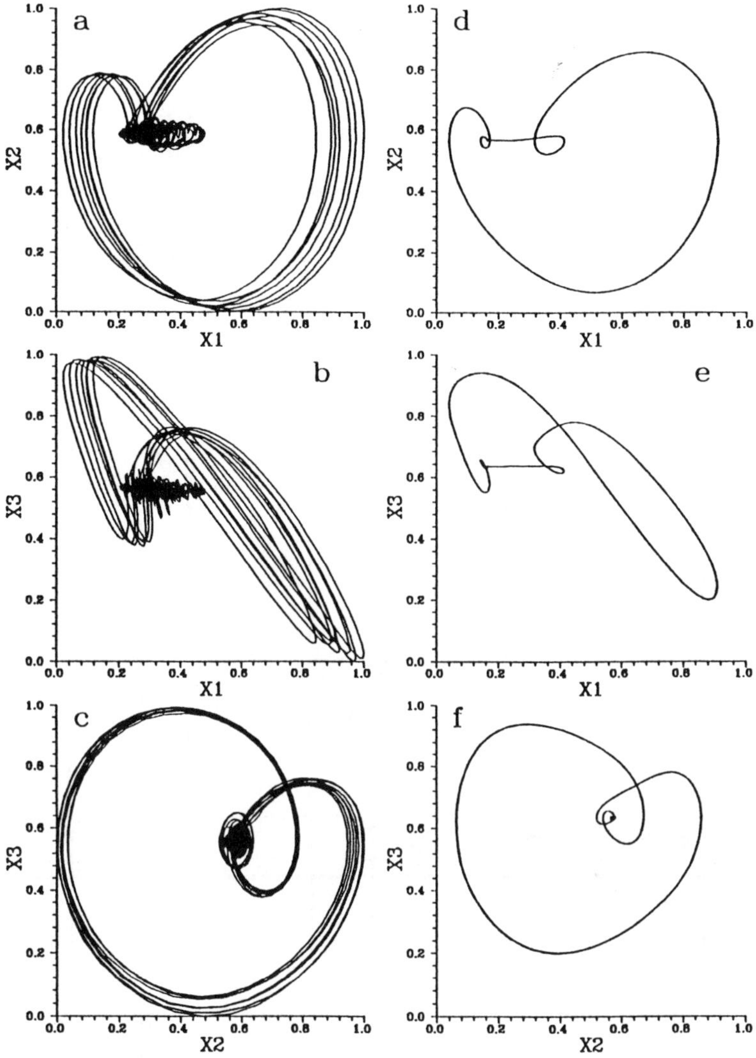

Fig. 7. a), b), c) projections of attractor reconstructed by differentiating the realization of HII; d), e), f) projections of solution of ODEs reconstructed on a), b) and c).

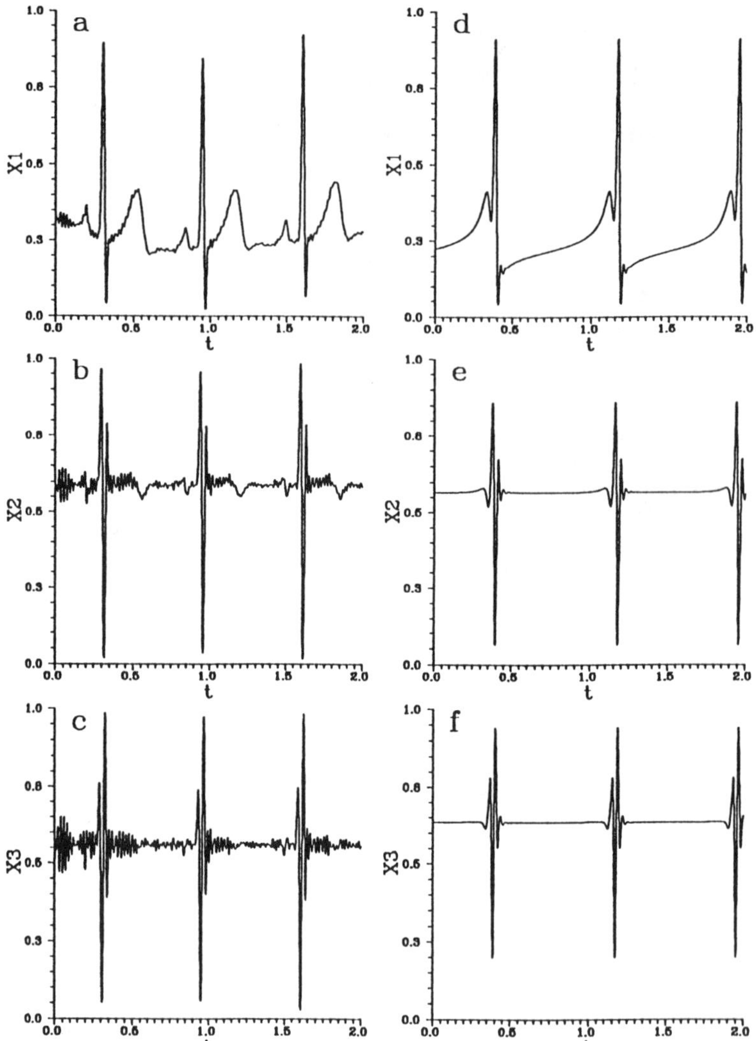

Fig. 8. a), b), c) time dependences of a, $\dot{a}$, $\ddot{a}$ of HH; d), e), f) time dependences of coordinates of solution of reconstructed ODEs.

of the reconstructed first coordinate (Fig. 8(d)) also differs from the initial ECG (Fig. 8(a)), though all the P, Q, R, S, T - peaks typical for ECG are present in it.

We also applied the reconstruction method to a system ABP. We did not succeed in reconstructing this system, as well as the system HH, using the phase portrait obtained by delay method. Its pahse portrait was restored by differentiating (Fig. 9(a,b,c)). For $s = 27, m = 15, \Delta t^* = 10^{-11}, \nu = 3 (N = 3)$ we got a system of equations whose attractor is shown in its projections in Fig. 9(d,e,f), and its coordinates are given in Fig. 10(d,e,f) together with the initial ones (Fig. 10 (a,b,c)). Both the initial and reconstructed attractors are chaotic, their shape being similar, but there is some difference in the behavior of their coordinates, especially of the third ones.

Properties of the Reconstructed Equations

We shall not write here the obtained differential equations because of their complexity. Their right-hand parts have the form (6), the number of coefficients $C_{j,l_1,l_2,\ldots,l_N}$ for $\nu = 3$ being equal to $20 \times 3 = 60$. The range of the coefficients was rather large ($\pm 10^4$).

A very important result is the roughness of the equations obtained, i.e., when one slightly varies their parameters, their solution is also varied slightly. When varying the parameters more considerably one could observe bifurcations of soluitons. Consider, for example, the equations modeling HH (corresponding to Fig. 7). When increasing the parameter $C_{1,012}$ one could see a sequence of period-doubling bifurcations of initial limit cycle followed by the birth of strange attractor, a sequence of inverse period-doubling bifurcations and transition to period-one cycle. When decreasing the same parameter a gradual increase of the period of oscillations could be noticed with subsequent abrupt transition of a trajectory to infinity, reminding of a bifurcation of a limit cycle's junction with separatrix loop (8). Note, that the same bifurcations could be observed when varying any of system's parameters, but there were a few parameters which could be varied in a rather great range, being suitable for investigation, while the variation of the other ones in a small range lead to very quick change of regimes.

Now, we must pay attention to a following disagreeable instant. In cases when we restored the phase trajectory by successively differentiating the coordinates we hoped to obtain the equations in the form of:

$$\dot{x} = y, \ \dot{y} = z, \ \dot{z} = F(x, y, z). \tag{9}$$

But, in spite of the qualitative correspondence of the phase portraits, we did not obtain the differntial equations of such a form, though when

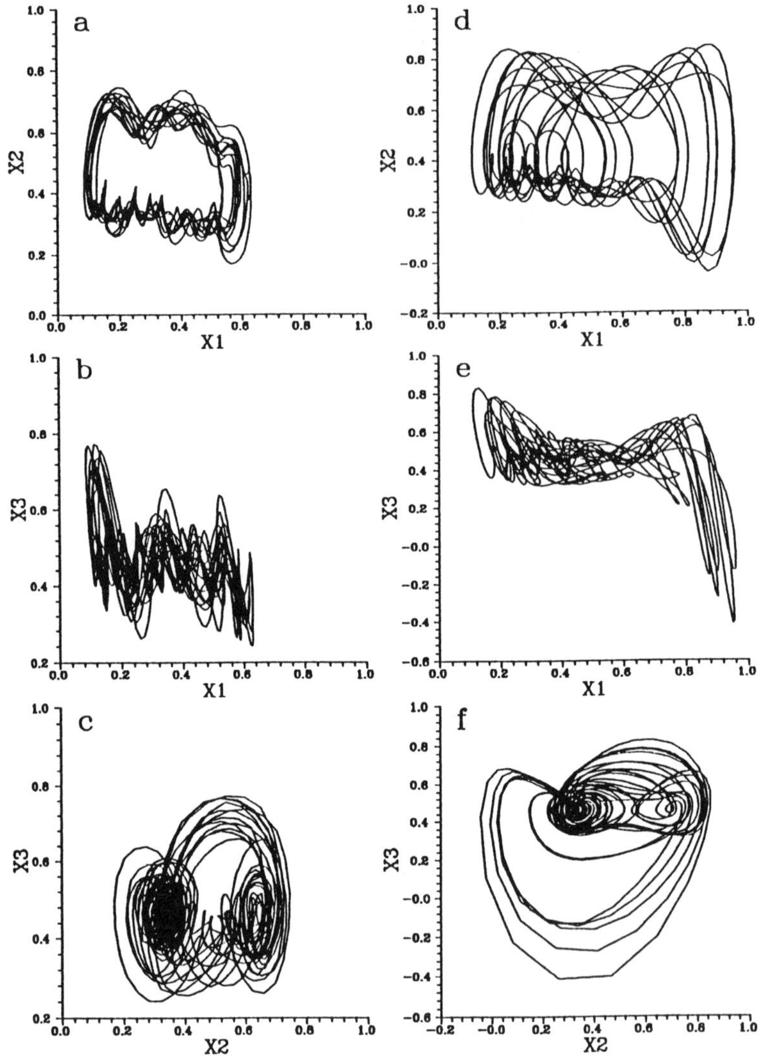

Fig. 9. a), b), c) projections of attractor reconstructed by differentiating the realization of ABP; d), e), f) projections of solution of ODEs reconstructed on a), b) and c

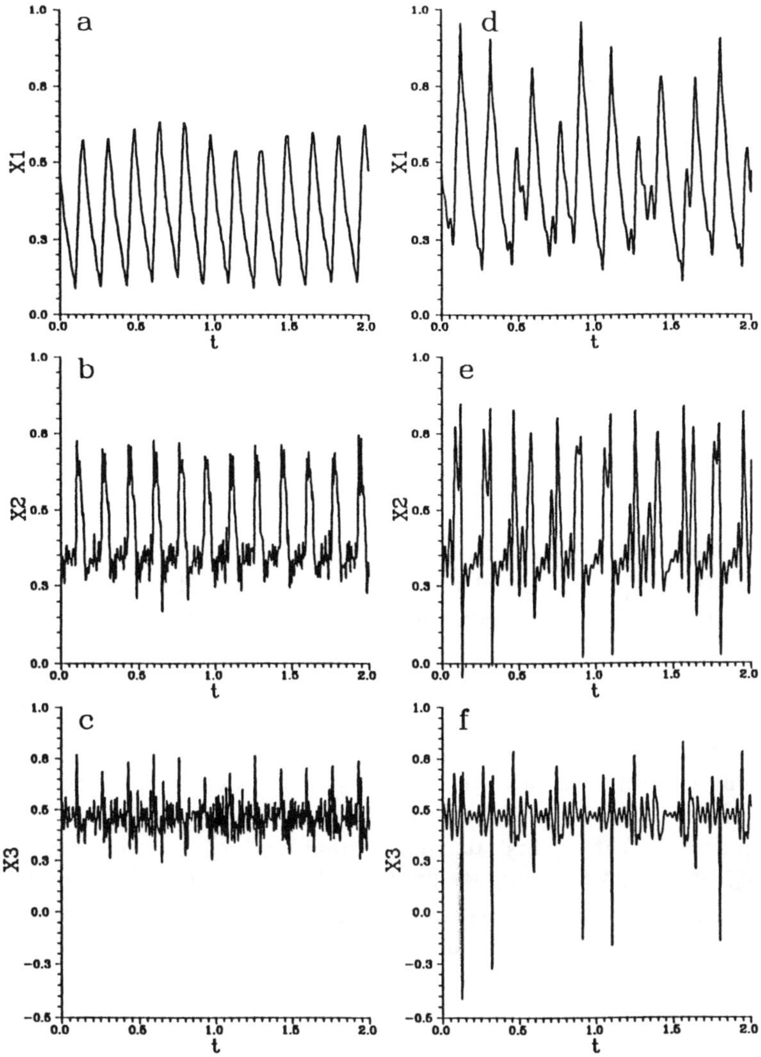

Fig. 10. a), b), c) time dependences of a, $\dot{a}$, $\ddot{a}$ of ABP; d), e), f) time dependences of coordinates of solution of reconstructed ODEs.

reconstructing the model systems with this method we had the expected result. All the three reconstructed equations had a complex form and no one coefficient could be neglected. This can be connected with that the embedding dimension to which we had to confine ourselves due to the lack of computer memory is not sufficient, and also with possible non-stationarity of initial data (i.e., the phase trajectory did not belong to the system's attractors due to influence of some collatoral factors when measuring the realization from alive organisms) or a too small sampling rate. As one more explanation of our failure, one can deliver a suggestion that the systems under study were not dynamical ones which may be described by finite-dimensional systems of ordinary differential equations. Therefore, their reconstruction by our method may be impossible in principle.

CONCLUSIONS

1. The algorithm of reconstruction of dynamical systems on one realization was tested on famous model systems, and that proved its workability.

2. Calculation of correlation dimensions of biological systems under study was performed.

3. An attempt is made to apply the reconstruction method to real medical and biological systems. As a result, some rough equations were obtained whose solutions qualitatively correspond to the initial data.

4. Some control parameters of reconstructed systems are found whose variations lead to bifurcations of solutions.

One must not treat our first attempt of creating the qualitative models of real biological systems as a final result. We only showed that the reconstruction method is applicable to noisy data obtained in the course of experiment (in our case, biological). The volume of this paper does not allow us to present the more detailed investigation of the obtained equations and, namely, the result of their solution with noise of different intensity. The prospect of further study in this field is the increase of embedding dimension when reconstructing the equations, the use of more accurate methods of finding the right-hand parts of equations, the use of other methods of embedding the data in the phase space.

ACKNOWLEDGEMENTS

We express our sincere acknowledgements to professor of Saratov Medical University G.Ye. Brill for kindly giving us the experimental realizations of

IFH, to P.I. Saparin and N.B. Igosheva for the given measurements of ECG, to professor of SSU T.G. Anishchenko and A.N. Murashov for the experimental realizations of arterial blood pressure of a rat. The work was partially supported by International Science Foundation (grant RNO 000) and Russian foundation of Natural Sciences (grant 93-8.2-10).

REFERENCES

1. Martin Casdagli, *Physica D*, **35**, 335 (1989).

2. J.D. Farmer and J.J. Sidorowich, *Phys. Rev. Lett.*, **59**, 845 (1987).

3. H.D.I. Arbanel, R. Brown, J.B. Kadtke, *Phys.Rev. A*, **41**, 1782 (1990).

4. G.Gouesbet and C.Letellier, *Physical Review E*, **49**, 4955 (1994).

5. R. Brown, N.F. Rulkov, E.R. Tracy, *Phys. Rev. E*,**49**, 3784 (1994).

6. N.H.Packard, J.P. Crutchfield, J.D.Farmer, and R.S. Shaw, *Phys. Rev. Lett.*, **45**, 712 (1980).

7. G.B. Mindlin, H.G. Solari, M.A. Natiello, R.Gilmore, X.-J. Hou, *J. Nonlinear Sci.*, **1**, 147 (1991).

8. Anishchenko, V.S., *Slozjnye kolebanija v prostykh sistemakh* (in Russian), Moscow: Nauka, 1990.

9. F.Takens, in: *Lecture Notes in Math.*, Springer, Berlin (1981).

10. T.S. Parker and L.O. Chua, *Practical Numerical Algorithms for Chaotic Systems*, New York: Springer-Verlag, 1989, ch.7, pp. 191-199.

11. A.Bransater, H.L. Swinney, *Phys. Rev. A*, **35**, 2207 (1987).

12. A.M. Fraser, H.L. Swinney, *Phys. Rev. A*, **33**, 1134 (1986).

13. W. Liebert, H.G. Schuster, *Phys. Lett. A*, **142**, 107 (1989).

14. K. Pawelski, H.G. Schuster, *Phys. Rev. A*, **35**, 481 (1987).

15. H. Atmanspacher, H. Schungraber, W. Voges, *Phys. Rev. A*, bf 37, 1314 (1988).

16. James Theiler, *J. Opt. Soc. Am. A*, **7**, 1055 (1990).

17. J.W. Havstad, C.L. Ehlers, *Phys. Rev. A*, **39**, 845 (1989).

18. J. Theiler, *Phys. Rev. A*, **34**, 2427 (1986).

19. M.A.H. Nerenberg, C. Essex, *Phys. Rev. A*, **42**, 7065 (1990).

20. V.S. Anishchenko, N.B. Smirnova,*SPIE*, **2098**, 137 (1993).

21. A. Destexhe, J.A. Sepulchre, A. Babloyantz, *Phys. Lett. A*, **132**, 101 (1988).

22. V.S. Anishchenko, M.A. Safonova, P.I. Saparin (in Russian), *Radiotechnika i electronika*, **37**, 467 (1992).

Exploring the Nonlinear Dynamics of a Physiologically Viable Model Neuron

John F. Lindner*, William L. Ditto**

*Applied Chaos Laboratory****
School of Physics
Georgia Institute of Technology
Atlanta GA 30332-0430

Abstract. We describe efforts underway to explore the nonlinear dynamics of the Pinsky-Rinzel model neuron. Via computer simulations, we seek to discover nonlinear phenomena in this physiologically accurate model, thereby complementing ongoing and future experiments. Here we describe the model in detail and analyze it using tools of nonlinear dynamics to demonstrate nontrivial behaviors.

INTRODUCTION

Recent experiments suggest the importance of nonlinear dynamics in biological systems, most notably in hearts and brains (1). There is a need for physiologically relevant computer simulations to complement and guide this experimental work. In the paper, we study a realistic model of a neuron.

Based upon extensive physiological data, Traub (2) developed a 120-variable 19-compartment model of a pyramidal cell from the CA3 region of the hippocampus of the brain. Subsequently, Pinsky and Rinzel (3) reduced this to a much more manageable 8-variable 2-compartment model while still preserving its physiological relevance. This is the model we have implemented and will discuss.

PINSKY-RINZEL NEURON: OVERVIEW

A typical neuron contains a soma, dendrites, and an axon (as well as presynaptic connections). The soma is a lipid bilayer membrane filled with salts (such as KCl and NaCl) ionized in solution. Concentration gradients and electrical gradients cause a potential difference (voltage) between the inside and the outside of the resting cell. Selectively permeable and voltage dependent ion channels (gates, pores) in the lipid membrane act as ion pumps. The dendrites gather electrical impulses that can raise the potential of the soma above a threshold beyond which ions are pumped in and out, thereby causing the neuron to "fire" by emitting a traveling electrical wave along the axon. Figure 1a depicts a neuron.

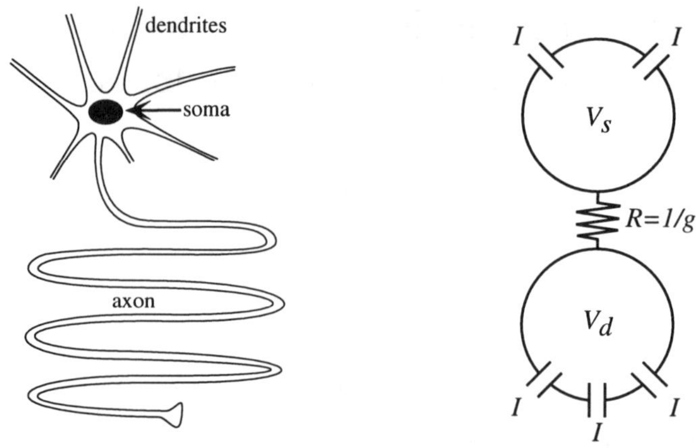

FIGURE 1. Schematic of a real neuron (left) and schematic of the Pinsky-Rinzel model neuron (right).

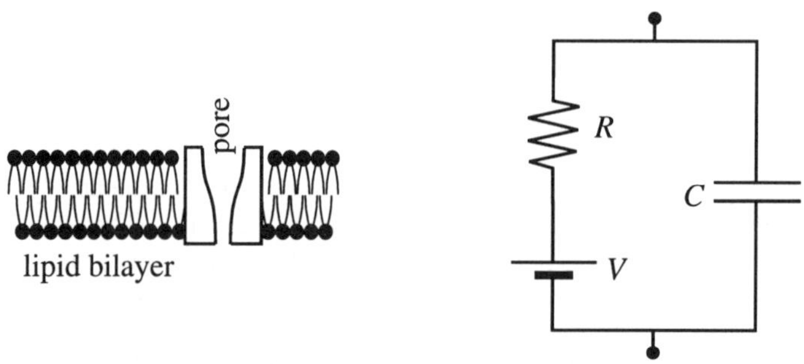

FIGURE 2. A patch of the cell membrane modeled as an equivalent electrical circuit.

The Pinsky-Rinzel model neuron, illustrated in Figure 1b, consists of somatic and dendritic compartments resistively coupled at different potentials. Currents due to the diffusion of K, Na, and Ca ions allow the potentials to vary. A patch of the cell membrane is modeled as an equivalent electrical circuit consisting of a capacitor and a resistor in parallel, as shown in Figure 2. Schematically, the current balance equations for the two compartments follow from differentiating the capacitance definition $CV = Q$:

$$CV_s' = \Sigma I, \tag{1a}$$

$$CV_d' = \Sigma I, \tag{1b}$$

where a typical current is obtained from

$$IR = r(V - V_{\mathrm{ref}}), \tag{2}$$

where the voltage-dependent gating variable r obeys

$$r' = (r_{inf} - r) / \tau \tag{3a}$$

or alternately

$$r' = \alpha(1 - r) - \beta r. \tag{3b}$$

The functions r_{inf} and τ (or alternately α and β) are phenomenologically determined from experimental data so as to mimic the opening and closing of membrane gates (see Figure 3).

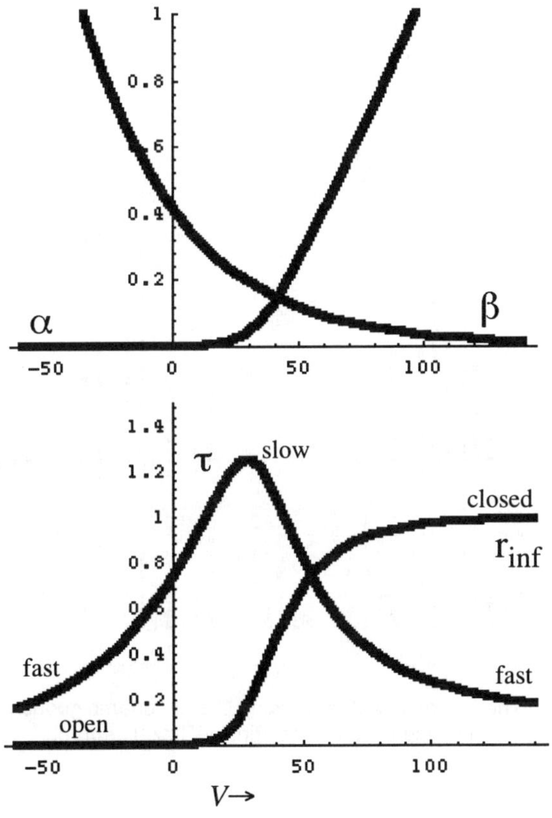

FIGURE 3. Phenomenological rate parameters determine the opening and closing of membrane gates and hence the modeling of the ion pumps.

```c
#include "math.h"
/*********************************************************************/
#define S_NMDA 250
#define SQR(x) ((x)*(x))
#define SMALLER(x,y) ( (x)<(y)? (x):(y) )
#define CHI(x) (SMALLER((x)/250.0, 1.0))
/*********************************************************************/
double h = 0.999;      // gating variables         // INITIALIZATION
double n = 0.001;
double s = 0.009;
double c = 0.007;
double q = 0.010;
double Ca = 0.2;       // calcium level
double v_s = -4.6;     // soma voltage
double v_d = -4.5;     // dendrite voltage
/*********************************************************************/
double p = 0.5;        // relative area of soma to dendrite
double g_c = 2.1;      // coupling conductance between soma & dendrite
double c_m = 3.0;      // membrane capacitance

double v_Na = 120.0;   // reversal potentials
double v_Ca = 140.0;
double v_K = -15.0;
double v_l = 0.0;
double v_syn = 60.0;

double g_l = 0.1;      // ionic conductances
double g_Na = 30.0;
double g_KDR = 15.0;
double g_Ca = 10.0;
double g_KAHP = 0.8;
double g_KCa = 15.0;

double i_se = 0.75;    // applied soma current
double i_de = 0.0;     // applied dendrite current

double g_NMDA = 0.0;   // synaptic conductances
double g_AMPA = 0.0;
/*********************************************************************/
double t = 0.0;        // times in milliseconds
double dt = 1E-5;
/*********************************************************************/
void main(void)        // EVOLVE PR NEURON ONE STEP FORWARD IN TIME
{
    double alpha_m,beta_m,tau_m,m_inf,dm_dt;
    double alpha_h,beta_h,dh_dt;
    double alpha_n,beta_n,dn_dt;
    double alpha_s,beta_s,ds_dt;
    double alpha_c,beta_c,dc_dt;
    double alpha_q,beta_q,dq_dt;
    double dCa_dt;
    double i_s,dVs_dt;
    double r_inf,i_syn,i_d,dVd_dt;                 // auxiliary variables
    /*********************************************************************/
    alpha_h = 0.128*exp((17.0-v_s)/18.0);
    beta_h = 4.0/(1.0+exp((40.0-v_s)/5.0));
    dh_dt = (1 - h)*alpha_h - h*beta_h;            // h gate equation
    /*********************************************************************/
alpha_n = 0.016*(35.1-v_s)/(exp((35.1-v_s)/5.0)-1.0);
beta_n = 0.25*exp(0.5-0.025*v_s);
dn_dt = (1 - n)*alpha_n - n*beta_n;             // n gate equation
/*********************************************************************/
alpha_s = 1.6/(1.0 + exp(-0.072*(v_d-65.0)));
beta_s = 0.02*(v_d-51.1)/((exp(v_d-51.1)/5.0)-1.0);
ds_dt = (1 - s)*alpha_s - s*beta_s;             // s gate equation
/*********************************************************************/
if(v_d<=50.0)
{
    alpha_c = exp((v_d-10.0)/11.0-(v_d-6.5)/27.0)/18.975;
    beta_c = 2.0*exp((6.5-v_d)/27.0)-alpha_c;
}
else//v_d>50.0
{
    alpha_c = 2.0*exp((6.5-v_d)/27.0);
    beta_c = 0.0;
}
dc_dt = (1 - c)*alpha_c - c*beta_c;             // c gate equation
/*********************************************************************/
alpha_q = SMALLER(0.00002*Ca, 0.01);
beta_q = 0.001;
dq_dt = (1 - q)*alpha_q - q*beta_q;             // q gate equation
/*********************************************************************/
alpha_m = 0.32*(13.1-v_s)/(exp((13.1-v_s)/4.0)-1.0);
beta_m = 0.28*(v_s-40.1)/(exp((v_s-40.1)/5.0)-1.0);
tau_m = 1/(alpha_m+beta_m);
m_inf = alpha_m*tau_m;                          // Na activates instantly
/*********************************************************************/
dCa_dt = -0.075*Ca-0.13*g_Ca*SQR(s)*(v_d-v_Ca); // calcium current
/*********************************************************************/
i_s = - g_c*(v_s-v_d)/p                 // electrotonic coupling
      - g_l*(v_s-v_l)                   // leak current
      - g_Na*SQR(m_inf)*h*(v_s-v_Na)    // inward Na current
      - g_KDR*n*(v_s-v_K)               // delayed-rectifier K current
      + i_se/p;                         // soma electrode current
dVs_dt = i_s/c_m;                       // CV = Q -> CV'= I
/*********************************************************************/
r_inf = 1.0/(1.0+0.28*exp(-0.062*(v_d-60.0)));
i_syn = g_NMDA*S_NMDA*r_inf*(v_d-v_syn);// NMDA synaptic current
i_d = - g_c*(v_d-v_s)/(1-p)             // electrotonic coupling
      - g_l*(v_d-v_l)                   // leak current
      - g_Ca*SQR(s)*(v_d-v_Ca)          // inward Ca current
      - g_KAHP*q*(v_d-v_K)              // K afterhyperpolarization current
      - g_KCa*CHI(Ca)*c*(v_d-v_K)       // Ca activated K current
      - i_syn/(1-p)                     // synaptic current
      + i_de/(1-p);                     // dendrite electrode current
dVd_dt = i_d/c_m;                       // CV = Q -> CV'= I
/*********************************************************************/
h += dt*dh_dt;                  // synchronous Euler update of 8 variables
n += dt*dn_dt;                  // (Euler updates NOT recommended!)
s += dt*ds_dt;
c += dt*dc_dt;
q += dt*dq_dt;
Ca += dt*dCa_dt;
v_s += dt*dVs_dt;
v_d += dt*dVd_dt;
```

FIGURE 4. The fine print: Actual working C code unambiguously defines our implementation of the Pinsky-Rinzel neuron.

PINSKY-RINZEL NEURON: IMPLEMENTATION

To avoid any possible confusion due to ambiguity or typographical error, we define our implementation of the Pinsky-Rinzel model by Figure 4, which contains actual working C code. Notice that there are 8 first-order differential equations in 8 unknowns, the current equations being supplemented by an equation governing the concentration of calcium ions.

In our actual programs, we can toggle between two different integration schemes: fixed time step 4th-order Runge-Kutta and variable time step 5th-order Runge-Kutta (4). The variable time step is ideal for the spiky time series generated by the model: giant steps between spikes/bursts accelerate the integration while tiny steps during the spikes/bursts preserve its accuracy.

To determine Inter-Spike Intervals (ISIs), we toggle between 2 different peak detection schemes. The first triggers on a threshold value of the somatic voltage. The second triggers on a threshold of the rate-of-change of the somatic voltage. The latter requires the user to supply a refractory time during which there is no triggering.

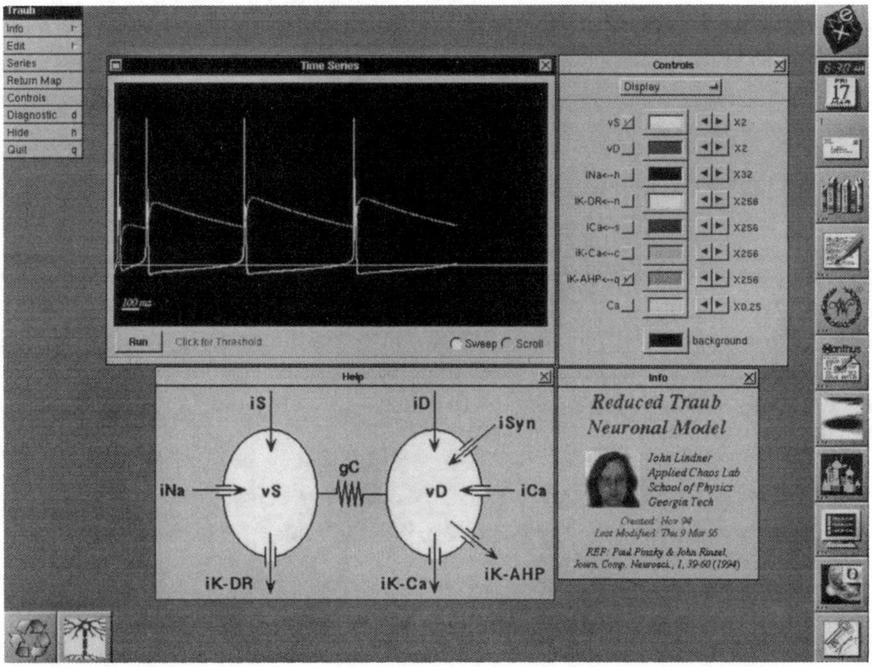

FIGURE 5. Screenshot of the NEXTSTEP application "Traub", an early version of our impementation of the Pinsky-Rinzel reduced Traub model neuron.

We first implemented the model on a 60 MHz Pentium (5) computer running the NEXTSTEP operating system. NEXTSTEP is a powerful and elegant operating system and a superb application development environment. Figure 5 is a screenshot of our NEXTSTEP application "Traub", a precursor to "PR Neuron". We have also ported the application to the Macintosh operating system using the Code Warrior development environment. On an 80 MHz Power PC Macintosh, the model neuron spikes at a rate of ~3Hz when duplicating Figure 2a of reference (3).

EXAMPLE ANALYSIS

We have begun to apply the tools of nonlinear dynamics to the Pinsky-Rinzel model neuron. We have obtained interesting time-one return maps for the interspike intervals. Some of these maps display period-2, period-6, and possible quasiperiodic behavior, as well as maps consistent with chaos. We are attempting to locate a period doubling route to chaos in the model.

Figure 6 is an example of these results. We stimulate the neuron with a constant external current and, in this parameter regime, its bursting is dominated by 2 frequency components. Note the structure in the ISI time-one return map and in the corresponding ISI histogram. The time series was sampled once every 100 points.

FUTURE WORK

We are currently using the simulations to complement experiments which seek a variety of nonlinear effects in neural tissue. We expect such phenomena as stochastic resonance, control of chaos, and chaotic encoding of information to be relevant to actual neural function. We are also extending these explorations to include networks of neurons. We plan to tune the model to make it as quantitatively accurate as possible, so that we can first test *in silico* what we later will try *in vitro,* and ultimately *in vivo*.

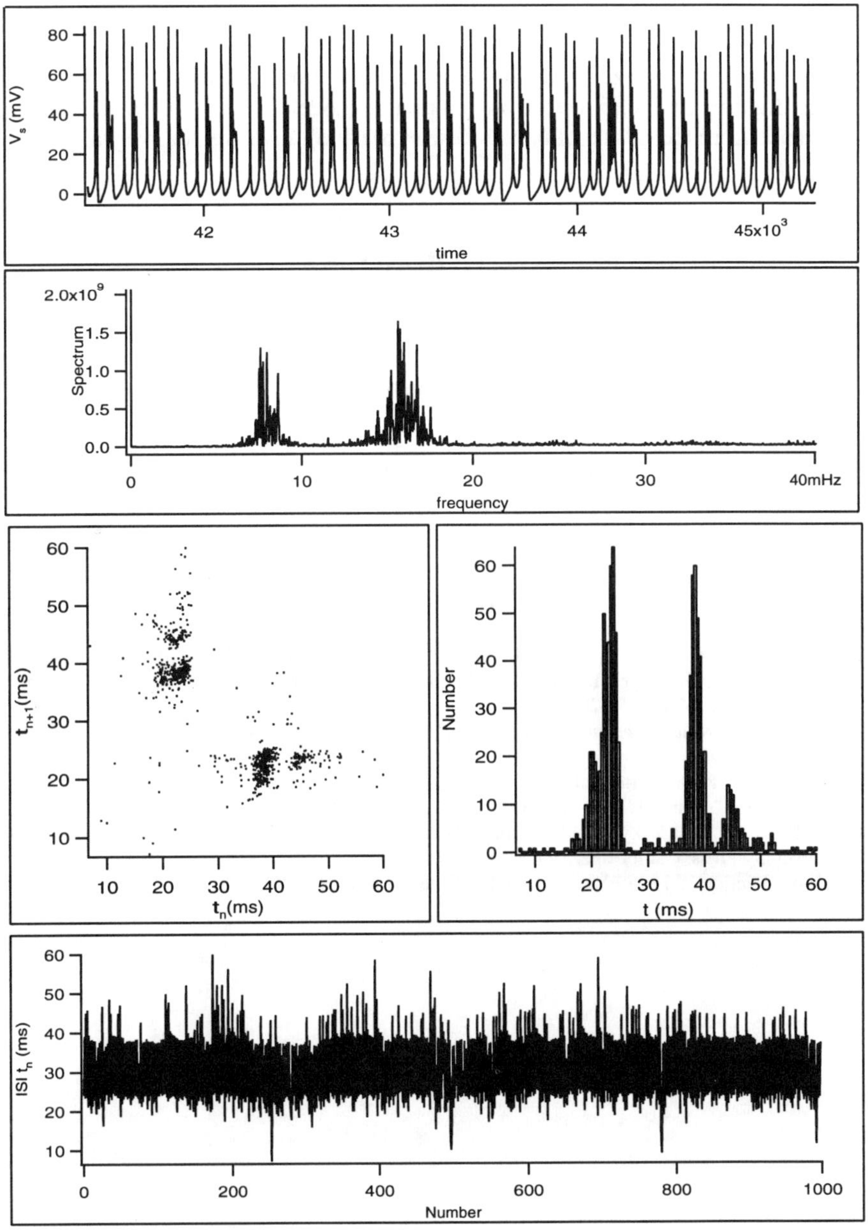

FIGURE 6. Time series and its spectrum (top). ISI return map , ISI histogram, and ISI sequence (bottom). The soma is being stimulated by an external current.

ACKNOWLEDGMENTS

We thank John Rinzel and Paul Pinsky for helpful discusssions. We thank James Aguirre for helping test our computer models. W. L. D. acknowledges assistance from the Office of Naval Research Young Investigator Program.

AUTHOR NOTES

*jlindner@acl1.physics.gatech.edu
**wditto@acl1.physics.gatech.edu
***http://acl1.physics.gatech.edu/aclhome.html

John F. Lindner's permanent address is:
The College of Wooster, Wooster, OH 44691.

REFERENCES

1. A. Garfinkel, M. Spano, W. Ditto, J. Weiss, "Controlling Cardiac Chaos", Science 257, 1230-1235; S. Schiff, K. Jerger, D. Duong, T. Chang, M. Spano, W. Ditto, "Controlling Chaos in the Brain", Nature, 370, 615-620 (1994).
2. R. Traub, R. Miles, *Neuronal Networks of the Hippocampus,* (Cambridge University Press; New York, 1991).
3. P. Pinsky, J. Rinzel, "Intrinsic and Network Rhythmogenesis in a Reduced Traub Model for CA3 Neuron", Journ. Comp. Neuroscience, 1, 39-60 (1994).
4. W. Press, S. Teukolsky, W. Vetterling, B. Flannery, *Numerical Recipes in C,* (Cambridge University Press; Cambridge, 1992).
5. All our Pentium chips are new ones, free of the notorious FDIV errror.

10. NEW METHODS FOR MODELING NONLINEAR SYSTEMS

Stochastic Nonlinear Modeling

Lisa Borland

Department of Physics
University of California at Berkeley, Berkeley, CA 94720

Abstract. The problem of simultaneously modeling the underlying nonlinear stochastic and deterministic dynamics of a time series is discussed.

INTRODUCTION

We present a short review of a procedure which we call SEQUIN (Stochastic EQUation INFerence) to estimate the underlying nonlinear deterministic and stochastic dynamics of an observed time series. This procedure has been developed extensively in previous papers [1-5], where we have shown that it is capable of analyzing Markov processes in both one and higher dimensions. It successfully reconstructs the underlying dynamics of systems where both the deterministic and stochastic dynamics are nonlinear functions of the observed state variable.

STOCHASTIC EQUATIONS OF MOTION

Assume that we have observations of one particular realization of a stochastic process. Due to the noise, a second realization will hardly give us the same observations. In trying to model the underlying dynamics of the process it is therefore useful to study properties of the evolution of the probability of realizations. We shall see that it is possible to derive the underlying equations of motion of the process from estimations of the joint temporal probability distribution of the process. This approach allows us to study both continuous and discrete time systems [6]. We shall in the following review how the technique works for one-dimensional continuous time systems. First, however, we remind the reader of the general theory of stochastic equations.

The particular class of processes which we have in mind to characterize are such which can be described by a stochastic Ito-Langevin equation [cf. 7] of the form

$$\dot{q} = K(q) + g(q)\eta(t) \tag{1}$$

In this equation, $K(q)$ represents a deterministic force, while $g(q)$ is a stochastic force modulated by the random variable $\eta(t)$ which has the properties

$$< \eta(t) >= 0 \tag{2}$$

and

$$< \eta(t)\eta(t') >= \delta(t - t') \tag{3}$$

Written in a differential form, it can be shown that eq(1) becomes

$$dq = K(q)dt + g(q)dw \tag{4}$$

where

$$< dw >= 0 \tag{5}$$

and

$$< dw^2 >= \sqrt{dt} \tag{6}$$

Both $K(q)$ and $g(q)$ may be nonlinear functions of q.

An equivalent representation of the process defined by eq(4) is given by the Fokker-Planck equation

$$\frac{d}{dt}P(q(t)) = -\frac{d}{dq}(K(q)P(q)) + \frac{1}{2}\frac{d^2}{dq^2}(G^-1(q)P(q)) \tag{7}$$

with

$$G^{-1}(q) = \frac{1}{2g^2(q)}, \tag{8}$$

which describes the temporal evolution of the probability distribution $P(q(t))$ of values of q.

SEQUIN - A REVIEW

The method SEQUIN which we present here is based on a methodology which we have previously formulated for continuous time processes in one and higher dimensions [1-5]. For simplicity, we restrict our current review to the one-dimensional case.

Starting from a stationary time series of a stochastic process, it is our goal to explicitly reconstruct the underlying dynamics in the form of an Ito-Langevin equation (4). We wish to emphasize that this is a different problem to the one of fitting a curve through the time series, which is the purpose of many modeling techniques such as ARMA and other regression methods [cf. 8] Our approach attempts to capture the structural form of a generating equation for the time series.

The time series $q(t)$ is sampled at time points $t = 1, 1+\tau, \cdots, j, j+\tau, \cdots, T$ such that it can be represented as a sequence $q_i, i = 1, \cdots, T$. The procedure SEQUIN starts out by estimating the joint probability distribution $P(q_{i+\tau}, q_i)$ which, for small but finite τ, corresponds to a short-time solution of eq(7). We then show that it is possible to extract explicit formulas for the underlying dynamics of eq(4) from $P(q_{i+\tau}, q_i)$. The way this is done consists of 3 steps, which are presented below.

Step 1: The Probability Distribution

First, an unbiased estimate of the joint short-time probability distribution $P(q_{i+\tau}, q_i)$ of the process is estimated using the Maximum Information Principle (MIP) of Jaynes [9]. To this end, we need to calculate certain short-time correlations of the type

$$< q_i q_{i+\tau} >= \frac{1}{T} \sum_{i=1}^{i=T} q_i q_{i+\tau} \tag{9}$$

from the data. These serve as constraints to be used in conjunction with the MIP, and simultaneously determine the model class with which we wish to represent the underlying dynamics. Using the model class of polynomials, constraints of the form

$$\begin{aligned} &< q_i^\nu > && \nu = 0 \cdots, N, && (10) \\ &< q_i^\kappa q_{i+\tau} > && \kappa = 0 \cdots, M, && (11) \\ &< q_i^\mu q_{i+\tau}^2 > && \mu = 0 \cdots, L, && (12) \end{aligned}$$

allow us to study processes with additive and weak multiplicative noise. We shall alternatively refer to the set of constraints eq(10) - eq(12) as

$$< \Omega_j >_T \qquad j = 0, \cdots, R \tag{13}$$

with $R = N + M + L + 3$. The subscript T indicates that they are calculated as temporal averages from the time series. The polynomial orders N,M and L correspond to the polynomial degree with which one chooses to represent the problem. Note, however, that we need not characterize the variable $q_{i+\tau}$ by more than its second moment. This choice of constraints can be extended to deal with other classes of processes, e.g. processes where strong multiplicative noise may occur [3],[5].

The formal expression which we obtain as the MIP estimate of $P(q_{i+\tau}, q_i)$ is given by the product of a conditional distribution $P(q_{i+\tau} \mid q_i)$ and a marginal distribution $P(q_i)$ according to

$$P(q_{i+\tau}, q_i) = P(q_{i+\tau} \mid q_i)P(q_i) \tag{14}$$

with

$$P(q_i) = \exp\{-\sum_{\nu=0}^{N} \lambda_\nu^{(1)} q_i^\nu\} \tag{15}$$

and

$$P(q_{i+\tau} \mid q_i) = N(q_i)\exp\{-\frac{1}{\mid C(q_i)\mid}\left(q_{i+\tau} - \frac{B(q_i)}{2C(q_i)}\right)^2\}. \tag{16}$$

Here,

$$B(q_i) = \sum_{\mu=0}^{M} \lambda_\mu^{(2)} q_i^\mu \tag{17}$$

$$C(q_i) = \sum_{\kappa=0}^{L} \lambda_\kappa^{(3)} q_i^\kappa \tag{18}$$

$N(q_i)$ must be chosen so that $P(q_{i+\tau} \mid q_i)$ is normalized to 1. For simplicity, we assume in this paper that $C(q_i) =\mid C(q_i) \mid > 0$. In these equations, the parameters $\{\lambda_j\}, j = 0, \cdots, R$ are Lagrange multipliers

$$\lambda_\nu^{(1)}, \quad \nu = 0, \cdots, N, \tag{19}$$

$$\lambda_\mu^{(2)}, \quad \mu = 0, \cdots, M, \tag{20}$$

$$\lambda_\kappa^{(3)}, \quad \kappa = 0, \cdots, L \tag{21}$$

which correspond respectively to the constraints $< \Omega_j >_T$ of eq(10) - eq(12).

Step 2: Estimating the Parameters

In the above section, we wrote down a formal estimate of $P(q_{i+\tau}, q_i)$ in eq(15) and eq(16). This estimate is a function of the unknown parameters $\{\lambda\}$. To find the parameters $\{\lambda\}$, we minimize the Kullback-Leibler distance between the "true" unknown distribution of the process and our MIP estimate P. This is done with a gradient strategy technique resulting in:

$$\dot{\lambda}_j = < \Omega_j >_P - < \Omega_j >_T \qquad j = 0, \cdots, R \tag{22}$$

where $< \Omega_j >_T$ represents the calculated values of eq(10) - eq(12) while $< \Omega_j >_P$ is of the type

$$< q_i q_{i+\tau} >_P = \int\int q_i q_{i+\tau} P(q_{i+\tau}, q_i) dq_{i+\tau} dq_i. \tag{23}$$

$P(q_{i+\tau}, q_i)$ is a function of the $\{\lambda\}$ as according to eq(15) and eq(16). The numerical values of the parameters $\{\lambda\}$ are obtained from the stationary solution of eq(22), which corresponds to the case when averages $< \Omega_j >_P$ over the estimated distribution P are equal to the calculated ones $< \Omega_j >_T$.

Step 3: The Equations of Motion

If the conditional probability distribution of eq(16) behaves as a δ-function with respect to the variable $q_{i+\tau}$ for $\tau \to 0$, it can be shown that it must be a short-time propagator to the Fokker-Planck equation of eq(7). This is the case if $P(q_{i+\tau} \mid q_i)$ takes on the form

$$P(q_{i+\tau} \mid q_i) = N \exp\{\frac{1}{G\tau}(q_{i+\tau} - q_i - \tau K(q_i))^2\} \tag{24}$$

By expanding the parameters $\{\lambda\}$ occurring in the expression of eq(16) for $P(q_{i+\tau} \mid q_i)$ into power series of τ we can identify eq(16) with eq(24). By then taking the limit $\tau \to 0$, we are able to extract the following equations for the functions $K(q)$ and $G(q)$:

$$G(q_i) = \tau \sum_{\kappa=0}^{M} \lambda_\kappa^{(3)} q_i^\kappa \tag{25}$$

and

$$K(q_i) = \frac{1}{2\tau} \frac{\lambda_0^{(2)} + \sum_{\mu=1}^{M} (\lambda_\mu^{(2)} - 2\lambda_{\mu-1}^{(3)}) q_i^\mu}{\sum_{\kappa=0}^{M} \lambda_\kappa^{(3)} q_i^\kappa} \tag{26}$$

Using the equality

$$g(q_i) = \frac{1}{\sqrt{2G(q_i)}} \tag{27}$$

according to eq(8), we have thereby obtained expressions for the deterministic force $K(q)$ and the stochastic force $g(q)$ occurring in the Ito-Langevin equation of eq(4).

Summary

We have presented, in three steps, the theory behind SEQUIN. In practice, one simply calculates the constraints $< \Omega_j >_T$ of eq(10) - eq(12) and inserts these values into the eq(22) to obtain numerical values for the parameters $\{\lambda\}$. Then, the parameters $\{\lambda\}$ together with the value of the sampling time τ are inserted into equations eq(26) and eq(25) resulting in explicit expressions for the deterministic and stochastic terms of the Ito-Langevin equation eq(4).

DISCUSSION

We have given a short review of the procedure SEQUIN for reconstructing the underlying dynamics of a stochastic time series. The discussion was limited to that of one-dimensional stationary processes. However, as shown in earlier papers, it is in principle possible to analyze processes of arbitrary dimension in a similar fashion. We have recently extended SEQUIN to deal with discrete time processes as well. A variety of simulated time series have been used to validate the method, and we are currently applying it to real experimental data.

REFERENCES

1. Haken, H., *Information and Self-Organization, A Macroscopic Approach to Complex Systems*, Springer Ser. Synergetics, Vol. 40, Springer Verlag, Berlin, Heidelberg, 1988.

2. Borland,L., and Haken, H., "Unbiased Determination of Forces Causing Observed Processes. The case of additive and weak multiplicative noise", *Z. Phys. B - Condensed Matter* **81**, 95, 1992.

3. Borland, L., and Haken, H., "Unbiased Estimate of Forces from Measured Correlation Functions, including the Case of Strong Multiplicative Noise", *Ann. Physik* 1, 452, 1992.

4. Borland, L., and Haken. H., "Learning the Dynamics of Two-Dimensional Stochastic Markov Processes", *Open Systems and Information Dynamics* **1** No. 3, 311, 1993.

5. Borland, L., Haken. H., " On the Constraints Necessary for Macroscopic Prediction of Stochastic Time-Dependent Processes", *ROMP* **33**, 35, 1993.

6. Borland L., in prep., 1995.

7. Gardiner, C.W., *Handbook of Stochastic Methods*, 3rd Printing, Springer Verlag, Berlin, Heidelberg, 1994.

8. Tjϕstheim, D., " Nonlinear Time Series: A Selective Review", *SFB 123*, Universit"at Heidelberg, 1992.

9. Jaynes, E. T., " Information Theory and Statistical Mechanics", I, II *Phys. Rev.* **106**, 620 (1957), **108**, 171, 1957.

An Oscillatory Neural Network Unit Model

M.Bazhenov, M.Rabinovich, and L.Rubchinsky

Institute of Applied Physics, Russian Academy of Science
46 Uljanov Str., 603600 Nizhny Novgorod, Russia

Abstract. A mathematical model of a neuron describing its oscillatory activity is constructed and investigated. The model in the form of a system of three ordinary differential equations demonstrates both regular behaviour of the time dependence of electric membrane potential and its chaotic dynamics. The absence of rigorous heteroclinic trajectories in the phase space of the system is proved and the transition to chaos is investigated. The processes of synchronization of the oscillators are discussed.

INTRODUCTION

The neuron models currently used for modelling neural networks can be classified into three groups. The first one comprises the models for which detailed information about ion membrane currents is needed. These include, in particular, the classical Hodgkin–Huxley model (1) as well as its numerous generalizations taking into account additional ion channels, the action of neuromediators, etc. To the second group belong simplified neuron models based on the Van der Pol or the FitzHugh–Nagumo (2,3) equations. Their simplest variant is the Hopfield's model (4) in which a neuron, like spin, may be only in two states (they may be associated with different branches on the curve of slow motions in the Van der Pol or FitzHugh–Nagumo equations). Such simplified models are highly promising for modelling neural networks with a very large number of elements, when topology of couplings is of major significance whereas intrinsic neuron dynamics may be modelled by combining several elements possessing trivial dynamics. Finally, the third group includes phenomenological models which are based on the simplest possible equations but, at the same time, give a qualitatively correct picture of the dynamics and typical bifurcations of the neuron. The most attractive of them are the Rose–Hindmarsh (5) and Chay (6,7) models. Their remarkable property is that they describe generation of spikes and bursts, slow oscillations and chaotic activity of a neuron.[1] The model we propose here belongs to the third group. Its novelty compared to

[1]Paraphrasing Einstein we can say: We need the simplest possible neuron model but not simpler than that.

other available models is that it gives a qualitatively correct description of fast oscillations of two different types which are observed in some neuron cells. We believe that such models are very promising for modelling the principles of operation of small neural networks, for example, the ones responsible for motory activity of living organisms. They may also enable one to answer very difficult questions such as: Why is collective dynamics of small nervous systems frequently simpler than the dynamics of a single neuron? How does the phenomenon of multistability occur in a small nervous system consisting of chaotic elements?

CONSTRUCTION OF THE MODEL

There exist numerous experimental data indicating that oscillatory activity is inherent in structural elements of nervous system. The time dependence of electric potentials is frequently oscillations in which one can clearly distinguish portions having different forms, frequencies, etc., i.e., different patterns (see, e.g., (8–10) as well as (11) where data of experiments on neurons from lobster stomach are presented).

From general considerations, the simplest mathematical ODE model may be three-dimensional. The form of oscillations observed in experiments enables us to make some suppositions about the structure of phase space of the system of interest, in particular, about the existence in this phase space of singular points of the saddle-focus type, with each saddle-focus corresponding to a pattern of its own.

We propose the following dynamical system consisting of three ODEs:

$$\begin{aligned} \dot{x} &= z - 2y^2 + (\delta - \alpha z)y + \gamma x \\ \dot{y} &= 2xy - (\delta - \alpha z)x \\ \dot{z} &= -2z(x+1) \end{aligned} \tag{1}$$

This system is a modification of the system proposed for description of the stochasticity emerging in decay processes (12). Similar systems are known in biology. For example, mathematical models for membrane regulation of cell cycle (13) and others. In a certain sense, our model resembles the known model of the Wilson–Cowan oscillator (14) but the two-dimensional phase space and simple form of influence function of the latter make description of complex (periodic or chaotic) dynamics impossible.

The solution $x(t)$ of the system (1) for the found values of the parameters $\alpha = \alpha_0 = 2.5$, $\delta = \delta_0 = 2.5$ and $\gamma = \gamma_0 = 0.25$ (see Fig. 1) coincides qualitativly with data of experiments (11) fairly well.

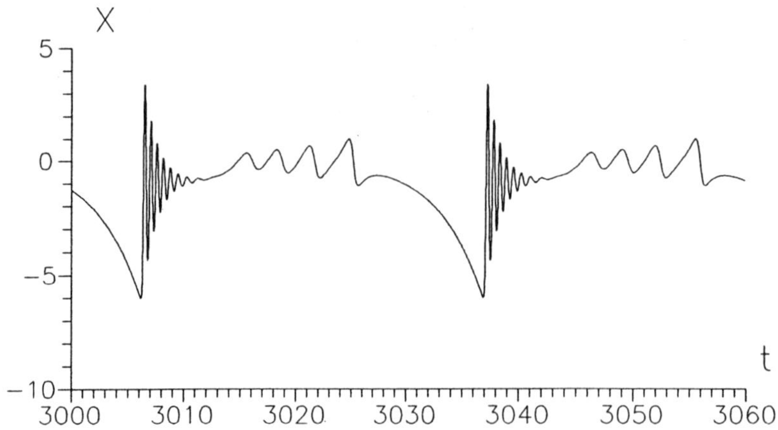

Figure 1: Solution $x(t)$ of system (1) for $\alpha = 2.5$, $\delta = 2.5$ and $\gamma = 0.25$. Periodic oscillations of complex form with two pronounced types of temporal patterns are observed.

THE PROPERTIES OF THE MODEL

Numerical investigation shows that one stable limit cycle of a rather complex form exists in the phase space of the system (1). An axonometric projection of this cycle is presented in Fig. 2. The limit cycle passes close to the singular points of the system and, consequently, its form depends significantly on the vector field structure in the neighborhood of the singular points, which can be considered in a linear approximation.

The system (1) possesses three equilibrium states:

1. S_1 $(x = 0,\ y = 0,\ z = 0)$ – unwinding saddle-focus with eigenvalues $\lambda_1 = -2$ and $\lambda_{2,3} = \gamma \pm \sqrt{\gamma^2 - 4\delta^2}/2$;

2. S_2 $(x = 0,\ y = \delta/2,\ z = 0)$ – saddle with eigenvalues $\lambda_1 = \gamma$, $\lambda_2 = 0$ and $\lambda_3 = -2$; and

3. S_3 $(x = -1,\ y = (\delta - \gamma\alpha)/2,\ z = \gamma)$ – winding saddle-focus.

The two saddle-foci possessing different imaginary parts of the eigenvalues (i.e., frequencies) correspond to different patterns depending on the function $x(t)$ (see Fig. 1).

There exists in the phase space of the system (1) an invariant set (surface $z = 0$). This means that any solution of (1) belonging to the surface $z = 0$ at $t = 0$ will belong to this set for any $t > 0$. Furthermore, it is easy to prove that $z(t) > 0$ for any $0 < t < \infty$ if $z(0) > 0$ and the condition $z = 0$ can be

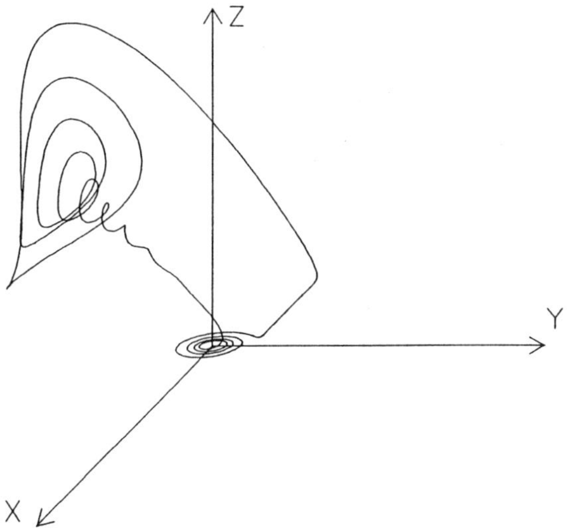

Figure 2: Limit cycle of system (1) for $\alpha = 2.5$, $\delta = 2.5$ and $\gamma = 0.25$.

fulfilled only when $t \to \infty$. The system (1) is reduced on the manifold $z = 0$ to the two-dimensional ODE system

$$\begin{aligned} \dot{x} &= -2y^2 + \delta y + \gamma x \\ \dot{y} &= 2xy - \delta x \end{aligned} \qquad (2)$$

It is easy to see that the function $V = x^2 + y^2$ as well as the derivative for the solutions of the system (1) $\dot{V} = \gamma x^2$ are positive definite if $\gamma > 0$ (the set $x = 0$ is not a phase trajectory of (2)) and consequently any solution of (2) belonging to the surface $z = 0$ tends to infinity when $t \to \infty$. One must note that the line $y = \delta/2$ is invariant to the system (2) and simple analysis of the phase space of (2) indicates that $y \to \delta/2$, $x \to -\infty$ when $t \to \infty$.

The considerations presented above enable us to describe qualitatively the behaviour of the phase trajectories of the system (1) as follows. Any trajectory for which the condition $1+x > 0$ is fulfilled (we consider the invariant halfspace $z \geq 0$) will tend asymptotically to the surface $z = 0$. It will occur rather fast due to the strong compression of the phase flow along the z-direction in the vicinity of the point S_1 ($\lambda_1 = -2$). Then the phase trajectory evolves slowly in the ϵ-vicinity ($0 < \epsilon \ll 1$) of the surface $z = 0$ (the value of ϵ will be less to the trajectories lie closer to the one-dimension manifold connecting S_1 and S_3). This portion of the trajectory may be approximately described by the two-dimensional system (2). However, as soon as the condition $1 + x > 0$ is violated (this always takes place as $x \to -\infty$ when $t \to \infty$ in the ϵ-vicinity of the manifold $z = 0$) the trajectory will leave the ϵ-vicinity of $z = 0$ and its

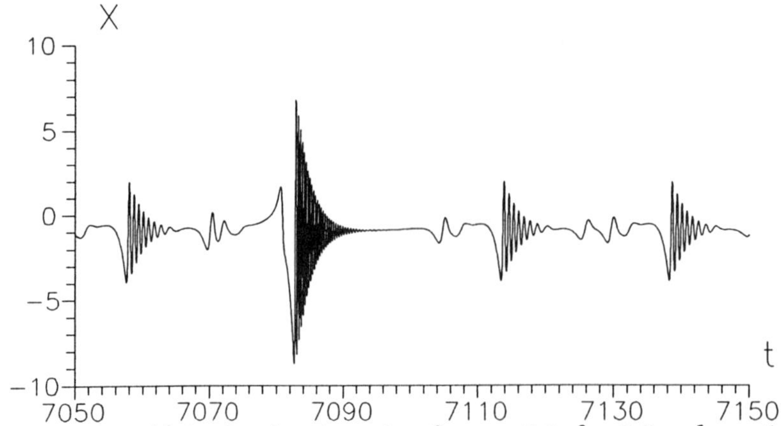

Figure 3: $x(t)$ in the chaotic regime for $\alpha = 2.5$, $\delta = 2.5$ and $\gamma = 1$.

further behaviour will be determined by the complete system (1). Numerical simulation demonstrates that, further, the trajectory returns to the vicinity of the point S_3 and, consequently, the condition $1+x > 0$ is fulfilled again. Thus, the trajectory remains bounded although it is difficult to prove this supposition analytically. In particular, it is easy to see that for any number $R > 0$ there exist the initial conditions inside the sphere of radius R $(x_0^2 + y_0^2 + z_0^2 < R^2)$ for which the condition $x(t)^2 + y(t)^2 + z(t)^2 > R^2$ is met when $t > t_1 > 0$ where t_1 depends from R and initial conditions.

It is interesting to investigate the bifurcations and the birth of chaos in our dynamical system with the change of the parameters. For $\alpha = 0.25$, $\gamma = 0.25$ and small δ (of the order of 0.2), a stable limit cycle of simple form exists in the system (1) and the oscillations are quasi-sinusoidal. As δ is increased up to 0.38, a period doubling bifurcation occurs and with a still further increase of δ up to δ_0 no bifurcations are observed.

Chaos may exist in the system (1). As γ grows up to $\gamma = 0.77$, a period doubling bifurcation takes place and further a strange attractor is formed through a cascade of period doublings as γ grows up to about 0.88. The projection of the attractor and the solution $x(t)$ in the chaotic regime are presented in Figs. 4, 5.

Of great significance is the existence of simple periodic, complex periodic and chaotic regimes in one system, for which at least a three-dimensional space is needed. We will speak about it in detail below in text.

By introducing into the system (1) additional parameters one can control location of the singular points of the system, i.e., the average values of different patterns in $x(t)$. Let us write the third equation of (1) in the form

$$\dot{z} = -2z(x + \beta) \tag{3}$$

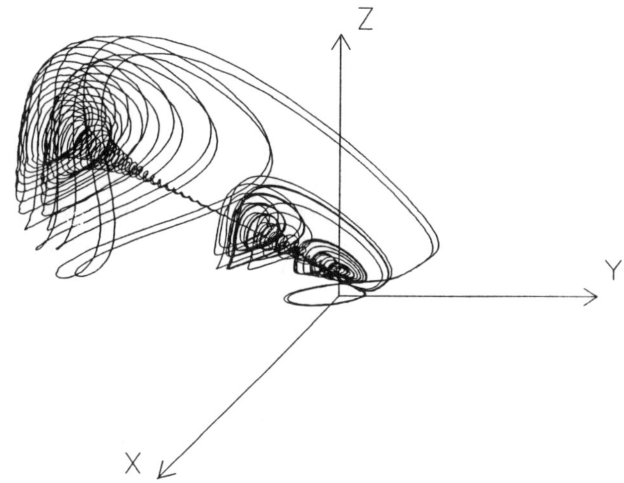

Figure 4: Strange attractor in system (1) for $\alpha = 2.5$, $\delta = 2.5$ and $\gamma = 1$.

By choosing β to be less than unity, for example $\beta = 0.6$, we find that the coordinate of the point S_3 and, hence, the average value near which x oscillates change to be $-\beta = -0.6$. However, this variation of β entails a change in the number of oscillations in the patterns and, with a great decrease of β, the structure of the phase space changes significantly because S_1 and S_2 have no negative eigenvalues at $\beta < 0$.

Let us note that, strictly speaking, system (1) is not structurally stable in the class of the general dynamical systems. This happens because of degeneration of the system (1): $z = 0$ plane is an invariant set. That is why structurally stable heteroclinic trajectories can not exit in the phase space of (1) and only one heteroclinic trajectory, which is not structurally stable, connects singular points S_3 and S_1. One can easily eliminate the degeneration by changing the last equation of (1) by $\dot{z} = -2(z - ax)(x+1)$ where $0 < a \ll 1$. Since a is fairly small the behaviour of the system doesn't change qualitatively but there is no any degeneration. It's worth to note that stable and unstable two-dimensional manifolds of the singular points S_1 and S_3 will intersect structurally stablly in a modified system. If we suppose that there exists, for some values of the parameters, a heteroclinic trajectory connecting S_3 and S_1 in the system (1) (with $a \neq 0$), we will have a heteroclinic circuit and its breaking down with the change of parameters may give birth to the stable periodic motion investigated in our paper.

DISCUSSION

Our investigations enable us to understand the mathematical images of complex and simple behavior of the cell and, which is no less important, the transitions to other regimes with the change of the parameters. Since we are concerned with structurally stable dynamical systems with the variation of parameters our system gives a qualitatively correct picture even if it does not describe the real process quantitatively correctly. This is what makes such phenomenological models valuable.

Apparently, periodic oscillations of complex form having different time scales in different portions, and the birth of chaos are possible in a single cell. (Perhaps, in view of this the formation of regular structures in neural networks and in the brain appears to be still more nontrivial.) A great amount of various experimental and theoretical investigations (this one including), as well as general mathematical considerations on the existence of strange attractors in dynamical systems verify that the chaos is quite typical in biology.

It appears highly promising to use models similar to the one described here as elements of neural networks. There exists broad literature on investigation of different regimes: synchronization, chaos, structures in chains, lattices and other systems of coupled oscillators and their extended analogs (e.g. (15–18)). Our numerical experiments on two systems of the form (1) with simple linear dissipative coupling show the existence of the phenomena such as synchronization and chaos formation when partial systems behave non-chaotically. An inverse situation may also occur. The motion of two chaotic neurons may become regular due to coupling. The coupling makes one of the saddle periodic motions belonging to the attractor stable. As the coupling parameters are changed, another saddle motion becomes stable, and so on. Investigation of two and more chaotic elements like the ones we considered in this paper is one of our primary tasks for the nearest future.

We believe that the use of rather simple oscillators possessing complex behavior as oscillatory elements of systems (which may correspond either to a single neuron cell or to a whole ensemble of neuron cells) will enable one to construct mathematical models for a great amount of phenomena in ensembles of coupled cells, in particular in the brain, and to get an insight into them.

ACKNOWLEDGEMENTS

We highly appreciate P. Rowat's information about experiments on electric activity of neurons (11). M.B. wishes to acknowledge the hospitality of the Niels Bohr Institute where some of this work was carried out. L.R. wishes to thank International Soros Scientific Educational Program for Soros student grant. This paper was made possible in part thanks to the Russian Foundation for Basic Research (project code N 94–02–03263-a) and the International Science Foundation (grant N NOU 000).

REFERENCES

1. A.L.Hodgkin and A.F.Huxley, *J. Physiol.* (London), **117**, 500 (1952).

2. R.FitzHugh, *Biophys. J.*, **1**, 445 (1961).

3. J.Nagumo, S.Arimoto and S.Yoshizawa, *Proc. IRE*, **50**, 2061 (1962).

4. J.J.Hopfield, *Proc. Nat. Acad. Sci. USA*, **79**, 2554 (1982).

5. J.L.Hindmarsh and R.M.Rose, *Proc. R. Soc. Lond.*, **B221**, 87 (1984).

6. T.R.Chay, *Physica D*, **16**, 223 (1985).

7. Y.S.Fan and T.R.Chay, *Phys. Rev. E*, **51**, 1012 (1993).

8. W.Singer, *Progress in Brain Research*, ed. by T.P.Hicks, S.Molotchnikoff and T.Ono, **95**, p.461 (1993).

9. R.R.Llinas, *Science*, **242**, 1654 (1988).

10. A.Goldbeter and F.Moran, *Eur. Biophys. J.*, **15**, 277 (1988).

11. P.Rowat, "Experiments with neurons lobster stomatogastric gangloin" (unpiblished data, 1994).

12. M.I.Rabinovich, *Sov. Phys. Usp.*, **21**, 443 (1978).

13. D.I.Chernavskii, A.A.Polezhaev, and E.I.Volkov, *Cell Biophys.*, **45**, 143 (1982).

14. H.Wilson and J.Cowan, *Kybernetik*, **13**, 55 (1973).

15. A.Selverston, *Ann. Rev. Neurosci.*, **16**, 531 (1993).

16. G.V.Osipov and M.M.Sushchik, *Phys.Lett.A*, **201**, 205 (1995).

17. M.Bazhenov, M.Rabinovich, and L.Rubchinsky, submitted to *J. Stat. Phys.*

18. H.D.I.Abarbanel, M.V.Bazhenov, R.Huerta, M.I.Rabinovich, L.L.Rubchinsky, and M.M.Sushchik, "Synchronization in neural ensembles" (being prepared for publication).

A New Method of Modelling and Numerical Simulation of Nonlinear Dynamical Systems

Tiberiu Coloşi* and Steliana Codreanu**

*Technical University of Cluj-Napoca
15, Daicoviciu St., RO-3400 Cluj-Napoca, Romania
**"Babeş-Bolyai" University, Department of Theoretical Physics,
1, Kogălniceanu St., RO-3400 Cluj-Napoca, Romania

Abstract. This work presents the most significant aspects of an original method of modelling and numerical simulation of nonlinear (linear) dynamical systems (1) it assures the local-iterative linearization (LIL) of nonlinear (linear) differential equations and transforms them, in the close proximity of a pivot moment, into algebraic equations. The use of this method is illustrated in the study of a particular nonlinear dynamical systems. The conclusions highlight the advantages of the proposed procedure.

INTRODUCTION

The proposed (LIL) method refers to the dynamical systems or process, mathematically described by the following differential equation:

$$\frac{d^n y}{dt^n} = F\left(u, \frac{du}{dt}, \ldots, \frac{d^m u}{dt^m}, y, \frac{dy}{dt}, \ldots, \frac{d^{n-1} y}{dt^{n-1}}\right), \tag{1}$$

where u=u(t) is an independent, known variable (the input signal) and y=y(t) is the dependent, unknown variable (the output signal). The initial conditions are known and always $n \geq m$ with $n \geq 1$. The right member of the equation (1) is a continuous nonlinear (linear) function. We observe that (1) represents the analytical model of numerous scientifical or technological dynamical processes.

By defining the state (phase) variables:

$$x_1 = y, \quad x_2 = \dot{x}_1, \quad \ldots, \quad x_n = \dot{}_{n-1}$$

and by using the notations:

$$u_1 = u, \quad u_2 = \frac{du}{dt}, \quad u_3 = \frac{d^2 u}{dt^2}, \quad \ldots, \quad u_{m+1} = \frac{d^m u}{dt^m},$$

the equation (1) becomes:

$$\dot{x}_1 = x_2 = F_1(x_2) = F_1, \tag{2}$$
$$\dot{x}_2 = x_3 = F_2(x_3) = F_2, \tag{3}$$

....

$$\dot{x}_{n-1} = x_n = F_{n-1}(x_n) = F_{n-1}, \tag{4}$$
$$\dot{x}_n = F_n(u_1, u_2, \ldots, u_{m+1}, x_1, x_2, \ldots, x_n) = F_n, \tag{5}$$

or in a vector form

$$\dot{\boldsymbol{x}} = \boldsymbol{F}(\boldsymbol{u}, \boldsymbol{x}) = \boldsymbol{F}. \tag{6}$$

We choose an arbitrary pivot moment t_k and in its close proximity we develop in a Taylor series $\dot{x}_k(\theta) = \dot{x}(t_k + \theta)$ and $F_k(\theta) = F(t_k + \theta)$. Then by integrating the equation (6) we have:

$$\int_{-\delta t}^{\delta t} \sum_{q=0}^{\infty} \frac{\theta^q}{q!} \cdot \boldsymbol{x}^{(q+1)} \cdot d\theta = \int_{-\delta t}^{\delta t} \sum_{q=0}^{\infty} \frac{\theta^q}{q!} \cdot \boldsymbol{F}_k^{(q)} \cdot d\theta, \tag{7}$$

which leads to:

$$\dot{\boldsymbol{x}}_k + \frac{\delta t^2}{3!} \cdot \dddot{\boldsymbol{x}}_k + \frac{\delta t^4}{5!} \cdot \boldsymbol{x}_k^{(5)} + \ldots = \boldsymbol{F}_k + \frac{\delta t^2}{3!} \cdot \ddot{\boldsymbol{F}}_k + \frac{\delta t^4}{5!} \cdot \boldsymbol{F}_k^{(4)} + \ldots \tag{8}$$

If we approximate the above derivatives, until three regressive sequences, by:

$$\left(\frac{d^q \boldsymbol{v}}{dt^q}\right)_{t_k} = \boldsymbol{v}_k^{(q)} \cong \frac{1}{\Delta t^q} \cdot \sum_{j=0}^{3} \delta_{qj} \cdot \boldsymbol{v}_{k-j}, \tag{9}$$

where $\mathbf{v}_{k-j} = \mathbf{v}(t_k - \Delta t)$ for $\Delta t = 2\delta t$, considered low enough, the integration of the equation (6) leads to the finite difference algebraic equation, in vectorial form:

$$\sum_{j=0}^{\omega} \sigma_{11j} \boldsymbol{x}_{k-j} \cong \Delta t \sum_{j=0}^{\omega} \sigma_{10j} \boldsymbol{F}_{k-j}, \tag{10}$$

The coefficients δ_{qj} and σ_{1pj} are presented in Table 1 and Δt is now the integration step.

TABLE 1. Coefficients needed for numerical calculations

q	δ_{q0}	δ_{q1}	δ_{q2}	δ_{q3}
1	11/6	-18/6	-9/6	-2/6
2	2	-5	4	-1
3	1	-3	3	-1

p	σ_{1p0}	σ_{1p1}	σ_{1p2}	σ_{1p3}
0	13/12	-5/24	1/6	-1/24
1	15/8	-25/8	13/8	-3/8

ϵ_{011}	ϵ_{02}	ϵ_{03}
3	-3	1

THE NUMERICAL MODELLING

The numerical solution of the equation (1) can be obtained, by using (10), in the general form of the state vector at sequence k, where $k=\frac{t_k}{\Delta t}$, as:

$$\boldsymbol{x}_k = g \cdot \boldsymbol{F}_k + \boldsymbol{h}_{xFk}, \tag{11}$$

where:

$$g = \Delta t \, \frac{\sigma_{100}}{\sigma_{110}} \tag{12}$$

and

$$\boldsymbol{h}_{xFk} = \frac{1}{\sigma_{110}} \sum_{j=1}^{3} \left(\Delta t \cdot \sigma_{10j} \cdot \boldsymbol{F}_{k-j} - \sigma_{11j} \cdot \boldsymbol{x}_{k-j} \right). \tag{13}$$

In the above relationship g and $\mathbf{h}_{xFk}$ represent the transfer coefficient, respectively the free response to the k-th sequence, while $g \cdot \mathbf{F}_k$ is interpreted as the forced response to the same sequence. The term $\mathbf{h}_{xFk}$ is called "the history of the process at the sequence k" because it contains only the regressive sequence $\mathbf{F}_{k-j}$ and $\mathbf{x}_{k-j}$, (j=1,2,3).

The vector $\mathbf{F}_k$ from (11) is approximated by:

$$\boldsymbol{F}_k \cong \boldsymbol{F}(\boldsymbol{u}_k, \boldsymbol{x}_{k\epsilon}), \tag{14}$$

where

$$x_{k\epsilon} = \sum_{j=1}^{3} \epsilon_{0j} \cdot x_{k-j}, \tag{15}$$

is the state vector extrapolated with the coefficients ϵ_{0j} presented in the Table 1.

The relationship (11) represents, in a local-iterative linearized form, the numerical modelling, with three regressive sequences, of the dynamical process analogical modelled by the equation (1).

THE NUMERICAL SIMULATION

As a result of numerous examples investigated, we have elaborated a general flow chart for the numerical simulation by L.I.L. (Fig. 1). This flow chart contain 9 program modules and a starting subroutine, as follows:

Module 1.

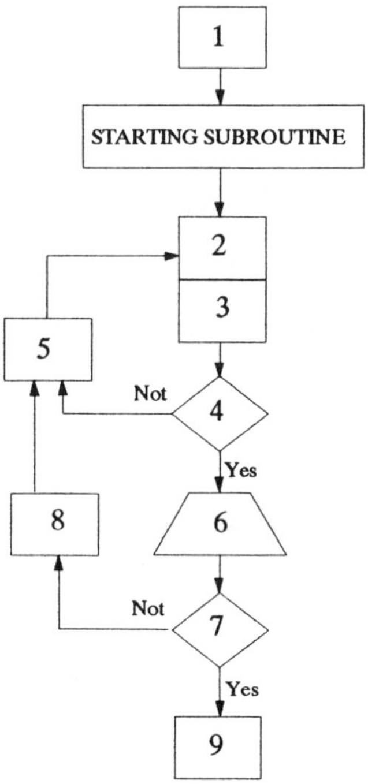

FIGURE 1. The flow chart

1.1. The program nomination and the definition of the components of the vectors **u**(t) and **F**[**u**(t),**x**(t)].

1.2. The declaration of the coefficients and structure parameters contained in 1.1.

1.3. The delimitation in time of the program: t_0 (the starting time); t_f (the final time), Δt (the integration step), $\Delta\theta$ (the extraction step of the results, as a multiple of Δt).

1.4. The declaration of the initial conditions $\mathbf{x}_0=\mathbf{x}(t_0)$ and $\mathbf{u}_0=\mathbf{u}(t_0)$, of the coefficients σ_{10j}, σ_{11j} (j=0,1,2,3) and ϵ_{0j} (j=1,2,3) (Table 1) and also g from (12).

The starting subroutine: Assures the computation of the fictitious starting vectors $\tilde{x}_{k-2}$, $\tilde{x}_{k-3}$ and $\tilde{F}_{k-2}$, $\tilde{F}_{k-3}$ with the observation that $\tilde{x}_{k-1}=x_0$ and $\tilde{F}_{k-1}=F(u_0, x_0)$.

We elaborated a starting subroutine by using a standard fourth order Runge-Kutta method.

Module 2. Assures the iterative increment of the pivot moment $t_k=t_{k-1}+\Delta t$.

Module 3. Calculates in the following order: $\mathbf{h}_{xFk}$ from (13); $x_{k\epsilon}$ from (15); $u_k=u(t_k)$ from module 1.1.; $F_k=F(u_k,x_{k\epsilon})$ from (14) and finally x_k from (11).

Module 4. Is a conditional jump instruction for the extraction, with the period $\Delta\theta$, of the calculation result: t_k, u_k, and x_k; if $t_k\geq\theta$ then transfer to Module 6; if not then transfer to Module 5.

Module 5. Realizes the following sequential advances:

5.1.: $t_{k-1} = t_k$;

5.2.: $\mathbf{x}_{k-3}=\mathbf{x}_{k-2}$; $\mathbf{x}_{k-2}=\mathbf{x}_{k-1}$; $\mathbf{x}_{k-1}=\mathbf{x}_k$;

5.3.: $\mathbf{F}_{k-3}=\mathbf{F}_{k-2}$; $\mathbf{F}_{k-2}=\mathbf{F}_{k-1}$; $\mathbf{F}_{k-1}=\mathbf{F}_k$;

Then transfer to Module 2.

Module 6. Displays with the period $\Delta\theta$, the calculated results: t_k, u_k and x_k.

Module 7. Is a conditional jump instruction: if $t_k\geq t_f$ then transfer to Module 9.

Module 8. Assures the incrementation with the step $\Delta\theta$ of the moment θ, for the extraction of the results need to the next stage (Module 4).

Module 9. Contains the instructions for the stopping of the calculations and for the finish of the program if $t_k\geq t_f$.

We remark the simplicity of the flow chart presented in the Figure 1, and also its flexibility to any modifications of the program.

AN APPLICATION OF L.I.L. METHOD

We used the L.I.L. procedure in the study of the dynamics of an irregular pulsating star (2), The analytical model of this nonlinear systems is the following differential equation:

$$\dddot{x}+(\eta-\beta-\dot{x})\,\ddot{x}+(\gamma-\alpha-\beta\eta+\alpha x+\beta\dot{x})\,\dot{x}-\alpha\eta x=0, \tag{16}$$

which for the modelling and numerical simulation is written in the form

$$\begin{aligned}\dot{x}&=y=F_1(y),\\ \dot{y}&=\alpha x+\beta y+z=F_2(\alpha,\beta,x,y,z),\\ \dot{z}&=-\gamma y-\eta z+yz=F_3(\gamma,\eta,y,z).\end{aligned} \tag{17}$$

We remark that the equation (17) correspond to the general vector form (6) particularized for three state variables. These equations describe a nonlinear oscillator, where x, y and z are the displacement, the velocity and respectively the force acting on the oscillator. The equations contain also the structure parameters α, β, γ and η, which play an important role in the system dynamics. For different sets of parameters, the orbits of the oscillator can be very different.

The system has three fixed points:

i) $x=0;\ y=0;\ z=0,\ \textit{if}\ \alpha\neq 0\ \textit{and}\ \eta\neq 0;$

ii) $x=-\dfrac{a}{\alpha};\ y=0;\ z=a,\ \textit{if}\ \alpha\neq 0\ \textit{and}\ \eta=0\ \ (a\in\mathbb{R});$

iii) $x=b;\ y=0;\ z=0,\ \textit{if}\ \alpha=0\ \textit{and}\ \eta\neq 0\ \ (b\in\mathbb{R}).$

and their stability was analyzed with the usual methods. Because the aim of this chapter is the application of L.I.L. procedure we do not insist neither on the phenomenological details nor on the mathematical formalism of the stability, which may be found in the references (3).

Following the simulation steps presented above we find the orbits in the phase space of the oscillator for different sets of parameters which was compared with the analytical results.

For example, if

$$\alpha\eta<0;\quad \eta-\beta>0;\quad \gamma-\alpha-\beta\eta>0;\quad (\eta-\beta)^2>3(\gamma-\alpha-\beta\eta). \tag{18}$$

the first fixed point is a stable one. But if

$$\alpha\eta>0;\quad \eta-\beta<0;\quad \gamma-\alpha-\beta\eta>0;\quad (\eta-\beta)^2>3(\gamma-\alpha-\beta\eta). \tag{19}$$

this fixed point is unstable.

If

$$\alpha\eta>0;\quad \eta-\beta>0;\quad \gamma-\alpha-\beta\eta<0;\quad 3|\gamma-\alpha-\beta\eta|>(\eta-\beta)^2. \tag{20}$$

we obtain the condition of Hopf bifurcation in the plane xOy from an unstable focus to a stable limit cycle.

If

$$\beta<0;\quad \gamma-a-\alpha>0;\quad \beta^2-4(\gamma-a-\alpha)<0, \tag{21}$$

the second fixed point is stable.

When

$$\beta=0,\quad \Lambda=\pm(\gamma-a-\alpha), \tag{22}$$

the trajectory must be a circle with Λ the circle ray and the origin of the circle in the point $\left(-\dfrac{a}{\alpha},0,a\right)$. The results are illustrated in Table 2 and Figures 2-7.

TABLE 2. Results of the considered example.

Figure	Final state	Parameters	Initial conditions	Δt	Graphical programming SCREEN 1:PSET()
2	Stable focus	α=1.0 β=-0.5 γ=-1.0 η=2.0	x_0=0.00 y_0=0.10 z_0=-0.01	0.1	$900 \cdot y_k+160$, $900 \cdot x_k+100$
3	Unstable focus	α=1.0 β=2.5 γ=5.0 η=1.5	x_0=0.00 y_0=0.01 z_0=0.02	0.1	$20 \cdot y_k+160$, $20 \cdot x_k+100$
4	Hopf bifurcation	α=1.0 β=0.1 γ=0.5 η=0.2	x_0=0.0 y_0=0.1 z_0=0.0	0.02	$10 \cdot y_k+160$, $10 \cdot x_k$
5	Stable focus	α=1.0 β=0.1 γ=2.0 η=1.0	x_0=1.0 y_0=1.0 z_0=1.0	0.02	$500 \cdot y_k+200$, $1000 \cdot x_k-700$
6	Limit cycle	α=-5.0 β=0.0 γ=2.0 η=0.0	x_0=1.0 y_0=1.0 z_0=1.0	0.02	$20 \cdot z_k+100$, $20 \cdot y_k-700$
7	Strange attractor	α=-0.5 β=0.5 γ=1.0 η=4.0	x_0=1.0 y_0=0.1 z_0=0.0	0.05	$10 \cdot z_k+180$, $10 \cdot y_k+100$

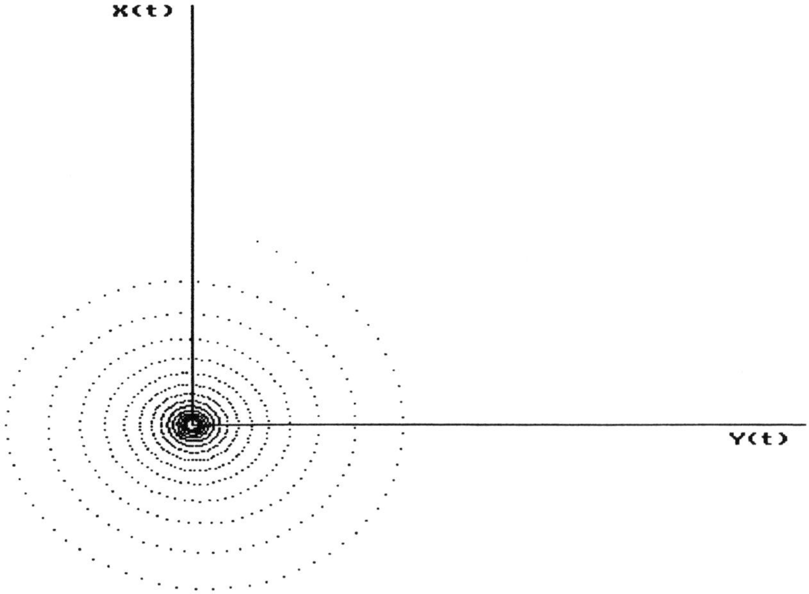

FIGURE 2. Stable focus

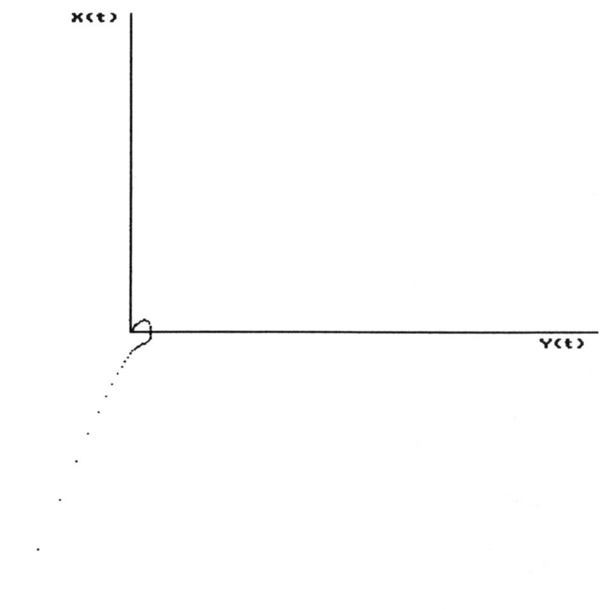

FIGURE 3. Unstable focus

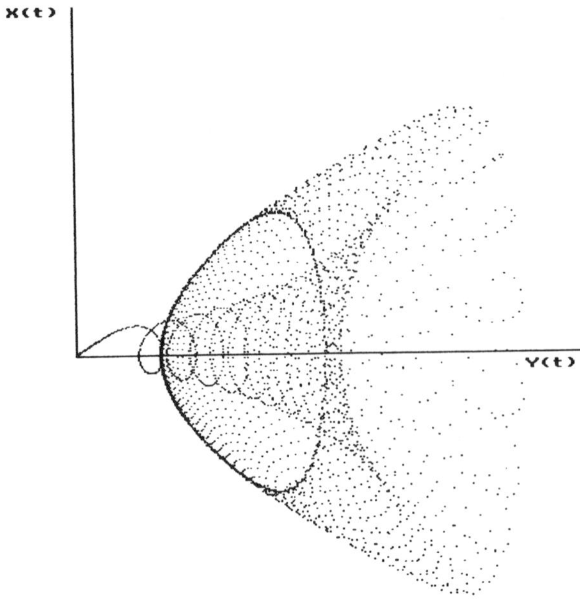

FIGURE 4. Hopf bifurcation

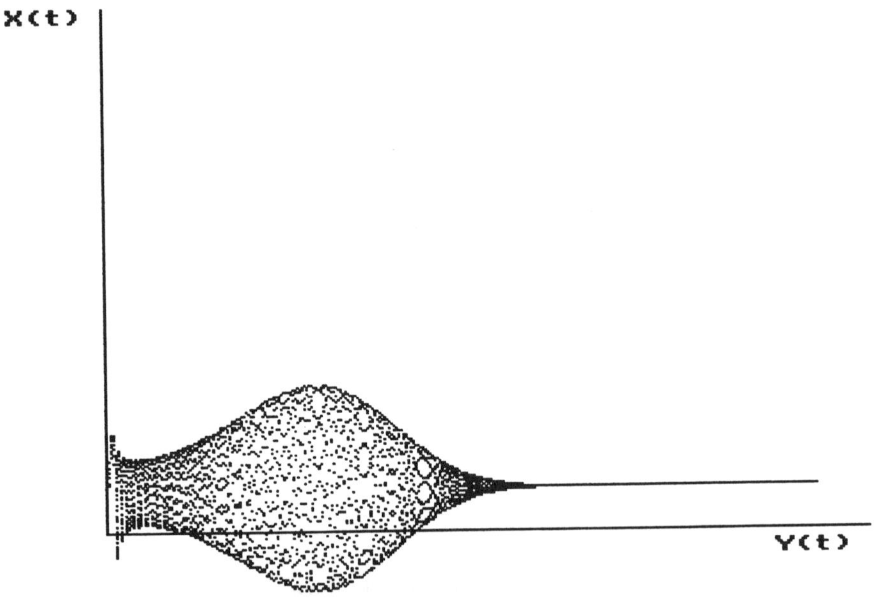

FIGURE 5. To a stable focus

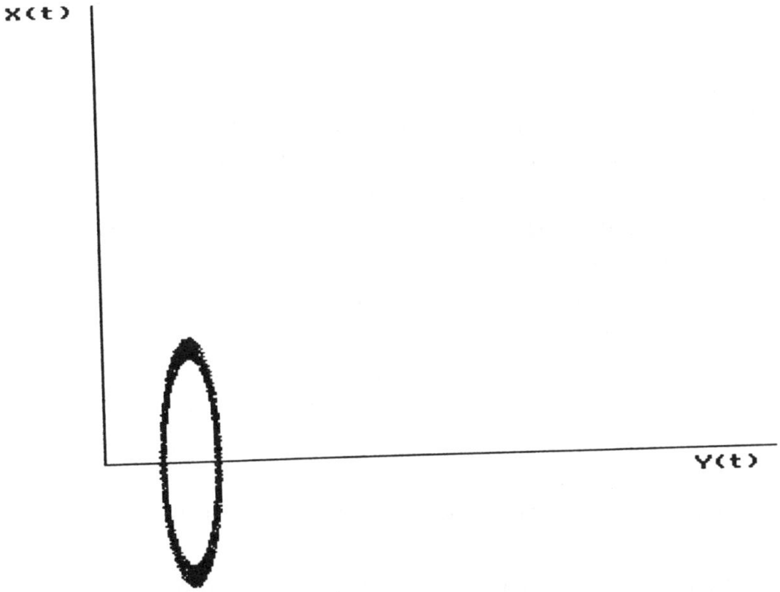

FIGURE 6. Limit cycle

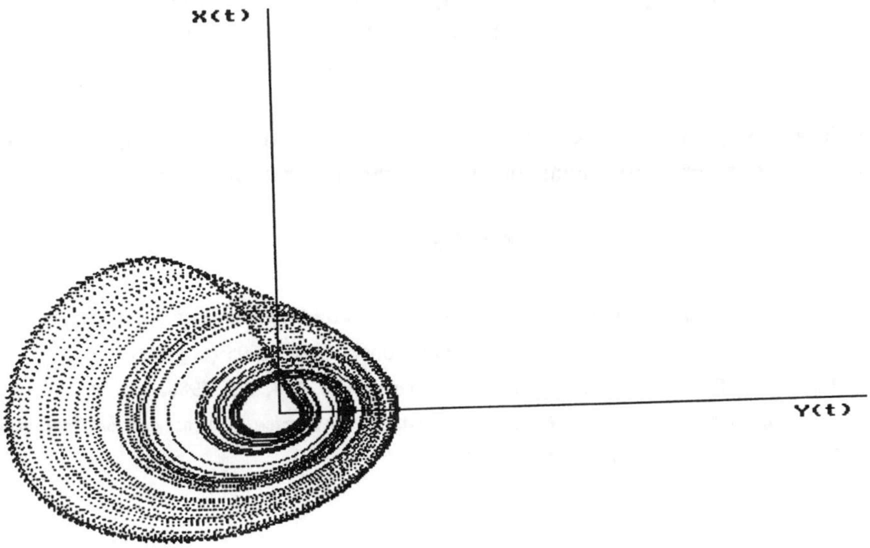

FIGURE 7. Strange attractor

CONCLUSIONS

The proposed method for the numerical modelling and simulation by local-iterative linearization of dynamical processes presents the following features:
a) Different from the other methods of linearization of nonlinear dynamical processes, which are valid only when the deviation around the static working point are small (with usually rough restrictions), the L.I.L. method localizes, with a remarkable precision, the linearization in the immediate proximity of the pivot moment t_k which is iteratively incremented with the integrating step Δt.
b) The integration precision is high. The relative cumulated errors (in %) of order (10^{-6}-10^{-2}) are obtained. These performances are due to:
- the multistep procedure, which uses three (or five) regressive sequences;
- the elimination of all odd order derivatives of F_k and of even order derivatives of x_k from the terms neglected in the Taylor series (8);
- the possibility to avoid the matrix inversion, which leads also to an augmentation of the calculation speed;
c) For the same initial conditions, t_0, Δt, t_f and in the same domain of errors, the computing time is smaller (at least with 30%) than those spent with usual fourth order Runge-Kutta method.
d) The numerical convergence (stability and consistency) is very good.
e) Being a multistep method of numerical integration, the L.I.L. procedure needs, for the calculus starting, a single step method like Runge-Kutta (standard, Gill) one.
f) The flow chart is simple, flexible to the modification of the input or of the structure and very convenient for the computer programming.

REFERENCES

1. Coloşi, T., Codreanu, S., Naşcu, I., *Numerical Modelling and Simulation of Dynamical Processes by Local Iterative Linearization*, Cluj: Scientific Book House, 1995 (in press).
2. Tanaka, Y., Takeuti, M., *Astrophysics and Space Science*, 148, 229, 1988.
3. Guckenheimer, J., Holmes, P, *Nonlinear Oscillations, Dynamical Systems and Bifurcations of Vector Fields*, Berlin: Springer-Verlag, 1990.

Jieke Theory and Logistic Model

Hong–Xing Cao* and Guo–Lin Feng+

* *Chinese Academy of Meteorological Sciences, 46 Baishiqiaolu, 100081, Beijing, China*

+ *Agriculture College, Yangzhou University, 12 Sunonglu, Yangzhou, China*

Abstract. What is a shell or a JIEKE (in Chinese) is introduced firstly, jieke is a sort of system boundary. From the concept of jieke theory, a new logistic model which takes account of the switch effect of the jieke is suggested. The model is analysed and nonlinear mapping of the model is made. The results show the feature of the switch logistic model far differ from the original logistic model.

INTRODUCTION

A system can be divided as two parts: a system inner and a system boundary. Thereby, a specific topic of systems sciences, which is to study the system boundary in general sense, has been suggested since 1988 (1). Several basic works have been acquired (2). Our objective is to study a sort of system boundary which plays a role in defending existence of the system and exchanging between the system and its environment. We call this sort of system boundary a JIEKE in Chinese or a shell in English. For example, in the human society a castle in Europe and the Great Wall in China defend an invasion from an enemy, in meteorology, an eye wall in a typhoon or a hurricane keeps clear and sunny weather within the eye, in biology, a cell membrane assures alive of a cell and exchangeable between the cell and its environment. A turtle has a hard, strong shell which defends itself and keep it living for a quite long time.

In respect to structure the jieke is divided two parts: the wall and the door or the pass. Thus, we define two functional elements of the jieke: an open ratio and an exchange rate. From the second law of thermodynamics, an entropy criterion of the lower limit of the exchange rate is derived (3); from the reaction diffusion equation, the upper limit of the open ratio is derived too.

As a similar consideration in the cybernetics, dynamics of jieke was established (4). One application of the dynamics is to set up an immigration-population model which can explain quantitatively interaction between population change and immigration control.

LOGISTIC MODEL WITH JIEKE EFFECT

A well-known logistic model is

$$x_{n+1} = \mu x_n(1-x_n) = \mu x_n - \mu x_n^2$$

where μ is a parmeter, n is the time of iteration. This is a formula which has been investigated thoroughly. Many nonlinear phenomena were found from the model. For example, period doubling, Feigenbaum number and so on are all come from the logistic model.

In view of jieke theory, a logistic model with jieke effect, which can describe an immigration-population dynamics with the customs control, is suggested in this paper. Actually, we set a switch into the model, i.e

$$x_{n+1} = (\mu + \delta) x_n (1-x_n) \qquad (1)$$

$$\delta = \left[\begin{array}{ll} 1 \text{ or } \nu & \text{if } (x_{n+1}-x_n) \leqslant c \\ 0 & \text{if } (x_{n+1}-x_n) > c \end{array} \right. \qquad (2)$$

where ν is a net immigration rate through the customs of a country, c is a given constant relevant to a population increase rate. For population, a value of x is positive, if $x_n < 0$ or $x_n > 1$ than $x_{n+1} < 0$, namely x_{n+1} becomes negative. So it follows

$$0 < x_n < 1$$

Take differential calculus and set it to be zero

$$dx_{n+1} / dx_n = (\mu + \nu) - 2(\mu + \delta)x_n = 0$$

then the maximum is obtained

$$x_n = 1/2$$

correspondingly $x_{n+1} = (\mu + \delta)/4$

suppose $x_{n+1} < 1$, then $x_{n+1} = (\mu + \delta)/4 < 1$

if $\delta = 0$ then $\mu < 4$

if $\delta = 1$ then $\mu < 3$.

ITERATIVE COMPUTATION

Taking $\nu = 1$ and $C = 0.02$ which is equivalent to high increase rate of population, (1) can be computed iteratively. δ in (2) means if population increment $\Delta x_n = x_{n+1} - x_n$ is greater than a given threshold c, the immigration come-in will be prohibited, otherwise the come-in will be permitted. Actually, the model (1) is a switch model whose concept is come from the jieke theory that studies in general sense the structure, the function and the behaviour of a system boundary. Main points of computational result are as follows:

If $\mu = 0.01$ then $\delta = 1$ and x_{n+1} approaches to 9.899548×10^{-3};

If $\mu = 0.1$ then $\delta = 1$ and x_{n+1} approaches to 9.090909×10^{-2};

it means that the population will contract and become a bit if $\mu \in [0, 0.1]$.

If $\mu = 0.2$ then $\delta = 1$ and x_{n+1} approaches to 0.1666666;

If $\mu = 0.3$ then $\delta = 1$ and x_{n+1} approaches to 0.23076923076923076;

If $\mu=0.32$ then $\delta=1$ and x_{n+1} approaches to 0.24242424;

If $\mu>0.32$ then $\delta=1$ or $\delta=0$ alternately and x_{n+1} varies periodically. This kind of periodic behaviour of the model (1) retains to

$$\mu_{\infty}=3.0239999999997042$$

If $\mu>\mu_{\infty}$ then the behaviour of the model becomes chaotic.

Similar to Feigenbaum (5) number, we calculate

$$\varepsilon_n=\lim_{n\to\infty}\frac{\mu_n-\mu_{n+1}}{\mu_{n+1}-\mu_{n+2}} \qquad (3)$$

the ε_n do not approach to a fixed value but to the special period as in Figure 1. which is composed of period length T=26 and its maximum being 1.25672 and its minimum being -14.5127.

Above points demonstrates that the periodic and chaotic behaviours of the logistic model with jieke effect are controlled by the switch δ. The control chaos of the model will be further investigated in the future.

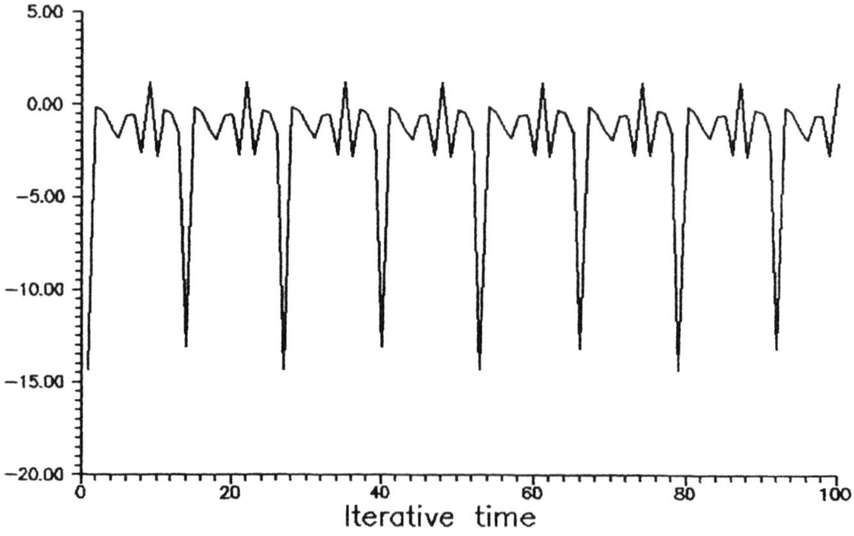

FIGURE 1. Periodic oscillation of Feigenbaum-number-like in the logistic model with jieke effect.

REFERENCE

1. Cao Hong-Xing, Jieke phenomenon and its scientific frame, Technical Report of Chinese Academy of Meteorological Sciences, No. 88005, 1-16 (1988), (in Chinese).
2. Cao Hong-Xing, Modelling of a system boundary, appear in ***Kybernetes***, Vol. 23, 1994.
3. Cao Hong-Xing, Entropy equilibrium equation with jieke effect and its application to earth sciences, in ***Comprehensive Advance on Astronomy, Earth Science and Biology***, Beijing, China Sciences Technology Press, 1989, pp. 122-126 (in Chinese).
4. Cao Hong-Xing, Dynamical equation of shell, in ***Proceedings of the Second International Conference on Systems Science and Systems Engineering***, Zheng Weimin (ed.), International Academic Publisher, 1993, pp. 538-541.
5. Feigenbaum, M. J., ***J. Statist. Phys.*** 19, 25-52 (1978).

Chaotic Dynamics of Weakly Nonlinear Systems

D.M.Vavriv

Radio Astronomy Institute of Ukrainian Academy of Sciences
4 Krasnoznamennaya Street, 310002 Kharkov, Ukraine

Abstract. A review is given on the recent results in studying chaotic phenomena in weakly nonlinear systems. We concern with the class of chaotic states that can arise in physical systems with any degree of nonlinearity however small. The conditions for, and the mechanisms of, the transitions to chaos are discussed. These findings are illustrated by the results of the stability analysis of practical microwave and optical devices.

INTRODUCTION

Progress made in recent years in the field of chaos study has shown that the chaotic states are typical not only for strongly nonlinear systems, but they can arise in weakly nonlinear dissipative systems as well [1-6]. There are a lot of typical situations where the chaotic states manifest themselves under the weakly nonlinear excitation conditions, including various, types of oscillators with a quasiperiodic forcing, multimode autonomous and nonautonomous systems, distributed systems, etc. The existence of the chaotic states in the quasilinear limit implies that the condition for the chaos to arise is not very stringent for many practical situations, and that the influence of the chaotic instabilities on the dynamics of real systems can be much more severe as compare with the recent thinking.

The weakly nonlinear systems are characterized by specific mechanisms of the transition to chaos as compare with strongly nonlinear systems. In the next section we discuss these differences, and then we consider the chaotic phenomena in some practical systems.

2. MECHANISMS FOR CHAOS

To illustrate the main differences in mechanisms of the transition to chaos in weakly and strongly nonlinear systems, let us consider the quasiperiodically forced Duffing-type oscillator

$$\ddot{x} + \omega_0^2 x = \varepsilon\left[-\left(\delta_0 + \delta_1 x^2\right)\dot{x} + \gamma x^3 + A_1 \cos\omega_1 t + A_2 \cos\omega_2 t\right]. \qquad (1)$$

The chaos origin in this oscillator is usually related with the existence of saddle points, in the unperturbed oscillator with $\gamma>0$ when the dissipation and amplitudes of the external forcing are equal zero. We call this type of singular points as the *original saddle points.* In the case of the periodically forced oscillator, chaotic motion arises because of the formation of a homoclinic structure in the phase space of the system due to the transverse intersections between the stable and unstable manifolds of hyperbolic periodic orbits associated with these points. The threshold for chaos with respect to the amplitude of the external force is proportional to $\varepsilon^{-3/2}$, and a strong nonlinearity is required for the chaos onset. The same conclusions can be made for the quasiperiodically forced oscillator when the chaos onset is again related with the original saddle points [7].

However, in the case of the quasiperiodically forced oscillator, there is also another way of the chaos arising that is not related with the existence of the original saddle points, and where the strong nonlinearity is not required. The alternative way of the chaos onset is associated with the occurrence of a new saddle orbit in the phase space of the system under the action of one of the external harmonic component. We call this orbits as *the induced saddle orbits.* With the availability of an additional incommensurate spectral component of low intensity in the external force, a normally hyperbolic invariant torus is formed in the phase space instead of the saddle orbit. The chaos results from the transverse intersection between the stable and unstable manifolds of this torus when the perturbation is increased. The formation of a homoclinic structure occurs here at much more lower values of the amplitudes of the perturbation as compared to the previous case. This mechanism of the transition to chaos is typical for single-degree-of-freedom systems with a quasiperiodic excitation. Multimode nonlinear systems acquire chaotic states in much similar way, however, induced saddle orbits and corresponding homoclinic structures can arise here also due to the interaction of modes without external forcing.

The are no analytical methods that can predict the chaotic states in weakly nonlinear oscillators by using the initial equation, like (1), directly. Several ways around this problem were proposed, which include: averaging methods [1,3], current Lyapunov exponents' technique [8], Melnikov's technique applied to the averaged equations [4,5,9], mapping [10], harmonic balance method [11]. The application of these methods enables to define quite accurately the regions of the control parameters where chaos can arise in the weakly nonlinear limit. For example, with respect to the oscillator (1), it was found [6,9] that the threshold value A_{thr} of the amplitudes of the external forcing is

$$A_{thr} = \frac{2\omega_0\delta_0}{3}\left(\frac{\sqrt{3}\delta_0}{\gamma}\right)^{1/2} \tag{2}$$

and for the chaos to arise the parameter of nonlinearity γ should exceed a critical value determined by the parameter of nonlinear dissipation δ_1, namely $\gamma > \sqrt{3}\delta_1$.

It should be noted that these conditions, unlike that for the strongly nonlinear oscillator, do not contain the parameter ε. It means that any degree of nonlinearity however small can initiate the chaos onset. Owing to this, the chaotic states can exert a strong influence on the dynamics of many practical systems, their output noise level, stability. For the purposes of illustration, we discuss in the next section results of theoretical and experimental investigations of various types of microwave and optical devices.

CHAOS AND STABILITY OF PRACTICAL SYSTEMS

An important motivation for the first investigations in this direction was result of experimental studies of resonant amplifiers which indicated that the amplifiers, being stable with a harmonic input signal, lose their stability when a multifrequency signal is applied [12, 13]. The theoretical studies have show that this is due to the chaotic states arising in the weakly nonlinear limit. It was predicted and proved experimentally that the chaotic instability presents a severe hazard to transistor and diode-type amplifiers because of the inherent reactive nonlinearity of semiconductor circuits. It was found that an electron beam can also introduce a large enough degree of reactive nonlinearity into an open resonator for the chaotic states arising in free-electron lasers [13]. For the later case the threshold for chaos is essentially decreased because of the high Q-factor of the resonator. It emerges that the conventional valve amplifiers have the highest immunity against the chaos onset due to the practical absence of reactive nonlinearities in their circuits.

The parametric amplifiers, both microwave and optical ones, should be considered as that with the highest susceptibility to the chaotic instability. It is worth noting that the factors which are responsible here for the chaos onset are precisely the same as that which, from the point of view of the classical theory of parametric devices, provide low noise amplification. Indeed, according to this theory, parametric amplifiers possess a low noise output level because they consist of a reactive circuit and they utilize an ac power supply rather then dc supply. Proceeding from the theoretical and experimental results [14,15], it is clear that exactly the combination of this type of nonlinearity and the ac power supply leads to the possibility of the chaos onset in the weakly nonlinear limit even with a harmonic input signal. These findings work also for Josephson-junction parametric amplifiers, SQUIDs [16], and other types of similar devices.

The investigations of the conditions for the chaos onset in the microwave generators show that the synchronized generators are subjected to the chaotic instability due to a variety of perturbations typical at practice. Such a perturbation can be a parasitic low-intensity component in the spectrum of the synchronizing signal, or low-frequency hindrance in a supply circuit which was experimentally demonstrated with Gunn-diode oscillators [17]. Multimode nature of the majority of microwave and optical systems creates a number of additional ways for the transition to chaos in weakly nonlinear systems. The chaotic states can arise, for example, as a result of the interaction of two active modes as well as active and passive ones [18], not to mention a variety of roads to chaos through the interaction of a grater number of modes. It was demonstrated that not only the resonant modes' interaction leads to the chaotic state arising, but the non resonant interaction can have a pronounced effect on the system dynamics [19].

CONCLUSION

The review given points to the fact that the investigations of the chaotic dynamics of weakly nonlinear systems constitute now a separate direction of the research in the field of chaos being important both for the basic study and applications. Chaotic states are typical for a variety of practical weakly nonlinear systems and many classical results concerning their dynamics and stability ought to be reconsidered from this point of view.

ACKNOWLEDGMENTS

The author wishes to thank O.A. Tretyakov, A.B. Belogortsev, I.Yu.Chernyshov, V.B. Ryabov, S.A. Sharapov, and Yu.A.Tsarin for useful discussions and collaborations. This work was partly supported by the International Science Foundation (grant # U33000).

REFERENCES

[1] Miles, J., *Proc. Nat.Acad.Sci., Phys.Sci.USA* **81**, pp.3919-3923 (1984).
[2] Belogortsev,A.B., Vavriv, D.M.,and Tretyakov, O.A., *Sov.Phys.JETP* **65**, pp.737-740 (1987).
[3] Belogortsev, A. B. et al, *Sov.Phys.Tech.Phys.* **33**, pp.174-179 (1988).
[4] Yagasaki, K., Sakata, M., and Kimura, K., *ASME J. of Appl. Mech.* **57**, pp.209-221 (1990).
[5] Yagasaki, K., *ASME Journal of Appl. Mech.* **59**, pp.161-167 (1992).
[6] Belogortsev, A.B. et al, *Appl.Mech.Rev.* **46**, pp.372- 384 (1993).
[7] Wiggins, S., *Phys.Lett.* **A124**, pp. 138-142 (1987).
[8] Vavriv, D. M., and Ryabov, V. B., *Comput. Maths & Math.Phys.* **32**, pp.1259-1269 (1992).
[9] Ryabov, V. B., and Vavriv, D. M., *Phys.Lett.* **A153**, pp.431-436 (1991).
[10] Belogortsev, A. B., *Phys.Lett.* **A161**, pp.352-356 (1992).
[11] Vavriv, D. M., and Ryabov, V. B., *Pis'ma v Zh.Tekh.Fiz.* **17** , pp.55-60 (1991).
[12] Belogortsev, A. B. et al., *Telecommun.Radio Eng.* **5**, pp.82-85 (1988).
[13] Vavriv, D. M., and Tretyakov, O. A., *Theory of Resonant Amplifiers with Distributed Interaction of O-type*, Kiev: Naukova Dumka Press, 1987.
[14] Vavriv, D. M. et al., *Sov. Phys. Tech. Phys.* **36**, pp.1325-1331 (1991).
[15] Chernyshov, I. Yu., and Vavriv, D. M., *Phys.Lett.* **A165**, pp. 117-123 (1992).
[16] Bulgakov, S. A. et al, *J.Low Temp.Phys.* **83** (5/6), pp. 241-255 (1991).
[17] Vavriv, D. M. et al., *Pis'ma v Zh.Tekh.Fiz.* **14**, pp.903-908 (1991).
[18] Belogortsev, A. B. et al, *Electronic Lett.* **26**, pp.1354-1355.
[19] Belogortsev, A. B. et al, *Sov.Phys.Tech. Phys* **36**, pp.380-385 (1991).

Fundamental Threshold of Chaos in Some Nonlinear Oscillators

V.B.Ryabov
Institute of Radio Astronomy, 4 Krasnoznamennaya St., 310002 Kharkov, Ukraine

Abstract

A technique for predicting chaos arising in a broad class of nonlinear oscillatory systems is proposed. It is based on the notion of running Lyapunov exponents and uses the local stability properties of trajectories for determining the "safe" areas in the phase space where any trajectory is regular and stable in the sense of Lyapunov. The combination of this approach with harmonic balance method permits to obtain the corresponding "safe" regions in the control parameter space. The borders of these regions may be considered as threshold lines delimiting the areas of possible chaotic instability. An example of the two-well Duffing oscillator demonstrates good agreement between theoretically predicted values of control parameters where chaos arises with those obtained numerically. The technique is especially effective for rather high dissipation levels when other known methods such as Melnikov's criterion or combination of harmonic balance with analysis of variational equations fail to provide correct results.

One of the principal problems in the dynamical chaos theory is to predict its appearance. Many efforts have been made in recent years [1-9] in order to develop a theoretical method that permitted to specify the domains in control parameter space of dynamical systems described by sets of ordinary differential equations where chaotic behavior is likely to occur. Despite some positive results obtained so far, now we, actually, have not many analytic tools available for solving the pointed out problem.

If the system under study is weakly dissipative and possesses a closed homoclinic or heteroclinic trajectory at zero dissipation then Melnikov's method can be applied. This approach has been successfully used for many one-and-a-half degrees of freedom oscillators [1, 2, 4, 5, 7] and also in some higher dimensional cases [2]. However, it should be stressed that this technique gives neither sufficient nor necessary conditions of chaos arising, since it merely predicts the appearance of transverse intersections of stable and unstable manifolds of a particular saddle-type periodic orbit. It is clear that, on the one hand, the homoclinic structure associated with the periodic orbit may be non attractive and in such a case it defines the chaotically looking transients to the regular final state, and on the other hand, there may exist other saddle orbits whose manifolds may intersect independently of the ones treated by Melnikov's theory. In addition, this approach, being in essence an asymptotic technique, as a rule is applicable only at low level of dissipation and becomes very inaccurate when the corresponding parameter is not small.

Another way to predict the appearance of chaos is to study the stability of existing periodic attractors. In such a manner, the combination of the well known harmonic balance method with approximate analysis of associated Hill's type equation in some situations leads to more precise predictions of chaos arising threshold [3, 8, 9] than can be derived by means of Melnikov's theory. However, the accuracy of this criterion is also strongly dependent on the dissipation level.

A method proposed by Shil'nikov [6] should be also mentioned as the one permitting to obtain chaos arising conditions for some very specific types of dynamical systems.

In Hamiltonian, i.e. zero dissipation case, the criterion of resonance overlap due to Chirikov [10] is frequently used. Sometimes it gives reasonable results, even at nonvanishing dissipation value [11], but, again, it is, as a rule, very inaccurate, especially at higher dissipation levels.

In the present letter we propose an approach that in some cases gives much better prediction compared to the methods mentioned so far. It is based on the notion of running (or local) Lyapunov exponents (RLE) [12 - 18] which was recently demonstrated as a very useful tool for deriving chaos arising threshold in some nonlinear oscillators [12, 13, 19]. Actually, it permits to obtain the maximal value of oscillation amplitude below which chaos can not occur at all. Accordingly, the larger value of the amplitude means the possibility for chaos to appear. From this view, this method may be considered as a one giving the necessary condition of strange attractor formation in the phase space. Here we generalize the results of [13] to the case of general type passive oscillators and obtain explicit formulae for chaos arising threshold in the particular example of Duffing-Holmes equation with harmonic excitation.

RLE are introduced in the following way [12]. Let us consider a dynamical system described by the set of n ordinary differential equations

$$\frac{d\mathbf{x}}{dt} = \mathbf{F}(\mathbf{x}, t), \qquad \mathbf{x} \in \Re^n \tag{1}$$

Stability of an arbitrary trajectory $\mathbf{x}^*(t)$ is defined by the variational system

$$\frac{d\mathbf{y}}{dt} = \hat{\mathbf{J}}(\mathbf{x}^*(t))\mathbf{y} \tag{2}$$

where $\hat{\mathbf{J}}(\mathbf{x}^*(t)) \equiv \partial \mathbf{F}(\mathbf{x}^*(t))/\partial \mathbf{x}$ is n×n Jacobi matrix, $\mathbf{y}$ is an arbitrary n-vector in tangent space. Let us now choose the set of mutually orthogonal vectors $\{\mathbf{y}_k\}$, k = 1, 2, ..., n. Then, the spectrum of RLE, which is denoted by $\mu_k(t)$, is introduced as the logarithmic derivative of the norms of these vectors

$$\mu_k(t) = \frac{d}{dt} \ln \|\mathbf{y}_k\| \tag{3}$$

The functions $\mu_k(t)$ characterize the local expansion and contraction properties of the phase flow in the vicinity of an arbitrary trajectory in the directions defined by vectors $\{\mathbf{y}_k\}$. The set of standard measures of motion, Lyapunov characteristic exponents (LCE) [20, 21], which is usually used for distinguishing chaos among other possible regimes, is directly related to RLE through the following formula

$$\lambda_k = \lim_{T\to\infty} \frac{1}{T} \ln \frac{\|y_k(T)\|}{\|y_k(0)\|} = \lim_{T\to\infty} \frac{1}{T} \int_0^T \mu_k(t)dt \tag{4}$$

So, LCE is simply the long time averages of corresponding RLE. Positive sign of the largest LCE λ_1 indicates the exponentially fast separation of nearby trajectories under the dynamics which is considered as a typical feature of the motion on a chaotic attractor.

The basic idea underlying the proposed approach originates from the relation (4). It is evident, that λ_1 may take a positive value only if μ_1 is also greater than zero during some time intervals. On the contrary, if the inequality

$$\mu_1(t) < 0 \tag{5}$$

holds at any time, the system is "superstable" and chaos is impossible. From the inequality (5), together with (1), (2), it turns out possible to obtain the position of the domain in the phase space, where the system is "safe", in the sense that no chaos appears, provided the trajectory never leaves the region where μ_1 is negative. So, we propose first to specify the stability area in the phase space and then obtain the estimate for the motion amplitude and derive the threshold of chaos arising in the parameter space.

In order to demonstrate how to implement this technique for a particular system, we take the following equation [1-3]

$$\frac{d^2x}{dt^2} + \delta\frac{dx}{dt} + N(x) = f(t) \tag{6}$$

which naturally arises in many problems of mechanics, electronics, optics, field theory, etc. Here, x is a generalized coordinate, δ is linear dissipation parameter, N(x) is potential function, f(t) is external force. The explicit expression for the largest RLE has the form

$$\mu_1 = \frac{1}{2}\left[-\delta + \delta\cos\Theta + \left(1 - \frac{dN(x)}{dx}\right)\sin\Theta\right] \tag{7}$$

where Θ stands for the direction of the vector $\mathbf{y}_1$. It is easy to see from (7) that, since Θ can take any value within the interval [0; 2π], μ_1 can take any sign and, hence, no conclusion about the stability is possible at this stage. However, if to make a linear coordinate transform

$$\mathbf{z} = \hat{\mathbf{B}}\mathbf{x} \tag{8}$$

which preserves the stability properties of trajectories and adjust the elements b_{ij} of the matrix $\hat{\mathbf{B}}$ in accordance with

$$2(j_{21}b_{12}b_{22} - j_{12}b_{11}b_{21}) + (b_{21}b_{12} + b_{11}b_{22})(j_{11} - j_{22}) = 0 \tag{9}$$

where j_{ij} are elements of the matrix $\hat{\mathbf{J}}$ from (2), then Θ is replaced by Ψ, the prefactor at the $\cos\Psi$ term vanishes, and we have instead of (7)

$$\mu_1 = \frac{1}{2}\left\{-\delta + \left[\frac{1}{\xi} - \xi\left(\frac{dN(x)}{dx} - \frac{\delta^2}{4}\right)\right]\sin\Psi\right\},\ \xi = \frac{b_{22}}{b_{11}} \tag{10}$$

which is more appropriate for the subsequent treatment. Notice, that, when deriving the relation (10), we set the value of the element b_{12} equal to zero in order to make the constant term in (10) independent from the coordinate x.

It is evident from (10) that if

$$\left|\frac{1}{\xi} - \xi\left(\frac{dN(x)}{dx} - \frac{\delta^2}{4}\right)\right| < \delta \tag{11}$$

μ_1 is always negative and, therefore, only regular behavior of solutions is possible for the system (6), when the control parameters and the amplitude of motion belong to the area defined by (11). We would like to stress here, that this result is valid for any form of the nonlinear potential function N(x) and for any type of external force. The value of ξ is arbitrary at the moment. In any particular problem it should be specified on the basis of the form of N(x) function. It seems reasonable to choose the value of ξ in such a way that the stability area in the phase space was maximal. This choice permits to obtain as accurate estimates for chaos arising threshold as possible in the framework of the present method.

At the final stage of the analysis, some different method should be used for getting the relation between control parameters and the location of motion in the phase space. For every particular form of the N(x) potential and type of external force this constitutes a separate problem, to be solved simultaneously with the task of parameter ξ choice.

Let us now consider the simplest situation and put

$$\begin{aligned} N(x) &= \kappa x + \alpha x^2 + \beta x^3 \\ f(t) &= P\cos(\nu t - \theta) \end{aligned} \tag{12}$$

So defined, the function N(x) corresponds to Duffing type equation instead of (6). The behavior of such type dynamical system is well documented. The choice of the sinusoidal external force facilitates considerably the final stage of the problem, allowing harmonic balance method to be used [3].

Further on, the inequality (11) transforms to the system

$$\begin{aligned} G(x) + 4\xi\delta &> 0 \\ G(x) - 4\xi\delta &< 0 \end{aligned} \tag{13}$$

where

$$G(x) = 12\beta\xi^2x^2 + 8\alpha\xi^2x + \xi^2(4\kappa - \delta^2) - 4. \qquad (14)$$

A typical solution of the system defined by (13, 14) is illustrated in Fig.1. The areas of stability $x \in [x_1;x_2]$ or $x \in [x_3;x_4]$ are placed between two parabolas whose positions are defined by all the control parameters and the value of ξ as well. So, we have

$$x_{1\div4} = -\frac{\alpha}{3\beta} \pm \sqrt{\frac{1}{3\beta}\left(\frac{\alpha^2}{3\beta} - \kappa + \frac{\delta^2}{4} + \frac{1}{\xi^2} \mp \frac{\delta}{\xi}\right)} \qquad (15)$$

and the largest Lyapunov exponent λ_1 is negative if the motion occurs within the crosshatched areas in Fig.1.

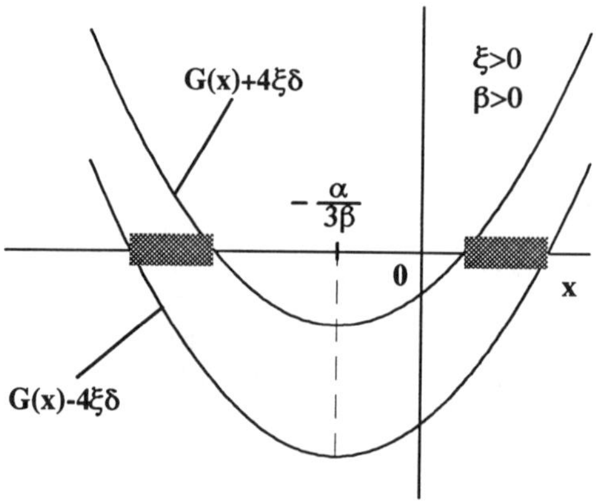

FIG.1. Solution of inequalities (13). Stability area is crosshatched

Now we are in a position to relate the parameters of the external force to the area where the motion takes place. We use here the results of applying the harmonic balance technique to the case of symmetric potential ($\alpha = 0$; $\kappa < 0$; $x_{1,2} = -x_{4,3}$). In addition, this choice of parameter values enables us to compare our results with both the predictions following from the Melnikov theory [7] and chaos arising criteria developed in [8, 9]. It should be stressed, however, that we restrict our consideration to the pointed out case only for the sake of clarity. The proposed technique may be applied in exactly the same way to the problems with other form of the nonlinear potential function, such as one-well Duffing oscillator with softening or hardening type of nonlinearity, or pendulum equation, etc.

Following [8], the solution is assigned the form

$$x(t) = x_0 + A\cos(\nu t) \qquad (16)$$

Substituting (16) into (6) with taking (12) into account we obtain the well-known formulae [8] which fix the position of the trajectory in the phase space

$$x_0^2 = -\frac{\kappa}{\beta} - \frac{3}{2}A^2; \tag{17a}$$

$$\left[\delta^2 \nu^2 + \left(2\kappa + \nu^2 + \frac{15}{4}\beta A^2\right)^2\right] A^2 = P^2 \tag{17b}$$

Combining (17a) with (15), the value of maximal amplitude assuring the stable motion can be derived. Then this magnitude has to be substituted to (17b) for obtaining the threshold of stability in P-ν parameter plane. It should be also taken into account that Eq. (17b) admits multiple solutions in a definite range of control parameters, known as a hysteresis phenomenon. Within this range at least two attractors with different oscillation amplitudes coexist in the phase space. Correspondingly, the chaos arising conditions for both attractors are not similar. Typically, the larger amplitude attractor transforms to a chaotic one, whereas the smaller orbit remains periodic under the variation of parameters.

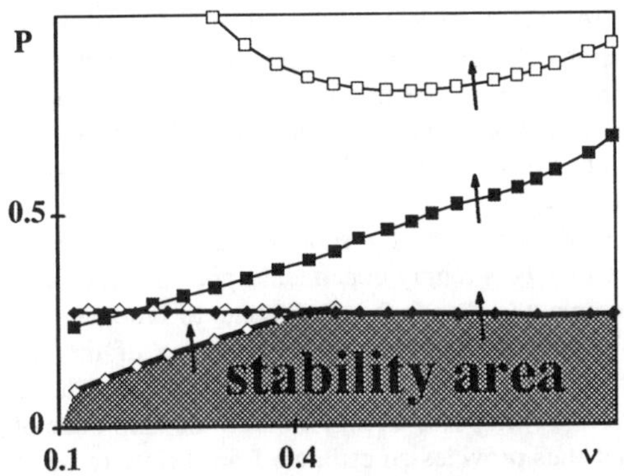

FIG. 2. Sate diagram of system (6) with potential and external force defined by (12) at $\alpha = 0$; $-\kappa = \beta = 0.5$; $\delta = 1$. —■— - actual threshold of chaos arising ; —□— - border of chaos from Melnikov's criterion (18); —♦— - border of the stability area defined from combining the RLE - criterion and harmonic balance method (Eqs. (15) and (17)); —◇— - hysteresis boundary from (17b). Arrows indicate the direction of change in the control parameters leading to chaos.

In Fig.2 an example of state diagram for the system (6), (12) is given. The crosshatched area delimited by heavy lines corresponds to the solution of (15), (17), with taking into consideration the hysteresis phenomenon. The line of actual chaos arising defined in numerical experiments is also shown. Comparison of mutual arrangement of these lines permits to conclude that the proposed technique gives a satisfactory estimate for chaos arising threshold.

It is interesting to compare the results with the predictions made on the basis of other techniques, such as Melnikov's method and the approach based on the combination of harmonic balance and variational equations analysis. In the case of Melnikov's criterion the chaos arising threshold is located above the line [7]

$$P > \frac{-\delta\kappa\sqrt{-2\kappa}}{3\nu\sqrt{\beta}}\cosh\left(\frac{\pi\nu}{2\sqrt{-\kappa}}\right) \tag{18}$$

Our experience indicates that at small values of dissipation parameter δ, the pointed out methods work better than ours, giving more accurate estimates for the border of chaotic areas. However, when dissipation becomes larger, both approaches fail and lead to absolutely wrong predictions. The threshold values obtained by our method seem never be inconsistent with the results from direct integration of governing equations, thus providing the necessary condition for chaos to arise.

We suppose that the proposed method may be used as a first step in investigating the possibility of the presence of chaos in passive nonlinear oscillatory systems. Although it strongly underestimates the threshold of chaos at low values of dissipation, at larger magnitudes of δ the predictions made are better then the ones provided by other existing techniques. Such an approach seems to be especially useful in the situations when chaotic behavior is an undesired effect. From this view, it provides the stability criterion for the system under study and thus guarantees the absence of possible additional noise source originating from the chaotic dynamics. (See [19] for an example of application to the problem of SQUID stability).

It was recognized recently that the threshold of chaos arising in nonlinearity parameter for oscillatory systems may be strongly dependent upon the type of external excitation. As shown in [22], for example, the change from harmonic to bifrequency resonant excitation in an equation of class (6) results in considerable lowering of the chaos onset. From this follows an important question, to what extent the thershold value may be lowered. We suppose that the relation (11) imposes a restriction on the amplitude of motion under which chaos is impossible and thus provides an estimate from below to this value. We would like to emphasize that the result (11), being independent of the type of external force, gives a fundamental limit for chaotic instability to appear in a broad class of nonlinear dynamical systems.

References

[1] J. Guckenheimer and P. Holmes, *Nonlinear Oscillations, Dynamical Systems, and Bifurcations of Vector Fields* (Springer-Verlag, New York, 1983).

[2] S. Wiggins, *Global Bifurcations and Chaos: Analytical Methods* (Springer, Berlin, 1988).

[3] Y. Ueda, *The Road to Chaos* (Aerial Press, Santa Cruz, 1994).

[4] V.B. Ryabov and D.M. Vavriv, Phys. Lett. A **153**, 431 (1991).

[5] V.K. Melnikov, Trans. Moscow Math. Soc. **12**, 1 (1963).

[6] L.P. Shil'nikov, Sov. Math. Dokl. **6**, 163 (1965).
[7] P. Holmes and J. Marsden, Arch. Rat. Mech. Anal. **76**, 135 (1981).
[8] W. Szemplinska-Stupnicka, Ingenieur-Archiv **58**, 354 (1988).
[9] T. Kapitaniak, Phys. Lett. A **144**, 322 (1990).
[10] B.V. Chirikov, Nonlinear Resonance (Novisobirsk, 1977) (in Russian)
[11] A.B. Belogortsev, D.M. Vavriv, and O.A. Tret'yakov, J. Sov. Phys. Tech. Phys. 32, 174 (1988).
[12] D.M.Vavriv and V.B.Ryabov, Dokl. Akad. Nauk UkrSSR A **2**, 50 (1990).
[13] D.M. Vavriv and V.B. Ryabov, Comput. Maths. Math. Phys. **32**, 1259 (1992).
[14] H. Herzel and Th. Schulmeister, in *Dynamical Systems and Environmental Models* (Akademie, Berlin, 1987).
[15] J.S. Nicolis, Rept. Progr. Phys. **49**, 1109 (1986).
[16] B. Eckhardt and D. Yao, Physica (Amsterdam) **65D**, 100 (1993).
[17] J.M. Nese, Physica (Amsterdam) **35D**, 237 (1989).
[18] P. Grassberger and I. Procaccia, Physica (Amsterdam) **13D**, 34 (1984).
[19] S.A. Bulgakov, V.B. Ryabov, V.I. Shnyrkov, and D.M. Vavriv, J. Low Temp. Phys. **83**, 241 (1991).
[20] G. Bennetin, L. Galgani, A. Giorgilli, and J.M. Strelcyn, Meccanica **15**, 9 (1980).
[21] A. Wolf, J.B. Swift, H.L. Swinney and J.A. Vastano, Physica (Amsterdam) **16D**, 258 (1985).
[22] D.M. Vavriv, V.B. Ryabov, and I.Yu. Chernyshov, Sov. Phys. Tech. Phys. **36**, 1325 (1991).

Nonlinear Dynamics and Cryptosystem

Ying Liu

Dept. of Math. and Computer Science
Savannah State College
Savannah, Georgia 31404

Abstract. In this paper, a new cryptosystem using dynamical systems is introduced, which features storing information in stable configurations of dynamical systems but representing the information in arbitrary configurations of dynamical systems. The message p in a plaintext can be considered as a stable configuration of a dynamical system. The encrypted string of p in a ciphertext is an arbitrary configuration which will lead the dynamical system to the attractor defined by the original string p. The encryption procedure, C = E(p), treats a string in a plaintext as an attractor of a finite dynamical system, and will generate a configuration C randomly, as long as the configuration leads to a correct attractor. The decryption procedure regenerates attractors of a dynamical system, p = D(C) = D(E(p)), from the configurations contained in the ciphertext. Three families of cryptosystems (simple, compound, and stochastic cryptosystems) are presented.

1. INTRODUCTION

Dynamical systems consist of a space together with a mapping rule. The space is called configuration space. Each point in the space is called a configuration. In this paper, the configuration is a subset of X = { 0, 1, 2, ... , N-1 }. All such configurations together form a space H(X), called a configuration space. An evolution of a system in its configuration space is specified by a set of production rules. A specification of a dynamical system includes a definition of a configuration space H(X) and a set of rules for motions in the configuration space, H(X) --> H(X). Most dynamical system evolutions are irreversible. An orbit of a system is a trajectory in its configuration space. Orbits merge with time, and after many time steps, orbits starting all possible initial configurations become concentrated onto "attractors". These attractors typically contain only a very small fraction of possible configurations.

A dynamical system is defined by a set of mapping rules in a configuration space, which is, in turn, specified by a set of parameters, t. All such sets of parameters together form a space, T, the parameter space. Depending on the position in a parameter space, $t \in T$, a dynamical system can exhibit extremely regular behaviors, like fractals, on one end, and Chaos on the other end. Such behavior can be easily observed. An example is Feigenbaum iteration (with only one parameter, a): $x_{n+1} = a\, x_n(1 - x_n)$ [1].

Dynamical systems can be classified by their attractor topology. In the following, we will classify the dynamical systems into nine classes[2]:

1. Systems with a single null attractor.
2. Systems with a single point attractor.

3. Systems with a finite number of point attractors.
4. Systems with an infinite number of point attractors.
5. Systems with a single periodic attractor.
6. Systems with a finite number of periodic attractors.
7. Systems with an infinite number of periodic attractors.
8. Systems with regular infinite attractors.
9. Strange attractors.

When we limit X to be a finite set, the classification is reduced to seven classes:

1. Systems with a single null attractor.
2. Systems with a single point attractor.
3. Systems with a finite number of point attractors.
5. Systems with a single periodic attractor.
6. Systems with a finite number of periodic attractors.
8. Systems with attractors that grow with the cardinality of X. The dependence of the attractor on the size of X can be formulated.
9. Systems with attractors that grow with the cardinality of X. The dependence of the attractor on the size of X can not be formulated and can be determined only through simulation.

Dynamical systems which generate simple fractals are in class 2 or 3 while those which generate Chaos are in class 9. By turning parameters, a dynamical system can across between classes. A dynamical system graph of a dynamical system is defined as follows: Let $|X| = N$, then there are 2^N nodes in the graph, one for each configuration of the dynamical system. If state i goes to state j, there will be an edge from node i to node j. The above classification naturally divided the configurations of a dynamical system into two categories: transient and persistent. Let a dynamical system have N configurations, if a vertex can return to itself within a N-step traversal, it is called persistent. If it does not return to itself within N steps, it is called transient.

Recently, information processing using stable configurations of nonlinear dynamical systems, or persistent node of a digraph, has been used intensively both in neural networks and Fractal information processing. In this paper, a new cryptosystem is developed based on a similar approach. We will use the ideas of storing information in stable configurations of dynamical systems. A string p in a plaintext can be considered as a stable configuration of a dynamical system. The encrypted string in a ciphertext is an arbitrary configuration which will lead the dynamical system to the attractor defined by the original string p. The encryption procedure treats a string in a plaintext as an attractor of a dynamical system, and will generate a configuration randomly, as long as the configuration leads to a correct attractor. The collection of such configurations forms a ciphertext. The process of decryption is to generate attractors of a dynamical system from the configurations contained in the ciphertext.

In the systems using keys, it is well known that to break a key of length L, if the decoding of the key can be reduced to a NP problem, then an exhaustive search has the time complexity of $O(2^L)$, where 2^L is the cardinality of the key space. For NP-problems, there is no better algorithms. In this paper, we will show that our new cryptosystems have a much larger key space. For binary dynamical systems, if the size of dynamical systems is L, the exhaustive search must deal with much more cases than 2^L; that is, it deals with the following number of cases:

$$N = 2^{L\,2^L}$$

There are several open problems here, while Key-algorithms are reducible to NP-problems, whether the algorithms proposed are NP-reducible is unknown.

Section 2 discusses the basic ideas. Section 3 discusses the possible advantages of the approaches. Section 4, 5 and 6 present three families of cryptosystems: simple, compound, and stochastic systems.

2. DATA ENCRYPTION AND DECRYPTION

Point attractors and cyclic attractors with low periods can be used as containers for data. Storing a string in an attractor of a dynamical system can achieve data encryption: the strings in plaintexts are attractors, any configurations of the dynamical systems which lead to correct attractors encrypt the attractors. It is possible to encrypt a string, p, in a plaintext, specified by an attractor of a dynamical system, by storing a configuration, q, of the dynamical systems which leads the dynamical system to p.

Let X be a complete metric space. A string is a compact subset of X. All possible strings together form a space, called a data space H(X). A dynamical system consists of a space H(X), together with a set of mappings. If a system has k attractors:

$$P = \{ p_1, p_2, \dots p_k \}.$$

the space H(X) is quantized into k sets by the system:

$$H(X) = D_1 \cup D_2 \cup \dots \cup D_k.$$

Each D_i is called an attractor basin. Let $q \in D_i$ be an arbitrary configuration, then the evolution of the system with q as an initial point will lead to the attractor p_i.

We can define a mapping F as

$$F\colon H(X) \dashrightarrow P \quad or \quad F(q) = p_i, \; q \in D_i$$

The decryption process is to retrieve a string $p \in P$ from any initial configuration of a dynamical system $q \in H(X)$. Given a string $p \in P$, a configuration of a dynamical system can be found such that: $q = F^{1}(p)$. The encryption process is to obtain a point $q \in H(X)$ from $p \in P$.

An encryption algorithm is given in Fig. 1. The decryption algorithm corresponding to the encryption algorithm in Fig.1 is given in Fig. 2.

1. Create a partial attractor list;
2. Assign each base string to one or more attractors;
3. Decompose a plaintext into base strings:
   $\{ p_1, p_2, ..., p_n \}$;
4. While (not all pi are encrypted) do
   randomly generate a configuration u;
   while (u does not lead the dynamical system to p_i) do
   randomly generate a configuration u;
   end while;
   assign u to q_i;
   end while;
5. The ciphertext is $\{ q_1, q_2, ..., q_n \}$.

Fig.1. An example of an encryption algorithm using dynamical systems.

1. While (not all q_i are decrypted) do
   trace the attractor p_i with q_i as an initial state;
   end while;
2. Convert $\{ p_1, p_2, ..., p_n \}$ into original plaintext.

Fig.2. The decryption algorithm corresponding to the encryption algorithm of Fig. 1.

Since a dynamical system can be visualized by a dynamical system graph, the above algorithms have a geometric picture. The configurations in attractors are persistent nodes in a digraph. Specially, the point attractors are sinks in digraphs. The plaintext is encrypted into persistent nodes, then transient nodes are used in ciphertext.

The cryptosystem used here should be in class 3 (See section 1) or in class 6 with low periods. They are classified into several types: simple system; compound systems; and stochastic cryptosystems. Examples of the simple dynamical cryptosystems are neural cryptosystems[3]; cellular cryptosystems[4,6]; iterated function system [5] cryptosystems (IFS cryptosystem); and automata network [7] cryptosystems. These systems are the foundations of compound systems and stochastic systems.

3. DISCUSSION

The simple and compound systems are "binary" dynamical systems while the stochastic dynamical systems are not. In this section, we will discuss some general properties of the approach.

3.1 What is being Kept Secret in Our System

The encryption function E and decryption function D are given in the last section. We assume the adversary knows the algorithms. What is kept secret is the dynamical systems used. The following theorem will immediately show the advantages of our approach.

Theorem Let $|X| = L$, that is, X has L element: $X = \{ 0, 1, 2, ..., L-1 \}$, then there exist

$$N = 2^{L2^L}$$

dynamical systems.

It is the specification of a dynamical system that is being kept secret. As the above theorem shows, there are quite a few of them, much more than the public key or common key systems, assuming that the length of a key and the cardinality of the space X are equal. Of course, the cardinality of X can be an arbitrary number, just as the length of a key can be any number.

3.2 A Better System ?

(1) The search space has 2^M points, where $M = L*2^L$, and L is the cardinality of a space X. In systems using keys, the search space has only $2^{L'}$ points, where L' is a length of a key. The cardinality of X can be any positive integer, just as the case of the length of a key. Therefore, the key space is larger. However, given a family of dynamical systems, it is still open to determine whether it is NP-reducible to find a specification of a dynamical system.

(2) The encryption mapping, $C = E (P)$, is not unique. There are many ways to map a plaintext to ciphertext after a dynamical system is chosen. The feature is summarized in the following theorem.

Theorem Let $X = \{ 0, 1, 2, ..., L-1\}$. Let a dynamical system be chosen and is used to encrypt n symbols. Then the maximum selections of a ciphertext for a plaintext is

$$\left(\frac{2^L}{n}\right)^m$$

where m is the number of strings in the plaintext.

This theorem indicates that once a dynamical system is chosen, the mapping from a plaintext P to ciphertext C is not unique. A same plaintext C can be sent through a channel many times. Each time the ciphertext is different.

3.3 Break the Systems by Search

We assume the algorithm is known. To break a system, one has to find the dynamical system used in the encryption mapping.

Let $|X| = N$. There is a systematic way to define all dynamical systems. This will take the space complexity of $O(2^N)$. All dynamical systems can be ordered by the systematic definition.

Definition All the dynamical systems together form a space, called dynamical system space, D. A point in D is a dynamical system [2].

Definition The canonical order[2] for dynamical systems with |X|=N is defined as follows. Each dynamical system is represented by an integer in radix 2^N and the length of the integer is 2^N. These integers can be, in turn, mapped to decimal integers: 0, 1, 2, ..., $2^m - 1$, where $m = N2^N$.

Example. Let N = 3, an integer with length 8 in radix 8 will represent a dynamical systems:

0 0 0 0 0 0 0 0,
0 0 0 0 0 0 0 1,
...
0 0 0 0 0 0 0 7,
0 0 0 0 0 0 1 0,
...
7 7 7 7 7 7 7 7.

The dynamical system 0000 0007 implies that states 001, 010, 011, 100, 101, 110, 111 go to state 000, and state 000 goes to state 111

Each dynamical system is represented by an integer in radix 2^N and the length of the integer is 2^N. These integers can be considered to form a 2^N-dimensional lattice with 2^N points in each dimension. The distance in D can be simply the Euclidean distance (L_2 distance) or the Manhattan distance (L_1 distance) in the 2^N-dimensional lattice.

An exhaustive search for breaking the cryptosystem will test each of the dynamical systems. Once a dynamical system is chosen, one has to test all possible ciphertexts for a given plaintext. The number of ciphertexts for a given plaintext grows exponentially as the length of the plaintext.

4. AN EXAMPLE OF SIMPLE SYSTEM

Take a neural cryptosystem as an example: a message, m, in a plaintext can be considered as a stable configuration in a neural network. The encrypted string of m in a ciphertext is an arbitrary configuration which will lead the neural network to the attractor defined by the original message m. The encryption procedure, C = E(m), treats a message m in a plaintext as an attractor of a neural network, and will generate a configuration C randomly, as long as the configuration, C, leads to a correct attractor, m. The decryption procedure regenerates the attractor, m = D(C) = D(E(m)), of the neural network from the configurations, C, contained in the ciphertext.

In the case of neural networks, only neural networks of class 3 and class 6 with low periods should be considered. The following theorems should be used to select these second order neural networks.

Theorem A neural network is stable [15] if

(1) the connection matrix satisfies

$$M = M^T$$

and (2) the operation mode satisfies

$$\Delta X M \Delta X \geq 0$$

Corollary

a. A neural network is stable if [16]
   (1) the operating mode of the neurons is asynchronous;
   (2) M is symmetric and the diagonal elements are zero.

b. A neural network is stable if [15]
   (1) the operating mode of the neurons is synchronous;
   (2) M is symmetric and can be written as

$$M = N^T N.$$

c. A neural network is stable if [15]
   (1) the operating mode of the neurons is synchronous;
   (2) N is symmetric and M can be written as

$$M = NN.$$

d. A neural network is stable if [15]
   (1) the operating mode of the neurons is synchronous;
   (2) A is antisymmetric and M can be written as

$$M = AAAA.$$

e. A neural network is stable if [8]
   (1) M can be written as

$$M = \begin{pmatrix} 0 & N \\ N^T & 0 \end{pmatrix}$$

   (2) X is divided into two parts X = (Y Z):

$$\begin{pmatrix} 0 & N \\ N^T & 0 \end{pmatrix} \begin{pmatrix} Y \\ Z \end{pmatrix} = \begin{pmatrix} NZ \\ N^T Y \end{pmatrix}$$

The updating rule is Y, Z, Y, Z, ...

   (3) the operating mode of the neurons for Y or Z is synchronous;

f. A neural network is stable if [15]

(1) M can be written as

$$M = \begin{pmatrix} 0 & M_{12} & M_{13} \\ M_{12}^T & 0 & M_{23} \\ M_{13}^T & M_{23}^T & 0 \end{pmatrix}$$

(2) X is divided into two parts X = (X1 X2 X3);
The updating rule is X1, X2, X3, X1 .... .

(3) the operating mode of the neurons for Xi is synchronous;

Theorem The maximum period of a neural network is 2 if [9]

(1) the connection matrix is symmetric; and

(2) the operation mode is synchronous.

The maximum period of a neural network is 4 if [9]

(1) the connection matrix is antisymmetric; and

(2) the operation mode is synchronous.

Similar results can be found for IFS in [5, 10].

5. COMPOUND SYSTEMS

Formal language theory [11] and Hutchinson operators[1] can be used as a control device in dynamical systems. Given a family of dynamical system, one can systematically add a control device to the system.

5.1 Finite Automata[11]

Definition: A finite automaton is a 5-tuple

$$M = (Q,\Sigma,\delta,q_0,F),$$

where

Q is a finite set of states;
$\sum$ is a finite input alphabet;
q_0 in Q is the initial state;
$F \subseteq Q$ is the set of final states; and
δ is the transition function mapping $Q\times\sum$ to Q.

Each finite automaton accepts a language. We define L(M), the language accepted by M to be all strings that lead to a final state. In particular, the language accepted by a finite automaton is called regular expression.

5.2 One State Automata Dynamical Systems

In this section, we introduce dynamical systems with more than one mapping rule. The configuration space H(X) consists of all subsets of X. Since points in H(X) are also sets, operations like union, intersection and negation make sense. The operator union will be used in the following definition.

Definition Let { w_i, i = 0, 1, ... K-1 } be K mapping rules of a dynamical system, the Hutchinson operator,

$$W = \bigcup w_i$$

is defined as

$$W(B) = \bigcup w_i(B), \quad B \in H(X)$$

Definition: A one-state dynamical system is a doublet

$$A = \{ H(X), \Sigma \},$$

where

(1) H(X) is a configuration space; and
(2) Σ is a set of mapping rules

$$\Sigma = \{w_0, w_1, \ldots, w_{K-1}\}.$$

The configurations evolve by iteration of the mapping

$$\Sigma(B^{(t)}) = \bigcup_{i=0}^{K-1} w_i(B^{(t-1)}), \qquad B \in H(X).$$

Consider H(X) as a state of a finite automata, then the above machine is a one state finite automaton. H(X) is the initial state and the final state. $\sum$ defines the alphabet. The transition function maps (H(X), $\sum$) to H(X).

5.3 Dynamical Systems with Internal Control

We now formally define the automata dynamical system:

Definition: An automata dynamical system, E, is a 5-tuple

$$E = \{ R; \Sigma; M; I; F \}$$

where

(1) R = { $H(X_1)$,..., $H(X_n)$ } is a finite set of states;
(2) Σ is a mapping alphabet;
(3) M is a n by n transition matrix. Each of the elements of this matrix, M_{ij}, is a subset of Σ. The transition rule is

$$(X_j, w) \to X_i, \quad if \ w \in M_{ij}.$$

The set M_{ij} can be empty.
(4) The initial state is $I = \{X_I\}$ which sets the initial configuration.
(5) The set of final states F is a subset of R which determines the configurations of the whole dynamical system.

The states that are not in F and I are hidden units of the system. F is responsible for representing the system. The attractors of F are the attractors of the system. I is responsible for setting the initial configuration. The above definition can be modified such that there is more than one state in I, however such a machine is no more powerful than the above one because the former can be simulated by the latter.

5.4 Other Types of Dynamical Systems with Internal Control

The above dynamical system with internal control is based on Hutchinson operator and finite automata. There are many operators like Hutchinson operators, and there are many machines like finite automata, therefore, there are several ways to construct other types of systems with internal control. This includes:

(1) replace finite automata by more powerful machine, for example, pushdown automata [11];
(2) replace Hutchinson operator by "generalized Hutchinson operator".

In the following, we will show a few examples of the second case.

Definition: An intersection Hutchinson operator for an alphabet

$$\Sigma = \{w_0, w_1, \ldots, w_{K-1}\}.$$

is

$$\Sigma(B^{(t)}) = \bigcap_{i=0}^{K-1} w_i(B^{(t-1)}), \qquad B \in H(X).$$

Definition: A mixed Hutchinson operator for an alphabet

$$\Sigma = \{w_0, w_1, \ldots, w_{K-1}\}.$$

is

$$\Sigma(B^{(t)}) = G(\, w_i(B^{(t-1)})), \quad i=0,\ldots,K-1 \quad B \in H(X).$$

For example,

$$G(a,b,c) = a \cup b \cap c,$$

then

$$\Sigma(B^{(t)}) = w_1(B^{(t-1)}) \cup w_2(B^{(t-1)}) \cap w_3(B^{(t-1)}).$$

5.5 Compound Cryptosystems

The simple dynamical cryptosystems are neural cryptosystems[3]; cellular cryptosystems[4]; iterated function system cryptosystems (IFS cryptosystem); and automata network cryptosystems. From these systems, using the theory of section 5.4, many compound systems can be generated [15,10,11]. Example are: (1) several networks together can form a one-state automata neural network. With an additional FA, a automata neural network can be made [15]; (2) similar theories exist for IFS [17, 11, 12]; (3) a set of neural networks, a set of IFS, and a finite automata can form an automata-neural-IFS system. Examples here are lengthy and the readers are referred to [17,10,12,13,15]. Once the compound systems are established, they can be used in the same way as simple systems. In some sense, a compound, A, uses a larger simple system, B. However, only part of the configurations of B are visible to outside. To be more specific, only part of X (therefore only part of H(X)) are visible to outside. A "projection" of B is visible, through the final states of A.

6. STOCHASTIC SYSTEMS

The simple systems and compound systems are binary systems. In this section, we will discuss grayscale systems. One of these types is a stochastic dynamical system. In simple systems, a string $p \in H(X)$ is a binary string. In this section, the string is no longer binary.

An grayscale string f(x) with G intensity levels can be written as

$$x = 0,1,2, \ldots, M;$$

$$f(x) \in \{0,1,2,\ldots, G-1\}, \qquad G>1.$$

Let's think of the above string as a one-dimensional image. Each pixel can be considered as a state in a Markov chain and each image is considered as an distribution [14] of the Markov chain. Specifically, the markov chain is defined by a stochastic dynamical system. In this section, we will establish this connection and we will discuss the limit-behavior of stochastic dynamical systems. Probabilistic Iterated Function Systems (PIFS) will be used to illustrate the idea.

6.1 Stochastic Dynamical Systems

Given a family of simple dynamical systems, one can systematically construct a family of stochastic dynamical systems. In the following, we will use IFS to illustrate the idea.

Definition An IFS [5] is, C = {X, Σ}, where X is a metric space and Σ is a set of contractive mappings.

The stochastic version of IFS is called probabilistic IFS (PIFS). A PIFS has a set of mappings. At each time step, only one mapping is chosen. The probability of choosing the mapping w_i is p_i.

Definition A PIFS [5] is, C = {X, Σ}, where X is a metric space. The mapping rule is

$$\Sigma = \{ (w_i, p_i), i=1,2,3 \ldots K \}$$

$$\sum p_i = 1.$$

Here, w_i is a mapping and p_i is the probability to choose the mapping w_i. Each time only one mapping is chosen.

The parameter space of PIFS, T, is defined by the set of free parameters used to specify PIFS. These parameters are the probabilities p_i, i = 1, 2, ..., K; and the parameters in the mapping w_i, i = 1, 2, 3, ..., K. Let X = { 0, 1, 2, 3, ... , N-1 }, where N is the number of pixels in the string. Each integer in X can be considered as a state in a Markov chain.

6.2 Invariant Distributions

Let X = { 0, 1, 2, ..., N-1 }. Let $i \in X$ be a state in a Markov chain. A mapping w_k will connect state i to state $w_k(i)$ with probability p_k. This will define a Markov chain [14].

Let P = { p_{ij} } be a stochastic matrix. In an **irreducible** finite Markov chain, the limits [14]

$$u_k = \lim p_{jk}^{(n)} \qquad n \longrightarrow \infty$$

exist and are independent of the initial state j. In other words, one can start with any initial state j, after infinite steps, the probability of being in state k is always the same. Furthermore

$$\sum u_k = 1$$

$$u_j = \sum_i u_i p_{ij} \quad or \quad U^T = U^T P \quad or \quad U = P^T U$$

where P^T stands for transpose of P.

Whatever the initial distribution, the probability of the state k is u_k. The initial distribution will be converted to a stationary distribution, the invariant distribution.

Example Let the initial distribution of a Markov chain be

$$a_k^{(0)} = a_k.$$

After one step, the distribution is

$$a_k^{(1)} = \sum_j a_j p_{jk}.$$

After n steps

$$a_K^{(n)} = \sum_j a_j p_{jk}^{(n)}.$$

After infinite steps, the distribution is the invariant distribution. Note that if k is a transient state, $u_k = 0$. In other words, $u_k > 0$ implies that k is a persistent state for a finite Markov chain.

Using the example of PIFS, it defines a Markov chain. Each pixel is a state of the Markov chain. The state of a finite Markov chain is divided into persistent and transient. It can be proved that each reducible Markov chain possesses an invariant distribution [14], therefore, PIFS has stable distributions.

6.3 Stochastic Cryptosystems

For each stochastic dynamical system, A, a dynamical system B can be generated as follows. The distribution space of A defines the configuration space of B. In other words, a distribution of A is a configuration of B, all such distributions of A together form the configuration space of B. The mapping of B is defined as follows: Let the initial distribution of a Markov chain be

$$a_k^{(0)} = a_k.$$

After one step, the distribution is

$$a_k^{(1)} = \sum_j a_j p_{jk}.$$

This defines the mapping rule of the dynamical systems, B. After n steps

$$a_K^{(n)} = \sum_j a_j p_{jk}^{(n)}.$$

This defines an orbit in the configuration space of B. After infinite steps, the distribution is the invariant distribution of A, reaching the attractor of B.

The difference between here and the simple system is that the strings in simple systems are binary, while here the strings are not binary ones. The dynamical system, B, generated from a stochastic dynamical system, A, can be used in a cryptosystem in the same way as simple systems.

Note that, if the mappings of an IFS are contractive, PIFS has only one invariant distribution (i.e., the PIFS is in class 2):

Theorem: If a PIFS is contractive[5], the PIFS has only one invariant distribution.

Our cryptosystem can not use class 2 dynamical systems. In order for a PIFS to be in class 3, the contraction condition must be removed. A family of local contractive IFS [10], for example, are indeed in class 3.

Theorem: If a PIFS is local contractive [10], the PIFS can have more than one invariant distribution.

The membership problems (i.e., which system is in which class ?) is the main topic of dynamical systems. Our scheme employs class 3 dynamical systems and class 6 dynamical systems with low periods. We listed a few results for neural network and PIFS. We will not address the membership problems of other systems here.

7. CONCLUSION

In conclusion, a new cryptosystem is developed in this paper. In this scheme, a plaintext is a collection of attractors of a dynamical system. They are encrypted into a collection of arbitrary configurations. The advantages of the new systems includes possible increasing data security and easy implementation. The idea of encryption is to store a string, p, in an attractor of a dynamical system. Then an encrypted string of p in the ciphertext is an arbitrary configuration of the dynamical system which leads the system to p. Several families, including several simple systems, several compound systems, and several stochastic systems, are introduced.

ACKNOWLEDGMENTS

I would like to thank my wife Gina Porter Liu for proof reading of the paper.

REFERENCES

1. H. O. Peitgen, H. Jurgens, and D. Saupe, Chaos and Fractals, Springer-Verlag, 1992.
2. Y. Liu, "Fractal, Neural Networks, Cellular Automata, Formal Languages and Coding Theory," Proc. of IEEE SMC '92, pp.1663 - 1669, October, 1992.
3. Ying Liu, " A Neural Cryptosystem," Proc. of SimTec'93, pp. 291 - 296, November, 1993.
4. Ying Liu, " A parallel algorithm for cryptography using cellular automata," Proceedings of Fifth ISMM International Conference on Parallel and Distributed Computing and Systems, Pittsburgh, pp. 155 - 159, 1992.
5. M.F. Barnsley, Fractals Everywhere, Academic Press, 1988.
6. S. Wolfram, "Universality and complexity in cellular automata", Physica 10D, (1984) 1-35.
7. G. Weisbuch and S. Ryckebusch, Complex Systems Dynamics, Addison-Wesley, 1991.
8. B. Kosko, Neural Networks and Fuzzy Systems, Prentice hall, 1991.

9. J. Bruck, "On the Convergence properties of the Hopfield Model," Proceedings of the IEEE, Vol. 78, No. 10, 1579-1585, 1990.
10. M.F. Barnsley and L. P. Hurd, Fractal Image Compression, AK Peter, Ltd. 1993.
11. J.E. Hopcroft and J.D. Ullman, Introduction to automata theory, language, and computation complexity, Addison-Wesley, 1979.
12. Ying Liu and Hede Ma, " ω-Orbit Finite Automata for Data compression," Proceeding of Data Compression Conference '91, IEEE Computer Press, Snowbird, Utah, 166 - 175, April, 1991.
13. Ying Liu and Hede Ma, "Pattern recognition using ω-Orbit Finite Automata, Proc SPIE 1606, Boston, pp. 226 - 240, November, 1991.
14. W. Feller, An introduction to probability theory and its application, John Wiley and Sons, 1968.
15. Y. Liu, "Stability of Automata Neural Networks," Proc. SPIE, Vol. 1904, pp.291-304, February, 1993.
16. J. J. Hopfield, "Neural networks and physical systems with emergent collective computation abilities," Proc. of Nat. Academy Sci., USA, vol. 81, pp.3088-92, 1984.
17. M.F. Barnsley, et. al., Construction Approximation, V5, p1, 1989.

Autosolitons in Applied Physics and Traffic Flow

Boris S. Kerner

Research Institute, Daimler-Benz AG, F1V/VT, HPC: G253, 70546 Stuttgart, Germany

Abstract. A review of investigations of autosolitons in nonlinear systems which are of interest for the applied physics and for the transportation research is presented. Autosolitons are solitary intrinsic states which can be formed in a broad class of physical, chemical, biological dissipative distributed media and in traffic flow. Properties of autosolitons which are general for physical systems and for traffic flow will be discussed. Based on results of recent investigations of traffic jams in traffic flow, a comparison of nonlinear characteristics of traffic jams and with nonlinear properties of autosolitons which can be formed in active systems with diffusion will be given. Forms, properties, processes of evolution of autosolitons in traffic flow, in semiconductors and in gas discharge plasma are considered.

1. INTRODUCTION

Autosolitons are macroscopic intrinsic, localized states in nonequilibrium distributed systems, elementary building blocks of more sophisticated inhomogeneous macroscopic states of dissipative systems. Autosolitons can be formed in a broad class of physical, chemical, biological dissipative nonequilibrium media and in traffic flow. A formation of an autosoliton in a local region of a medium can be considered as a process of self-organization in the medium. Further evolution of autosolitons and possible sophisticated interactions between autosolitons determine other numerous scenarios of self-organisation. A turbulence in a medium may be regarded as complex autosolitons which parameters (width, amplitude, velocity, ets) are randomly change in time or as a pattern of autosolitons which appear and disappear at random in different parts of the medium.

A new approach to problems of self-organization and turbulence, treating these phenomena as a result of spontaneous formation and subsequent evolution of autosolitons has been presented in the book by Kerner and Osipov (1), where theory and physics of autosolitons which may be formed in a wide class of active systems with diffusion have also been considered. The nature of autosolitons in such kind of physical, chemical, biological nonequilibrium active media is very diverse. For example, an autosoliton in a form of a impulse running without attenuation may be realized in nerve fibre and in other excitable media (2) - (10). In semiconductors devices, autosolitons are often appeared as bright spots - regions of high temperature of carriers (11) - (13). Autosolitons in the form of a solitary stratum or a sequence of strata may appear in a gas discharge (14), (15). Autosolitons may also be formed in composite superconductors, nonlinear interferometers and in a lot of other active systems with diffusion which are of interest for the applied physics. A history and results of a general nonlinear theory of autosolitons in a wide class of active systems with diffusion have been summarised in (1), (16) - (18).

A formation of localized structures, - autosolitons, is not an exceptional property of dissipative physical, chemical or biological systems. Different kind of localized structures may also be self-formed in social nonlinear distributed systems, for example, in traffic flow, where autosolitons in the form of traffic jams can appear (19), (20).

The dissipative physical systems, in particular, active systems with diffusion, belong to the inanimate world where the well-known laws of nature are valid (1), (16) - (18). In traffic flow, on the contrary, the human effort to move with acceptable risk as fast as possible is important. During the formation of traffic jams, this human effort plays the decisive role (19), (20).

On the one hand, as a result of the mentioned difference between active systems with diffusion and traffic flow, a nature, forms, processes of an evolution and a lot of other characteristics of autosolitons in active systems with diffusion (1), (16) - (18) and of traffic jams in traffic flow (19), (20) it is found to be completely different. On the other hand, from a general point of view, the competition between safety requirements and risk readiness of drivers in traffic flow can be interpreted as a competition between active and damping processes which causes a self-organisation process in traffic flow - the formation of a traffic jam (19), (20). A competition between different kind of active and damping processes in active systems with diffusion is also decisive during the self-organisation process leading to the formation of autosolitons (1), (16) - (18). Apparently for this reason, some properties of autosolitons in active systems with diffusion (1), (16) - (18), as it has been found in (19), (20), are qualitatively similar to the behavior of traffic jams.

In this article, properties of autosolitons which are general for active systems with diffusion and for traffic flow will be investigated. Besides, based on results of works by Kerner and Konhäuser (19), (20), as well as on results of new investigations, nonlinear characteristics of traffic jams and a comparison of their characteristics with nonlinear properties of autosolitons which can be formed in active systems with diffusion will be given.

In Sect. 2, general properties of autosolitons will be considered. A process of a formation of traffic jams, the form of traffic jams, their characteristics and main parameters will be explained in Sect. 3. In Sect. 4, based on results of the book (1) and on the articles (18), (21), a brief discussion of general properties of autosolitons in active systems with diffusions as well as a brief consideration of diffusion autosolitons in semiconductor and in a gas discharge systems which are examples of active systems with diffusion will be made.

2. SOME GENERAL PROPERTIES OF AUTOSOLITONS

2.1. Local "Active" and "Damping" Processes

A possibility of an existence of an autosoliton in a medium is linked with a local instability which may be realized in some range of a controlling parameter of a system. This local instability in the medium is caused by a space-time competition between some local "active" and local "damping" processes in the medium.

Although nonlinear characteristics and the physics of systems, where autosolitons may be formed, can be entirely different, a lot of nonlinear properties of autosolitons in these systems may be qualitative the same. In particular, it occurs if the nonlinear properties of local active and local damping processes causing an appearance of autosolitons have a qualitatively similar character. This situation for example is realized in active systems with diffusion (9), (10), (1), (16) - (18). In these systems it often occurs that a local active process is due to a positive feedback in one of the variables of a system. Such variable is called an activator. The local damping process which controls a local change of the activator can also be fulfilled by some other variable (or a function of variables) of a system. This variable (or a function) is called an inhibitor. In different active systems with diffusion roles both of the activator and of the inhibitor can play entirely different variables. Nevertheless, because a space-time behaviour of the activator and the inhibitor in entirely different active systems with diffusion may be qualitative similar, nonlinear properties of autosolitons in these different active systems are, as a rule, qualitative the same (9), (10), (1), (16) - (18).

For example, charge carriers and, consequently, a current in semiconductor structures and in gas discharge systems can often appear due to an avalanche collision ionisation. The rate of this collision ionisation is, as a rule, an increasing function both the concentration of the carriers and the electric field. An increase in the concentration of the carriers causes an increase in the rate of the collision ionisation. Therefore, the concentration of the carriers may further increase, and so on. In other words, in this case the active process is the avalanche collision ionisation. The concentration of the carriers n plays here a role of the "activator". On the other hand, an increase in the concentration of the carriers can lead to a decrease in the local electric field and, consequently, to an decrease in the rate of the collision ionisation. This damping process competes with the active process considered above. Therefore, a role of an inhibitor plays the electric field, exactly, a voltage drop across an active layer of a system V_i (see Sect. 4). The characteristic length of a spatial change of the activator n is considerable more than the characteristic length of a spatial change of the inhibitor V_i. Therefore, a local increase in the concentration causes an increase in the rate of the collision ionisation which cannot be suppressed by the decrease in the voltage drop V_i, if the local region under consideration is small enough. As a result, a local solitary state - an autosoliton - in the form of a current filament, where the concentration of the carriers and the rate of the impact ionisation are higher than on the periphery of the system, may spontaneously be formed in the semiconductor or in the gas discharge systems (Fig. 1) (Sect. 4).

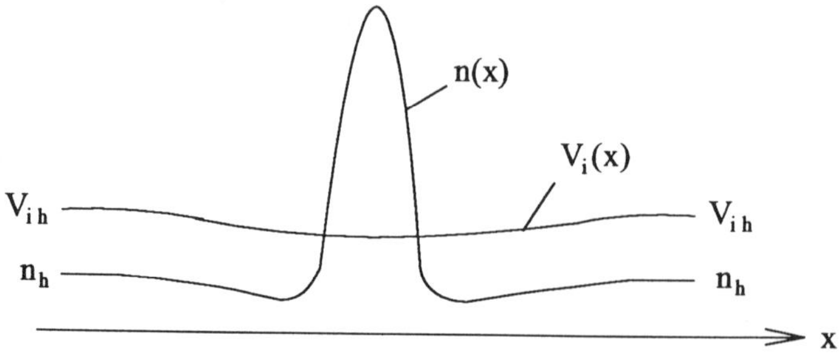

FIGURE 1. Form of a static autosoliton in semiconductor and gas discharge systems (see explanations in Sect. 4). n_h and V_{ih} are values of concentration of the carriers n and the voltage drop V_i on the periphery of the autosoliton (Ref. (1)).

In traffic flow, where a non-stationary moving autosoliton can be self-formed (Fig. 2), it is also possible to distinguish local active and damping processes (20).

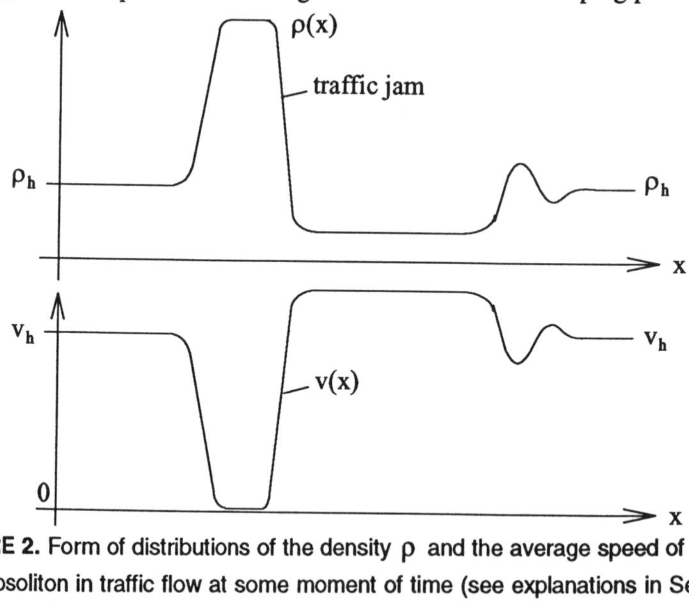

FIGURE 2. Form of distributions of the density ρ and the average speed of vehicles v in an autosoliton in traffic flow at some moment of time (see explanations in Sect. 3.2). ρ_h and v_h are the density and the average speed of vehicles in an initially homogeneous traffic flow (Ref. (20)).

Active processes appear in traffic flow due to interactions between vehicles which are realized when the density of vehicles is high enough. Indeed, drivers should slow down in a local region of a road, if the density of vehicles in this local region increases. In other words, the average speed of vehicles decreases in the local region, where the density is increased. The decrease in the density of vehicles in the local region of the road under consideration can cause the further slowing down of vehicles, and so on. Therefore, a local region of high density, - traffic jam - where vehicles cannot move at all, may be self-formed in an initially homogeneous traffic flow (Fig. 2) (20). Naturally, damping processes in traffic flow can suppress this active process. In this case a traffic jams is not formed. The conditions for a traffic jam formation in traffic flow will be discussed in Sect. 3.

Therefore, a role of an activator in traffic flow plays the average speed of vehicles v or the density of vehicles ρ. Damping processes can be linked with a carefully behaviour of drivers. Formally, it may be possible to determine that a role of local "inhibitor" in traffic flow plays a local flux of vehicles q^* in a system of co-ordinate moving with some velocity v_{tr}:

$$q^* = \rho(v - v_{tr}) \tag{1}$$

However, the velocity v_{tr} in the formula (1) itself depends on parameters of traffic flow and on distributions of the density and the average speed of vehicles in a local region of the road, where the flux q^* in the formula (1) should be calculated (Sect. 3.6).

In particular, due to this complex character of local damping processes in traffic flow, autosolitons in the form of traffic jams (Fig. 2) (Sect. 3) have a lot of qualitatively completely different nonlinear properties (20) in comparison with properties of autosolitons in active systems with diffusion (Fig. 1) (9), (10), (1), (16) - (18). Nevertheless, because the autosoliton in the form of the traffic jam can also exist owing to a competition between some local active and damping processes, it is possible to formulate some nonlinear properties of autosolitons which may be general for both traffic flow and active systems with diffusion.

2.2. Spontaneous Formation of Autosolitons

We can distinguish the following two types of self-organization phenomena connecting with a spontaneous appearance of an autosoliton in a distributed system (1), (17), (18), (20), (21):

1. Local instability of a distributed system: Spontaneous occurrence of a localized critical fluctuation acting as "nucleation centre" for a formation of an autosoliton.

2. The "local breakdown" effect. This effect results in a dynamical local restructuring of an initial state of a system. It causes a spontaneous appearance of an autosolitons in a deterministic way, i.e., even when fluctuations in a system may be neglected.

Systems, where the first type of self-organization phenomena, i.e., the local instability is realized, may be formally considered as "metastable" distributed systems in which small localized perturbations fade out. If a critical localized fluctuation in such system occurs, which amplitude exceeds some finite critical value, an autosoliton can spontaneously be formed in the system. In other words, it is possible to consider this problem as an escape problem from a homogeneous state of a system into a state of the system with an autosoliton. The critical localized fluctuation plays here a role of a "nucleation center" for the formation of the autosoliton. We will distinguish two different types of escape problems and correspondingly two types of localized fluctuations which may act as the "nucleation center" for the formation of the autosoliton.

2.2.1. Bistability and "First" Escape Problem

The first type of an escape from a homogeneous state of a system into a state of the system with an autosoliton is the well-known escape problem from a stable state of a system which have at least two stable stationary states (*a* and *c* in Fig. 3(a)) and one unstable stationary solution (*b* in Fig. 3(a)) (e.g., (22)).

Let us assume that the stable state *a* in Fig. 3 corresponds to a homogeneous state of the system and the other stable stationary state *c* corresponds to an autosoliton. We will only be interest for the situation, when an unstable stationary solution *b* in Fig. 3(a) has a finite but a small enough amplitude. In this case for a lot of distributed systems the homogeneous state *a* and also inhomogeneous solutions both between points *a, b* and in a small vicinity of the point *b* in Fig. 3(a) can be approximately described by the well-known "general" Ginzburg-Landau equation of the form (e.g., (23) - (25)):

$$\frac{\partial u}{\partial t} = -\frac{\delta \Phi}{\delta u^*} + \zeta(\mathbf{r}, t), \tag{2}$$

where $\zeta(\mathbf{r}, t)$ is a stochastic function and the functional

$$\Phi[u(\mathbf{r}, t)] = \int_V U(u, \nabla u, \Delta u) d^3 r \tag{3}$$

has the meaning of "free energy"; the stochastic noise sources are assumed to be delta-correlated:

$$\overline{\zeta(r_1, t_1)\zeta(r_2, t_2)} = S\delta(t_1 - t_2)\delta(r_1 - r_2). \tag{4}$$

The extrema of the functional Φ (a and b in Fig. 3(a)) correspond to possible stationary distributions, both stable (a) and unstable (b) (Ref. (22), (23)).

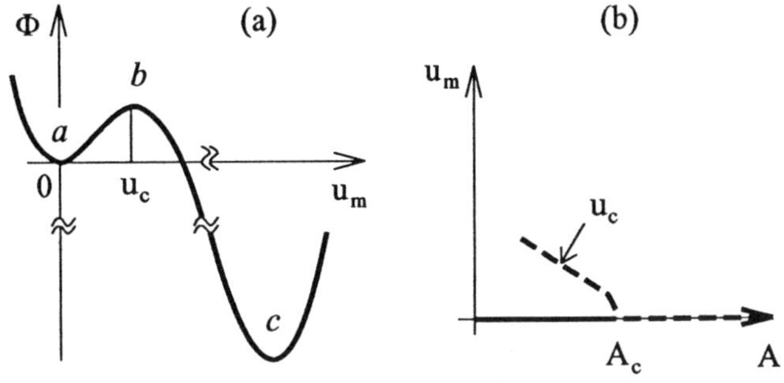

FIGURE 3. Form of the potential Φ for bistable systems (a) and a dependence of the amplitude u_c of an unstable inhomogeneous solution by a subcritical bifurcation (b) (Ref. (7) - (9), (22)). Wavy lines in Fig. (a) indicate that to the contrary the stationary state *b*, the stationary state *c* corresponds to an inhomoheneous solution of a large amplitude.

We will also consider only cases, when at some critical value A_c of the controlling parameter A of systems under consideration a subcritical bifurcation is realized (Fig. 3(b)). Besides, let us assume that an instability in this critical point occurs with respect to the growth of global nonhomogeneous fluctuations with some characteristic value of a wave number $k_c \neq 0$. Notice that the words "global fluctuations", in comparison with localized fluctuations, mean that the global fluctuations cover all the system and do not localized somewhere in the system.

Due to the subcritical bifurcation in the critical point $A = A_c$, stable low-amplitude states may not appear in a small vicinity of this point, but a sharp transition into high-amplitude inhomogeneous states can usually occur. Emphasize that large-amplitude inhomogeneous states, in particular, large-amplitude autosolitons, are, as a rule, impossible to describe within the frame work of Eq. (2) (Ref. (1)). These large-amplitude states call for the initial fundamental equations which describe a distributed nonlinear system. We will discuss the results of investigations of the initial fundamental equations describing traffic jams in traffic flow and autosolitons in active systems with diffusion in Sect. 3 and Sect. 4, correspondingly.

However, Eq. (2) can be used in some cases to find the probability for an autosoliton to arise in stable system with respect to fluctuations of very small amplitude, at least for the controlling parameter A although less, but close to A_c. As it has already been mentioned, it can be done because the amplitude u_c of critical fluctuation whose growth leads to the formation of an autosoliton is relatively small, though finite quantity, in the case, when $A < A_c$, but A is close to

A_c. Under these conditions, the dependence of the potential Φ of the amplitude u_m of the fluctuation $u(\mathbf{r}, t)$ looks like a potential well (Fig. 3(a)) and the point *b* in Fig. 3(a) corresponds to one of the points on the branch which describes an unstable inhomogeneous solution at $A < A_c$ (dotted line in Fig. 3(b)). In other words, in the case under consideration this unstable stationary solution (point *b* in Fig. 3(a) and dotted line in Fig. 3(b)) corresponds to a critical fluctuation $\Delta\theta_c(\mathbf{r})$, which has a critical amplitude u_c.

In the examples of semiconductor and gas discharge structures mentioned above as well as in a lot of other active systems with diffusion a critical fluctuation $\Delta\theta_c(\mathbf{r})$ at $A < A_c$ has a form of a localized function which is shown in Fig. 4. A critical amplitude u_c of this localized fluctuation has been calculated in (26) (Ref. (21)).

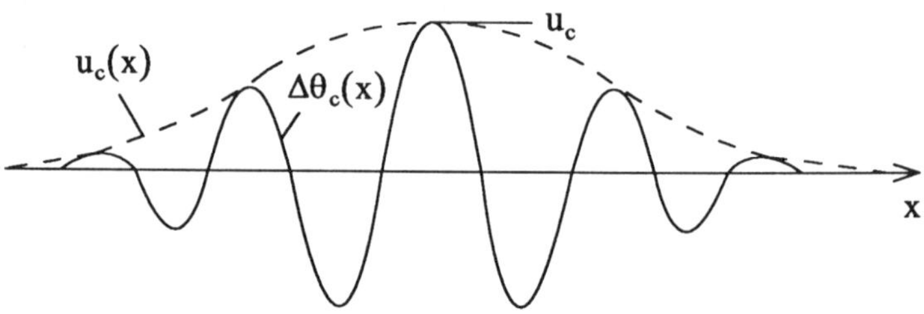

FIGURE 4. Form of a critical fluctuation in active systems with diffusion.

In traffic flow on a circular road of some circumference L the controlling parameter is a total number of vehicles N on the whole road, i.e., in this case $N \equiv A$ and a critical value of the controlling parameter $N_c \equiv A_c$. The value N determines the density of vehicles in a homogeneous state of traffic flow: $\rho_h = N/L$. A critical value of the controlling parameter N_c, therefore, corresponds to the critical density ρ_{c1} (Ref. (20)) (Sect. 3.1.2): $N_c = \rho_{c1} L$. As it follows from the investigations of stationary inhomogeneous solutions in traffic flow on a circular road made in Ref. (20), a critical fluctuation at $N < N_c$, i.e., at $\rho_h < \rho_{c1}$, has a form of a function $\Delta\rho(x)$ which is shown in Fig. 5(a).

This function $\Delta\rho(x)$ describes a local increase in the density of vehicles and a corresponding decrease in the density of vehicles in the rest of the road. Indeed, due to the same total number of vehicles on the whole road N for both an initially homogeneous stable state ρ_h (solid line at $A < A_c$ in Fig. 3(b)) and for an inhomogeneous unstable solution $\rho_h + \Delta\rho(x)$ (dotted line at $A < A_c$ in Fig. 3(b)), the function $\Delta\rho(x)$ (Fig. 5(a)) should satisfy the obvious integral condition (Ref. (20)):

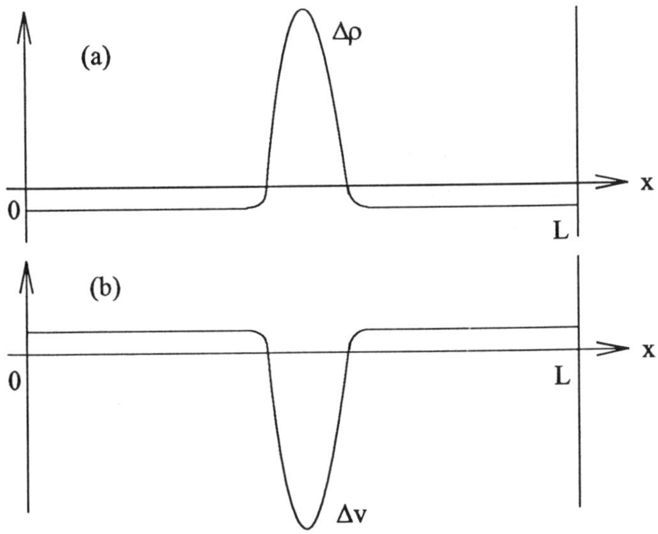

FIGURE 5. Form of critical fluctuations of the density (a) and of the average speed of vehicles (b) for a circular road of some circumference L.

$$\int_0^L \Delta\rho(x)dx = 0. \tag{5}$$

The corresponding solution for the critical fluctuation of the average speed of vehicles is obviously a function $\Delta v(x)$ with one minimum (Fig. 5(b). A critical amplitude u_c of the fluctuation $\Delta\rho(x)$, i.e., the amplitude of the stationary solution, corresponding to the point *b* in Fig. 3(a), has been calculated in (27).

The discussed critical fluctuations correspond to the solution $u(\mathbf{r}) = u_c(\mathbf{r})$ of Eq. (2) which has the smallest maximum value of the functional Φ (e.g., Ref. (22)). An average time T_{es} of the escaping from the stable homogeneous state (*a* in Fig. 3(a)) corresponds to the average time it takes the amplitude of the fluctuation $u(\mathbf{r})$ to reach its critical value u_c (Fig. 3(a)) (e.g., (22)):

$$T_{es} \approx \tau_{ch} \exp\left[2\Phi[u_c]/S\right], \tag{6}$$

where τ_{ch} is the characteristic changing time of variables of the systems; S corresponds to the dispersion of fluctuations in the system.

The formula (6) as well as the consideration made above, strictly speaking, is valid, when fluctuations in the system have a small enough amplitude. Critical fluctuations (Fig. 4 and 5) which have been considered above are a result of a development of these fluctuations in a vicinity of the critical points A_c (or N_c). An example of these small amplitude fluctuations in semiconductor and gas discharge systems are fluctuations of the recombination of charge carriers. In traffic flow, small amplitude fluctuations, which may lead to the development of the critical fluctuation $\Delta\rho(x)$ (Fig. 5), can be linked with the fact that a driver cannot fulfill exactly the same acceleration or slowing down in interactions with other drivers, even if this driver repeats the travel process with the same initial conditions.

2.2.2. Local Instability and "Second" Escape Problem

Besides "usual" fluctuations which may be responsible for the development of the critical perturbations shown in Fig. 4 and Fig. 5, localized critical fluctuations of the different type ("second" type) may occur in distributed nonlinear systems: Relative large amplitude strong localized fluctuations. Such localized fluctuation can be formed due to single random localized events in the system which arise from different physical reasons in comparison with the fluctuations considered in Sect. 2.2.1.

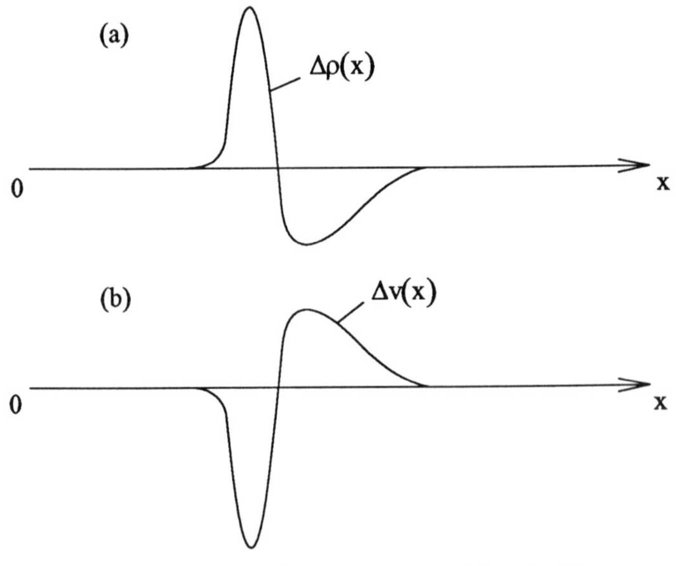

FIGURE 6. Form of a localized fluctuation of the density (a) and of the average speed of vehicles (b) in traffic flow.

For example, in traffic flow a localized relative large fluctuation can appear due to a such changing of a lane of one of the drivers which is an unexpected event for other drivers or to a random solitary too sharp slowing down of one of the drivers which is not be provoked of a present distributions of the density and of the average speed of vehicles. In these cases drivers which move direct up-stream must sharp slow down too. Therefore, up-stream from the driver, which has changed a lane unexpected for other drivers or slowed down too sharp, the density of vehicles becomes higher. Besides a local region of lower density down-stream from this driver is formed. In other words, the localized random processes discussed above can cause the spontaneous appearance of the macroscopic localized perturbation of the density $\Delta\rho$ and of the average speed of vehicles Δv shown in Fig. 6.

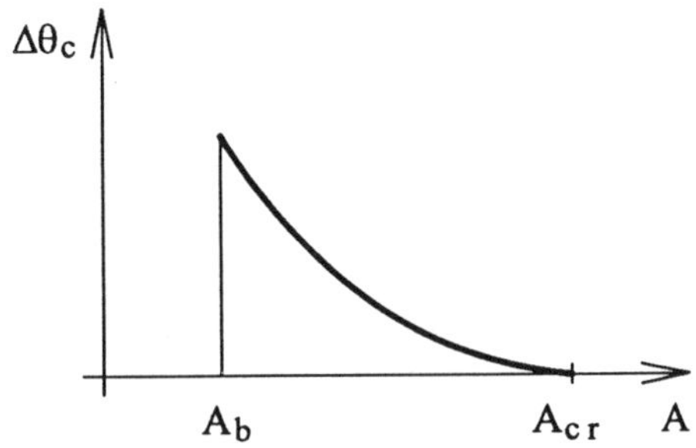

FIGURE 7. A dependence of the amplitude of the critical localized fluctuation $\Delta\theta_c$ of the value of the controlling parameter of a system A.

In general, we will use for a localized fluctuation and for an amplitude of this fluctuation, which may spontaneously be formed in nonlinear distributed systems under consideration, the designations $\Delta\theta$ and $\Delta\theta_m$, correspondingly. When the amplitude $\Delta\theta_m$ of the localized fluctuation exceeds some critical value $\Delta\theta_c$, which depends on the controlling parameter A of a system (Fig. 7), then the amplitude of this perturbation self-increases and finally an autosoliton can be formed in the system. Contrarily, if $\Delta\theta_m < \Delta\theta_c$, this localized fluctuation fades out. In other words, the homogeneous state of the system is a "metastable" state with respect to localized fluctuations of the "second" type. Emphasize that this "second" escape problem, i.e., the escape from an initially stable state due to an occurrence of the localized fluctuation under consideration has at least the following differences in comparison with the usual ("first") escape problem which has been discussed in Sect. 2.2.1:

1. The critical localized fluctuation, i.e., a localized fluctuation which amplitude $\Delta\theta_m$ is equal to the critical value $\Delta\theta_c$, may not correspond any stationary solution of the system.
2. When the controlling parameter of a system A tends to the critical value A_c, contrarily with the case of the "first" escape problem (Fig. 3(b)), the critical value $\Delta\theta_c$ for an amplitude of the localized fluctuation under consideration, which due to a single random localized event occurs somewhere in the system, may not tend to zero.

3. The latter means that the average time of the escaping from the stable homogeneous state (*a* in Fig. 3(a)) may not correspond now to the formula (6). The average time of the escaping from the stable homogeneous state (*a* in Fig. 3(a)) in the case under consideration can be calculated as the average time it takes a localized fluctuation with an amplitude $\Delta\theta_m > \Delta\theta_c$ due to a single random localized event in the system to occur somewhere in the system.

A critical value of the controlling parameter A_{cr}, at which any amplitude localized fluctuations of the "second" type under consideration can grow, may be in some systems higher than the critical value A_c, where a homogeneous and inhomogeneous solutions are branching (Fig. 3(b)). In this case, in the range of the controlling parameter $A_c < A < A_{cr}$ the homogeneous state of the system is unstable with respect to global fluctuations in the whole system. For this reason, in this range of values A a state of a system can only be considered as a "metastable" state with respect to the "second" type of localized fluctuations, if an occurrence of localized critical fluctuations of the "second" type considerable faster leads to an escape from the unstable homogeneous state than the growth of global fluctuations.

2.2.3. "Local Breakdown" Effect. Dynamical Appearance of Autosoliton

A homogeneous state of a system is a hypothetical state of a distributed system. Distributions of variables in real systems are always inhomogeneous. In semiconductor and gas discharge systems a reason for an inhomogeneous state may be recombination and ionisation processes near the boundaries of the systems which are the additional process to those in the volume of the system. Traffic flow on a highway is always inhomogeneous due to entering and exiting traffic to on- and off-ramps or an existence of vehicles in traffic flow which have different "free" speeds.

A process of a formation of an autosoliton in a distributed system shows qualitatively new peculiarities even if distributions of variables are only slightly inhomogeneous: An autosoliton in such inhomogeneous systems can spontaneously be self-formed in *a deterministic way*, i.e., even if an influence of any fluctuations on the processes of a formation of autosolitons may be neglected (1), (16), (17).

Such deterministic self-formation of autosolitons is caused by the "local breakdown" effect in a system (1). The "local breakdown" effect results in a dynamical local restructuring of slightly inhomogeneous distributions of variables of the distributed system. This effect starts, when the controlling parameter of a system A reaches some critical value A_c^-. This critical value of the controlling parameter is less than the critical values A_c (Fig. 3(b)) or A_{cr} (Fig. 7) which correspond to the growth of global (Sect. 2.2.1) or localized (Sect. 2.2.2) fluctuations of any amplitude in a homogeneous state of a system. Fluctuations may cause a formation of an autosoliton earlier than the controlling parameter of the system A reaches the critical point A_c^-, where the "local breakdown" effect starts.

A realization of the "local breakdown" effect in a presence of slightly inhomogeneous distributions of variables of a system may be explained with the help of the S-shaped functions $\theta(\eta)$ shown in Fig. 8.

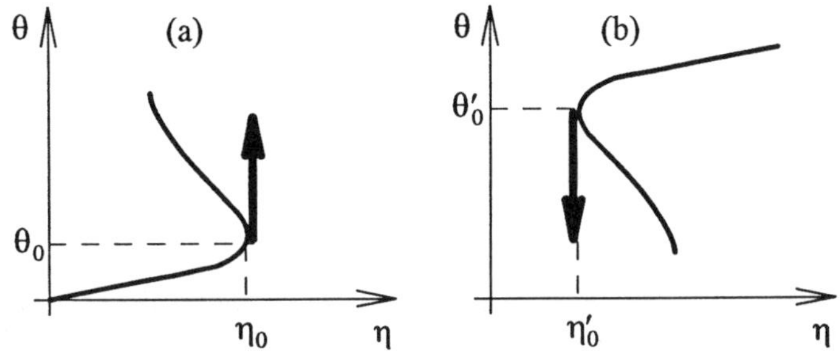

FIGURE 8. To the explanation of the "local breakdown" effect. Arrows indicate the "local breakdown" in a vicinity of an inhomogeneity in distributions of variables of the system.

For these functions $\theta(\eta)$ at $\eta=\eta_0$ (Fig. 8(a)) or at $\eta=\eta_0'$ (Fig. 8(b)) we have $d\theta/d\eta=\infty$. Here θ is a variable of a system which describes the active process and η is a variable (of a function of variables) which describes the damping process in a system. When $A \to A_c^-$, a true function $\theta(\eta)$, which corresponds to the real functions $\theta(\mathbf{r})$ and $\eta(\mathbf{r})$ in a small vicinity of an inhomogeneity in distributions of variables of the system, abuts on the limiting point $\eta=\eta_0$ (Fig. 8(a)) or on the limiting point $\eta=\eta_0'$ (Fig. 8(b)). In the first case, for $A > A_c^-$ the value of "activator" θ must increase abruptly from $\theta \approx \theta_0$ to higher values (the arrow in Fig. 8(a)). In the second case, the value of "activator" θ must decrease abruptly from $\theta \approx \theta_0'$ to lower values (the arrow in Fig. 8(b)) (1). These processes of a abruptly increase or decrease in a value of the "activator" are deterministic processes, i.e., they can also occur when fluctuations are negligible.

Emphasize that the discussed true function $\theta(\eta)$ of real distributions of variables in the system only in *local* regions of the whole system abuts on the limiting points $\eta=\eta_0$ or $\eta=\eta_0'$ of the S-shaped functions $\theta(\eta)$ shown in Fig. 8. For this reason the "breakdown" effect considered is a local effect.

In the examples of semiconductor or gas discharge systems, as it has already been mentioned, the role of the activator plays the concentration of the carriers: $\theta \equiv n$. The "local breakdown" effect can be realized in this case in a local region of the deterministic perturbation of the activator (Fig. 9) which appears, if the rate of the ionisation in a vicinity of some point $x=0$ is higher than in the rest of the system. In this case, for $A > A_c^-$ (here A is a voltage drop V across the system: $A \equiv V$) the value of activator θ must increase abruptly to higher values (the arrows in Fig. 8(a) and Fig. 9). Such "local breakdown" causes the deterministic formation of the autosoliton shown in Fig. 1 (1).

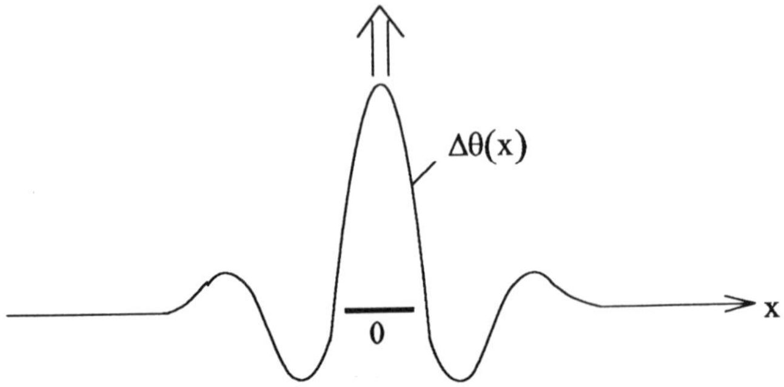

FIGURE 9. Form of the deterministic perturbation of the activator near a local inhomogeneity in semiconductor or in gas discharge systems (1). Arrow indicates the "local breakdown" effect.

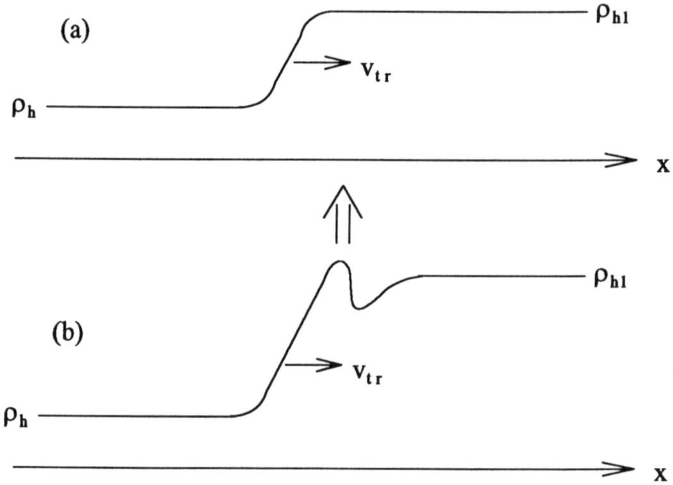

FIGURE 10. Form of a moving transition layer. The density ρ_{h1} in Fig. (a) is lower than in Fig. (b). Top arrow in Fig. (b) indicates the "local breakdown" effect.

In traffic flow, the role of "activator" plays the density (or the average speed of vehicles) and the role of the damping parameter plays the function q^* given by the formula (1). The "local breakdown" effect can start in traffic flow for example in a vicinity of a moving transition layer which is formed owing to an existence of different levels of the density of vehicles in an inhomogeneous traffic flow (28), (29).

Let us consider the transition layer between lower density ρ_h up-stream and some higher density ρ_{h1} of vehicles down-stream which moves with some positive velocity v_{tr} (Fig. 10). If the density ρ_{h1} is low, then the distribution of the density in the transition layer is a monotonous function (Fig. 10(a)). When the density ρ_{h1} is increased, a deterministic perturbation can appear direct down-stream the transition layer (Fig. 10(b)). To explain the reason of an appearance of the deterministic perturbation in the moving transition layer, notice that drivers which reach the moving transition layer with high speed v_h must slow down inside the transition layer. Indeed, the density of vehicles is increasing there. To satisfy the safety requirements, drivers, meeting a hindrance (the transition layer), slow down more than they make in a "quasi-stationary" case, i.e., when the density is would very smoothly change in space. This effect can lead to the occurrence of the deterministic perturbation of the density in the moving transition layer (Fig. 10(b)).

The form of this deterministic perturbation is similar to those which has been shown in Fig. 6. If the density ρ_{h1} is further increased, the amplitude of this perturbation increases too and, beginning at some critical value of the density, an avalanche-like increase in the amplitude of the perturbation, i.e., the "local breakdown" process, starts. Such "local breakdown" effect can also be explained based on the consideration of Fig. 8(a) made above, where we have now $\theta \equiv \rho$ and $\eta \equiv q^*$ (1). As a result of the "local breakdown" effect, a traffic jam spontaneously appears in traffic flow (Sect. 3.6.5).

2.2.4. Evolution of Autosolitons

After an autosoliton has appeared in a distributed system, a lot of new self-organization processes or an occurrence of a space-time chaos can spontaneously be realized there. Some of these processes will be briefly discussed in Sect. 3 and 4. Conditions of a realization of these complex nonlinear effects depend on a range of the controlling parameter A, where autosoliton can exist in the system. In general, effects of a transformation of the autosoliton into other complex states of the systems and, especially, the types of scenarios of self-organization and of a space-time chaos depend on the peculiarities of the nonlinear system. Nevertheless, there are some general properties of an evolution of autosolitons in different systems:

1. There is the boundary (threshold) value of the controlling parameter A_b (Fig. 7): At $A < A_b$ an autosoliton cannot exist. This means that if an autosoliton exists in the system, it will disappear, if the value A becomes lower than A_b. At $A = A_b$ an autosoliton has a minimum width and/or a minimum amplitude.

2. The critical value of a local perturbation which can cause a spontaneous formation of the autosoliton is maximal at $A = A_b$ (Fig. 7).
3. There are at least two types of distributed systems under consideration. In systems which are unstable at $A > A_{cr}$ (Fig. 7), an amplitude of autosolitons can increase monotonous with the value of the controlling parameter A. Contrarily, in systems, which are unstable in a limited range $A_{cr} < A < A'_{cr}$ of the controlling parameter A (Fig. 11), an amplitude of an autosoliton is limited. Besides, two different kind of autosolitons can be exited in such systems in the different ranges of the controlling parameter $A_b < A < A_{cr}$ and $A'_{cr} < A < A'_b$, correspondingly. The critical value of a local perturbation which can cause a spontaneous formation of the autosoliton of the second type is maximal at $A = A'_b$ (Fig. 11), but a width and/or an amplitude of this autosoliton are minimal in the second boundary (threshold) point A'_b.

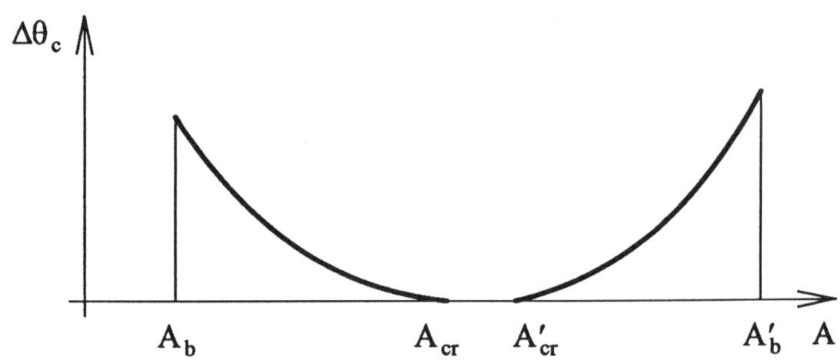

FIGURE 11. Dependencies of the amplitude of the critical localized fluctuation $\Delta\theta_c$ of the value of the controlling parameter of a system A.

3. TRAFFIC JAMS

The phenomenon of a traffic jam without obvious reason is one of the frequently observable collective effects. Almost every driver some time met this phenomenon. First, the drivers, which were up-stream far from the traffic jam, could move with some relatively high speed in an average homogeneous traffic flow. Suddenly, drivers found themselves in the traffic jam - a region with high density of vehicles, where they almost could not move at all. After some time the traffic jam passed by and drivers could move again at a high speed in a new almost homogeneous traffic flow, but, to their great surprise, nothing had happened on the road which could have caused the traffic jam they met. This experience of drivers is confirmed by experimental investigations (see results, e.g., in (30), (31)).

From many observations it can be assumed that the appearance of a traffic jam without obvious reason may be a local effect in an initially average homogeneous traffic flow on a long road. This supposition has recently been confirmed by the theory of cluster-effect in an initially homogeneous traffic flow developed by Kerner and Konhäuser (19), (20). It has been found there that a traffic jam, i. e. a cluster of vehicles, where the density is high and the average speed of vehicles is low (up to a standstill), can be self-formed in traffic flow. A traffic jam even can appear spontaneously in a homogeneous traffic flow formed by a source and a sink of vehicles, which are situated so far one from another that the connecting road can be considered as being of infinite length. The formation of a traffic jam on such a long road is a purely local non-linear collective effect.

In this section, based on results of the original works (19), (20), (28), (29), a review of theoretical investigations of physics, structure and properties of traffic jams which can spontaneously be self-formed in traffic flow will be presented. Besides, characteristics of traffic jams which are general for autosolitons (Sect. 2) will be discussed.

3.1. Kinetic Model of Traffic Flow

3.1.1. Basic Equations

If the number of vehicles moving on a road is sufficiently large and only average characteristic of the motion are interesting, traffic flow can be considered as a one-dimensional compressible flow. In this kinetic approach (Ref. (31) - (35)), the kinetic model of traffic flow is described by the well-known continuity equation:

$$\frac{\partial \rho}{\partial t}+\frac{\partial q}{\partial x}=0, \tag{7}$$

the equation of motion of vehicles (Ref. (19)):

$$\rho\left[\frac{\partial v}{\partial t}+v\frac{\partial v}{\partial x}\right]=\rho[V(\rho)-v]-c_0^2\frac{\partial \rho}{\partial x}+\frac{\partial^2 v}{\partial x^2} \tag{8}$$

and periodical boundary conditions:

$$q(0,t)=q(L,t), \tag{9}$$

$$v(0,t)=v(L,t),\ w(0,t)=w(L,t). \tag{10}$$

In Eqs. (7) - (10) $\rho(x,t)$ is the density ($0<\rho\leq 1$) and $v(x,t)$ is the average speed of vehicles ($v\geq 0$),

$$w=\frac{\partial v}{\partial x}; \tag{11}$$

$$q(x,t) = \rho(x,t)v(x,t) \tag{12}$$

is the flux of vehicles in traffic flow; V is a safe ("maximal and out of danger") speed which is achieved in a both time-independent and homogeneous traffic flow, $V(\rho)$ is a monotonous decreasing function of ρ, i.e., its derivative (Ref. (31) - (35))

$$\xi(\rho) = \frac{dV(\rho)}{d\rho} < 0; \tag{13}$$

$c_0 = \text{const}$ (Ref. (34), (35)); L is the length of the road. In equations (7) - (13), and as of now v, V and c_0 are measured in units of ℓ/τ, the length in units of ℓ, the time in units of τ, where τ is the characteristic time relaxation of the average speed v to the "maximal and out of danger" speed V, $\tau = \text{const}$ (Ref. (34), (35)); the density of vehicles ρ is measured in units $\hat{\rho}$, where $\hat{\rho}$ is the maximal possible density of vehicles on the road, for n-lanes road $\hat{\rho} = n/\hat{a}$, where $\hat{a}$ is an average length of vehicles; the value w is measured in units of $1/\tau$, the flux q in units of $\hat{\rho}\ell/\tau$, the value ξ in units of $\ell/(\tau\hat{\rho})$; $\ell = \sqrt{\mu\hat{\rho}^{-1}\tau}$; $\mu = \text{const}$ (Ref. (32)).

The equation of motion (8) formally follows (Ref. (20)) from the well-known Navier-Stokes equations, more precisely, the "Navier-Stokes-like" equation for traffic flow, which for a one-dimensional compressible flow reads

$$\rho\left[\frac{\partial v}{\partial t} + v\frac{\partial v}{\partial x}\right] = X - \frac{\partial p}{\partial x} + \frac{\partial}{\partial x}\left(\mu\frac{\partial v}{\partial x}\right), \tag{14}$$

where $p = \rho c_0^2$ is the local "pressure"; for the given case the sum of all inner forces X, which appear due to interactions between individual vehicles, is given by

$$X = \rho\frac{V(\rho) - v}{\tau}. \tag{15}$$

The force X (15) in the "Navier-Stokes-like" equation for traffic flow (14) takes into account the relaxation process of the average speed of vehicles v to a safe ("maximal and out of danger") speed $V = V(\rho)$, which is achieved in a both time-independent and homogeneous traffic flow. The effect of X in Eq. (14) can best be understood, if its isolated action on the acceleration of an average driver in homogeneous traffic flow is studied. Under these artificial conditions, equation (14) reduces to:

$$\frac{dv}{dt} = \frac{V(\rho) - v}{\tau} \tag{16}$$

and it is apparent from this equation, that for a given density ρ, the average driver is decreasing his speed $(dv/dt < 0)$ if his instant speed v is higher than $V(\rho)$ and vice versa. The term $-\partial p/\partial x$ in the "Navier-Stokes-like" equation for traffic flow (14) describes an anticipation factor, i.e., drivers accelerate to a decrease in the density ρ or they slow down when the density increases (e.g., Ref. (32)). This means that a gradient of the pressure $\partial p/\partial x$ should have the same sign that a gradient of the density: $\partial p/\partial x = c_0^2 \partial\rho/\partial x$. Because a possible dependence c_0^2 of the density, as it follows from numerical investigations made in Ref. (29), is not important for qualitative results discussed below, one can use in Eq. (8) $c_0 = \text{const}$. In other words, for investigations of a traffic jam formation it is possible to express the local "pressure" p in Eq. (14) by $p = \rho c_0^2$.

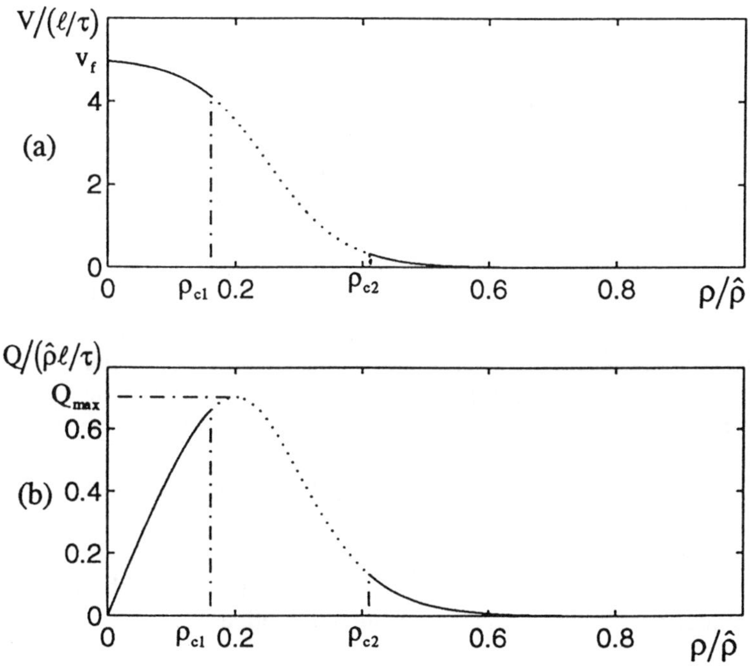

FIGURE 12. The example of the function $V(\rho)$ (a) and of the corresponding fundamental diagram $Q(\rho)$ (b). Results of the numerical computations for $V(\rho) = 5.0461\left\{\left[1+\exp(((\rho/\hat{\rho})-0.25)/0.06)\right]^{-1} - 3.72\cdot 10^{-6}\right\}\ell/\tau$, $c_0 = 2.0497\,\ell/\tau$. The found critical values of density for L=800ℓ are $\rho_{c1} \cong 0.1613\hat{\rho}$, $\rho_{c2} \cong 0.4118\hat{\rho}$ (Ref. (19), (20)).

The function $V(\rho)$ in Eq. (8) is determined by the average balance between safety requirement and risk readiness of the drivers as well as legal traffic regulations, average vehicle technology and road conditions, i.e. $V(\rho)$ is a phenomenological function (Ref. (31) - (35)). Despite the variety of quantities that determine the function $V(\rho)$, it strongly depends on ρ and some important statements with respect to the general shape of the function $V(\rho)$ can be made (e.g., Ref. (19)):

1. At not too small values ρ the value V should decrease as ρ increases, because drivers decrease their average speed if the headway to the vehicle in front of them is reduced. At the limit where $\rho \to \hat{\rho}$ vehicles cannot move at all and for this reason $V(\rho)|_{\rho\to\hat{\rho}} \to 0$.
2. On the other hand, at small enough values of ρ there is almost no interaction between vehicles and they can move with some average "free" speed v_f which depends on road conditions and average vehicle technology.

The assumption, that there is a continuous and smooth overlapping between the domains of small and large densities and that $V(\rho)$ approaches its maximal value v_f as ρ tends to zero, seems reasonable.

Therefore, $V(\rho)$ is a monotonous decreasing function of ρ (Fig. 12(a)), i.e. their derivative $\xi(\rho)<0$ (13). The conclusion that this relationship of the average speed of vehicles in a *homogeneous and time-independent* traffic flow V on the density of vehicles ρ has the decreasing character (Fig. 12(a)) were confirmed by a lot of experimental investigations of traffic flow (e.g., (31) - (35)). The related dependence on ρ of the traffic volume $Q(\rho) = \rho V(\rho)$ is obviously a function with only one maximum (Fig. 12(b)). Traffic engineers call this relationship *fundamental diagram* (e.g., (31) - (35)). As it has been mentioned this fundamental diagram (Fig. 12(b)) has sense only for a homogeneous and time-independent traffic flow.

3.1.2. Stochastic Macroscopic Traffic Flow Model

To investigate processes of growth of the critical fluctuations in traffic flow, it is possible to use stochastic macroscopic traffic model which includes the continuity equation (7), the stochastic equation of motion

$$\rho\left[\frac{\partial v}{\partial t}+v\frac{\partial v}{\partial x}\right]=\rho[V(\rho)-v]-c_0^2\frac{\partial \rho}{\partial x}+\frac{\partial^2 v}{\partial x^2}+\rho\zeta(x,t) \qquad (17)$$

and the boundary conditions (9), (10). In (17) a stochastic function $\zeta(x,t)$ describes a random component of an acceleration or a slowing down of vehicles in traffic flow (Ref. (28)). A value ζ in (17) is measured in unites of ℓ/τ^2.

A correlation length χ of the stochastic function $\zeta(x,t)$ can not be less than the value $1/\rho$, corresponding to an average distance between vehicles (including the length of a vehicle). On the other hand, fluctuations of an acceleration or of a slowing down of vehicles being situated far enough from one another may be not correlate. A correlation time ψ of the stochastic function $\zeta(x,t)$ may depend both on the density and on the average speed of vehicles. In particular, an acceleration or a slowing down for small time-intervals partially correlate and only for large time-intervals does not correlate.

These properties of the stochastic function $\zeta(x,t)$ have been taken into account in Ref. (28), where numerical investigations of the instability of traffic flow in a vicinity of the critical density ρ_{c1} have been made. It has been found there that the random component of an acceleration or a slowing down of vehicles in traffic flow described by the stochastic function $\zeta(x,t)$ in Eq. (17) causes the corresponding change in the density and in the average speed of vehicles in a vicinity of an initially homogeneous state of traffic flow. In some time, one or a few traffic jams are self-formed from this random process. If the density ρ_h is further increased, a very complicated picture of a lot of traffic jams, which merge under collisions, have different amplitudes, different widths, different velocities and are not situated periodically in space, can spontaneously appear (see Sect. 3.5) as a result of the growth of fluctuations in traffic flow (28).

3.1.3. Instability of Homogeneous Traffic Flow

For given values of a total number of vehicles on the road N and the length of the road L, there is only one homogeneous and time-independent state ρ_h, v_h for traffic flow under consideration:

$$\rho_h = \frac{N}{L}, \quad v_h = V(\rho_h). \tag{18}$$

The corresponding value of the flux (traffic volume) in the homogeneous and time-independent traffic flow is $q_h = v_h \rho_h$. As it follows from formulas (18), the given values N and L determine *one point* on the fundamental diagram $Q(\rho)$: $q_h = Q(\rho_h)$ and one point on the speed density relationship $V(\rho)$: $v_h = V(\rho_h)$.

In Ref. (20) it has been shown that the given value N can represent not only the homogeneous solution (18) but also the non-homogeneous solutions in the form of a local clusters of vehicles. An appearance of the local cluster corresponds to a self-formation of a traffic jam in an initially homogeneously traffic flow.

A local cluster of vehicles can formally spontaneously appear owing to a *global* instability of the initially homogeneous traffic flow which is realized if the density ρ_h exceeds some characteristic value ρ_{c1} (20). Indeed, as it follows from the linear analysis of an instability of the initially homogeneous traffic flow made in Ref. (19), the homogeneous state of traffic flow loses its stability with respect to a growth of the non-homogeneous global fluctuation of the density of vehicles. On the threshold of this global instability $\rho_h = \rho_{c1}$ a critical perturbation is:

$$\delta\rho(x) = \delta\rho_0 \exp(2\pi ix/L), \quad \delta v(x) = \delta v_0 \exp(2\pi ix/L). \tag{19}$$

It represents the most long-wave non-homogeneous perturbation in a homogeneous traffic flow, which moves with the phase velocity

$$v_p = v_h - c_0. \tag{20}$$

The global instability of the initially homogeneous flow occurs in some interval of value ρ_h:

$$\rho_{c1} < \rho_h < \rho_{c2}, \tag{21}$$

if inside this interval the value ξ (13) satisfies the condition:

$$\xi < -\left[1 + (2\pi/L)^2 \rho_h^{-1}\right](c_0/\rho_h). \tag{22}$$

As it follows from (22), one can find the critical values $\rho_h = \rho_{c1}$ and $\rho_h = \rho_{c2}$ (21) from the equation:

$$\left[-1 - (\rho_{ci}/c_0)\xi(\rho_{ci})\right]\rho_{ci} = (2\pi/L)^2, \ (i = 1,2). \tag{23}$$

In the range of the density (21) there are, obviously, a lot of other points of bifurcations, where more short-wave global perturbations

$$\delta\rho(x) = \delta\rho_0 \exp(2\pi imx/L), \quad \delta v(x) = \delta v_0 \exp(2\pi imx/L), \ m=2,3, \tag{24}$$

may grow. The critical values of density $\rho_h = \rho_{c1}^{(m)}$, m=2,3,..., of the growth of such global fluctuations (24), for example, in a vicinity of the density ρ_{c1}, can be found from the obvious formulas

$$\left[-1 - (\rho_{cm}/c_0)\xi(\rho_{cm})\right]\rho_{cm} = (2\pi m/L)^2, \ m=2,3,... \tag{25}$$

As it follows from Ref. (20), the kinetics of a growth of the global perturbation (19) at the density ρ_h, which exceed the critical value ρ_{c1} (23) but is rather close to it, shows that during the growth of an amplitude of the perturbation its previous form (19) is gradually deformed: The moving transition layer between some lower and higher values of the density appears in traffic flow. At some time in the vicinity of the maximum of this transition layer, a local perturbation of density and average speed of vehicles is formed by itself. This local perturbation occurs in a deterministic way in the transition layer. The reason for the appearance of the deterministic perturbation has been discussed in Sect. 2.2.3 (Fig. 10(b)). As a result of the appearance of the localized perturbation, the "local breakdown" starts in a vicinity of the maximum of this perturbation (Sect. 2.2.3). For this reason, the further development of the distribution of the density and of the average speed of vehicles on the whole road, which leads to a formation of a traffic jam, is almost entirely governed by the growth of the amplitude of this self-formed deterministic local perturbation.

It is found in Ref. (20) that this formation of the traffic jam on the road is qualitatively similar to a formation of a traffic jam which occurs, if instead of the global fluctuation (19) only a *local* fluctuation with finite amplitude, which has a form similar to the form of the self-formed deterministic perturbation in traffic flow discussed above, appears. Such *local* instability of the homogeneous traffic flow with respect to local fluctuations of a finite amplitude (Sect. 2.2.2; Fig. 6) may also occur if the density $\rho_h < \rho_{c1}$, i.e., in a stable traffic flow with respect to global fluctuations of small amplitude.

In the next Sect. 3.2, based on results derived in Ref. (20), a formation of a traffic jam caused by an occurrence of a local fluctuation in traffic flow will be considered.

3.2. Structure of Traffic Jams

3.2.1. Kinetics of traffic jam formation

Let us consider the kinetics of the formation of a traffic jam in an initially homogeneous traffic flow (Fig. 13) which is only slightly disturbed at t=0 by a rather small but finite amplitude local perturbation $\Delta\rho(x)$ localized at $\approx 150\ell$ from the beginning of the road (Ref. (20), (21)). The used initial local perturbation of density of vehicle at t=0

$$\Delta\rho(x) = \Delta\rho_m\left\{\cosh^{-2}(0.2(x-x_0)) - 0.25\cosh^{-2}(0.05(x-25\ell-x_0))\right\} \quad (26)$$

represents two local regions, following down-stream, one after another: A region of large and a region of low density of vehicles. This form of the initial perturbation (Fig. 13(b)) is similar to those shown in Fig. 6. It corresponds to the demand following from Eq. (7) that the initial total quantity of vehicles in the whole region of the localisation of the localized perturbation cannot be changed (see also the formula (5)). This local initial perturbation of the density of vehicles first moves down-stream with only slightly increasing amplitude (Fig. 13(c)).

After some time ($\approx 36\tau$), the local perturbation under consideration comes to a stop and its amplitude begins to grow rapidly (Fig. 13(d)), forming in time a cluster of vehicles of large amplitude, i. e., the traffic jam, where the average speed of vehicles sharply decreases in time (Fig. 13(e)). The developing traffic jam moves up-stream.

By the time $\approx 200\tau$, the density and the average speed of vehicles inside this up-steam moving cluster reaches the almost invariable values ρ_{max} and $v_{min} \approx 0$, i.e., inside the cluster there is nearly a standstill (Fig. 13(f)). The width of the cluster L_s, i. e., the distance between the cluster's fronts, where the density and the average speed of vehicles sharply changes in space, monotonously increases in time. It occurs because the up-stream front of the cluster moves with a higher negative velocity v_{gl} compared to the velocity v_{gr} of the down-stream front. Therefore, the derivative

$$\frac{dL_s}{dt} = v_{gr} - v_{gl} \tag{27}$$

is positive. In the case under consideration (Fig. 13(a)) the width of the traffic jam L_s increases with a rate of $\approx 0{,}1274\ell/\tau$.

The increase in L_s is linked to the fact that the flux of vehicles through the up-stream front of the cluster, i. e., the flux into the cluster q_{in}, is stronger than the flux through the down-stream front, i.e., the flux out of the cluster q_{out} (Fig. 13(f)):

$$q_{in} > q_{out}, \tag{28}$$

where q_{in} equals to the flux in the initially homogeneous flow q_h.

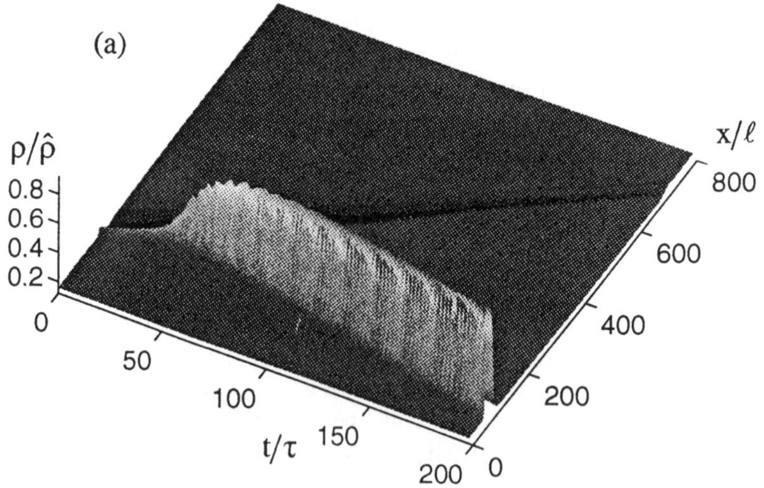

FIGURE 13(a) See caption on page 802.

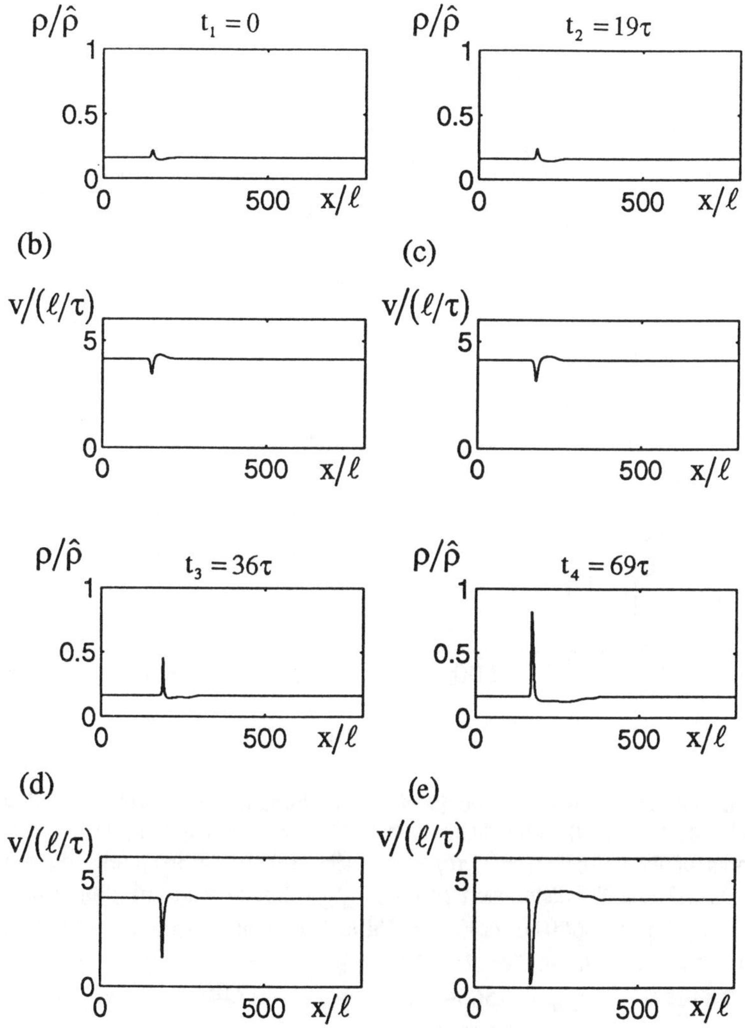

FIGURE 13(b) - (e) See caption on Page 802.

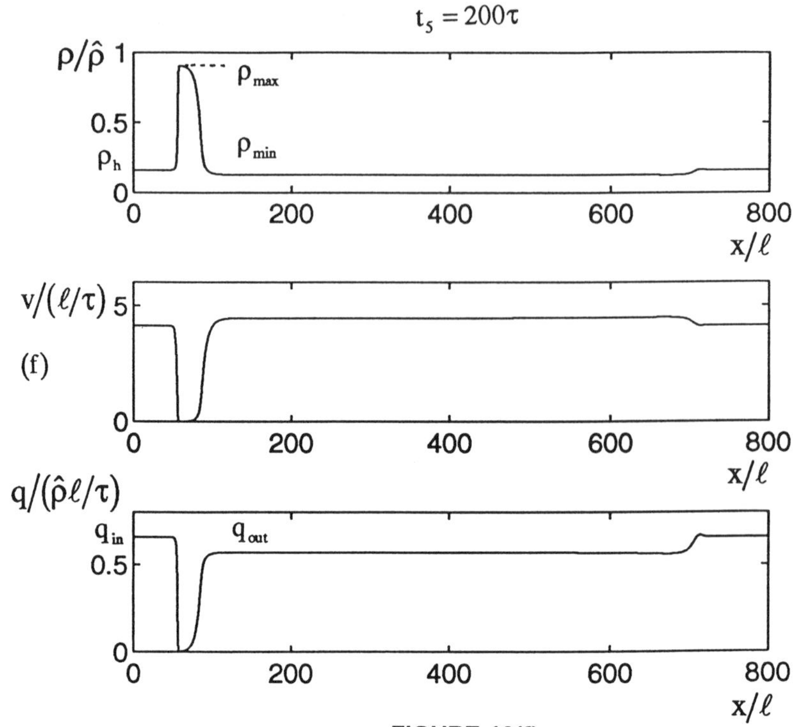

FIGURE 13(f)

FIGURE 13. The kinetics of the local cluster formation (20): (a) - the dependence $\rho(x,t)$, (b, c, d, e, f) - the distributions $\rho(x)$, v(x) and q(x) (f) in the intermediate moments of time ((b) - $t_1 = 0$, (c) - $t_2 = 19\tau$, (d) - $t_3 = 36\tau$, (e) - $t_4 = 69\tau$; (f) - $t_5 = 200\tau$. The initial distribution $\rho(x,0) = \rho_h + \Delta\rho(x)$ with $\Delta\rho$ (26), $\Delta\rho_m = 0.06\hat{\rho}$, $\rho_h = 0.16\hat{\rho}$ ($q_h \cong 0.6601\,\hat{\rho}\ell/\tau$), $x_0 = 150\ell$. The other parameters are the same as in Fig. 12 (Ref. (20), (21)). The found values for $t_5 = 200\tau$ (f) are $\rho_{min} \approx 0.1251\hat{\rho}$, $v_w \approx 2.8424\,\ell/\tau$, $q_{out} \approx 0.5609\,\hat{\rho}\ell/\tau$, $v_{gl} \approx -0.8448\,\ell/\tau$, $v_{gr} \approx -0.7174\,\ell/\tau$, $v_{min} \cong 0.0024\,\ell/\tau$, $\rho_{max} \approx 0.9039\hat{\rho}$.

The condition (28) means that the cluster of vehicles - a traffic jam - acts as some new local source of vehicles on the road moving up-stream which "stores" more and more vehicles in the course of time.

This up-stream moving local source of vehicles forms down-stream it a new almost homogeneous traffic flow with smaller density ρ_{min} (Fig. 13(f)). For this reason, far down-stream from the traffic jam, there moves a transition layer with positive velocity

$$v_w \cong \frac{q_h - q_{out}}{\rho_h - \rho_{min}} \tag{29}$$

between the initially homogeneous traffic flow and the new homogeneous traffic flow formed by the traffic jam.

Therefore, the local cluster of vehicles is a non-stationary localized structure which consists of:

1. The proper cluster of vehicles, i.e., the traffic jam.
2. The new homogeneous traffic flow formed by the traffic jam down-stream.
3. The transition layer between this new and the initially homogeneous flow moving down-steam (Fig. 13(a), (f)).

3.2.2. Average Vehicle Trajectories

Furthermore, it is instructive to consider the average vehicle trajectories on the t-x plane (Fig. 14), corresponding to the local cluster shown in Fig. 13(a).

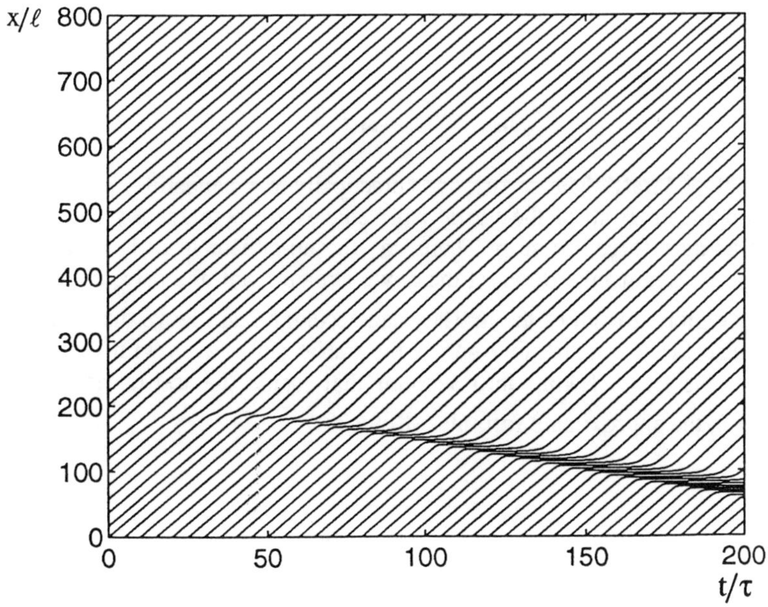

FIGURE 14. The average vehicle trajectories corresponding to Fig. 13(a) (20).

One can see in Fig. 14 that the driver which has caused the local perturbation at t=0 and at the distance $\approx 150\ell$ from the beginning of the road can always travel with high speed in the traffic flow. Indeed, at time $\approx 36\tau$, when the local perturbation he has caused stops at position $\approx 189\ell$ and its amplitude begins to grow rapidly forming the traffic jam, this driver is already at position $\approx 298\ell$. Therefore, this driver is far away from the traffic jam he is responsible for. Consequently, this driver does not know anything about the traffic jam which he has "prepared" for the other drivers moving at some distance behind him.

Indeed, one can see from Fig. 14 that drivers which have been at $t \approx 75\tau$ at the beginning of the road must reduce their speed sharply at the time, when they reach the traffic jam. As it has been mentioned, the width of the traffic jam L_s is increasing during the course of time. For this reason, the later a driver approaches the traffic jam, the longer he has to wait in it until the traffic jam, due to its negative velocity, has passed him. This fact can be seen in Fig. 14, where the vehicle trajectories are considerably denser in the upstream moving traffic jam and the slope of the vehicle trajectories is nearly parallel to the x-axis. This indicates that the speed of the vehicles is approximately zero. Therefore, inside the traffic jam there is nearly a standstill.

3.2.3. Velocity and Width of Traffic Jam

When the width L_s of the traffic jam is large enough ($t \geq 200\tau$, Fig. 13(a)), the average speed of vehicles in the traffic jam approaches $v_{min} \cong 0$ and consequently the flux

$$q_{min} = v_{min}\rho_{max} \cong 0, \tag{30}$$

i.e., the developed traffic jam corresponds practically to a standstill.

At the up-stream front of the traffic jam vehicles are permanently entering the traffic jam and are forced to stop. Therefore, the up-stream front is moving with the velocity v_{gl} against the flow. At the down-stream front, the drivers realize the low density region in front of them. They can accelerate and leave the traffic jam. Therefore also the down-stream front of the traffic jam moves with the velocity v_{gr} against the flow. Taking into account that inside the traffic jam the flux $q_{min} \cong 0$ (30), the approximate formulas (Ref. (20))

$$v_{gr} = -\frac{q_{out}}{\rho_{max} - \rho_{min}}, \tag{31}$$

$$v_{gl} = -\frac{q_h}{\rho_{max} - \rho_h}. \tag{32}$$

can be derived. In these formulas

$$q_{out} = v_{max}\rho_{min};$$ (33)

ρ_{min} and $v_{max} \cong V(\rho_{min})$ are the density and the average speed of vehicles directly down-stream of the traffic jam, correspondingly; ρ_{max} is the density of vehicles inside the traffic jam. As $\rho_h > \rho_{min}$ (Fig. 13(f)) and consequently $q_h > q_{out}$, from formulas (31) and (32) it follows that $|v_{gl}| > |v_{gr}|$, corresponding to the results discussed in Sect. 3.2.1.

The flux q_{out} (33) is determined by the possible rate of vehicles escaping from a standstill inside the traffic jam. In other words, the flux q_{out}, or else the density ρ_{min} (as $q_{out} \equiv V(\rho_{min})\rho_{min}$), does not depend upon the density in the homogeneous flow ρ_h. The same conclusion is valid for the density inside the traffic jam ρ_{max}. Therefore, the velocity v_{gr} (31) does not also depend on ρ_h. On the contrary, the velocity v_{gl} (32) becomes more and more negative the higher the density ρ_h is. For this reason, as it follows from the formula (27), the width of the traffic jam

$$L_s(t) = L_s(t_0) + \left(\frac{q_h}{(\rho_{max} - \rho_h)} - \frac{q_{out}}{(\rho_{max} - \rho_{min})}\right)(t - t_0)$$ (34)

is the stronger increasing in time, the higher the density ρ_h is. In (34) t_0 is a fixed moment of time.

3.2.4. Characteristic Parameters of Traffic Jams

Emphasize that if the width of the cluster is large enough (Fig. 13(a), $t \approx 200\tau$), parameters of the local cluster $\rho_{min}, \rho_{max}, v_{max}, v_{gr}$ are the characteristic parameters of traffic flow: They are independent of the density ρ_h and of the length of the road L (Ref. (20)). On the one hand, the values $\rho_{min}, \rho_{max}, v_{max}, v_{gr}$ determine the main other characteristics of a traffic jam $q_{out}, L_s, q_b, \rho_b$ and they are also linked with the parameters of the kinetic model of traffic flow (Fig. 12). On the other hand, if we would consider a transition layer which is formed by vehicles escaping from a long queue of motionless vehicles with the density ρ_{max}, the values $v_{gr}, \rho_{min}, v_{max}$ would be approximately equal to the velocity of this transition layer, the density and the average speed of vehicles down-stream it, correspondingly. Therefore, experimental investigations of an escaping of vehicles from a such long queue may allow to estimate both parameters of kinetic model of traffic flow and the values q_{out}, ρ_{min}, v_{max}, v_{gr}, L_s, q_b, ρ_b discussed.

3.3. Physics and Properties of Traffic Jams

3.3.1. Active and Damping Processes in Traffic Flow

In order to understand the physics of the self-formation of the local cluster discussed above, let us now consider qualitatively the behavior of a local increase $\Delta\rho(x) > 0$ of the density in time. In this local region, the value V decreases: $\Delta V(x) = (dV(\rho)/d\rho)\Delta\rho(x) < 0$, as $dV/d\rho < 0$ (13). This decrease of the "maximum and out of danger" speed V forces drivers to decrease their average speed v sharply if the value $|dV/d\rho|$ is large enough. On the other hand, from the equation of continuity, it follows that a local decrease in average speed of vehicles v is accompanied by a local increase in the density of vehicles ρ and vice versa. Therefore, the decrease in v in the local region under consideration causes a further increase in the local value of ρ in this region and, therefore, leads to further subsequent decreases in the values V and v and also to an increase in the value ρ in this local region. This avalanche-like process of the decrease in the average speed and the corresponding increase in the density of vehicles in the local region of traffic flow one can consider as a some kind of active process, which causes the spontaneous appearance of a traffic jam in traffic flow.

On the other hand, the local increase in the density and the corresponding local decrease in the average speed considered above could disappear due to the diffusion ("viscosity") process, the influence of the gradient of "pressure" and also to the relaxation process of the average speed v to the "maximum and out of danger" speed V. Therefore, there is a competition between these two kind of processes: the *active processes*, trying to increase the amplitude of the local perturbation, and the *damping processes*, acting in the opposite direction (Sect. 2.1). If the active processes win this competition, a traffic jam appears on a road (Fig. 13(a)).

3.3.2. Boundary (Threshold) Density and Boundary Flux of Traffic Jam's Existence

The active process becomes weaker if the density ρ_h decreases. Indeed, for low density the interactions between vehicles seldom occur and the value $|dV/d\rho|$, which characterizes the strength of the active process, is small (Fig. 12(a)). For this reason, there should be some boundary (threshold) value of the density ρ_b for an excitation of a traffic jam, i.e., in a homogeneous flow there should be a boundary flux $q_b = V(\rho_b)\rho_b$: At

$$\rho_h < \rho_b, \tag{35}$$

i.e., at

$$q_h < q_b, \tag{36}$$

a local perturbation of any amplitude fades in time, i.e. a traffic jam cannot develop. In other words, a homogeneous state of traffic flow of low density $\rho_h < \rho_b$ (35) is a stable state.

3.3.3. Disappearance of Traffic Jam

On the other hand, from the latter qualitative consideration it follows that if a traffic jam would exist in the flow but up-stream of the traffic jam the density ρ in less than the density direct down-stream from the cluster ρ_{min}:

$$\rho < \rho_{min}, \tag{37}$$

this traffic jam should disappear in the course of time. Indeed, in this case the flux into the cluster $q_{in} \cong \rho V(\rho)$ is less than the flux out the cluster q_{out} (the formula (33)):

$$q_{in} < q_{out} \tag{38}$$

and the width of the traffic jam

$$L_s(t) = L_s(t_0) + \left(\frac{q_{in}}{(\rho_{max} - \rho)} - \frac{q_{out}}{(\rho_{max} - \rho_{min})}\right)(t - t_0) \tag{39}$$

should monotonously decrease.

From these qualitative considerations it follows an important conclusion that the boundary (threshold) density ρ_b for a traffic jam existence should be close to the density of vehicle formed direct down-stream from the traffic jam ρ_{min}; consequently, the boundary (threshold) flux q_b for an excitation of the traffic jam is close to the flux q_{out} (Ref. (20)):

$$\rho_b \cong \rho_{min}, \tag{40}$$

$$q_b \cong q_{out}. \tag{41}$$

Therefore, the condition of a disappearance of a traffic jam (35) or (36), using (40), (41), can be formulated as the following: A traffic jam will gradually disappear, if the density of vehicles up-stream it becomes lower than the boundary (threshold) density of vehicles:

$$\rho < \rho_b , \tag{42}$$

because in this case the condition (38) is fulfilled and the width of traffic jam (39) is decreasing.

3.4. Amplitude of Critical Perturbation

3.4.1. "Metastable" State of Traffic Flow

Discussing the kinetics of traffic jam formation in Sect. 3.3.1, the question about the amplitude $\Delta\rho_m$ of the initial local perturbation in the formula (26), which is needed for the traffic jam to be formed, has not been considered. In fact, the amplitude of a local perturbation should exceed some critical value $\Delta\rho_c$ for the traffic jam to be excited (Fig. 15) (Ref. (20)).

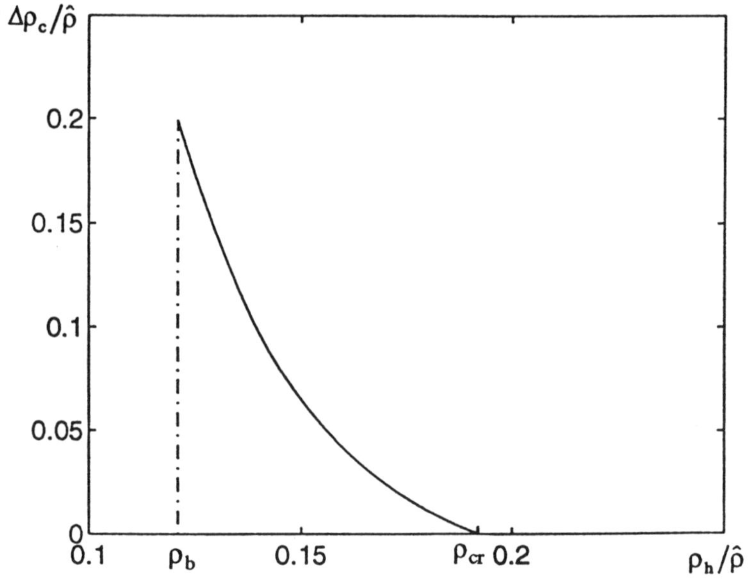

FIGURE 15. The critical amplitude of the initial local perturbation $\Delta\rho_c$ (Ref. (20)). The parameters of the model of traffic fllow are the same as in Fig. 12.

After the traffic jam has been formed its shape and properties do not depend on the amplitude of this local perturbation any more. Contrarily, if the amplitude of the initial local perturbation $\Delta\rho_m$ in the formula (26) is lower than $\Delta\rho_c$, this local perturbation fades in time. Emphasize that a qualitative form of the dependence $\Delta\rho_c(\rho_h)$ (Fig. 15) is the same that it has already been discussed in Sect. 2.2.4 (Fig. 7).

As it has been noticed above, there is the boundary density of vehicles in the initially homogeneous flow ρ_b for the existence of a local cluster in this flow. The critical amplitude of a local perturbation $\Delta\rho_c$ is maximal at this boundary point (Fig. 15) (Ref. (20)).

There is also some critical value of the density of vehicles ρ_{cr} which is higher than ρ_b. At $\rho_h > \rho_{cr}$ any amplitude of the initial local perturbation, i.e., which has even an arbitrary small initial amplitude, grows on the long road in time. This means that the value $\Delta\rho_c$ is a monotonously falling function of ρ_h in the interval $\rho_b \le \rho_h \le \rho_{cr}$ and it tends to zero at $\rho_h \to \rho_{cr}$ (Fig. 15) (20).

Therefore, in the range of density of vehicles

$$\rho_b < \rho_h < \rho_{cr} \tag{43}$$

the homogeneous state of traffic flow is a "metastable" state: A random appearance of a localized critical fluctuation, which amplitude exceed a critical value $\Delta\rho_c$ (Fig. 15) causes an excitation of a traffic jam in the homogeneous traffic flow (20). A local critical perturbation which growth leads to a self-formation of traffic jams in traffic flow can be considered as a "nucleation center" in traffic flow.

A random appearance of a fluctuation with an amplitude being large enough, in a vicinity of the traffic jam, can cause a disappearance of this traffic jam and a returning of traffic flow in the homogeneous state, as the latter is stable in the interval of density (43) with respect to local fluctuations with small amplitude. In other words, if the dispersion of localized fluctuations in traffic flow is large enough, one can suppose that there is a considerable probability to find in this density interval a process of random appearance and of disappearance of different traffic jams on the road (20). Emphasize that the critical density ρ_{cr} of the growth localized fluctuations is obviously higher than the critical density ρ_{c1} of the growth of the global fluctuation (19) in traffic flow (Ref. (20)):

$$\rho_{cr} > \rho_{c1}. \tag{44}$$

Therefore, in the range of the density

$$\rho_{c1} < \rho_h < \rho_{cr} \tag{45}$$

the homogeneous traffic flow is unstable with respect to global fluctuations on the whole road. For this reason, in this range of the density (45) traffic flow can only be considered as a "metastable" state, if an occurrence of localized critical fluctuations considerable faster leads to the escape from the unstable homogeneous state of traffic flow than the growth of global fluctuations.

3.4.2. Narrow Traffic Jams

If the density ρ_h equals ρ_b (or only slightly exceeds it) , narrow traffic jams, as a rule, are formed (20). Inside such traffic jams the average speed of vehicles v_{min} is considerably higher than zero. As a result the values ρ_{min} and q_{out} which should be used in the formula (34) are a little bit lower for narrow traffic jams than the corresponding values for the wide traffic jam (Fig. 13(a), $t \approx 200\tau$) discussed above. For the chosen model parameters (Fig. 12) $\rho_b \cong 0.1209\hat{\rho}$, $q_b \cong 0.5465\,\hat{\rho}\ell/\tau$. For narrow traffic jams, where $v_{min} > 0$ and $q_{min} = v_{min} \cdot \rho_{max} > 0$, the formulas (31), (32) should be replaced correspondingly by

$$v_{gr} = \frac{q_{min} - q_{out}}{\rho_{max} - \rho_{min}}, \tag{46}$$

$$v_{gl} = \frac{q_{min} - q_h}{\rho_{max} - \rho_h}. \tag{47}$$

It follows from the formulas (46), (47) that the velocities v_{gr}, v_{gl} for narrow traffic jams are higher than for the wide traffic jam (Ref. (19), (20)) discussed. They can even become positive as the density ρ_h decreases.

3.5. Reasons of Appearance of Multiple Traffic Jams

3.5.1. Structure of Transition Layer. Complex Local Cluster of Vehicles

When the structure of the local cluster has been discussed (Sect. 3.1), for a simplification of the consideration the space oscillations of the density in the transition layer which can appear there have not been taken into account. The largest peak of density ρ in this transition layer usually changes its position in the transition layer in time. Nevertheless, the peaks of ρ in the space oscillations finally fade down-stream (Ref. (20)).

The higher the density ρ_h is, the larger the amplitude of the peaks in the oscillations in the transition layer will be. It is clear from now on that, at some value ρ_h, the amplitude of the largest peak can reach a critical value in the course of time, which is large enough to start the avalanche-like process discussed in Sect. 3.3. This process leads to the spontaneous self-formation of a complex localized structure which consists of two or more traffic jams (Ref. (20)).

If the value ρ_h is increased further, the avalanche-like process of traffic jam formation can start in many peaks in the distribution of the density in the transition layer. Therefore, a sequence of traffic jams, which appears subsequently in space and time, is created. The traffic jams in this sequence have, as a rule, different amplitudes, different widths, different velocities and are not situated periodically in space (20).

3.5.2. Sequence of Traffic Jams in Traffic Flow

The local clusters shown in Fig. 13(a) appears due to the occurrence of only one local perturbation in the traffic flow. As a lot of local perturbations in a traffic flow on a road long enough can arise independently, a lot of local clusters can almost simultaneously appear. The possibility of this process of an appearance of a lot of traffic jams, which can completely cover the flow, is especially large, when the value ρ_h approaches the value ρ_{cr}, because the value of the critical amplitude of a local perturbation is decreased (Fig. 15). As in the latter case the traffic jams can have different amplitudes, different widths, different velocities and are not situated periodically in space (19), (20), the sequence of such traffic jams on the ρ, q phase plane can represent, as it follows from the numerical investigations made (20), a very complicated picture. The corresponding functions $\rho(x_0, t)$ and $q(x_0, t)$, i.e., the density and the flux at a fixed point $x = x_0$ on the road, can be very complicated functions of time. This agrees with the experimental investigations of traffic flow (e.g., (36)), where for a relative high density of vehicles the large scattering of values of the flux, which have been measured at a fixed point on a road, have been found in the ρ, q phase plane. In other words, a chaotic behavior of traffic flow which is often observed in experimental investigations may be linked to the spontaneous appearance of a lot of interacting traffic jams with different parameters in a traffic flow (20).

3.6. "Dipole Layer" Effect in Traffic Flow

Up to now we have considered the range of density $\rho_b \le \rho_h < \rho_{cr}$, where different kind of traffic jams can be self-formed. In a range of higher density of vehicles

$$\rho'_{cr} < \rho_h < \rho'_b, \qquad (48)$$

another non-linear effect can be realized in an initially homogeneous traffic flow: If a localized fluctuation, which amplitude exceeds some critical value $\Delta\rho_c'$, occurs than a localized structure in traffic flow in the form of a "dipole-layer" can spontaneously be self-formed (Fig. 16) (Ref. (28)).

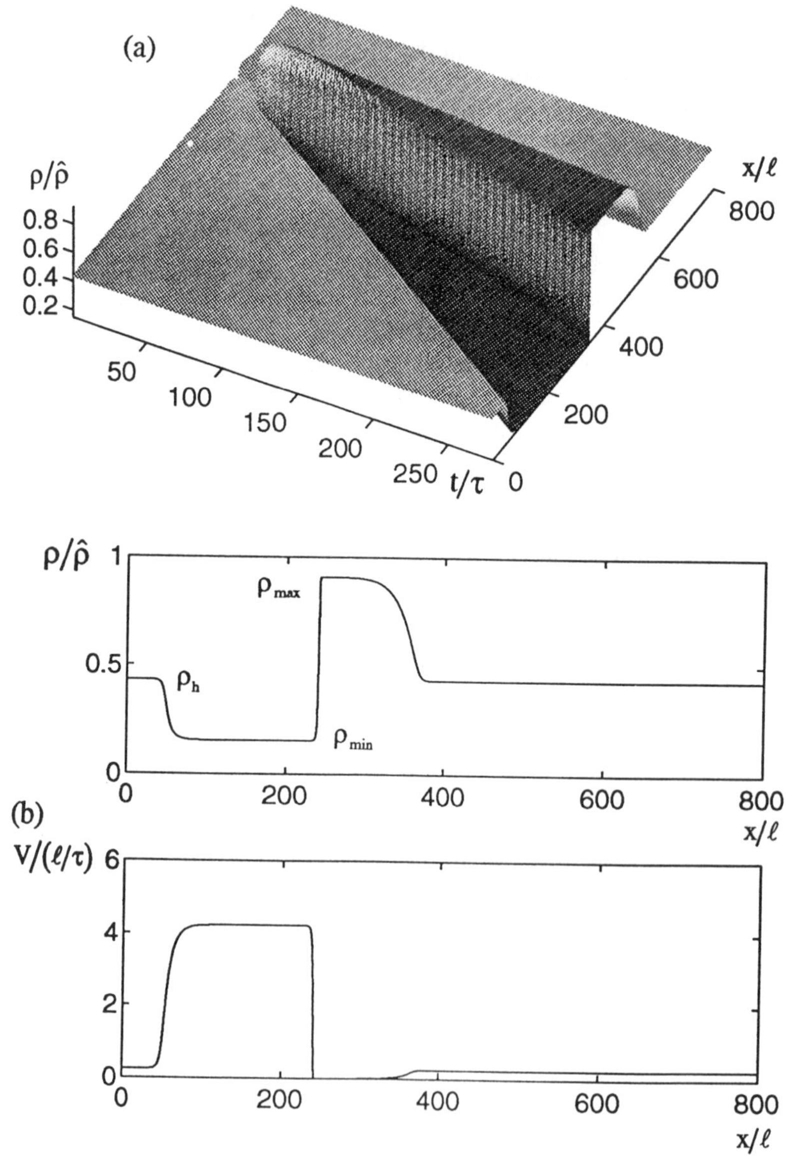

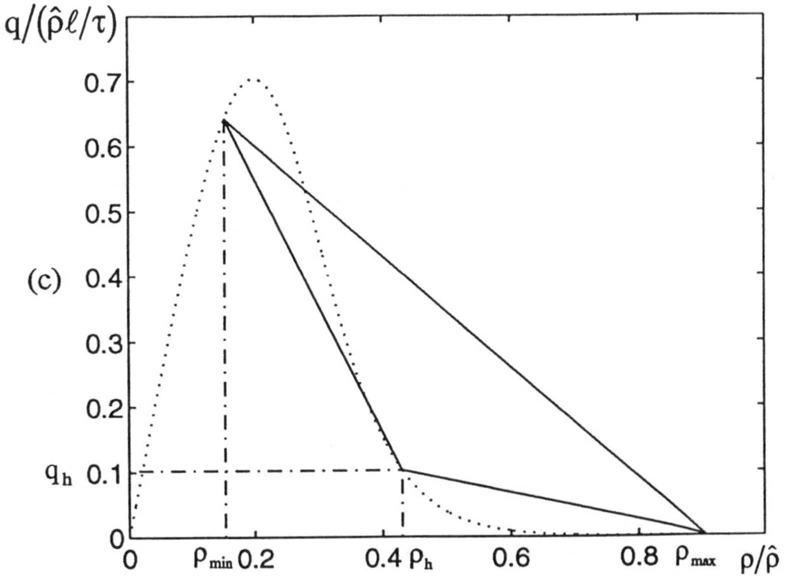

FIGURE 16. Kinetics of the formation of the "dipole layer" in traffic flow (Ref. (28)): (a) - the dependence $\rho(x,t)$; (b), (c) - the distributions $\rho(x)$ and v(x) (b) and the "dipole layer" in the ρ,q phase plane (c) at $t=180\tau$. Dotted lines in (c) represent the fundamental diagram (Fig. 12(b)). The initial distribution $\rho(x,0)=\rho_h-\Delta\rho(x)$ with $\Delta\rho(x)$ (26), $\Delta\rho_m=0.12\hat{\rho}$, $\rho_h=0.43\hat{\rho}$, $x_0=600\ell$. The other parameters are the same as in Fig. 12.

In (48) the boundary density $\rho_b'\cong\rho_{max}$, where ρ_{max} is the maximal density in a traffic jam (Fig. 13(f)), and the amplitude of the critical perturbation $\Delta\rho_c'$ is an increasing function of the density ρ_h: It tends to zero at $\rho_h\to\rho_{cr}'$ and it is maximal at $\rho_h=\rho_b'$. In other words, the range of density (48) is qualitatively analogous to the second range of the controlling parameter in Fig. 11 (in the case under consideration the role of the controlling parameter plays the density ρ_h: $A\equiv\rho_h$), where autosolitons in nonlinear systems can be excited.

3.6. Traffic Jams in Slightly Inhomogeneous Traffic flow

A homogeneous traffic flow is obviously a hypothetical state of traffic flow: A real traffic flow on a highway is always inhomogeneous due to entering and exiting traffic to on-ramps and off-ramps. If an initially traffic flow is inhomogeneous, an effect of a random appearance and disappearance of local clusters on a long road may be realized (Ref. (20)). Besides, in an inhomogeneous traffic flow a traffic jam can spontaneously be self-formed in a *deterministic way*, i.e., even when fluctuations in traffic flow are negligible and they may be neglected (Ref. (28), (29)). After a traffic jam due to its moving up-stream has left a source of a local inhomogeneity of traffic flow, its properties may become qualitatively the same as well as in a homogeneous traffic flow discussed above. In this section based on the results of the work by Kerner, Konhäuser and Schilke (29) a deterministic spontaneous formation of traffic jams will be considered.

3.6.1. Processes of Random Appearance and Disappearance of Traffic Jams

An effect of a random appearance and disappearance of local clusters on a long road can be observed if the density in the regions, where the traffic flow is dense, is higher than ρ_b and the density in the diluted regions is lower than ρ_b (Ref. (20)). If the extension of these regions of different density is large enough to consider traffic flow inside each of them as almost homogeneous, local clusters, as it has been described above, can spontaneously appear in regions of high density, where $\rho > \rho_b$. When a cluster, which moves against the flow (Fig. 13(a)), finds itself in a region of low density, where $\rho < \rho_b$, it will disappear, if the extension of the region of a low density is large enough. Indeed, the flux of vehicles q_{out}, formed by the cluster down-stream, exceeds the flux in front of cluster q_{in} in the region with $\rho < \rho_b$ and therefore the width of the cluster L_s gradually decreases.

Qualitatively the same process of random appearance and disappearance of local clusters can be realized on a multi-lanes road even if the flow is initially homogeneous on each lane, but the density on one lane is $\rho > \rho_b$ and on the other lanes the densities are $\rho < \rho_b$. Under those circumstances a local cluster can spontaneously appear on the lane with high density $\rho > \rho_b$ and then can disappear in the course of time due to lane changing of vehicles to the other lanes with low density up-stream of the traffic jam or/and inside the traffic jam.

3.6.2. Macroscopic Traffic Flow Model for Inhomogeneous Traffic Flow

Due to entering and exiting traffic to on-ramps and off-ramps traffic flow is, as a rule, inhomogeneous. The kinetic model of traffic flow on a long road with I on and J off-ramps includes the continuity equation

$$\partial\rho/\partial t + \partial q/\partial x = \sum_{i=1}^{I} g_i(x - x_i, t) - \sum_{j=1}^{J} r_j(x - x_j, t), \tag{49}$$

the equation of motion (8) and the boundary conditions (9), (10). The terms on the right-hand side of Eq. (49) express the fact of entering traffic g_i to the on-ramp "i" situated at $x = x_i$, and exiting traffic r_j to the off-ramp "j" situated at $x = x_j$. For a road long enough a length of sections, where vehicles are entering or exiting the road may be considered as being negligible short. Therefore, it can be assumed that $g_i(x - x_i, t) = g_i^0(t)\varphi_i(x - x_i)$ and $r_j(x - x_j, t) = r_j^0(t)\phi_j(x - x_j)$, where functions $\varphi_i(x - x_i)$ and $\phi_j(x - x_j)$ are strongly localized near the corresponding on- and off-ramps. To be more precisely, the regions of their localization should be negligible short in comparison with the distances between on- and off-ramps. In this case, as it follows from numerical investigations made in Ref. (29), a choice of the later functions is not important for the qualitative characteristics of effects which will be considered below. Values of integrals $q_a^{(i)} = g_i^0(t)\int_0^L \varphi_i(x - x_i)dx$ or $q_d^{(j)} = r_j^0(t)\int_0^L \phi_j(x - x_j)dx$ correspond to fluxes of vehicles from $q_a^{(i)}$ and $q_d^{(j)}$ to the on-ramp "i" or to the off-ramp "j", respectively.

3.6.3. Deterministic Spontaneous Appearance of Traffic Jams. "Local Breakdown" Effect

Processes of a deterministic spontaneous appearance of traffic jams has been found in Ref. (29) from numerical investigations of model (49), (8), (9), (10). These processes can be realized in a vicinity of transition layers which are formed between different levels of the density of vehicles in an inhomogeneous traffic flow (Sect. 2.2.3). The deterministic spontaneous appearance of traffic jams is linked with the "local breakdown" effect which is a deterministic effect and it results in a dynamical local restructuring of traffic flow. As a result of the "local breakdown" effect, a traffic jam is self-formed in traffic flow.

We will distinguish two different cases:

1. Motionless transition layers in traffic flow.
2. Moving transition layers in traffic flow.

The character of the "local breakdown" effect and the conditions of a traffic jam formation are in these two cases considerably different.

3.6.4. "Local Breakdown" Effect in Motionless Transition Layers of Inhomogeneous Traffic Flow

Let us first propose that at time $t < t_0 = 0$ there is an initially homogeneous traffic flow with some density ρ_h, average speed of vehicles v_h and a flux $q_h = v_h\rho_h$. Obviously that when, beginning at the time $t = 0$, some additional flux $q_a = q_a^{(1)}$ of vehicles appears from an on-ramp, the initially homogeneous traffic flow on a road becomes inhomogeneous: The total flux $q_t = q_h + q_a$ builds on the road directly down-stream from the on-ramp traffic flow with higher density $\rho_{h1} > \rho_h$. If the total flux q_t is hold a constant, the transition layer between the initially value of density ρ_h and the new higher value of density ρ_{h1} does not move (Fig. 17(a)).

When the flux q_a is slightly increased, the density of vehicles ρ_{h1} is slightly increased too. This gradually behaviour of a change of the density down-stream holds as long as the total flux q_t is less than a value Q_{max} which corresponds to the maximum point on the fundamental diagram (Fig. 12(b)). When the flux q_a is increased further, so that the total flux $q_t = q_h + q_a$ exceeds the value Q_{max}, a local avalanche-like change in traffic flow occurs (Fig. 17(b)): The density of vehicles increases and the average speed of vehicles decreases in an avalanche-like manner in the localized region which is situated down-stream from the on-ramp.

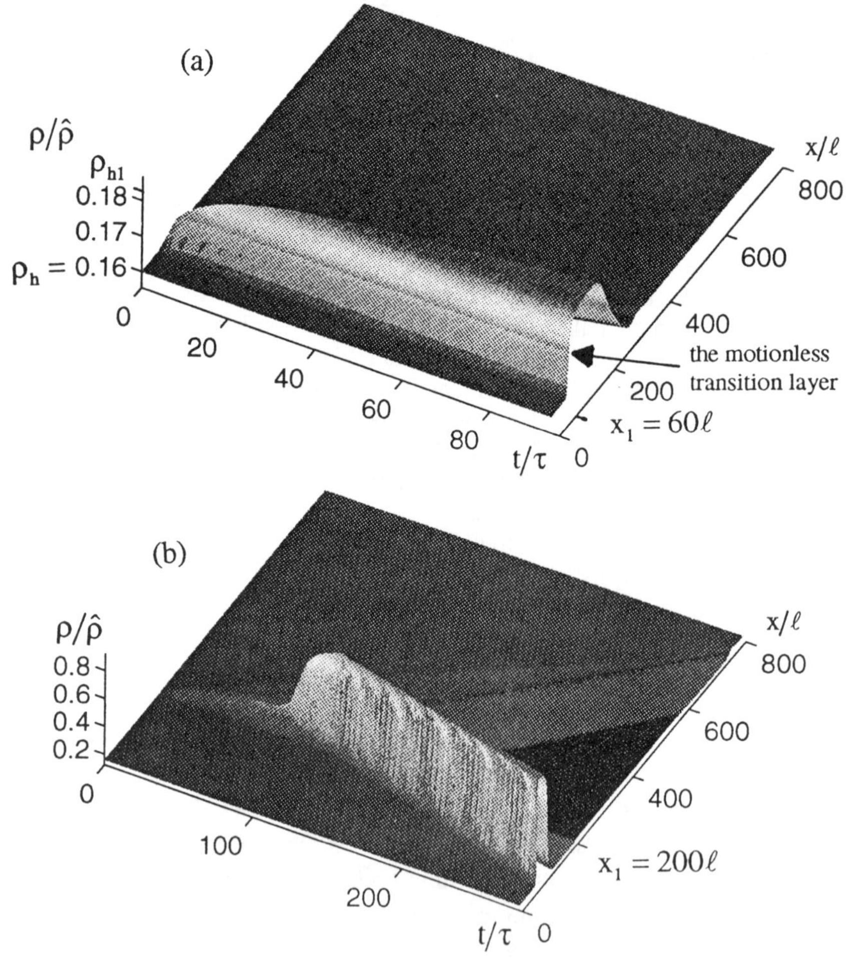

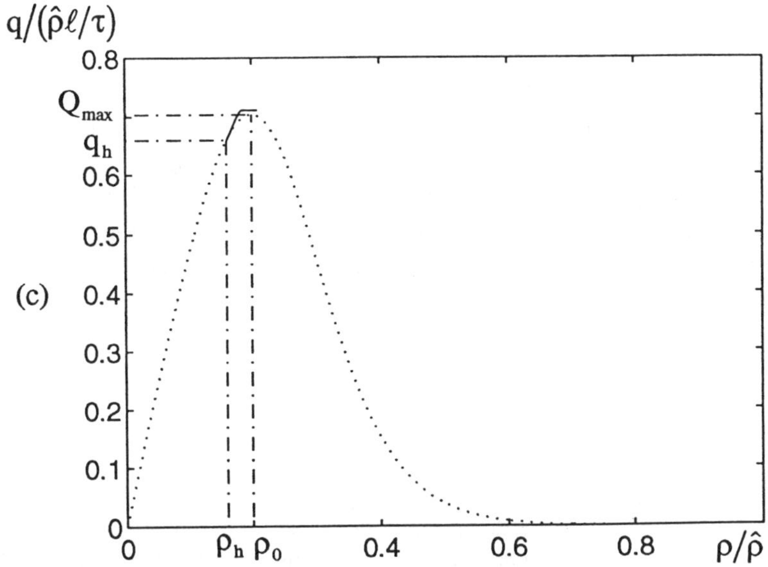

FIGURE 17. A deterministic formation of a traffic jam in motionless transition layers (Ref. (29)): (a) the dependence $\rho(x,t)$ for the case, when a traffic jam is not formed because a total flux down-stream from the on-ramp $q_t < Q_{max}$ ($q_t = 0.6906\hat{\rho}\ell/\tau$, $\rho_{h1} = 0.1833\hat{\rho}$, $Q_{max} = 0.7035\hat{\rho}\ell/\tau$, $\rho_0 \cong 0.199\hat{\rho}$); (b) $\rho(x,t)$ for the case, when a traffic jam is formed due to $q_t > Q_{max}$ ($q_t = 0.7111\hat{\rho}\ell/\tau$); (c) the fundamental diagram (dotted line) and the true function $q(\rho)$ (solid line), corresponding to the transition layer at $t = 69\tau$, for the case $q_t > Q_{max}$ shown at (b). $\rho_h = 0.16\hat{\rho}$. The on-ramp in Fig. (a) is situated at $x_1 = 60\ell$ and in Fig. (b) it is situated at $x_1 = 200\ell$. The other parameters are the same as in Fig. 12.

Due to the "local breakdown" effect, a traffic jam is formed in traffic flow (Ref. (29)). The form and the properties of this traffic jam are similar to those, when a traffic jam in an initially homogeneous traffic flow has appeared (Fig. 13). In contrary, the considered "local breakdown" effect in traffic flow results in a dynamical local restructuring of traffic flow, which is not associated with the presence of fluctuations in traffic flow.

The "local breakdown" effect in the motionless transition layers in traffic flow can be explain from the general considerations made in Sect. 2.2.3. Indeed, let us notice that an analogue of the "damping" parameter η in a nonlinear distributed system, in the case under consideration plays the flux q and a role of parameter θ plays the density ρ (or the average speed of vehicles). On the other hand, the function $\rho_h(q_h)$ which corresponds to the fundamental diagram, is S-shaped, and at $q_h = Q_{max}$ we have $d\rho/dq = \infty$ (dotted line in Fig. 17(c)). Recall that the function $\rho_h(q_h)$ practically defines the connection between values ρ_{h1} and q_t directly down-stream from the on-ramp. When $q_t = q_{h1} + q_a \to Q_{max}$, the true function $\rho_{h1}(q_t)$ in the transition layer directly down-stream from the on-ramp abuts on this limiting point $q_h = Q_{max}$. Therefore, for $q_t > Q_{max}$ (solid line in Fig. 17(c)) the density of vehicles should increase abruptly directly down-stream from the on-ramp from $\rho_{h1} \approx \rho_0$, where ρ_0 is the density in the point of the maximum on the fundamental diagram (Fig. 17(c)), to higher values. This local breakdown, that is, the abrupt local increase of the density of vehicles, naturally results in an appearance of a traffic jam (Fig. 17(b)).

3.6.5. "Local Breakdown" Effect in Moving Transition Layers of Inhomogeneous Traffic Flow

When a flux is higher than Q_{max} (solid line in Fig. 17(c)), it is quite expected some kind of "catastrophic" processes to appear in traffic flow. A "catastrophic" process due to the local breakdown effect described above really occurs in traffic flow leading to a self-formation of a traffic jam (Fig. 17(b)). However, as we will show below, a deterministic occurrence of a traffic jam can usually be realized on a highway, even when the flux in the traffic flow in any regions of the road is considerably lower than the characteristic value Q_{max}.

To show this effect (Ref. (29)), let us return to the case considered in Fig. 17(a), where the total flux $q_t < Q_{max}$ and, therefore, the local breakdown effect has not been occurred. Because in Fig. 17(a) during the time $0 \le t \le 95\tau$ the flux q_a was a constant, the transition layer could not move. In real traffic flow, a flux q_a is not usually a constant during a long period of time. For example let us consider the case when $q_a = \begin{cases} q_a, & 0 \le t \le 95\tau \\ 0, & t > 95\tau \end{cases}$, i.e., for $t > 95\tau$ the flux q_a from the on-ramp becomes zero (Fig. 18(a)). In this case the transition layer, which has been caused by the entering traffic to the on-ramp during the time $0 \le t \le 95\tau$ (Fig. 17(a)), at $t > 95\tau$ begins to move (Fig. 18(a)) with a positive velocity which is approximately equal:

$$v_{tr} \approx \frac{q_t - q_h}{\rho_{h1} - \rho_h}, \tag{50}$$

Emphasize once more that in the case under consideration the total flux $q_t = q_h + q_a$ for any time is considerably lower than the value Q_{max}.

Although the total flux in traffic flow is chosen (Fig. 17(a) and Fig. 18) to be considerably lower than the characteristic value Q_{max} (Fig. 17(c)), a self-formation of traffic jam in a deterministic way starts after the transition layer begins to move (Fig. 18(a)). To understand this effect, let us introduce two new variables (29):

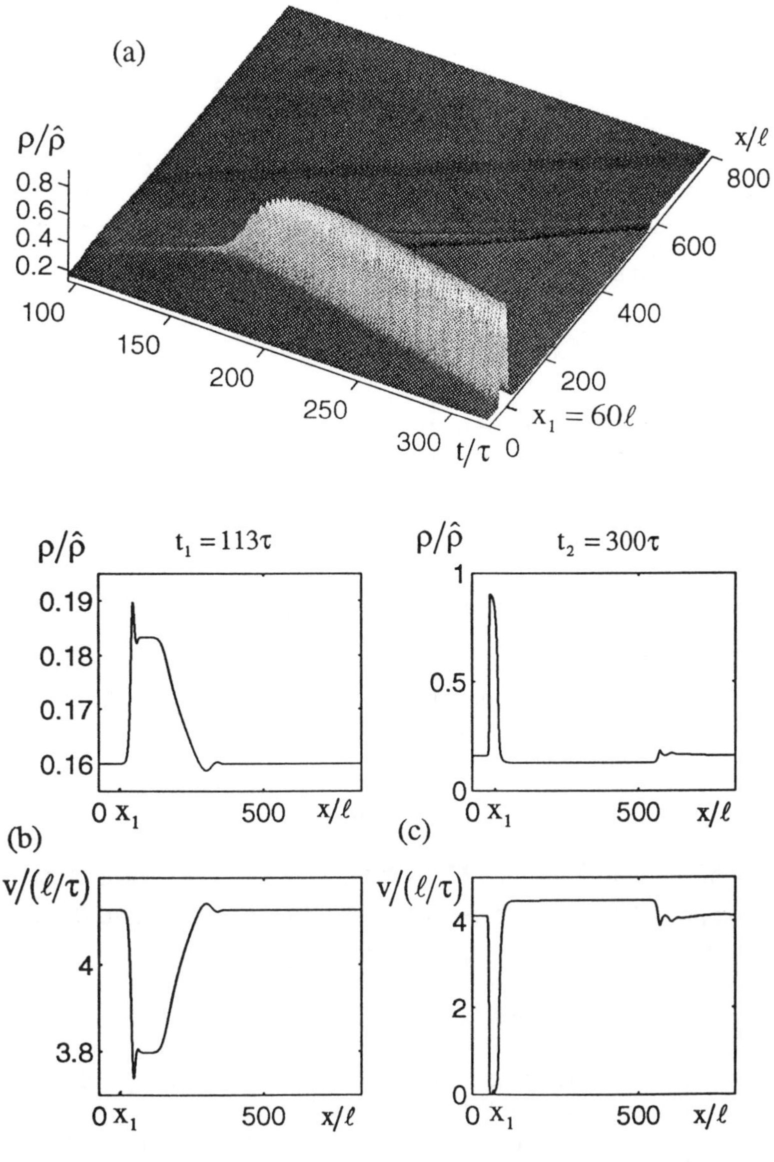

FIGURE 18. A deterministic formation of a traffic jam in a moving transition layer (Ref. (29)): (a) the dependence $\rho(x,t)$ for the time $t > 95\tau$ (this time is later than for the distribution $\rho(x,t)$ shown in Fig. 17(a)), when due to $q_a|_{t>95\tau} = 0$ the transition layer moves; (b) - (c) the corresponding distributions $\rho(x)$ and v(x) in two intermediate moments of time $t = 113\tau$ (b) and $t = 300\tau$ (c). The other parameters are the same as in Fig. 17(a).

$$q^* = \rho(v - v_{tr}), \tag{51}$$

$$Q^* = \rho(V(\rho) - v_{tr}), \tag{52}$$

where q^* and Q^*, correspondingly, are the "true" flux in traffic flow and the flux in a homogeneous traffic flow in a system of co-ordinate which moves with the velocity of the transition layer v_{tr}. The function $Q^*(\rho)$ is S-shaped, exactly, it is a curve with one maximum point (Q^*_{max}, ρ^*_0) which corresponds to the condition:

$$dQ^*/d\rho\Big|_{\rho_h} = v_h + \rho_h \xi(\rho_h) - v_{tr} = 0, \quad \rho_h = \rho^*_0, \tag{53}$$

where the value ξ is given by the formula (13). The flux q^* in the transition layer is practically a constant because the transition layer moves with an almost constant velocity v_{tr} (50) (Ref. (20)). If the difference $\rho_{h1} - \rho_h$ in the transition layer is small enough, it follows from (50) that for the velocity v_{tr} we can write the formula:

$$v_{tr} \approx v_h + \rho_h \xi(\rho_h), \tag{54}$$

where it has been taken into account that $q_t \cong v_{h1}\rho_{h1}$ and $q_h = v_h \rho_h$. From comparison Eq.(53) with Eq.(54) it follows that

$$q^* \cong Q^*_{max}. \tag{55}$$

In other words, the true function $q^*(\rho)$ in the transition layer practically abuts on the limiting point Q^*_{max} of the function $\rho_h(Q^*)$ which has S-shaped (Sect. 2.2.3). Therefore, the density of vehicles has a tendency to increase abruptly directly down-stream the transition layer causing the "local breakdown" in the moving transition layer and, finally, causing a deterministic appearance of a traffic jam.

In other word, the "local breakdown" effect in a moving transition layer in traffic flow can qualitatively be described, based on the concept of a local S-shaped characteristics considered in Sect. 2.2.3. In this case the S-shaped dependence is really realized in a small vicinity of the maximum of the local perturbation between the density (or the average speed of vehicles) and a parameter q^* (51). In other words, the density (or the average speed of vehicles) plays a role of "activator" and the parameter q^* (51) plays a role of a "damping" parameter in moving transition layers in traffic flow (see also Sect. 2.2.3).

Let us explain the effect of the deterministic traffic jam formation shown in Fig. 18 in more detail (Ref. (29)). First notice that, as it has already been explained in Sect. 2.2.3, directly down-stream of the moving transition layer a deterministic local perturbation (an increase) of the density of vehicle usually occurs (Fig. 18(b)). Such local perturbation can only be seen directly down-stream the transition layer, if a velocity v_q of this perturbation is nearly to the velocity of the transition layer v_{tr}, i.e., when

$$v_{tr} \approx v_q. \tag{56}$$

Indeed, if for example v_{tr} were considerably more than v_q, the local perturbation could not be localized directly down-stream the transition layer and, therefore, the local breakdown effect could not start.

The velocity v_q of small amplitude perturbations in traffic flow can be estimated from the formula (20), i.e.,

$$v_q = v_h - c_0. \tag{57}$$

At low density of vehicles, when the derivative $|dV/d\rho|$, i.e., the value $|\xi|$ in Eq. (54), is very small, we have $v_{tr} \approx v_h$. In this case the velocity v_{tr} of the transition layer, as it follows from the formula (57), is considerably more than the velocity v_q of perturbations in traffic flow. Therefore, at low density of vehicles the condition (56) cannot be fulfilled and the "local breakdown" effect cannot start. When the density is increased, the derivative $|dV/d\rho|$ increases too. Consequently, the velocity of the transition layer v_{tr} (54) decreases and it can approach the velocity v_q (57).

Substituting the formulas (54) and (57) into the condition (56) we can find that the characteristic density of vehicles, which corresponds to the condition (56), satisfies the formula

$$-(\rho_h/c_0)\xi(\rho_h) \approx 1. \tag{58}$$

From the comparison of the formula (58) with the formula (23) can be seen that the density of vehicles which follows from (58) approximately coincides, when L is large enough, with the characteristic value ρ_{c1} which has been found in the theory of the global instability of a homogeneous traffic flow.

As it follows from numerical investigations made in Ref. (29), for a starting of the "local breakdown" effect, besides the carrying out the condition (56), it is also necessary that an amplitude of the deterministic perturbation under consideration exceeds some critical value. The amplitude of this perturbation is increased if either the density ρ_h or the density ρ_{h1} are increased.

For this reason the occurrence of the "local breakdown" effect depends both on the density ρ_h and the density ρ_{h1}. To start the "local breakdown" effect, the density ρ_{h1} should be higher that ρ_{c1}, but, as it has already been mentioned, the density ρ_{h1} can be considerably lower than the density ρ_0 which corresponds to the maximum point on the fundamental diagram (Fig. 17(c)). The density ρ_h can be lower than ρ_{c1} but it should be higher than the boundary density ρ_b.

A real traffic flow on the highway can consist of a lot of moving transitions layers between different levels of the density of vehicles. If conditions which have been considered above are fulfilled in a vicinity of one of the moving transition layers, a traffic jam in the deterministic way will be formed in such a flow. Because the traffic jam moves up-stream, it may approach in some time one of the off-ramps, where due to exiting traffic down-stream from this off-ramp the density ρ and, consequently the flux q_{in} on the highway are lower than the boundary (threshold) values of the density $\rho_b \cong \rho_{min}$ and the flux $q_b \cong q_{out}$, correspondingly. In this case, as it follows from the formula (39), the width of traffic jam decreases and the traffic jam gradually disappear. Both effects of the deterministic spontaneous appearance of traffic jams down-stream from on-ramps and then a disappearance of this traffic jam down-stream from off-ramps have been observed in numerical calculations made in Ref. (29).

4. DIFFUSION AUTOSOLITONS

In this section, based on the general theory of autosolitons in active systems with diffusion (1), two examples of such systems will be considered:

1. Back-biased *pn* structure.
2. Gas discharge system.

Although these two distributed active systems with diffusion belong to different kind of physical systems, we will see that the basic equations, describing distributions of the variables of these systems, are identical in their structure. As a results, a form and a lot of properties of autosolitons which can spontaneously appear in these systems are very similar.

4.1. Avalanche Current Filaments in Back-Biased *pn* Structure

The effect of a stratification of an initially homogeneous distribution of the density of an avalanche current and of a spontaneous formation of multi-filaments states have experimentally been found out and explained in the investigation of the back-biased *pn* structure made of α-SiC by Kerner, Litvin and Sankin (37). In this section, based on this work, the physics, model and some properties of autosolitons in the form of current filaments in *pn* structure made of α-SiC will be given.

4.1.1. Physics of Appearance of Current Filaments

It has been found in (37) that, if an orientation of the crystal of *pn* junction has been chosen on such a way that the electric field directed parallel to the natural superlattice of α-SiC, an initially homogeneous distribution of an avalanche current, beginning at current densities $j > j_c = 10 - 20\,A/cm^2$, has stratified. The stratification has been realised at a positive differential conductivity of the sample. It is manifested as a change, as the current increases, from a breakdown pattern which has been uniform over the area to a nonuniform breakdown. This nonuniform breakdown has corresponded to a two-dimensional dissipative structure in the form of numerous and brightly luminous avalanche-current filaments of different dimensions which have been not periodically located over the area of the sample. The results of the theory of this phenomena developed in (37) and of the numerical investigations (38) have been in good agreements with the experiment.

Let us consider the physics of this effect (37). A natural superlattice exists along the C crystallographic axis in α-SiC crystals. The existence of such a superlattice is known to cause a qualitative difference between the heating of electrons in an electric field directed parallel to the superlattice axis and that in an electric field directed perpendicular to this axis. The difference is that the heating of electrons in a field parallel to the axis of the superlattice is hindered by the miniband nature of the electron spectrum. In other words, in the case of a semiconductor with a superlattice the Bragg reflection from the miniband edge hinders heating of the electron gas by a field parallel to the superlattice axis and, consequently, hinders the impact ionization process. When an electron acquires energy, the transformation of the electron from one miniband to another is possible only by virtue of a quasi-elastic scattering of electrons by phonons. Such a situation is the characteristic for a low current density.

As the current density is raised, the transformation of the electron from one miniband to another may be possible also due to electron-electron collisions. It may occur, because the rate of electron-electron collisions is proportional to the concentration of the electrons n, i.e., this rate is an increasing function of the density of the avalanche current. The latter effect may be responsible for a rising dependence of the average rate of ionization ν_i in the space charge region of a *pn* junction on the carrier density n. This rising dependence ν_i of n can cause a growth of a local perturbation of the current density in the structure. The latter can lead to a spontaneous appearance of current filaments. Indeed, the avalanche current density is:

$$j = env_e \tag{59}$$

i.e., it is proportional to the concentration of electrons n in the space charge region of a *pn* junction, because the drift velocity of electrons v_e in this region is practically constant. The local increase in the density of current implies the local increase in the electron concentration in the space charge region of a *pn* junction and, therefore, it implies the increase in the average rate of ionization ν_i. The latter causes the further local increase in the previous local perturbation of current density and so it causes the further local growth of the average rate of ionization ν_i and so on.

Considering this avalanchelike growth of the local current density in the *pn* junction, the voltage drop across quasineutral parts, i.e. the resistive layers of the structure (Fig. 19(a)), has not been taken into account. For the simplicity, taking into account the specific resistivity ρ_s of the only one of the qusineutral regions n region in Fig. 19(a), we get the formula for the total voltage drop across the structure V:

$$V = j\rho_s + V_i \tag{60}$$

where V_i is the voltage drop across the *pn* junction. The formula (60) is valued, if the distribution of current density over the area of the structure is homogeneous.

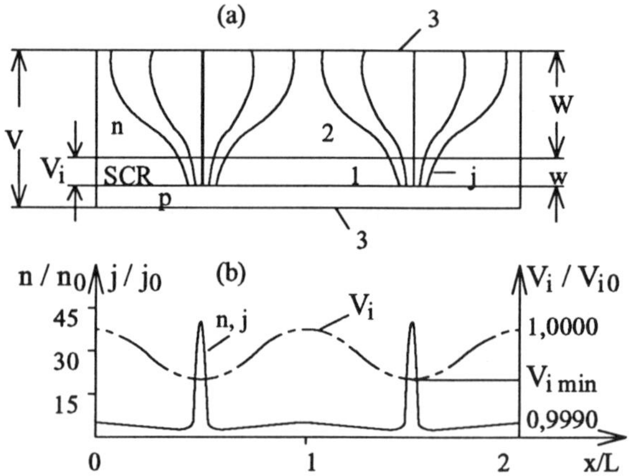

FIGURE. 19. Formation of current filaments in back-biased *pn* structure: schematic representation of the structure (a) (lines indicate the current lines within two periods of a current filaments; space charge region (SCR) of the *pn* junction (1), resistive layer (2), metallic electrode (3)); numerically calculated distribution of n, j, and V_i (b) (Ref. (38)).

From the formula (60) it follows that, if a homogeneous increase in the current density δj occurs, we get that the voltage drop across the *pn* junction V_i should decrease, if the value of the total voltage drop V=const:

$$\delta V_i = -\delta j \rho_s . \tag{61}$$

On the other hand, it is well-known that the average rate of ionization ν_i is a strong increasing function of the voltage drop across the *pn* junction V_i. Therefore, if the value ρ_s is not very small, the decrease in V_i (61) suppresses the growth of the homogeneous perturbations of the current, inducing by the rising dependence of the average rate of ionization ν_i on the carrier density n. In other words, the homogeneous perturbations of the current density cannot grow.

A different situation is encountered when the growth of the current density is nonhomogeneous. To understand it, let us consider the growth of current density in a small local regions of a size ℓ of the area of the *pn* junction (see current lines in the space charge region (SCR) in Fig. 19(a)). If the thickness of the resistive layer of the structure W is large enough:

$$W >> \ell, \tag{62}$$

the current density spreads over the quasineutral n type region of the structure, as shown by lines in Fig. 19(a). Because of that, the change in the voltage drop across the n type region, i.e., the value $V_\rho = V - V_i$, produced by a local increase in the avalanche current in the small neighbourhood of size ℓ, will be much less, in proportion to $\ell/W << 1$, than that in the above case of a uniform increase in the avalanche current. In other words, a local increase in the density of the current causes a very small change in the voltage drop across the resistance layer: $V_\rho << \delta j \rho_s$. As a result, taking into account that $\delta V_i = -\delta V_\rho$, instead of the formula (61), we have

$$|\delta V_i| << \delta j \rho_s \tag{63}$$

Corresponding to (63), the decrease in the value of the voltage drop across the *pn* junction δV_i can be so small that it cannot suppress the effect of the growth of the current density causing by the rising dependence of the average rate of ionization of the current density in the local region under consideration. Therefore, if the density of current increases in local regions of the very small size ℓ (62), the voltage drop in the resistive layer is not enough to prevent the further growth of current density in these small local regions of the structure. So, local solitary regions of high density of current may arise in a homogeneous *pn* structure (Fig. 19(b)).

Notice that the value ℓ, where a local growth in the avalanche current density can be realized, is defined by the length of a diffusion spreading of free carriers during the time $\tau_n = w/v_e$ of their passage across the space charge region of the *pn* junction. In other words, $\ell \approx (D\tau)^{1/2}$, where D is the coefficient of diffusion of charge carriers. In the case under consideration Ref. (37) we have $\ell \sim 10^{-6}$ cm and $W \sim 10^{-1}$ cm. Therefore, the condition (62) which is important for the formation of multi-filaments current states in *pn* structure is really fulfilled.

4.1.2. Basic Equations

The basic equations which describe the distributions of the concentration n and of the voltage drop V_i in the model of the *pn* structure (Fig. 19) have been proposed in Ref. (37) and they have further been substantiated in (38). The model includes the balance equation for the concentration in the space charge region of the *pn* junction:

$$\tau_n \frac{\partial n}{\partial t} = \ell^2 \Delta_\perp n + n v_i(n, V_i) \tau_n - n + G\tau_n \tag{64}$$

and the equation for the distribution of the voltage drop V_i which is linked with the spreading of the current density in the quaineutral region of the structure:

$$\tau_V \frac{\partial V_i}{\partial t} = L^2 \Delta_\perp V_i - j\rho_s + V - V_i, \tag{65}$$

where $\tau_V = \rho_s C$, C is the capacitance per unit area of the junction; $L \approx W$; G is the rate of thermal and/or tunnel generation of the carriers in the space charge region of the *pn* junction; the other designations have been explained in Sect. 4.2.1. Equations (64) and (65) can be added by the Kirchhoff's equation:

$$E_{ex} = V + R_L \int_S j d\mathbf{r}_\perp , \tag{66}$$

where R_L is the load resistor, E_{ex} is the electromotoric force of the voltage source.

In one-dimensional case the model (64), (65) have been solved numerically in (38) (Fig. 19(b)). It has been found that the autosoliton in the form of current filament (Fig. 1) can really be formed in the semiconductor structure under consideration.

The model (64), (65) can be written in the basic form for active systems with diffusion as a set of two diffusion nonlinear equations describing the distributions of the activator θ and of the inhibitor η. For one-dimensional case we can write these equations in the form (1):

$$\tau_\theta \frac{\partial \theta}{\partial t} = \ell^2 \frac{\partial^2 \theta}{\partial x^2} - q(\theta, \eta, A), \tag{67}$$

$$\tau_\eta \frac{\partial \eta}{\partial t} = L^2 \frac{\partial^2 \eta}{\partial x^2} - Q(\theta, \eta, A), \tag{68}$$

where τ_θ, τ_η, ℓ and L are the characteristic times and lengths of a variation of, respectively, the activator θ and the inhibitor η; A is the controlling parameter of the system; q and Q are the nonlinear functions. The general theory of autosolitons describing by the model (67), (68) has been reviewed in (1). As it follows from (1), the form and parameters of autosolitons as well as the scenarios of a self-organization and turbulence in active systems with diffusion depend on the nonlinear character of functions q, Q and on the values

$$\varepsilon = \frac{\ell}{L}, \quad \alpha = \frac{\tau_\theta}{\tau_\eta}. \tag{69}$$

In the semiconductor structures under consideration $n \equiv \theta$, $V_i \equiv \eta$, $\tau_n \equiv \tau_\theta$, $\tau_V \equiv \tau_\eta$, $V \equiv A$,

$$q(\theta, \eta, A) \equiv q(n, V_i) = -n\nu_i(n, V_i)\tau_n + n - G\tau_n, \tag{70}$$

$$Q(\theta, \eta, A) \equiv Q(n, V_i, V) = j\rho_s - V + V_i, \tag{71}$$

the value $\varepsilon \ll 1$ and, as a rule, the value $\alpha > 1$. Correspondingly to the results of the general theory of autosolitons in active systems with diffusion (1), the *pn* structure under consideration corresponds to the class of so called "K-system", where, as a rule, static autosolitons are formed. This conclusion corresponds to the results of the numerical investigations, where static autosolitons (Fig. 1 and Fig. 19(b)) have been found (38).

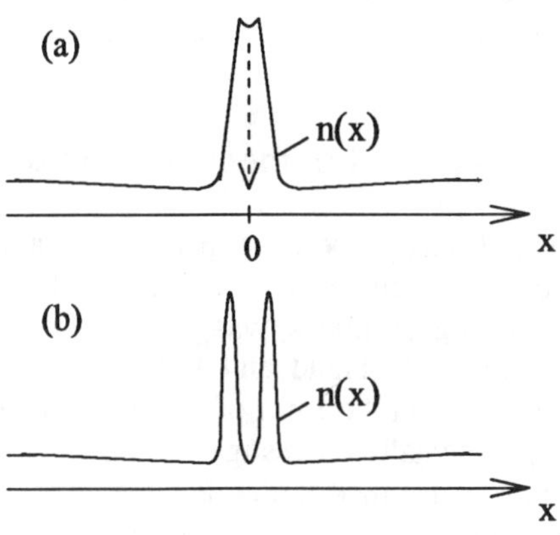

FIGURE 20. Schematic representation of the splitting of a static autosoliton caused by the "local breakdown" effect in the center of the autosoliton: The distribution of the activator (the concentration n of the carriers) at the initial moment time (a) and at once after local breakdown in the center of autosoliton (b).

Notice that the scenarios of an evolution of autosolitons in K-system for one-dimensional case are mainly determined by two nonlinear effects (1):

1. The "local breakdown" effect in the center of an autosoliton (Fig. 20) or in oscillating tails of an autosoliton.
2. The "pumping" of activator between autosolitons (Fig. 21).

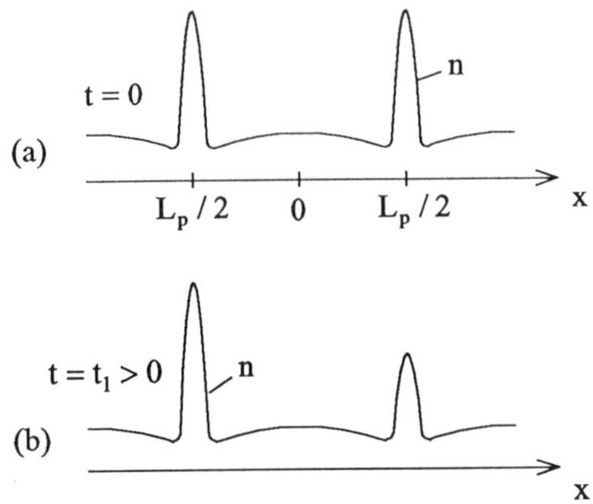

FIGURE 21. Illustrating the "pumping" of activator between two autosolitons which are situated at some distance L_p (a): (b) - some intermediate distribution of the activator (the concentration n of the carriers) realized in the process of the "pumping" activator effect.

Both the local breakdown effect and the "pumping" of activator effect have been found in the numerical investigations of the model of the semiconductor *pn* structure (38).

4.2. Avalanche Current Filaments in Gas Discharge System

A structure with gas discharge between a metallic cathode and a bulky high-resistance semiconductor anode, where a gas discharge region besides the electrodes is limited by two glass plates, was proposed and realized experimentally by Radehaus, Dirksmeyer, Willebrand and Purwins (Fig. 22) (39). In this gas discharge system autosolitons in the form of current filaments have experimentally been found and experimentally investigated in detail (39) - (41). These experimental results have confirmed the earlier theoretical predictions of a lot of nonlinear effects in active systems with diffusion connected with an appearance and evolution of autosolitons in these systems (42) - (44), (1), (16), (17).

In this section we will consider a physical model of this gas discharge system as it has been proposed in (45). Besides we will briefly discuss some properties of autosolitons which can appear there.

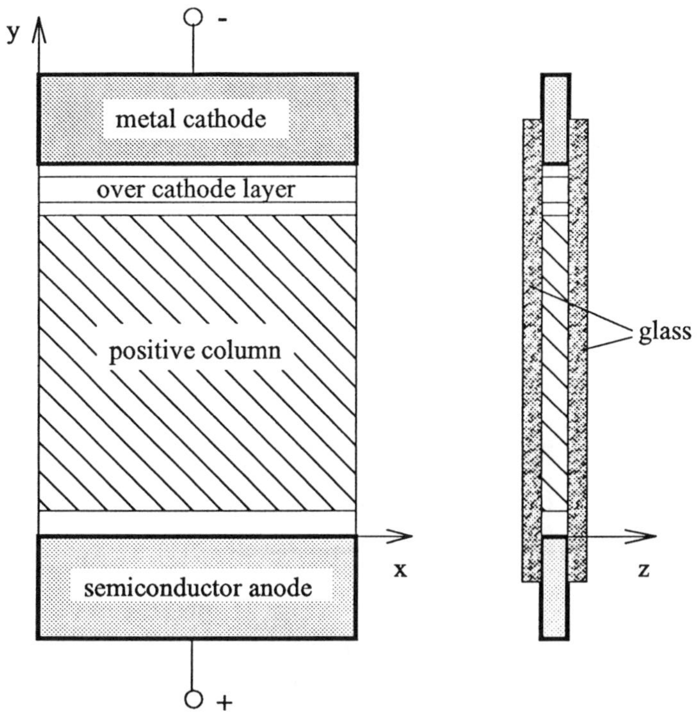

FIGURE 22. Model of the gas discharge system (39).

4.2.1. Model and Basic Equations

For simplification of a model we will neglect effects in an over cathode layer (Fig. 22) and also other boundary effects near the semiconductor anode (Fig. 23). Besides we will propose that on the one hand, the concentration of electrons n in the gas discharge is in many order of magnitude less than the density of neutral molecules of a gas, on the other hand, it is large enough to use a quasi-hydrodynamic approach to the gas plasma and also to use the consideration of this plasma in the quasi-neutral approximation. In additional, we will consider only the case of small enough distance between the glass plates of the system. In this case the recombination of electrons and ions near the glass plates plays the decisive role in the balance of the concentration of the carriers. In other words, the ambipolar diffusion of electrons and ions to the glass plates in such "diffusion type" of a gas discharge (46) determines mainly the concentration n of electrons and ions which appear in the system due to the process of impact ionisation in the gas.

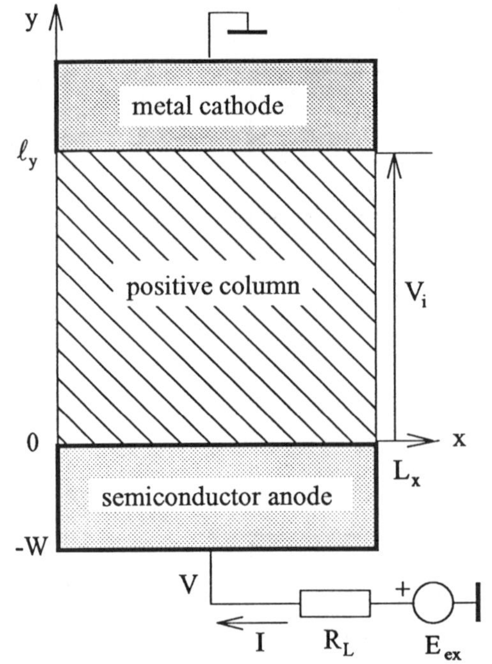

FIGURE 23. Simplified model of the gas discharge system (39).

In this model the equations which describe the balance of the concentration and of the average energy of electrons in the gas discharge (46):

$$\frac{\partial n}{\partial t} = e^{-1}\operatorname{div}\mathbf{j}_e + n\nu_i(n,T) + G_0 - \gamma(n,T)\,n^2 \tag{72}$$

$$\frac{3}{2}\frac{\partial(n\,T)}{\partial t} = -\operatorname{div}\mathbf{j}_\varepsilon + \mathbf{j}_e\mathbf{E} - n\frac{T - T_i}{\tau_\varepsilon}, \tag{73}$$

after taking into account the quasi-neutral approximation, can be averaged in the direction of z-axis of the system (Fig. 22) and the diffusion time of recombination can be introduced (46):

$$\tau_D = \frac{\ell_z^2}{\beta^2 D}. \tag{74}$$

Notice that in the equations (72), (73) and in the formula (74) the following designations are used: $\nu_i(n,T)$ is the average rate of the impact ionisation per one electron, T is the electron temperature, G_0 is the rate of ionisation (photos, thermal, ets) which does not depend on the electron concentration, γ is a coefficient of a volume recombination of the plasma (for example, Auger-recombination), **E** is the electric field, T_i is the temperature of ions and neutral molecules (T_i=const), τ_ε is the characteristic time of energy relaxation of the electrons, ℓ_z is the distance between the glass plates, $\beta \sim 1$, D is the ambipolar coefficient of diffusion,

$$\mathbf{j}_e = e\mu_e n\mathbf{E} + \mu_e T\nabla n + (1+\alpha_T)\mu_e n\nabla T \tag{75}$$

is the density of the electron current,

$$\mathbf{j}_\varepsilon = -e^{-1}\left(\frac{5}{2}+\alpha_T\right)(\mu_e nT\nabla T + T\mathbf{j}_e) \tag{76}$$

is the density of the energy flux of electrons, μ_e is the mobility of electrons, $\alpha_T = \partial \ln \mu_e / \partial \ln T$.

As a result of the quasi-neutral approximation and mentioned averaging of the equations in the direction of z-axis (Fig. 22), we get the following equation of the balance of the concentrations of electrons (Ref. (45)):

$$\begin{aligned}\tau_D^0 \frac{\partial n}{\partial t} &= \ell^2 \nabla_\perp (T\nabla_\perp n) + (1+\alpha_T)\ell^2 \nabla_\perp (n\nabla_\perp T) - \frac{\tau_D^0 \mu_i}{e} \mathbf{j}\nabla_\perp \mu_e^{-1} + \\ &+ n\nu_i(n,T)\tau_D^0 + G_0\tau_D^0 - nT - \tau_D^0\gamma(n,T)n^2,\end{aligned} \tag{77}$$

and of the balance of energy of electrons

$$\begin{aligned}\tau_\varepsilon^0 \frac{3}{2}\frac{\partial (nT)}{\partial t} &= (\tilde{L}^0)^2 \nabla_\perp \left(\frac{n\mu_e T}{\mu_e^0}\nabla_\perp n\right) + \frac{3}{2}\frac{\tau_\varepsilon^0}{e}\mathbf{j}\nabla_\perp T - \\ &- \frac{\tau_\varepsilon^0 T}{en}\mathbf{j}\nabla_\perp n + \frac{j^2\tau_\varepsilon^0}{en\mu_e} - \frac{\tau_\varepsilon^0 nT}{\tau_\varepsilon},\end{aligned} \tag{78}$$

where μ_i is the mobility of ions, $\ell = \ell_z/\beta = (\tau_D^0 D^0)^{1/2}$, $\tau_D^0 = \tau_D(T_i)$, $D^0 = D(T_i) = \mu_i T_i e^{-1}$, $\tilde{L}^0 = (\mu_e^0 T_i \tau_\varepsilon^0 e^{-1})^{1/2}$, $\tau_\varepsilon^0 = \tau_\varepsilon(T_i)$, $\mu_e^0 = \mu_e(T_e)$, $\mathbf{j} = \mathbf{j}_e$.

Here we will restrict the discussion of autosolitons in the form of current filaments. In other words, distributions of variables only in the direction of x-axis (Fig. 23) will be considered and the possible nonlinear effects which are linked with the nonhomogeneous distributions of variables in the direction of y-axis of the structure will be neglected. In this approximation $n = n(x)$, $T = T(x)$ and $j = j(x)$.

In addition, the characteristic length L of a spatial change in the voltage drop V_i across the gas plasma, which appears due to the spreading of the current in the bulk semiconductor electrode (Fig. 23), and the characteristic time of the relaxation τ_V of this voltage drop V_i will be proposed to be considerably more than corresponding values $\tilde{L}$ and τ_ε for the temperature of electrons in the gas plasma.

Taking these approximations into account, we can write the equation of the balance of the concentrations of carriers (77) in the form (Ref. (45)):

$$\tau_D^0 \frac{\partial n}{\partial t} = \ell^2 \frac{\partial}{\partial x}\left(T\frac{\partial n}{\partial x}\right) + (1+\alpha_T)\ell^2 \frac{\partial}{\partial x}\left(n\frac{\partial T}{\partial x}\right) + \tag{79}$$
$$+ n\nu_i(n,T)\tau_D^0 + G_0\tau_D^0 - nT - \tau_D^0\gamma(n,T)n^2$$

and the equation (78) allows us to write the local connection between the voltage drop V_i across the gas plasma and the temperature of electrons T:

$$V_i^2 = \frac{\ell_y^2 T}{e\tau_\varepsilon \mu_e}, \tag{80}$$

where ℓ_y is the length of the region of the gas plasma in y-direction (Fig. 23). The equations (79), (80) (Ref. (45)) should be completed by two equations:

- by the equation (65) which was derived for this system in Ref. (47) and which in one-dimensional case under consideration has the form:

$$\tau_V \frac{\partial V_i}{\partial t} = L^2 \frac{\partial^2 V_i}{\partial x^2} - j\rho_s + V - V_i, \tag{81}$$

- by the equation (66) which in this case takes the form:

$$E_{ex} = V + R_L \int_0^{L_x} j(x)dx. \tag{82}$$

The set of the equations (79) - (82), completed by boundary conditions which may be taken as the following:

$$\left.\frac{\partial n}{\partial x}\right|_{x=0} = \left.\frac{\partial n}{\partial x}\right|_{x=L_x}, \quad \left.\frac{\partial V_i}{\partial x}\right|_{x=0} = \left.\frac{\partial V_i}{\partial x}\right|_{x=L_x}, \tag{83}$$

makes a physical model of the gas discharge system (Fig. 23). Notice that in the equation (81) for the gas discharge system $\tau_V = \rho_s C$, C is the capacitance per unit length (in x-direction) of the system, ρ_s and W are the specific resistivity and the width of the semiconductor anode of the gas discharge system (Fig. 23).

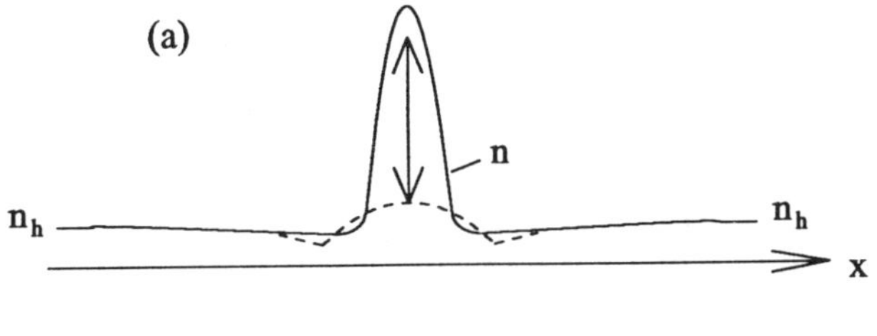

FIGURE 24. Form of spike pulsating (a), rocking (b) and travelling (c) current filaments (1), (16).

As well as in the *pn* structure considered in Sect. 4.1, the average rate of the impact ionisation in the gas discharge plasma $\nu_i(n, T(V_i))$ is an increasing function of both the voltage drop across the gas discharge V_i and the concentration of the carriers n. As a result, a role of the activator in the gas discharge systems plays the concentrations of the carriers n and the role of the inhibitor plays the voltage drop V_i. The equation (79) is qualitative very similar to the equation (64), describing the distribution of the concentration of carriers in the space charge region of the *pn* junction. Therefore, an autosoliton of the same form (Fig. 1) can appear in the *pn* structure and in the gas discharge system. The same nonlinear effects, as it is shown in Fig. 20 and Fig. 21, can also be important for the processes of self-organization and turbulence in the gas discharge systems.

However, owing to a changing in some parameters of the gas discharge system, as the gas pressure, the distance between electrodes, the resistivity of the semiconductor anode, it was possible to find out experimentally nonlinear effects (Ref. (41)) which are characteristic for active systems with diffusion which correspond to the class of so called "KΩ-systems" (Ref. (1), (16), (43), (44)). To this class of active systems with diffusion belong systems for which the inhibitor η is slow and long-range compared with the activator θ: $\varepsilon = \ell/L << 1$ and $\alpha = \tau_\theta/\tau_\eta < 1$.

In particularly, besides static autosolitons (Fig. 1), another types of autosolitons: pulsating, rocking and travelling autosolitons in the form of current filaments can be formed in KΩ-systems (Fig. 24) (1), (16), (43), (44). These types of autosolitons (Fig. 24) have experimentally been found out in the gas discharge system (Ref. (41)). It may be linked with a possibility that at some parameters of the gas discharge system the both conditions $\varepsilon = \ell/L << 1$ and $\alpha = \tau_D/\tau_V < 1$ which are characteristic for KΩ-systems (1), (16), (43), (44), exactly, that the voltage drop across the gas discharge V_i is slow and long-range compared with the concentration of the carriers n in the gas discharge, can be fulfilled.

Emphasize that the scenarios of self-organisation and turbulence may be very diverse in KΩ-systems (1). In particular, it is linked with two effects (1), (16), (43), (44):

1. Static, pulsating, rocking and travelling autosolitons may coexist in KΩ-systems.
2. Effects of a spontaneous transformation of one type of autosolitons into another, for example due to interactions between autosolitons, may be realized in KΩ-systems (Fig. 25).

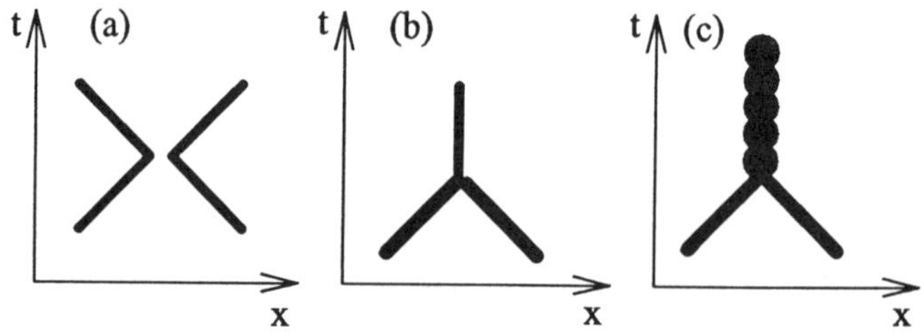

FIGURE 25. Transformation of one type of autosoliton into another upon collision: repulsion of two travelling autosolitons (a), transformation of two travelling autosolitons into a static (b) or pulsating autosoliton (c).

The both effects of the coexistence of different types of autosolitons and of the transformation of one type of autosoliton into another (Fig. 25(a)), predicted in the general theory of autosolitons (1), (16), (43), (44), have been found out in experimental investigations of the gas discharge system (41), (48).

4.2.2. Physics of Appearance of Current Filaments

The physics of an appearance and of an existence of autosolitons in the form of current filaments in the gas discharge system is very similar to those which is responsible for the occurrence of avalanche current filaments in the semiconductor *pn* structure (Sect. 4.1.1).

When the current density in the gas discharge j is relative large, the average rate of ionization ν_i in the gas plasma becomes an increasing function of the concentration of electrons n (46). It is linked with a rising dependence of the value ν_i of the rate of electron-electron collisions which is proportional to the concentration of the electrons.

This rising dependence ν_i of n can cause a growth of a local perturbation of the avalanche current density in the structure. The latter can lead to a spontaneous appearance of current filaments. Indeed, the avalanche current density is $j = env_e$, i.e., it is proportional to the concentration of electrons n in the gas discharge. The local increase in the density of current implies the local increase in the electron concentration in the gas discharge. Therefore, it implies the increase in the average rate of ionization ν_i. The latter causes the further local increase in the previous local perturbation of the current density and so it causes the further local growth of the average rate of ionization ν_i and so on.

Considering this avalanchelike growth of the local current density in the gas discharge, the voltage drop across the semiconductor anode (Fig. 23) with the specific resistivity ρ_s, has not been taken into account. First, let us consider a homogeneous (in x-direction) growth of the current density. In this case we get the obvious formula for the total voltage drop across the structure V: $V = j\rho_s + V_i$, where V_i is the voltage drop across the region of gas discharge (Fig. 23). From this formula it follows that if a homogeneous increase in the current density δj occurs, we get that the voltage drop across the gas discharge region V_i should decrease, if the value of the total voltage drop V=const: $\delta V_i = -\delta j \rho_s$.

On the other hand, the average rate of ionization $\nu_i(n, T(V_i))$ is a strong increasing function of the temperature of electrons, i.e., correspondingly to the formula (80), it is also a strong increasing function of the voltage drop V_i. Therefore, if the value ρ_s is not very small, the decrease in V_i suppresses the growth of the homogeneous perturbations of the current, inducing by the rising dependence of the average rate of ionization ν_i on the carrier density n. In other words, the homogeneous perturbations of the current density can be suppressed.

A different situation is encountered when the growth of the current density is nonhomogeneous. To understand it, let us consider the growth of a perturbation of the current density $\delta j(x)$ in a small local region of a size ℓ of the gas discharge. If the thickness of the semiconductor anode of the gas discharge system W is large enough, $W \gg \ell$, the current density spreads over the semiconductor anode of the structure. Because of that, the change in the voltage drop across the semiconductor anode, i.e., the value $V_\rho = V - V_i$, produced by a local increase in the avalanche current in the small neighbourhood of size ℓ, will be much less, in proportion to $\ell/W << 1$, than that in the above case of a uniform increase in the avalanche current. In other words, a local increase in the density of the current causes a very small change in the voltage drop across the semiconductor anode and, therefore, it causes a very small change in the value V_i: $|\delta V_i| << \delta j \rho_s$.

This decrease in the value of the voltage drop across the gas discharge region V_i can be so small that it cannot suppress the effect of the growth of the current density causing by the rising dependence of the average rate of ionization of the current density in the local region under consideration. Therefore, if the density of current increases in the local regions of the very small size, the voltage drop in the semiconductor anode is not enough to prevent the further growth of current density in these small local regions of the structure. So, local solitary regions of high density of current may arise in the gas discharge system (Fig. 1 and Fig. 24)).

The stability of static current filaments (Fig. 1) in some range of the voltage drop across the system V is linked with the damping of the critical fluctuations of current density which are localized in the filament (1), (16). This damping effect is realized due to a corresponding spatial change in the local value of the voltage drop across the gas discharge region V_i, when the localized fluctuation of the current density appears.

However, at some parameters of the gas discharge system the characteristic time τ_V of the change in the voltage drop V_i becomes larger than the time of a recombination of the carriers τ_D which determines the time of a change in the density of current in the gas discharge. As a result, a localized fluctuation of the current density in the static current filament, which changes in time with some frequency ω: $\tau_V^{-1} < \omega < \tau_D^{-1}$, causes a change in the voltage drop V_i which may be too slow to damp the critical fluctuations of the current density in the static current filament. In this case, as it follows from the general theory of autosolitons in active systems with diffusion (1), (16), pulsating, rocking, travelling autosolitons (Fig. 24), or complex space-time structures, or else a chaos can spontaneously appear in the gas discharge system. The latter behaviour of current filaments have really been found out in experimental investigations of the gas discharge system (41).

CONCLUSIONS

Autosolitons in traffic flow, i.e., traffic jams, have a lot of completely different properties and characteristics compared with autosolitons in active systems with diffusion. Indeed, let us compare for example a travelling autosoliton in active systems with diffusion with a local cluster of vehicles in traffic flow. The travelling autosoliton (pulse) can occur in active systems with diffusion, where the process of the inhibition is slow enough compared with the process of the activation. The velocity of such travelling autosoliton is determined by the velocity of the diffusion propagation of an "excitation front" in the active system. Contrarily, the local cluster of vehicles is characterised by three different velocities:

1. The velocity of the up-stream front.
2. The velocity of the down-stream front.
3. The velocity of the transition layer between the initially state of traffic flow and the new homogeneous state of traffic flow formed by the traffic jam down-stream.

The first velocity is determined by the flux of vehicles in the initially state of traffic flow up-stream from the traffic jam. The second velocity depends on the possible rate of an "escaping" of vehicles from the traffic jam. The velocity of the transition layer depends on the fluxes and on the density of vehicles in both the initially state of traffic flow and the new homogeneous state of traffic flow formed by the traffic jam down-stream.

However, we could see that autosolitons both in active systems with diffusion and in traffic flow have also some general nonlinear properties:

- Autosolitons appear due to a competition between some local active and local damping processes in distributed systems.
- Autosolitons can spontaneously be formed owing to an occurrence of a localized fluctuation or due to the "local breakdown" effect in a distributed system.
- There is the boundary (threshold) value of the controlling parameter, where autosolitons have the minimum amplitude and/or the minimum width.
- An occurrence of either autosolitons in active systems with diffusion or traffic jams in traffic flow is a quite natural general property of the corresponding nonlinear systems and can be considered as some kind of self-organisation.
- An appearance and an evolution of autosolitons in active systems with diffusion determine the main scenarios of self-organization and turbulence which can be realized in these systems. Consequently, the main scenarios of self-organization and turbulence which may be realized in traffic flow are linked with the nonlinear properties of different traffic jams which can appear in this flow. Besides, as the traffic jam moves, as a rule, up-stream, it will reach in some time a source of the road. In other words, the spontaneous appearance of the traffic jam somewhere on a long road can cause a sharp change in the whole road network which includes the road under consideration and can also be responsible for very complex nonlinear processes in the road network.

Because the processes of an occurrence and of an evolution of autosolitons determine the behaviour of a lot of nonlinear distributed systems, both general and specific properties of autosolitons may be an interesting and important subject of future investigations.

ACKNOWLEDGEMENTS

I would like to thank Dr. P. Konhäuser for fruitful discussions of this review.

REFERENCES

1. Kerner, B.S., and Osipov, V.V., *Autosolitons: A New Approach to Problems of Self-Organization and Turbulence*, Dordrecht: Kluwer Academic Publishers, 1994.
2. Wiener, N., and Rosenblueth, A., *Arch. Inst. Cardiol. Mex.* **16**, 205-265 (1946).
3. FitzHugh, R., *Biophys. J.* **2**, 11-21 (1962).
4. Nagumo, I., Arimoto, S., Yoshizawa, S., *Proc. IRE*, **50**, 2061-2070 (1962).
5. Zhabotinskii, A.M., and Zaikin, A.N., *J. Theor. Biol.* **40**, 45-61 (1973).
6. Winfree, A.T., *Sci. Amer.* **230**, 82-95 (1974).
7. Nicolis, G., and Prigogine, I., *Self-organization in Nonequilibrium Systems*, New York: Wiley, 1977.
8. Haken, H., *Introduction to Synergetics*, Berlin: Springer, 1977.
9. Mikhailov, A.S., *Foundations of Synergetics I*, Berlin: Springer, 1990.
10. Meron, E., *Phys. Rep.* **218**, 1- 66 (1992).
11. Kerner, B.S., and Osipov, V.V., *Sov. Phys.-JETP* **47**, 874-885 (1978).
12. Kerner, B.S., and Osipov, V.V., *Sov.Phys.-Semiconductors* **13**, 424-431 (1979).
13. Kerner, B.S., and Sinkevich, V.F., *JETP Lett.* **36**, 436-439 (1982).
14. Nedospasov, A.V., *Sov. Phys.-Uspekhi* **11**, 174-187 (1968).
15. Kerner, B.S., and Osipov, V.V., *Radio Eng. Electr. Phys.* **27**, 122-132 (1982).
16. Kerner, B.S., and Osipov, V.V., *Sov. Phys.-Uspekhi* **32**, 101-138 (1989).
17. Kerner, B.S., and Osipov, V.V., *Sov. Phys.-Uspekhi* **33**, 679-719 (1990).
18. Kerner, B.S., "Autosolitons in Form of Current Filaments and Electric Field Domains in Semiconductors and Devices", in: *Nonlinear Dynamics and Pattern Formation in Semiconductors and Devices* ed. by F.-J Niedernostheide, Springer Proceedings in Physics, Berlin: Springer, 1995, pp. 70-111.
19. Kerner, B.S., and Konhäuser, P., *Phys. Rev. E* **48**, 2335-2338 (1993).
20. Kerner, B.S., and Konhäuser, P., *Phys. Rev. E* **50**, 54-83 (1994).
21. Kerner, B.S., "Mechanism of Local Chaos in Active Distributed Media", in *Experimental and Theoretical Advances in Biological Pattern Formation* eds. by H.G. Othmer, P.K. Maini and J.D. Murray, NATO ASI Series A: Life Sciences, New York and London: Plenum Press, 1993, pp. 191-210.
22. Gardiner, C.W., *Handbook of Stochastic Methods for Physics, Chemistry and the Natural Sciences*, Berlin: Springer, 1994.
23. Haken, H., *Instability Hierarchies of Self-organizing Systems and Devises*, Berlin: Springer, 1983.
24. Kuramoto, V., and Tsuzuki, T., *Prog. Theor. Phys.* **54**, 687-699 (1975).
25. Swift, J., and Hohenberg, R.C., *Phys. Rev. A.* **15**, 319-328 (1977).
26. Kerner, B.S., and Klenov, S.L., (unpublished) (1992).
27. Kurtze, D.A., and Hong, D.C., to be published.
28. Kerner, B.S., Konhäuser, P., and Schilke, M., presented at the 7ht World Conference on Transportation Research, Sydney, Australia, 1995.
29. Kerner, B.S., Konhäuser, P., and Schilke, M., *Phys. Rev. E*, **51**, N. 6 (1995).
30. Treiterer, J., Ohio State University Technical Report PB 246 094, Columbus, Ohio (unpublished) (1975).
31. Leutzbach, W., *Introduction to the Theory of Traffic Flow* Berlin: Springer, 1988.
32. Kühne, R., "Traffic Patterns in Unstable Traffic Flow on Freeways", in *Highway Capacity and Level of Service*, edited by U. Brannolte, Rotterdam: A.A.Balkema, 1991, pp. 211-223.
33. Prigogine, I., and Herman, R., *Kinetic Theory of Vehicular Traffic,* New York: American Elsevier publishing Co.,1971.
34. Payne, H.J., "Models of Freeway Traffic and Control", in *Mathematical Models of Public Systems*, Simulation Council Proc., 1971, Vol. 1, No. 1, pp. 51-61.
35. Whitham, G.B., *Linear and Nonlinear Waves,* New York: John Wiley & Sons, 1974.

36. Hall, F.L., Brian, L.A., and Gunter, M.A., *Transpn. Res. A* **20A**, 197-210 (1986).
37. Kerner, B.S., Litvin, D.P., and Sankin, V.I., *Sov. Tech. Phys. Lett.* **13**, 342-343 (1987).
38. Gafiichuk, V.V., Datsko, B.I., Kerner, B.S., and Osipov, V.V., *Sov. Phys. Semicond.* **24**, 806-811 (1990).
39. Radechaus, Ch., Dirksmeyer, T., Willebrand, H., and Purwins, H.-G., *Phys. Lett. A* **125**, 92-94 (1987).
40. Willebrand, H., Niedernostheide, F.-J., Ammelt, E., Domen, R., and Purwins, H.-G., *Phys. Lett. A* **153**, 437-445 (1991).
41. Willebrand, H., Hünteler, T., Niedernostheide, F.-J., Domen, R., and Purwins, H.-G., *Phys. Rev. A* **45**, 8766-8775 (1992).
42. Kerner, B.S., and Osipov, V.V., *Sov. Phys.-JETP* **52**, 1122-1132 (1980).
43. Kerner, B.S., and Osipov, V.V., *Sov. Phys.-JETP* **56**, 1275-1282 (1982).
44. Kerner, B.S., and Osipov, V.V., *Mikroelectronika* **12**, 512-529 (1983) (in Russian).
45. Kerner, B.S. (unpublished) (1991).
46. Golant, V.E., Zhilinskii, A., and Sakharov, S.A., *Fundamentals of Plasma Physics*, New York: Wiley, 1980.
47. Radechaus, Ch., Kardell, K., Baumann, H., Jäger, D., and Purwins, H.-G., *Z. Phys. B*, **65**, 515-525 (1987).
48. Willebrand, H., Matthiessen, K., Niedernostheide, F.-J., Domen, R., and Purwins, H.-G., *Contrib. Plasma Phys.* **32**, 57-68 (1992).

DR. HENRY D. I. ABARBANEL
INST FOR NONLINEAR SCIENCE
U OF CAL, S/D
9600 GILMAN DRIVE
LA JOLLA, CA 92093
Phone 619 534 5590
Fax
E-mail hdia@gibbs.ucsd.edu

DR. MAURICE ADAMS
CASE WEST. RES.
GLENNAN BLDG 417
CLEVELAND, OH 12345
Phone 216 368 6449
Fax 216 368 6445
E-mail mla5@po.cwru.edu

PROF. OLUFEMI ADEYEMI
KELLEY HALL
ELEC ENGRG DEPT
UNIV. OF R.I.
PROVIDENCE, RI 02881
Phone 401 792 5872
Fax 401 782 6422
E-mail adeyemi@ele.uri.edu

DR. THEODORE ANDERSON
NUWC, CODE 3433
39 SMITH STREET
NEW LONDON, CONN 06320
Phone 203 440 5558
Fax 203 440 4624
E-mail

HELENE ANDERSON
NUWC, CODE 3422
39 SMITH STREET
NEW LONDON, CT 06320
Phone 203 440 4864
Fax 203 440 6725
E-mail helene@aol.com

PROF. VADIM ANISHCHENKO
SARATOV STATE UNIVERSITY
ASTRAKHANSKAYA STR 83
SARATOV, RUSSIA, 410071
Phone 8452 51 2876
Fax 8452 24 0446
E-mail wadim@chaos.saratov.su

PROF. VLADIMIR ASTAKHOV
SARATOV STATE UNIVERSITY
ASTRAKHANSKAYA STR 83
SARATOV, RUSSIA, 410071
Phone 8452 51 2876
Fax 8452 24 0446
E-mail wadim@chaos.saratov.su

DR. WILLIAM AVERA
NRL, CODE 7442
BLDG 2437
STENNIS SP. CTR, MI. 39529
Phone 601 688 4778
Fax 601 688 4476
E-mail avera@will.nrlssc.navy.mil

BRUCE BATES
NUWC, CODE 2211
1176 HOWELL STREET
NEWPORT, R.I. 02841
Phone 401 841 3354
Fax 401 841 4749
E-mail

PROF. MAXIM BAZHENOV
RUSSIAN ACAD OF SCIENCES
46 ULJANOV STR
NIZHNY NOVGOROD, RUSSIA
Phone
Fax 783 123 67291
E-mail bazh@app. nnov.su

DR. LISA BORLAND
U OF CAL, BERKELEY
DEPT OF PHYSICS
BERKELEY
LOS ANGELES, CA 94720
Phone
Fax
E-mail lisa@gojira.berkeley.edu

DR. MARIBETH BOZEK-KUZMICKI
THE MITRE CORP.
7525 COLSHIRE DRIVE, W662
MCLEAN, VA 22102
Phone 703 883 5584
Fax 703 883 6708
E-mail maribeth@mitre.org

MICHELLE BRIGHT
NASA LEWIS
21000 BROOKPARK RD.
CLEVELAND, OH 44135
Phone 216 433 2304
Fax 216 433 8643
E-mail mbright@limspop.lerc.nasa.gov

DR. DAVID BROOMHEAD
DRA(MALVERN)
ST. ANDREWS ROAD
GREAT MALVERN, UNITED KINGDOM
Phone
Fax
E-mail dsb@signal.dra.hmg.gb

JEFFREY S. BRUSH
CAMBRIDGE RESEARCH ASSOCIATES
1430 SPRING HILL ROAD, SUITE 200
MCLEAN, VA 22101
Phone 703 790 0505
Fax 703 790 0370
E-mail jbrush@cambridge.com

DR. ADI BULSARA
NCCOSC, CODE 573
SAN DIEGO, CA 92122
Phone 619 553 1595
Fax 619 553 1269
E-mail bulsara@manta.nosc.mil

DR. THOMAS CARR
NRL
SP.PROJ. IN NONLINEAR SCIENCES
WASH., D.C. 20375
Phone 202 767 3195
Fax 202 404 8357
E-mail carr@nls7.nrl.navy.mil

DR. ROBERT CAWLEY
NSWC, CODE B-44
10901 NEW HAMPSHIRE AVENUE
SILVER SPRING, MD 20903-5640
Phone 301-394-2256
Fax 301-394-5135
E-mail rcawley@nswc.navy.mil

PROF. CHRISTINE CHAN
GSO/URI
BOX 200, SOUTH FERRY ROAD
NARRAGANSETT, RI 02882
Phone 401 792 6676
Fax 401 792 6728
E-mail chan@gsosun1.gso.uri.edu

SHALABH CHANDRA
CLEVELAND CLINIC FOUNDATION
DEPT OF CARDIOLOGY, EUCLID AVE.
CLEVELAND, OH 44196
Phone 216 445 7287
Fax 216 445 7306
E-mail schandra@cardioth.ccf.org

PROF. J. J. COLLINS
BOSTON UNIVERSITY
44 CUMMINGTON STREET
BOSTON, MA 02215
Phone 617 353 9635
Fax 617 353 5737
E-mail collins@enga.bu.edu

PROF. R. F. COSTANTINO
DEPARTMENT OF ZOOLOGY
U.R.I.
KINGSTON, R.I. 02881
Phone 401 792 2469
Fax 401 792 4256
E-mail byg101@uriacc.uri.edu

DR. JOSEPH CUSUMANO
DEPT OF ENGINEERING SCIENCE & MECHANICS
PENN STATE UNIVERSITY
227 HAMMOND BLDG
STATE COLLEGE, PA 16802
Phone 814 865 3179
Fax 814 863 7967
E-mail jpc@hydr06.psu.edu

MR. ROY DEAVENPORT
NUWC, CODE 3112
39 SMITH STREET
NEW LONDON, CT 06320
Phone 203 440 4779
Fax 203 440 6696
E-mail n/a

DR. WILLIAM DITTO
GEORGIA INST OF TECH
SCHOOL OF PHYSICS
ATLANTA, GA 30332
Phone 404 894 5216
Fax 404 853 9958
E-mail
wditto@acl1.physics.gatech.edu

DR. DANIEL DUCHARDT
THE MITRE CORP.
428 LONG HILL ROAD
GROTON, CT 06340
Phone 203 449 1851
Fax 203 449 0559
E-mail dduchard@mitre.org

DR. RICHARD EVANS
SAIC
SUITE 201, 21 MONTAUK
AVENUE
NEW LONDON, CT 06320
Phone 203 442 0564
Fax 203 442 4162
E-mail
ndxdrab@rrznuser.unihannover.de

DR. JAMES F. FORREN
PSI
7923 JONES BRANCH DRIVE
MCLEAN, VA 22102
Phone 703 734 3424
Fax 703 734 3456
E-mail jforren@plansys.com

DR. TED FRISON
RANDLE, INC.
P. O. BOX 1010
GREAT FALLS, VA 22066
Phone 703-759-5257
Fax 703 757 7639
E-mail ted@chaotic.com

PROF. LUCA GAMMAITONI
UNIV OF PERUGIA
PHYSICS DIVISION
PERUGIA, ITALY
Phone 3907 5585 3033
Fax 3907 544 666
E-mail gammiatoni@perugia.infin.it

JOHN GEARY
NUWC, CODE 3111
39 SMITH STREET
NEW LONDON, CT 06320
Phone 203 440 5816
Fax 203 440 6696
E-mail n/a

DR. ARI GOLDBERGER
HARVARD MED/BETH ISRAEL
HOSPITAL, ROOM GZ435
330 BROOKLINE STREET
BOSTON, MA 02215
Phone 617-735-4199
Fax 617 735 4833
E-mail ary@astro.bih.harvard.edu

STEPHEN GREINEDER
NUWC, CODE 2121
39 SMITH STREET
NEW LONDON, CT 06320
Phone 203 440 5329
Fax 203 440 4010
E-mail
stephen_greineder@code20.nl.navy.m
il

PROF. DMITRI GRIBKOV
UNIVERSITY OF MOSCOW
PHYSICS DEPARTMENT
MOSCOW, RUSSIA, 123362
Phone
Fax
E-mail gribkov@cam.cornell.edu

DR. RICHARD GRIMM
CLEVELAND CLINIC
FOUNDATION
DEPT OF CARDIOLOGY
EUCLID AVENUE
CLEVELAND, OH 44196
Phone 216 445 7287
Fax 216 445 7306
E-mail

DR. JOHN GUCKENHEIMER
CORNELL
MATHEMATICS
ITHACA, N.Y. 14853
Phone 607 255 4336
Fax 607 255 9860
E-mail gucken@cam.cornell.edu

DR. STEPHEN HAMMEL
IRAMMP PROGRAM MANAGER
NAVSWC, B44
SILVER SPRING, MD 20903
Phone 301 394 2797
Fax 301 394 4645
E-mail shammel@relay.nswc.navy.mil

DR. THOMAS HAYWOOD
NRL, CODE 7121
4555 OVERLOOK AVE SW
WASH., D.C. 20375
Phone 202 767 2196
Fax 202 404 781
E-mail haywood@wave.nrl.navy.mil

PROF. HANS-PETER HERZEL
TECHNICAL UNIV AT BERLIN
DARMSTADT, GERMANY
Phone +49 30 3142 2088
Fax +49 30 3142 1130
E-mail hertzel@itp1.physik.tu-berlin.de

PROF. JOACHIM HOLZFUSS
INSTITUTE FOR APPLIED
PHYSICS
DARMSTADT, GERMANY
Phone 496 151 16-2884
Fax 491 151 16-4534
E-mail
joachim.holzfuss@physik.thdarmstadt.de

DR. GUAN-HSONG HSU
NAVAL SURFACE WARFARE
CENTER, B-44
10901 NEW HAMPSHIRE AVENUE
SILVER SPRING, MD 20903
Phone 301-394-5289
Fax 301 394 5289
E-mail ghsu@relay.nswc.navy.mil

DR. RONALD HUGHES
1600 TANYARD HILL RD
GAITHERSBURG, MD 20879
Phone 301 548 0188
Fax 301 548 0189
E-mail chaos@telcoinfo.com

DR. MARIO INCHIOSA
NCCOSC, CODE 573
SAN DIEGO, CA 92152
Phone 619-553-1595
Fax 619 553 1269
E-mail inchiosa@nosc.mil

DR. GARY JACYNA
THE MITRE CORP.
7525 COLSHIRE DRIVE, W525
MCLEAN, VA 22102
Phone 703 883 6972
Fax 703 883 7852
E-mail gjacyna@mitre.org

PROF. NATALIA JANSON
SARATOV STATE UNIV
ASTRAKHANSKAYA STR 83
SARATOV, RUSSIA, 410071
Phone 8452 51 2876
Fax 8452 24 0446
E-mail wadim@chaos.saratov.su

DR. CHRISTOPHER JONES
BROWN UNIVERSITY
DEPT OF APPLIED MATH
PROVIDENCE, RI 02912
Phone 401 863 2044
Fax 401 863 1355
E-mail pmiller@cfm.brown.edu

DR. JAMES KADTKE
UNV OF CALIF, SAN DIEGO.
9500 GILMAN DRIVE/IPAPS-360
LA JOLLA, CA 92093
Phone 619-453-5243
Fax 619 453 3938
E-mail jbkpla@sdphv1.ucsd.edu

PROF. TOMASZ KAPITANIAK
TECH. UNIV.OF LODZ
STEFANOWSKIEGO
LODZ, POLAND
Phone 484 231 2231
Fax 484 236 5646
E-mail tomaszka@lodz1.p.lodx.pl

DR. RICHARD KATZ
NUWC, CODE 33A
39 SMITH STREET
NEW LONDON, CT 06320
Phone 203 440 5986
Fax 203 440 5420
E-mail katzra@nl.nuwc.navy.mil

PROF. BORIS KERNER
DAIMLER-BENG
HEUSCH/BOESEFELD GMBH,
LIEBIGSTRAPS 20
AACHEN, GERMANY 52070
Phone 0241 168 0575
Fax 0241 168 0577
E-mail

DR. DAVID KIL
LOCKHEED-SANDERS
P.O. BOX 868
NASHUA, NH 03061
Phone 603 885 3373
Fax 603 885 0631
E-mail carlos@sanders.com

PROF. LASLO KISS
JATE UNIV. INST. FOR
EXPERIMENTAL PHYSICS
DOM TER 9
SZEGED, H6720, HUNGARY
Phone 36 62 311 154
Fax 36 62 311 154
E-mail laszlo@sol.cc.u-szeged.hu

PROF. YURI KRAVTSOV
MOSCOW ST UNIV
SPACE RES. INST.
MOSCOW, RUSSIA
Phone
Fax
E-mail kravtsov@asp.iki.rssi.ru

PROF. JURGEN KURTHS
UNIV OF POTSDAM
POTSDAM, GERMANY
Phone +49 331 977 1611
Fax +49 331 977 1142
E-mail jkurths@agnld.uni-potsdam.de

PROF. CLAUDIA LAINSCSEK
INST. FUER THEORETISCHE
PHYSIK
TECHNISCHE UNIVERSITAET
GRAZ
PETERSGASSE 16, A-8010
GRAZ, AUSTRIA
Phone +43 316 873
8195
Fax +43 316 814741

DR. P.C. LALL
NUWC, CODE 000C
NEWPORT DIVISION
NEWPORT, R.I. 02841
Phone 401-841-4736
Fax 401 841 1231
E-mail lallpc@nuwc.navy.mil

PROF. WERNER LAUTERBORN
DRITTES PHYSIKALISCHES
INSTITUT
UNIVERSITAET GOETTINGEN,
BUERGERSTR. 42-44
D-37073 GOETTINGEN,
GERMANY
Phone +551 39 7713
Fax +551-39-7720
E-mail lb@physik3.gwdg.de

DR. JEFFERY LEADER
USMA,WEST POINT
ARL/2800 POWDER MILL ROAD
ADELPHI, MD
Phone 301 394 2002
Fax 301 394 2002
E-mail leader@iris35s.arl.mil

DR. PATRICK MILLER
BROWN UNIVERSITY
APPLIED MATH DEPT
PROVIDENCE, R.I. 02912
Phone 401 863 2044
Fax 401 863 1355
E-mail pmiller@cfm.brown.edu

DR. FRANK C. MOON
CORNELL UNIVERSITY
UPSON HALL
MECHANICAL & AEROSPACE
SYSTEMS
ITHACA, NY 14853
Phone 607 255 7146
Fax
E-mail fcm3@cornell.edu

DR. EDWARD C. MOZLEY
NAVAL RESEARCH LAB
CODE 7442, BLDG 2438
STENNIS C
Phone 601 688 4855
Fax 601 688 4476
E-mail emozley@nrlssc.navy.mil

COMMODORE CHARLES MUNNS
COMSUBDEVRON 12
COMMANDER, SUBMARINE
DEVELOPMENT SQUADRON 12
NEW LONDON, CT 06349
Phone 203 449 3166
Fax 203 449 3167
E-mail n/a

DR. RICHARD NADOLINK
NUWC, CODE 10
NEWPORT DIVISION
NEWPORT, R.I. 02841
Phone 401-841-6760
Fax 401 841 4660
E-mail

CHUNG NGUYEN
NUWC, CODE 2211
NEWPORT DIVISION
NEWPORT, R.I. 02841
Phone 401 841 3354
Fax 401 841 4749
E-mail

DR. ALBERT NUTTALL
NUWC, CODE 302
39 SMITH STREET
NEW LONDON, CT 06320
Phone 203 440 4618
Fax 203 440 6841
E-mail
nuttall@al.vsdec.nl.nuwc.navy.mil

DR. ALAN V. OPPENHEIM
MIT
RM. 36-615
CAMBRIDGE, MA 02173
Phone 617 253 4177
Fax 617 253 8495
E-mail avo@mit.edu

DR. LEV OSTROVSKY
U OF COLORADO
325 BROADWAY,
R/E/ETI
BOULDER, CO 80303
Phone 303 497 6138
Fax 303 497 3577
E-mail lostrovsky@et1.noaa.gov

PROF. EDWARD OTT
UNIV OF MARYLAND
INST. FOR PLASMA RESEARCH
COLLEGE PARK, MD 20742
Phone 301 405 5033
Fax 301 405 1678
E-mail n/a

MR. STANLEY PAWLUKIEWICZ
THE MITRE CORPORATION
428 LONG HILL ROAD
GROTON, CT 06340
Phone 203 449 1851
Fax 203 449 0559
E-mail stanp@mitre.org

DR. STEVEN PINCUS
NATIONAL ACADEMY OF
SCIENCE
900 MOOSE HILL ROAD
GUILFORD, CT 06437
Phone 203 458 3455
Fax 203 458 9063
E-mail n/a

DR. NATHAN PLATT
NSWC, WHITE OAK, CODE B-44
10901 NEW HAMPSHIRE AVE.
SILVER SPRING, MD 20903-5640
Phone 301 394 3899
Fax 301 394 4645
E-mail nxp@critical.nswc.navy.mil

DR. KLAUS PRANK
MED SCHL/HANNOVER
HANNOVER, GERMANY 30625
Phone 4951 1532 3827
Fax 4951 1532 3825
E-mail sirprank@dhvmhh1.bitnet

MR. GEORGE J. QUART
JOHNS HOPKINS UNIVERSITY
ROOM 7-358
LAUREL, MD 20723
Phone 301 953 6000
Fax 301 953 5950
E-mail

DR. AZIZUL H. QUAZI
NUWC, CODE 3122
39 SMITH STREET
NEW LONDON, CT 06320
Phone 203 440 4726
Fax 203 440 6228
E-mail
quazi@a1.vsdec.nl.nuwc.navy.mil

DR. FRANCES SHIN
LOCKHEED-SANDERS, INC.
P.O. BOX 868
NASHUA, NH 03061-0868
Phone 603 885 3373
Fax 603 885 0631
E-mail carlos@sanders.com

DR. MICHAEL SHLESINGER
CODE 1112
OFFICE OF NAVAL RESEARCH
800 N. QUINCY STREET,
ARLINGTON, VA 22217
Phone 703 696 4220
Fax 703 696 6881
E-mail shlesim@onrhq.onr.navy.mil

DR. LARRY SIROVICH
THE ROCKEFELLER UNIVERSITY
1230 YORK AVENUE
NEW YORK, NY 10021-6399
Phone 212 327 8031
Fax
E-mail
chico@camelot.rockefeller.edu

DR. KATEPALLI SREENIVASAN
YALE UNIVERSITY
MASON LABORATORY
P.O. BOX 2159
NEW HAVEN , CT 06520
Phone 203 432 4771
Fax
E-mail

DR. TAMAS TEL
EOTVOS UNIVERSITY
INST FOR THEORET. PHYSICS
PUSHKIN UNIV.
1088 BUDAPEST, HUNGARY
Phone
Fax
E-mail ethpoo@ursus.bke.hu

PROF. JURAS ULBIKAS
SEMICONDUCTOR PHYSICS
INST.
VILNIUS, LITHUANIA
Phone 370 2 616 433
Fax 370 2 627 123
E-mail root@jur.pfi.lt

DR. D.M. VAVRIV
UKRANIAN ACAD OF SCIENCES.
4 KRASNOZNAMENAYA STREET
KHARKOV, UKRAINE, 310002
Phone
Fax
E-mail rai@ira.kharkov.ua

MS. KARA WAGNER
BB&N
UNION STATION
NEW LONDON, CT 06320
Phone 203 447 3261
Fax 203 447 1132
E-mail kwagner@bbn.com

ROGER L. WOODALL
NUWC, CODE 2113
DETACHMENT NEW LONDON
NEW LONDON, CT 06320
Phone 203 440 5230
Fax 203 440 5235
E-mail n/a

MR. JIANGUO YAN
FLORIDA INT. UNIV.
ELECTRICAL ENGINEERING
MIAMI, FL 33199
Phone 305 348 4124
Fax 305 348 3707
E-mail yan@eng.fiu.edu

DR. ROLF ZOLLER
CARL SCHENCK AG
LAUDWEHRSTR. 55
DARMSTADT, GERMANY 64273
Phone 0049 6151 32 3681
Fax 00496151 32 1256
E-mail zoeller@csd.de

AUTHOR INDEX

A

B

C

D

E

F

G

H

I

J

K

L

M

N

O

P

Q

R

S

T

U

V

W

Y

Z

AIP Conference Proceedings

	Title	L.C. Number	ISBN
No. 210	Production and Neutralization of Negative Ions and Beams (Brookhaven, NY, 1990)	90-55316	0-88318-786-8
No. 211	High-Energy Astrophysics in the 21st Century (Taos, NM, 1989)	90-55644	0-88318-803-1
No. 212	Accelerator Instrumentation (Brookhaven, NY, 1989)	90-55838	0-88318-645-4
No. 213	Frontiers in Condensed Matter Theory (New York, NY, 1989)	90-6421	0-88318-771-X 0-88318-772-8 (pbk.)
No. 214	Beam Dynamics Issues of High-Luminosity Asymmetric Collider Rings (Berkeley, CA, 1990)	90-55857	0-88318-767-1
No. 215	X-Ray and Inner-Shell Processes (Knoxville, TN 1990)	90-84700	0-88318-790-6
No. 216	Spectral Line Shapes, Vol. 6 (Austin, TX 1990)	90-06278	0-88318-791-4
No. 217	Space Nuclear Power Systems (Albuquerque, NM 1991)	90-56220	0-88318-838-4
No. 218	Positron Beams for Solids and Surfaces (London, Canada 1990)	90-56407	0-88318-842-2
No. 219	Superconductivity and Its Applications (Buffalo, NY 1990)	91-55020	0-88318-835-X
No. 220	High Energy Gamma-Ray Astronomy (Ann Arbor, MI 1990)	91-70876	0-88318-812-0
No. 221	Particle Production Near Threshold (Nashville, IN 1990)	91-55134	0-88318-829-5
No. 222	After the First Three Minutes (College Park, MD 1990)	91-55214	0-88318-828-7
No. 223	Polarized Collider Workshop (University Park, PA 1990)	91-71303	0-88318-826-0
No. 224	LAMPF Workshop on (π, K) Physics (Los Alamos, NM 1990)	91-71304	0-88318-825-2
No. 225	Half Collision Resonance Phenomena in Molecules (Caracas, Venezuela 1990)	91-55210	0-88318-840-6
No. 226	The Living Cell in Four Dimensions (Gif sur Yvette, France 1990)	91-55209	0-88318-794-9
No. 227	Advanced Processing and Characterization Technologies (Clearwater, FL 1991)	91-55194	0-88318-910-0

	Title	L.C. Number	ISBN
No. 228	Anomalous Nuclear Effects in Deuterium/ Solid Systems (Provo, UT 1990)	91-55245	0-88318-833-3
No. 229	Accelerator Instrumentation (Batavia, IL 1990)	91-55347	0-88318-832-1
No. 230	Nonlinear Dynamics and Particle Acceleration (Tsukuba, Japan 1990)	91-55348	0-88318-824-4
No. 231	Boron-Rich Solids (Albuquerque, NM 1990)	91-53024	0-88318-793-4
No. 232	Gamma-Ray Line Astrophysics (Paris-Saclay, France 1990)	91-55492	0-88318-875-9
No. 233	Atomic Physics 12 (Ann Arbor, MI 1990)	91-55595	088318-811-2
No. 234	Amorphous Silicon Materials and Solar Cells (Denver, CO 1991)	91-55575	088318-831-7
No. 235	Physics and Chemistry of MCT and Novel IR Detector Materials (San Francisco, CA 1990)	91-55493	0-88318-931-3
No. 236	Vacuum Design of Synchrotron Light Sources (Argonne, IL 1990)	91-55527	0-88318-873-2
No. 237	Kent M. Terwilliger Memorial Symposium (Ann Arbor, MI 1989)	91-55576	0-88318-788-4
No. 238	Capture Gamma-Ray Spectroscopy (Pacific Grove, CA 1990)	91-57923	0-88318-830-9
No. 239	Advances in Biomolecular Simulations (Obernai, France 1991)	91-58106	0-88318-940-2
No. 240	Joint Soviet-American Workshop on the Physics of Semiconductor Lasers (Leningrad, USSR 1991)	91-58537	0-88318-936-4
No. 241	Scanned Probe Microscopy (Santa Barbara, CA 1991)	91-76758	0-88318-816-3
No. 242	Strong, Weak, and Electromagnetic Interactions in Nuclei, Atoms, and Astrophysics: A Workshop in Honor of Stewart D. Bloom's Retirement (Livermore, CA 1991)	91-76876	0-88318-943-7
No. 243	Intersections Between Particle and Nuclear Physics (Tucson, AZ 1991)	91-77580	0-88318-950-X
No. 244	Radio Frequency Power in Plasmas (Charleston, SC 1991)	91-77853	0-88318-937-2
No. 245	Basic Space Science (Bangalore, India 1991)	91-78379	0-88318-951-8

	Title	L.C. Number	ISBN
No. 246	Space Nuclear Power Systems (Albuquerque, NM 1992)	91-58793	1-56396-027-3 1-56396-026-5 (pbk.)
No. 247	Global Warming: Physics and Facts (Washington, DC 1991)	91-78423	0-88318-932-1
No. 248	Computer-Aided Statistical Physics (Taipei, Taiwan 1991)	91-78378	0-88318-942-9
No. 249	The Physics of Particle Accelerators (Upton, NY 1989, 1990)	92-52843	0-88318-789-2
No. 250	Towards a Unified Picture of Nuclear Dynamics (Nikko, Japan 1991)	92-70143	0-88318-951-8
No. 251	Superconductivity and its Applications (Buffalo, NY 1991)	92-52726	1-56396-016-8
No. 252	Accelerator Instrumentation (Newport News, VA 1991)	92-70356	0-88318-934-8
No. 253	High-Brightness Beams for Advanced Accelerator Applications (College Park, MD 1991)	92-52705	0-88318-947-X
No. 254	Testing the AGN Paradigm (College Park, MD 1991)	92-52780	1-56396-009-5
No. 255	Advanced Beam Dynamics Workshop on Effects of Errors in Accelerators, Their Diagnosis and Corrections (Corpus Christi, TX 1991)	92-52842	1-56396-006-0
No. 256	Slow Dynamics in Condensed Matter (Fukuoka, Japan 1991)	92-53120	0-88318-938-0
No. 257	Atomic Processes in Plasmas (Portland, ME 1991)	91-08105	0-88318-939-9
No. 258	Synchrotron Radiation and Dynamic Phenomena (Grenoble, France 1991)	92-53790	1-56396-008-7
No. 259	Future Directions in Nuclear Physics with 4π Gamma Detection Systems of the New Generation (Strasbourg, France 1991)	92-53222	0-88318-952-6
No. 260	Computational Quantum Physics (Nashville, TN 1991)	92-71777	0-88318-933-X
No. 261	Rare and Exclusive B&K Decays and Novel Flavor Factories (Santa Monica, CA 1991)	92-71873	1-56396-055-9
No. 262	Molecular Electronics—Science and Technology (St. Thomas, Virgin Islands 1991)	92-72210	1-56396-041-9
No. 263	Stress-Induced Phenomena in Metallization: First International Workshop (Ithaca, NY 1991)	92-72292	1-56396-082-6

	Title	L.C. Number	ISBN
No. 264	Particle Acceleration in Cosmic Plasmas (Newark, DE 1991)	92-73316	0-88318-948-8
No. 265	Gamma-Ray Bursts (Huntsville, AL 1991)	92-73456	1-56396-018-4
No. 266	Group Theory in Physics (Cocoyoc, Morelos, Mexico 1991)	92-73457	1-56396-101-6
No. 267	Electromechanical Coupling of the Solar Atmosphere (Capri, Italy 1991)	92-82717	1-56396-110-5
No. 268	Photovoltaic Advanced Research & Development Project (Denver, CO 1992)	92-74159	1-56396-056-7
No. 269	CEBAF 1992 Summer Workshop (Newport News, VA 1992)	92-75403	1-56396-067-2
No. 270	Time Reversal—The Arthur Rich Memorial Symposium (Ann Arbor, MI 1991)	92-83852	1-56396-105-9
No. 271	Tenth Symposium Space Nuclear Power and Propulsion (Vols. I–III) (Albuquerque, NM 1993)	92-75162	1-56396-137-7 (set)
No. 272	Proceedings of the XXVI International Conference on High Energy Physics (Vols. I and II) (Dallas, TX 1992)	93-70412	1-56396-127-X (set)
No. 273	Superconductivity and Its Applications (Buffalo, NY 1992)	93-70502	1-56396-189-X
No. 274	VIth International Conference on the Physics of Highly Charged Ions (Manhattan, KS 1992)	93-70577	1-56396-102-4
No. 275	Atomic Physics 13 (Munich, Germany 1992)	93-70826	1-56396-057-5
No. 276	Very High Energy Cosmic-Ray Interactions: VIIth International Symposium (Ann Arbor, MI 1992)	93-71342	1-56396-038-9
No. 277	The World at Risk: Natural Hazards and Climate Change (Cambridge, MA 1992)	93-71333	1-56396-066-4
No. 278	Back to the Galaxy (College Park, MD 1992)	93-71543	1-56396-227-6
No. 279	Advanced Accelerator Concepts (Port Jefferson, NY 1992)	93-71773	1-56396-191-1
No. 280	Compton Gamma-Ray Observatory (St. Louis, MO 1992)	93-71830	1-56396-104-0

	Title	L.C. Number	ISBN
No. 281	Accelerator Instrumentation Fourth Annual Workshop (Berkeley, CA 1992)	93-072110	1-56396-190-3
No. 282	Quantum 1/f Noise & Other Low Frequency Fluctuations in Electronic Devices (St. Louis, MO 1992)	93-072366	1-56396-252-7
No. 283	Earth and Space Science Information Systems (Pasadena, CA 1992)	93-072360	1-56396-094-X
No. 284	US-Japan Workshop on Ion Temperature Gradient-Driven Turbulent Transport (Austin, TX 1993)	93-72460	1-56396-221-7
No. 285	Noise in Physical Systems and 1/f Fluctuations (St. Louis, MO 1993)	93-72575	1-56396-270-5
No. 286	Ordering Disorder: Prospect and Retrospect in Condensed Matter Physics: Proceedings of the Indo-U.S. Workshop (Hyderabad, India 1993)	93-072549	1-56396-255-1
No. 287	Production and Neutralization of Negative Ions and Beams: Sixth International Symposium (Upton, NY 1992)	93-72821	1-56396-103-2
No. 288	Laser Ablation: Mechanismas and Applications-II: Second International Conference (Knoxville, TN 1993)	93-73040	1-56396-226-8
No. 289	Radio Frequency Power in Plasmas: Tenth Topical Conference (Boston, MA 1993)	93-72964	1-56396-264-0
No. 290	Laser Spectroscopy: XIth International Conference (Hot Springs, VA 1993)	93-73050	1-56396-262-4
No. 291	Prairie View Summer Science Academy (Prairie View, TX 1992)	93-73081	1-56396-133-4
No. 292	Stability of Particle Motion in Storage Rings (Upton, NY 1992)	93-73534	1-56396-225-X
No. 293	Polarized Ion Sources and Polarized Gas Targets (Madison, WI 1993)	93-74102	1-56396-220-9
No. 294	High-Energy Solar Phenomena: A New Era of Spacecraft Measurements (Waterville Valley, NH 1993)	93-74147	1-56396-291-8
No. 295	The Physics of Electronic and Atomic Collisions: XVIII International Conference (Aarhus, Denmark, 1993)	93-74103	1-56396-290-X
No. 296	The Chaos Paradigm: Developments an Applications in Engineering and Science (Mystic, CT 1993)	93-74146	1-56396-254-3

	Title	L.C. Number	ISBN
No. 297	Computational Accelerator Physics (Los Alamos, NM 1993)	93-74205	1-56396-222-5
No. 298	Ultrafast Reaction Dynamics and Solvent Effects (Royaumont, France 1993)	93-074354	1-56396-280-2
No. 299	Dense Z-Pinches: Third International Conference (London, 1993)	93-074569	1-56396-297-7
No. 300	Discovery of Weak Neutral Currents: The Weak Interaction Before and After (Santa Monica, CA 1993)	94-70515	1-56396-306-X
No. 301	Eleventh Symposium Space Nuclear Power and Propulsion (3 Vols.) (Albuquerque, NM 1994)	92-75162	1-56396-305-1 (Set) 156396-301-9 (pbk. set)
No. 302	Lepton and Photon Interactions/ XVI International Symposium (Ithaca, NY 1993)	94-70079	1-56396-106-7
No. 303	Slow Positron Beam Techniques for Solids and Surfaces Fifth International Workshop (Jackson Hole, WY 1992)	94-71036	1-56396-267-5
No. 304	The Second Compton Symposium (College Park, MD 1993)	94-70742	1-56396-261-6
No. 305	Stress-Induced Phenomena in Metallization Second International Workshop (Austin, TX 1993)	94-70650	1-56396-251-9
No. 306	12th NREL Photovoltaic Program Review (Denver, CO 1993)	94-70748	1-56396-315-9
No. 307	Gamma-Ray Bursts Second Workshop (Huntsville, AL 1993)	94-71317	1-56396-336-1
No. 308	The Evolution of X-Ray Binaries (College Park, MD 1993)	94-76853	1-56396-329-9
No. 309	High-Pressure Science and Technology—1993 (Colorado Springs, CO 1993)	93-72821	1-56396-219-5 (Set)
No. 310	Analysis of Interplanetary Dust (Houston, TX 1993)	94-71292	1-56396-341-8
No. 311	Physics of High Energy Particles in Toroidal Systems (Irvine, CA 1993)	94-72098	1-56396-364-7
No. 312	Molecules and Grains in Space (Mont Sainte-Odile, France 1993)	94-72615	1-56396-355-8

	Title	L.C. Number	ISBN
No. 313	The Soft X-Ray Cosmos ROSAT Science Symposium (College Park, MD 1993)	94-72499	1-56396-327-2
No. 314	Advances in Plasma Physics Thomas H. Stix Symposium (Princeton, NJ 1992)	94-72721	1-56396-372-8
No. 315	Orbit Correction and Analysis in Circular Accelerators (Upton, NY 1993)	94-72257	1-56396-373-6
No. 316	Thirteenth International Conference on Thermoelectrics (Kansas City, Missouri 1994)	95-75634	1-56396-444-9
No. 317	Fifth Mexican School of Particles and Fields (Guanajuato, Mexico 1992)	94-72720	1-56396-378-7
No. 318	Laser Interaction and Related Plasma Phenomena 11th International Workshop (Monterey, CA 1993)	94-78097	1-56396-324-8
No. 319	Beam Instrumentation Workshop (Santa Fe, NM 1993)	94-78279	1-56396-389-2
No. 320	Basic Space Science (Lagos, Nigeria 1993)	94-79350	1-56396-328-0
No. 321	The First NREL Conference on Thermophotovoltaic Generation of Electricity (Copper Mountain, CO 1994)	94-72792	1-56396-353-1
No. 322	Atomic Processes in Plasmas Ninth APS Topical Conference (San Antonio, TX)	94-72923	1-56396-411-2
No. 323	Atomic Physics 14 Fourteenth International Conference on Atomic Physics (Boulder, CO 1994)	94-73219	1-56396-348-5
No. 324	Twelfth Symposium on Space Nuclear Power and Propulsion (Albuquerque, NM 1995)	94-73603	1-56396-427-9
No. 325	Conference on NASA Centers for Commercial Development of Space (Albuquerque, NM 1995)	94-73604	1-56396-431-7
No. 326	Accelerator Physics at the Superconducting Super Collider (Dallas, TX 1992-1993)	94-73609	1-56396-354-X
No. 327	Nuclei in the Cosmos III Third International Symposium on Nuclear Astrophysics (Assergi, Italy 1994)	95-75492	1-56396-436-8
No. 328	Spectral Line Shapes, Volume 8 12th ICSLS (Toronto, Canada 1994)	94-74309	1-56396-326-4

	Title	L.C. Number	ISBN
No. 329	Resonance Ionization Spectroscopy 1994 Seventh International Symposium (Bernkastel-Kues, Germany 1994)	95-75077	1-56396-437-6
No. 330	E.C.C.C. 1 Computational Chemistry F.E.C.S. Conference (Nancy, France 1994)	95-75843	1-56396-457-0
No. 331	Non-Neutral Plasma Physics II (Berkeley, CA 1994)	95-79630	1-56396-441-4
No. 332	X-Ray Lasers 1994 Fourth International Colloquium (Williamsburg, VA 1994)	95-76067	1-56396-375-2
No. 333	Beam Instrumentation Workshop (Vancouver, B. C., Canada 1994)	95-79635	1-56396-352-3
No. 334	Few-Body Problems in Physics (Williamsburg, VA 1994)	95-76481	1-56396-325-6
No. 335	Advanced Accelerator Concepts (Fontana, WI 1994)	95-78225	1-56396-476-7 (Set) 1-56396-474-0 (Book) 1-56396-475-9 (CD-Rom)
No. 336	Dark Matter (College Park, MD 1994)	95-76538	1-56396-438-4
No. 337	Pulsed RF Sources for Linear Colliders (Montauk, NY 1994)	95-76814	1-56396-408-2
No. 338	Intersections Between Particle and Nuclear Physics 5th Conference (St. Petersburg, FL 1994)	95-77076	1-56396-335-3
No. 339	Polarization Phenomena in Nuclear Physics Eighth International Symposium (Bloomington, IN 1994)	95-77216	1-56396-482-1
No. 340	Strangeness in Hadronic Matter (Tucson, AZ 1995)	95-77477	1-56396-489-9
No. 341	Volatiles in the Earth and Solar System (Pasadena, CA 1994)	95-77911	1-56396-409-0
No. 342	CAM -94 Physics Meeting (Cacun, Mexico 1994)	95-77851	1-56396-491-0
No. 343	High Energy Spin Physics Eleventh International Symposium (Bloomington, IN 1994)	95-78431	1-56396-374-4
No. 344	Nonlinear Dynamics in Particle Accelerators: Theory and Experiments (Arcidosso, Italy 1994)	95-78135	1-56396-446-5

	Title	L.C. Number	ISBN
No. 345	International Conference on Plasma Physics ICPP 1994 (Foz do Iguaçu, Brazil 1994)	95-78438	1-56396-496-1
No. 346	International Conference on Accelerator-Driven Transmutation Technologies and Applications (Las Vegas, NV 1994)	95-78691	1-56396-505-4
No. 347	Atomic Collisions: A Symposium in Honor of Christopher Bottcher (1945-1993) (Oak Ridge, TN 1994)	95-78689	1-56396-322-1
No. 348	Unveiling the Cosmic Infrared Background (College Park, MD, 1995)	95-83477	1-56396-508-9
No. 349	Workshop on the Tau/Charm Factory (Argonne, IL, 1995)	95-81467	1-56396-523-2
No. 350	International Symposium on Vector Boson Self-Interactions (Los Angeles, CA 1995)	95-79865	1-56396-520-8
No. 351	The Physics of Beams Andrew Sessler Symposium (Los Angeles, CA 1993)	95-80479	1-56396-376-0
No. 352	Physics Potential and Development of $\mu^+ \mu^-$ Colliders: Second Workshop (Sausalito, CA 1994)	95-81413	1-56396-506-2
No. 353	13th NREL Photovoltaic Program Review (Lakewood, CO 1995)	95-80662	1-56396-510-0
No. 354	Organic Coatings (Paris, France, 1995)	96-83019	1-56396-535-6
No. 355	Eleventh Topical Conference on Radio Frequency Power in Plasmas (Palm Springs, CA 1995)	95-80867	1-56396-536-4
No. 356	The Future of Accelerator Physics (Austin, TX 1994)	96-83292	1-56396-541-0
No. 357	10th Topical Workshop on Proton-Antiproton Collider Physics (Batavia, IL 1995)	95-83078	1-56396-543-7
No. 358	The Second NREL Conference on Thermophotovoltaic Generation of Electricity	95-83335	1-56396-509-7
No. 360	The Physics of Electronic and Atomic Collisions XIX International Conference (Whistler, Canada, 1995)	95-83671	1-56396-440-6
No. 361	Space Technology and Applications International Forum (Albuquerque, NM 1996)	95-83440	1-56396-568-2

	Title	L.C. Number	ISBN
No. 362	Two-Center Effects in Ion-Atom Collisions (Lincoln, NE 1994)	96-83379	1-56396-342-6
No. 363	Phenomena in Ionized Gases XXII ICPIG (Hoboken, NJ, 1995)	96-83294	1-56396-550-X
No. 364	Fast Elementary Processes in Chemical and Biological Systems (Villeneuve d'Ascq, France, 1995)	96-83624	1-56396-564-X
No. 365	Latin-American School of Physics XXX ELAF Group Theory and Its Applications (México City, México, 1995)	96-83489	1-56396-567-4
No. 366	High Velocity Neutron Stars and Gamma-Ray Bursts (La Jolla, CA 1995)	96-84067	1-56396-593-3
No. 367	Micro Bunches Workshop (Upton, NY, 1995)	96-83482	1-56396-555-0
No. 368	Acoustic Particle Velocity Sensors: Design, Performance and Applications (Mystic, CT, 1995)	96-83548	1-56396-549-6
No. 369	Laser Interaction and Related Plasma Phenomena (Osaka, Japan 1995)	96-85009	1-56396-445-7
No. 370	Shock Compression of Condensed Matter-1995 (Seattle, WA 1995)	96-84595	1-56396-566-6
No. 371	Sixth Quantum 1/f Noise and Other Low Frequency Fluctuations in Electronic Devices Symposium (St. Louis, MO, 1994)	96-84200	1-56396-410-4
No. 372	Beam Dynamics and Technology Issues for + - Colliders 9th Advanced ICFA Beam Dynamics Workshop (Montauk, NY, 1995)	96-84189	1-56396-554-2
No. 373	Stress-Induced Phenomena in Metallization (Palo Alto, CA 1995)	96-84949	1-56396-439-2
No. 374	High Energy Solar Physics (Greenbelt, MD 1995)	96-84513	1-56396-542-9
No. 375	Chaotic, Fractal, and Nonlinear Signal Processing (Mystic, CT 1995)	96-85356	1-56396-443-0
No. 376	Chaos and the Changing Nature of Science and Medicine: An Introduction (Mobile, AL 1995)	96-85220	1-56396-442-2
No. 377	Space Charge Dominated Beams and Applications of High Brightness Beams (Bloomington, IN 1995)	96-85165	1-56396-625-7